Environmental Science and Engineering

Environmental Science

For further volumes:
http://www.springer.com/series/7487

Viorel Badescu · Richard B. Cathcart
Editors

# Macro-engineering Seawater in Unique Environments

## Arid Lowlands and Water Bodies Rehabilitation

Springer

*Editors*
Prof. Viorel Badescu
Candida Oancea Institute
Polytechnic University Bucuresti
Spl. Independentei 313
060042 Bucuresti, Romania
e-mail: badescu@theta.termo.pub.ro

Richard B. Cathcart
Geographos
W. Olive Avenue 1300
Burbank, CA 91506-2225, USA
e-mail: rbcathcart@gmail.com

ISSN 1863-5520

ISBN 978-3-642-14778-4 e-ISBN 978-3-642-14779-1

DOI 10.1007/978-3-642-14779-1

Springer Heidelberg Dordrecht London New York



*Cover design:* deblik, Berlin

Printed on acid-free paper

Springer is part of Springer Science+Business Media (www.springer.com)

*This book is not dedicated to those complaining about the wind neither to those expecting it to change but to those who adjust the sails.*

# Foreword

Humans have for long had an impact on their environment. Very often the impacts were accidental and unintended consequences of human action—for example, vegetation loss due to fires getting out of control, or soil erosion produced by vegetation removal. However, as technological power developed humans have been capable of an increasing range of deliberate modifications of the environment as is made evident when one considers such features as the polders of the Netherlands, the huge irrigation systems of northern India, or the great levees of the Mississippi River. Today as our technological power increases still further, but also because our natural resources (such as water and fuel) are under increasing pressure, and because we are having a range of increasingly deleterious environmental impacts (such as climate change) that need to be halted or reversed, there is an opportunity and need to consider, in an imaginative and innovative way, possible means by which deliberate environmental modifications can be achieved by macro-engineering schemes.

One area of particular recent concern has been the application of geo-engineering to help cope with the threat of climate change. Suggested techniques have included changing the planetary albedo to reflect incoming solar radiation, the planting of huge plantations to sequester carbon dioxide, the fertilization of the oceans with iron, nitrates or phosphates to encourage the growth of marine organisms whose carbonate skeleton will then fall to the ocean floor, the pumping of cold nutrient-rich ocean water upwards to encourage algal blooms that would serve to pump down $CO_2$, the production of a human volcano by injecting sulfate aerosols into the atmosphere, and the placing of solar reflectors into the atmosphere.

However, an area where there may be scope for macro-engineering activities is in the world's drylands. These are important regions in terms of the proportion they make up of the Earth's land surface (somewhere between 30 and 40%), and the number of people who inhabit them (perhaps 2 billion). Some proposed schemes to ameliorate the severity of dryland climates by flooding lake basins have a long history. In the 1920s for example, Schwarcz proposed that the climate of southern Africa could be improved if the Kalahari basin were flooded. Similarly the possibility of linking those desert basins with have their floors below sea-level to the oceans in order to generate power was proposed for the Qattara basin in Egypt by John Ball at much the same time.

One of the big problems of arid lands is salinity. Natural levels of salinity are often high, but they are currently being enhanced by such processes the spread of irrigation and the removal of native vegetation. Another environment where salt poses a problem is in the oceans. Here also there are possibilities for schemes that control currents, permit desalination, provide protection against storm surges, and enable modification of sea water chemistry.

The case studies presented in this book provide an exciting range of possibilities, though, as always, one needs to be aware that macro-engineering schemes will in all probability have a range of unexpected and unwelcome environmental consequences. In addition they should not be used in the place of efforts to reduce carbon emissions, to conserve water, and to limit our energy consumption. Moreover, because of their huge costs, potential global impacts and great areal extent, they are bound to require a large degree of international collaboration. Institutions do not yet exist with the expertise or authority to assess or enforce responsible use of the global conmons and of macro-engineering techniques. Notwithstanding these serious caveats it may be time to re-evaluate the importance and feasibility of macro-engineering schemes.

Oxford, UK, June 2010 Andrew Goudie

# Preface

As professionalized, practising macro-engineers, we assert the emerging key-factor of our complex world's future geo-economic development success equation must consist of macro-project proposals that become widely recognized, generously explored and soundly implemented. For continued and increasing sustainable human survival and prosperity, the existing Earth-biosphere and our Earth-world's seven billion living persons must, ever more frequently, utilize financially feasible macro-projects, vetted environmentally and made implemental by established and organized macro-engineering, simply because both realms (i.e., the Earth-biosphere and Earth's anthropocentric zone) nearly overlap in duties and tangible results.

In other words, herein, we assume macro-engineering to be a needed and wanted geosozology tool that humankind has employed anciently and recently, and will continue to employ as long as our species exists, to protect conservatively and to save (preserve) the whole Earth. Interestingly, during the late twentieth century and early twenty-first century, the technical means by which geoscientists approach the exploration and exploitation of our world's ocean and the nearest region outer space enveloping Earth have tended to mesh synergistically, especially with the development of grand networks of monitoring systems (various orbiting satellites and seabed, deep-sea and ocean–atmosphere interface sensors). Geoscientists have discovered that coastal groundwater, which once flowed into the ocean unseen, is being depleted by growing human populations resident on the world's coastline. Indeed, dammed reservoirs have, since circa 1900, reduced the world-ocean's rise by ~30 mm—in other words, retarding global sea-level rise. Mirroring aircraft contrails in the sky over the ocean, bottom trawler trails rapidly gouged into the seabed by moving ships are visible in images garnered by Earth-orbiting satellites; in 2005, the General Fisheries Commission for the Mediterranean banned trawling in that segment of the world-ocean below depths of 1000 m because it is likely that unknown, possibly valuable, undersea ecosystems were being damaged as well as sunken historical time-capsules. Sound generated by shipping following heavily used sea-lanes is known to adversely affect marine mammals and air pollution emitted by ships affects human health (in seaports) and adds contaminating fallout materials to the turbulent seawater.

Swimming marine life, rivaling the seawater stirring caused by global winds, is now proved to mix the ocean fluid on a spatially large scale. The editors surmise that today's "fish mega-shoals" may one day be equaled in their churning of seawater by awesome man-made machines, big and numerous operational artifacts such as huge future Mega-Float mobile artificial islands, existing shipping fleets and future gigantic super-ships with many exterior motor-pods and even, perhaps, future vast multi-unit sailing wind-power farms. Such technology-induced seawater turbulence could affect any future human choices and internationalized establishment of particular pelagic protection regions; indeed, high-seas "Ocean Parks" may already be infeasible for the conservation of marine species because the habitat is open to unlimited seawater transformation caused by worldwide shipping, the cheapest of all means of cargo and passenger transportation.

Plastic trash dumped into the ocean releases hazardous chemicals, such as bisphenol A (BPA) and PS oligomer, upon natural decomposition; such pollutants disrupt the functioning of hormones in impinged animals, affecting their reproduction. Yet, during 2010, industrialized coast-sited aquaculture produced about 50% of the total fish and shellfish consumed by Earth's humans; commercial high-seas aquaculture is still being tested and developed for further widespread oceanic deployment. Coastal seawater bodies and the underlying seafloor, exploited by humans for tens of thousands of years, will probably be ever more intensively exploited during the coming 90 years of the twenty-first century. Noteworthy for our textbook's macro-engineering purposes, of the 2010 United Nations Organization-recognized 192 countries, there are 33 ecosystem-nations with land below current global sea-level.

Nowadays ecosystem-nations must strive to preserve visible coastal seascapes from intrusive floating billboards towed offshore by tugboats. Semi-enclosed marine systems everywhere are formed by the combination of coastal seawater bodies and river catchments on the bordering world-ocean land. Soon, fish resources and dissolved oxygen may become anthropically regulated in both watery segments. Sometimes, in circumstances of unusual and special geophysical and societal conditions such as those present in the Netherlands and at Hong Kong, China, freshwater reservoirs have been created by closing off from the world-ocean a suitable large bay or several strategic gulfs and coves. In those specific instances an ocean segment has been replaced by river or pipeline reservoir. "Coastal Zone Management" has economic costs, of course, as with the prevalent early decade of the twenty-first century global economic malaise and yet, so far, "the end of the world" phrase cheekily applies accurately only to a temporarily halted commercial macro-project in Dubai, "The World" collection of artificial islands in the Red Sea.

Herein, we treat many unique facets—potential factors in the postulated equation of widespread human land-ocean usage—of our world's present-day macro-engineered and to-be-macro-engineered coastlines. By 2050, our Earth's human population is predicted to exceed 9 billion persons. Progressive interdisciplinary understanding of the Earth-biosphere articulates an ever-encompassing network of digitized data and enveloping idea relationships responding to human beliefs and thoughts about Earth's prospective nature-influenced future, projected food, materials and energy consumer demands, as well as superimposed natural and unnatural global change. Seawater aquaculture has, therefore, the near-term future potential of becoming the second most prolific means of food production for human consumption after the very ancient honorable activity of agriculture on land.

We view positively the macro-engineering mind-set as a social and technological enterprise and expertise that benefits humanity by permitting exceptional technological and social innovations to be explored during a period of obvious and world-public noticed Earth-biosphere changes. In short, we foresee macro-engineers as the crew, not "mere passengers", of Spaceship Earth since adaptive biosphere management on a self-altering (nature) and anthropogenically impacted terrestrial planet requires timely advancements in our understanding of Earth-surface dynamics in order to influence it in a purposeful manner, including imaginative mitigation and curative macro-engineering project proposals. The more individual persons comprehending and evaluating macro-projects, the better the chance flaws and dangers in macro-project proposals become exposed publicly and the bigger the collection of organized knowledge from which to spin-off appropriate macro-engineering solutions.

Seacoast-focused willful macro-engineers always may advance, and sometimes retreat, with a certain degree of uncertainty, not only because humanity's future and ultimate end is thankfully still undefined conceptually, but also because useful knowledge of most of the natural and social-economic oceanic and coastal processes are still, and probably will always will remain, incomplete.

So, it is our intent herein, to merely suggest, visionary mega-projection style, some technically possible new useful infrastructures, built as intriguing and attractive macro-projects in the coastal region, some of which are inland depressions and/or adjacent world-ocean arid landscapes, which can be used to adapt or to rigorously control ongoing and initiating Earth-surface geodynamical processes. Certainly beyond unnecessary pollution and contamination, the world-ocean and the coast are physically malleable and may already be trending to future abrupt, probably unpredictable, nature-instigated changes. Accordingly, we refuse to offer miniscule solutions with, and based on, our inquisitional macro-problem equation. We do support any and all expert efforts to advance the ideas and technical cures necessary for the prediction of, and adjustments to, dynamic seacoast change—the enveloping "Coastal Zone Management"—at the all-important macro-engineering scale which allow the pervasive restoration, preservation and creation of valuable and beautiful landscapes immediately adjoining the Earth's ocean.

Earth's surface material and energy disequilibrium is almost strictly a response mainly to internal radiogenic heat and external solar radiation. Soon, large-scale human exploration, exploitation and manipulative settlement activities in drylands—extraction of freshwater, food, fodder and fiber production, tapping of geothermal and solar energy resources and waste disposal will add a powerful new anthropogenic forcing additional to nature's tectonic and climatic forcings.

This book takes account of the simple fact that if the welfare of our species is to be preserved for a long time to come, macro-engineering is certain to become a more widespread profession since its practitioners are in the forefront of the remaking, unmaking and extension of our extant infrastructure. Macro-projects with a regional or global impact demand collaboration and negotiation among many participating parties. Omnipotence or omni-benevolent characteristics are not claimed by macro-engineers; we do claim that macro-engineering aims to design Earth-biosphere benignity into our spatially large-scale creations so that they can be constructed with minimal energy and raw materials, few hazardous macro-project design-life moments, and little (if any) harmful by-products and discarded waste that require costly disposal such as entombment.

The goal of this edited book is to gather together a number of macro-projects involving seawater management, arid lowlands and salt-water bodies rehabilitation, that were already studied and reported in the literature as well as new, original macro-engineering concepts. A secondary purpose of this book is to inform persons with imagination and personal drive to solve macro-problems related to humanity's striving for long-term survival, a topic of considerable interest worldwide (engineering of Earth's climate, for example). Another purpose is to provide a history of the literature related to macro-engineering, which is a purpose-driven meld of geo-engineering ideas, so that our book's readers will have a solid base upon which to commence their own life-long advancements to change for the better the inhabitable Earth-biosphere. Also, the book is intended to foster a needed professional spirit because macro-engineering's creative and scientific activity transforms global nature and can transmogrify any global nature to serve the needs and wants of large numbers of persons—we create what did not exist.

This book is structured along logical lines of progressive thought on the topic. The first part of the book contains three chapters and deals with the histories of ancient and modern use of seacoasts. After the introductory Chap. 1presenting the Earth's past global sea-level change and its connection with human settlements, the first in this series, Chap. 2, examines the prototypic history of a man-made Mediterranean Sea port's construction (Ischia Harbour in southern Italy). Chapter 3 documents ongoing anthropogenic coastal development promoted by extant great wealth, accumulated mostly since World War II, in the United Arab Emirates.

The series of informative macro-engineering thinking forming the second part of the book (Chaps. 4–7) outline possible macro-projects capable of massive changes in the coastlines of the Dead Sea, Red Sea and Persian Gulf caused by canal and massively scaled hydropower dam installations.

Part 3 is entirely focused on the Black Sea with a series of seven histories and proposals (Chaps. 8–14) which examine virtually every relevant macro-engineering aspect of that famous body of seawater known worldwide today and mentioned often in humanity's recorded history.

Part 4 is constituted of a single, yet important contribution (Chap. 15) that examines the possibility of artificially stirring a northern Europe seawater body (Baltic Sea).

Part 5 is composed of essays (Chaps. 16–18) describing the twenty-first century possibilities of refreshment of the Aral Sea and Iran's Lake Uremia with seawater or river freshwater importation macro-projects.

Part 6, a series of three statements (Chaps. 19–21), are strictly focused on the potential rehabilitation of some vital arid zone regions now dominated by moving or movable surface granular materials using unique and unusual macro-projects. These offered installations, however, are preceded by a biographical story of a pioneer of sand movement understanding (Ralph Bagnold).

Part 7 deals with the seawater flooding of land regions situated below present-day global sea-level. A series, consisting of four inspiring articles (Chaps. 22–25) soundly document the difficulties, and probable rewards, of submerging some near-coast wastelands.

Part 8, is a five-part series (Chaps. 26–30) devoted entirely to harnessing energy and obtaining freshwater from the world's salt-laden ocean by various modern industrial means.

Part 9 (Chaps. 31–33) address various macro-projects designed specifically for the protection (reduction of vulnerability) of particular Earth geographical regions.

More details about the 33 chapters of the book are given below.

Chapter 1, by Geoffrey N. Bailey and Geoffrey C.P. King, shows that for most of human existence on the Earth over the past 2 million years, sea levels have oscillated with amplitude of more than 100 m in response to the expansion and contraction of the continental ice sheets. At the termination of the most recent Ice Age, global sea level rose from a low-stand of −130 m (about 16,000 years ago) to reach its modern-day level about 6,000 years ago. Vertical changes of this scale have been a continuous accompaniment to the growth and expansion of human societies throughout pre-history, creating profound changes to the landscapes in which people made and earned their livings. Alongside these effects in coastal regions, tectonic changes associated with tectonic plate motions, Earth-crust rifting, earthquakes and volcanic activity have produced equally profound changes of the physical landscape in many of our world's regions. Paradoxically, these regions of geological instability and dynamic landscape change appear to have attracted concentrations of human populations from the earliest period of our history as a biological species. One explores the geophysical processes associated with these changes and their effects on past human societies. One shows that these dynamic processes of geological change, although they can have destructive consequences in the short term, often create and sustain a complex topography with attractive ecological conditions for human settlement and may have exercised powerful selection pressures in favour of the human evolutionary trajectory. One shows how a long-term perspective made possible by an archaeological focus on human-landscape interactions can help to shed new light on the nature of the physical changes themselves, unravel the way in which different sorts of processes operating on different time scales interact, and provide insights into the likely longer term consequences of our present-day and future actions.

Chapter 2 authored by Stefano Carlino, Elena Cubellis, Ilia Delizia and Giuseppe Luongo examines recent and historical sources with a view to reconstructing the circumstances leading in 1854 to the opening of the natural harbour of Ischia, the execution phases of the works and the morphological changes arising.

An active volcanic island, Ischia has since the late seventeenth century been a major European destination for spa treatment. It underwent a period of change after the harbour was opened up, which represented not only an outlet towards the mainland but also an important factor of social and cultural aggregation for the island. The historical analysis is followed by the description of the geology of Ischia Harbour and by observations of the current state of the island as well as the main issues concerning the increase in volcanic and seismic risk resulting from urban expansion and the increase in tourism since the first half of the twentieth century. Some of the well-known risks to the people of Naples from Vesuvius are comparable with the risks all of the permanent residents of Ischia must endure.

Chapter 3, penned by Pernilla Ouis explores the recent development in the United Arab Emirates regarding the creation of artificial islands and other mega development projects on islands. This development is discussed in terms of a recent "islomania" in the region, as a part of the strategy to diversify the economy towards tourism and international finance. Luxury housing, luxurious hotels, constructed beaches, and facilities for amusement and leisure are constructed on such man-made isolated places which are interestingly exemplified with the world-famous artificial archipelago "The World" and the three Palm islands in Dubai and the reclamation of Saadiyat and Sowwah Island in Abu Dhabi, and similar macro-projects. These islands can be seen as important urban flagships for the country on the global tourism and financial industries market, but also as cultural icons legitimizing modernity with reference to tradition. Hybrid urbanism is discussed, arguing that the islands create segregation rather than promoting a cultural and social mix. The islands are interpreted as colonial, utopian spaces for the global elite. Further, the artificial islands create an inversion between the map and nature. Such islands are important symbols viewed from aircraft and from Earth-orbit, and a new kind of tourist, the Google Earth tourist, is identified. Ecological as well as economic sustainability issues are discussed in terms of economic dependency, and hence vulnerability even in the alleged post-oil era.

Chapter 4 by Shahrazad Abu Ghazleh, Abed-Alqader Abed and Stephen Kempe shows that the Dead Sea (DS) is a hyper-saline lake occupying one of a series of pull apart-basins formed in the Earth-crust along the Dead Sea Transform Fault. It represents the lowest terrestrial point with a briny water level of 422 m b.s.l (as of June, 2009). Since 1932, the lake level has experienced a dramatic drop amounting to 0.4 m/a on average due to the accelerating freshwater consumption in the DS Basin. The analysis of the terrain model of the DS and shoreline altimetry shows that the Dead Sea has dropped vertically by 1 m/a during the last 14 years and lost 0.65 $km^3/a$ and 4.6 $km^2$ from its volume and area, respectively. The polynomial function derived from the terrain model used can predict the near-future changes in the lake's volume and area. Therefore, in 2020, the lake is expected to lose 7.7 $km^2$ and 50.6 $km^2/a$ from its current volume and area, respectively. The continuous decline of the DS level caused severe environmental consequences for the lake and for its surroundings, threatening the limited natural resources and the natural habitat of the area. Although the rehabilitation of the Jordan River is the safest ecological alternative for maintaining the DS and the river flow, it would not solve the macro-problem entirely however. The Red-Dead Canal seems to be a sustainable long-term alternative with several attractive benefits for the riparian countries. The volume-level model and the previous investigations of the groundwater discharge and evaporation rate from the salt-ponds suggests that the projected Red-Dead Canal must have a carrying capacity of 1.5 $km^3/a$ in order to halt lake level decrease (stabilization). The canal will also utilize the altitude difference of $\sim$400 m to generate hydropower and produce freshwater by desalinization. Moreover, such a canal will maintain tourism and the profitable potash industry of the Dead Sea and reduce the natural hazard caused by the lake's shoreline advance. The possible environmental impacts of the canal could be minimized or avoided using a comprehensive plan of environmental hazards assessment and mitigation.

Chapter 5 by Damien Closson, Herbert Hansen, François Halgand, Nada Milisavljevic, Frédéric Hallot and Marc Acheroy refers to the Red Sea-Dead Sea canal, showing the origins of this project and the challenges it faces. The opening segment of the chapter deals with the freshwater shortage problem in Jordan, Israel and Palestine. The information is taken from pioneering studies which were conducted during the 1990s. Once the general background for understanding the issues behind the need for the Red Sea-Dead Sea Canal has been drawn, a comprehensive technical description of the macro-project's major physical elements is provided on the basis of a pre-feasibility study conducted in the mid-1990s. The work of the feasibility study for the canal project is then presented. It started only six months before the writing of this chapter and will last approximately one year and a half, perhaps until 2012–2013. This part enumerates the major geological and engineering challenges that will require complete technical answers.

Chapter 6 by Roelof Dirk Schuiling, Viorel Badescu, Richard B. Cathcart, Jihan Seoud and Jaapchan C. Hanekamp shows that artificial closure of the Red Sea at its southern entrance (the Bab-al-Mandab Strait) could well lead to the world's largest hydropower generation, of the order of 50,000 MW. The cost and time-scales involved are, obviously, beyond normal twenty-first century economical considerations. Macro-engineering projects of this size cause a massive destruction of existing ecologies. On the positive side of the environmental assessment scale, however, are the vast reductions of aerial greenhouse gas emissions, and the reduced pace of exhaustion of mined fossil hydrocarbon resources. This chapter examines the ethical and environmental dilemmas and some of the political implications of macro-engineering, as exemplified by the Bab-al-Mandab Macro-project.

Chapter 7 by Roelof Dirk Schuiling, Viorel Badescu, Richard B. Cathcart and Piet van Overveld shows that compartmented ocean gulfs offer a means of artificially creating a water depression, which can be used for a regionally significant hydroelectric macro-project. One examines the case for a dam at the Strait of Hormuz that blocks a large gulf situated in an arid region. A 35 m evaporation of this concentration basin will reduce its seawater surface area by 53% and allow generation of 2,050 MW (or possibly 2,500 MW) of electricity. The conclusion is that the proposed Electricity Development Infrastructure Node (EDIN), echoing the Biblical Eden, is a feasible and desirable macro-project. If the macro-project starts in the near-term future, it would require a significant change in the logistics of oil and gas transport from this region. Alternatively, it can be considered as an attractive future solution for the energy requirements of the region after exhaustion of its underground oil and gas reserves.

Chapter 8 by Yukinobu Oda, Kazunori Ito, Takahide Honda and Solomon Yim discusses construction techniques for deep-water an immersed tunnel under strong seawater current conditions. The Bosphorus Tube Tunnel, submerged beneath the heavily trafficked Bosphorus Strait, a part of the upgraded railroad system in Istanbul, Turkey (the Marmaray Project), is constructed using the tunnel element immersed method. The maximum installation depth of 60 m makes it for the time being the world's deepest immersed tunnel. The Bosphorus Strait is very narrow horizontally and winding, akin to some rivers on land. The seawater current speed changes rapidly and sometimes runs >2 m/s. To overcome such severe current conditions various techniques have been applied. Some of those are reported here. In particular, a seawater current forecast system, developed to maintain quality and safety of tunnel construction, is presented and discussed. The flow of the Bosphorus Strait changes under the influence of not tide but, rather, ambient weather conditions, mainly air-pressure and wind. The forecast system estimates the seawater levels at both ends of the strait on the basis of weather information and predicts the seawater current by taking the density interface effect into account. The predictive model is statistically developed based on a year-long observation data set. The current forecast gives sufficiently accurate information for the everyday marine construction operation.

Chapter 9 by Mircea Dimitrie Cazacu and Sergiu Nicolaie presents relevant research of a joint effort by the Power Engineering Faculty from Politehnica University Bucharest and the National Research Institute in Electrical Engineering, also from Bucharest. In the first part one presents the contributions of theoretical and experimental researches, concerning the viscous fluid flow through the axial rotors of hydraulic and wind turbo machines, by approaching aspects regarding the novelty of the mathematical apparatus (the method of maximal power extracted from the kinetic energy of current fluid flow for turbines) and, also, the special technical applications obtained (e.g., lighting buoys with energetic autonomy on rivers). Also one presents the experimental tests and by validating the previous values obtained in theoretical way, for the both axial rotors (with three and four blades). The tests were performed on a glassed-in hydrostatic channel, and also in a small open wind stand, obtaining a maximal power coefficient CP = 0.56. In the second part of the chapter the author-team presents relevant research related with Ecological River traffic in the biosphere reservation of the Danube River Delta, using electric boats based on renewable energies. Related with this application they present the method to obtain the maximal propulsion force for the propeller of the boat.

Chapter 10 by Eugen Rusu refers to the implementation of the spectral phase averaging models in the Black Sea and to the usage of the wave prediction system developed in two important directions. These are to evaluate the wave energy resources and the most relevant interactions that may occur between sea-waves and seawater currents. The study is focused on the western part of the sea which is traditionally considered as being more energetic. The wave prediction system implemented for the Black Sea is based entirely on the SWAN model (acronym for Simulating WAves Nearshore), which is used both for wave generation and near-shore transformation. This methodology has the advantage that a single model covers the full scale of the modeling process. Various tests were performed considering data measured at three different locations. Special attention was paid to the white-capping sea-wave process that is still widely considered the weak link in deep-water wave models. Comparisons carried out against measured data show that the wave prediction system generally provides reliable results, especially in terms of significant sea-wave heights and mean periods. After the model system calibration by increasing the resolution in geographical space, the field distributions of wave energy were analyzed for both high and average sea-wave conditions. The second application of the wave prediction system is related with the evaluation of the wave current interactions at the entrance of the Danube Delta. The wave conditions in this coastal sector are usually significant from energetic point of view and relatively strong currents, induced there by the Danube River outflow, lead to interactions between waves and currents. This process modifies considerably both sea-waves' magnitude and direction affecting also the coastal navigation and the sediment transport patterns. Five case studies corresponding to the most relevant patterns of the environmental matrix were analyzed. Finally, in order to assess the current effect for a longer time scale an analysis concerning the variation of the main wave parameters was performed for a three-month period considering some reference points. The results show that the currents produce considerable changes in the wave field especially as regards the significant wave heights, mean wave directions and wavelengths. The Benjamin-Feir Index was also estimated. The analysis of the variation induced by the current over this spectral shape parameter indicates that, in certain conditions in the target area the sea-wave heights can not be considered Rayleigh distributed and freak waves may also occur.

Chapter 11 by Mircea Dimitrie Cazacu and Dan Aurel Machita shows that considering the velocity potential expression, corresponding to Gerstner's traveller wave motion, one obtains the relation that gives the value of the constant C, determined by applying of Daniel Bernoulli's relation on the wave free surface, for a same time moment, as function of wave height. One determines the link between the wave kinetic parameters and water mean deep in the conveying channel, to obtain the real values of this constant. Developing the multiple utilities idea for the hydropower arrangements, introduced in Romania by the famous Romanian scientist, professor and engineer Dorin Pavel (1900–1979), they present in this chapter a special hydro-energetic building, placed in front of the Black Sea shoreline at the submarine isobaths between 4 and 10 m, which realizes the following utilities: (i) captation of nonpolluted and inexhaustible sea-wave energy, which on the short Romanian littoral of $\sim$200 km has the same size order as that of the state's interior river potential energy, (ii) shore protection against destructive sea-wave action in the conditions of diminishing of Danube River alluvium transport, as a result of the extant hydropower dams inland and of dispositions to struggle against inland soil erosions, (iii) captation of nonpolluted and inexhaustible energy of wind, by laying a row of wind turbines atop the proposed Sea Highway macro-project, (iv) high speed sea link by means of the slider ships on air cushion in this near-shore zone without large sea-waves, between the littoral and the energy collectors captation devices, (v) terrestrial high-speed transport on this Sea Highway parallel to the litorral, (vi) transport of gaseous or liquid hydrocarbons by pipelines mounted atop and/or within the Sea Highway infrastructure.

Chapter 12 by Roelof Dirk Schuiling shows that metal wastes are produced in large quantities by a number of industries. Their disposal in isolated waste deposits is certain to cause many subsequent problems, because the sealing of disposal sites usually starts to leak, often within a short time period after the disposal site has been filled, capped and legally closed. The contained heavy metals are leached from the disposal site and will ultimately contaminate the surrounding soil profile and the groundwater. Storage as metal sulfides in an anoxic environment is the safest method to handle these detestable metal wastes. It is proposed to use the world's largest anoxic basin of seawater, the Black Sea, as a georeactor. Metal wastes will be transformed, sustainably, into harmless and immobile sulfides, which are incorporated in the lifeless black-colored sea-bottom mud, where they are likely to be safely stored at no further economic cost for millions of years.

Chapter 13 by Mircea Dimitrie Cazacu and Raducu Viorel Iancu deals with the advantageous naturalization of the Black Sea waters. After a short history of the Black Sea is presented, different researches concerning the extraction of hydrogen sulphide in some patented methods and installations with the aim-point goal of Black Sea deep-water naturalization, because in few decades the accumulation of hydrogen sulphide will transform the place into an marine dead zone. In the frame of the patent one claims more possibilities of environment-friendly technologies, as: the energy of the emitted gases from the deep-water extraction using a turbo-exhauster, the energy due to the hydrogen sulphide burning using a hot gas turbine, as well as the production of the high purity sulphuric acid and also of the deuterium, necessary to the heavy water fabrication in the present atomic epoch and for the producing of the future energy by deuterium fusion reaction. The advantages of the proposed original method and installation for the naturalization of the deep water of freshwater lakes, seas or oceans are as follows: (a) the easy progressive starting of the accelerated unsteady flow of the biphasic fluid from the vertical pipe of very great length, (b) the elimination of the eruptive dangerous phenomenon, created by the accelerated lifting of the gas bubbles by their expansion in the vertical pipe, (c) the deposition of the water without dissolved gases in a superior reservoir, from where these (maybe afterwards) could be restored periodically to the Black Sea, (d) the utilization of the compressed gases detention energy by their elimination from the deep water, (e) the use of energy, given by the hydrogen sulfide burning, through a turbo exhauster, (f) the utilization of the eliminated gases from the deep-water by sulfuric acid production and (g) the eventual separation of the deuterium contained in the deep-sea waters and the gases from depth, necessary for the present heavy-water production and for the anticipated future sustained fusion nuclear reactors generating electrical power.

Chapter 14 by Salah Naman, Engin Ture and Nejat Veziroglu refers to an industrial extraction pilot-plant for stripping $H_2S$ gas from Black Sea water. The results from the laboratory scale extraction pilot-plant unit for the separation of $H_2S$ gas from Black Sea seawater led the authors to build a novel industrial extraction pilot-plant to concentrate $H_2S$ gas from 10 ppm concentration to above 10,000 ppm concentration. A technology for extraction and concentration of $H_2S$ gas is essential. The conceptual pilot-plant proposed in this chapter is, in principle, similar to the laboratory pilot-plant developed at the University of Duhok (Iraq), and it could work to pump seawater directly from Black Sea. It contains a screen with electrical heater to fix the temperature of stripping, a water chiller at the top to separate any water droplet or vapor. The research on industrial pilot-plant has shown that, this unit could operate both on as well as underneath the Black Sea's surface.

Chapter 15 by Anders Stigebrandt and Bengt Liljebladh shows that, in the future, it might be interesting to oxygenate large natural seawater basins for different reasons such as nutrient management or promotion of preferred ecological states. Before such enterprises in mega-engineering should be undertaken, the basin's biogeochemical and ecological response upon oxygenation must be well characterized. The present chapter describes an experiment in the seawater of Byfjorden fjord located on Sweden's West Coast which has restricted exchange due to a submarine sill. The basin usually contains high concentrations of hydrogen sulfide but during the experiment it will be kept oxic by pumping down oxygenated surface seawater into the lowest part of the marine geologic basin. This may improve the oxygen conditions in the basin in two ways—by direct addition of 02-saturated seawater and by decreasing the density of the basin's seawater which will increase the frequency of exchange with the marine area outside the fjord and, thus, bring more water (and oxygen) into the basin. The authors develop a model that computes the evolution of the oxygen concentration in similar isolated basins when adding a flow of surface seawater. Knowing the rate of oxygen consumption, the model may be used to estimate the required pump capacity to obtain specified oxygen conditions in a wide range of marine basins.

Chapter 16 by Richard B. Cathcart and Viorel Badescu proposes a comprehensive control strategy to partially recreate the Aral Sea which directly involves several hydrological factors: (i) overland pipeline conveyance of Caspian Sea water deposited into the Aral Sea Basin; (ii) overland pipeline conveyance of Aral Sea brine deposited into the Caspian Sea and (iii) compensating overland pipeline importation of Black Sea water to the Caspian Sea. Effects of hydrological management of the Syr Darya and the Aral Sea's regulated Western Basin as well as a technical/economic model are offered. Three operational scenarios produce water levels stabilized at 28.9, 27.2 and 25.7 m, respectively. A macro-engineering solution is considered. Increased runoff from Syr Darya and other freshwater sources raises the Aral Sea level by 1.4–3.5 m. Depending on the scenario, the Aral

Sea returns to its 2005 elevation. Caspian Sea water importation, at flow rates of 14 $km^3$/year leads to Aral Sea stabilization at $\sim$32 m. Some considerations of the macro-project's ecological, cultural and social consequences are examined.

Chapter 17 by Roelof Dirk Schuiling and Viorel Badescu shows that the Aral Sea has shrunk and became a large salt-pan, because the freshwater from the two rivers that used to feed the lake (Amu Darya and Syr Darya) is almost entirely used for irrigation. In this chapter some possibilities to return to the original (1960) situation are studied. After discussing some of the alternatives, it is proposed to construct a canal along a more southerly route than the original Sibaral Canal, starting from the Zaisan Lake along the Irtysh River. This solution requires the expensive construction of a major tunnel through the Khrebet Tarbagataj mountain range. Thereafter, it will flow through the Balkash Lake, saving several hundred kilometers of canal construction, and discharge its water in the lower reaches of the Syr Darya. From here it will flow into Aral Sea, slowly restoring it towards its original (1960) level. Several flanking water-saving measures are considered. Most of the drive to restore the Aral Sea is for ecological reasons. There may also be a serious climatic threat to avoid, although this is a matter of debate. It is found that the river discharge from Siberia into the Arctic Ocean is on the increase, and this may affect the great world-ocean conveyor belt. This would have dire consequences for the climate in western and northern Europe. This could be avoided by diverting part of the water towards the Aral Sea. A restoration of the Aral Sea will have beneficial effects on climate, human health, fishery and ecology in general.

Chapter 18, by Hossein Golabian, starts by showing that accelerating demand for freshwater to be used in agriculture, industry and human consumption is a global phenomenon. Pressures on freshwater sources, i.e. rivers and underground reservoirs are causing gradual but unavoidable desiccation of the endoric water bodies. Many lakes have already dried up and been converted to anthropic deserts all over the world. Many of them are in remote Earth regions and forgotten, but some other are under news spotlight. The Aral Sea is the most famous case. The next one in this series of environmental disasters is Urumia Lake in the northwest of Iran, our world's second largest salty water lake and an internationally important wetland. Urumia Lake is located in a densely populated area and contains $\sim$8 billion MT of salt and other minerals which upon drying of the lake will become a gigantic "salt bomb" owning to wind deflation and will endanger the population in an surrounding area of >500 km radius. Its salt content must be contained and this goal cannot be achieved except that the lake is kept filled by enough water. There is no other reliable, accessible, and sustainable water resource other than Caspian Sea within some 300 km distance which can act as a helping elder sister to Urumia Lake. This chapter describes the idea of supplying severely needed water quantity for the stabilization of Urumia Lake by intentional transference from the nearby Caspian Sea.

Chapter 19 by Michael Welland deals with the sediment transported by wind and moving water in arid regions, focusing on the pioneering work of Ralph Bagnold. Ralph Alger Bagnold pursued two highly distinguished careers, one military and the other scientific. His early pioneering exploratory expeditions through the Eastern Sahara demonstrated the means by which motorized travel in such landscapes could be achieved and inspired his life-long love of our world's largest desert (contiguous area). His enduring scientific legacy lies in his groundbreaking work on sediment transport by both wind and flowing water, work that began in the 1930s and continued for the rest of his life. This work, and its subsequent refinement, continues to underpin the multidisciplinary design and evaluation of essentially all macro-engineering projects. Bagnold as a scientist is difficult to categorise. He described himself as an amateur who had never held an academic position "or had any professional status." But, as he wrote in his autobiography, he felt that this gave him "the rather unusual advantage of considering problems with an open mind, unbiased by traditional textbook ideas that had remained untested against [field] facts". He was, at heart, an engineer, but one with an acute and perceptive capacity to apply other disciplines—physics, mathematics, geology—to his identification and analysis of any problem. It was Bagnold's intuitive and radical interdisciplinary approach to sediment transport activated by wind and water, over the course of more than fifty years and fifty scientific papers, that enabled today's Earth scientists and mega-project engineers to plan and pursue research and projects equipped with a deep, if still imperfect, understanding of these critical natural processes. Bagnold's classic book, The Physics of Blown Sand and Desert Dunes, continues to hold the position of the second most-cited academic publication of any kind in the field of geomorphology. And modern textbooks, at least in the Earth Sciences, will contain wording such as uttered by M.R. Leeder: "In the absence of any modern readable text solely dedicated to the principles of sediment transport, a return to Bagnold's 1954 classic (for wind) and to his 1966 (for water) should provide the necessary inspiration."

Chapter 20, by Magnus Larsson, shows that by deliberate injection of dune sand with prepared cultures of the bacterium *Bacillus pasteurii*, an aerobic bacterium pervasive in natural soil deposits, they can solidify such granular accumulations into sandstone structures. A network of such infrastructures could mitigate the threat of desertification, while providing adaptive shelter spaces from the shifting sands. Or, as characterized by a July 2010 magazine article published in Popular Science (pages 48–49), "Ecotopia" might be established in the Sahara. Discontinuous dryland regions cover more than one-third of our Earth's land surface, and desertification—"the diminution or destruction of the biological potential of the land"—is certainly a major threat to human welfare on all continents, affecting more than 100 countries in the world. Some estimates suggest that 35% of Earth's land surface, as well as the livelihoods of ~850 million people, are at risk nowadays. More than 80% of Africa's drylands are moderately or severely desertified, a figure that equals more than one-third of all desertified land in the world. Sand dunes cover only about one-fifth of our existing deserts, but those extreme regions are good places to introduce a barrier of greenery in order to halt the shifting sands and stop the dunes from migrating. The idea of a 'Green Wall for the Sahara' was first proposed by a former President of Nigeria (Olusegun Obasanjo) in 2005. The initiative originally called for twenty-three African countries to cooperate in order to plant trees across a 7,000 km by 15-km shelter belt just south of the Sahara. The macro-project presented here would turn 6,000 km of sand into a pan-African sandstone city, a linear urbanization, and support the Great Green Wall for the Sahara and Sahel Initiative through a localised cementation of the desert sand via microbe-induced carbonate precipitation (MICP). The spatial pockets would help retain scarce freshwater and mineral resources, while also serving as habitable and programmable spaces for a nodal network converging with the planned Sahara Railway. The result would be a habitable wall-like infrastructure straddling latitudinally an entire continent, binding villages, people, and ecosystem-countries together. All design is fundamentally about aggregation and erosion. One adds and one takes away. The novel process of engineered architectural lithification, creating from a pile of loose sand a solid sedimentary rock structure, a sandstone building, effectively involves gluing one grain of sand to the next on a microscopic level. The building speaks of the chronology of the sand, the vast rhythms of Earthly geological history, the evolution of human villages and burgeoning cities covered in shimmering grains, forgotten in a sea of sand.

Chapter 21 by Viorel Badescu and Richard B. Cathcart proposes macro-engineering using tactical technologies that stabilize and promote the vegetation of barren near-coast sand dune fields with imported seawater. Extracted seawater that would otherwise, as commonly postulated, increase the Earth–ocean volume. Anthropogenic saturation of the ground with pumped seawater should fix widespread active sand dune fields in deserts (such as the westernmost Sahara). Seawater extraction from the ocean, and its deposition on dune sand, is made via solar-powered pipeline. Stabilization of one major erg in Mauritania is evaluated as a pre-implementation case study. The financial cost of the macro-project is estimated as a few billion USA\$—less than about 0.1% of the USA's 2010 gross domestic product (GDP). The initial investment may be between 0.6 and 1.1 billions USA\$.

Chapter 22 by Nicola M. Pugno, Richard B. Cathcart and Joseph J. Friedlander proposes the connection and maintenance of an artificially created Mediterranean Sea gulf. Dug, maintained and enlarged by nuclear-powered dredgers, the CATS (Chotts Algeria-Tunisia Scheme) could, eventually, result in the development of a North Africa-version of Europe's Mainport Rotterdam. A CATS-style harbor is a sheltered part of seawater deep enough to provide anchorage for ships or a place of refuge. If built, the CATS could offer shipping some protection from both long- and short-period Mediterranean Sea waves, easy safe access to the Mediterranean Sea in all types of weather, adequate depth and maneuvering room within the CATS harbor, shelter from storm winds and minimal navigational channel maintenance dredging. Of course, the environmental impact of pollutants generated by routine shipping operations in the CATS harbor is unavoidable. The authors propose to control the Mediterranean Sea's future level rise, limiting it to just a 1-meter rise within the Basin, with a low-cost barrier at Gibraltar Strait. Simultaneously, protective infrastructure investments at Venice, Italy, and other historic locales could be stopped and funds may be diverted or partially re-directed to realization of the CATS, especially because Europe now views North Africa as a site for energy generation that can be transmitted to Europe via undersea electric cables. The authors outline, in significant economic, geographical and social details, the entire macro-engineering development process for the CATS. The history of proposals for the chotts development, including nuclear excavation, is also discussed.

Chapter 23 by Ragab A. Hafiez reveals that the Qattara Depression, in Egypt's Western Desert, is the largest land depression—located entirely below the mean global sea level—in the Eastern Sahara. In this chapter, the available SRTM-90 meter digital elevation data have been used to delineate the real periphery of the depression, to generate an accurate contour map, and to develop a new terrain model of the volume and surface area for each level of the entire Qattara Depression. Regardless of the physical and chemical factors that may affect the contained volume of imported water, at each contour level the volume is found to be increased in an ascending, but irregular, order from $<1.0\ km^3$ at the dent's floor to more than 190 $km^3$ near the top. The dominance of evaporites near the bottom, and fractured limestone at the top, of the depression are, also, governing factors of the arbitrary mechanically transported seawater volume. The role of such resulted inland seawater lake due to the seawater transportation in changing the local climate is expected to be considerable by the increase of relative humidity and decrease of ambient temperature. The change of the region's climate, from a hyper-arid to an arid or semi-arid climate, is expected and desirable.

Chapter 24 by Pandora Hope, Andrew B. Watkins and Robert L. Backway elucidates that drought and the growing demand for freshwater are increasing problems in many regions of the world. The extent of the regions and the number of people affected by freshwater stress is projected to increase in the twenty-first century. One proposed macro-engineered solution is the extensive flooding of key locations with seawater, with the aim to increase downwind rainfall. This chapter examines the theoretical impact, and hence potential benefit to an potential agricultural region, of such an inundation macro-project; namely the seawater flooding of Lake Eyre in central Australia and the impact this may have upon the Murray Darling Basin (MDB), which lies eastwards of Lake Eyre and is often termed Australia's 'food basket'. The first method of analysis uses periods when Lake Eyre has been naturally flooded with freshwater (it is an ephemeral lake, and often has a surface of salt crust) as a proxy for an artificial seawater inundation. An estimate of the monthly time-series of Lake Eyre water volume was created and compared with MDB rainfall. Results suggest, at best, only a small contemporaneous correlation, explaining about 10% of the variance. When Lake Eyre volume leads MDB rainfall, correlations drop close to zero. This simple analysis suggests the climate of the MDB is unlikely to be driven in any significant fashion by flooding at Lake Eyre. The second method to assess the impact of such inundation uses climate models to artificially flood Lake Eyre, and then to compare the response with identical simulations using a dry lake. Climate models for this purpose range from very high resolution, meso-scale weather models that simulate only a small region but include many surface processes, to global models that capture all the interactions with the broad-scale Earth-atmosphere circulation. In between are regional models, which can cover a broader region of interest and draw upon information provided at their boundaries. Studies using both mesoscale and regional models indicate that enhanced localized air cooling is evident in the region of an artificially flooded lake. The rainfall response however, is less clear, with regions further afield, such as the MDB relative to Lake Eyre, showing no consistent climate response. As computers become faster and computer time cheaper, higher resolution studies can be performed. Such experiments may become routine for all proposed macro-engineering projects which involve inundating large regions infused with the very human hope of changing the climate favorably 'down-wind'.

Chapter 25 by Viorel Badescu, Richard B. Cathcart, Marius Paulescu, Paul Gravila and Alexander A. Bolonkin proposes a macro-engineering project which exploits technologies that have the potential to quickly enliven the arid region surrounding Lake Eyre. The plan is focused on biosaline agriculture. The distinctive macro-project components are: Lake Eyre is gradually filled to a higher level by controlling evaporation and by pumping seawater from the nearby Southern Ocean using cheap tensioned textile tubes. Most of the necessary energy to power the pumps and other devices could be produced by photovoltaics, in a very attractive application without electricity storage requirements. Eventually, Lake Eyre is to be lidded with a floating impermeable plastic cap and/or with buoyant hollow plastic balls that reduce evaporation. Filled with seawater, Lake Eyre can be used as a reservoir for irrigation in biosaline agriculture. Elementary mathematical models for the proposed tasks and rough macro-economical estimations are presented. Based on overall results, one concluded that the macro-project is feasible with existing twenty-first century technologies and could become profitable in a few decades time. But beyond paying off the capital investment, the greatest implementation benefit is the prospect of creating beneficial conditions for human settlement in the low-fertility soils of the Lake Eyre Basin. Additional research directions are briefly presented.

Chapter 26, written by Peter Flynn and Jason Songjian Zhou, shows that down-welling ocean currents move carbon into the deep ocean and play a major role in controlling the content level of atmospheric carbon. The formation of North Atlantic Deep Water (NADW) also releases heat to the atmosphere, which is a major contributor to a mild climate in Europe. Modification of down-welling seawater currents by either an increase in carbon concentration or an increase in volume is a possible response to the increase in anthropogenic carbon in our planet's atmosphere and to the possible weakening of the NADW. Seven possible methods of modifying down-welling currents were screened, including using existing industrial techniques for exchange of heat between seawater and the superior air. Increasing carbon gas concentration in down-welling currents is not practical due to the existing high degree of saturation of high latitude surface seawater. Two of the methods for increasing the volume of down-welling seawater currents were found to be impractical, and four were too expensive to warrant further consideration. Formation of thicker annual sea ice by pumping seawater onto the surface of ice sheets is the least expensive of the seven methods for increasing the volume of down-welling currents. One Sverdrup of incremental flow ($10^6$ $m^3$ $s^{-1}$) would have an estimated capital cost of USA$45G and an annual operating and maintenance cost of $1.3G. This macro-engineering method could be employed to avoid a repeat of a mini Ice Age in Europe associated with interruption in the flow of NADW.

Chapter 27 by Roger H. Charlier shows that, broadly-speaking, implementation remains hesitant in the field of harnessing ocean energies. Some research is still in process. Ocean sources of energy can and should be put to work; they are non-polluting, and minimally environment impacting. Their extraction is impeded less by technology than be the onerous capital intensiveness of the necessary investment. Tides, waves, marine winds, have been tapped, other possibilities remain currently more macro-engineers' dreams; they include marine currents, seawater salinity differential, conversion of biological products, marine geothermal energy. Ocean thermal energy conversion is being used in some modest way, but economically it remains unattractive. Before 2010, more than twenty five countries were involved in ocean renewable energy technology development activities. Lack of targeted national priorities and policies remains a major developmental and structural barrier. Offshore marine wind turbines dot today many landscapes, even if environment-related objections are raised (bird kills, weakened air movement down-wind). Wind tapping for electricity generation is still a "young" industry and surrounded by unresolved financial and technical macro-problems. OTEC uses the difference of temperature prevailing between different ocean seawaters layers to produce electrical power. Following the 1973 Oil Crisis, there was a renewed flurry of public interest for OTEC facilities. An OTEC system requires a temperature difference of at least 20–25°C to operate, limiting, in principle, its use to tropical regions. The closed-circuit system is far more ecologically benign than the open-circuit system. Often discussed, ocean currents hold immense potential. However, they have come repeatedly under severe criticism by concerned ecologists and dedicated environmentalists. Harnessing the Gulf Stream near the State of Florida, or placing a barrage within the English Channel, is viewed as unconceivable macro-projects. The ocean currents are driven by wind and solar heating of the seawaters near the Earth's equator, though some ocean currents result from density and salinity variations of water. Harnessing marine currents' energy could be achieved by using submerged water turbines similar to wind turbines that would have rotor blades, a generator for converting the rotational energy into electricity, and a means of transporting the electric current to shore to connect with the nearest electrical grid.

Chapter 28 by Steven Czitrom, Iván Penié Rodríguez and Guadalupe de la Lanza Espino shows that the buffering capacity and low hydrologic dynamics in seacoast water bodies, not necessarily blue-water lagoons, are characteristics that favor a high economic and biological productivity, also rendering them vulnerable to the accumulation of sediments and exogenous substances. The ecological rehabilitation of these bodies of naturally variable salinity water requires decreasing the input of pollutants and increasing the hydrologic dynamics and biodiversity.

With these process objectives in mind, a proposal to install a SIBEO (their Spanish acronym for Wave Energy-Driven Seawater Pump) in each of two nominated coastal lagoons and one seaport in Mexico is made, as the starting point for a national eco-coastal management macro-plan based on the use of renewable energy derived from natural ocean sea-waves. The analyses carried out show that the SIBEO can help diminish the eutrophication process in the coastal zone where it is applied, thus favoring a biodiversity recovery and increased fisheries at relatively low economic costs.

Chapter 29 by Stephen Salter, Joao Cruz, Jorge Lucas and Remy Pascal shows that the electrical energy needed to drive desalination plants can be produced from renewable energy sources if their erratic energy outputs can be made to match the strict specifications of an electrical network (grid). However, energy from sea-waves can also be used directly, using the vapor-compression desalination process. This inherently provides the correct impedance match to seawater wave inputs and does not suffer from wild swings of power outputs. This chapter extends previous work by the authors, by the inclusion of numerical predictions and proposals for a device ocean mooring design.

Chapter 30 by Alexander A. Bolonkin, Shmuel Neumann and Joseph J. Friedlander refers to a unique method for the extraction of freshwater (desalination of seawater) by using solar energy. It entails the following: (1) installation of the desalination system occurs on the sea and, as such, does not require expensive coastal land surface; (2) Desalinization uses solar energy which produces freshwater at a cost close to zero; (3) the device is made from thin film which is far less costly than the conventional reverse osmotic or multi-flash plant materials; (4) ecologically sound, the desalinization device uses sea-waves for water and wetted air circulation; (5) highly transportable, the desalination device may be delivered anywhere by vessels through available shipping sea-lanes; (6) preassembled or disassembled, the desalination device is very compact and easily stored in conventional storehouses; (7) the desalination device also functions as a rain-water collector; (8) environmentally friendly, the desalination system does not dump brine, which is hazardous to some marine life, into the seawater; (9) a portable, inexpensive version of the seawater distiller is available to help travelers produce freshwater when it may be urgently needed.

Chapter 31 by William J. Merrell, Lyssa Graham Reynolds, Andres Cardenas, Joshua R. Gunn and Amie J. Hufton conveys an appreciation of the devastation caused by the 2009 Hurricane Ike and its massive storm surge has forced the Houston/Galveston region to consider means of significantly suppressing storm surge near the Texas shoreline. Due to the significant concentrations of population and the multitude of nationally important industries (such as petroleum refineries) in the region, as well as the difficulties in highway evacuation, attention has focused on the possible construction of a coastal barrier—the Ike Dike—that would help protect the entire Galveston Bay region from hurricane-induced storm surge. Although the devastation caused by Hurricane Ike was the catalyst for action, the realization that other hurricanes could potentially have a much higher surge and the fact that the region is hit by a major hurricane about every 15 years are the factors driving the Ike Dike macro-project concept. The Ike Dike concept envisions a coastal spine composed of the existing Galveston Seawall, extensions of the protection afforded by the Seawall to west Galveston Island and the Bolivar Peninsula and storm surge gates at Bolivar Roads and San Luis Pass.

Chapter 32 by Subba Rao V. Durvasula explains that the Gulf of Khambhat is a high tidal energy environment on the western coast of India and is located in an intensely active seismic zone. The Gulf receives annually 38 $km^3$ freshwater and $74 \times 10^9$ kg of sediment from 12 tributary rivers. The State of Gujarat currently suffers from an acute shortage of freshwater and hydroelectric power, and to alleviate these macro-problems a planned mega-engineering project, 'Kalpasar', is outlined. ["Kalpasar" originates from the Hindu mythological "Kalpa Vriksha" (wishing tree) and means a freshwater lake that "fulfills all human wishes".] The mega-project plan includes damming the Gulf and its tripartite division into a) freshwater reservoir, b) a brackish water tidal basin, and c) the open ocean. A highly stressed environment the Gulf of Khambhat acts as a sink to nutrients from agricultural runoff, and for industrial pollutants and its carrying capacity is in excess of its limit, causing resource conflicts. The proposed tripartite division would have strong environmental impact on the fluvial ecosystem as well as on the inter-basin estuarine and marine ecosystems and may lead to overall deterioration of the Gulf environment. Based on a review of similar mega-engineering projects elsewhere, a comprehensive baseline survey over multi-site, multi-year is recommended to provide a better understanding of the structure, functioning and response and recovery of the Gulf ecosystem to perturbations. A follow up by a thorough evaluation of the impending environmental impacts, similar to those addressed by a Before/After and Control/Impact (BACI model) is emphasized. Other recommendations include construction of several small dams on the various inland rivers to relieve the freshwater shortage, to consider alternative sources of energy, to provide assurance to the public and to develop integrated preparedness systems for management of disasters.

Chapter 33 by Richard B. Cathcart, Alexander A. Bolonkin and Radu D. Rugescu refers to a Bering Strait macro-project. Since great quantities of soil organic carbon previously retained by Arctic permafrost is becoming increasingly unstable because of Arctic Ocean warming, a Bering Strait Seawater Deflector has been devised to stall or even reverse further sea-ice melting. Sea-ice removal by regional climate change, of course, causes the onset and advancement of solar-driven warming of the Arctic Ocean and defrosting of Arctic Ocean Basin permafrost. A sudden and deliberate twenty-first century closure of the Bering Strait is likely to cause an abrupt decrease of the mean temperature of the seawater in the adjacent Arctic Ocean, thus helping to re-establish a greater-than-today sea-ice cover floating on the Arctic Ocean. P. Borisov (1901–1973) was the first to elucidate a costly macro-engineering concept of a concrete dike-like barrier placed within the Bering Strait. Cathcart, Bolonkin and Rugescu, instead, offer a low-cost screen—they call it a textile curtain—separating the Bering Sea from the Arctic Ocean. Their purpose is to significantly reduce the amount of heat in North Pacific Ocean seawater currently entering the Arctic Ocean in order to, ultimately, instigate ground re-freezing of some Arctic Basin permafrost. Successful re-freezing of permafrost would, thus, slow or possibly halt the twenty-first century release of massive quantities of methane locked in that warming soil type because methane in twenty-one times more potent than carbon dioxide gas in warming the Earth-atmosphere.

The principal audience for this book consists of researchers (engineers, physicists, biologists) involved or interested in macro-engineering. The book may become useful for industry developers interested in joining national or international geo-engineering programs. Finally, it may be used for undergraduate, postgraduate and doctoral teaching in faculties of engineering and natural sciences.

The Editors

# Acknowledgments

A critical part of writing any book is the review process, and the authors and editors are very much obliged to the following researchers who patiently helped them read through subsequent chapters and who made valuable suggestions: Prof. Salah Abdelel-Moghith (Desert Research Center, Cairo, Egypt), Prof. Najib Abou Karaki (University of Jordan, Amman, Jordan), Dr. Naser Agh (Urmia University, Iran), Dr. Claudia Agraz (University of Campeche, México), Prof. Salah M. Aliwi Al-Abbasi (Iraqi Cultural Attache', Berlin, Germany), Prof. Gheorghe Baran (Polytechnic University of Bucharest, Romania), Dr. Mariana Bernardino (Technical University of Lisbon, Portugal), Mr. Richard B. Cathcart (Geographos, Burbank, CA, USA), Prof. Stefano Ciurli (University of Bologna, Italy), Dr. Justin Costelloe (University of Melbourne, Australia), Prof. Jason DeJong (University of California, Davis, CA, USA), Dr. Ravi V. Durvasula (University of New Mexico, Albuquerque, NM, USA), Prof. Götz Ebhardt (Rossdorf, Germany), Prof. Billy Edge (North Carolina State University, USA), Prof. David Evans (Bristol University, UK), Dr. Charles Finkl (Coastal Planning & Engineering, Boca Raton, FL, USA), Mr. Joseph Friedlander (Shave Shomron, Israel), Dr. Peter F. Friend (Cambridge University, UK), Dr. Peter Gluck (UPC-Romania, Cluj, Romania), Prof. S.M. Golabi (University of Tabriz, Iran), Prof. Steven Hardy (University of Nebraska, USA), Dr. Robert Harriss (Houston Advanced Research Center, TX, USA), Dr. J. Marvin Herndon (Transdyne Corporation, San Diego, CA, USA), Dr. Christopher Kilburn (University College London, UK), Dr. Boris Kruglyak (Brooklin, NY, USA), Dr. Mark Krinker (New York, USA), Dr. Amit Kumar (University of Alberta, Edmonton, Canada), Dr. Michael Lipsett (University of Alberta, Edmonton, Canada), Dr. Silvio Guido Marinone (Centro de Investigación Científica y de Educación Superior, de Ensenada, Ensenada, Baja California, Mexico), Prof. Victor A. Mazur (Odessa State Academy of Refrigeration, Ukraine), Prof. Norimi Mizutani (Nagoya University, Japan), Prof. Alexandru Morega (Polytechnic University of Bucharest, Romania), Dr. Jörg F.W. Negendank (German Research Centre for Geosciences, Potsdam, Germny), Prof. Shmuel Neumann (Michlala Jerusalem College, Israel), Dr. Gheorghe Oaie (National Institute of Marine Geology and Geo-Ecology,

Bucharest, Romania), Dr. Fanel-Viorel Panaitescu (Constanta Maritime University, Romania), Prof. Mariana Panaitescu (Constanta Maritime University, Romania), Prof. Lars Rahm (University of Linköping, Sweden), Prof. Eugen Rusu (Galati University Dunarea de Jos, Romania), Prof. Elias Salameh (University of Jordan, Amman, Jordan), Dr. Brian Sadler (Water Policy Services, Australia), Dr. Kafia Mawlood Shareef (Hawler Medical University, Arbil, Iraq), Prof. Anne K. Soper (German University of Technology in Oman, Muscat, Sultanate of Oman), Prof. Ioan Strat (Galati University Dunarea de Jos, Romania), Dr. James E. Stewart (Bedford Institute of Oceanography, Dartmouth, Canada), Prof. Vatche P. Tchakerian (Texas A&M University, USA), Prof. Huib P. van Heel (Delft Technical University, the Netherlands), Prof. Claudio Vita-Finzi (University College London, UK), Prof. Erik Wiren (Royal Institute of Technology, Stockholm, Sweden), Prof. Donald Zeigler (Old Dominion University, Norfolk, VA, USA).

The editors, furthermore, owe a debt of gratitude to all authors. Collaborating with these stimulating colleagues has been a privilege and a very satisfying experience.

# Contents

# List of Contributors

**Abdulkader M. Abed**, Department of Geology, University of Jordan, Amman, 11942, Jordan, e-mail: aabed@ju.edu.jo

**Marc Acheroy**, Royal Military Academy, Avenue de la Renaissance, 30, Brussels, 1000, Belgium, e-mail: Marc.Acheroy@elec.rma.ac.be

**Robert L. Backway**, Commodore of the Lake Eyre Yacht Club, Belgrave Heights, Victoria, Australia, e-mail: bob_backway@hotmail.com

**Viorel Badescu**, Candida Oancea Institute, Polytechnic University of Bucharest, Spl. Independentei 313, Bucharest, 060042, Romania, e-mail: badescu@theta.termo.pub.ro

**Geoffrey N. Bailey**, Department of Archaeology, King's Manor, University of York, York, YO1 7EP, UK, e-mail: gb502@york.ac.uk

**Alexander A. Bolonkin**, CR & C, 1310 Avenue R, #6-F, Brooklyn, 11229, NY, USA, e-mail: aBolonkin@juno.com; e-mail: aBolonkin@gmail.com

**Stefano Carlino**, Istituto Nazionale di Geofisica e Vulcanologia, Sezione di Napoli Osservatorio Vesuviano, Via Diocleziano 328, Napoli, 80124, Italy, e-mail: stefano.carlino@ov.ingv.it

**Andres Cardenas**, Department of Marine Sciences, Texas A&M University at Galveston, Galveston, P.O. Box 1675, 77553, TX, USA, e-mail: cardenasa@tamug.edu

**Richard B. Cathcart**, Geographos, 1300 West Olive Avenue, Suite M, Burbank, 91506-2225, CA, USA, e-mail: rbcathcart@gmail.com

**Mircea Dimitrie Cazacu**, str. Cpt. Aviator Nicolae Drossu nr. 11, Bucuresti sector 1, cod postal, 012071, Romania, e-mail: cazacumircea@yahoo.com

**Roger H. Charlier**, Faculty of Sciences, Vrije Universiteit Brussel-VUB 2 Pleinlaan, Brussels, 1050, Belgium, e-mail: rocharli@vub.ac.be; e-mail: rogerhcharlier@yahoo.com, Faculty of Florida Atlantic University, Boca Raton, USA

**Damien Closson**, Royal Military Academy, Avenue de la Renaissance, 30, Brussels, 1000, Belgium, e-mail: damien.closson@yahoo.fr

**Joao M. B. P. Cruz**, GL Garrad Hassan, Rua Nova do Almada, N 59, 2nd Floor, Lisboa, 1200-288, Portugal, e-mail: joao.Cruz@gl-garradhassan.com

**Elena Cubellis**, Istituto Nazionale di Geofisica e Vulcanologia, Sezione di Napoli Osservatorio Vesuviano, Via Diocleziano, Napoli, 328-80124, Italy, e-mail: elena.cubellis@ov.ingv.it

**Steven Czitrom Baus**, Instituto de Ciencias del Mar y Limnología, Unidad Académica de Geologa Marina y Ambiental, Circuito Exterior S/N, Ciudad Universitaria, UNAM, México D.F., 04510, Mexico, e-mail: stevenczitrom@gmail.com

**Guadalupe De laLanza Espino**, Instituto de Biología, Circuito Exterior S/N, Ciudad Universitaria, UNAM, México D.F., 04510, Mexico, e-mail: gdlle@servidor.unam.mx

**Ilia Delizia**, Dipartimento di Storia dell'Architettura e Restauro, Università di Napoli Federico II, Via Monteoliveto 3, Napoli, 80134, Italy, e-mail: ilia.delizia@unina.it

**Subba Rao V. Durvasula**, 12 The Horseshoe, Manor Park, Dartmouth, Nova Scotia, B2Y 4E5 902-463-7804, Canada, e-mail: seshu35@gmail.com

**Peter Flynn**, Department of Mechanical Engineering, University of Alberta, Edmonton, AB, T6H 5K7, Canada, e-mail: peter.flynn@ualberta.ca

**Joseph J. Friedlander**, P.O. Box 12, Shavei Shomron, 44858, Israel, e-mail: jjfriedlander@gmail.com; e-mail: shahrazad@geo.tu-darmstadt.de

**Shahrazad Abu Ghazleh**, Institute for Applied Geosciences Physical Geology and Global Cycles, Darmstadt University of Technology, Schnittspahnstr. 9, Darmstadt, 64287, Germany

**Lyssa Graham Reynolds**, Green Light Multimedia, Inc., P.O. Box 2039, Galveston, 77553, TX, USA, e-mail: Lyssa@LyssaGraham.com

**Paul Gravila**, Physics Department, West University of Timisoara, V. Parvan Str., Nr. 4, Timisoara, 300223, Romania, e-mail: gravila@physics.uvt.ro

**Hossein Golabian**, Center of Tehran University Architects, #141 Shariati Street, Tehran, Iran, e-mail: hossein@golabian.com

**Joshua R. Gunn**, Department of Marine Sciences, Texas A&M University at Galveston, P.O. Box 1675, Galveston, 77553, TX, USA, e-mail: joshuagunn1151@yahoo.com

**Ragab A. Hafiez**, Al-Mokattam, PO Box 3, Cairo, 11571, Egypt, e-mail: ragabr@gmail.com

**François Halgand**, Tractebel Engineering S.A, Le Delage, rue du 19 mars 1962, 5, Gennevilliers Cedex, 92622, France, e-mail: francois.halgand@gdfsuez.com

**Frédéric Hallot**, Royal Military Academy, Avenue de la Renaissance, 30, Brussels, 1000, Belgium, e-mail: frederic.hallot@rma.ac.be

**Jaapchan C. Hanekamp**, CEO, HAN, Zoetermeer, The Netherlands, e-mail: hjaap@xs4all.nl

**Herbert Hansen**, Keyobs SA, Zoning Industriel des Hauts Sarts, Rue d'Abooz, 31, Herstal, 4040, Belgium, e-mail: hhansen@keyobs.com

**Takahide Honda**, Hydraulic and Environmental Engineering Research Section, Civil Engineering Research Institute Technology Center, Taisei Corporation, 344-1 Nase-cho, Totsuka-ku, Yokohama, 245-0051, Japan, e-mail: hndtkh01@pub.taisei.co.jp

**Pandora Hope**, Centre for Australian Weather and Climate Research, Australian Bureau of Meteorology, GPO Box 1289, Melbourne, 3001, Victoria, Australia, e-mail: p.hope@bom.gov.au

**Amie J. Hufton**, Department of Marine Sciences, Texas A&M University at Galveston, P.O. Box 1675, Galveston, 77553, TX, USA, e-mail: amie.hufton@gmail.com

**Raducu Viorel Iancu**, str. Ceahlau, nr. 19, bloc 73, scara 5, apart. 69, Bucuresti, sector 6, 060374, Romania, e-mail: radu.iancu@czone.ro; e-mail: raducu.iancu@gmail.com

**Kazunori Ito**, Hydraulic and Environmental Engineering Research Section, Civil Engineering Research Institute Technology Center, Taisei Corporation, 344-1 Nase-cho, Totsuka-ku, Yokohama, 245-0051, Japan, e-mail: kazunori.ito@sakura.taisei.co.jp

**Stephan Kempe**, Institute for Applied GeosciencesPhysical, Geology and Global Cycles, Darmstadt University of Technology, Schnittspahnstr. 9, Darmstadt, 64287, Germany, e-mail: kempe@geo.tu-darmstadt.de

**Geoffrey C. P. King**, Institut de Physique du Globe de Paris, 1 rue Jussieu, 75005, Paris cedex 05, France, e-mail: king@ipgp.fr

**Magnus Larsson**, 192 Hoxton Street, London, N15LH, UK, e-mail: magnus@magnuslarsson.com

**Bengt Liljebladh**, Department of Earth Sciences, University of Gothenburg, Box 460, Gothenburg, 40530, Sweden, e-mail: beli@gvc.gu.se

**Jorge A. A. Lucas**, Institute of Energy Systems, School of Engineering and Electronics, University of Edinburgh, Mayfield Road, Edinburgh, EH93JL, Scotland, e-mail: J.Lucas@ed.ac.uk

**Giuseppe Luongo**, Dipartimento di Scienze della Terra, Università di Napoli Federico II, Largo San Marcellino 10, Napoli, 80123, Italy, e-mail: giuseppe.luongo@unina.it

**Dan Aurel Machita**, str. Hrisovului nr. 18, bl. A12, sc. B, et. 8, ap 73, Bucuresti, sector 1, Romania, e-mail: mediu2@mt.ro; e-mail: flota@mt.ro

**William Merrell**, Department of Marine Sciences, Texas A&M University at Galveston, P.O. Box 1675, Galveston, 77553, TX, USA, e-mail: merrellw@tamug.edu

**Nada Milisavljevic**, Royal Military Academy, Avenue de la Renaissance, 30, Brussels, 1000, Belgium, e-mail: nada@elec.rma.ac.be

**Salah A. Naman**, College of Education, Department of Chemistry, University of Duhok, Kurdistan Region, Duhok, Iraq, e-mail: salah.naman@yahoo.com

**Shmuel Neumann**, Michlala Jerusalem College, Jerusalem, Israel, e-mail: drneumann@netvision.net.il; e-mail: domemountain@gmail.com

**Sergiu Nicolaie**, Head Department of Efficiency in Conversion and Consumption of Energy from National Institute for Research & Development in Electrical Engineering ICPE –CA, Splaiul Unirii, Nr. 313, District 3, 030138, Bucharesti, Romania, e-mail: ecosergio@icpe-ca.ro

**Yukinobu Oda**, Hydraulic and Environmental Engineering Research Section, Civil Engineering Research Institute Technology Center, Taisei Corporation, 344-1 Nase-cho, Totsuka-ku, Yokohama, 245-0051, Japan, e-mail: od-ykn00@pub.taisei.co.jp

**Pernilla Ouis**, Faculty of Health and Society, Malmö University, Malmö, 205 06, Sweden, e-mail: Pernilla.Ouis@mah.se

**Piet van Overveld**, Volker Wessels, Woerden, The Netherlands, e-mail: pvanoverveld@vhb-vsce.nl

**Remy C. R. Pascal**, Institute of Energy Systems, School of Engineering and Electronics, University of Edinburgh, Mayfield Road, Edinburgh, EH93JL, Scotland, e-mail: v1rpasca@staffmail.ed.ac.uk

**Nicola M. Pugno**, Department of Structural Engineering and Geotechnics, Politecnico di Torino, Laboratory of Bio-Inspired Nanomechanics "Giuseppe Maria Pugno", Corso Duca degli Abruzzi 24, Torino, 10129, Italy, e-mail: nicola.pugno@polito.it

**Marius Paulescu**, Physics Department, West University of Timisoara, V. Parvan Str., Nr. 4, Timisoara, 300223, Romania, e-mail: marius@physics.uvt.ro

**Iván Penié Rodríguez**, Instituto de Ciencias del Mar y Limnología, Unidad Académica de Geología Marina y Ambiental, Circuito Exterior S/N, Ciudad Universitaria, UNAM, México D.F., 04510, Mexico, e-mail: ivan.penie @gmail.com

**Radu D. Rugescu**, University "Politehnica" of Bucharest, Splaiul Independentei 313 sector 6, Bucharest, 060042, Romania, e-mail: rugescu@yahoo.com

**Eugen Rusu**, University Dunarea de Jos of Galati, 55 Domneasca. Str., Galati, 800008, Romania, e-mail: erusu@ugal.ro

**Stephen H. Salter**, Institute of Energy Systems, School of Engineering and Electronics, University of Edinburgh, Mayfield Road, Edinburgh, EH93JL, Scotland, e-mail: S.Salter@ed.ac.uk

**Roelof D. Schuiling**, Faculty of Geosciences, Utrecht University, 80.021, Utrecht, 3508 TA, The Netherlands, e-mail: schuiling@geo.uu.nl

**Jihan Seoud**, Environmental Consultant UNDP, Beirut, Lebanon, e-mail: jihan.seoud@undp.org

**Anders Stigebrandt**, Department of Earth Sciences, University of Gothenburg, Box 460, Gothenburg, 40530, Sweden, e-mail: anst@gvc.gu.se

**I. Engin Ture**, UNIDO-ICHET, Sabri Ulker Sk. 38/4 Cevizlibag, Istanbul, Turkey, e-mail: enginture@halic.edu.tr

**T. Nejat Veziroglu**, UNIDO-ICHET, Sabri Ulker Sk. 38/4 Cevizlibag, Istanbul, Turkey, e-mail: veziroglu@iahe.org

**Andrew B. Watkins**, National Climate Centre Australian Bureau of Meteorology, GPO Box 1289, Melbourne, Victoria, 3001, Australia, e-mail: a.watkins@bom.gov.au

**Michael Welland**, Orogen Limited, 175 Southwark Bridge Road, London, SE1 0ED, UK, e-mail: mw@throughthesandglass.com

**Solomon Yim**, School of Civil and Construction Engineering, Oregon State University, Corvallis, OR, 97331, USA, e-mail: solomon.yim@oregonstate.edu

**Jason Songjian Zhou**, Weatherford Canada, 2801 84 Ave., Edmonton, AB T6P 1K1, Canada, e-mail: songjian.zhou@ca.weatherford.com

# Living with Sea Level Change and Dynamic Landscapes: An Archaeological Perspective

**Geoffrey N. Bailey and Geoffrey C. P. King**

## 1 Introduction

For most of human existence on this planet over the past 2 million years, sea level has been substantially lower than the present and has swung through changes of more than 100 m in response to the glacial–interglacial climatic cycle. At a time when modern society is increasingly concerned about the potentially destructive impact over the coming decades of a sea-level rise of 3 m or so, it is sobering to realize that prehistoric societies across the world faced a sea-level rise between about 16,000 and 6000 years ago of 130 m (Fig. 1). That change of course was spread over many human generations and many millennia, so that the full effects would not have been experienced within a single human lifetime. Nevertheless, the long-term cumulative effect of sea level rise and loss of territory would have been dramatic. On a world scale, substantial areas of continental shelf were successively exposed, creating potentially attractive territories for human settlement and migration and land connections between major land masses, and then removed again by sea level rise (Fig. 2). In Europe, during the last glacial period, the total land mass of the continent was extended by as much as 40% at the maximum marine regression (Fig. 3), with a corresponding loss of land when sea levels rose as the continental glaciers melted into the oceans. In some parts of the sea-level cycle, and especially in regions where the slope of the continental shelf is shallow, the effects would have been noticeable and sometimes dramatic within the lifetimes and memories of the people affected. Moreover, these changes have taken place repeatedly over the long Pleistocene history of human existence.

G. N. Bailey (✉)
University of York, York, UK
e-mail: gb502@york.ac.uk

G. C. P. King
Institut de Physique du Globe, Paris, France

V. Badescu and R. B. Cathcart (eds.), *Macro-engineering Seawater in Unique Environments*, Environmental Science and Engineering,
DOI: 10.1007/978-3-642-14779-1_1, © Springer-Verlag Berlin Heidelberg 2011

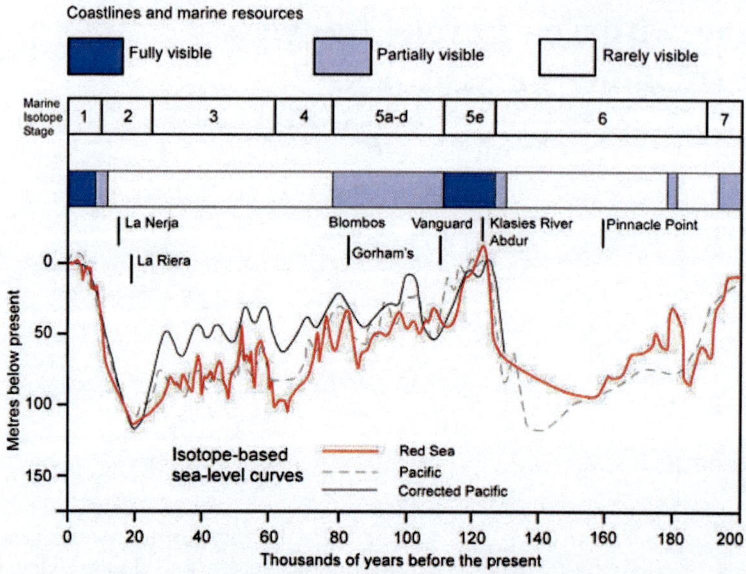


**Fig. 1** Smoothed curve of sea-level change over the past 200,000 years, showing amplitude of sea-level change and impact on the visibility of palaeoshorelines and archaeological sites with evidence of coastal and marine exploitation. Site names refer to coastal sites in Africa and Mediterranean Europe, mostly caves sites, with long stratified sequences of archaeological material, associated with steeply shelving coastlines or periods of high sea level, and showing some evidence of marine food remains (shells of marine molluscs and marine vertebrates). Similar cycles of sea-level change repeated at similar intervals back to about 0.8 million years, with long periods of low sea level punctuated by short episodes of high sea level as high as or higher than today. Before that the amplitude of sea-level change appears from the deep-sea isotope record to have been less, but sea levels lower than present were the norm back to at least 2 million years ago. Sea-level data from Lambeck and Chappell (2001), Shackleton (1987), Siddall et al. (2003), Van Andel (1989), Waelbroeck et al. (2002). For information on archaeological sites see Bailey et al. (2008a, b), © G. Bailey

Sea level change is only one component of an unstable land surface. Changes of the Earth's crust have been continuously remolding many of the landscapes in which human populations have made their living, whether through tectonic plate motions and rifting, or isostatic adjustments in response to varying loads of ice and water associated with sea level change and high latitude glaciation. Superimposed on these geological changes, of course, were the major climatic changes associated with the glacial cycle, sometimes involving dramatic changes on a time scale of decades and even years. In short, the world in which we have evolved as a species is highly unstable, geologically speaking, and will continue to be so.

What impact did these changes have on prehistoric societies, how did people cope with such changes in the deeper past, and what can we learn from this long-term archaeological perspective about the challenges which confront our own society and civilization? We cannot hope to cover the full range of environmental changes associated with human development in the space of a short chapter.

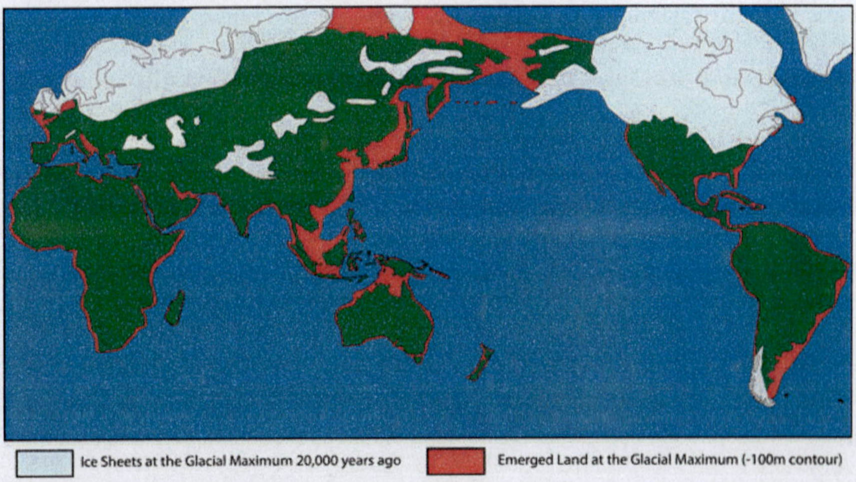


**Fig. 2** World map showing maximum extent of the continental sheets 20,000 years ago (*pale blue*), and the extent of new land created by a corresponding drop in sea level of 100 m (*red*)

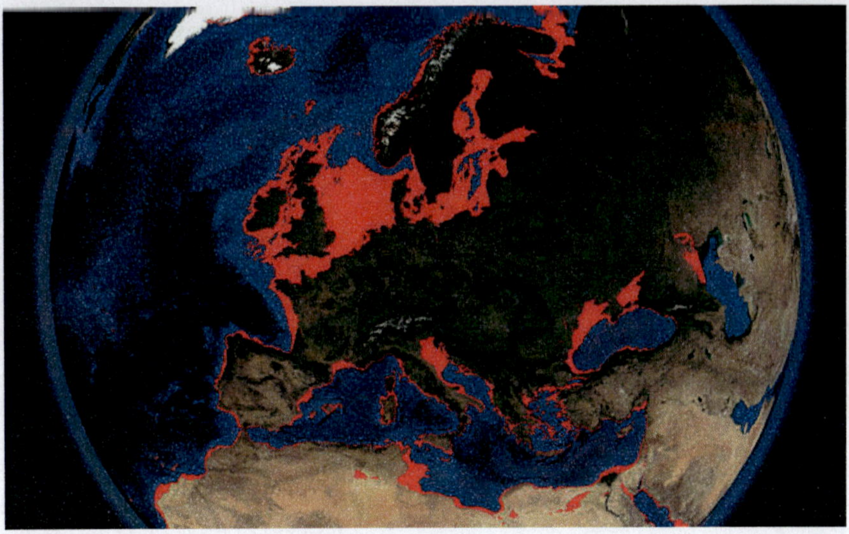

**Fig. 3** Map of Europe showing the extra land (in *red*) exposed by the maximum lowering of sea level at the last glacial maximum, 20,000 years ago. Picture supplied courtesy of Simon Fitch and Ben Geary, University of Birmingham, with data from USGS NED and ETOPO2

Long-term climate change in particular is a vast topic, and is well covered elsewhere (e.g., Elias 2006; Maslin and Christensen 2007), though we touch on climate factors to the extent that they interact with or are modified by geological changes. What we will concentrate on here are the major geological processes that have created instabilities in the human landscape, particularly those associated with active tectonics and sea level change, and the likely impacts both positive and negative that such changes have had on prehistoric societies. We will highlight two themes.

The first is the counter-intuitive proposition that regions of high geological instability, despite their potential hazards and high risks of destructive impact, have repeatedly created ecologically attractive conditions for human settlement, and have played a major selective role in our long-term evolution, dispersal and social development. The second is that very nearly all of the environmental changes that are happening around us today or are likely to happen in the future, including those resulting from human intervention, whether deliberate or unintended, have happened at some time in the past. The archaeological record of deep time provides a magnifying lens for observing more about the nature and causes of past changes in natural systems, for understanding something of their human significance and their impact on human society; at the same time it offers a sort of large-scale laboratory for observing their long-term consequences and disentangling the relative contributions of different processes operating on different time scales.

## 2 Active Tectonics

Tectonic processes such as plate motions, mountain uplift and basin submergence have generally been assumed to operate too slowly to be of relevance to the scale of human activity or human history. Over the past 25 years, however, detailed studies of earthquake activity, beginning with the El Asnam earthquake of 1980 in Algeria (King and Vita-Finzi 1981; King and Yielding 1984; Philip and Meghraoui 1983), and the Coalinga earthquake of 1983 in California (King et al. 1988; Stein and King 1984; Stein et al. 1988), have demonstrated the intimate relationship between earthquakes and changes in the surface morphology of the physical landscape. The fault displacement or folding episode caused by a single earthquake, or a short series of earthquakes repeating on the same fault zone, can modify topography over relatively short time spans ($10–10^3$ years), with the creation of localized barriers, disruption of surface drainage, and damming back of stream flows and sediment. Over longer time spans ($10^4–10^8$ years) the cumulative effect is major fault scarps and depressions, and ultimately large-scale basins and mountain ranges.

In the El Asnam case, the earthquake was one of the largest recorded in Algeria, and its immediate effects were widespread damage to buildings, the deaths of over 2000 people, and homelessness for a further 300,000 (Ambraseys 1981). The earthquake caused surface faulting that extended horizontally over a distance of 24 km, raised the local Ser el Maarouf ridge by 5 m, and by partially damming back the flow of the Chelif River, re-created a lake in the neighboring basin that had been recorded as a wetland environment on historical maps but had

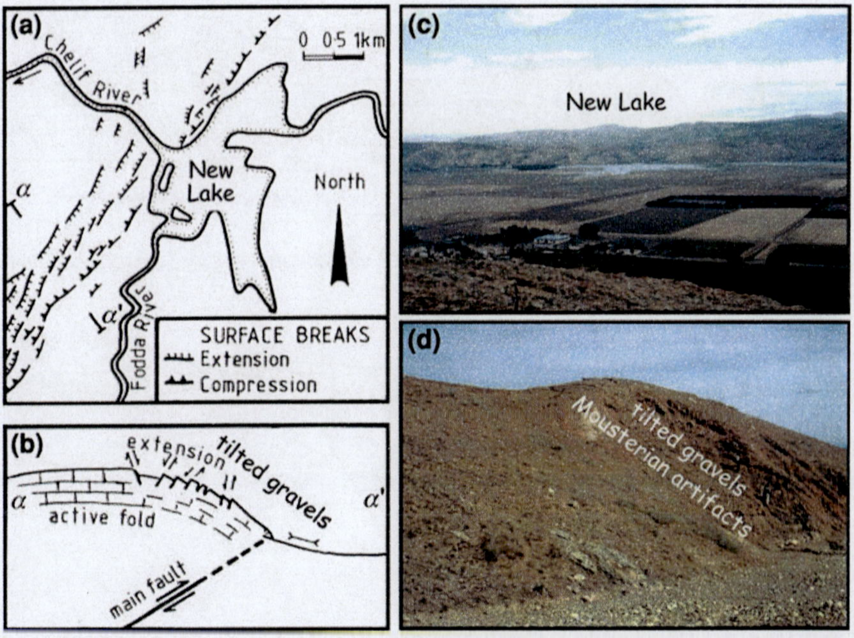


**Fig. 4** Map of the El Asnam area, showing zones of faulting, uplifted ridge and local lake basin, © G. Bailey

subsequently dried out (Fig. 4). Progressively tilted gravels making up the local ridge, the earliest with Mousterian artifacts dating back at least 30,000 years, indicated that the ridge had undergone progressive uplift as a result of repeating earthquakes on the same fault over a long period, each earthquake accentuating or rejuvenating a complex topography of barrier and basin. The short-term effect of the earthquake was highly destructive, but the longer-term effects in the wider landscape could be regarded as beneficial, creating a region of localized topographic complexity and a well-watered basin that was clearly attractive to the prehistoric hunter-gatherers of the region and would have offered ecological benefits to local populations at any period whatever their technology and mode of subsistence.

These effects are especially marked near plate boundaries. Where terrestrial plates are converging, the dominant large-scale effect is crustal contraction, uplift and mountain building, and the local effect is reverse faulting and folding of the surface topography (Fig. 5a). Where plates are diverging, the dominant large-scale effect is crustal stretching, subsidence and formation of marine basins with some uplift at the margins, and the local effect is normal faulting (Fig. 5b). Where plates are moving sideways with respect to each other (strike-slip), both effects may be observed locally (e.g., Bilham and King 1989). Earthquakes are active in all these examples and can be accompanied by volcanic activity in regions of crustal stretching, strike-slip or rifting. In some regions of the world, notably in the

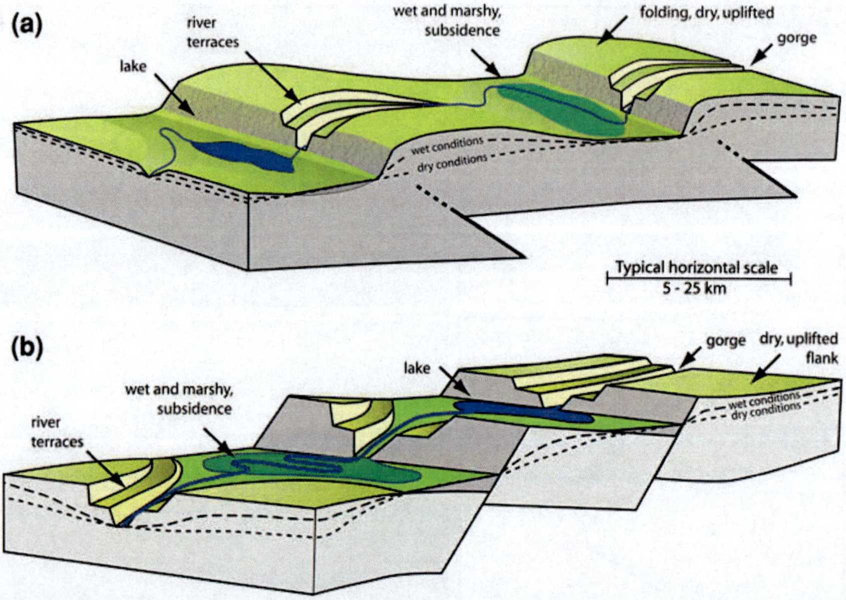


**Fig. 5** Schematic illustration of the effects of changing water table in a tectonically active landscape: **a** contractional (reverse-faulting); **b** extensional (normal-faulting) environments. In **a** the surface expression of faulting appears in the form of folds rather than faults that cut the surface. In **b** faults usually fully cut the surface. The *scale bar* is approximate. Both figures show how faulting and folding can trap water and sediments to sustain intermittent areas of well-watered and fertile conditions even when the water table changes because of climate change. In each figure two faults are shown that could host repeated earthquakes with Richter magnitudes between 6 and 7. Smaller events in the same region with magnitudes between 3 and 4 will occur 10,000 times more frequently. In regions of Greece or Western Turkey major events repeat every 200 years or less and create both major and minor features (e.g. Armijo et al. 1996). Secondary effects are important, creating deep river valleys and gorges and faulting that produces small private valleys or secluded bays at a range of scales. Greece and Western Turkey are very active, but tiny fractions of that rate can create and maintain features. Overall rates for the East African Rift are a factor of one hundred smaller, and less is known historically, but motion can still generate spectacular features within a human life time. Fault scarps tens of metres high can be created by repeated movement on such fault lines in less than 100 years, © G. Bailey

Eastern Mediterranean, overall convergence of major plates is mediated by smaller platelet motions, resulting in a complex interplay of extension, contraction and strike-slip (e.g. Flerit et al. 2004).

These processes can be rapid in human terms. The Gulf of Corinth, now nearly 1 km deep has been created within the past 1 Ma, well within the time span of human interest (Armijo et al. 1996). On shorter time scales the Gulf of Corinth has dropped 25 m since 500 BC, and the Byzantine chapel of Kenchreai was partially submerged by earthquake activity (Vita-Finzi and King 1985).

At first sight the main effect of such processes appears disruptive or destructive, especially at the smallest scale of the spatio-temporal spectrum, that of the individual earthquake or volcanic eruption. However, over longer time scales, the effects are potentially beneficial, creating a complex topography with ecological diversity, rejuvenation of local water supplies and fertile sediments. Uplifted barriers and ridges afford opportunities for hunter-gatherer populations dependent on the hunting of elusive prey animals to use the topography to tactical advantage, and similar advantages to pastoralist populations concerned to protect their livestock from theft or predation. This may explain the apparent paradox that many of the largest concentrations of archaeological sites and evidence of past human activity are concentrated in tectonically active zones close to plate boundaries or rift zones, whether we are talking about our earliest ancestors in the African Rift, the advanced Palaeolithic hunters of Mediterranean Europe, the earliest Old World civilizations, or the population centers of more recent times (Jackson 2006; King and Bailey 2006; Force 2008). As long as activity is maintained over the longer run of centuries, millennia, and longer, these advantageous features are constantly maintained or rejuvenated. Without repeated activity resulting from regional tectonic activity, the long-term effect would be erosion and smoothing of the land surface and a drop in the water table (Fig. 6). Without active tectonics there would be no topographic complexity, and without complex topography human evolution and social development might have taken a very different turn.

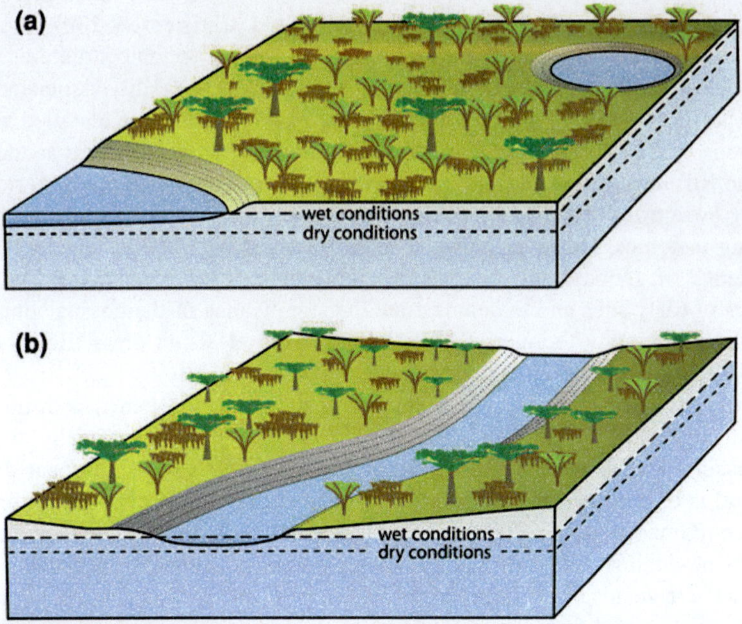


**Fig. 6** Schematic illustration of effects of changing water table in a flat landscape: **a** a region with water-holes and lakes; **b** a region with a river flowing through it. In these conditions, the water table is sensitive to climatically induced changes in precipitation, © G. Bailey

## 3 The African Rift and Human Evolution

According to current consensus, Africa is the cradle of early human evolution, with the earliest dated and most numerous finds of human fossils and stone-tool assemblages, documenting the transition from a tree-dwelling, plant-eating ape-like ancestor to omnivorous ground-dwelling humans. The first steps on this evolutionary trajectory were towards some form of bipedalism combined with continued tree-climbing abilities, and occurred at least 4.5 million years ago with *Ardipithecus ramidus*, evolving through the Australopithecines, and leading to the evolution of the genus *Homo* after 2.5 Ma. The emergence of *Homo ergaster* and *Homo erectus* ~1.8 million years ago saw a fully developed bipedal capacity to range widely over open terrain, increased brain capacity, the widespread use of stone tools, a greater dependency on animal protein whether by scavenging or hunting, and dispersal more widely within and beyond Africa (Cachel and Harris 1998, Delson et al. 2000, Klein 1999). The longest sequences, and the most complete record of successive biological and cultural assemblages occur in the East African Rift and in South Africa, and the evolution of anatomically modern humans (*H. sapiens sapiens*) also appears to have taken place in Africa some time after about 150,000 years ago both on grounds of early dated fossil finds and the phylogenetic mapping of modern DNA lineages (McDougall et al. 2005; Stringer and Andrews 1988; Torroni et al. 2006).

Climate change has most often been invoked as a significant external driver of evolutionary change (Potts 1996a, b; Maslin and Christensen 2007), but the association of early finds with one of the world's largest tectonic structures in the East African Rift also points to the impact of geological instability associated with active tectonics as a significant factor (Fig. 7). Until recently, detailed reconstructions of the original topography in such a highly active structure as the East African Rift have proved difficult to achieve, especially for the earliest periods of human evolution, because so much of the landscape has been transformed by ongoing tectonics. However, new techniques employing satellite imagery and a knowledge of geophysical processes are now being used to map the landscape settings of early sites and to demonstrate the significance of their association with complex topography (King and Bailey 2006, King et al., in press; Bailey et al. 2010; Bailey and King 2010).

In the rift setting, progressive widening of the rift is associated with uplift of the rift flanks and down dropping of the rift floor to produce internally draining lake basins. The active part of the rift at any given time is usually near the rift axis and is only a few kilometers wide, though it may move over longer periods of time (Grandin et al. 2009). This is accompanied by repeated earthquake activity producing near-vertical fault escarpments and volcanic lava flows, both of which create localized barriers to easy movement by fast-moving predators. Water supplies and fertile sediments are sustained and rejuvenated by repeated movement on fault zones, leading to ecological richness and habitat diversity. Moreover this process can maintain favorable and stable conditions of food and

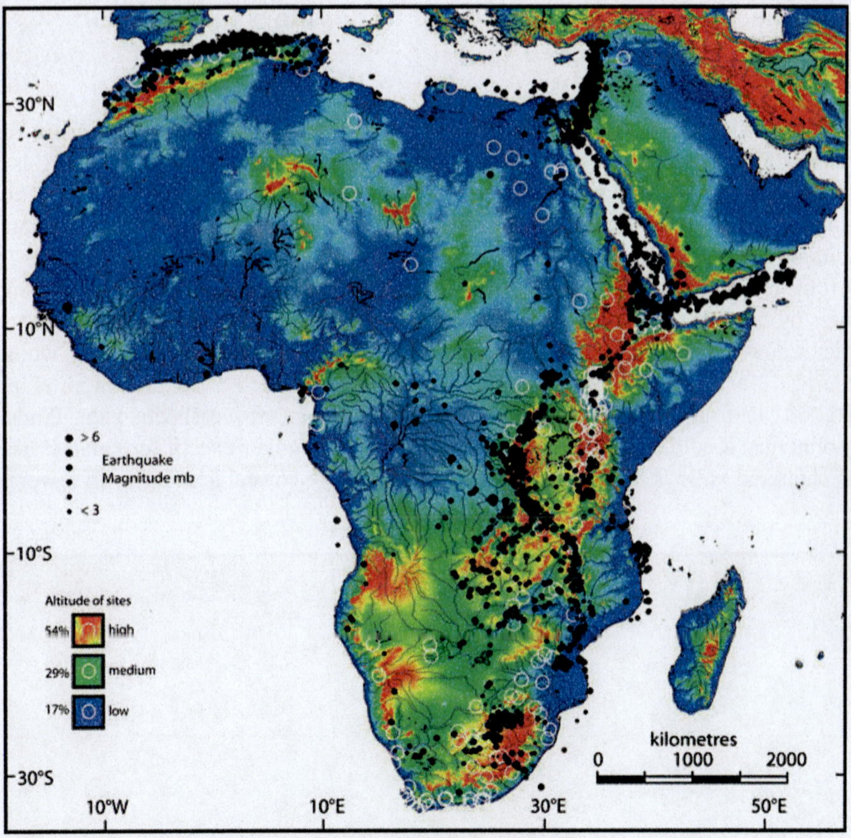


**Fig. 7** Map of Africa showing distribution of fossil and archaeological sites (*white circles*), earthquake activity (*black circles*) and topography, © G. Bailey

water supplies independently of climate change, offering a measure of buffering against climatic aridification. The combination of localized barriers and partial enclosures created and sustained by active faults and lava flows forms a complex topography which an intelligent hominin can use to tactical advantage to avoid predators, find safety for vulnerable young, and improve access to mobile or elusive prey animals. These features provide a powerful set of external selection pressure favoring the human evolutionary trajectory towards the development of an unspecialized omnivore with a significant dietary component of meat to feed an enlarged brain, a requirement for abundant water supplies, an extended period of infant dependency, and locomotory adaptations to the negotiation of broken topography (King and Bailey 2006). In South Africa, where many early finds have also been found, the tectonic style is different but tectonic activity is present and produces many of the same topographic features present in the Rift (Bailey et al. 2010).

## 4 Palaeolithic Hunters, Tectonic Landscapes and Erosion in Northwest Greece

Further insight into the role of a dynamically active landscape comes from the archaeological record of the Epirus region of Northwest Greece, where there is a sequence of human occupation extending over at least 100,000 years in a landscape that has undergone repeated physical modification because of tectonics and climate change (King and Bailey 1985; Bailey et al. 1993; Bailey 1997). Archaeological deposits comprise open air sites and rock shelters used as camp sites by hunters moving over large territories between the coastal lowlands and the mountainous hinterland (Fig. 8). Most of the archaeological evidence falls within the last glacial period (110,000–10,000 years ago), which at its maximum about 20,000 years ago saw the establishment of a permanent ice sheet in the Pindus mountains, a colder and drier climate with little tree cover except for small stands in sheltered valleys, and the creation of an extensive coastal lowland with lowered

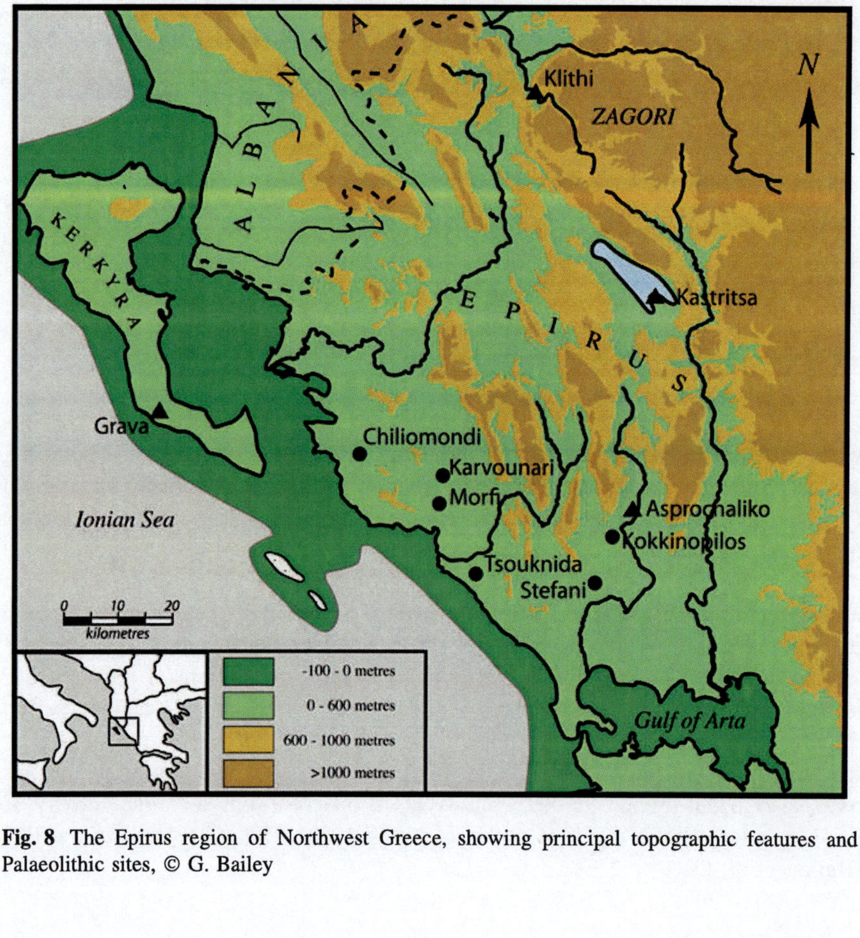


**Fig. 8** The Epirus region of Northwest Greece, showing principal topographic features and Palaeolithic sites, © G. Bailey

sea level. Large herds of red deer (*Cervus elaphus*), wild ass (*Equus hydruntinus*), wild cattle (*Bos primigenius*), ibex (*Capra ibex*), and chamois (*Rupicapra rupicapra*) formed the main supplies of animal food for the Palaeolithic hunters.

The region is prone to earthquakes and the dominant tectonic style is compression and uplift, with some strike-slip and normal faulting because of the interaction of smaller plates in the eastern Mediterranean region. The surface expression of these geophysical processes is a complex topography with Northwest–Southeast trending mountain ranges alternating with broad basins, some, such as the Ioannina basin, with long-standing lakes. Some of these topographic features have persisted and been accentuated throughout the past 1 million years and more, notably the Ioannina lake basin and its impressive surrounding limestone ridges. However, other parts of the landscape have been transformed by long-term tectonic deformation, and areas that were formerly basins with accumulations of fluvial and lacustrine sediments have now been uplifted locally and transformed into barren badlands with slopes stripped bare of vegetation by soil erosion, most famously the red beds of Kokkinopilos, where the uplift and erosion of sediments originally formed in a well-watered basin have exposed some of the earliest stone age artifacts of the region.

The dominant bedrock geology is limestone, and this bedrock-type tends to support the most attractive grazing for herd animals. There are also substantial areas of a younger, softer rock, known as flysch, comprising laminated siliceous sandstones and siltstones, which produce soils more quickly than the limestone but with lower fertility and more susceptibility to erosion. Long-term tectonic processes of compression have uplifted and disturbed the younger flysch, so that it tends to form extensive areas of eroded topography on the flanks of the limestone ridges (Fig. 9). The result is extensive tracts of inaccessible country, which even

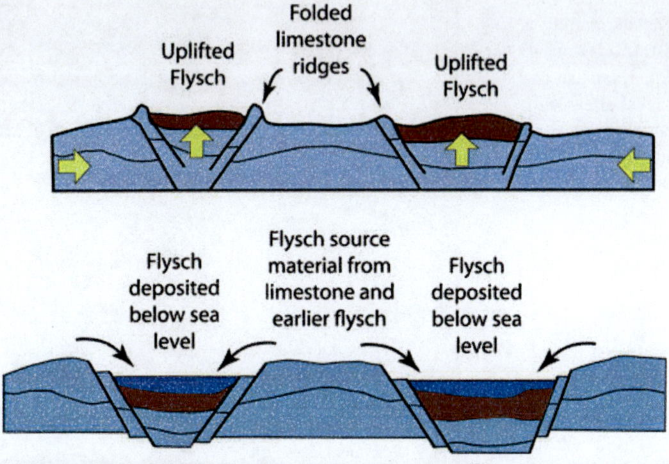


**Fig. 9** Block diagram showing the effects of tectonic compression on the configuration of limestone and flysch terrain. *Lower figure* shows earlier phase of deposition, upper figure shows a later phase after compression, © G. Bailey

today have little productive capacity and low population densities, and in the prehistoric past would have formed barriers to the movement of animals, circumscribing basins and valleys of more attractive terrain.

It is clear from the distribution of Palaeolithic sites close to the entry and exit points of these major grazing basins that prehistoric hunters were capable of using these topographic features as a means of monitoring and controlling animal movements on a regional scale (Fig. 10). Locally, sites are located on or close to fault zones and are well placed to take advantage of local fertility and local barriers to trap animals during the course of their migratory movements (Fig. 11).

Much of the erosion that creates badlands landscapes on the flysch geology and more localized areas of erosion on the limestone is typical of the eroded landscapes visible more widely in the Mediterranean (Van der Leeuw 1998; Hordern and Purcell 2000; Grove and Rackham 2001), and has usually been attributed to over exploitation in recent millennia, because of overgrazing by goats, cutting down of trees for timber and firewood, and overexploitation by mechanized agriculture.

However, if we consider the archaeological time scale, it is clear that erosion has been going on in the Epirus landscape long before the agricultural and technological developments of recent millennia. During glacial periods erosion took

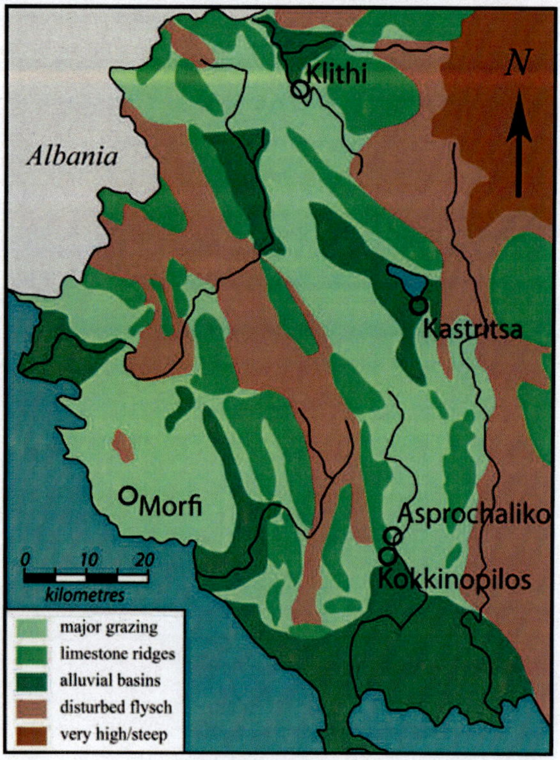


**Fig. 10** Regional map showing the relationship between major grazing basins, geological barriers and archaeological site distributions. Kokkinopilos and Morfi are open air sites with Mousterian and Upper Palaeolithic artefacts spanning at least the past 150,000 years. Asprochaliko, Kastritsa and Klithi are rockshelters with stratified sequences that together span most of the Last Glacial period from 100,000 years to 10,000 years ago, © G. Bailey

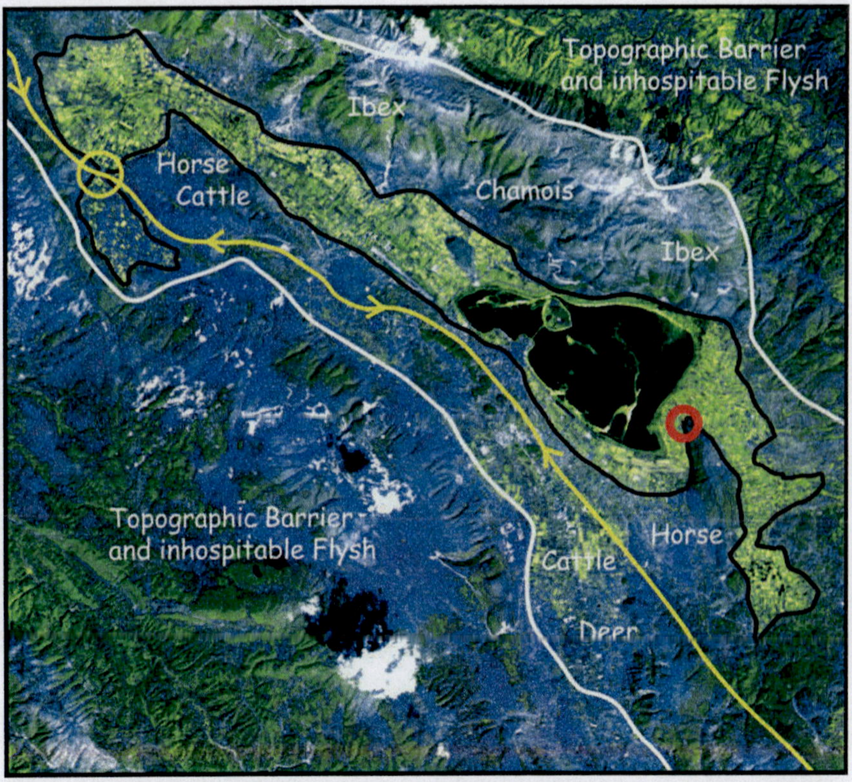


**Fig. 11** Local topographic features in the vicinity of the Palaeolithic rockshelter of Kastritsa. This site was used repeatedly as a hunting base from at least 26,000 years ago to 12,000 years ago, a period spanning climatic conditions of maximum cold and aridity at the Last Glacial Maximum. The site is well placed to monitor the movements of animals in the region without disturbing them, while also combining the local topographic features of limestone ridges and the lake edge to divert the animals into a topographic cul-de-sac where they could easily be trapped and killed, © G. Bailey

place on a massive scale, creating huge fans of sediment at the foot of the higher mountains, and the accumulation of thick alluvial deposits in river and lake basins (Macklin et al. 1997). This erosion far outweighs the supposed impact of recent human intervention and most of it took place long before the appearance of domestic animals or the increasing demands of agricultural and urban populations.

Undoubtedly some of the erosion has been caused by recent land-use practices, but there are at least two other processes at work in this landscape, operating on different time scales. On a time scale of tens of thousands to hundreds of thousands of years, repeated cycles of cold and dry glacial climates periodically removed most of the tree cover and accelerated the breakdown of soil and bedrock through freeze–thaw effects. On a time scale of millions to tens of millions of years, the Epirus landscape has been subjected to progressive tectonic compression and uplift, in which offshore sediments created by earlier cycles of erosion have been compacted and uplifted to produce the hard-rock geology we see today. Thus the underlying tectonic instability has made the land surface especially susceptible to disturbance, whether triggered by earthquakes, climatic effects or human intervention (King et al. 1997).

The humanly-induced erosion of recent millennia, far from appearing to be an exceptional effect, turns out to be a relatively minor continuation of processes that have been operating in this landscape over a much longer period of time and at a much larger scale. Since erosion has a much longer history than the domestic goat, it becomes hard to pin all the blame on the latter. On the contrary, from this longer-term perspective, it seems more likely that goat husbandry, so far from being the cause of erosion, represents a successful adaptation to a chronically degraded landscape that was in existence long before human settlement and cannot be made productive for human benefit in any other way. Whereas on the shorter time scale of the recent historical period, it appears that goats cause erosion, on the longer time scale of the Pleistocene the roles of cause and effect are reversed, so that it would more appropriate to say that erosion 'causes' or 'selects for' goats (Green and King 1996; Bailey 2007).

Moreover, if we expand the spatial scale of observation, it is clear that erosion in one place results in the accumulation of sediment somewhere else. The massive Pleistocene fans of sediment that form low hills at the foot of the more prominent mountain ridges are often the focus of modern village settlements because of their attractive soils and water supplies. In a complex topography, then, erosion may have a more generally beneficial effect in bringing together soil that is thinly distributed over hill and mountain slopes, and concentrating it in basins and lowland river valleys and coastal plains, where it provides some of the most important agricultural land for the modern economy. Thus erosion, which seemed at a local scale to be largely negative, turns out at a larger spatial scale to be positively beneficial. Such observations have serious implications for modern conservation policies and technological interventions (Van der Leeuw 1998; Van der Leeuw and Redman 2002).

## 5 Sea Level Change and Coastal Prehistory

We return to the issue of sea-level change, with which we began. During the past decade, archaeologists have become more acutely aware that the periodic drop in sea level during glacial periods exposed large tracts of coastal territory that are now submerged, and which probably represented some of the most attractive territory for prehistoric populations. This was already apparent in the Epirus example discussed above (see Fig. 8). However, it is only very recently that archaeologists have begun to recognize that the surviving archaeological record on dry land is most probably a severely truncated fragment of the original picture, that some of the most important evidence for the earliest developments in human prehistory may be missing because of submerged of earlier landscapes by sea-level change, and that there are realistic possibilities of acquiring new information from this hidden world (Flemming 2004; Bailey et al. 2008a, b; Gaffney et al. 2009; Benjamin et al. 2011, in press).

Coastal territories are often more attractive than their hinterland counterparts, because of better groundwater supplies, more fertile sediments, more equable climates, greater ecological diversity, and the addition of marine resources at the shore edge. They are often tectonically active, especially at plate boundaries where oceanic material is colliding with the margin of a continental plate and being subducted beneath it, and are subject to varying process of long-term uplift or subsidence, resulting both from tectonic and isostatic movements (Inman 1983). Even when sea level is relatively stable, reworking of material at the coast edge by erosion and accumulation of sediments creates an ecologically and geologically dynamic environment. It follows that coastlines are likely to pose many of the same sorts of characteristics of geological instability combined with topographic complexity and ecological attractiveness that are typical of tectonically active regions. Moreover, some of the largest concentrations of population and the largest settlements are likely to have been concentrated in coastal regions in the deeper prehistoric past, as they are today.

However, the impact of Pleistocene sea level change has also been to remove such evidence for most of human prehistory. Occasionally, ancient shorelines that have been uplifted by isostatic rebound in high latitudes, or by subduction and tectonic uplift at plate boundaries, are preserved above sea level and give some insight into the coastal adaptations of much earlier human societies (Bailey and Flemming 2008). Deeply stratified coastal caves also give some indication of coastal activities in short-lived periods of earlier high sea level (see Fig. 1). However, these conditions are exceptional, and most of the evidence is now submerged and will require underwater exploration. Despite some early speculations to the contrary (Sauer 1962), it has usually been assumed by archaeologists from the general absence of evidence, that a serious human interest in coastlines and marine resources only began some time after about 7000 years ago, because it is from that time on that we see abundant evidence in the archaeological record of shell mounds, fishing gear, the use of boats, and intensive exploitation of shellfish, fish and sea mammals—indeed a worldwide explosion of such evidence. However, it now seems more likely that this apparent pattern simply reflects the fact that sea level stabilized at about the present position from about 7000 years ago onwards, and that earlier expressions of human interest in maritime activities and marine resources have a much deeper history that has been lost beneath the ocean. The fact that Australia and New Guinea are now known to have been first colonized by human populations 50,000 years ago (Hiscock 2008), requiring sea journeys over distances of at least 60 km, is a very significant indicator in support of such a proposition.

It now seems likely that if we are to obtain a fuller picture of early patterns of human migration and dispersal, the extinction of the Neanderthals and their replacement by modern humans spreading out of Africa, early developments in seafaring, fishing and the exploitation of marine resources, the earliest dispersal of agriculture and even the earliest roots of development of the great Old World civilizations, and of course the social impact of sea level change and past human response to it, we will have to look for evidence underwater. All these processes, representing some of the most significant developments in human prehistory, took place when sea levels were lower than present.

The steady accumulation of chance archaeological finds demonstrates that archaeological settlements and whole cultural landscapes can survive inundation by rising sea levels, often with excellent conditions of preservation. New technologies, stimulated by industrial developments on the sea bed, are beginning to offer the prospect of realistic underwater exploration where previously this was regarded as impractical, and new collaborations are under way to develop systematic programs of research (SPLASHCOS 2010). Moreover, these hidden underwater archives of evidence do not simply have a bearing on archaeological issues of human development but include geological and sedimentary records that are likely to refine very considerably our understanding of the processes of long-term sea-level change and the regional and local expression of climate change as it affected these now submerged coastal territories. We are on the threshold of a new phase of investigation that is likely to transform, in the coming decades, our knowledge and understanding of all these interlinked processes. At present, there is a relatively limited base of evidence, but we describe below two examples that show how this new field of research is likely to impact on a broader understanding.

## 6 The Red Sea and the Arabian Peninsula

The Red Sea region and the Arabian Peninsula (Fig. 12) are currently at the centre of interest about the likely pattern of dispersal of early human populations out of Africa and their adaptations, particularly with regard to the possible role of sea crossings and marine resources in facilitating early patterns of movement and habitat expansion (Petraglia and Rose 2009). It has generally been assumed that the most likely pathway, particularly for the earliest expansion of archaic human populations out of Africa at about 1.8 Ma, but also for the dispersal of anatomically modern humans after about 150 ka, was via the Nile Valley and the Sinai Peninsula into the Near East, mainly on the grounds that this is the only dry land route and that a pathway across the southern end of the Red Sea would have been blocked by a sea crossing of some 30 km and by predominantly arid landscapes in the Arabian Peninsula. More recent discussions and new work have challenged this long-standing assumption, suggesting that a land connection might have existed across the southern end of the Red Sea when sea levels were very low, or that early inhabitants of the region might have already been exploiting marine foods and developing simple craft for crossing sea barriers, perhaps providing an impetus to dispersal out of Africa and around the rim of the Indian Ocean, at least in the case

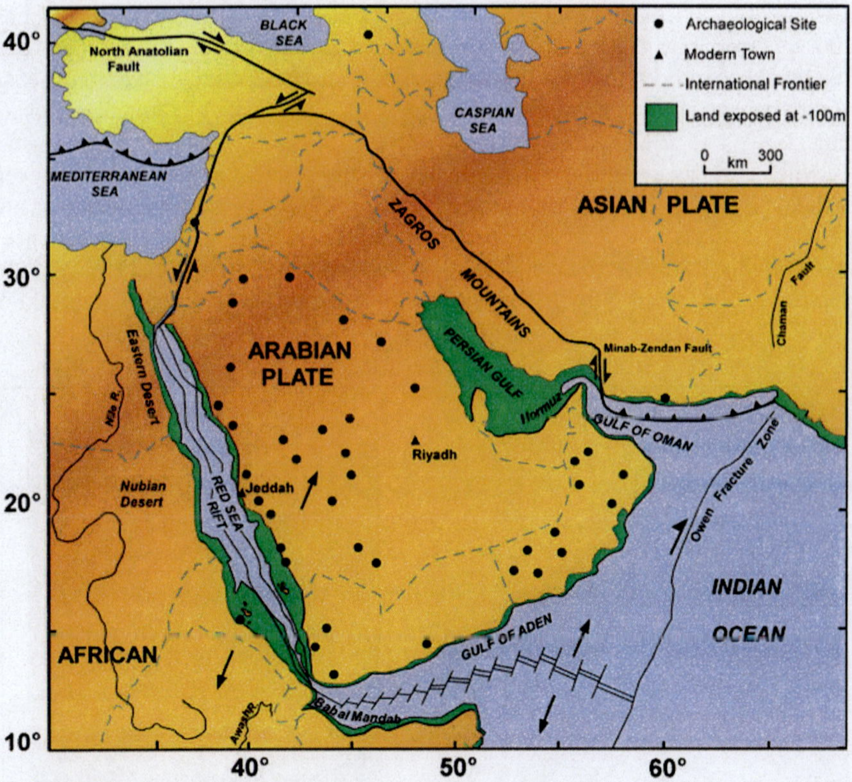


**Fig. 12** General map of the Red Sea and the Arabian Peninsula. Archaeological sites are mostly surface finds of Early Stone Age and Middle Stone Age tools that could cover a range of dates from >1 million years to <40,000 years. When sea level was very low, the southern end of the Red Sea would have been reduced to a narrow channel running through extensive coastal lowland territory, and the Persian Gulf would also have been a well-watered lowland basin, © G. Bailey

of anatomically modern humans some time after 150,000 years ago (Stringer 2000; Walter et al. 2000).

Abundant finds of stone tools demonstrate that there has been a human presence, at least intermittently in the southern Arabian Peninsula, extending far back into the prehistoric past, probably as far back as 1 million years and perhaps earlier. Moreover, some of these finds are in regions that today are too dry for habitation, which implies a wetter climate than today (Parker 2009). Detailed analysis of the isotope composition of marine organisms in deep sea cores with sediments extending back over the past 350,000 years (Siddall et al. 2003) demonstrates that at no time in that period, even when sea level dropped to its lowest level at glacial maxima, was the Red Sea cut off from the Indian Ocean at its shallow southern end. Had that happened, evaporation within a closed basin would have taken isotope values to much more extreme values than are in fact recorded. Independent measurements of the position of the palaeoshoreline, taking account of local isostatic adjustments of the crust and tectonic effects (Bailey et al. 2007), corroborate this pattern, showing that at glacial maxima the southern Red Sea would have been reduced to a long, narrow channel extending over a distance of 100 km and no more than a few kilometers wide, but without permanent closure (Fig. 13).

Whether this narrow channel would have been sufficient to deter crossing, even without seafaring skills, is open to question. It is clear that a dry crossing was not possible, at least not within the past half million years, but the short distances involved during quite a long period of the sea-level cycle would have been relatively easy to cross with simple rafts or even by swimming (Bailey 2009).

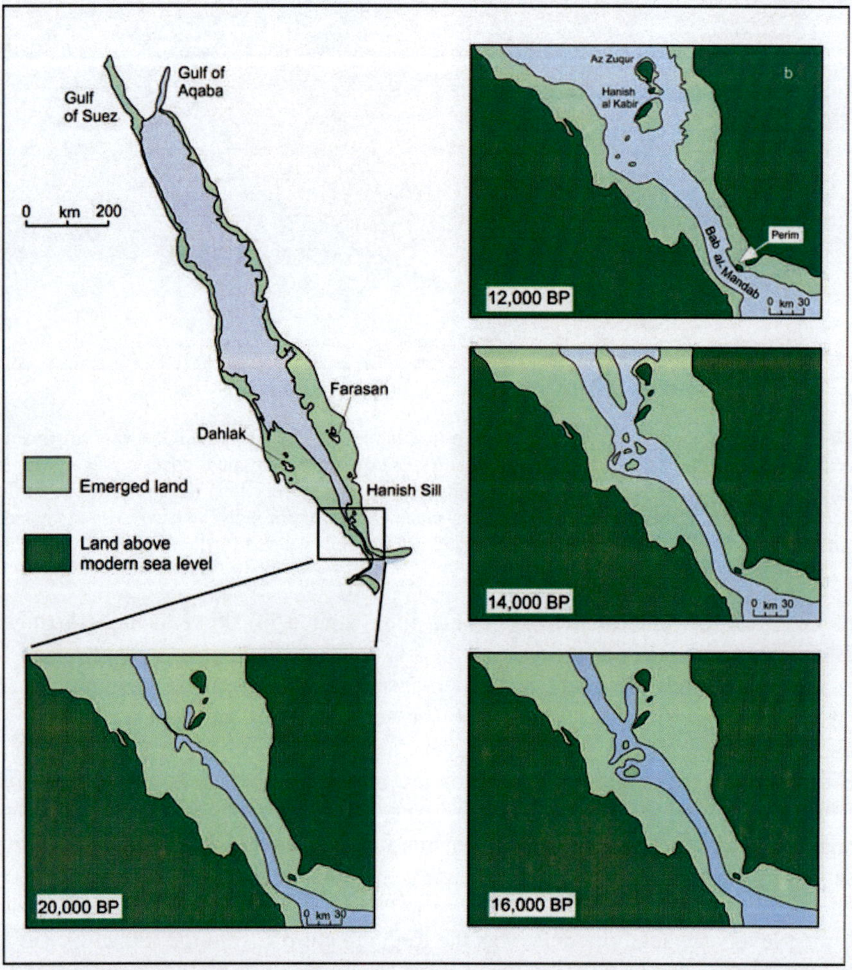


**Fig. 13** Palaeoshorelines at the southern end of the Red Sea at different dates in the glacial sea-level cycle. Data supplied courtesy of Kurt Lambeck, © G. Bailey

More relevant to the question of human habitation in the region is the extensive areas of now submerged landscape, as much as 100 km wide, that would have been available during periods of low sea level at the southern end of the Red Sea.

Studies of climate change suggest that wetter climates in the region tended to coincide with periods when sea level was close to or only slightly lower than the present level. This suggests that the greater ease of crossing the Red Sea when sea levels were low would have been offset by a generally arid landscape. However, preliminary investigations of the submerged shelf region indicate a surprisingly complex topography, due to the effect of salt tectonics, with uplift of evaporite deposits accompanied locally by salt withdrawal to form deep depressions, as well as to more widespread processes of rift propagation. The result is a landscape with localized barriers and basins that would have been very familiar to human populations adapted to conditions in the African Rift, with all the advantages for human settlement described earlier. Some of this exposed region would most likely have been covered in linear sand dunes accumulated from deposits picked up by wind action from the exposed sea floor, as is the case in parts of the modern coastal plain (Munro and Wilkinson 2007). But this was most likely only one element in a complex mosaic of surface conditions. In such topography we would expect to find local basins of greater fertility with traps for sediment and water, and it has been hypothesized that ground water movements associated with a drop in sea level would have created better-watered conditions than is the case today (Faure et al. 2002). This would have been of great significance to human settlement, offering an environmental refugium during periods when climatic conditions are likely to have been at their most arid in the adjacent hinterlands.

Conversely, the fertility of the marine environment in this region is at its greatest when sea level is high and relatively stable, as is the case in present-day conditions, with vast quantities of marine molluscs in shallow bays and productive inshore fisheries. These have clearly been exploited throughout the period of modern sea level over the past 6000 years, as is attested by the numerous large shell mounds representing the material expression of former settlements on the Red Sea coastline and its offshore islands. When sea level dropped, particularly to very low levels, restriction of marine inflow from the Indian Ocean would have resulted in increased salinities, injurious to some marine species and possibly high enough to suppress plankton production in some parts of the Red Sea (Siddall et al. 2003). We cannot be sure that similar shell mounds do not exist on the palaeoshorelines that are now submerged, and underwater explorations are under way to search for their remains. But, rapidly shifting shorelines during a period of sea-level change may have inhibited the establishment of large shell beds on the scale that can be observed in the modern environment.

Thus, the complex cycle of changes in sea-level conditions, climate and local productivity on land and at sea would have created a different balance of advantages and disadvantages for human populations established in the region. Any human populations with the flexibility to switch between terrestrial and marine resources, and between hinterland and coastal territories, would have been able to maintain a permanent presence in the region over long periods in the face of the many shifts in local conditions. Hence this region could have played a key bridging role in facilitating population movements and cultural contact between Africa and southern Asia from the very earliest periods of human development, rather than representing a barren and little visited cul-de-sac that acted as a barrier to such interchange.

## 7 Sinking Coastlines in Northern Europe

One of the regions of Europe that was more dramatically affected than any other by the sea-level rise at the end of the last glacial period was the Northwest region around the Baltic and the North Sea. The relatively shallow bathymetry of these basins resulted in a very extensive coastal landscape during periods of low sea level (Fig. 14), with correspondingly dramatic changes in palaeogeographic configuration when sea level rose. The melting of the ice sheets that sat over northern Britain and Scandinavia also resulted in isostatic rebound of the Earth's crust, amounting to more than 100 m in northern Norway, and submergence around the southern rim of the North Sea and the Baltic, creating a complex interplay between

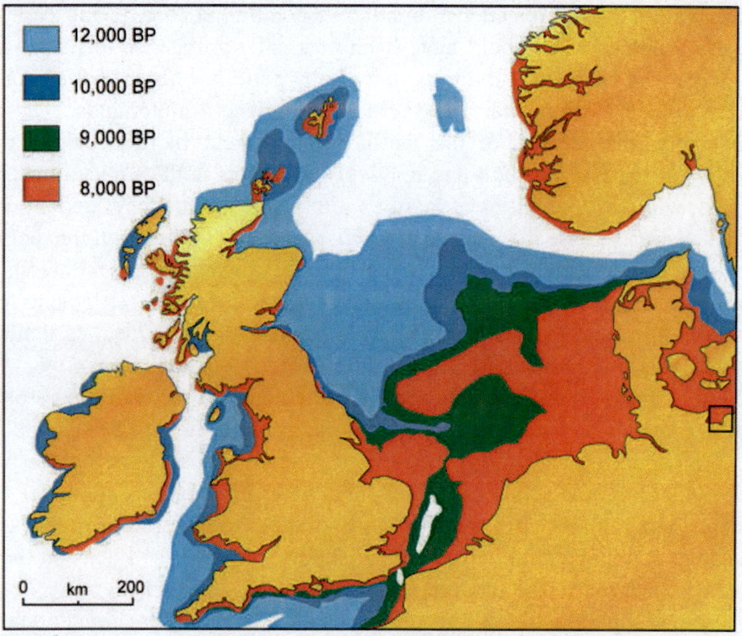


**Fig. 14** Map of the North Sea basin and western Baltic showing the successive change of coastline position with progressive sea level rise at the end of the Last Glacial period between 12,000 years ago and the establishment of the modern coastline position about 6000 years ago. The square on the far right marks the position of the Wismar Bay (see Fig. 15). Data from Bailey et al. (2008a), © G. Bailey

sea level rise and vertical crustal movements that varied in different regions and that is still only understood to a general level of approximation by a combination of theoretical models and a limited number of dated benchmarks for the position of the shoreline at different periods and in different regions.

This region has also provided some of the most abundant and well preserved evidence of submerged archaeological sites anywhere in the world, particularly in the shallow waters around Denmark and along the Baltic shoreline of Germany. In this region, over 2,000 underwater archaeological sites are known, mostly dating from the Mesolithic period from about 9000 years ago onwards (Fischer 2004, 2007; Skaarup and Grøn 2004; Harff et al. 2007). These submerged finds include a remarkable wealth of detailed evidence including a substantial body of organic materials and wooden artefacts preserved in anaerobic marine peats, including communal fish traps, dugout canoes, and human burials. Most of this material lies within a water depth of about 10 m, easily accessible to scuba divers. Earlier material may exist in deeper water, particularly on the North Sea side, but investigations here have so far been limited to mapping of the underwater landscape using acoustic records from the oil and gas industries, and the dredging up of artefacts and animal bones by fishing trawlers (Flemming 2004; Gaffney et al. 2009).

The archaeological material is not only important in its own right as a source of information about past human settlement and material culture, but it also provides

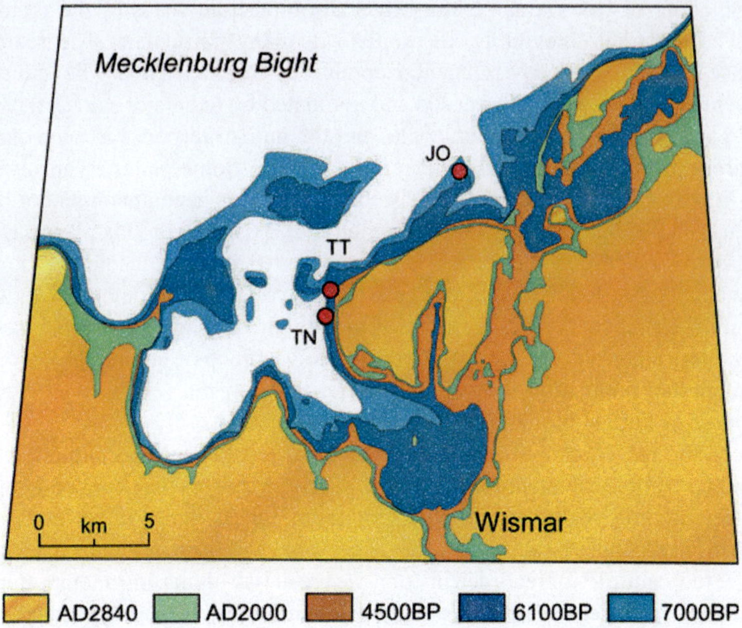


**Fig. 15** The Wismar Bay of northern Germany, showing the changing position of the shoreline at successive positions during the past 7000 years and the predicted position of the shoreline at AD 2840. *Red circles* indicate submerged archaeological settlements: *JO* Jäcklegrund-Orth, *TT* Timmendorf-Tonnenhaken, *TN* Timmendorf-Nordmole. Data from Harff et al. (2007)

precisely located and easily dateable materials and environmental indicators in the form of exploited plants, animals and marine fauna. Such information can be used to give precision to patterns of sea level and environmental change, which can be fed into more general models with predictive value. One example of this approach comes from the Wismar Bay of northern Germany (Fig. 15). This region has been sinking slowly as a result of isostatic effects following the ice retreat, and coastlines with Mesolithic and Neolithic settlements are now underwater. Excavation of some of these sites has helped to track in detail the shift from freshwater to marine water conditions as sea level rose, and to model the pattern of relative sea level change. Using these data, projections have been made of what will happen to the position of the coastline between now and AD 2840. The data show that there will be a continued encroachment of the sea on the present-day coastal region, with the inundation of important areas of modern settlement and industrial activity, requiring either relocation of major facilities or the building of defensive structures.

## 8 Conclusion

The examples considered above provide a range of examples of long-term geological change, their effects on human settlement, and the implications for a deeper understanding of past human history over the long time spans of the prehistoric past. The record is, inevitably, incomplete in many particulars. The search for evidence of the varied environmental conditions that existed on the submerged continental shelf, repeatedly exposed and inundated by the major sea level changes of the glacial-interglacial climate cycle, and the human response to such changes, has barely begun. Many details will remain elusive. Some patterns, however, are clear. First, geological instability has been a continuous accompaniment to human evolution and social development throughout our existence on this planet. Despite the potentially destructive effects of earthquakes, tsunamis, volcanic eruptions and sea-level rise that dominate our attention as observers of our contemporary world, these short-term events are themselves symptomatic of longer term processes that, in the long run, have often created ecologically attractive conditions for human settlement and population growth, and exercised powerful selection pressures on the course of human biological and social development.

Secondly, the combined effects of geological, ecological and climatic change represent a complex interaction of many different processes that frequently operate on different time scales and at different temporal rhythms. If we confine our attention only to those processes that are observable within an individual life time, or over the few centuries of recorded history, we are at risk of missing an important part of the overall picture. As the example of erosion in the Epirus landscape makes clear, changing the time span of observation changes our understanding of cause and effect, and may actually lead to a completely erroneous interpretation of causal factors. In such a complex natural system, causation is unlikely to be a simple matter of linear cause and effect, but a more complex interaction of variables involving varying contributions of proximate triggers and underlying boundary conditions. The latter in particular may appear fixed, until we expand our time span to observe the changes they undergo over longer periods of time.

Finally, the archaeological record is testament to the fact that we have evolved as an adaptive species with flexible behavior patterns capable of responding to changes in our environmental circumstances, or of circumventing them. Many of the geological changes discussed above that appear destructive or damaging from the point of view of a particular place or a particular time, can be seen to have beneficial effects if we expand our geographic or our temporal perspective. One person's soil erosion in one part of the physical landscape is another person's sediment accumulation. The loss of territory as sea level rose at the end of the last ice age was compensated for, in many regions, by the establishment of productive inshore fisheries and intertidal mollusk beds as sea level stabilized at about its present position. Cyclical or repeated geological change, whether caused by tectonics or changes in sea-level, often rejuvenated water supplies and sediment fertility in the longer run, or opened up entirely new landscapes for human settlement, rather than destroying previously productive ones. Mobility and migration were important keys to finding solutions to past environmental problems, and the development of adaptive strategies for avoiding or minimizing their effects. Paradoxically, the geological changes that seem to us so destructive in our perceptions of day-to-day existence were the stimulus to creating the problem-solving species that we have evolved into. In the modern era we have acquired immense capacities for coping with the many threats to our existence, but we have also acquired an equally powerful ability to create new and often unforeseen ones, with consequences that may extend far into the future. Of course, the past record of human interaction with geological or environmental change cannot provide a precise template for predicting the future, but the long-term perspective supplied by such a record can provide many insights that should help to inform our understanding of our present condition and the likely consequences of our present and future actions.

**Acknowledgments** We acknowledge support from our primary funding agencies, the British Academy, The Natural Environment Research Council (NERC, UK) through its EFCHED programme (Environmental Factors in Human Evolution and Dispersal), the Leverhulme Trust, the !Khure France-South Africa project, and the EU through its COST (Cooperation in Science and Technology) scheme in support of COST Action TD0902, which have funded the field investigations and stimulated the international and interdisciplinary collaborations from which the ideas in this chapter have benefited. This is IPGP paper number 3047.

## References

Ambraseys NN (1981) The El Asnam earthquake of 1980; conclusions drawn from a field study. Q J Eng Geol Hydrogeol 14:143–148

Armijo R, Meyer B, King GCP, Rigo A, Papanastassiou D (1996) Quaternary evolution of the Corinth rift and its implications for the Late Cenozoic evolution of the Aegean. Geophys J Int 126:11–53

Bailey GN (ed) (1997) Klithi: palaeolithic settlement and quaternary landscapes in northwest Greece, vol 2. McDonald Institute for Archaeological Research, Cambridge

Bailey GN (2007) Time perspectives, palimpsests and the archaeology of time. J Anthropol Archaeol 26:198–223

Bailey GN (2009) The Red Sea, coastal landscapes and hominin dispersals. In: Petraglia M, Rose J (eds) The evolution of human populations in Arabia. Springer, Dordrecht, The Netherlands, pp 15–37

Bailey GN, Flemming N (2008) Archaeology of the continental shelf: marine resources, submerged landscapes and underwater archaeology. Quat Sci Rev 27:2153–2165

Bailey G, King GCP (2010) Dynamic landscapes and human evolution: tectonics, coastlines and the reconstruction of human habitats. Quat Sci Rev doi: 10.1016/j.quascirev.2010.06.019

Bailey GN, King GCP, Sturdy DA (1993) Active tectonics and land-use strategies: a Palaeolithic example from northwest Greece. Antiquity 67:292–312

Bailey GN, King GCP, Flemming NC, Lambeck K, Momber G, Moran LJ, Al-Sharekh A, Vita-Finzi C (2007) Coastlines, submerged landscapes and human evolution: the Red Sea Basin and the Farasan Islands. J Isl Coast Archaeol 2:127–160

Bailey G, Barrett J, Craig O, Milner N (2008a) Historical ecology of the North Sea basin: an archaeological perspective and some problems of methodology. In: Rick T, Erlandson J (eds) Human impacts on ancient marine ecosystems. University of California Press, Berkeley, pp 215–242

Bailey G, Carrión JS, Fa DA, Finlayson C, Finlayson G, Rodríguez-Vidal J (eds) (2008b) The coastal shelf of the Mediterranean and beyond: corridor and refugium for human populations in the Pleistocene. Quat Sci Rev 28:23–24

Bailey G, Reynolds S, King GCP (2010) Landscapes of human evolution: models and methods of tectonic geomorphology and the reconstruction of hominin landscapes. J Hum Evol. doi: 10.1016/j.jhevol.2010.01.004

Benjamin J, Bonsall C, Pickard K, Fischer A (eds) (2011, in press) Underwater archaeology and the submerged prehistory of Europe. Oxbow, Oxford

Bilham RG, King GCP (1989) The morphology of strike slip faults; examples from the San Andreas, California. J Geophys Res 94:10204–10216

Cachel S, Harris JWK (1998) The lifeways of *Homo erectus* inferred from archaeology and evolutionary ecology: a perspective from East Africa. In: Petraglia MD, Korisettar R (eds) Early human behaviour in global context. Routledge, London, pp 108–132

Delson E, Tattersall I, Van Couvering JA, Brooks AS (eds) (2000) Encyclopedia of human evolution and prehistory, 2nd edn. Garland, New York

Elias SA (ed) (2006) Encyclopedia of quaternary science. Elsevier, Oxford

Faure H, Walter RC, Grant DR (2002) The coastal oasis: ice age springs on emerged continental shelves. Global Planet Change 33:47–56

Fischer A (2004) Submerged stone age—danish examples and north sea potential. In: Flemming NC (ed) Submarine prehistoric archaeology of the North Sea: research priorities and collaboration with industry. CBA Research Report 141, London, pp 23–36

Fischer A (2007) Coastal fishing in Stone Age Denmark—evidence from below and above the present sea level and from human bones. In: Milner N, Craig OE, Bailey GN (eds) Shell niddens in Atlantic Europe. Oxbow, Oxford, pp 54–69

Flemming NC (ed) (2004) Submarine prehistoric archaeology of the North Sea: research priorities and collaboration with industry. CBA Research Report 141, London

Flerit F, Armijo R, King GCP, Meyer B (2004) The mechanical interaction between the propagating North Anatolian fault and the back-arc extension in the Aegean. Earth Planet Sci Lett 224:347–362

Force E (2008) Tectonic environments of ancient civilizations in the eastern hemisphere. Geoarchaeology 23(5):644–653

Gaffney V, Fitch S, Smith D (2009) Europe's lost world: the rediscovery of Doggerland. CBA Research Report 160. Council for British Archaeology, York

Grandin R, Socquet A, Binet R, Klinger Y, Jacques E, de Chabalier JB, King GCP, Lasserre C, Tait S, Tapponnier P, Delorme A, Pinzuti P (2009) September 2005 Manda Hararo-Dabbahu rifting event, Afar (Ethiopia): constraints provided by geodetic data. J Geophy Res 114:B08404. doi:10.1029/2008JB005843

Green S, King GCP (1996) The importance of goats to a natural environment: a case study from Epirus (Greece) and southern Albania. Terra Nova 8(6):655–658

Grove AT, Rackham O (2001) The nature of Mediterranean Europe: an ecological history. Yale University Press, New Haven

Harff J, Lemke W, Lampe R, Lüth F, Lübke H, Meyer M, Tauber F, Schmölcke U (2007) The Baltic Sea coast—a model of interrelations among geosphere, climate, and anthroposphere. Special Paper 426. Geological Society of America, New York, pp 133–142

Hiscock P (2008) Archaeology of ancient Australia. Routledge, London

Hordern P, Purcell N (2000) The corrupting sea: a study of Mediterranean history. Blackwell, Oxford

Inman DL (1983) Application of coastal dynamics to the reconstruction of paleocoastlines in the vicinity of La Jolla, California. In: Masters PM, Flemming NC (eds) Quaternary coastlines and marine archaeology. Academic Press, London, pp 1–49

Jackson J (2006) Fatal attraction: living with earthquakes, the growth of villages into megacities, and earthquake vulnerability in the modern world. Philos Trans R Soc A 364:1911–1925

King GCP, Bailey GN (1985) The palaeoenvironment of some archaeological sites in Greece: the influence of accumulated uplift in a seismically active region. Proc Prehist Soc 51:273–282

King GCP, Bailey GN (2006) Tectonics and human evolution. Antiquity 80:1–22

King GCP, Vita-Finzi C (1981) Active folding in the Algerian earthquake of 10 October, 1980. Nature 292:22–26

King GCP, Yielding G (1984) The evolution of a thrust fault system: processes of rupture initiation, propagation and termination in the 1980 El Asnam (Algeria) earthquake. Geophys J R Astron Soc 77:915–933

King GCP, Stein R, Rundle J (1988) The growth of geological structures by repeated earthquakes: 1. Conceptual framework. J Geophys Res 93:13307–13318

King GCP, Sturdy DA, Bailey GN (1997) The tectonic background to the Epirus landscape. In: Bailey GN (ed) Klithi: Palaeolithic settlement and Quaternary landscapes in Northwest Greece. Volume 2: Klithi in its local and regional setting. McDonald Institute for Archaeological Research, Cambridge, pp 541–559

Klein RG (1999) The human career, 2nd edn. University of Chicago Press, Chicago

Lambeck K, Chappell J (2001) Sea level change through the last glacial cycle. Science 292: 679–686

Macklin M, Lewin J, Woodward J (1997) Quaternary river sedimentary sequences of the Voïdomatis basin. In: Bailey GN (ed) Klithi: Palaeolithic settlement and Quaternary landscapes in Northwest Greece. Volume 2: Klithi in its local and regional setting. McDonald Institute for Archaeological Research, Cambridge, pp 347–360

Maslin MA, Christensen B (2007) Tectonics, orbital forcing, global climate change, and human evolution in Africa. J Hum Evol 53(5):443–634

McDougall I, Brown FH, Fleagle JC (2005) Stratigraphic placement and age of modern humans from Kibish, Ethiopia. Nature 433:733–736

Munro RN, Wilkinson TJ (2007) Environment, landscapes and archaeology of the Yemeni Tihāmah. In: Starkey J, Starkey P, Wilkinson T (eds) Natural resources and cultural connections of the Red Sea. British Archaeological Reports International Series 1661. Archaeopress, Oxford, pp 13–33

Parker AG (2009) Pleistocene climate change in Arabia: developing a framework for hominin dispersal over the last 350 ka. In: Petraglia MD, Rose JI (eds) The evolution of human populations in Arabia. Springer, Dordrecht, The Netherlands, pp 39–49

Petraglia MD, Rose JI (eds) (2009) The evolution of human populations in Arabia. Springer, Dordrecht, Netherlands

Philip H, Meghraoui M (1983) Structural analysis and interpretation of the surface deformation of the El Asnam earthquake of October 10, 1980. Tectonics 2(1):17–49
Potts R (1996a) Humanity's descent: the consequences of ecological instability. William Morrow, New York
Potts R (1996b) Evolution and climate variability. Science 273:922–923
Sauer CO (1962) Seashore—primitive home of man? Proc Am Phil Soc 106:41–47
Shackleton NJ (1987) Oxygen isotopes, ice volume and sea level. Quat Sci Rev 6:183–190
Siddall M, Rohling EJ, Almogi-Labin A, Hemleben C, Meischner D, Schmelzer I, Smeed DA (2003) Sea-level fluctuations during the last glacial cycle. Nature 423:853–858
Skaarup J, Grøn O (2004) Møllegabet II: a submerged Mesolithic settlement in southern Denmark. British Archaeological Reports International Series 1328. Archaeopress, Oxford
SPLASHCOS (2010) Submerged prehistoric archaeology and landscapes of the continental shelf. EU COST Action TD0902. http://php.york.ac.uk/projects/splashcos/. Accessed 10 June 2010
Stein R, King GCP (1984) Seismic potential revealed by surface folding: the 1983 Coalinga, California, Earthquake. Science 224:869–872
Stein R, King GCP, Rundle J (1988) The growth of geological structures by repeated earthquakes: 2 field examples of continental dip-slip faults. J Geophys Res 93:13319–13331
Stringer CB (2000) Coasting out of Africa. Nature 405:53–55
Stringer CB, Andrews P (1988) Genetic and fossil evidence for the origin of modern humans. Science 239:1263–1268
Torroni A, Achilli A, Macaulay V, Richards M, Bandelt H-J (2006) Harvesting the fruit of the mtDNA tree. Trends Genet 22:339–345
Van Andel T (1989) Late Quaternary sea-level changes and archaeology. Antiquity 63:733–745
Van der Leeuw SE (ed) (1998) The Archaeomedes Project—understanding the natural and anthropogenic causes of land degradation and desertification in the Mediterranean Basin. Office for the Official Publications of the European Community, Luxembourg
Van der Leeuw S, Redman CL (2002) Placing archaeology at the center of socio-natural studies. Am Antiquity 67(4):597–605
Vita-Finzi C, King GCP (1985) The seismicity, geomorphology and structural evolution of the Corinth area of Greece. Philos Trans R Soc Lond A 314:379–407
Waelbroeck C, Labeyrie L, Michel E, Duplessy JC, McManus JF, Lambeck K, Balbon E, Labracherie M (2002) Sea-level and deep water temperature changes derived from benthic foraminifera isotopic records. Quat Sci Rev 21:295–305
Walter RC, Buffler RT, Bruggemann JJ, Guillaume MMM, Berhe SM, Negassi B, Libsekal Y, Cheng H, Edwards RL, von Gosel R, Neraudeau D, Gagnon M (2000) Early human occupation of the Red Sea coast of Eritrea during the Last Interglacial. Nature 405:65–69

# History of Ischia Harbour (Southern Italy)

**Stefano Carlino, Elena Cubellis, Ilia Delizia and Giuseppe Luongo**

## 1 Introduction

On 17 September 1854, under the initiative of the Bourbon Ferdinand II (1830–1859), King of the Two Sicilies, the opening of the new harbour of Ischia was celebrated. It has since become a major maritime port and marina in the Bay of Naples (Fig. 1).

The port constituted the fundamental marine transporation element connecting the island and the mainland, providing easier access to the island and promoting the progressive growth of the local economy. At the time, Ischia showed great diversity between one zone and another, and much of the island was almost inaccessible and sparsely inhabited. Noted for its active volcanism from the early fourteenth century and persistent seismicity until 1883, "before 1853 this island was almost impracticable, … to the detriment of many natural advantages that it has over others, it had a wild appearance, to say the least" (Annali Civili del Regno delle Due Sicilie—Annals of the Kingdom of the Two Sicilies 1855).

However, due to its morphological characteristics and nature of its settlements, the northern side held out more attractions: the coastline was low and from here it was easier to reach Ischia Castello, the administrative centre of the island. In particular, occupying the site of the present-day harbour of Ischia was a lake close to the coastline, surrounded by low rises generated by recent volcanic activity. On the most southerly hill rose the house of chief Court Physician Francesco Buonocore that had been built during 1735 on family-owned land at the back of the central section of the lake nearly opposite today's harbour entrance. It was a

S. Carlino (✉) and E. Cubellis
Istituto Nazionale di Geofisica e Vulcanologia, Naples Italy
e-mail: stefano.carlino@ov.ingv.it

I. Delizia and G. Luongo
Università di Napoli Federico II, Naples Italy

V. Badescu and R. B. Cathcart (eds.), *Macro-engineering Seawater in Unique Environments*, Environmental Science and Engineering,
DOI: 10.1007/978-3-642-14779-1_2, 

**Fig. 1** The Bay of Naples and the island of Ischia

country villa which captured the spa waters bubbling up from underground and had thus become a house of health and well-being for nobles and dignitaries attached to the court of Naples who needed treatment.

The site had already been chosen by an earlier Bourbon king, Ferdinand IV, as a "royal delight" (1784), insofar as it was suitable for sating his passion for nature and for fishing in the lake. Later, in 1854, it was to come within the town-planning programmes of the northern side, carried forward by Ferdinand II, which led to the establishment and rapid development of Villa de'Bagni, the original name of the first settlement of Ischia Porto (Quaranta 1855). (The numbering of Ferdinand's royal titles is somewhat confusing. He was, in fact, Ferdinand IV of Naples and became, at a later stage in his reign, Ferdinand I of the Two Sicilies.)

An ancient volcanic formation dating to the fourth century BC, the lake was perfectly suited to being converted into a harbour. Known as the *Lago del Bagno* or *de' Bagni* due to the presence of hot springs on its perimeter, it was almost circular and deep enough to provide keel clearance for small fishing vessels. Moreover, it was separated from the sea by a small isthmus: a narrow sandy dune on which marine vegetation was deposited during rough seas. Since time immemorial, the lake had been used for fish-farming, constituting one of the few sources of revenue for the local administration.

The lacustrine basin, just like the rest of the island, was the product of volcanic activity which generated an extraordinarily complex landscape in continuous evolution (Figs. 2a, b, 3). This is due to the island's geological history which has

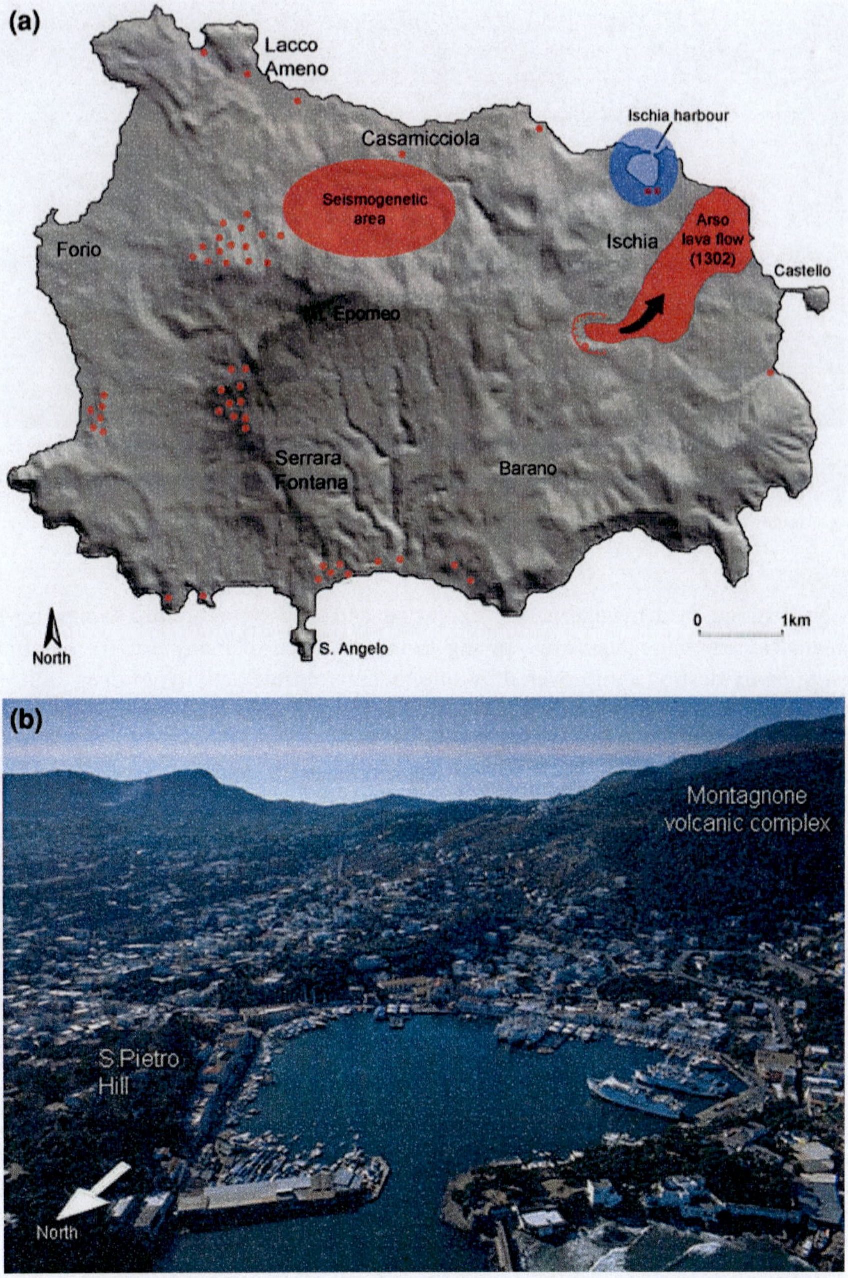


**Fig. 2** The island of Ischia and main features related to its recent dynamic; *dots* represent the main hot springs and fumaroles fields (**a**) (modified after Luongo et al. 2006). The *circle* highlights the Ischia Harbour zone (**b**)

**Fig. 3** The Ischia island as appears from the ferryboat which comes from Naples. On the right side (north), the top of Mt. Epomeo. It is also marked the Ischia harbour zone and the Castello lava dome (photo S. Carlino, 2008)

been distinguished by alternating explosive and effusive eruptions which have created a very variegated area, owing to the presence of many eruptive centres being partly destroyed or covered by subsequent volcanic activity, of deep valleys produced by erosion of pyroclastic top layers, of marine terraces and hills with sub-vertical walls, which testify to volcanic and tectonic processes on the island, especially active in the last 10,000 years, and with intense seismic activity recorded historically in the past 800 years. The most recent major volcanic eruption occurred in 1302, emitting a lava flow that affected the eastern side of Ischia, partly invading the area east of the harbour (Fig. 2a) (Vezzoli 1988; Civetta et al. 1991; Cubellis et al. 2004; Carlino et al. 2006; Luongo et al. 2006).

The main historical sources for information on the opening of the harbour, an operation which was to change the lives of the islanders, the morphology of the area, and the roles and hierarchy of the island's settlements are the Annali Civili del Regno delle Due Sicilie (1855), which supply technical and descriptive elements mixed with notes of praise and celebration for the king. The technicians entrusted with the task of cutting the isthmus to join the lake with the sea had, in the nearby Campi Flegrei, known historically as the Phlegraean Fields, the example of the work of Lucius Cocceius Auctus in 37 BC, who cut the isthmuses separating the Lucrine Lake from the sea and from Lake Avernus to build *Portus Iulius* (Fig. 4a, b) for the Roman fleet.

A further important source is the constant flow of correspondence between Camillo Quaranta, appointed commissioner for harbour works, and the king and the various ministers of the Bourbon Royal House, varyingly involved in operations (Naples State Archive, Ministry of Public Works). Recent years have seen some critical reviews of this source (Delizia 1987, Delizia and Delizia 2006; Rispoli 2007). Archaeological reconstruction of the area (Buchner 1946; Rittmann

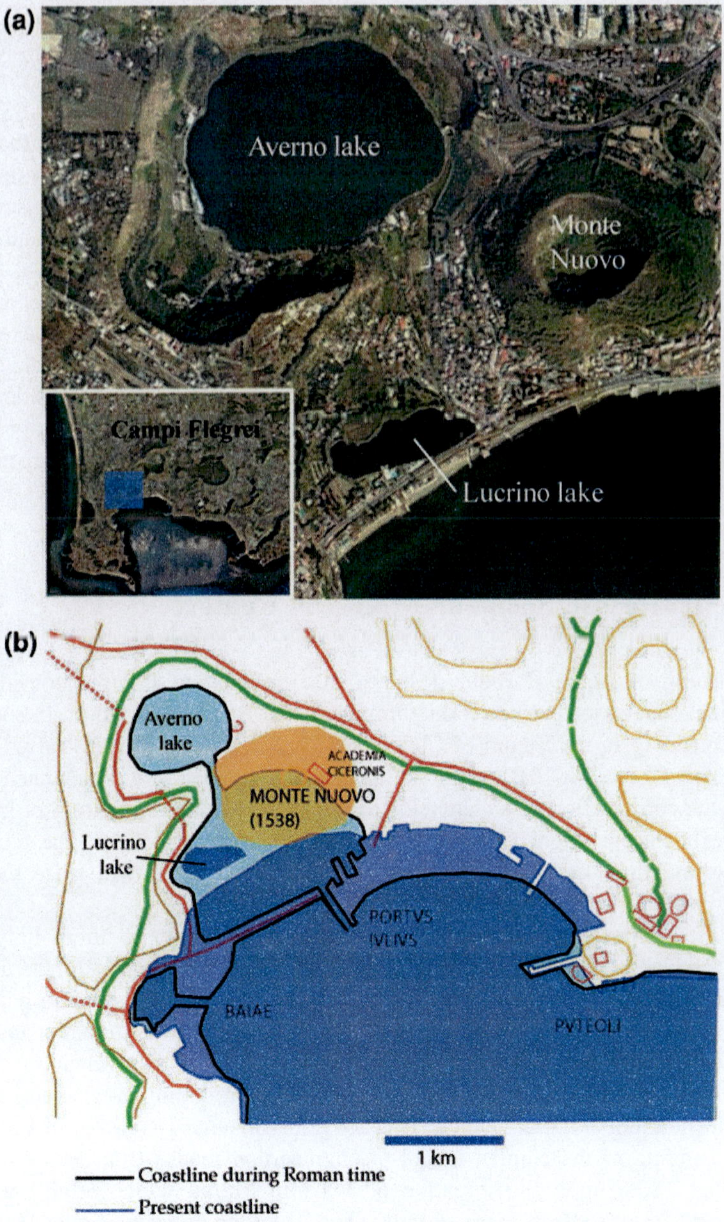


**Fig. 4** Campi Flegrei Caldera - Avemus and Lucrine lakes and Mt. Nuovo tuff cone (1538) (**a**). During Roman times the isthmus which separated the two lakes was removed in order to obtain a channel for the passageway of the Roman fleet (Portus Iilius). The products of Mt. Nuovo eruption covered a large part of Lucrine lake (**b**). The submerged archaeological ruins indicate that the ground level of the Roman times is lower than the present-day one.this result is due to subsidence and resurgence processes occurred in the last 2000 years (Castagnoli 1977; Morhange et al. 2006; Pappalardo 2006)

and Buchner 1948), together with volcanological studies, has sketched out a reconstruction of the relations between prehistoric settlements and eruptions, as well as variations in the coastline in the last 2000 years.

No less important for our analysis is the maps (Alisio and Valerio 1983) and iconographic representations (Caputo 2000) produced either side of the great conversion of the lake into a harbour. Towards the mid-nineteenth century, the island's geology and natural history were the subject of extensive scientific inquiry, as ably illustrated in the geological maps of Fonseca (1847) and Fuchs (1873).

The set of available data, studies and reports are analysed herein to reconstruct the circumstances that led to the opening of Ischia Harbour, the macro-execution phases and the resulting change in the island's morphology. Our historical analysis is followed by the description of the geology of the harbour, some considerations on the current state of the island and on issues relating to the increase in volcanic and seismic risk resulting from urban expansion and the increase in tourism since the early twentieth century.

## 2 The Historical, Social and Cultural Context

The opening of Ischia Harbour occurred in a period that saw the considerable reforming impetus of public works throughout the Bourbon kingdom in southern Italy, especially in the region of Campania. This climate of reform started under the Spanish viceroy in 1610 when the first major intervention was launched in the region, the construction of a channel across the Campanian Plain north of Naples, the so-called Regi Lagni (Fiengo 1988), whose aim was to avoid the recurrent floods tormenting the local people and preventing urban growth since the pre-Roman period.

As often happens in the history of great monarchies, in southern Italy both during the Bourbon and French dominations (1737–1860), the sovereigns used to prove their greatness and benevolence by carrying out public works on a "grand scale", at the same time providing essential contributions for regional change and improvement. Modern macro-engineering endeavours are equivalent to "grand scale" in olden times. The city of Naples and its surrounding areas were protagonists of such modernizing change. This active involvement would ultimately lead to the founding of the Naples school of engineering, thanks to a decree of 18th November 1808, upon the initiative of Joachim Murat, which established the *Engineers Corps for Bridges and Roads*. Thus, arose the imposing figure of the all-directing State engineer, whose career was founded on a mostly meritocratic basis, in the interests of more rational land use.

With the arrival of the Bourbon Ferdinand II in 1830, the city of Naples and its surrounding areas experienced a period of considerable economic development, which gave it a modern, advanced image. In this period the first gas-lights were installed in the streets of Naples, roads and communication networks in general were improved and built ex novo, and in 1839 the first railway line established in Italy was built, connecting the city of Naples with the town of Portici situated on the slopes of Vesuvius. Also the island of Ischia was included in this vast programme of public works, some of which never progressed beyond the project planning phase. However, it cannot be denied that the Bourbons did much for the island (D'Ascia 1867): new highway networks were opened up, hugely improving road communications over very rugged terrain; a new ~4.8 km-long section of the freshwater aqueduct was built, submarine telegraphic links were established between Ischia and the mainland; the Portosalvo church was built and the harbour was created, opening up new horizons for developing the island's economy. The presence of a safe port, whose natural morphology made it sheltered from the westerly storms that create havoc in the Bay of Naples (Fig. 5), continues to be an essential function. Since Bourbon times, it has made the island easily accessible from the ports in Naples, Pozzuoli and Capo Miseno.

It was precisely in the early nineteenth century that systematic navigation in the Tyrrhenian sea began with the aid of large steamships, designed in the late eighteenth century in the USA, then in the UK and Italy. The invention of the steamship was so well received in the Kingdom of the Two Sicilies that King Ferdinand II, on the proposal of the Minister of Finance, decided to start up a shipbuilding industry. The Kingdom of the Two Sicilies became the third largest steamship producer in the world and the first in the Mediterranean.

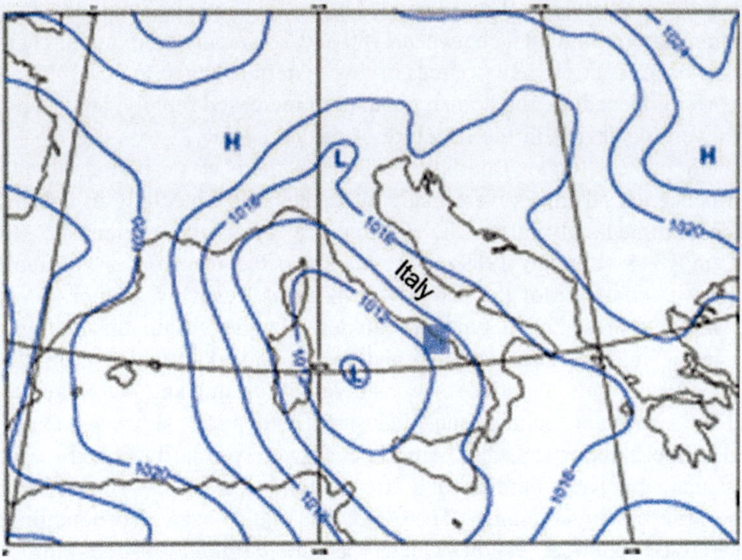


**Fig. 5** Typical low pressure over the Tyrrhenian Sea which produces NW and SW winds and waves in the western and eastern sector respectively. The Bay of Naples is highlighted with a *square* (wetteronline.de)

However, the cultural climate of that time and region was not characterised only by the spirit of innovation and industrial development: the late eighteenth century saw the start of the first scientific studies which were to lead to the birth of modern geology and for which the island of Ischia would play an important role in interpreting geological phenomena. The publication of *Principles of Geology* (1830) by Charles Lyell (1797–1875) who followed the Scottish James Hutton's (1726–1797), *Theory of the Earth with Proofs and Illustrations* (1795) gave a new impetus to studying the Earth's dynamics (Fischer and Garrison 2009). Lyell was in Naples in October 1828, attracted by the minute description of the rocks of the island of Ischia which the geologist Gian Battista Brocchi had made during his stay in Naples between 1811 and 1812. For Lyell the volcanic areas and recently-formed soils assumed considerable importance for his theory of gradualism, because he felt that in the Earth's history the causes which today slowly change the Earth's surface had always acted. The proof of this came from recent soils, where major changes were still under way. Lyell headed for Ischia, where he found fossil shells of marine origin on Mt. Epomeo as high as 600 m, with which he was able to demonstrate, with the aid of the Neapolitan naturalist Oronzo Gabriele Costa (1787–1867) who identified the fossils, that the island had recently undergone substantial tectonic uplift.

The drive towards research in actual geology also came from the need to identify and harness natural resources, especially energy resources needed to develop the industries that were arising in the technologically more advanced European countries. The positivist movement was to dominate much of European scientific and literary culture in the nineteenth century. Empirical facts were considered as underlying all real knowledge. In this social climate the Industrial Revolution started: in the UK between 1780 and 1800 there were major changes in the means of production, new sources of raw materials began to be exploited, new markets were opened up, the human population increased rapidly, and deep-rooted changes were occurring in the structure of settled society.

In southern Italy, under Bourbon domination there was a period, albeit short-lived, when great advances were made in science and especially in volcanology, concluded immediately after the Conference of Italian Scientists, held in Naples, in 1845 (Luongo 1989). For scientists, the theatre for volcanological developments consisted of the active volcanoes of Vesuvius, Campi Flegrei and Ischia, which represented a type of "golden triangle", with the metropolis of Naples in the middle. The historical and social context in which Ischia harbour was opened was thus a period of considerable cultural and scientific ferment, infused by a spirit of technical and industrial innovation. However, the reasons that led to the harbour at Ischia being opened were primarily linked to personal choices made by King Ferdinand II, to his fondness for the site as a holiday residence for the royal family. That said, the choice was also conditioned by the far-sighted political vision which was the hallmark of the king's main undertakings: this was still virgin territory, susceptible to developmental changes that would eventually leave the unmistakable mark of the Bourbon monarchy.

## 2.1 First Settlements and Development of the Island

Once known by the name of *Pithecusa,* the island of Ischia was the site of the earliest known Greek settlement in Italy. It was founded by Euboean Greeks from Eretria and Chalcis *in* ca. 770 BC who established a flourishing trading post in the present-day bay of San Montano, followed in the fourth century BC by a small settlement on the north-eastern coast, now the site of the port of Ischia (Buchner 1971; Monti 1968, 1980). At a later date, a violent cataclysm is reported, which not only drove away the first Greek peoples, but also changed the configuration of the landscape: it brought about the depression of a coastal stretch with the formation of the lake and the small volcanic edifices of Montagnone and Rotaro. We hear of this event from the period's great polymath, Pliny the Elder, who died at Stabiae following the eruption of Vesuvius in 79 AD. In his *Naturalis Historia* (23–79 AD), he writes that on the island of Ischia, then called Aenaria, the earth swallowed up a now-forgotten unnamed town—*oppidum haustum profondo*—and that after this catastrophe a lake was formed—*alioque motu terrae stagnum emersisse*. Indeed, below the products of the eruption that led to the formation of the *Lago del Bagno* the remains of sixth and fifth century BC pottery were found, besides roof tiles from a Greek temple of the same era, now conserved in the island's museum (Buchner and Gialanella 1994; Buchner 2004).

Volcanic activity was such that the cultures that followed did not leave very significant traces of their presence on the island. The "rediscovery" of Ischia only occurred from the mid sixteenth century onward, when Giulio Iasolino (1538–1622), a medical doctor from Calabria, professor of anatomy at Naples University, started a systematic, painstaking study of all the hot springs on the island, which he introduced into curative practice. The long task of analysing and recognising the hot springs culminated in 1588 with the publication of *De' rimedi naturali che sono nell'isola di Pithecusa, hoggi detta Ischia* (On the natural remedies on the island of Pithecusa, today called Ischia) (Iasolino 1588), a work of great importance and editorial success, which was to boost Ischia's fame, thanks in part to the useful map of sites included, engraved by the mapmaker from Viterbo, Mario Cartaro, which was later used in the most important European atlases.

Interest in Ischia may also have grown due to the eruption of the nearby Campi Flegrei in 1538, which made the more popular thermal baths of Pozzuoli and Baia impracticable (Buchner 1958). From this time on, visitors to Ischia would be lured and connected to the development of the bathing and spa treatment industry which the pioneering construction of the Pio Monte della Misericordia establishment at Casamicciola (1601–1603), on the Gurgitello springs, contributed to inaugurate, develop and spread.

In the mid nineteenth century, there were already 12 spas/bathing establishments on the island where "the infirm flock from every part of Europe to try out the beneficial effects of these gifts that nature appears to have bestowed generously on this island" (Annali Civili del Regno delle Due Sicilie 1855). The island's ancient rural vocation, chiefly linked to wine production, would gradually give way to an economy based on spa and recreational tourism. As a result, there was an influx of new monetary capital that increased hotel capacity. The island's economy would progressively grow, but the inhabitants would continually have to come to terms with the forces of nature: in 1881 and 1883 the island was shaken by two large earthquakes which devastated island's northern sector, with the epicentres between the Casamicciola and Lacco Ameno municipalities. The earthquake of 28 July 1883 was especially destructive: it caused almost all the buildings in Casamicciola to collapse, heavy damage in almost the whole island and led to a death toll of 2,333 persons, many of whom were tourists. In the worst-hit municipalities there ensued a period of economic recession and a population decrease that would only begin to recover in the early twentieth century (Fig. 6) (Cubellis and Luongo 1998; Cubellis et al. 2004; Luongo et al. 2006; Carlino et al. 2009).

After the Second World War, starting ca. 1950, the tourist flow would increase, with a consequent chaotic, undisciplined boom in construction on the land, due to the lack of authoritative land-use planning regulation. This led to an aggravation, besides other things and events, of the volcanic and seismic risk to people and infrastructure. Today the island has six lawful municipalities (Ischia: 18,253 inhabitants; Casamicciola: 7,374; Lacco Ameno: 4,273; Forio: 14,554; Serrara Fontana: 3,060; Barano: 8,591, ISTAT data) making up a total population of >56,000 persons. The island's three main municipalities have boat and yacht marinas. Only that of Ischia is a natural ship harbour.

Since the opening of Ischia Harbour in 1854, the island has undergone radical economic, social and environmental changes. New shipping routes have permitted a rapid increase in daily commuter flows to and from the island. However, nowadays, more than four million tourist visitations impact the island every year, most people visit during May to September. Natural human population increase, tourism and urban region expansions have brought about an exponential increase in risk

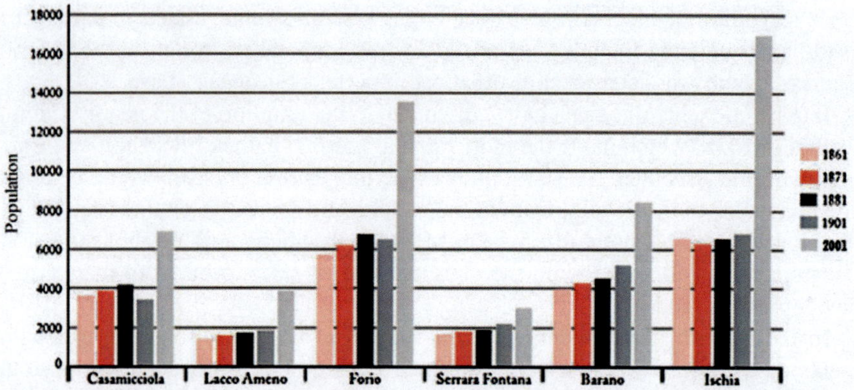


**Fig. 6** Population growth on the island from 1861 to present (ISTAT). After the 1883 earthquake there was a slight population decrease in the northern and western sectors (Casamicciola and Forio). A rapid increase has occurred in recent times (from Luongo et al. 2006)

correlated with seismic and volcanic activity. Besides, the island has been stricken in recent times (1910, 2006, 2009) by floods that have caused serious damage and dozens of victims, especially in the northern sector (Cubellis et al. 2008, 2009; Luongo 2009; Carlino et al. 2010). Attention is thus also laid on all those macro-management problems linked to the region's vulnerability and the consequent increase in risk (Alberico et al. 2008) which has now reached levels that are no longer publicly acceptable, compromising the island's basic vocation as a "place of well-being" in which the adventurous, tired, sickly and healthy alike can enjoy the unique and picturesque beauty of "civilized" volcanic landscapes. There have been a few seawater pollution episodes caused by industrial activity and shipping in the bay region nearby Ischia Harbour.

## 3 The Lake of Bagno Before the Harbour's Construction: Descriptions and Representations

"Only a sandbank, about 50 feet wide, separates it from the sea: it is a small version of a Dead Sea, with the difference that the lacustrine basin three-quarters of a mile in circumference is the bed of an ancient volcanic crater, formed by the small promontory of lava of S. Pietro a Pantanello to the east by the volcanic hills of Sant' Alessandro to the west and north. This lake has been given the inappropriate name of Pantanello which means little marshy pond; it communicates with the sea via a channel dug at the end of the sandbank. Hence the water is continuously changed in the basin, which has a sandy bed and resembles a pond full of exquisite fish, mussels and other crustaceans. At the centre of the lake rises a lava rock on which there is a small hut for fishing tackle which is let, reaping revenue for the town of Ischia. On the western shore of the lake there is an estate endowed with almost everything that constitutes, in this place, a good rural economy. The residence is small but clean. Suited to the needs of its owner it is situated at the centre of the vegetable garden that extends across the lower part of the estate, almost on the water's edge" (Haller 1822). Thus Conrad Haller, a French traveller to the Bay of Naples, described what would be the future harbour of Ischia, a circular lake, the so-called Lago del Bagno, formed within an ancient volcanic crater whose circumference measured 1.2 km, separated from the sea by a sandbank 15–20 m wide and about 250 metres long. Albeit relatively small, about 115,000 $m^2$, the stretch of brackish water was more similar to a lake than a marshy pond (pantano), since it communicated with the sea thanks to a small man-made channel, called Bocca Vecchia (old mouth), which allowed the continuous change of water and, closed off by a lock, favoured the proliferation of fish and crustaceans. The small lake was, thus, an ideal place for fishing, for the farming of molluscs and crustaceans, but also for the hunting of coot and other waterfowl.

With the iconographic and cartographic representations of the *Lago del Bagno* before its conversion into a harbour, the morphology of the site—and the sentiment it inspired—can be fairly faithfully reconstructed. Such was the enchantment of the site that the Bourbon kings, Ferdinand IV, and later Francis I and Ferdinand II, succumbed to its appeal, so as to make it become a work of total transformation which led to the creation of the royal residence on the island. It was precisely with this in mind that Ferdinand IV commissioned the court artist, Philip Hackert, in 1792, to paint a view of the area. Exhibited in the Palace of Caserta in the Room of Ferdinand IV, the painting depicts the whole stretch of the lake with the Aragonese Castle in the background and on the eastern shore the Buonocore lodge, now residence of the Bourbons. The lodge became the property of the royal household following a donation made by the descendants of Francesco Buonocore, even though it would appear to have been an appropriation rather than a spontaneous donation. Besides, Hackert's painting depicts in the foreground fishermen intent on casting their nets from small boats, a fairly profitable activity for the local community (Fig. 7). In the central sector of the lake the small lava rock described by Haller can be seen, with the storehouse for fishing tackle on top. This block of lava represents the remainder of a magmatic body which had fuelled the eruption, later solidified within the volcanic conduit and exposed after the crater collapse.

The morphology of the crater prior to the opening of the harbour is masterfully depicted in the topographic plan drawn by Carlo Vanvitelli around 1792 and in the historic map of 1840 (Reale Ufficio Topografico di Napoli 1840, source IGMI). The latter also supplies some land use features. At that time the whole area was sparsely inhabited, except for the small village of *Villa dei Bagni e della Casina*

**Fig. 7** View of Ischia lake, Philip Hackert (1792). Palazzo Reale Caserta (from De Seta 2005)

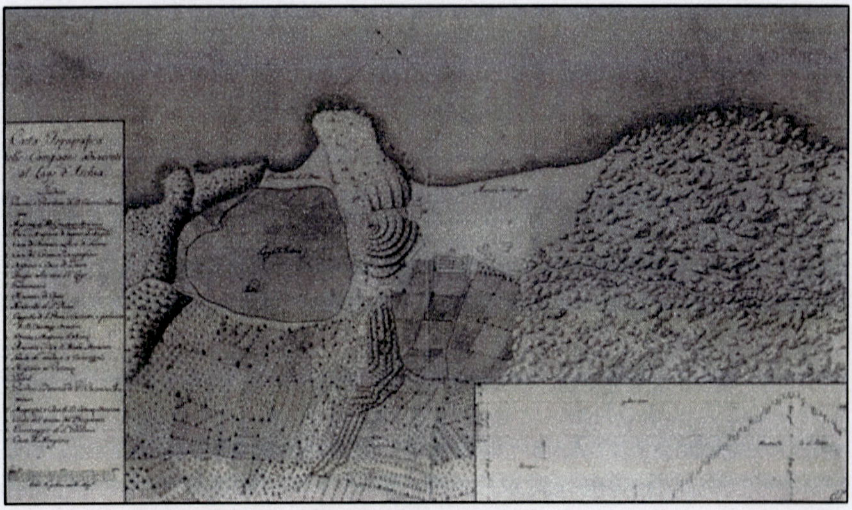

**Fig. 8** An antique topographic map of the *Lago del Bagno* and neighbouring countryside. A sandy isthmus separates the lake from the sea in the north. A little channel in the western side allowed the water to circulate between the lake and the sea. In the southern sector of the lake a small lava block crops out (Vanvitelli 1739–1821) (from Delizia and Delizia 2006)

**Fig. 9** Topographic map of the Royal Topographic Office of Naples, scale 1:25,000, 1840 (IGMI source). On the left sector of the lake (sea side) the little channel for water circulation (*arrow*), in the eastern and southern sector the Villa dei Bagni settlements and the Royal Lodge (Casino Reale), respectively. The whole area around the future harbour was formed by cultivated fields and woods

*Reale*. The lake was surrounded by woods and plots for growing fruit and vegetables. In both maps the small channel communicating with the sea can be recognised, situated westward (Figs. 8, 9). This channel was opened in 1670 to prevent the lakeside shores from becoming marshland due to the presence of hot-water springs and the depositing of current-borne marine vegetation during storms at sea (Quaranta 1855; Delizia 1987; Minervini 2004). Indeed, winter storms carried beyond the narrow sand dune a large quantity of *posidonia* (sea-grass) which rotted during the summer, producing a typical smell of sulphur. The channel mouth was kept closed by a series of reed gates, which permitted a change of water, while the retained fish quietly prospered in the lake.

Towards the mid nineteenth century the island became a site of extreme interest for scholars of volcanology and seismology, in a cultural climate in which scientific debate became lively. Of the products generated by this period of scientific enquiry mention should be made of the production of the first geological map of the island of Ischia, at a scale of 1:25,000, by Ferdinando Fonseca (1847), associated to the French Geological Society (Fig. 10). It is not a real geological map in the modern sense of such cartography, as it lacks some elements on the succession of eruptive events on a stratigraphic and structural basis. Rather, it is an areal subdivision of the most widespread lithotypes on the island. Fonseca distinguished four types of outcrops: trachytes; tuff and incoherent volcanic material; sub-Apenninic marl; quaternary shell aggregates.

In 1873, Fuchs proposed a geological map of the island of Ischia, again at the scale of 1:25,000 (Fig. 11) (Fuchs 1873). There are two curiosities here. First, the

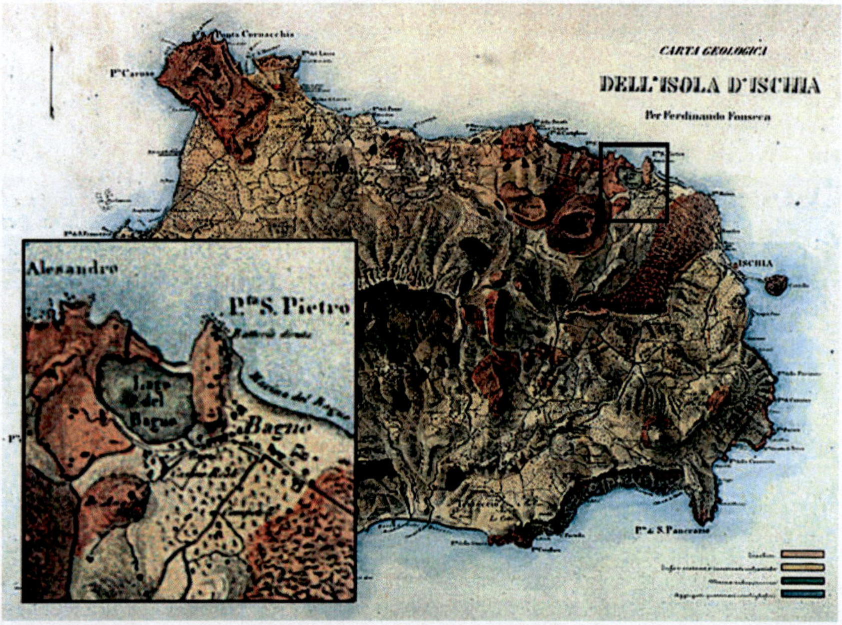


**Fig. 10** First geological map of Ischia produced by Fonseca (1847), scale 1:25,000. The most recent lava flow and lava and scoria cones are represented in red-brown. Inset with a close-up of the *Lago del Bagno* zone

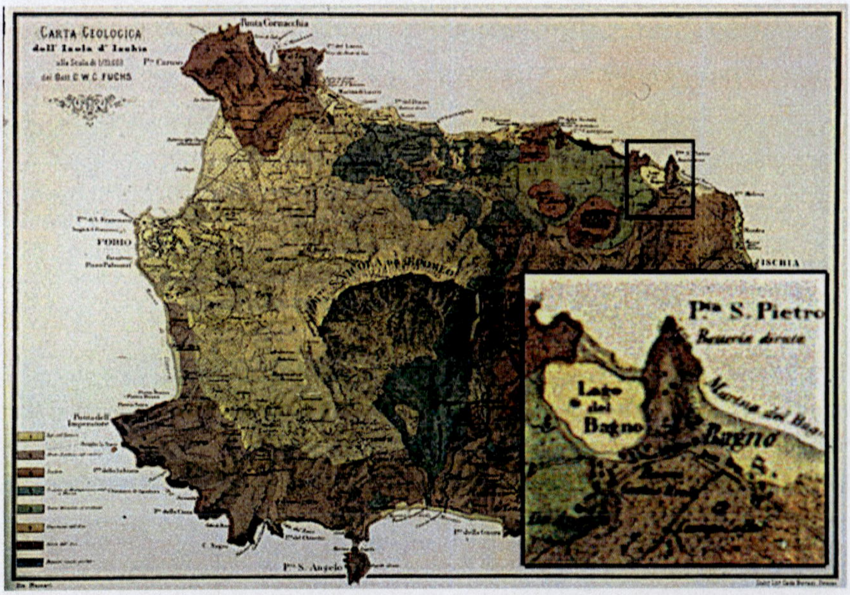


**Fig. 11** Geological map of the island produced by C. W. C. Fuchs, scale 1:25,000. The map is dated after the harbour opening (1873). As highlighted in the square, the isthmus separating the lake from the sea is still present. Thus it is likely that the topographic map utilized by Fuchs is the same as that used by Fonseca (1847)

town called *Ischia* corresponds to the present-day Ischia Ponte close to the castle; secondly, the map does not show the cut in the isthmus which occurred in 1854 with the opening of the harbour in *Lago di Bagno*. It may be deduced that the basis for Fuchs's map was the same as that used by Fonseca. Fuchs's geological map was published in Volume II of "Memories serving to describe the geological map of Italy". This may be considered the first geological map published by the geological Royal Committee established by decree of Victor Emanuel II on 15th December 1867 and according to standards laid down by the Royal Decree issued 15th June 1873.

For Fuchs, the considerations made by Fonseca only partly apply, in that he distinguishes the temporal succession of the various outcropping products: in this new map Fuchs believes that the Epomeo tuff is the most ancient outcrop, above which rest layers of pumice, tuff and lava flows produced by the various eruptive centres on the island. Fuchs distinguishes eight types of outcrops: Epomeo tuff; flows, domes and cones of trachytes; products of a similar composition to marl; scoria of trachyte and obsidian; the Arso layer; Arso scoria; recent fossiliferous deposits. In both the geological maps of Fonseca and Fuchs, there is clear documentation of the sand bank that separated *Lago del Bagno* from the sea, the volcanic edifices of Montagnone and Rotaro, the Arso lava flow eastward and the morphology of the whole future harbour area.

## 4 The Opening of Ischia Harbour

Ferdinand II was personally very fond of the island of Ischia. Together with his second wife, Maria Theresa, and their many offspring, he spent several months there every year, relaxing by the *Lago dei Bagni*, in the Royal Lodge that dominated its shoreline from the hill nearby (Minervini 2004). When the King arrived, he said: "...of the lake we shall make a harbour, it will be the lifeblood of Ischia". As often happened during the Bourbon period, the vacation macro-project idea swiftly became reality. Its purpose was twofold: increase the consensus for Bourbon initiatives to renew the kingdom and transfer their own representative on the island from the castle to those delightful, unspoilt places. This was the necessary step to carry out the "Royal Delights" project which the same Ferdinand II completed with the conversion of the Buonocore residence into a Royal Lodge, with the construction of the Church of Portosalvo and the restoration of the royal space adjacent to the whole area of the future port (Delizia 1990).

The Provincial Council of Naples turned down the planning application from the Council Authority of Ischia to build the harbour, since the intervention was not deemed useful and the island *not susceptible to development* (Rispoli 2007). On 13th March 1853 the far-sighted King Ferdinand II, convinced of the importance of the work, overrode the decision and passed a decree to make the lake into a harbour for Ischia. Although the initial estimate was 50,000 ducats, modifications when the project was under way made the cost climb to over 126,000 ducats, a considerable sum in those times. By 25th July of the same year the works had already begun, and they were completed in only 14 months, a relatively short period of construction time given the resources and technology at their disposal (Annali Civili del Regno delle Due Sicilie 1855).

From the preliminary analysis of the lake's characteristics, commissioned by the king, we have the first information on the nature and economic quantity of the works to be undertaken. The latter involved the removal of part of the sandbank separating the lake from the sea to the north, with an entrance of ~500 palms (~130 m). A jetty was also to be constructed to protect the harbour mouth from the strong NNW winter winds, about 700 palms long (~182 m) consisting of about 541 cubic *canne* (~10,000 $m^3$) of rock obtained from a cliff beyond the small river mouth to the west, and the bed was to be dredged to allow access for large vessels as well (Archivio di Stato di Napoli, Ministero dei Lavori Pubblici 1853–1854). The whole lake-bed was excavated, removing material about one metre deep, amounting to 115,000 $m^3$. To enable these operations, on 31st May 1854 the King commanded a small specialised fleet to be transferred, consisting of two steam dredgers, a small steamship, and a four-boat tug to protect the harbour entrance (Quaranta 1855; Carelli 1858; Rispoli 2007). In addition, an ancient water/debris collection tank situated near the Buonocore residence that was blocked would be emptied, so as to prevent storm-related flooding caused by streams to the south invading the lake-bed, making safe navigation more difficult.

The macro-project for constructing the harbour was assigned to the Inspector of Bridges and Roads Luigi Oberty and Lieutenant Domenico Milo of the Engineering Corps, while the project's director was Camillo Quaranta, a commissioner of the Royal Navy. The latter left behind copious documents concerning the state of the works, which point out the difficulties encountered in the work, the mishaps and some technical details on intervention topologies (Delizia 1987). Before embarking on the sand removal works, Quaranta first strengthened the lake shores with walls supported on robust iron frames. He then built further embankments to flank the short channel entrance opening to the sea and raised the quay for moorings (Quaranta 1855; Minervini 2004). The cutting and removal of the sand bank to open up the channel to the sea was completed 4 months later. The works in progress are beautifully depicted in a painting by Francesco Mancini (1830–1903), which shows the building site with the channel to the sea already open (Fig. 12). King Ferdinand II was not ever present during these operations, but waited for the work to be completed, on 31st July 1854, before entering for the very first time aboard the Royal Steamship *Il Delfino*. It was a sort of technical test and verification of the works prior to the official inauguration held on 17th September. The news was reported in the *Giornale delle Due Sicilie* as an event of worldwide importance and enthusiastically feted by the population of the whole island. The inauguration occurred amidst the boom of discharging artillery, the lyrical notes of several musical bands, the charitable sounds of the excited throng of pleased islanders decked out in party costumes and about 200 ships and boats. The parade

**Fig. 12** The gouache by F. Mancini (1853) reproducing the works during the opening of the isthmus separating the *Lago del Bagno* from the sea (Private Coll.) (from Caputo 2000)

of vessels was preceded by the Royal Launch, followed by the warships *Tancredi*, *Saetta*, *Delfino* and *Antilope*. Thus began a new era for Ischia, in which the harbour played a key role in changing both the island itself and relations between the island and the mainland (Quaranta 1855; Mirabella 1913).

Works to improve the structure of the harbour continued for about 2 years. At the end of the protective breakwater on 15th December 1856 a fifth-order lighthouse was first lit. For ships arriving from the northeast, Ischia Harbour lighthouse, together with the smaller warning light from the nearby island of Procida, represented an important reference point for coastal navigation in the channel that separated the two islands from the Italian mainland. Southward, instead, Ischia lighthouse with that of Capri, about 21 miles to the south, would be very useful for large vessels sailing to and from the islands of Sicily and Malta. Works to redesign the harbour area were completed with the construction of the church of Santa Maria di Portosalvo in 1856, crowning the programme of Ferdinand II of the Royal Delights of Ischia in the last few years of his reign (Rispoli 2007; Delizia 2007a, b).

Ischia Harbour, therefore, played a fundamental part in opening the island towards the sea and the mainland, allowing new trading and cultural links, further increasing spa and recreational tourism, supplying a new structure to the island in which the harbour became the pivot for the island's social and demographic growth (Fazio 1896).

From 1943 to 1945 the harbour played a role of vital importance as an Allied naval base in the Mediterranean Sea Basin, taking its cue from what was happening on the mainland in the regions of Campania and Lazio, a few miles away from the island (Silvestri 2004). In the years after the Second World War the

**Fig. 13** Ischia Harbour in the 1950s with the old cableway (postcard of the island)

Harbour of Ischia still retained all its splendour (Fig. 13) (Taddeo 1954). However, the post-war subsequent economic boom and increase in tourism were to change the area progressively, resulting from excessive population pressure. This was to lead to a gradual covering/sealing of the transformed natural landscape by the distinctly man-made landscape. However, the volcanic structure of this region is still predominant; its natural evolution must be taken into account when assessing the island's future economic and population development scenarios. Ischia Harbour is deep enough today to navigationally accommodate rather large motorized and sail-driven ships.

## 5 Recent Geological Studies of Ischia

The island of Ischia, located at the margin of the Bay of Naples, has an area of 47 $km^2$ with 34 km of coastline. It is formed by volcanic rocks (pyroclastic flow and fall, lava domes and lava flows) from eruptive centres largely destroyed or covered by subsequent activity. The oldest volcanic activity dates back about 150 ka BP while the most recent eruption occurred in 1301–1302 (Fig. 14).

In recent times geophysical, geological and volcanological studies have been performed to gain insights into the volcano-tectonic dynamic of the Campanian volcanic district, also including the island of Ischia (Cubellis and Luongo 1998; Judenherc and Zollo 2004; Carlino et al. 2006; Berrino et al. 2008; Paoletti et al. 2009; Sbrana et al. 2009; Vezzoli et al. 2009; de Vita et al. 2010). Many of these studies agree with the hypothesis, first proposed by Rittmann (1930), that volcano-tectonic processes of the island are linked to the occurrence of a shallow magmatic intrusion (about 2 km deep) beneath its central sector. This intrusion probably took place after the large ignimbrite eruption which occurred about 55 ka ago, with emission of 20–45 $km^3$ of pyroclastic flow, pumice flow and ash-fall deposits, called Mt. Epomeo Green Tuff (MEGT). This eruption produced a sub-circular caldera collapse, about 10 km × 7 km wide, which underwent a process of resurgence, starting about 33 ka ago (Vezzoli 1988), and formed the uplifted structure of Mt. Epomeo (787 m above sea-level). The edges of the uplifted Mt. Epomeo block are marked by a system of faults and fractures (Vezzoli 1988; Fusi et al. 1990; Orsi et al. 1991; Luongo et al. 1995; Cubellis and Luongo 1998; Tibaldi and Vezzoli 1998, 2004; Carlino et al. 2006; Sbrana et al. 2009). The average uplift of ~800 m, deduced by the present-day height of marine deposits on Mt. Epomeo, occurred as a discontinuous process, until 5 ka BP, at an average velocity of about 3.3 cm $year^{-1}$. During the last 10 ka the geological history of the island was also punctuated by the occurrence of large-scale avalanching processes, producing the dismantlement of the southern slope of Mt. Epomeo between 8 and 5 ka (Vezzoli 1988; Tibaldi and Vezzoli 2004; Luongo et al. 1995; Carlino and Cubellis 2005; Carlino et al. 2006, 2008; Luongo et al. 2008, 2009).

The seafloor around the island shows an articulate morphology due to volcanic, tectonic and erosional marine geomorphologic processes. Many shallow banks,

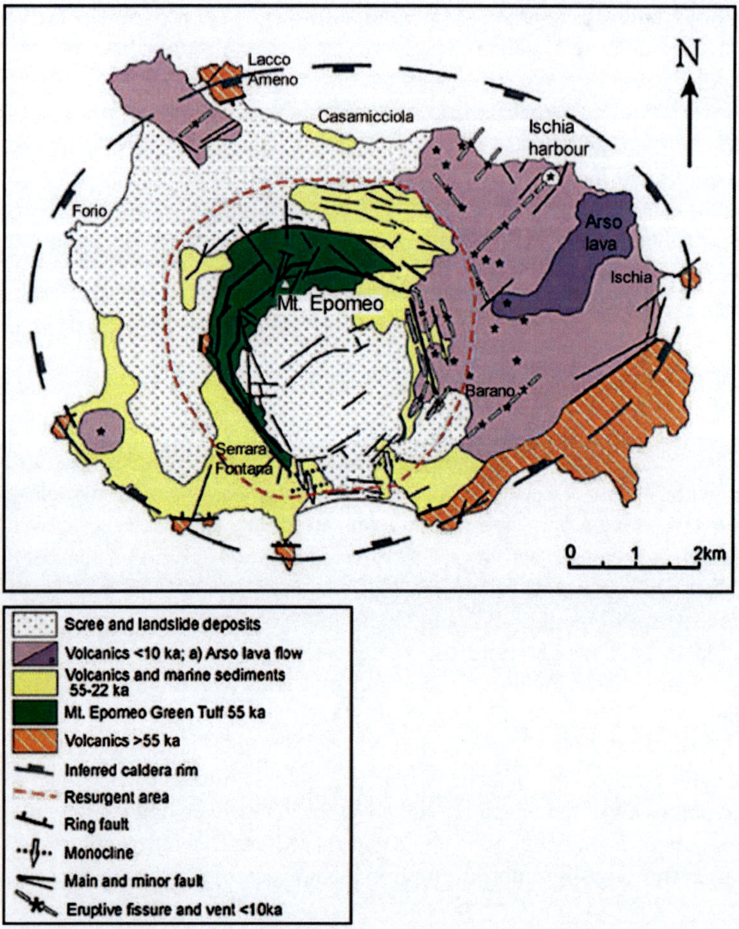


**Fig. 14** Geological map of the island with the main tectonic features. Two relevant structures are drawn: the Green Tuff caldera and the resurgent area of Mt. Epomeo (modified after Vezzoli et al. 2009)

which are the result of submarine volcanic activity, are located around the island at shallow depth. The southern part of the sea-bottom is riven by different canyons, the largest of which lies on the main tectonic alignments of the area. The continental shelf is well developed in the northern part of the island with a slight descending slope, while the southern part has a marked shelf break very close to the coastline. Another remarkable feature of the sea-bottom around the island is the presence of large debris deposits resulting from the avalanching processes which involved the slope of Mount Epomeo (Chiocci and De Alteriis 2006). The largest of these underwater debris deposits is located in the southern part of the island where a typical “hummocky topography” is observed at a depth of 600

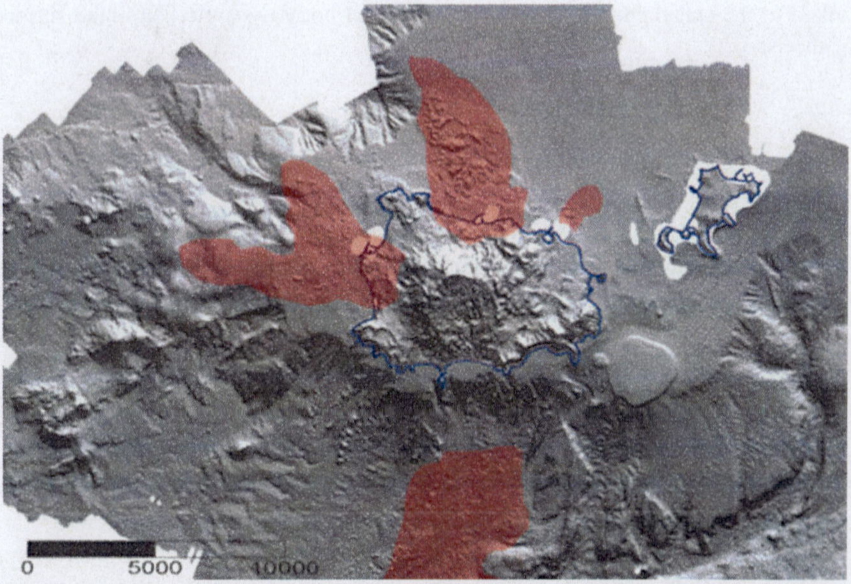


**Fig. 15** Digital Terrain Model (DTM) of emerged and submerged area of Ischia volcanic field. The different slope of the scarps in the northern and southern sector is evident. The red zones represent the main debris avalanche deposits whose source has been identified on the island. To the right the island of Procida (modified after de Alteriis et al. 2006)

to 100 m below sea-level, while only minor landslides have involved the continental shelf in near the entrance to Ischia Harbour (Fig. 15).

## *5.1 The Recent Sub-surface Dynamic*

The sequential time history of eruptions in the island can be obtained from 150 ka BP up to the last great eruption in AD 1301–1302 with lava flow in the eastern sector. Instead, the data on the earthquakes are usually reliable only from the thirteenth century onwards. From that time, many earthquakes devastated the northern part of the island; the strongest, and latest, major earthquake occurred in 1883 (Imax = XI MCS degree), producing severe damage and 2,333 victims. Afterwards, a low level of seismicity was observed on Ischia (Cubellis and Luongo 1998; Cubellis et al. 2004; Luongo et al. 2006; Carlino et al. 2009).

Presently, a diffuse hydrothermal system, with fumaroles and hot-springs with maximum surface temperatures of ~100°C, is recognizable around the resurgent block of Mt. Epomeo, while submerged archaeological ruins near the existing coastline, and levelling data since the early twentieth century testify to a general subsidence of the island (Luongo et al. 1987; Pingue et al. 2005; Manzo et al. 2006; De Martino et al. 2007). A low subsidence rate (few millimetres/year) can be linked to the stasis phase of the shallow magma body dynamic below the surface of the island.

### 5.1.1 The Geology of Ischia Harbour

The crater of Ischia Harbour was formed several centuries BC. The volcanic products in the eastern sector (San Pietro Hill) overlie a palaeosol developed on top of an older trachyte containing pottery remains from the fifth century BC and roof tiles of the sixth to fifth centuries BC (Buchner 1986; Vezzoli et al. 2009) (Fig. 16). The Ischia Harbour crater was formed by a phreatomagmatic eruption, during which the explosive energy increased, followed by a magmatic phase with a strombolian activity (Fig. 17a, b). A small positive gravity Bouguer anomaly (Maino and Tribalto 1971), close to the harbour, highlights the presence of a shallow dense mass which can be interpreted as due to the solidified lava lake formed during the last phase of the eruption (Fig. 18). This interpretation is supported by the presence of a rock block emerging from the lake surface.

What remains of the Ischia Harbour eruptive centre is a sub-circular crater ~400 m across in diameter, located along a NE–SW eruptive fissure. The related products are composed, from bottom to top, of stratified tuffs, black scoriae,

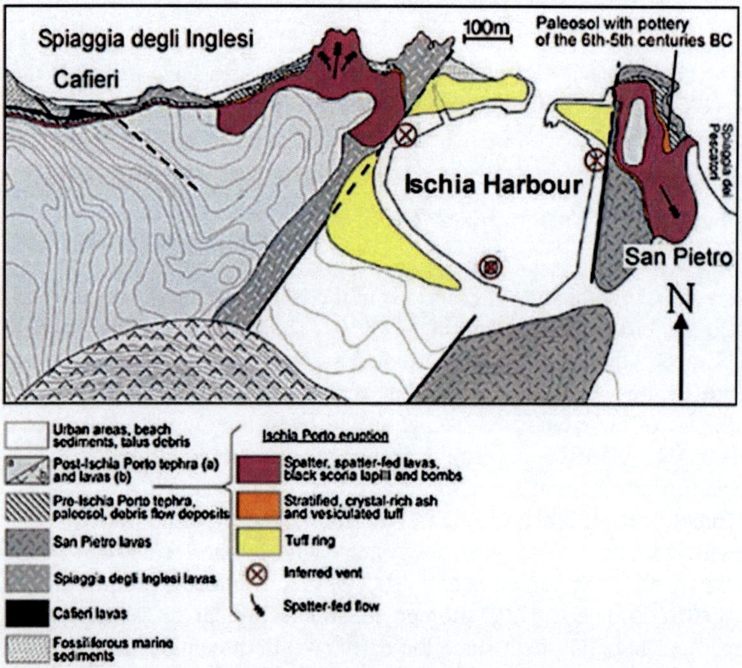


**Fig. 16** Geological map of Ischia harbour (modified after Vezzoli et al. 2009)

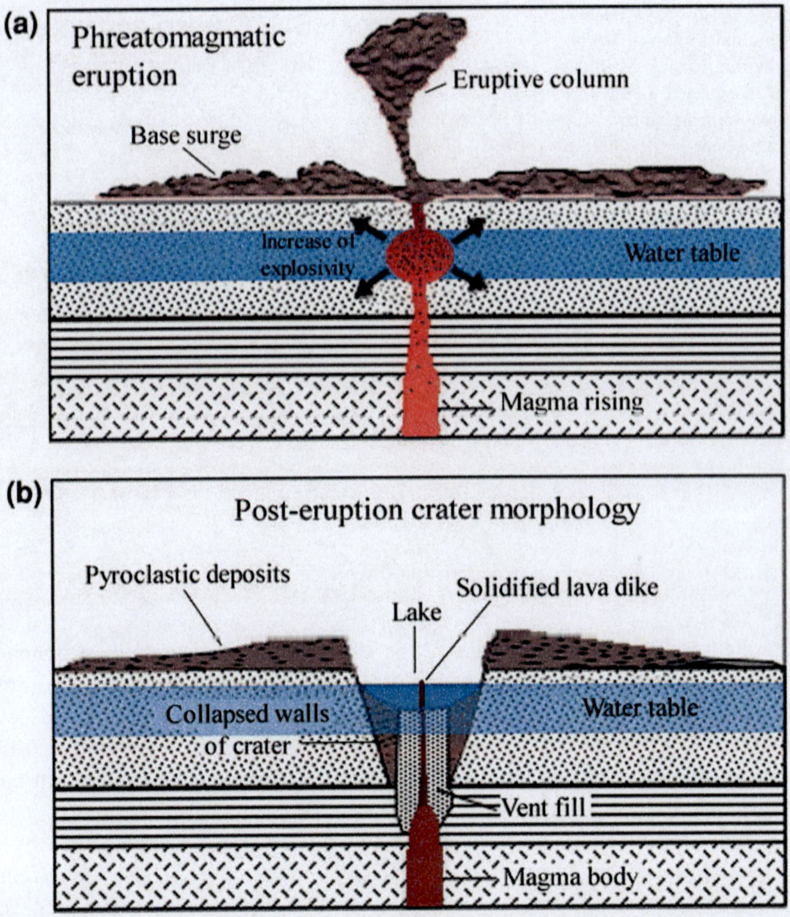


**Fig. 17** Sketch of the formation of a crater like Ischia Harbour (maar-type). This is formed after small phreatic eruptions during which the magma comes into contact with water and increases its explosivity (**a**). After the eruption a moderate collapse of the crater occurs, while the lava which has feed the eruption solidifies forming a dike; the volcanic depression is refilled by the water (**b**)

spatters and spatter-fed lavas. The basal unit shows cross laminations and abundant lithic content, which suggests a pyroclastic surge origin due to phreatomagmatic explosions. During the eruption, fire-fountains probably generated the scoriae and spatter deposits located along N–S and NE–SW-striking fracture within the Ischia Harbour crater. The internal and external slopes of San Pietro hill, to the east, and the Spiaggia degli Inglesi marine cliff, to the west, are mantled by stratified tuffs, scoriae, and spatter deposits. The external slopes are characterised by the presence of, spatter bombs welded together and flowed, ballistic lithic blocks and accretionary lava balls (Vezzoli et al. 2009) (Fig. 16).

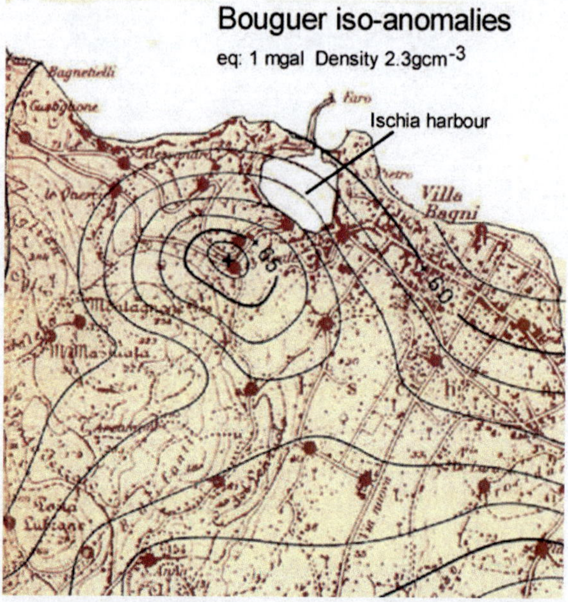


**Fig. 18** The positive Bouguer anomaly inferred from the gravimetric survey of the island, interpreted as the occurrence of a shallow magmatic body very close to the harbour area (modified after Maino and Tribalto 1971)

On the west side of the harbour, the oldest unit comprises volcanogenetic sediments composed of stratified, fossiliferous sand and ashy silt. The fossil content indicates a palaeo-bathymetry between 50 and 80 m below sea level, dating from 9,800 to about 6,400 years ago (Barra et al. 1992; Vezzoli et al. 2009). At present, these sediments are uplifted up to 50 m, testifying to a minimum uplift of about 100 m during the last 10 ka.

The most recent coastline variation of the harbour area is exemplified by the presence of archaeological ruins on the land close to the crater. These finds are also located along the coastline in the eastern sector of the island (Castello d'Ischia), in the north-western and western part (Lacco Ameno and Forio) and in the southern as well (Sant'Angelo). The finds have been dated between the fourth century BC and second AD, and consist of Greek and Roman walls, and the remains of thermal baths and pools (Buchner 1965). They provide evidence for the general subsidence of the island since they are commonly located 1.5–2 m below the present sea level; thus an inversion of the ground movements occurred, later than the Roman period; Friedlander (1938) using a tide-gauge measured a subsidence velocity of 3.4 mm year$^{-1}$. Taking into account this value and the depth of the ruins two hypothesis are possible on the subsidence: in the first the subsidence started in the first century AD, but its velocity during the last 2000 years was far less than that observed by Friedlander; alternatively, in the second the subsidence started a long time later than the Roman period.

Nowadays two tide-gauges, which belong to a wider network for sea-level monitoring in the Bay of Naples, operate on the island, at Ischia and Forio harbours. The data are also utilised for slow ground deformation monitoring related to the dynamic of Neapolitan volcanoes (Capuano et al. 2004).

## 6 The Island of Ischia: Development and Natural Risks

Since 1854, when the harbour was opened up, the island has undergone far-reaching changes in terms of its economy, society and land use. The Harbour at Ischia has become a major tourist destination besides the main point of access to the island: its shores have been radically changed by human intervention and by a development policy chiefly focusing on increasing the building meterage (Luongo et al. 2006) (Figs. 19a, b, 20). In the Ischia municipality the resident population has more than doubled since the early 1800s (ISTAT 1861–2001) and the relative urban expansion has failed to take account of the knowledge acquired of natural phenomena occurring on the island and especially their possible effects upon the region (Luongo et al. 2006). After all, the island has an active volcanic dynamic, and is affected by seismicity, chiefly in the northern sector, and has a high predisposition to landslides and floods on land (Luongo 2009; Carlino et al. 2010).

In this context, the remarkable recent urbanisation on the island and the lack of planning that accentuates the vulnerability of the region has produced an exponential increase in geological risk. Previous visible eruptions, felt earthquakes and uncontrolled freshwater flooding on the island of Ischia have produced a wealth of literature on such catastrophic natural hazard events. In general, these events are recorded in such disparate fashion, as in newspapers, in reports of property boundary and damage disputes, historical and sociological analysis, poetry and artistic works and some published scientific analysis (Cubellis and Luongo 1998, and references therein).

Eruptions, earthquakes and tsunamis have been recorded in myths, legends, demonstrated by archaeological finds, described in historical documents and the results of recent scientific and engineering surveys. Documented descriptions of

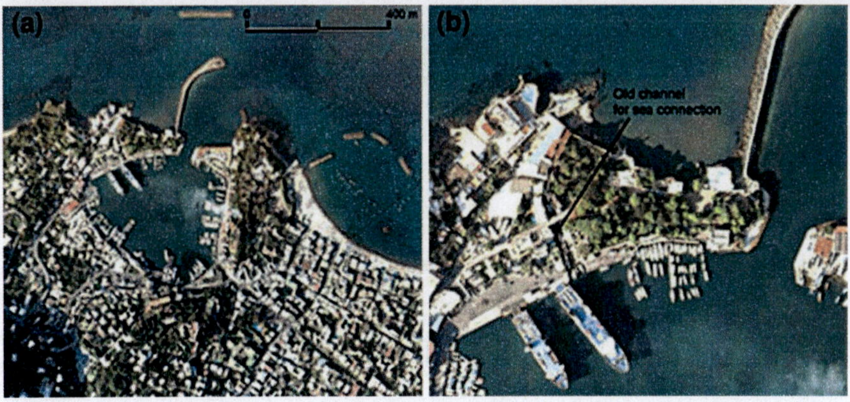


**Fig. 19** Ischia Harbour today (**a**); the considerable urbanization of the area is evident. The old channel for the circulation of the water can still be made out, though it has turned into a cement structure (**b**) (from Google Earth)

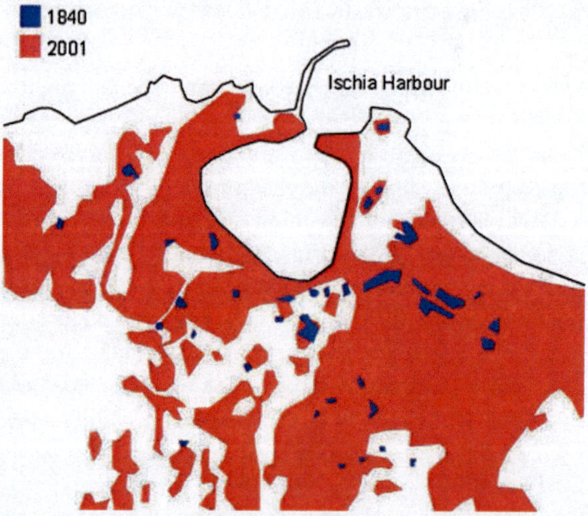


**Fig. 20** A comparison between settlements around Ischia Harbour in 1840 and 2001. A huge increase in built-up areas has occurred, with a consequent increase in exposure and volcanic and seismic risk

historical eruptive events are only available for the last eruption of 1301–1302, while there are records for eruptive events in the early centuries AD. More comprehensive accounts are available about historical seismicity. Information and documentation have been available since the 1228 earthquake. However, more detailed, useful interpretations of earthquakes concern events from the end of the eighteenth century. There is an almost inexhaustible literature related to the 1881 and 1883 earthquakes, pointing out the relationship between seismicity and the volcanic history of the island. These two seismic events occurred during an intense period of production of geological maps and fieldwork study on the island's physical characteristics. The effects of the earthquakes were classified using the former scale of intensity, and different mechanisms of the seismic source were suggested (Luongo et al. 1987; Cubellis and Luongo 1998; Luongo et al. 2006).

The earliest information on hydrogeological disasters on the island dates back to the sixteenth century when major floods affected the island during extreme local weather conditions. An archetypal phenomenon was the 1910 flood which caused serious property damage and remarkable changes in morphology of the areas directly affected by the sudden inundations. In recent times (for example, AD 2006 and 2009) floods scourged the island again, producing debris flows which devastated its northern sector, resulting in human injuries, some few fatalities and remarkably heavy damage to the island's infrastructure.

Analysis of catastrophic events on the island occurring during historical times shows, in synthesis, that its northern sector was the region affected by the heaviest earthquake and flood damage. This condition is caused by the region's tectonics and morphology, as highlighted by the integrated analysis of structural data and seismicity.

Although in recent times our knowledge of the island's geological phenomena has improved, and special attention has been laid on the effects of their most intense manifestations and related risks, regrettably among the unwary communities exposed to environmental risk few people have adopted the culture of preventive action. In light of this fact, the level of attention to extreme geological phenomena apparently only grows in concomitance with catastrophic events, while political decision-makers limit themselves to intervening only in the emergency phase, without any long-term programming of responsible land-use management policies to ensure the future reduction of geological risk. This situation becomes hard to macro-manage when, faced with the recurrence of entirely natural phenomena capable of instigating very significant damage to property and inducing tragic loss of human life, there is still the dogged continuation of illegal building, which has during the early twenty first century produced a level of risk to residents and visitors that is no longer really acceptable by a truly modern society.

## 7 Conclusions

The opening of Ischia Harbour in 1854 was undoubtedly an intervention which radically changed the hierarchical relations between the island and the mainland. It is also a fine example of the art of nineteenth century macro-engineering (Rispoli 2007). It was an operation which, in terms of techniques and execution times, was certainly ahead of its time. Although the reasons that drove King Ferdinand II to open up the pre-existing Bagno Lake, to make it the island's main port, were chiefly personal, what lay behind the choice was the spirit of renewal in regional policies, which was one of the hallmarks of the Bourbon government. At that time the island was already well known for treatments with its thermal waters, especially in Casamicciola, where since the late sixteenth century many visitors had gone to enjoy their therapeutic effects. However, the lack of safe harbours throughout the island made access difficult, especially in winter months when the frequency of sea storms increases considerably. With the opening-up of Ischia Harbour, the situation radically changed: maritime traffic and the flow of visitors shifted towards the eastern side of the island which thus became easily accessible, also for large steamships. The new centre of life on Ischia would become endowed with further improvement works, especially on the road network, which—together with the very presence of the harbour—was to generate a population increase in the *comune* of Ischia, rather than elsewhere on the island, and lead to new urban settlements.

The natural features of the harbour, obtained from an ancient crater lake, are certainly singular. However, its so well-defined shape has prevented its expansion (Rispoli 2007), which would seem necessary to handle the substantial increase in maritime traffic. Hence the morphology of the ancient lake poses a constraint to expanding today's port. Today, its navigational depth is <10 m. Maritime traffic can only be regularised on Ischia through a strategy to improve the efficiency of port services and passenger vessels. During the peak tourist season more than 20 large vessels dock and sail daily, handling about 15,000 passengers, and over 35 fast-ferry watercraft handling 8,000 passengers, making a total flow of 23,000 passengers per day. It is some surprise that, in 1853, the Provincial Council of Naples failed to approve the construction of the new harbour of Ischia, which it considered a site "unsusceptible to development". Clearly, for such regions endowed with natural, untamed landscape and cultural attractions, interventions resulting in better accessibility and fruition produce not only an increase in tourist traffic, but also proliferation in the resident human population. Moreover, as the island economy's organisational centre, the Harbour brings about the transition from settlement of the coastal zone to that of the inland zone. The trade taking place in the market-harbour promotes a deep-rooted conversion of production, from a closed, internalized intra-consumption economy to an inter-regional trade economy (Vella and Barbera 2001). In this sense, Ischia Harbour represented not only a window looking to Italy's mainland but also a key factor in the island's social and cultural aggregation.

## References

Alberico I, Lirer L, Petrosino P, Scandone R (2008) Volcanic hazard and risk assessment from pyroclastic flows at Ischia island (southern Italy). J Volcanol Geotherm Res 171:118–136

Alisio G, Valerio W (1983) La cartografia napoletana dal 1871 al 1889. Napoli

Annali Civili del Regno delle Due Sicilie (1855) Archivio di Stato di Napoli

Archivio di Stato di Napoli, Ministero dei Lavori Pubblici (1853–1954), fascio 317, Napoli

Barra D, Cinque A, Italiano A, Scorziello R (1992) Il pleistocene superiore marino di Ischia: paleoecologia e rappórti con l'evoluzione tettonica recente. Studi Geologici Camerti 1:231–243 (Special Volume)

Berrino G, Corrado G, Riccardi U (2008) Sea gravity data in the Gulf of Naples. A contribution to delineating the structural pattern of Phlegraean Volcani District. J Volcanol Geotherm Res 175:241–252

Buchner P (1946) Il protomedico Francesco Buonocore (1689–1768) ed il suo casino sopra l'odierno porto d'Ischia. In: Ricerche Contribuit e Memorie. Centro Studi Isola d'Ischia, Italy

Buchner P (1958) Giulio Isolino medico calabrese che dette nuova vita ai bagni d'Ischia. Milano

Buchner Niola D (1965) L'isola d'Ischia. Studio geografico. Memorie di Geografia Economica e Antropica, n.s.3. Napoli

Buchner G (1971) Gli Scavi di Pithecusa. Centro di Studi su l'isola d'Ischia. Ricerche, contributi, memorie, 2, atti relativi al periodo 1944–70, Napoli, pp 515–531

Buchner G (1986) Eruzioni vulcaniche e fenomeni vulcano-tettonici di età preistorica e storica nell'isola d'Ischia. In: Tremblements de Terre, eruptions volcaniques et vie des hommes dans la Campanie antique. Bibliotheque de l'Institut Francais de Naples, Deuxième serie, vol VII. Pubblications du Centre Jean Berard, Naples, pp 145–188

Buchner P (2004) Quanto mi sei caro, piccolo grazioso porticcioulo! In: La Rassegna d'Ischia, Anno XXV, no. 6, pp 43–46

Buchner G, Gialanella C (1994) Museo Archeologico di Pithecusa Isola d'Ischia, Istituto Poligrafico dello Stato- Libreria dello Stato, Roma

Capuano P, Buoncore B, Tammaro U, Obrizzo F, La Rocca A, Pinto S, Russo A, Di Sena F, Pingue F (2004) Caratteristiche spettrali delle variazioni del livello marino delle baie di Napoli e Pozzuoli. ASITA, Roma, pp 615–620

Caputo R (2000) Ischia nelle vedute romantiche dell'Ottocento nelle collezioni private. Comune di Ischia, G&E Sarnelli

Carelli G (1858) Porto d'Ischia. Napoli

Carlino S, Cubellis E (2005) The potential causes of Mt. Epomeo flank failure, Ischia Island (Southern Italy). Geophys Res Abstr 7:04171

Carlino S, Cubellis E, Luongo G, Obrizzo F (2006) On the mechanics of caldera resurgence of Ischia Island (southern Italy). In: Troise C, De Natale G, Kilburn CRJ (eds) Mechanisms of activity and unrest at large calderas. Geological Society, London, Special Publications no. 269, pp 181–193

Carlino S, Cubellis E, Luongo G, Obrizzo F (2008) Resurgence and flank failure of Mt.Epomeo, Ischia Island (Southern Italy). In: Oggiano G, Carmignani L, Funedda A, Conti P (eds) Riassunti estesi del 84° Congresso Nazionale della Società Geologica Italiana, Sassari 15–17 settembre 2008, vol 3, pp 176–177. http://hdl.handle.net/2122/4811

Carlino S, Cubelis E, Maturano A (2009) The catastrophic 1883 earthquake at the island of Ischia (southern Italy): macroseismic data and the role of geological conditions. Nat Hazards. doi: 10.1007/s11069-009-9367-2

Carlino S, Cubellis E, Iannuzzi R, Luongo G (2010) A conceptual model of geological risk in the Ischia Island (Italy): highlights on volcanic history, seismicity and flooding. Geophysical Research Abstracts, vol 12, EGU2010-0, 2010 EGU General Assembly 2010

Castagnoli F (1977) Topografia dei Campi Flegrei. Atti Convegno Internazionale "I Campi Flegrei nell' Archeologia e nella Storia". Accademia Nazionale dei Lincei, Roma, pp 41-79

Chiocci FL, de Alteriis G (2006) The Ischia debris avalanche: first clear submarine evidence in the Mediterranean of a volcanic island prehistorical collapse. Terra Nova 18(3):202–209

Civetta L, Gallo G, Orsi G (1991) Sr and Nd isotope and trace element constraints on the chemical evolution of the magmatic system of Ischia (Italy) in the last 55.000 ka. J Volcanol Geotherm Res 46:213–320

Cubellis E, Luongo G (1998) Il Terremoto del 28 luglio 1883 a Casamicciola nell'Isola d'Ischia 'Il contesto fisico', Monografia N.1 – Servizio Sismico Nazionale. Istituto Poligrafico e Zecca dello Stato, Rome, pp 49–123

Cubellis E, Carlino S, Iannuzzi R, Luongo G, Obrizzo F (2004) Management of historical seismic data using GIS: the island of Ischia (Southern Italy). Nat Hazards 33:379–393

Cubellis E, Mazzarella A, Scagliola L (2008) 1910 - L'alluvione nell'isola d'Ischia. Ambiente Territorio, n 1 – febbraio 2008, pp 48–53

Cubellis E, Carlino S, Iannuzzi R, Luongo G (2009) Eruptions, Earthquakes and flooding damage at Ischia Island. GEOITALIA 2009, VII Forum Italiano di Scienze della Terra. Rimini 9–11 settembre 2009. Epitome 3:461

D'Ascia G (1867) Storia dell'isola d'Ischia. Tipografia Di Gabriele Argenio 1867, Napoli

De Alteriis G, Tonielli R, Passaro S, De Lauro M (2006) Isole flegree (Ischia e Procida). Liguori editore, Napoli, p 73

De Martino P, Tammaro U, Brandi G, D'Alessandro A, Dolce M, Esposito T, Malaspina S, Obrizzo F, Pingue F, Serio C (2007) Area vulcanica napoletana: 10 anni di osservazioni GPS. Atti Conferenza ASITA, Torino 2007, pp 925–930

De Seta C (2005) Hackert. Electa, Napoli

de Vita S, Sansivero F, Orsi G, Marotta E, Piochi M (2010) Volcanological and structural evolution of the Ischia resurgent caldera (Italy) over the past 10 k.y. The Geological Society of America special paper 464, pp 193–241

Delizia I (1990) Ischia d'altri tempi. Electa, Napoli, p 289

Delizia I (1987) Ischia l'identità negata. Edizioni Scientifiche Italiane, Napoli, p 296

Delizia I (2007a) La costruzione della Chiesa nel quadro delle iniziative e dei propositi borbonici. In: Santa Maria di Porto Salvo a Ischia 150/75. Tota Tua 75/150. Deltastudio, Ischia

Delizia I (2007b) La fabbrica oggi. In: Santa Maria di Porto Salvo a Ischia 150/75. Tota Tua 75/ 150. Deltastudio, Ischia

Delizia I, Delizia F (2006) Ischia e la modernità "Case da re e strutture pubbliche" - Progetti e interventi borbonici 1783-1854. Massa Editore, Napoli, pp 25–38

Fazio E (1896) Porto d'Ischia, grande stazione termale e climatica. Napoli

Fiengo G (1988) I Regi Lagni e la bonifica della «Campania felix» durante il viceregno spagnolo. Olschki, collana Biblioteca dell'Arch. storico italiano, Napoli

Fischer AG, Garrison RE (2009) The role of the Mediterranean region in the development of sedimentary geology: a historical overview. Sedimentology 56:3–41

Fonseca F (1847) Descrizione e carta geologica dell'Isola d'Ischia. Ann Acc Aspir Nat Napoli 1:163–200

Friedlander I (1938) Sui bradisismi dell'isola d'Ischia e sulla Grotta del Mago. Boll Soc Geogr It 7(3):44–54

Fuchs CWC (1873) Monografia geologica dell'Isola d'Ischia. Mem. R. Com. Ital. 2, pag 60 e carta geologica 1:25.000

Fusi N, Tibaldi A, Vezzoli L (1990) Vulcanismo, risorgenza calderica e relazioni con la tettonica regionale nell'isola d'Ischia. Memorie della Società Geologica Italiana 45:971–980

Haller C (1822) Tableau topographique et historique des Isles d'Ischia, de Ponza, Ventotena, de Procida et de Nisida, du Cap de Misene et du Mont Pausilipe, vol 8. Naples, 216 pp

Hutton J (1795) Theory of the earth with proofs and illustrations, 2 vols. Cadell and Devies, London

Iasolino G (1588) De' rimedi naturali che sono nell'isola di Pithecusa oggi detta Ischia. Napoli

ISTAT (1861–2001) Censimento della popolazione isola d'Ischia. Istituto nazionale di statistica, Roma. http://www.istat.it

Judenherc S, Zollo A (2004) The Bay of Naples (southern Italy): constraints on the volcanic structures inferred from dense seismic survey. J Geophys Res 109:B10312. doi: 10.1029/2003JB002876

Luongo G (1989) Il ruolo della Scuola napoletana nella Ricerca Vulcanologica. Atti del Convegno "Rischio vulcanico e programmazione territoriale. Ricordo di A Rittmann (1893–1980) a cura di G Binni, F Obrizzo, A Vista, pp 131–142

Luongo G (2009) Highlights on seismicity, volcanic history and flooding of Ischia Island (Italy) for a conceptual model of geological risk. "Volcanoes, landscapes and cultures" Conference Catania 2009

Luongo G, Cubellis E, Obrizzo F (1987) Ischia Storia di un'isola vulcanica. Liguori, Napoli, 164 pp

Luongo G, Cubellis E, Di Vito MA, Cascone E (1995) L'isola d'Ischia: dinamica e struttura del M. Epomeo. In: Cinquanta Anni di Attività Didattica e Scientifica del Prof. F. Ippolito. Liguori, Naples, p 64

Luongo G, Carlino S, Cubellis E, Delizia I, Iannuzzi R, Obrizzo F (2006) Il terremoto di Casamicciola del 1883: una ricostruzione mancata. Alfa tipografia, Napoli, p 64

Luongo G, Carlino S, Cubellis E, Obrizzo F (2008) Ischia Island (Sothern Italy): a model of caldera resurgence. In: Oggiano G, Carmignani L, Funedda A, Conti P (eds) Riassunti Estesi del 84° Congresso Nazionale della Società Geologica Italiana, Sassari 15–17 settembre 2008, vol 3, pp 499–500. http://hdl.handle.net/2122/4812

Luongo G, Carlino S, Cubellis E, Obrizzo F (2009) Resurgent and Avalanche Caldera at Ischia island (Southern Italy). GEOITALIA 2009, VII Forum Italiano di scienze della Terra. Rimini 9–11 settembre 2009. Epitome 3:276

Lyell C (1830–1833) Principles of geology, 3 vols. John Murray, London. Republished in 1990 by University of Chicago Press, Chicago, 1399 pp

Maino A, Tribalto G (1971) Rilevamento gravimetrico di dettaglio dellìisola d'Ischia (Napoli). Bollettino Serv. Geologico d'Italia, vol XCII, Roma

Manzo M, Ricciardi GP, Casu F, Ventura G, Zeni G, Borgstrom S, Berardino P, Del Gaudio C, Lanari R (2006) Surface deformation analysis in the Ischia island (Italy) based on spaceborne radar interferometry. J Volcanol Geotherm Res 151:399–416

Minervini R (2004) 17 settembre 1854 Il centenario del porto. In: La Rassegna d'Ischia, Anno XXV, no. 6, pp 29–34

Mirabella V (1913) Cenni storici e guida dell'isola d'Ischia. Napoli. Tipografia Fr. Tramontano, p 200

Monti P (1968) Ischia preistorica greco romana paleocristiana. Napoli

Monti P (1980) Ischia archeologia e storia. Tipografia Porzio, Napoli

Morhange C, Marriner N, Laborel J, Todesco M, Oberlin C (2006) Rapid sea-level movements and noneruptive crustal deformations in the Phlaegrean Fields caldera, Italy. Geology 34(2):93–96

Orsi G, Gallo G, Zanchi A (1991) Simpleshearing block resurgence in caldera depressions. A model from Pantelleria and Ischia. J Volcanol Geotherm Res 47:1–11

Paoletti V, Di Maio R, Cella F, Florio G, Motschka K, Roberti N, Secomandi M, Supper R, Fedi M, Rapolla A (2009) The Ischia volcanic island (Southern Italy): inferences from potential field data interpretation. J Volcanol Geotherm Res 179(1–2):69–86

Pappalardo U (2006) Il Golfo di Napoli, Archeologia e Storia di una terra antica. Arsenale editrice, Napoli

Plinio il Vecchio (23–79 AD) Naturalis Historia

Pingue F, Berrino G, Borgstrom SEP, Brandi G, Capuano P, Cecere G, D'alessandro A, De Martino P, Del Gaudio C, d'Errico V, La Rocca A, Malaspina S, Obrizzo F, Pinto S, Ricciardi GP, Ricco C, Russo A, Sepe V, Serio C, Siniscalchi V, Tammaro U, Aquino I, Dolce M (2005) Geodesia (Vesuvio, Campi Flegrei, Isola di Ischia; Vulcano-Isole Eolie). In: Attività di Sorveglianza dell'Osservatorio Vesuviano – Rendiconto 2003. A cura di G Macedonio e U Tammaro, pp 59–170. http://www.ingv.it

Quaranta B (1855) Del nuovo porto d'Ischia aperto per comando di Sua Maestà Ferdinando II re del Regno delle Due Sicilie. Annali Civili del Regno delle Due sicilie,vol LIII, pp 13–17

Reale Ufficio Topografico di Napoli (1840) Fonte IGMI. Cartografia storica dell'sola d'Ischia alla scala 1:25.000. Firenze

Rispoli F (2007) Oltre il Sagrato. In: Santa Maria di Porto Salvo a Ischia 150/75. Tota Tua 75/ 150. Deltastudio, Ischia

Rittmann A (1930) Geologie der Insel Ischia (Berlino). Zeitschrift für Vulkanologie VI:268

Rittmann A, Buchner G (1948) Origine e passato dell'isola d'Ischia. Gaetano Macchiaroli editore, Napoli

Sbrana A, Fulignati P, Marianelli P, Boyce AJ, Cecchetti A (2009) Exhumation of an active magmatic-hydrothermal system in a resurgent caldera environment: the example of Ischia (Italy). J Geol Soc London 166:1016–1073

Silvestri G (2004) Nel porto la base navale inglese. In: La Rassegna d'Ischia, Anoo XXV, no. 6, pp 47–50

Taddeo G (1954) L'Isola Verde. Napoli

Tibaldi A, Vezzoli L (1998) The space problem of caldera resurgence: an example from Ischia Island, Italy. Geologische Rundschaw 87:53–66

Tibaldi A, Vezzoli L (2004) A new type of volcano flank failure: the resurgent caldera sector collapse, Ischia, Italy. Geophys Res Lett 31:L14605. doi:10.1029/2004GL020419

Vella A, Barbera F (2001) Il territorio storico della città Vesuviana, struttura urbana e sviluppo della fascia costiera. Laboratorio Ricerche e Studi Vesuviani, Napoli

Vezzoli L (ed) (1988) Island of Ischia. Quaderni de La Ricerca Scientifica, 114, vol 10. CNR, Roma

Vezzoli L, Principe C, Malfatti J, Arrighi S, Tanguy JC, Le Goff M (2009) Modes and times of caldera resurgence: the <10 ka evolution of Ischia Caldera, Italy, from high-precision archaeomagnetic dating. J Volcanol Geotherm Res 186:305–319

# "And an Island Never Cries": Cultural and Societal Perspectives on the Mega Development of Islands in the United Arab Emirates

**Pernilla Ouis**

## 1 Why Islands? Islomania in the United Arab Emirates

This chapter aims at revealing cultural and societal aspects of the megaengineering of artificial islands and other island development projects in the United Arab Emirates (UAE). The UAE is a federation established in 1971 of seven traditional sheikhdoms or emirates, located on the Arabian coast of the Persian Gulf. The largest emirate Abu Dhabi; and also the name of the capital, but perhaps most famous is the city and emirate of Dubai for its many spectacular and extravagant megaprojects. The number and size of these new island projects can be understood by studying the elaborate chapter "Infrastructure" in the latest official *United Arab Emirates Yearbook 2009*. This chapter recounts for several projects of this kind that will be the starting point for my article and explained below in detail: In Abu Dhabi city the yearbook mentions that the islands named Sowwah, Al Reem, Saadiyat, Yas and Lulu are undergoing reclamation. In Dubai the world's largest artificial archipelago called The World has been finalized and the three large man-made Palm Islands are under construction with one recently being completed. In the coastal areas of the smaller and poorer northern emirates, islands are being developed as well, following the pattern of the larger emirates.

These islands are all under development, and facts regarding them can not be found in literature yet, but rather on the developing companies' homepages. The presentations found on such homepages are often expressed in visions and fantastic models, while most of these islands, at the time of writing this chapter (spring 2010), are nothing more than large construction sites. Only the future can

P. Ouis (✉)
Faculty of Health and Society, Malmö University, Malmö Sweden
e-mail: Pernilla.ouis@mah.se

V. Badescu and R. B. Cathcart (eds.), *Macro-engineering Seawater in Unique Environments*, Environmental Science and Engineering,
DOI: 10.1007/978-3-642-14779-1_3, 

tell if all these great projects will be realized as they are planned particularly given the current economic crisis. This chapter builds on secondary literature sources, Internet and my previous research (Ouis 2002a, b, 2007 and forthcoming) rather than conventional ethnographic fieldwork in the UAE. Moreover, the net-encyclopedia Wikipedia has been consulted, not as a reliable source for facts, but rather as identifying certain "rumors" regarding some projects, in cases where information no longer is available from alternative sources on the Internet.

The aim of the chapter is to discuss some forgotten aspects of cultural and societal aspects of such megaprojects (i.e. the artificial islands) from a human ecological view. A human ecologist approaches nature as if there were humans in nature. To state it simply, in human ecology the actions and interactions of humans in nature are included in the study of ecology. Humans are not disregarded as alien "anthropogenic" factors in nature.

With regard to the many island projects in the emirates, it could be correct to say that there is an "islomania" going on the UAE, a concept used to denote a strong affection or obsession with islands. The term was introduced by the writer Lawrence Durrell in his book *Reflections on a Marine Venus* (1953). Wikipedia, under the entry "islomania", quotes a letter by Durell, in which he writes: "Islomania is a rare affliction of spirit. There are people who find islands somehow irresistible. The mere knowledge that they are in a little world surrounded by sea fills them with an indescribable intoxication." John Gillis (2004) further explores the theme of islomania stating: "Islands evoke a greater range of emotions than any other land form. We project onto them our most intense desires, but they are also the locus of our greatest fears. We feel extraordinary free there, but also trapped. Associated with pleasure, islands also harbor pain, for they are prisons as frequently as escapes or refuges."

The enormous efforts and resources invested in island projects in the UAE can be interpreted as a way to profit on the mystery and lure attached to islands. The country is struggling to transform its economy from fossil fuel export into tourism and finance. Islomania is one strategy to create attractiveness in a rather barren environment, somehow empty both in terms of natural wonders and historical sites. The new islands fulfill many needs in Emirati society: as new urban flagships necessary for economic diversification, as icons of the traditional past, and as concrete islands enabling economic, religious, social and cultural segregation. A flagship denotes spectacular grand buildings or other centers to stimulate urban development and attract investors, tourists and residents on the global market (see for instance Smyth 1994). The boom in islands is also encouraged by a recent change in the law making it possible for foreigner to buy property on freehold ownership, something hitherto forbidden for non-UAE nationals. Furthermore, being a human ecologist I will attempt to touch upon some ecological aspects of these islands, as well as discussing how artificial islands create a nature-map inversion. Furthermore, a new type of traveler is being identified: the Google Earth tourist. Finally, I will discuss these islands from a perspective of dependency.

## 2 Island Projects in Abu Dhabi: The Responsible Sister

Abu Dhabi is the capital of the United Arab Emirates, but also the name of the largest (87% of the total territory of the country) and wealthiest emirate of the federation. This is the only emirate that truly fulfills the conditions of being a classical oil state, and the whole federation's expenses are mainly financed by Abu Dhabi's oil revenues. This emirate is invoking philanthropy and the tradition of "playing the role of the more responsible sister" (Easterling 2007), perhaps in relation to its more spoilt little sister Dubai. Still, the emirate invests to meet the post-oil era, which is reflected in its new island mega-projects. The development of the Sowwah Island and Al Reem Island in Abu Dhabi will be the core of the new Central Business District of the city. The district will be about 105 hectares of "commercial, residential and retail infrastructure supported by an extensive network of 13 bridges, public transport links and pedestrian facilities" (*United Arab Emirates Yearbook* 2009) to connect the area to the existing city centre. The development is associated with economic diversification. The chairman of the Sowwah Island's developer Mubadala, a government-owned company, comments:

> The work of Mubadala as a catalyst for diversification is particularly important. Through partnerships with local and international organizations, Mubadala builds new businesses and in some cases entire new sectors within the economy of the Emirate of Abu Dhabi. These initiatives expand the economic base of the Emirate and in doing so promise to deliver greater stability and sustainability to its economic performance over time. (Message by His Highness Mohamed Bin Zayed Al Nahyan Crown Prince of Abu Dhabi and Chairman of the Board, Mubadala 2010a).

Abu Dhabi has a green, environment-friendly profile on the global market (Ouis 2002a). However, the emirate struggles with the paradox of maintaining an ecological image while selling fossil fuels (that are seen as causing the most acute threat to the global environment due to green house gasses), and still trying to develop their economy. Development companies are well aware of the criticism from the environmental perspective regarding these megaprojects, and always try to address ecological issues. The developer's home-page promises that, although the Sowwah Island project will encompass "a diverse mix of attributes that are vital in the world's leading CBDs, including commercial, hotel, retail and residential clusters as well as community facilities, public amenities, parks and open spaces", it is designed as "a sustainable urban environment". The island will promote alternative means of transport, and reduce the use of energy and water due to its innovative architecture and landscape (Mubadala 2010b).

Sowwah Island, just as Al Reem Island, are both natural islands that are being developed. This is rather new in the region and is considered to be more "eco-friendly" than the creation of new artificial islands, which is the course taken by Dubai. Furthermore, it should be noted that Abu Dhabi since 2005 only accept members in the Gulf Cooperation Council (GCC) to own freehold property, while non-GCC members may have an arrangement with 99 year contracts (Easterling 2007). This suits well a more cautious image in dealing with foreign cultural influx. Al Reem will be developed for the same use as Sowwah and is marketed to new residents on the homepage in the package of "a city in the city" as so many islands, in addition to the value of islomania:

> Consider your dream home come tucked away in a picturesque island; verdant, mysterious and inspiring. Away from the stressful chaos in the city and its life; yet, with every convenience of the metropolitan lifestyle, that integrates work, home and recreation onto one rostrum, freeing up much valuable time for you to spend doing the things you enjoy. (Al Reem Island 2010).

Yet another island in Abu Dhabi, Saadiyat Island, will be developed and integrated with the city's Cultural District, which includes large museums such as Guggenheim Abu Dhabi (GAD) or Abu Dhabi Louvre that surely will become future flagships for the emirate. Abu Dhabi has a reputation of promoting itself as a center of culture and education, and the Saadiyat development plan should be viewed in this perspective. Hence, the island will be "home to one of the most contagious global spatial products of architourism and culture: the Guggenheim organization" (Easterling 2007). GAD will be 25 times larger than Bilbao and will be completed in 2012. It is expected to be "Abu Dhabi's most iconic masterpieces" (Tourism Development and Investment Company (TDIC) 2010a). The island will host 160,000 inhabitants, but will also include some nature reserves as well. The green image of Abu Dhabi is important in the global marketing of Abu Dhabi as previously discussed, as well as a way of preserving the legacy of the politics of the late Sheikh Zayed bin Sultan Al Nahyan, the "father" of modern UAE who ruled Abu Dhabi from 1966, and through federation from 1971 and until his death in 2004. The developer TDIC explains the project's vision: "Occupying 27 km$^2$ the entire project is due for completion by 2020 and is created around an eco-sensitive philosophy with a special low density masterplan" (TDIC 2010b). The island will have pedestrian access as well via a ten-lane highway bridge. It has a budget for more than 22 billion US dollar, and will be completed in 2018. The island will include 19 km of new beachfront, 29 hotels, 8,000 villas and 38,000 apartments, according to the UAE Yearbook 2009.

The yearbook also mentions the Lulu Island of Abu Dhabi, a man-made reclaimed island completed in 1992, but not opened to the public until 2007, designed mainly for leisure purposes. Emirates News Agency (2007) reported upon the opening of the island that the objective of the 10-km island development was to create a special world of experiences, combining entertainment with the cultural, recreational and leisure elements. Amenities include marinas, gardens, canals, botanical gardens and children water parks, an equestrian riding centre and a variety of resort accommodation.

But still there are more projects of this kind in Abu Dhabi; the Yas Island will be developed. This island is 25 km$^2$ and will create a 32 km of "new" waterfront and its development will include the creation of three large theme parks: Ferrari World Abu Dhabi, Warner Brothers Theme Park and Yas Island Water Park (UAE Yearbook 2009). The development of the island will include the building of new signature hotels, theme parks as mentioned, marinas, shopping malls and golf courses, just as the rest of the island projects. The "extra" on this island will perhaps be its car racing circuits (Yas Marina Circuit 2010). The island is expected to attract permanent residents as well as tourists.

This is a very short description of the Abu Dhabi island projects, but it has to be kept in mind that these projects include a massive infrastructure development of housing, towers, offices, roads, bridges, commercial and touristic sites. And apart from these island projects, other coastal projects are being developed in the country—the islands are just one scene for the megaprojects.

## 3 Duabi: The world of Artificiality

Dubai, the second emirate of the federation, long ago transformed its economy from fossil fuel into tourism. It is famous for its outstanding artificial island projects. The creation of the flagship of Dubai, the first "seven-star hotel" in the world, Burj al-Arab, was preceded with the construction of a large artificial island to build the hotel upon. The island is located 280 m from the beach and took 3 years to build; the construction of the hotel took 3 years as well. The hotel can only be reached by a gated bridge or helicopter.

Another spectacular coastal development project is Dubai Marina, a canal city in the "Venetian tradition" according to marketing, with a large central waterway that has been excavated from the desert with a length of 3 km. It was completed in 2004 and the area can accommodate more than 120,000 people (Design Build Network 2010). The marina is not an island as such, but is somehow a forerunner for the island projects in respect to architecture and master plan.

Dubai allows all foreigners to own property, a change in law from 1997 (in special areas, but after 2006 in the whole emirate) that probably has fuelled Dubai's boom in individual freehold property on the global market (Easterling 2007). The artificial archipelago The World of 300 islands together forming the shape of the world, was finalized in January 2008 and the building of infrastructure could commence. Each island can be purchased individually and the UAE yearbook 2009 reports that 50% of the islands have been sold, but it is uncertain if this number holds true after the financial crisis of 2008. The World has attained much international attention and is being promoted by the developing national company Nakheel as: "A journey. A saga. A legend. The World is today's great development epic. An engineering odessy to create an island paradise of sea, sand and sky, a destination has arrived that allows investors to chart their own course and make the world their own." (Nakheel 2010a). The uniqueness, mystic and originality of each island are used in the marketing. The World further described as "the most innovative real estate on Earth", while still "flourishing marine life", according to Nakheel (2010b).

Perhaps the most spectacular of all artificial islands are the Palm trilogy, three islands each in the shape of a palm tree also constructed by Nakheel. The construction of the first island in a 24/7 procedure, Palm Jumeirah began in June

2001 and was finalized in 2006. The island measures about 5 km by 5 km and residents have started to move in. The procedure for creating this kind of artificial islands (just as The World by the way) is to use sand dredged from the bottom of the sea and which is sprayed onto the designated area. Shortly after Palm Jumeirah was finished a new palm island, Palm Jebel Ali, was announced and reclamation work began. This island will be twice the size of Palm Jumeriah and calculated to be able to accommodate 250,000 residents by the year of 2020 according to the UAE Yearbook 2009. The buildings of the island are mainly private residences owned by foreigners in addition to luxury hotels, and the usual facilities for tourists: beaches, bars, clubs, marinas, restaurants, parks, towers etc.

The development of the third island, the largest of them all Palm Deira commenced in October 2004, an island that will have the incredible size of 46.35 $km^2$. However, it is difficult to get exact information regarding the development of this third palm that has just begun. According to Wikipedia, its "first announced design was 8 times larger than the Palm Jumeirah, and 5 times larger than the Palm Jebel Ali, and was intended to house one million people. Originally, the design called for a 14 km (8.7 mile) by 8.5 km (5.3 mile) island with 41 fronds. Due to a substantial change in depth in the Persian Gulf the farther out the island goes, the island was redesigned in May 2007. The project then became a 12.5 km (7.76 mile) by 7.5 km (4.66 mile) island with 18 larger fronds." It should be clarified that this source has been consulted, since the official site of the Palm Triology (by Nakheel) does not provide any information regarding the planned size for Palm Deira.

After the autumn 2008 and the financial crisis, neither the *Yearbook 2009* nor Nakheel's homepage is particularly explicit about the Palm Deira plans. The *Yearbook* expresses rather vaguely that Nakheel announced in December 2008 that some of its development schedule in the Gulf "is expected to slow down". Also the excavation project Arabian Canal, the 75 km long canal that will link Palm Jumeriah and Palm Jebel Ali, developed by Limitless announced that the company will review its construction schedule for the project. Still, the Yearbook 2009 mentions the new plans by Nakheel: The Universe, a cluster of coral-shaped islands that will emanate from Palm Jumeria to the other Palm Deira. These new islands will according to images found on the Internet be located between The World and Dubai coast. Hence, the whole coastal area will be filled with artificial islands. However, the Yearbook mentioned that the plans for The Universe will take 15–20 years to realize.

Another important urban flagship relating to Dubai's coast is the Hydropolis, a large underwater hotel supposed to become a "10-star hotel". This hotel should have opened already in 2006, but due to technical, ecological and perhaps financial, problems its opening was delayed. The Yearbook 2009 does notably not even mention this megaproject at all and the project seems to be in limbo. Again Wikipedia, under the entry Hydropolis, is one of the few sites giving any information, states that:

> Architects working on Hydropolis have had some difficulty selecting a suitable position for the complex, as concerns have been repeatedly raised about the displacement effect of

building a 260-hectare underwater structure. If their calculations are incorrect, tides and sea levels off the coast of Dubai could be severely affected, rendering vast tracts of coast uninhabitable due to chaotic tides, flooding, unpredictable wave patterns and a high possibility of whale and other sea-mammal beaching.

## 4 The Smaller Emirates: The Poorer Cousins from the Countryside

It is somehow easy to forget that the UAE is a federation of *seven* independent emirates, since Abu Dhabi and Dubai tend to dominate the whole country. However, the other smaller emirates in the north and east are much more reminiscents of the emirates before the boom in megaprojects. These emirates, Sharjah, Ajman, Umm al-Qaiwan, Ra's al-Khaimah and Fujairah, could be perceived as the poorer cousins from the countryside compared to urban and fashionable Abu Dhabi and Dubai. These poorer emirates are getting help to develop from national finances, and here islands are being developed as well. Al Nujoom Islands are developed in the Sharjah emirate, mentions the Yearbook 2009. This complex of ten islands are to be created by excavating an intricate system of waterways from the coast, rather than creating man-made islands by spraying sand as in Dubai. The Saudi based company Al Hanoo Holding Company is the developing company. The building of infrastructure and residences has just begun on this complex, and the islands will accommodate over 40,000 new residents when finalized.

In the emirate Ra's al-Khaimah a number of artificial islands are underway. Dana Island, designed as a main stem in three parts and ten leaf-shaped islands and Al Marjan Island, five coral-shaped islands, are also designed as multi-use developments, according to the Yearbook 2009. In this emirate the natural Saraya Islands are being developed as well. This project consists of the islands Al-Boum, Al-Marsa, Al-Sahab and Al-Wahat and they runs parallel to Ras Al Khaimah's coastline, creating a natural lagoon between the islands and the mainland (Sayara Ras Al-Khaimah 2010). It seems like these projects are repetitions of the grand touristic projects of Abu Dhabi and Dubai, but on a more moderate scale. However, all these projects, as well as those in Abu Dhabi and Dubai, are all hit by the decline in the economy.

## 5 On Cultural Icons: Legitimizing Modernity by Referring to Tradition

I have in my PhD thesis in Human Ecology from 2002 promoted the idea that legitimization of power in the UAE is dependent on reference to *both* tradition and modernity (Ouis 2002a). Another way of expressing the same thing is to say that the rulers' ambitions towards modernity are sanctioned by re-articulating important references to the past. Important icons from the traditional lifestyle are frequent in the modern emirates. Cultural symbols such as camels, falcons, tents, boats, pearls, coffee pots, palms and other references to the traditional Bedouin heritage and the former pearling and fishing economies are of importance in the modernization process. Throughout the emirates, these national icons are visible, in various flagship developments, as roundabout monuments, in buildings, landscaping and in advertisement. An example of these referents is the Burj al-Arab hotel that is a modern price of architecture, but based upon the shape of sails from a *dhow* (traditional boat), enabling the space between the sails to create the world's biggest atrium of 180 m in height. The hotel has almost literally become the ultimate flagship for Dubai; at least it was until the Burj Khalifa opened (see discussion below).

It is easy to speak ironically of the indulgence in "traditions" in this extremely ultra-modern society as found in the emirates, at least on the surface and in respect to technology and infrastructure. Still, traditions are important to Emirati culture and social structure. The "traditional" is somehow perceived as unauthentic by outsiders. The traditions referred to are definitely so-called *invented traditions*—modern attempts to establish continuity with a suitable historic past (Hobsbawn 1983). Still, this is not to say that these traditions are not genuine. Rather they are "articulations of tradition in a modern society" (Ouis 2002a). We have to keep in mind MacCannell's (1976) words that the victory of modernity is not the disappearance of the traditional society as such, but rather the ambitions to *preserve* and *conserve* it. Therefore, it makes no sense to be ironic over this constructed "folklore" in the United Arab Emirates, that has been criticized by different researchers as "folklure" (Khalaf 1992) or "fakelore" (Brorman Jensen 2007). The phenomenon of invented traditions appears in all societies during modernization processes, but are much easier identified in "The Other". I would however, argue that what is specific in the UAE is that the legitimization of power and the introduction of modernity are dependent on references to traditions, probably because of the tribal social fabric of the emirates and the speed of change.

The palm islands are typical examples of the successful combining of modernity and tradition. The very shape of the islands—as palm fronds—is ingenious in the perspective of creating long beaches, but also as an iconic shape referring to the traditional date palm. This tree has been planted in millions in the emirates, and is the base for the greening processes of the country (Ouis 2002b). Dates have been a prerequisite for survival in this harsh environment, and the very social structure in traditional Emirati society had the organization of the date farming at its core. The cover of the 1996 official UAE Yearbook 1996, at the official 25 anniversary of the birth of the nation, was the date palms' abundant greenery photographed from above, thus making the date palm perhaps the most important cultural icon in the UAE. On the official Palm Trilogy site it is said that the shape of the palm was chosen to solve the problem of Dubai's beach shortage: "The Ruler of Dubai drew a sketch of a palm tree, realising its fronds would provide more beach frontage than a traditional circular island. From this insight, the idea of Palm Jumeirah was born and the trilogy of islands envisioned." (Nakheel 2010c).

The statement above is interesting, since it attributes the genius shape to the ruler Sheikh Mohammed bin Rashid Al Maktoum. The idea of a palm shaped island is both rational, and hence modern, in the same time as it is a reference to the traditional. However, another source states that it was Warren Pickering, an architect based in Australia that came up with the palm idea to solve the beach shortage in Dubai, not Sheikh Mohammed (Elsheshtawy 2004). In official discourse it is however always the ruler that is portrayed as being personally responsible for the articulation of traditional values and symbols. By referring to tradition, the ruler's position is secured.

Yet another arrangement referring to tradition and the legitimization of power of Dubai ruling sheikh, also only viewed from air, are the planned housing typologies on Palm Jebel Ali forming a poem by the same ruler of Dubai. The translation of the poem in English goes:

> Take wisdom only from the wise, Not everyone who rides a horse is a jockey. It takes a man of vision to write on water, Great men rise to great challenges. (Translation at Eikongraphia 2010a)

It is an obvious interpretation of the poem that this is a eulogy to himself, however, what most outsiders do not understand is the importance of poetry for the sheikh's secure position and leadership. On his personal homepage his role as a poet is presented as an important aspect of his persona. His poetry is defined as classical Nabati poetry. It is explained as: "Nabati poetry shows the natural creativity of the Gulf's inhabitants and represents their roots in this land. Its form and content, literary significance, social function and historical value make it one of the Arabian Peninsula's great literary treasures" (Sheikh Mohammed 2010). To underline his genuine talent for poetry it is explained that his first published poetry, was published under pseudonym as he wanted to secure that he was not published only because he belonged to the ruling family. His talent in poetry is an asset evoking feelings of authenticity, rather than the artificiality Dubai is famous for embracing.

The tallest building in the world, Burj Khalifa (referred to as Burj Dubai during the building process), is also relating to the traditional and specific, by resembling a desert flower, with the tower's wings extend from a central core. On the tower's homepage it is stated: "No stranger to Middle Eastern design, architect Adrian Smith incorporated patterns from traditional Islamic architecture. But his most inspiring muse was a regional desert flower, the Hymenocallis, whose harmonious structure is one of the organizing principles of the tower's design." (Emaar 2010).

The example of Burj Khalifa reveals another interesting feature of these urban flagships: their vulnerability in addition to the competition between rulers of the different emirates. This tower—the tallest building in the world—that was formerly known as Burj Dubai, was supposed to become the pride of Dubai. However, the financial crisis hit Dubai severely, and it had to turn to Abu Dhabi for financial rescue. In return, it had to rename the tower after the ruler of

Abu Dhabi (and, it should be noted, the president of the UAE) Sheikh Khalifa bin Zayed bin Sultan al-Nahyan. The price tag for this move is said to be $1.5 billion (Tomlinson and Robertson 2010). They report from the top of the tower at the opening ceremony after the "humiliating" re-naming of the tower: "Emerging near the summit after an ear-popping lift ride to its 124th floor, one is reminded of how far Dubai has to go, seeing how much of the city remains incomplete." The traditional sheikdom's power and their long historical internal competition within the federation is actually manifested in the many and large megaproject such as in the various islands project within the country. The UAE as a whole exposes an over-capacity in international airports, hotels, shopping malls, flagships—you name it—due to its traditional competitive ruling structure, and one may add, the lack of central planning.

## 6 Hybrid Urbanism: Islands of Segregation

The word "hybrid" is often used to describe urban development in the UAE. Katodrytis defines Dubai in terms of hybrid urbanism, a term referring to identity discourses articulated in architecture and the built environment (AlSayyad 2001). Katodrytis (2007) describes the city to be everywhere and nowhere, a park-oriented cityscape, a city with no core but consisting of mind-zone spaces, where everything "seems to point to the twin towers of consumerism and tourism". Elsheshtawy (2004) problematizes the view of Dubai as a place of hybrid culture. He says that the official image propagated is that of a happy merging of cultures, while the reality is quite different. In the UAE each cultural, religious and/or ethnic group can create and maintain its set of institutions; they are even requested to do so. The Emirati society organizes differences by segregation.

My observation in the UAE is that it is an extremely segregated society, divided into isolated symbolic "islands" according to ethnic, cultural, economic, religious or other social belongings, opposed to all kind of hybrid ideals. Perhaps most segregated from the rest of society, is the large population of immigrant labour. Elsheshtawy (2004) discusses how no effort is made to resolve social problems in the UAE or to address problems of the immigrant workers or even to improve their conditions.

Marx believed that we have to demystify the social relations behind the consumer goods (i.e. to reveal the exploitation of the labour class), and I think it is necessary to identify the input of labour behind the islands. Basar et al. (2007) notes ironically that literature on Dubai are either "neo-liberal coffee-table gloss-dross or neo-left moralizing". With the risk of being associated to the neo-left, I think it is important to understand the associated injustices towards the majority of the population: the non-citizen, low-paid, low-skilled working class living in secluded areas. It is crucial to emphasize that the grand and fabulous megaprojects such as these artificial islands are dependent on the non-stop 24/7 labour of the guest workers, work performed under what most societies would consider unacceptable conditions. However, what they create with their work are places they will never have access to once these islands are finalized.

The islands themselves, once finalized, are example of Emirati socio-cultural-economic segregation. Who are they for? It is definitely not for the majority of the people already living in the emirates, not even the middle-class; instead the ultra-rich global elite is the target group for such constructs. Planners are expecting hundreds of thousands of this clientele to move to the UAE after the foreign freehold property boom. One may wonder if these plans and astronomical numbers of residents are realistic, but what characterizes these new environments? These islands are utopias, imaginary spaces, dreamscapes or fantasy enclaves, but above all they are colonial spaces, according to Jackson and della Dora (2009). They comment:

> Fantasy and luxury spaces—as the Dubai and post-Dubai islands are often deemed to be—normalise continuities of structural imbalances in the same way as do perceived 'failures' of urban space (slums, ruins, wastelands, exhausted spaces, etc.). Fantasies and failures *together* constitute the middle ground of so-called normal expectations of contemporary commodity-driven urban dwelling. Fantasy constructs the imaginary utopic lure as a commodity. Failure constructs the space to be overcome through technical mediation and consumption. (ibid).

As colonial spaces, the new islands have become gated communities for the global and local elite. Elsheshtawy (2004) comments regarding the Palm islands: "Their exclusive nature is further highlighted by locating them offshore, and controlling access through a variety of means: security guards, entrance fees, and by their sheer extravagance, prohibiting (or inhibiting) ordinary folk from entering." Brorman Jensen (2007) argues along similar lines saying that the Palm islands as gated communities opposes the idea of utopian spaces as the "antithesis of exclusion", saying: "The Palm has therefore become an inverse utopia and nothing less than an inversion of the Enlightenment's leitmotif." Social interaction in the Emirati society is built on the dichotomy exclusion and inclusion. It is very clear to each individual living there, with whom one can socialize and in which environment. For instance, Yas Island is presented to future residents:

> An island of dreams is taking shape, and for some this will be the home of their dreams. The Yas Island project is all about balance. Balance between thrills and relaxation, man and nature, leisure and residential environments. Across the island there will be a variety of modern low-rise and mixed-use residential areas nestled within landscaped public gardens and alongside parks that incorporate walking/jogging tracks and sporting fields. Imagine taking in the panoramic view of the island of dreams everyday from your luxurious living space. (Aldar 2010).

In reviewing many sites selling these new residents of islands, I have become aware that one often repeated marketing slogan for many of these islands is that the island will constitute "a city within the city". This alludes to the isolation of islands, and thus segregation. The statement by Gillis quoted in the introduction of this article speaks of islands as places where one can feel *both* free and trapped. The lure and attraction of the Emirati islands on the global market must surely be less if they can be perceived as segregated and guarded containers for the elite, almost like prisons.

The islands' closure towards the poorer majority of the population is a strategic closure against traditional and Islamic norms prevalent in the rest of the UAE. The islands will offer an escape for the local elite if they wish to challenge the Islamic morality, while physically still remain within their own country. Judging from the advertisement of the islands, it is Western norms and standards that will prevail in these environments. On the other hand, these islands can welcome Westerners to the UAE without a need for the UAE to liberalize the Emirati society. The islands provide a "free zone within" both for Westerners as well as the Emirati citizens for living according to liberal moral codes with for example access to alcohol. It is almost a paradisian space presented in the marketing of these islands:

> Visitors will come to enjoy the magnificent sunsets, unwind in the spas, pools and gardens and eat out at the fine selection of restaurants. There's a wide range of daytime activities, including water sports, aquariums, theme parks and of course miles of beautiful beaches. And at night the bars, clubs and shows will offer plenty of fun and amusement. So why not come and explore the endless possibilities Palm Jumeirah has to offer you. (Nakheel 2010d).

However, my claim that the island is an escape from Islamic moral codes, should be clarified. It is perhaps only a possibility for national Muslim *men* to escape the religious norms, since it is totally unacceptable socially for Emirati women to visit night clubs and bars or to swim (i.e. undress and take off their traditional, completely covering, black clothes, although some may use the so-called burkini or have access to gender segregated swimming pools). In all respects of segregation be it based on gender, religion, culture, ethnicity or economic status, it is easy to agree with Elsheshtawy (2004) conclusion that the effect of globalization is segregation in respect to the development in the UAE:

> Globalization has thus become another form of domination—primarily catering to those who have the necessary resources. While the imagery associated with the production of spatial forms is linked to cultural elements—an essential post-modern condition—thus suggesting hybridity, it ultimately caters to elite elements, thereby excluding ordinary citizens. Unfortunately, hybridity becomes a 'conspiratorial' actor by concealing its real intentions, cloaking itself in the mantle of tradition and (manufactured) heritage.

## 7 Artificial Islands: The Nature/Map Inversion

Mapping is a human activity in the encounter with nature. However, in relation to artificial islands, "this alliance has been turned upside down", notices Brorman Jensen (2007). The map—the human drawing—becomes nature, and the only limit for nature is the human fantasy. The Palm trilogy is considered to be the 8th wonder in the world, "a miracle that is singled out precisely by being something that cannot be imagined". In the press material for the island it is said to be "beyond even nature's vivid imagination" (ibid). Man has thus become the creator of nature. This fact has perhaps inspired Erandi de Silva to elaborate on a "logopelago" of islands in the shape of famous brands such as Disney-Isle-Land (in the shape of Mickey Mouse), Nike's Swoosh symbol, Ralph Lauren Polo Island or why not a Calvin Klein logo Island? (Basar 2004). The island is the symbol, the logo is represented with a particular exclusive lifestyle on the island itself. It is commented: "With projects such as The Palm and The World, Dubai has resuscitated the desire for one-liner icons that speak directly to the sky." (ibid).

Jackson and della Dora (2009) argue that the islands are constructed with the purpose to be recognized by their very artificiality, while they are also built to be invisible from the ground or sea level. They are built to been seen only by a "God-like viewer" such as a millionaire from his helicopter, or from the Burj al-Arab or Burj Khalifa or by the tourist flying into UAE by air. The islands as icons can also be viewed by the "global population of Google Earth armchair travellers" (ibid). The authors suggest a new Google Earth urbanism. I would say the islands open up for a kind of Google Earth *tourism* between fantasy and reality. I am strengthened in the view that the islands are more important as icons than real built environment, since most islands are not completed yet and may never be. At the moment, most islands are only visions and models on the developers' homepages or constructions sites which contours and forms are only vaguely seen by Google Earth. Still, they are powerful cultural icons and the fact that they are not yet realized seems to be less important than their power to allude to fantasy and the imaginary. However, the interpretation of those symbols is of course dependent of the context of the observer. For instance, one armchair tourist viewing the palm islands interprets the common phallus symbol in their shape:

> The iconography of the islands is not a direct copy of the image of the palm tree, but a more abstract representation of it and therefore leaves room for alternative readings. I tend to see in it also an ejaculating penis, although that reading is perhaps a bit perverse and simplistic. (Eikongraphia 2010b).

I previously stated that "the only limit for nature is the human fantasy". However, as a human ecologist, I do not really believe in that. Human experiences throughout history reveal that nature seems to have the "last word" in what we really can do towards nature. So how is the environment affected by the construction of artificial islands and the exploration of the already existing ones in the UAE? One environmental problem reported is the disruption of tidal flows of nutrients from the deeper parts of the ocean to the shallower coastal areas. All the developers present ecological concerns on their homepages, as mentioned earlier. For instance, the developers of The World, one of the few projects finalized and thus possible to be environmentally evaluated, claim on their homepage:

> Various sea life, such as coral anemones and sea squirts, thrive along and in between the breakwaters surrounding the islands, accepting it as part of their natural environment. The shoreline areas comprise diverse marine life such as sea cucumbers, oysters, sponges, plankton, butterfly fish and jellyfish. Marine life in and around The World is flourishing.

> Initiatives, such as the water treatment plan to reuse water and recycle waste, or the incentives provided for 'Green' developers, will aid in the continued harmony of The World, its customers and its environment. (Nakheel 2010e).

Reading such statements, one can even get the impression that artificial islands are good for the environment. I have in previous publications argued that the discourse of ecological modernization mystifies the true environmental conditions in the UAE (Ouis 2005, forthcoming). The conflict between exploitation and economic development versus the environment is concealed in this discourse, since it is argued that economic growth is a prerequisite for taking care of nature, not a cause for its destruction. Critics have argued that this new environmental discourse celebrating "sustainable development", protects the sustainability of *economic* development, rather than protecting nature. This attitude is seen in the green marketing of the islands. For instance, the development of Saadiyat Island off Abu Dhabi coast will include a nature reserve, but also a golf course. On the homepage this combination is not problematized as potential conflicting interests: "Home to flourishing natural wetlands, lush mangroves, free-flowing waterways, and the region's first ever tidal golf course. An unforgettable place where man and environment become one." (TDIC 2010c) The developers promise that they have "worked closely with trusted scientific partners in order to ensure that their development do not impinge upon the surrounding environment" and that not a single mangrove plant has been removed. (TDIC 2010d).

But perhaps the most ironic ecological consequences for these islands are the green house effect and the expected climate change and rise of sea water level. The petro dollar is ultimately the explanation of how these islands could be created in the first place, while the very same consumption of fossil fuels may cause the islands' destruction. The future sinking of the Maldives Islands is feared as a result of climate change, and such sinking rumors have hit the newly finalized archipelago The World, but for other reasons. Such speculations are strongly dismissed as false, since the developers argue that the technique of vibro-compaction make the islands having twice the load-bearing capacity of that of mainland Dubai (UAE Interact 2010). However, constructors of artificial islands have considered even the eventuality of sinking islands. New megaprojects with floating islands are being planned, a floating "eco-polis" rescuing peoples' homes from rising sea levels (Jackson and della Dora 2009).

Even if the islands are not threatening the local flora and fauna, neither that the islands will sink nor be affected by climate change, the most severe environmental problem still is not addressed at all in the discussions of the islands: the environmental problem related to lifestyle and consumption issues. This is a part of the mystification of true environmental problems in the discourse of ecological modernization. However, it is a well-known fact that it is the rich countries with high levels of consumption of energy, water and consumer goods, not the poor countries that are depleting the world's resources and affecting ecosystem services negatively, despite the fact that the former, richer countries have access to environmental up-to-date techniques. Hundreds, perhaps millions, new residents will come to this extremely sensitive Emirati environment, and they will require water, energy and consumer goods in abundance to sustain their elite lifestyle. For instance, the production of fresh water in the emirates is a very energy consuming and delicate issue, since the UAE locally has very little natural fresh water or ground water left to exploit. Water is produced by desalination of sea water. Local agricultural production in the emirates is also complicated, and many kinds of food stuff must be imported.

## 8 Issues of Dependency: No Island is an Island

The bard John Donne (1572–1631) has written: “No man is an island, entire of itself, every man is a piece of the continent, a part of the main”. These influential stanzas, popularized in a protest song in the 1960s, have often been quoted to point out that humans are social beings, dependent on each other. The ethnologist Saltzman (2001) alludes to this poem in her study about a Swedish island. A translation of the Swedish title of her book into English would be “No landscape is an island”. She argues that no landscape, not even the landscape of the island she studies, is an island—meaning an independent isolation—in space or time, or free from social issues (Saltzman 2001). Every landscape is interconnected in a web of social and environmental relations. This leads me to the conclusion in regard to the Emirati islands, paraphrasing Donne: *No island is an island.*

In this chapter I have shown a number of dependency relations that constitute the islands. The construction of these islands are utterly dependent on the sales of fossil fuels in exchange for cheap labour that can materialize these megaengineering projects. Following the dependency school of economic thought (see Frank 1966), exporters of natural resources are always in a dependent relation to the metropolis producing consumer goods. I think this vulnerability still holds true. Even though the demand for fossil fuels is still high, the global shift in energy sources that is slowly taking place due to climate change, will ultimately lessen the demand for fossil fuels. The leadership in the UAE is well aware of this fact, and tries to diversify its economy towards tourism, finance and consumerism. Here is the role for the islands, as explained earlier.

However, I think it is important to pose the question if these elite residents can exist in a post-oil era? Is it really possible to run business as usual? Will not sustainability issues and other ecological concerns put an end to consumerist, luxury lifestyles? Will it, for instance, be possible for millions of tourists to travel by air in a presumed post-oil future? Can nature provide for all these residents? Even if nature “manages” the loads of millions of tourists, the economy of these islands’ complexes are still vulnerable to global finance, as experienced in the 2008 crisis. Hence, can these islands be sustainable from an economic or ecological point of view? No island is an island, means that these islands are dependent of different ecological, economic and social relations. They are interwoven in a complex web of relations beyond their control. Such dependent relations can pose threats to the sustainability of the islands, or even the chances that they ever be finalized, if the state of the world changes. And the only thing that is constant, is change. Let us hope for the sustainability and stability for these islands projects by wishing that the lines in another song, this time by Simon and Garfunkel, holds true: "I am a rock, I am an island. And a rock feels no pain; And an island never cries."

## References

Aldar (2010) http://www.yasisland.ae/Residences_en_gb.html. Accessed 24 March 2010

Al Reem Island (2010) http://www.reemisland.com/. Accessed 10 March 2010

AlSayyad N (ed) (2001) Hybrid urbanism: on the identity discourse and the built environment. Greenwood Press, Westport

Basar S (2007) Logopelago: Erandi de Silva's ultra luxury lifestyle islands. In: Basar S, Carver A, Miessen M (eds) With/without: spatial products, practices & politics in the middle east. Bidoun & Moutamarat, Dubai, pp 144–152

Basar S, Carver A, Miessen M (eds) (2007) With/without: spatial products, practices & politics in the middle east. Bidoun & Moutamarat, Dubai

Brorman Jensen B (2007) Dubai: dynamics of bingo urbanism. The Architectural Publisher, Copenhagen

Design Build Network (2010) http://www.designbuild-network.com/projects/dubai-marina/. Accessed 16 March 2010

Durrell L (1953) Reflections on a marine venus: a companion to the landscape of rhodes. Faber & Faber, London

Easterling K (2007) Abu Dhabi: Extrastatecraft. In: Re_Urbanism: Transforming Capitals, Agrawal K et al. (ed) Perspecta 39: The Yale Architectual Journal, MIT Press, pp 4–16

Eikongraphia (2010a) http://www.eikongraphia.com/?p=223. Accessed 19 March 2010

Eikongraphia (2010b) http://www.eikongraphia.com/?p=217. Accessed 19 March 2010

Elsheshtawy Y (2004) Redrawing boundaries: Dubai, an emerging global city. In: Elsheshtawy Y (ed) Planning middle eastern cities: an urban kaleidoscope in a globalizing world. Routledge, New York, pp 169–199

Emaar (2010) http://www.burjkhalifa.ae/language/en-us/the-tower/design.aspx. Accessed 19 March 2010

Emirates News Agency (2007) Lulu island opens for public, http://www.uaeinteract.com/docs/Lulu_Island_opens_for_public/24757.htm. Accessed 16 March 2010

Frank AG (1966) The development of underdevelopment. Monthly Review, September 1966, pp 17–31

Gillis J (2004) Islands of the mind: how the human imagination created the atlantic world. Palgrave Macmillan, New York

Hobsbawn E (1983) Introduction: inventing traditions. In: Hobsbawn E, Ranger T (eds) The invention of tradition. Cambridge University Press, Cambridge, pp 1–14

Jackson M, della Dora V (2009) Dreams so big only the sea can hold them: man-made islands as anxious spaces, cultural icons, and travelling visions. Environ Plann A 41:2086–2104

Katodrytis G (2007) Metropolitan Dubai and the rise of architectural fantasy. In: Basar S, Carver A, Miessen M (eds) With/without: spatial products, practices & politics in the middle east. Bidoun & Moutamarat, Dubai, pp 244–249

Khalaf S (1992) Gulf socities and the image of unlimited good. Dialect Anthropol 17:53–84

MacCannell D (1976) The tourist: a new theory of the leisure class. Schocken Books, New York

Mubadala (2010a) http://www.mubadala.ae/en/category/about-mubadala/chairmans-message/. Accessed 10 March 2010
Mubadala (2010b) http://www.mubadala.ae/en/about-mubadala/investments/sowwah-island-project.html. Accessed 10 March 2010
Nakheel (2010a) http://www.theworld.ae/au_overview.html. Accessed 16 March 2010
Nakheel (2010b) http://www.theworld.ae/. Accessed 16 March 2010
Nakheel (2010c) http://www.thepalm.ae/vision.html. Accessed 16 March 2010
Nakheel (2010d) http://www.palmjumeirah.ae/about-palm-jumeirah.php. Accessed 24 March 2010
Nakheel (2010e) http://www.theworld.ae/ev_takingcare.html. Accessed 25 March 2010
Ouis P (2002a) Power, person, and place: tradition, modernity, and environment in the United Arab Emirates. Lund: Lund Studies in Human Ecology no 4
Ouis P (2002b) Greening the emirates: the modern construction of nature in the United Arab Emirates. Cultural Geographies 9(3):334–347
Ouis P (2007) Miljövänliga arabemiraten: Hur världens största resursförbrukare kan upprätthålla en miljövänlig image. Babylon: Tidskrift om Midtösten og Nord-Afrika 5(2):90–99
Ouis P (2010) Engineering the emirates: the evolution of a new environment. In: Brunn SD (ed) Engineering earth: the impacts of megaengineering projects. Kluwer (under publication)
Saltzman K (2001) Inget landskap är en ö: Dialektik och praktik i öländska landskap. Nordic Academic Press, Malmö
Sayara RA (2010) http://www.sarayarak.com/#1. Accessed 29 March 2010
Sheikh M (2010) http://www.sheikhmohammed.co.ae/vgn-ext-templating/v/index.jsp? vgnextoid=6d1d5c1090cc4110VgnVCM1000007064 a8c0RCRD. Accessed 22 March 2010
Smyth H (1994) Marketing the city: the role of flagship developments in the urban regeneration. E & FN Spon, London
Tomlinson H, Robertson D (2010) Burj Dubai becomes Burj Khalifa as Emirate loses out on crowning glory, Times on-line, published January 5, 2010, http://www.timesonline.co.uk/tol/news/world/middle_east/article6976011.ece. Accessed 22 March 2010
Tourism Development and Investment Company (TDIC) (2010a) http://www.saadiyat.ae/en/Content/About_Our_Partners.aspx. Accessed 16 March 2010
Tourism Development and Investment Company (TDIC) (2010b) http://www.tdic.ae/Projects/Saadiyat_Island/Overview.aspx. Accessed 16 March 2010
Tourism Development and Investment Company (TDIC) (2010c) http://www.saadiyat.ae/en/reserve/. Accessed 25 March 2010
Tourism Development and Investment Company (TDIC) (2010d) http://www.saadiyat.ae/en/Content/reserve/Overview.aspx. Accessed 25 March 2010
UAE Interact (2010) http://www.uaeinteract.com/docs/The_World_islands_not_sinking_Nakheel/39491.htm. Accessed 30 March 2010
Vine P (ed) (2009) United Arab Emirates Yearbook 2009. Trident Press, London. http://www.uaeinteract.com/uaeint_misc/pdf_2009/
Wikipedia: Entry: Hydropolis (2010) http://en.wikipedia.org/wiki/Hydropolis. Accessed 29 March 2010
Wikipedia: Entry: Islomania (2010) http://en.wikipedia.org/wiki/Islomania. Accessed 31 March 2010
Wikipedia: Entry: Palm Islands (2010) http://en.wikipedia.org/wiki/Palm_Islands. Accessed 19 March 2010
Yas Marina Circuit (2010) http://www.yasmarinacircuit.ae/. Accessed 16 March 2010

# The Dramatic Drop of the Dead Sea: Background, Rates, Impacts and Solutions

**Shahrazad Abu Ghazleh, Abdulkader M. Abed and Stephan Kempe**

## 1 Introduction

The Dead Sea (DS), the lowest continental point on Earth at 422.36 m below sea level (b.s.l.) (measurements as of June 1st, 2009; Talafeha 2009), is a unique ecosystem with distinctive geological, geographical and biological characteristics. It is composed of two basins: (a) the deep northern basin with the deepest submarine point of 770 m b.s.l. (Neev and Emery 1967) and (b) the shallow southern basin that dried and became exposed to the air in 1978 and is currently used by the commercial potash industry. The two basins are divided by the Lisan Peninsula at an altitude of ~400 m b.s.l. (Fig. 1). The climate in the DS basin varies from snow-capped Mount Hermon (Jabel El Sheik), with an annual precipitation of >1,200 mm to the arid regions of the Wadi Araba with an annual rainfall average of < 50 mm (Salameh and El-Naser 1990). Over the DS itself, average annual rainfall is ~70 mm while the annual evaporation ranges from ~1,300 to 2,000 mm (Abed 1985; Salameh and El-Naser 2000). The average temperature in the DS area is ~40°C in summer and ~15°C in winter.

Due to the water's high salinity only one green algae species and halophilic bacteria truly thrive in the DS waters. However, freshwater springs along the shores sustained a rich fauna, including rare species such as the leopard, the Nubian Ibex, the Rock Hyrax, and hundreds of birds.

Up to the 1970s, freshwater input in the DS was able to maintain a less salty topmost water layer (epilimnion) over-riding a salt-saturated bottom water layer (hypolimnion). The epilimnion was ~40 m deep, had a seasonal temperature

S. Abu Ghazleh (✉) and S. Kempe
Darmstadt University of Technology, Darmstadt Germany
e-mail: shahrazad@geo.tu-darmstadt.de

A. M. Abed
University of Jordan, Amman Jordan

V. Badescu and R. B. Cathcart (eds.), *Macro-engineering Seawater in Unique Environments*, Environmental Science and Engineering,
DOI: 10.1007/978-3-642-14779-1_4, © Springer-Verlag Berlin Heidelberg 2011

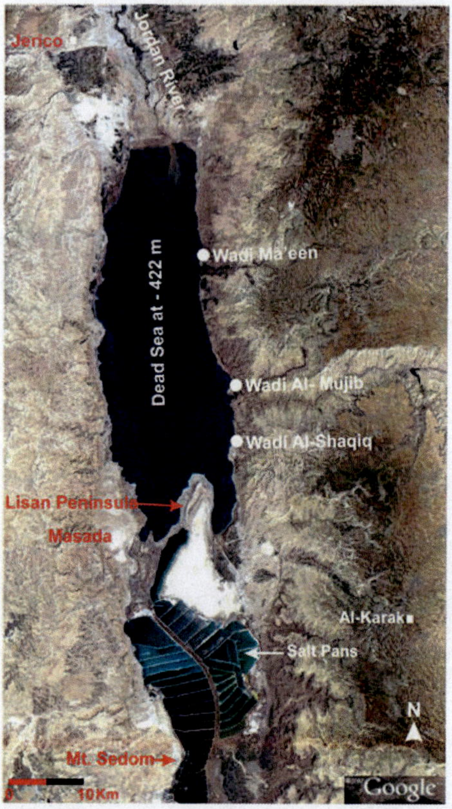


**Fig. 1** The location of the Dead Sea. The *red points* indicate the location of the measured terrace profiles. (The image was taken from Google Earth 2010)

variation of between 19 and 37°C and a salinity of about 30%. The water was particularly rich in sulphate and bicarbonate. Below the density interface (pycnocline, 40–100 m of depth) the hypolimnion was characterized by a uniform temperature of ~22°C and a salinity of >34%; it contained hydrogen sulphide and high concentrations of magnesium, potassium, chlorine, and bromine. The hypolimnion was unmixed for a long time. It is saturated with sodium chloride that precipitates as halite on the bottom (Abed 1985). The unique mineral composition of the DS water and its high salt concentration make it suitable for the production of potassium and magnesium. In addition, the high salt concentration provides natural cures against various skin sicknesses and the high oxygen content and low UVB exposure make the DS a prime area for therapeutic tourism (Abdel-Fattah and Pingitore 2009; Charlier and Chaineux 2009).

The DS area has also played a major role in shaping human history since early times. It is a cradle of the first civilizations and religions, and is located on the central land bridge connecting Asia, Africa and Europe. Several prominent archaeological and cultural sites are located all around the famous endorheic lake, forming a cultural landscape very important not only for the riparian countries but also world-wide.

The increasing freshwater demand in the DS basin—as one of the most arid regions on Earth—puts a big strain on the limited resources and has caused a significant change in the DS water balance. The diversion of the Jordan River and its tributaries in addition to the extraction of saltwater by the potash industry reduced the input from ~2 km$^3$/a before 1950 to 0.33 km$^3$/a in the late 70s (Salameh and El-Naser 1999; Al-Weshah 2000). As a result, the level of the DS dropped by more than 30 m in the last 70 years (Abu Ghazleh et al. 2009), and it is still dropping with severe environmental impacts. This chapter therefore aims to: (a) review the geological processes that led to the formation of the tectonic basin and its water bodies throughout geological history; (b) to investigate the most recent changes in the hydrological balance of the DS and the associated loss of volume and surface area; and (c) to identify the environmental consequences of the lake level lowering and discuss alternative macro-engineering solutions for the problem.

## 2 Geology of the Dead Sea

### *2.1 Dead Sea Transform*

The northwards movement of the African Plate (including Arabia) during the Cretaceous and Paleogene had pronounced geological effects on the Eastern Mediterranean. These effects are the formation of the Syrian Arc Fold belt (Turonian-Maastrichtian) (Krenkel 1924; Bowen and Jux 1987), the epeirogenic uplift that led to the withdrawal of the Tethys Ocean (Stow 2010) from Belad Al-sham (Levant) (Bender 1974), and its uplift and exposure to erosion.

By mid-Miocene, about 15 Ma (million years) ago, the Arabian Plate separated from the African Plate and began to move northwards (e.g., Abed 2000). At this time the Syrian-East African rift system formed, including the Gulf of Aden, the Red Sea (divergent boundary) and the Dead Sea transform boundary (DST).

The DST extends for about 1,100 km from the southern tip of the Gulf of Aqaba in the south to the East Anatolian Fault in southern Turkey in the north (Fig. 2). Along its rifted paths several pull-apart basins formed, including (from south to north): Gulf of Aqaba, Wadi Araba, DS, lower Jordan valley, upper Jordan valley, Bekka'h and Ghab basins. The annual rate of the NNE movement of the Arabian Plate (Jordan) relative to the Sinai-Palestine micro-tectonic plate is around 4.5 mm/a (Klinger et al. 2000). Since the Miocene, the total movement along the southern DST amounts to 107 km (Quennell 1958). the Gulf of Aqaba, which is a narrow, up to 1,850 m deep basin (the Gulf of Suez is, for comparison, only several tens of metres deep). From the northern end of the Gulf of Aqaba, the DST runs NNE along the eastern side of Wadi Araba-Dead Sea (Wadi Araba Fault, WAF) to peter out at the NE tip of the DS into the pre-existing Amman-Hallabat structure. Along the southern basin of the DS, the WAF is locally called Al-Safi Fault which is equivalent to the western Sedom Fault. The thin Miocene clastic

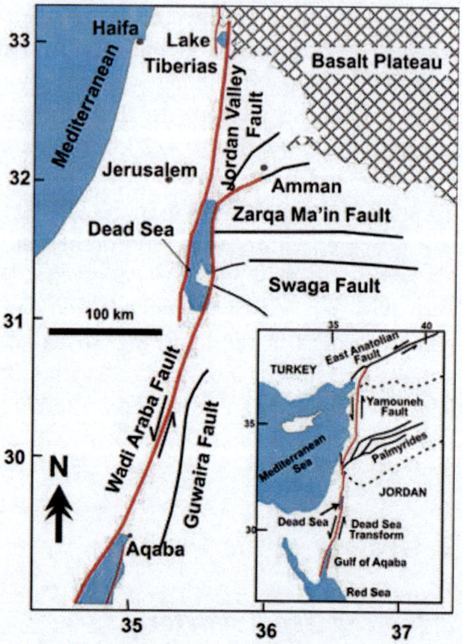


**Fig. 2** The Dead Sea Transform in Jordan, its two segments Wadi Araba Fault and Jordan Valley Fault, and their overlap in the DS basin. *Inset* shows the entire length of the DST from the Gulf of Aqaba in the south to the East Anatolian Fault in the north

sediment cover and the lack of Pliocene salt in the eastern block, suggests that the Ghor–Safi Fault was initiated earlier than the Sedom Fault (Al-Zoubi et al. 2002).

The Jordan Valley Fault starts at the SW corner of the DS basin and forms its western boundary and then diagonally crosses the River Jordan east up to the N end of the Tiberias-Hula basin. From there, the Yamouna Fault takes the movement through the Bekka'h and Ghab basins in Lebanon and Syria respectively, until it meets the East Anatolian Fault (Fig. 2). The DST, therefore, is not a single continuous transform fault, but consists of several segments (e.g. Galli 1999; Brew et al. 2001).

The floor of the DS rift has been subsiding since its formation in the Miocene. The subsidence, most probably is not a continuous process and it does not occur at the same rate throughout the various parts of the basins. For example, the DS basin has received more than 5,000 m of sediments since the early Miocene (around 25 Ma). This represents a minimum subsidence rate of 0.2 mm/a, or about 200 m/Ma. Subsidence is still continuing. Also, both shoulders of the DS Rift have been rising since its formation. The rate of uplift is certainly higher along its eastern shoulders (Jordan) compared with its western shoulders (Palestine). One of the important consequences of the uplift process is the isolation of the DS Rift from the Mediterranean Sea since the early Pleistocene. Both geological processes, subsidence and uplift, shaped the DS Rift. Subsidence and uplift are much faster than erosion and sedimentation, thus a huge accommodation space for sediments was created, the modern DS-Jordan depression. Due to the northwards movement of Jordan (relative to Palestine), the strata outcroppings on both sides of the DS are

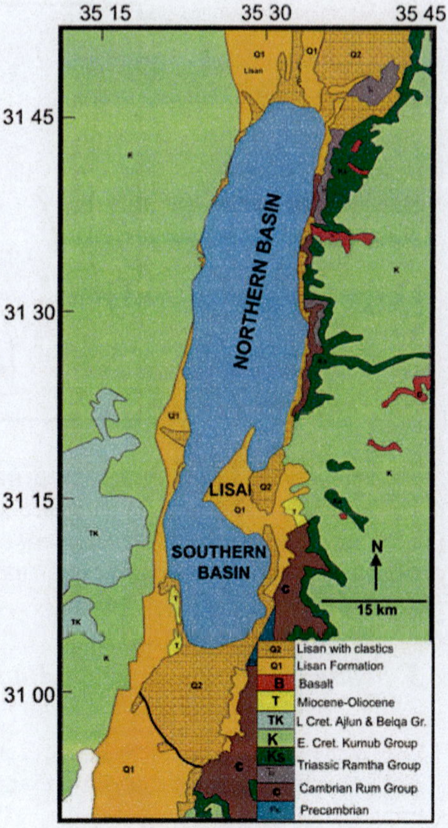


**Fig. 3** Generalized geological map of the DS area. Note how strata in the eastern shoulder are much older than those in the western shoulder. DS area and shore-lines as in the early 1960s at −397 m. Modified from Bender (1974) and Powell (1988)

not the same. The rocks along the eastern shoulder are much older geologically, involving units from the Precambrian to the Eocene, while along the western rim Precambrian and Paleozoic rocks are absent (Fig. 3).

## *2.2 Structural Geology*

Figure 2 illustrates that the DS basin is bordered by two transform faults; along the southern Dead Sea basin (SDSB), either Jordan moves N along the Wadi Araba Fault (WAF), while Palestine is moving south along the Jordan Valley Fault (JVF) (Quennell 1958), or Jordan moves N along the WAF faster than the northwards movement of Palestine along the JVF (Klinger et al. 2000). In both cases, the DS region is pulled in two opposite directions to produce a rhomb-shape pull-apart basin (Fig. 4).

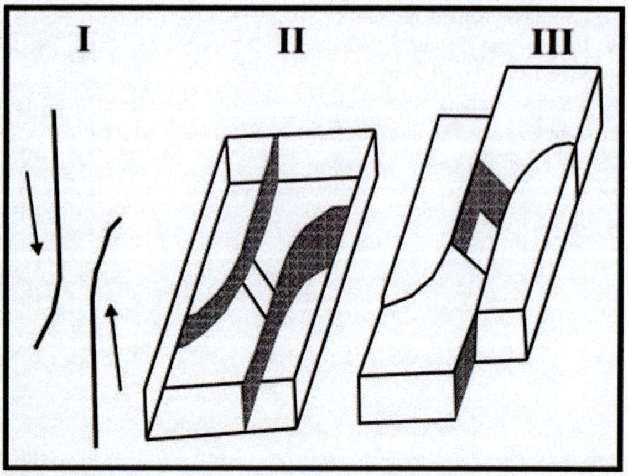


**Fig. 4** Stages for the formation of the DS basin. *I* The initiation of the two transforms along both sides of the basin. *II* Formation of transverse faults. *III* During the movement, a rhomboidal-shaped basin was formed. Modified after Quennell (1958)

The pull-apart process also created transverse faults formed in a NNE-SSW direction across the basin. Most conspicuous amongst them is the Khnaizereh Fault (KF), some 20 km south of the Southern Basin (Fig. 3). The KF is a normal listric or growth fault, its north-facing scarp marks the southern boundary of the DS basin.

## *2.3 Sedimentation and Water Bodies in the Dead Sea Basin*

Towards the end of the Eocene, the Tethys Ocean migrated to the west. The shoreline was slightly east of the present-day Mediterranean. Prior to that, and starting from the Cenomanian, a ca. 1,000 m thick sediment column—mostly carbonates—was deposited by the Tethys Ocean in and around the DS basin. The Oligocene was a time with enhanced epeirogenic movements (Bender 1974). By the Miocene time, the scene was set for the formation of the DS Rift. From this time onwards, the DS basin and the Jordan Valley has been the site for successive lakes of varying sizes and salinity. Full account of these lakes and their sediments is beyond the scope of this work. However, a few of the more prominent ones are briefly described.

### 2.3.1 Dana Conglomerate Formation

From the early Miocene onwards, the DS basin was topographically lower than its surroundings. Erosion and re-deposition of the older marine strata led to the formation of the Miocene Dana Conglomerate Formation. Its thickness at the shoulders of the DS basin is ca. 450 m, while it attains a thickness of about 2,000 m in the DS Basin (Powell 1988). The presence of micro fauna and the absence of evaporates suggests an open connection between the Mediterranean Sea and this infant basin. The Dana Formation consists of several fining upwards cycles made of conglomerates, calcarenite, calcisiltite and calcilutites.

### 2.3.2 Usdom (Sedom) Formation

The Usdom Formation is of Pliocene age. It is characterized by rock-salt with minor other evaporites. Its thickness at the centre of the DS basin is around 4,000 m below the Lisan Peninsula (Bender 1974). To the E and W, the salt beds decrease in thickness and begin to intercalate with fluvial sediments transgressing from the basin shoulders. The huge thickness of rock-salt can be explained by the DS basin being still connected to the Mediterranean Sea as a lagoon to provide the salt, subsidence to accommodate this thickness, and evaporation to adjust the salinity. Consequently, the Usdom Lake was a highly saline marine lagoon, similar in salinity to the present day DS where halite is now precipitated.

### 2.3.3 Quaternary Formations

By the end of the Pliocene, the western mountain range became elevated enough to severe the last connection between the Dead Sea-Jordan Valley and the Mediterranean Sea. The former area became, depending on precipitation, and, thus, on climate a source for freshwater throughout the Quaternary.

In general, precipitation was higher during the warm stages of the Pleistocene and the lakes in the region were larger and fresher. In Jordan, the Azraq-Umari Lake at 331 ka, marine isotope stage 9 (MIS9) (Abed et al. 2008), the Mudawwara Lake at the Saudi border at around 135 ka, MIS5 (Abed et al. 2000), and the Samra Lake 135–70 ka, MIS5-4 support this generalization. At around 27–25 ka BP a 1,000–1,800 $km^2$ large freshwater lake existed in the Jafr Basin, southern Jordan. It dried up during the last glacial maximum (LGM) and its basin is now exposed to wind erosion (Huckriede and Wiesemann 1968). Contemporary to the Jafr Lake, the Jordan Valley was filled with a deep water body, Lake Lisan at ~70–14 ka BP) (e.g., Waldmann et al. 2007). It reached a high level of −148 to −150 m at 30–23 ka BP (Abu Ghazleh and Kempe 2009). After the LGM, Lake Lisan experienced a sharp regression leaving behind prominent sequences of shoreline terraces and deep water deposits of aragonite and gypsum (Landmann et al. 2002; Haase-Schramm et al. 2003; Abu Ghazleh and Kempe 2009) now exposed on the Lisan Peninsula and along the shores of the DS. The following post-glacial freshwater, Damya Lake, also disappeared during the Younger Dryas (Abed and Yaghan 2000) followed by the current hyper-saline Dead Sea at ~10 ka BP. Its lake level was unstable during the Holocene. It experienced several risings and drops due to climatic changes. Two major wet phases were recorded from a DS sediment core at 10–8.6 and at 5.6–3.5 cal ka BP interrupted by arid events at 8.6, 8.2, 4.2 and 3.5 cal ka BP (Migowski et al. 2006). These events are thought to be associated with the collapse of agriculture and the major interruptions in the Near East cultural evolution (Migowski et al. 2006; Neuman et al. 2007). Table 1 summarizes the rock units reported during the Pleistocene.

**Table 1** Nomenclature of the post-Tethys formations in the DS-Jordan Valley (ka = 1,000 year, Ma = Million year)

| Age | | Formation | Lithology |
|---|---|---|---|
| Quaternary | | | |
| 10.5 ka | Holocene | Damya | Mainly laminated, freshwater |
| 128 ka | Upper Pleistocene | Lisan = Himma | Mainly laminated, fresh and saline |
| | | Samra | Mixed carbonates and clastics, fresh water |
| 700 ka | Middle Pleistocene | Naharim | |
| | | Kufranja | Gravel, fresh water |
| | | Tabaqat Fahl | Conglomerate, fresh |
| | | Abu Habil | Conglomerate, freshwater |
| 1.8 Ma | Lower Pleistocene | Ghor El-Katar | Mixed clastics in cycles |
| | | Shaghour | Conglomerate |
| Tertiary | | | |
| 5.3 Ma | Pliocene | Usdom (Sedom) | Rock salt, saline |
| 23.8 Ma | Miocene | Dana | Mixed clastics in cycles: conglomerate, calcarenite, calcilutite |

# 3 Hydrology of the Dead Sea

## *3.1 Water Balance and Lake Level Changes*

The DS is a terminal lake draining one of the largest hydrological basins in the Levant (Neev and Emery 1967) with a total area of 40,650 km$^2$ (Salameh and El-Naser 1999). The Jordan River is the main source of the water feeding the DS (Fig. 1) with an original long-term mean discharge of 1.21 km$^3$/a (Salameh and El-Naser 1999). Several tributaries of the DS from the eastern, western and southern sides contributed also with a discharge of ~0.5 km$^3$/a. In addition, groundwater delivered an annual discharge of ~0.2 km$^3$.

Nowadays, the discharge of the Jordan River has been reduced to ~0.1 km$^3$/a (Salameh and El-Naser 1999; Al-Weshah 2000) mainly due to the diversion of its water by the Israel Water Carrier and by consumption of other riparian countries. It is estimated that ~47% of the Jordan River is diverted by Israel, 22% by Jordan, 16% by Syria and 2% by Lebanon.

Meanwhile, the surface run-off and the groundwater inflow decreased to ~0.2 and 0.1 km$^3$/a, respectively (Table 2). An additional water amount of ~0.3 km$^3$/a is normally lost from the DS due to aerial evaporation from the commercial salt-pans south of the DS. Consequently, the level of the DS has dropped in the last 30 years with an alarming rate of 0.7 m/a (Abu Ghazleh et al. 2009).

The rapid drop of the DS level activated the head-ward erosion of the west- and east-draining wadis. The eroded material was deposited below the water, forming sub-lacustrine fan-deltas that are now exposed due to continued lowering of the lake level. The top-sets of these deltas were eroded by wave action forming a unique sequence of shoreline terraces (Abu Ghazleh et al. 2009) (Fig. 5) that

**Table 2** Water input into the DS prior and after freshwater resource development

| Freshwater resource | Amount km$^3$/a (prior to freshwater resources development) | Amount km$^3$/a (after freshwater resources development) |
|---|---|---|
| Jordan River | 1.207 | 0.100 |
| DS tributaries | 0.463 | 0.180 |
| Groundwater | 0.220 | 0.140 |
| Total | 1.89 | 0.547 |

Modified after Salameh and El-Naser (2000)

**Fig. 5** Sequence of shoreline terraces north of Wadi Al-Shaqiq fan-delta

represents annual lake level standstills. These terraces were surveyed in the field by Abu Ghazleh et al. (2009) and dated according to the lake level records (Figs. 6, 7). These data show that the DS dropped from −389 m in 1932 to −419 m in 2006–2007. Recent measurements, done by the Arab Potash company and by Talafeha (2009), show that the lake continued its drop to −420 m in 2008 and to −422 m in 2009. This amounts to a total drop of 33 m in the last 77 years with an average of 0.4 m/a.

However, the rate of the lake recession changed during the last 77 years. Four periods with different elevation reduction rates can be identified based on shoreline altimetry (Figs. 6, 7).

(1) From 1932 to 1955; the lake level dropped relatively slowly from −389 to −392 m with a drop rate of 0.1 m/a. This recession was associated with the foundation of the Palestine Potash Company in the early 1930s south-west of the DS that extracted only a small volume of lake water. However, before 1930 the DS level stayed stable because the fresh water inflow was equal to the evaporation rate. Therefore, the −389 m terrace documents the onset of regression of the DS caused mainly by human activity.
(2) From 1955 to 1978, the lake receded fast from −392 to −400 m with a drop rate of 0.35 m/a. This was caused by the diversion of the Upper Jordan by the

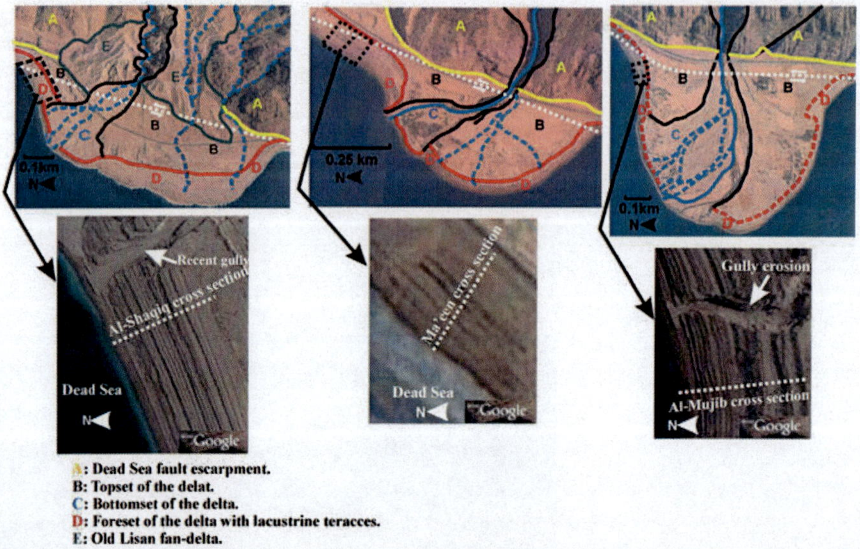


**Fig. 6** Fan deltas of the Dead Sea and sequences of lacustrine shoreline terraces north of Wadi Al-Shaqiq (*upper left*), north of Wadi Ma'een (*upper middle*), and north of Wadi Al-Mujib (*upper right*)

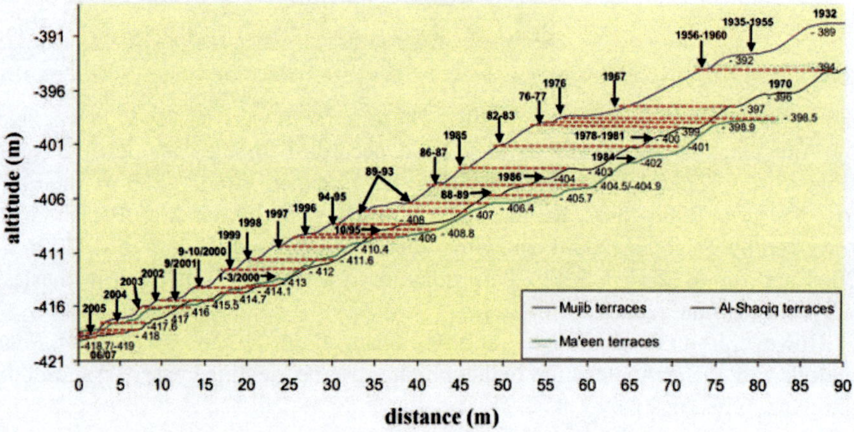


**Fig. 7** Three profiles of the DS terraces surveyed by DGPS and dated according to the recorded DS levels. After Abu Ghazleh et al. (2009)

Israel National Carrier between 1950 and 1960 in addition to diversion of the Yarmouk River by the same carrier and by water extraction by Jordan and Syria in the early 1970s. The average diversion of the Jordan and the Yarmouk water was estimated to be 575 Mio $m^3$/a by the Israel National Carrier, 135 Mio $m^3$/a by Syria and 110 Mio $m^3$/a by Jordan (Salameh and El-Naser 2009).

**Table 3** Water volume and surface area losses related to the lake level changes since 1932 based on the terrain model of the DS rift

| Period | Level change (b.s.l.) | Water volume loss ($km^3$) | Surface area loss $km^2$ | Drop rate m/a |
|---|---|---|---|---|
| 1932–1955 | 389–399 | 3 (0.13 $km^3/a$) | 78.1 (4 $km^2/a$) | 0.1 |
| 1955–1978 | 399–400 | 6.7 (0.29 $km^3/a$) | 286.8 (11.7 $km^2/a$) | 0.35 |
| 1978–1995 | 400–409 | 6 (0.36 $km^3/a$) | 56,3 (3.3 $km^2/a$) | 0.5 |
| 1995–2009 | 409–422 | 6.9 (0.49 $km^3/a$) | 72.5 (5.2 $km^2/a$) | 1 |
| 2010–2020 (expected) | 423–433 | 3.4 (0.34 $km^3/a$) (Eq. 1) | 55.8 (5.6 $km^2/a$) (Eq. 3) | 1 |

(3) From 1978 to 1995, the lake level declined from −400 to −409 m with a drop rate of 0.5 m/a. During this period, the southern basin completely dried up, and the Arab Potash Company was established on the eastern side, the salt ponds were expanded and pumps were used since then to fill them.

(4) During the last 14 years, the DS level declined with the alarming rate of 1 m/a from −408 to −422 m (Table 3). This fast regression resulted from expansion of the salt pans that are now consuming an average of ~0.3 $km^3/a$ (pumping minus return flows) (Salameh and El-Naser 1999). The peace treaty signed between Jordan and Israel in 1994 (The Harza JRV Group 1996) has entitled Jordan to use an additional amount of ~0.1 $km^3/a$ from the Upper Jordan and the Yarmouk River (Al-Weshah 2000) which also contributed to speeding up DS level decline.

## 3.2 Surface Area and Water Volume

Prior to human impact, the closed hydrological basin of the DS served as a natural rain gauge of the surrounding region and its natural level, area and volume changes were driven by climate. However the most recent lowering of the DS is mainly anthropogenic, due to human water consumption in the region and not a consequence of regional climate change.

Based on the USA's Space Shuttle Radar Topography Mission (SRTM), a terrain model of the DS rift volume and surface area was developed using ArcGIS (3D)-Analyst (ESRI)-"Surface Volume" tool-functionality (Abu Ghazleh et al. 2009). These data were used to derive functions describing the morphology of the rift basin above the water level and allow calculation of volume and area changes with elevation (Figs. 8, 9, 10). For volume changes a square-polynomial function Eq. 1 proved to be the best fit (Fig. 8). It was calculated using 30 decimals in order to get a precise fit (EXCEL):

$$\begin{aligned} WV = {} & 765249.552011049 \times 10^{-8} x^2 + 689054.133871516 \times 10^{-5} x \\ & + 168883.174228981 \times 10^{-2} \end{aligned} \tag{1}$$

**Fig. 8** Water volume-altitude model of the Dead Sea rift. Modified after Abu Ghazleh et al. (2009)

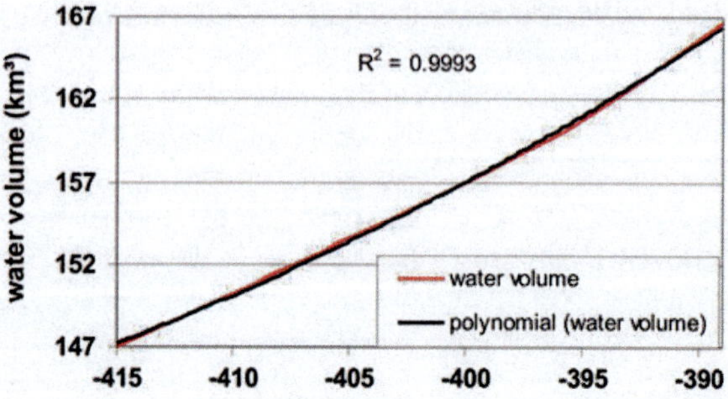


**Fig. 9** Surface area-altitude model of the entire Dead Sea rift

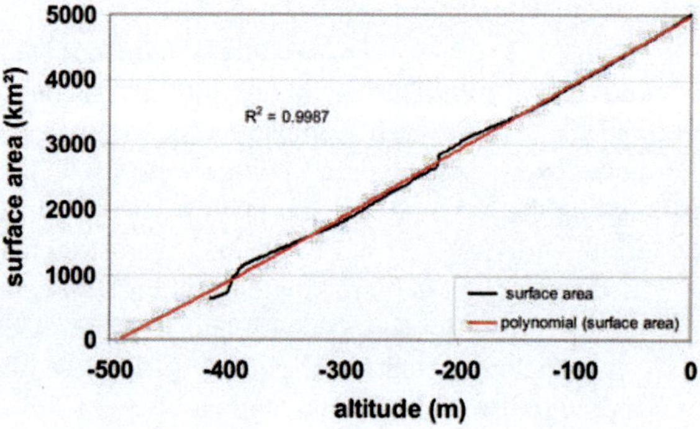


**Fig. 10** Linear surface area of the Dead Sea at an altitude of −399 to −434 m

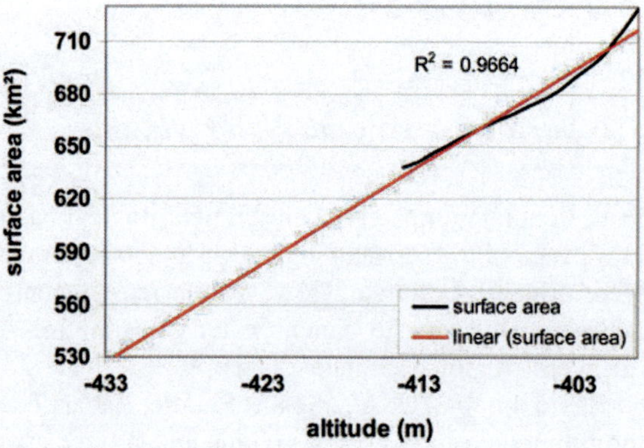


Likewise a cubic-polynomial function of the lake surface area Eq. 2 of the entire DS rift (zero to −434 m) proved to be the best fit (Sigma Plot) (Fig. 9). Although this function correlates highly with the lake levels ($R^2 = 0.998$), reflecting the bathymetry of the flanks of the former Dead Sea and the morphology of the rift valley, it is not suitable for calculating the rapid changes of DS area caused mainly by human activities during the last decades. Therefore, a linear function of the surface area during the last 55 years was used Eq. 3 (Fig. 10) in order to describe these changes and to estimate what changes are expected in the near-term future (Table 3).

$$SA = -0.00000373092x^3 - 0.00173775x^2 + 10.2099x + 4990.43 \tag{2}$$

$$SA = 5.5795x + 2943 \tag{3}$$

where: WV = water volume ($km^3$), SA = surface area ($km^2$) and $x$ = level in m below sea level.

The terrain model of the water volume and surface area shows that during the initial lowering of the DS (1932–1955), 3 $km^3$ of its volume and 78 $km^2$ of its area were lost with rates of 0.13 $km^3$/a and 4 $km^2$/a, respectively. Between 1955 and 1978 the DS lost 6.7 $km^3$ with a rate of 0.29 $km^3$/a, twice as fast as during the preceding period. Meanwhile, the lake area has significantly shrunk by 287 $km^2$ with a rate of ~12 $km^2$/a (Table 3). The large shrinkage of the surface area during this period was a result of the drying up of the shallow southern basin (Fig. 11). During the period 1978–1995, the lake lost a water volume virtually equivalent to that of the previous period. However, the loss in the surface area (56 $km^2$) was small compared to that of the previous recorded period. This indicates a change in the lake basin morphology from shallow to steep. The largest volume loss took place during the last 14 years, with a total amount of 6.9 $km^3$ and a rate of 0.49 $km^3$/a associated with the dramatic drop of the DS level of 1 m/a. At the same time, the surface area loss was small with only 72.5 $km^2$, again reflecting the steep morphology of the northern basin (Fig. 11). Salameh and El-Naser (1999; 2009) reported that the seaward migration of the interface between groundwater and DS water induced a groundwater discharge of ~0.5 $km^3$/a. This amount plus the volume-loss rate of the DS during the last 14 years suggest that the lake is currently loosing ~1 $km^3$/a directly from its surface.

This amount should be equal to the evaporation from the lake. Based on the evaporation rate of 1,300–2,000 mm/a reported by Abed (1985), Salameh and El-Naser (2000), Khlaifat (2008), and Khlaifat et al (2010) the total evaporation of the DS at present (667 $km^2$) is calculated to be 1 $km^3$, i.e. the same as the total volume loss. An additional water amount of ~0.3 $km^3$/a is lost from the DS due to evaporation from the industrial salt-pans (Salameh and El-Naser 1999). The volume loss of the DS (~1 $km^3$/a) in addition to that of the salt pans (0.3 $km^3$/a) suggests that the DS needs an amount of 1.3 $km^3$/a in order just to stop its continuous lowering. This amount is higher than that suggested by us originally (Abu Ghazleh et al. 2009) since it depends on the current volume loss of the DS (0.49 $km^3$/a) and not on the averaged volume loss of the last 30 years (0.47 $km^3$/a) employed previously. Moreover, the water volume loss from the salt-pans by evaporation was not considered by Abu Ghazleh et al. (2009).

Under the assumption that the morphology of the rift valley will not change drastically during the near-term future (Abu Ghazleh et al. 2009), and that the rate of DS water consumption remains the same, the volume- and area-altitude functions can be used to calculate the future water loss. Therefore, in the next decade [to 2020], the lake is expected to lose an additional amount of 3.4 $km^3$ and shrink by ~56 $km^2$ (Table 3).

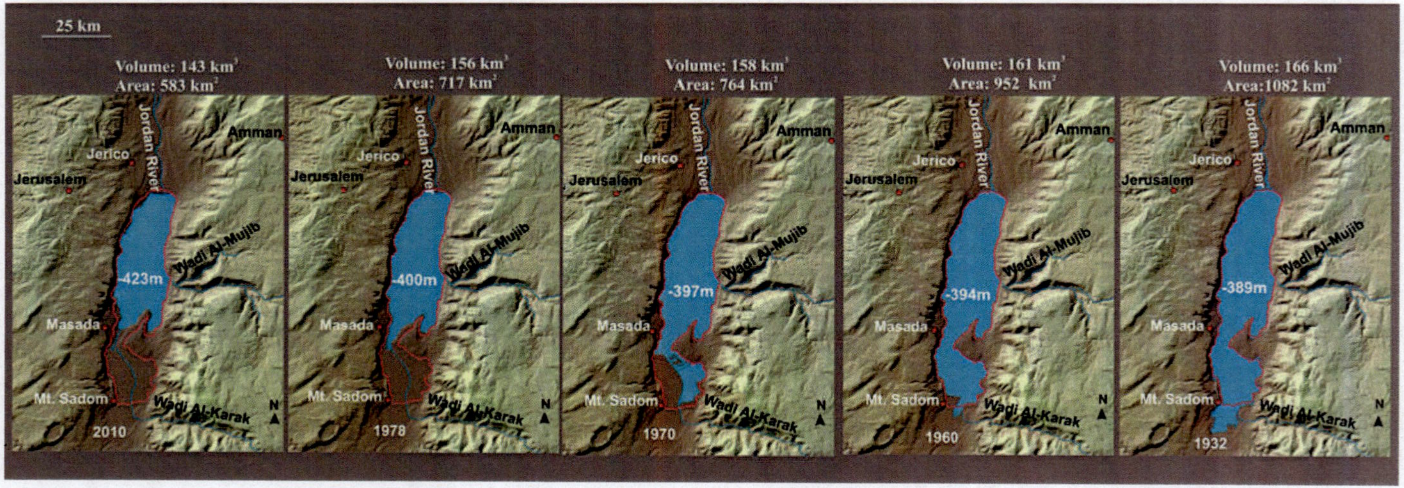


**Fig. 11** The changes in the Dead Sea level, volume and area over the last 77 years. (The animated map was taken from Riddle and Parker, http://deadseachange.webs.com/map.html). The levels, volumes and areas of the DS were calculated based on the terrain model. Note the large loss of the surface area between 1970 and 1978 and the small change in the area after 1978

Although the water level will continue to decline in the future, a complete demise of the DS is not expected. The water level will most probably approach a stable situation within 200 years. At a level of −550 m, the lake's maximal depth will be ~180–200 m and its surface area will have shrunk to 450 $km^2$. The decrease in the surface area and the increase of the salinity of the remaining brine will reduce the volume of annual evaporation. Subsequently, when evaporation will equal inflow, the lake level will stabilize (The Jerusalem Institute Report 2006). Thermodynamic computer simulation predicts that under current conditions, DS water may continue to evaporate to a level of −550 m. Simulated evaporation of DS water to a concentration factor of 50 yielded a 90 m thick column of chloride minerals containing halite, carnallite, bischofite, tachyhydrite, and $CaCl_2{\cdot}4H_2O$ (Katz and Starinsky 2009).

# 4 Environmental Degradation of the Dead Sea

The continuous lowering of the DS level during the last 77 years caused an ecological imbalance in the DS and in the surrounding coastal areas. Consequently, severe environmental problems arose, threatening the existence of the DS as a natural resource as well as the natural habitats of its shores and the livelihood of its people. The major environmental consequences are as follows.

## *4.1 Groundwater Outflow*

The rapid drop of the DS level led to a drop in the groundwater level and therefore to a seaward migration of the salt-/freshwater interface. This process resulted in groundwater outflow of ~0.5 $km^3$ per meter drop in the level of the DS in the period 1994–1998 (Salameh and El-Naser 2000). Consequently, several natural springs dried up affecting natural habitats and tourism industry of the lake shores. Also fresh ground water, potential drinking water, was irreplaceably lost.

## *4.2 Sinkholes*

Over the last several years, the coastal area around the declining DS has undergone a catastrophic collapse with a sudden appearance of hundreds of sinkholes (Fig. 12). This phenomenon is mostly developed due to the dissolution of the underlying salts layers by ground water outflow. Consequently, large caves were formed followed by surface collapses. Closson et al. (2005) and Abelson et al. (2006) suggested that the formation of sinkholes seems to be tectonically controlled, since most of them associated with subsidence zones and buried faults that

**Fig. 12** Sinkhole along the eastern coast of the Dead Sea (taken by Sawariah, Natural Resource Authority, Jordan)

allowed the ground water to penetrate salt layers. Previous studies reported more than 3,000 sinkholes occurred in the DS shores destroying fields, houses, roads and even the salt ponds and thus making the area unsafe for living and development. One of the most dangerous sinkholes occurred in 2000 in the major commercial salt pond number 19, SE of the DS. The dyke collapsed over almost 2 km and 56 million $m^3$ of brine poured into the northern part of the DS increasing the water level by 10 cm in less than 1 h (Closson et al. 2005).

## *4.3 Landslides*

The rapid emergence of delta bodies with deposits of mud and gravels can cause their instability because of the loss of buoyancy. Landslides could therefore occur spontaneously (failure of slopes) or could be triggered by earthquakes with the thread of tsunamis in the DS basin. The emerged delta bodies are also prone to quick erosion and wadi downcutting.

It was shown that half of the tsunami events that occurred in the Mediterranean Sea region were caused by landslides following earthquakes in the vicinity of the Dead Ssea (http://www.haaretz.com/hasen/ spages/1118864.html). Simulation of the DS water oscillation resulting from known earthquakes show clearly that waves of 4–6 m height could be generated from earthquakes of 7.3 Mw with agitated wave action duration of $\sim$1 h (Ichinose and Begin 2004).

Abou Karaki et al. (2005) pointed out that the unconsolidated deposits of the DS fan-deltas need a long time to settle until reaching a stable equilibrium. They explained that the exposure of such deposits to ground water flow causes read-justment that could cause landslides along the unstable margins of the fan-deltas. During the last 10 years, several treacherous landslides occurred along the Jordanian coast of the DS associated with the rapid drop of lake level (1 m/a). These landslides destroyed four coastal areas with intense tourism and industrial developments (Closson et al. 2010). One of these events occurred in the northeast of the DS in 1999 with the speed of 3–5 m/s (Turner and Schuster 1996), destroying

**Fig. 13** The 1999 landslide northeast of the Dead Sea

**Table 4** The status of the landslides along the eastern shore of the Dead Sea

| Land slide | Location | State | Velocity | Style |
|---|---|---|---|---|
| 1 | NE of the DS | Active (+) | Slow (1.6–156 m/a) | Complex (slump, earth flow) |
| 2 | NE of the DS | Active (−) | Very rapid in sep 99 (3 m/min–5 m/s) then very slow to inactive | Single |
| 3 | Ghour Hadithah | Active (++) | Slow (1.6–156 m/a) | Multiple (repeated slump) |
| 4 | Lisan Peninsula | Active (+) | Very slow (16 mm/a–1.6 m/a) | Single |

Modified after Closson et al. (2010)

~100 m of a beach rest house (Field observation, 1999) (Fig. 13). The status of the main landslides along the eastern shore of the DS was reported by Closson et al. (2010) (Table 4).

## *4.4 Halite Precipitation and Mushroom Problem*

Under continuous conditions of negative water balance of the DS, the salinity of the upper water increased and consequently an overturn of the water column occurred associated with the drying up of the southern basin in 1978/1979. As a result, the upper water mass of the DS became saturated with respect to aragonite ($CaCO_3$), anhydrite ($CaSO_4$) and halite (NaCl). However, before the overturn, only the lower water mass was saturated with respect to these minerals while the upper layer was saturated with respect to aragonite and anhydrite (Neev and Emery 1967). The halite began to precipitate from the DS in 1982 (Steinhorn 1983) and since then it has continued to precipitate almost uninterruptedly. Increasing salinity of the upper layer due to the continuous shrinkage of the lake could lead to massive salt incrustations along the entire lake. In the mid 1980s a natural

**Fig. 14** Mushroom like-shape of halite deposits in the salt pans south east of the Dead Sea

phenomenon was observed in the artificial ponds south of the DS. The NaCl started to crystallize in the form of "mushrooms" in various parts of the ponds. Columns of salt started to grow from the shallow bottom and once they reached the surface, they started to grow at an accelerated speed at the sides creating forms similar to mushrooms (Fig. 14).These deposits limited the surface area and the evaporation capability of the pond and disturb the mechanical managing of salt ponds.

## 4.5 Headward Erosion

The continuous lowering of the erosion base-level and the low consolidation of the alluvial fan deposits allowed rapid headward erosion of the west-draining wadis. This threatens bridges built on the fan deltas exposing their bases to erosion. The surface flow of some DS tributaries was noticed to disappear in the fan-delta bodies. Apparently, they penetrate the permeable layer of sand and gravels at the surface causing mobilization of the underlying deposits and consequently abrupt subsidence of parts of these deltas. This has been observed in the Wadi Al-Mukhairas delta, NE of the DS. Salameh (1997) reported that several bridges built on wadi gorges of the DS failed under the instability and subsidence of loose, coarse sediments upon which they were constructed. For instance, the Wadi Al-Mujib Bridge partially collapsed when its foundation was built on such sediments.

## 4.6 Salt Dust

Salt minerals such as halite (NaCl) and gypsum ($CaSO_42H_2O$) are precipitated from the evaporated water mass of the highly saline DS. Aerosols and salty dust can be transported over the surrounding areas by the NW–SE prevailing winds. This causes air pollution and can harm fertile lands, as well as people. XRD analysis of dust samples from Al-Karak city, east of the DS revealed the existence of halite and gypsum as minor mineral phases (El-Hasan et al. 2008).

# 5 Alternative Macro-Engineering Solutions

In order to restore the DS to its original level of ~30 years ago and reduce the environmental impacts caused by its lowering, as well as to find a reliable source of freshwater in the region that already suffers from severe water scarcity, three alternative macro-project proposals were suggested over the last years:

## *5.1 The Med-Dead Canal*

William Allen 1855 was the first to propose a connection of the Mediterranean Sea with the Red Sea via the DS. His canal was, however, to be used for shipping because he assumed that it would be cheaper than the projected Suez Canal (Vardi 1990). For a Mediterranean Sea-Dead Sea Canal (MDC) constructed to stabilize the water level of the DS, two alternative paths were suggested: those of the Northern Project alignments and the Qatif southern alignment (Fig. 15). The northern two alignments were planned to conduct Mediterranean water to Lake Tiberias or to the Jordan River through the Carmel Mountain with 80 km of pipeline plus 20 km of tunnel. It was also suggested to join this project with the National Water Carrier near Lake Tiberias. The Qatif alignment is to divert Mediterranean seawater from Gaza to Masada at the DS via a 80 km tunnel and a 20 km canal (The Harza JRV Group 1996). Although both of these routes are shorter and cheaper than the canal from the Red Sea, the macro-project was abandoned by Israel because it is not feasible economically and due to international objection to the construction of such a macro-project on a unilateral basis.

## *5.2 The Red-Dead Canal*

The Red Sea-Dead Sea Canal (RDC) is proposed to convey seawater from the Red Sea (Gulf of Aqaba) to the DS through the Wadi Araba with a total distance of ~180 km (Fig. 15). The principle of this macro-project is to utilize the altitude difference of >400 m between Wadi Araba and the DS in order to produce hydroelectric power that could be then used mostly for seawater desalinization. The rejected brines should be used to refill the DS to a target level that will be most probably not higher than −400 m. This represents the altitude of the sill dividing the northern basin from the southern one. Therefore, raising the DS level above this elevation will flood the salt pans in the southern basin of the DS and

**Fig. 15** The routes of the Med-Dead and Red-Dead Canals. (Redrawn after the Harza JRV Group 1996)

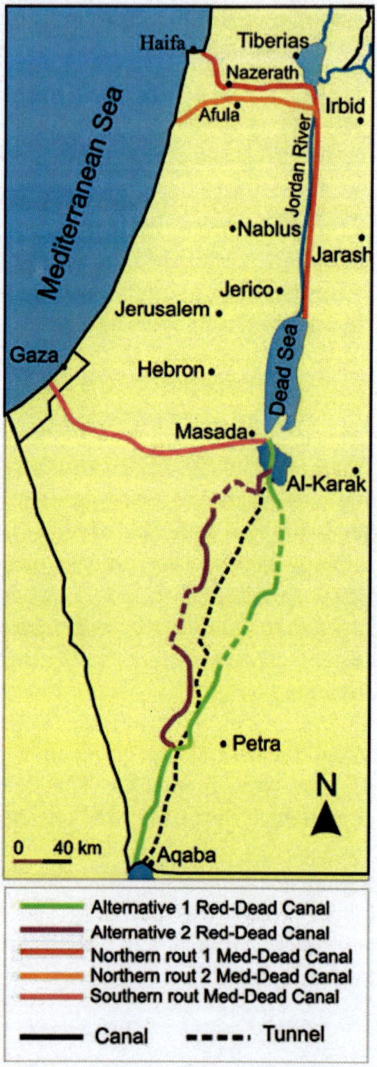


consequently destroy the salt industry (Gavrieli et al. 2005). Moreover, this will threaten several investments built at about −400 m. Gavrieli et al. (2005) suggested that an additional buffer of at least 2 m below the target level is required in order to accommodate abrupt, unpredictable lake level rises following intermittent rainy seasons. Abu Ghazleh et al. (2009) reported a lake level rise of 2 m in 1991/1992 due to extraordinary high precipitation of >1,000 mm/a in the eastern mountains of the DS.

In spite of the expected benefits of this macro-project, several questions related to environmental consequences arose:

### 5.2.1 Groundwater Aquifers

Four groundwater aquifers are located along the margin of Wadi Araba basin containing small amounts of water that are partially non-renewable. Two aquifers are in Cretaceous rocks and the other two are in Neogene and Quaternary alluvium deposits. Continuous infiltration of the seawater from the projected canal mainly threatens the shallow alluvium aquifer (100 m depth) that extends over a large area of Wadi Araba. The hydrological link between the alluvium and the other groundwater aquifers would cause a continuous leak of seawater to reach all groundwater resources in the area (Schirva-Schwartz et al. 2006). Abu Ghazleh et al. (2009) suggested that the canal should be built with an impermeable bed in order to avoid seawater pollution of the ground water and to avoid diminishing the energy output.

### 5.2.2 Earthquake Hazard

The Wadi Araba Fault (WAF) is a main part of the DSTF boundary separating the Arabian plate from the Sinai-Palestine sub-plate. It is characterized by the lowest seismic activity in between all segments of the DSTF based on the data of the Jordan Seismological Observatory Catalogue (JSO) (personal communication). This is also clear from the earthquake recorded along WAF that does not include any earthquake of magnitude >5 during the last 80 years (Fig 16). According to the GFZ-GPS annual monitoring of more than 40 GPS stations in Jordan, the seismic activity of the DSTF shows a tendency to migrate northward especially after the November 22, 1995 Gulf of Aqaba earthquake (Fig. 16). This is manifested by the 2004 DS, the 2008 Sour and the Palmyra earthquakes (JSO, Personal communication). Migowski et al. (2004) concluded that from A.D. 1600 to recent time, the earthquake epicentres are all located on the northern segment of the DST, while prior to 1000 B.C. and between A.D. 1000 and 1600 they appear to scatter along several segments of the DST.

Abou Eleanean et al. (2009) pointed out that the area of the 2004 earthquake that occurred along the NE of the DS is a source of low to moderate seismicity, whereas the area of the 2007 earthquake that occurred along WAF (Fig. 16) is a source of relatively low seismicity. The majority of earthquakes recorded by local and regional seismic networks along the DSF and ARF are too small to be included in the Global CMT, except the 2004 earthquake ($M = 5.2$), which was the largest in the DS during the last half-century.

Based on the reported earthquake damage in the DS area, Ken-Tor et al. (2001) estimated that the recurrence time of strong earthquakes (e.g., $M = 7.3$) in the southern half of the DSTF is $\sim$1,500 years. Therefore, Amit et al. (2002) suggested that the decrease in earthquake magnitudes and in the seismic activity for tens of years along the WAF emphasizes the potential of a major earthquake in the near future. However, it is difficult to predict any future seismicity simply because each earthquake activity will have different characteristics relative to other activities even along the same fault (JSO, personal communication).

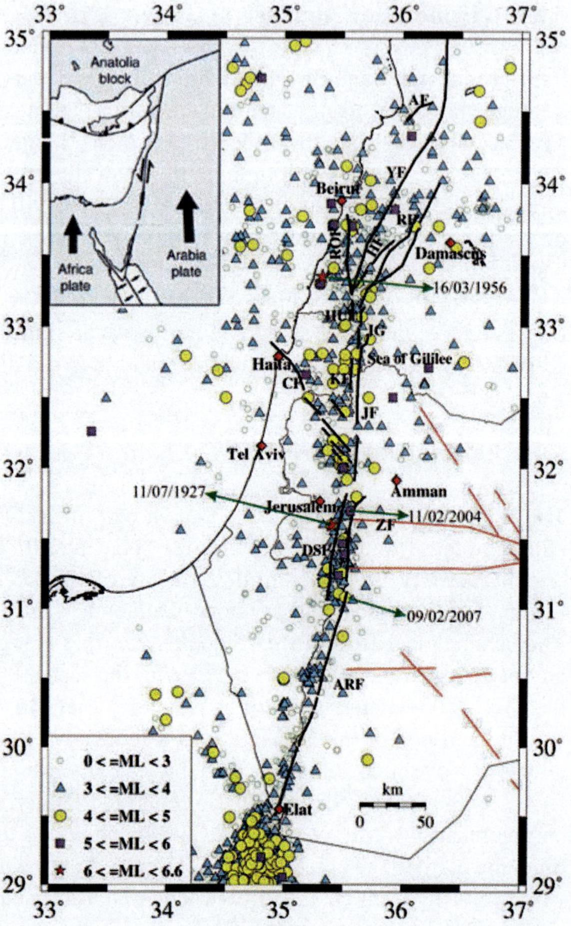


**Fig. 16** Recorded seismicity of the Wadi Araba and the Dead Sea Faults during the period 1900–2007. After Abou Eleanean et al. (2009)

However, the presence of numerous intersection points of any alternative alignment of the canal with potential active faults and the widespread alluvium and sand deposits with high amplification potential as well as the possibility of sudden surface displacement of 1–2 m, mandates a detailed study of these aspects at an early stage of the canal planning (Schirva-Schwartz et al. 2006). This should help to choose the most save route of the canal and minimize hazard consequences. Abu Ghazleh et al. (2009) suggested that possible ruptures of the canal bed by earthquakes would not be such a risk since the canal would be segmented by pumping and turbine stations, thus parts of it could be emptied and repaired, as in pipeline ruptures.

### 5.2.3 Lake Water Stratification

A large volume of seawater inflow (density ~1.03 kg/l) could dilute the surface water of the DS (density ~1.24 kg/l) and consequently cause a stratification of the water column. Experiences from the DS show that the high freshwater inflow during the rainy winter of 1991/1992 has diluted the surface water of the DS by up to 30% (Beyth et al. 1993). However, the stratification depends mainly on the rate at which the lake level will be raised. A slow and gradual pumping rate reduces the possibility of surface water dilution. Once the target level is attained, the volume of inflowing seawater will be adjusted to maintain a constant level. Incoming seawater will evaporate and the seawater-derived salts will accumulate in the upper water column. Thus the salinity and density of the surface water will continuously increase and reach that of the lower one. Consequently, the mixing of the water layers will occur (Gavrieli et al. 2005). However, if the rejected brine from the desalinization process should be drained to the DS, the scenario would be completely different. The rejected brine has almost the composition of seawater but about twice its concentration (Gavrieli et al. 2005). Thus a mixture of both rejected brine and Red Sea water would be denser and possibly no stable stratification would arise. It would be the ideal alternative to refill the lake and to avoid environmental consequences on the lake water.

### 5.2.4 Gypsum Precipitation

The mixing of sulphate-rich seawater (3,000 mg/l) with calcium-rich DS water (18,000 mg/l) could lead to gypsum precipitation. This was shown in laboratory and field experiments (Katz et al. 1977; Levy 1984). The major consequence of gypsum precipitation is the possibility that the gypsum crystals could remain suspended in the upper water mass causing whitening of the surface layer (Katz et al. 1977). This might affect the evaporation of the DS and consequently its climate. However, there is no certainty regarding the occurrence of whitening and most researchers agree that a prolonged event of whitening is unlikely (Gavrieli et al. 2005). Whitening of the DS would like serve to reflect sunlight, at least temporarily. Therefore more laboratory and field experiments are required to get reliable results regarding this phenomenon.

### 5.2.5 Blooming of Algae and Bacteria

Previous outdoor and laboratory experiments show that blooming is possible when the DS brine is diluted with seawater or freshwater by 10% or more and if phosphate is supplied (Oren and Shilo 1985). Therefore, blooming would presumably occur during the filling stage after diluting the surface water by inflowing seawater. This is also evident from the microbial blooming that occurred in the DS following the dilution of surface water during rainy winters of 1980 and 1992 (Oren 1983; Oren et al. 1995). Prolonged periods of blooming are not a desired outcome of the canal. It would change the colour of the DS and a high turbidity could lead to a higher evaporation rate which in turn would cause the necessity to introduce more seawater to keep the DS at the desired level (Gavrieli et al. 2005).

This phenomenon could only occur if the Red Sea water would be directly drained to the DS and not as a mixture of seawater and rejected brine. Since the possible dilution of the DS water will be limited to the filling period, the blooming is expected to be a minor and temporary problem. Some research is needed in order to understand the phosphate amounts involved and to study the likely organisms that could occur in blooms.

### 5.2.6 Development of Anoxic DS Lowest Water Mass

If stratification of the DS water occurs, the lowest water mass will be isolated from the atmosphere. After consumption of the dissolved oxygen through the oxidation of the organic matter sinking from the uppermost layer, bacterial sulphate reduction will occur and $H_2S$ will be produced. Under such conditions, iron becomes more mobile through its reduction to $Fe^{2+}$ (Gavrieli et al. 2005). The development of such anoxic lower water mass will not have a direct environmental impact on the surrounding of the DS. However, it may have a major impact on the potash industry, since the potash works would prefer to pump brine from the more concentrated lowest water body. Most of the $H_2S$ will be emitted to the atmosphere during its flowing to the evaporation pond and the rest will be chemically or bacterially oxidized. There will surely be an aerial stink associated with this emission. The permanent poisonous and dislikeable smell of $H_2S$ may become an environmental nuisance. Precipitation of iron and other trace metals that may occur in the carnallite ponds have to be removed from the salt products (Gavrieli et al. 2005).

### 5.2.7 Suggestions to Minimize the Environmental Impacts of the Canal Macro-Project

Nanofiltration (NF) of the Red Sea Water

The principle of this desalinization process is to utilize membranes with a higher molecular weight cut-off point than seawater reverse osmosis membranes (SWRO). NF rejects bivalent elements such as calcium, magnesium and sulphate, almost as well as (SWRO) but to a much lesser degree monovalent ions such as sodium and chloride. The rejected ions are 98% sulphate, 85–90% calcium and 94–98% magnesium. Moreover, this process blocks the passage of heavy metals, nutrients and other contaminations resulting in the following benefits (Hoffman 2008):

(a) Reducing the quantities of sulphate introduced into the DS and thus reducing precipitation of gypsum;
(b) Increasing the recovery ratio of the desalination process to 65% of the Red Sea water flow (1.86 $km^3$/a, as suggested by The Harza JRV Group 1996). This will result in a highly concentrated brine (10.2% TDS) that reduces algae and bacterial growth at the surface of the DS. Moreover, the low level of calcium,

magnesium and sulphate in this concentrated brine will improve the economics and purity of table salt production of the DS; and

(c) Reducing the concentration of organic matter and nutrients in the seawater increases the efficiency of chlorination. This will diminish the growth of marine organisms on all conveyance structures as well as algae and bacteria blooms in the DS.

Designing the Seawater Intake

A submerged intake system of the Red Sea water, including coarse filters that prevent suction of large floating objects, should be applied instead of a surface-sited intake. Some submerged intake points should be placed at specific sites where they could suck up clean water (Hoffman 2008). Any pollution would probably be small compared with the total required seawater intake. Therefore, it can be separated from the non-polluted seawater and removed using suitable facilities. The Gulf seawater flow pattern of the submerged intake points have to be modelled and optimized to replace suspended matter and stagnant water by a steady inflow of fresh seawater. If this clean-up process proves to be feasible, its benefits will outweigh the added cost of the submerged intake system (Hoffman 2008).

Utilizing the Dead Sea's Clean Solar Energy

The undesired possible result of density stratification could be used to create "solar ponds" to generate energy and producing desalinated water. In a "solar pond", a thin, less dense surface water insulates a bottom layer of brine. The brine is heated up to 95°C by the sun. Depending on pool depth and size the energy can be used continuously also during the nighttimes and during cloudy weather (Hoffman 2008).

## *5.3 Rehabilitation of the Jordan*

The abstraction of ~95% of the Jordan River's freshwater is the main cause of the dramatic shrinking of the DS. Nowadays, the water flow of the river is only 100–200 × $10^6$ $m^3/a$ (Al-Weshah 2000). The freshwater diversion is so large that in some summers, the Jordan River no longer reaches the DS. This has not only devastated the DS and its environment but also transformed the culturally and historically important Jordan River into little more than an open canal of agricultural run-off, diverted saline waters and untreated sewage (FoEME report 2005).

The amount of energy consumed by Israel to move freshwater from Lake Tiberias to the Mediterranean seacoast and Al-Negeb desert could be used to desalinate

Mediterranean Sea water, allowing an amount of 0.6 $km^3$ of the Jordan water to flow into the DS as it did previously. Moreover, a sufficient water management policy of the freshwater resource in the DS region (i.e. planting crops with minimal freshwater needs, recycling of waste water for agriculture, stopping freshwater pipe leakage, etc.) will also save a significant amount in each sector of the water economy in all of the riparian countries. The recent study of the FoEME (2005) suggested that these countries could restore 0.4–0.6 $km^3$/a, which could be transferred back into the Jordan River. Therefore, a total amount of $\sim$1.2 $km^3$ could be saved adopting this macro-engineering alternative. This will be enough to rehabilitate the Jordan River as well as protect its unique ecosystem. Meanwhile, it will slow down the rapid regression of the DS significantly, allowing time for macro-engineers and others to properly consider all long-term planning alternatives.

## 6 Summary and Conclusion

The DS is a pull-apart basin developed along the DST since the Miocene. After its disconnection from the Mediterranean Sea at the beginning of the Pleistocene, the basin was occupied by several water bodies of various sizes and salinities; the last one is the DS that is $\sim$10,000 years old. The DS has experienced several fluctuations during the Holocene caused by climatic changes. However, the most recent lowering of the lake since 1932 (0.4 m/a) is mainly due to impact of upstream freshwater diversion macro-projects together with mineral extraction industries in the southern basin of the DS. The terrain model of the DS rift yielded functions calculating water volume and surface area change with elevation. Moreover, the accurate altimetry survey and dating of the shoreline terraces helped to examine the lake level changes in details during the last decades. The over-exploitation of the freshwater resources in the DS basin during the last 14 years caused an alarming decrease in the water level, volume and surface area amounting to 1 m, 0.49 $km^3$, and 5.2 $km^2$ per year, respectively. The model functions can be used to predict the near future changes in the lake volume and area. Therefore, the lake is expected to drop to −433 m in 2020 and to lose 3.4 $km^3$ and 55.8 $km^2$ of its current volume and area, respectively.

The rapid decline of the lake level not only threatens the DS as a unique natural and historical resource but also caused severe environmental hazards all around the lake (i.e., sinkholes and landslides, among others). Due to these impacts and due to the mounting public freshwater demand in the lake basin, saving the DS seems to be mandatory. The rehabilitation of the Jordan River is a safe alternative from the environmental point of view. Although it will slow down the lake level drop significantly and preserve the ecological system of the river and the DS, it will not solve the problem entirely. The long-term alternative of the Red Sea-Dead Canal macro-project seems to be of very high potential with several positive implications. It will not only save the Dead Sea, but could also utilize the altitude difference of $\sim$400 m to produce hydroelectric power and, subsequently, freshwater by desalinization. In spite of possible environmental consequences of the Canal, it is still better than to take no action. However, a comprehensive approach of environmental assessment and impact analysis of the macro-project can address and find ways to mitigate the possible risks and minimize the negative environmental impacts. Moreover, additional micro-projects developed along the several possible Canal routes could enhance economical and social benefits for the riparian countries. Based on the average volume loss of the DS (0.49 $km^3/a$) and that of the commercial salt-pans (0.3 $km^3/a$) during the last 14 years and on the groundwater discharge to the DS (0.5 $km^3/a$), we suggest that the planned Red Sea-Dead Canal has to be assumed to have a seawater flow capacity of at least 1.3 $km^3/a$. However, a higher volume of seawater flow is required to raise the lake to the target level of about 400 m b.s.l. during the re-filling 21st Century period.

## References


Abdel-Fattah A, Pingitore E (2009) Low levels of toxic elements in Dead Sea black mud and mud-derived cosmetic products. Environ Geochem Health 31:487–492

Abed AM (1985) The geology of the Dead Sea. Dar Al-Urqam, Amman (in Arabic)

Abed AM (2000) Geology of Jordan, its water and environments. Jordanian Geologists Association, Amman (In Arabic)

Abed AM, Yaghan R (2000) On the paleoclimate of Jordan during the last glacial maximum. Palaeogeogr Palaeoclimatol Palaeoecol 160:23–33

Abed AM, Carbonel C, Collina-Girard J, Fontugne M, Petit-Maire N, Jean-Claude R, Yasin S (2000) Un paleolac du dernier interglaciaire Pleistocene dans l'extreme-sud hyperarid de la Jordanie. C R Acad Sci Paris Sciences de la Terrauet des Planetes/Earth Planet Sci 330:259–264

Abed AM, Yasin S, Sadaqah R (2008) The paleoclimate of the eastern Desert of Jordan during the marine isotope stage 9. Quat Res 69:458–468

Abelson M, Yechieli Y, Crouvi O, Baer G, Wachs D, Bein A (2006) Evolution of the Dead Sea sinkholes. In: Enzel Y, Agnon A, Stein M (eds) New frontier in Dead Sea paleoenvironmental research. Geological Society of America, Special Paper 401, pp 241–253

Abou Eleanean KM, Aldamegh KS, Zharan HM, Hussein HM (2009) Regional waveform inversion of 2004 February 11 and 2007 February 09 Dead Sea earthquakes. Geophy J Int 176:185–199

Abou Karaki N, Closson D, Salameh E, de Tervaent M, Barjous M (2005) Natural, induced and environmental hazards along the Dead Sea coast, Jordan. Hydrogeologie und Umwelt 14:1–25

Abu Ghazleh S, Kempe S (2009) Geomorphology of Lake Lisan terraces along the eastern coast of the Dead Sea, Jordan. Geomorphology 108:246–263

Abu Ghazleh S, Hartmann J, Jansen N, Kempe S (2009) Water input requirements of the rapidly shrinking Dead Sea. Naturwissenschaften 96:637–643

Al-Weshah RA (2000) The water balance of the Dead Sea: an integrated approach. Hydrol Process 14:145–154

Al-Zoubi A, Shulman A, Ben-Avraham Z (2002) Seismic ref lection profiles across the southern Dead Sea basin. Tectonophysics 346:61–69

Amit R, Zilberman E, Enzel Y, Porat N (2002) Paleoseismic evidence for time dependency of seismic response on a fault system in the southern Arava Valley, Dead Sea rift, Israel. Geol Surv Israel 114(2):192–206

Bender F (1974) Geology of Jordan. Borntraeger, Berlin

Beyth M, Gavrieli I, Anati D, Katz O (1993) Effects of the December 1991–May 1992 floods on the Dead Sea vertical structure. Israel J Earth Sci 42:45–47

Bowen R, Jux U (1987) Afro-Arabian geology. Chapman and Hall, London

Brew G, Brarazangi M, Al-Maleh K, Sawaf T (2001) Tectonic and geologic evolution of Syria. Geoarabia 6:573–616

Charlier RH, Chaineux M-CP (2009) The healing sea: a sustainable coastal ocean resource: thalasssotherpy. J Coast Res 25:838–856

Closson D, Abou Karaki N, Klinger Y, Hussein MJ (2005) Subsidence and sinkhole hazard assessment in the southern Dead Sea area, Jordan. Pure Appl Geophys 162:221–248

Closson D, Abou Karaki N, Hallot F (2010) Landslides along the Jordanian Dead Sea coast triggered by the lake level lowering. Environ Earth Sci 59:1417–1430

El-Hasan T, Momani K, Al-Nawayseh J, Al-Nawayseh K (2008) Dead Sea influence on the chemical and mineralogical characteristics of the dry deposition fallen over the eastern highland of central Jordan. Env Geol 54:103–110

FoEME (2005) Crossing the Jordan, Concept document to rehabilitate and promote prosperity to the Lower Jordan River Valley, 32 pp

Galli P (1999) Active tectonics along the Wadi Araba-Jordan Valley transform fault. J Geophys Res 104:2777–2796

Gavrieli A, Bein A, Oren A (2005) The expected impact of the Peace Conduct Project (The Red Sea-Dead Sea pipeline) on the Dead Sea. Mitig Adapt Strateg Glob Change 10:3–22

Haase-Schramm A, Goldstein SL, Stein M (2003) U-Th dating of Lake Lisan (late Pleistocene dead sea) aragonite and implications for glacial east Mediterranean climate change. Geochim Cosmochim Acta 68:985–1005

Hoffman D (2008) Greening the water Dead Sea-Red Sea conveyance project, ADAN Technical and economic services

Huckriede R, Wiesemann G (1968) Der jungpleistozane Pluvial-See von El Jafr und weitere Daten zum Quartar Jordaniens. Geologica et Palaeontologica 2:73–95

Ichinose GA, Begin ZB (2004) Simulation of Tsunamis and Lake Seiches for the Late Pleistocene Lake Lisan and the Dead Sea. Geol Surv Israel GSI/7/04: 1–50

Katz A, Starinsky A (2009) Geochemical history of the Dead Sea. Aquat Geochem 15:159–194

Katz A, Kolodny Y, Nissenbaum A (1977) The geochemical evolution of the Pleistocene Lake Lisan–Dead Sea system. Geochim Cosmochim Acta 41:1609–1626

Ken-Tor R, Agnon A, Enzel Y, Stein M, Marco S, Negendank WFG (2001) High-resolution geological record of historic earthquakes in the Dead Sea basin. J Geophys Res 106:2221–2234

Khlaifat A (2008) Dead Sea rate of evaporation. American Journal of Applied Sciences 5(8): 934–942

Khlaifat A, Hogan M, Phillip G, Nawayseh K, Amira J, Talafeha E (2010) Long–term monitoring of the dead sea level and brine physico-chemical parameters "from 1987 to 2008". J Mar Syst 81:207–212

Klinger Y, Avouac J, Abou Karaki N, Dorbath L, Bourles D, Reyss J (2000) Slip rate on the Dead Sea transform fault in northern Araba valley, Jordan. Geophys J Int 142:755–768

Krenkel E (1924) Der Syrische Bogen. Zentralblatt für Mineralogie Geologie und Palaontologie 9: 274–281, and 10: 301–313

Landmann G, Abu Qudaira GM, Shawabkeh K, Wrede V, Kempe S (2002) Geochemistry of Lisan and Damya Formation in Jordan and implications on palaeoclimate. Quat Int 89:45–57

Levy Y (1984) The influence of the admixture rate of partly evaporated Mediterranean water to the Dead Sea on the properties of gypsum that is formed in the brine. Mediterranean – Dead Sea projects, Summary of Research and Surveys. Mediterranean – Dead Sea Company 5: 279–282 (in Hebrew)

Migowski C, Agnon A, Bookman R, Negendank JFW, Stein M (2004) Recurrence pattern of Holocene earthquakes along the Dead Sea transform revealed by varve-counting and radiocarbon dating of lacustrine sediments. Earth Planet Sci Lett 222:301–314

Migowski C, Stein M, Prasad S, Negendank JFW, Agnon A (2006) Holocene climate variability and cultural evolution in the Near East from the Dead Sea sedimentary record. Quat Res 66(3):421–431

Neev D, Emery KO (1967) The Dead Sea, depositional processes and environments of evaporates. In: Bulletin, vol 41. Geological Survey of Israel, 147 pp

Neuman HF, Kagan JE, Schwab JM, Stein M (2007) Palynology, sedimetology and palaeoecology of the late Holocene Dead Sea. Quat Sci Rev 26:1476–1498

Oren A (1983) Population dynamics of halobacteria in the Dead Sea water column. Limnol Oceanogr 28:1094–1103

Oren A, Shilo M (1985) Factors determining the development of algal and bacterial blooms in the Dead Sea: a study of simulation experiments in outdoor ponds. FEMS Microbiol Ecol 31:229–237

Oren A, Gurevich P, Anati DA, Barkan E, Luz B (1995) A bloom of Dunaliella parva in the Dead Sea in 1992: biological and biogeochemical aspects. Hydrobiologia 297:173–185

Powell JH (1988) The geology of the Karak area, map sheet No. 3152 III. In: Geology directorate, vol 8. Natural Resources Authority, Amman, Jordan, 171 pp

Quennell AM (1958) The structures and geomorphic evolution of the Dead Sea rift. Geol Soc Quart Jour Lond 14:1–24

Salameh HR (1997) Geomorphology of the eastern coast of the Dead Sea. Geo J 41(3):255–266

Salameh E, El-Naser H (1990) Physical interpretation of the discharge coefficient of an aquifer, a maillet equation. Hydrogeologie und Umwelt, Heft 1:129–136

Salameh E, El-Naser H (1999) Does the actual drop in Dead Sea level reflect the development of water sources within its drainage basin? Acta Hydrochimica et Hydrobiologica 27:5–11

Salameh E, El-Naser H (2000) Changes in the Dead Sea level and their impacts on the surrounding groundwater bodies. Acta Hydrochimica et Hydrobiologica 28:24–33

Salameh E, El-Naser H (2009) Retreat of the Dead Sea and its effect on the surrounding groundwater resources and the stability of its coastal deposits. In: Möller P, Rosenthal E, Hötzel H (eds) The water of the Jordan valley. Springer, Berlin, pp 247–264

Schirva-Schwartz M, Calvo R, Bein A, Burg A, Nof R, Baer G (2006) Red Sea-Dead Sea Conduit: geo-environmental study along the Arava Valley. Geol Surv Israel 29:1–39

Steinhorn I (1983) In situ salt precipitation at the Dead Sea. Limnol Oceanogr 28:580–583

Stow D (2010) Vanished ocean: how tethys reshaped the world. Oxford University Press, USA, 288 pp

Talafeha E (2009) Ponds superintendent, Arab Potash Company, Personal interview, Safi, Jordan, June 4th, 2009

The Harza JRV Group (1996) Red Sea-Dead Sea Canal Project, Draft Prefeasibility Report, Main Report. Jordan Rift Valley Steering Committee of the Trilateral Economic Committee

The Jerusalem Institute for Israel studies (2006) The Dead Sea Basin: Assessment of current situation and prospects for the future under continued Dead Sea Water-Level Decline

Turner AK, Schuster RL (eds) (1996) 1996 Landslides—investigation and mitigation. National Academy Press, Washington, DC, Transportation Research Board Special Report 247, 673 pp

Vardi J (1990) Mediterranean-Dead Sea Project, Historical Review; Geological Survey of Israel, GSI/9/90: 31–50

Waldmann N, Starinsky A, Stein M (2007) Primary carbonates and Ca-chloride brines as monitors of a paleo-hydrological regime in the Dead Sea basin. Quat Sci Rev 26:2219–2228

Allen W (1855) The Dead Sea, A new route to India; with other fragments. Brown ad Co, London, pp 374

# The Red Sea–Dead Sea Canal: Its Origin and the Challenges it Faces

**Damien Closson, Herbert Hansen, François Halgand, Nada Milisavljevic, Frédéric Hallot and Marc Acheroy**

## 1 Introduction

The purpose of this chapter is to allow uninformed readers to understand the situation in which the Red Sea–Dead Sea canal arose. The water policy in the Middle East is complex and some key elements are provided so that the reader understands (1) why Jordan is currently actively seeking a fresh water source based on the desalination of seawater of the Red Sea, (2) why the Palestinians (West Bank), who have large quantity of freshwater in their ground, are also looking for freshwater from the canal project, (3) why Israel is involved in this project while its water reserves allow it to do without that channel.

The first part deals with the water problem in Jordan, Israel and Palestine. The information is taken from pioneering studies which were conducted during the 1990s. Thus, all mentioned sources are related to relatively old work but conclusions of which are still relevant in 2010. It is worth mentioning that the analytical concepts developed by these studies have since been used successfully throughout the world, proving their strength and continued relevance of the results and conclusions.

Once the general background for understanding the issues behind the need of the Red Sea–Dead Sea canal has been drawn, a technical description of the major elements is provided on the basis of a pre-feasibility study conducted in the mid 1990s. Again the references cited relate to the 1990s.

D. Closson (✉), N. Milisavljevic, F. Hallot and M. Acheroy
Royal Military Academy, Brussels, Belgium
e-mail: dclosson@mail.com

H. Hansen
Zoning Industriel des Hauts Sarts, Herstal, Belgium

F. Halgand
Tractebel Engineering S.A, Gennevilliers, France

V. Badescu and R. B. Cathcart (eds.), *Macro-engineering Seawater in Unique Environments*, Environmental Science and Engineering,
DOI: 10.1007/978-3-642-14779-1_5, © Springer-Verlag Berlin Heidelberg 2011 

Then, the work of the feasibility study for the canal project is presented. The feasibility study started only 6 months before the writing of this chapter and will last approximately one year and a half. Few novelties are thus presented as representatives of Coyne & Bellier, the company selected to perform this study, have to inform firstly funders of the project (World Bank) and the authorities of the countries concerned by this project. This part enumerates the major challenges that will require technical answers.

## 2 Water Management in the Dead Sea Riparian Countries of Israel and Jordan

The Dead Sea has been a centerpiece in the history of many cultures and religions for centuries. The region around the lake is considered by some as the cradle of human culture and civilization. It features numerous archeological and historic sites. The Dead Sea is the lowest spot on Earth: about 423 m below sea level in 2010. Its water is ten times more saline than ocean water, making it one of the saltiest water bodies in the world. Its distinctive chemical composition and fresh/salt water interface have created a unique ecology of international importance.

This lake and its environment are changing, as the water level is dropping due to a sharp decrease in inflow (Salameh and El-Naser 1999). The water level has fallen from 394 m below sea level in the 1960s to 423 m below sea level in 2010. As a result, its surface has been reduced by one third: from roughly 950 to 637 $km^2$ today. The level continues to drop at an alarming pace of more than one meter per year, and its surface is shrinking accordingly. The significant decline over the past 30 years is due to diversion of water from the Jordan River and from the Dead Sea itself. The Jordan River is the main feeder. The extracted water is of vital importance for the population and economy in the region. Environmental damage has already been incurred in the Dead Sea area (Salameh and El-Naser 2000a, b). Current damage includes loss of freshwater springs, river bed erosion, subsidence, landslides (Closson and Abou Karaki 2010), and occurrence of more than three thousand sinkholes (Closson and Abou Karaki 2009). If no action is taken, the further decline of the Sea is likely to cause more severe environmental, cultural, and economic damage. It is estimated that, if left unattended, the Dead Sea will reach a new equilibrium at an elevation that is about 100 m below the current level.

Water is a geographically limited resource, fundamental to both the societal and economic development of a state. Population Action International estimates that currently 25 countries containing 600 million people face either "water stress" or "scarcity". By 2025, according to projections, between 2.7 and 3.2 billion people in some 40 countries will be facing water stress or scarcity. It is estimated that by 2050, the number of water short countries will soar to 54, affecting 4 billion people, or 40% of the projected global population.

The 20 countries of the Near East and North Africa face the worst prospects. The Near East is the most water-short region in the world. In fact, the entire Near

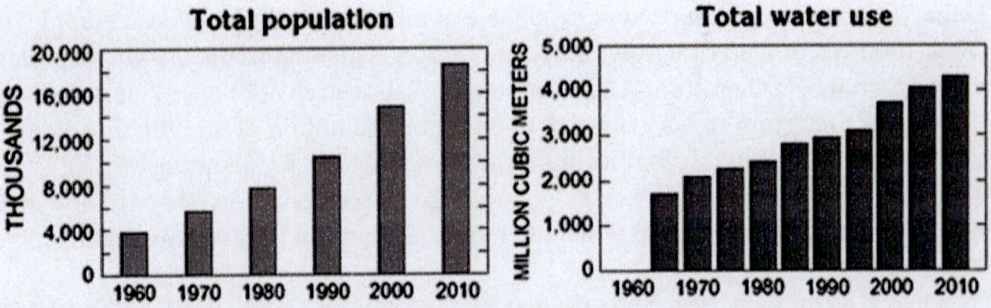


**Fig. 1** Increasing Israeli, Jordanian, and Palestinian population (*left*) and water use (*right*)

East "ran out of water" in 1972, when the region's total population was 122 million (Mitchell 1998). Since then, the region has withdrawn more water from its rivers and aquifer every year than has been replenished. For example, during the 1990s, Jordan withdrew 30% more water from groundwater aquifers every year than was replenished (Engelman and Leroy 1993).

Of 14 countries in the Near East, 11 are already facing water scarcity (Postel 1992). In five of these countries the populations are projected to double in 2020. Water is one of the major political issues confronting the region's leaders (Mitchell 1998). Since virtually all rivers in the Near East are shared by several nations, tensions over water rights could escalate into outright conflicts, driven by population growth and rising demand for an increasingly scarce resource.

In the initial stages of socio-economic development, water is considered as a free gift and access to shared water is uncomplicated. As population pressures increase, the socio-economic development of the state necessitates more active water exploitation, where the development of dams and inter-basin transfers is necessary for development in drier regions. When all the available water has been mobilized, the costs of further development and management of resources increase rapidly. This constant pressure to secure new sources of water can result in high competition at both the intra-state and inter-state level. When access is denied to alternative sources, a stronger politicization of water will occur.

Jordan River feeds the Dead Sea. It is by far the main tributary. This river Basin, some 18,300 km$^2$, is the most disputed area between Arabs and Israeli. It straddles the territories of four separate Arab political entities, and a Jewish one. These are respectively: Lebanon, the Syrian Arab Republic, the Palestinian Authority, the Hashemite Kingdom of Jordan, and Israel.

During the last 45 years, water resources in the Dead Sea watershed have been intensively developed to meet growing demands for this precious resource (Fig. 1).

## 3 "Virtual Water" Concept and Water Management Policies

Allan (1996a, b, c, 1997) analyzed the water management policies in the Middle East with the concept of "Virtual Water" as an organizing principle. Virtual Water is the volume of water needed to produce a commodity or service. For example, it takes 1000 $m^3$ of water to grow one tone of grain. This represents the virtual water value of grain. Furthermore, Allan defined the "allocative efficiency" as a method to increase the return to water, which is based on the notion of a rational choice as to which activity would bring the highest return to water. By adopting a strategy of allocative efficiency, a state makes possible the implementation of a virtual water strategy, through the importation rather than the production of low value, high water content foodstuffs.

Allan (1997) notes that, "more water moves into the Middle East as food trade than flows down the Nile and the Jordan in a year". Substantial volumes of water can be imported at a relatively low cost, rather than utilized at a relatively high cost. The upward trend in cereal imports reflects the capacity of the political economies within the region to meet their own strategic food needs.

Allan (1996a, b, c) also analyzed the changes that need to take place within a society in order to meet the challenge of increased water scarcity. This change is generally some form of "second phase water management, namely "demand side management"…(which) comes into play at a point in time when the first phase of supply-side management faces a crisis and is unable to mobilize more water by the application of "traditional" supply-side solutions" (Turton 1999a, b) (i.e., pumpage, aqueducts, pipelines).

Israel has a severely water deficit economy. This country has the third lowest level of renewable resources in the region. It provides the most advanced example of a transition to "water demand management". Israel has already moved from high water, low value agricultural produce (for example melons) to low water, high value produce (for example flowers, wine) to high value, low water industry and finally to virtual water importation. Allan (1996a, b, c) believes that, "demand side management is primarily essential in order to bring the allocation and management of the scarce water resources, into an environmentally and economically sustainable system", under which, "the economic activities being supported will be much more effective and the returns to water will be progressively enhanced".

Consequently, there was no escape from the politically stressful water relocation that Israel experienced and that its neighboring countries will be forced to address in the near future. It is evident that Israel has adopted a virtual water development strategy to balance its water budget. Despite needing up to four times more water than it is available, the political economy of Israel has been able to avoid conflict over access to scarce water resources because of its easy access to water that is embedded in the cereals, imported from rich water states. It is important to note that Israel, with the help of aid from the USA, was able to develop a diversified political economy by 1986. This enabled it to purchase its water entitlements on the international cereal market, therefore allowing it to alter its water policy. While Israel was able to implement a more sustainable water policy, Jordan, Gaza and the West Bank's less diversified economies have been unable to alter their development trajectories (with regard to water) and, as a result, face serious consequences if policy intervention does not take place. It is apparent that Jordan still has to undergo the difficult task of reallocating water resources between sectors, but out of sheer necessity it has already moved towards a virtual water strategy, with food imports rising sharply from 967 million USD in 1996 (EIU 1997) to 2120 million USD in 2009 (Jordan News Agency—Petra 2010).

## 4 The Karshenas' Theoretical Modeling

Karshenas' model illustrates the tendency of economies to deplete their environmental capital during the pre-industrial stage of economic and social development. The degree of over-exploitation that may occur is not necessarily irreversible provided that policy interventions are made and implemented (Turton 1999a).

In the case of water, countries, which develop and grow in their diversity and strength, will be able to pursue new and more efficient water allocating strategies, including the adoption of "demand side management" and allocative efficient strategies. Figure 2 outlines the trajectory regarded by Karshenas as the "normal"

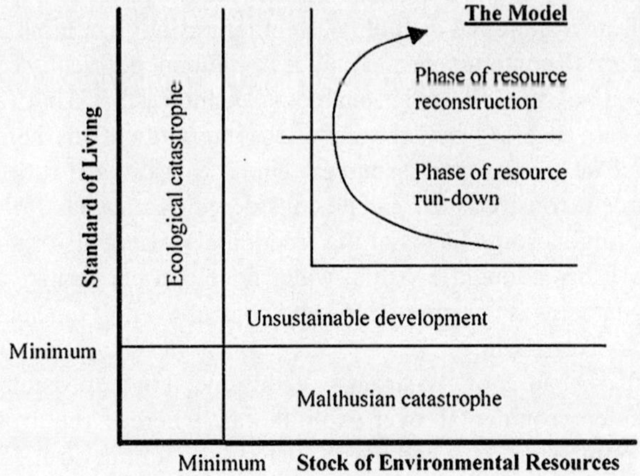


**Fig. 2** Attempt to conceptualize the relationship between economic development and the use of environmental capital such as water. The portion at the *top right hand* corner of the graph is the space where development can take place in a sustainable manner. The area to the *left* and *below* represents the depletion of resources in an unsustainable manner. The zones adjacent to the axes are those of ecological and Malthusian catastrophe, where environmental, and possibly even economic circumstances, become irreversibly degraded. The illustrated development trajectory is the one identified by Karshenas as reflecting a "normal" pattern for the use of environmental capital in which there are phases of resource use. Initially environmental capital is used to develop the economy and improve the welfare of the citizens. This may be beyond the threshold of sustainability in the short-term, resembling a degree of "over-exploitation". As the economy develops via the adaptation of new technologies, as well as by the integration of the national economy with larger regional or global economies, a new series of resource options begins to become available. With the increasing strength and diversity of the developing economy, policy makers can begin to contemplate the introduction of politically costly "demand side management" options, specifically by re-allocating water from one economic sector to another. This has been described by Karshenas as "natural resource reconstruction"

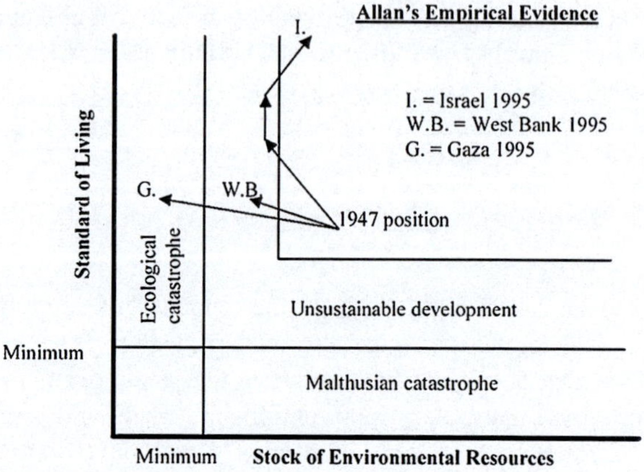


**Fig. 3** Trajectory produced by Allan, in an attempt to compare the trajectories of Israel, the West Bank and Gaza. Allan (1996a, b, c) notes that Israel's trajectory conforms to that which would be expected, of a diversified industrialized economy, stating: "By the second half of the 1980s, it was evident that water using policies were unsustainable and had to be modified"

pattern for the use of environmental capital. Under this trajectory, environmental capital is used to develop the economy, and improve the welfare of the citizens, even if this results in overexploitation of the resource base.

When the economy grows and becomes integrated with others, it opens doors to new sources of environmental capital. With other options available, policy makers are given the space to contemplate and maybe implement politically costly "demand side management" options. This then enables an economy that is both expanding and diversifying to gain access to virtual water on the international market, which lessens the pressure put on local water resources. Under this diversification, a policy of "natural resource reconstruction" can be considered (Karshenas 1994).

Figure 3 highlights some facets of the recent and present history. In 1967—two decades before it has adopted a virtual water development strategy to balance its water budget—Israel was determined to keep control over all the headwaters of the Jordan–Yarmuk river basin.

In the 1980s, because of its stronger economy, it was possible to make the transition from environmental over-exploitation, whereas the weaker and less diversified economies of Gaza and the West Bank could not do it (Turton 1999a). Instead, these economies remain on a trajectory heading towards environmental catastrophe, unless policy intervention takes place. Turton (1999a) suggests that this "in turn cannot be done unless Israel finally relinquishes control over the West Bank as it has done with Gaza". However, Israel is unlikely to let go of this source, which provides a significant part of its overall water supply.

# 5 The Predicament of Israel and Jordan

## *5.1 Israel*

Between the 1950s and 1970s, Israel has a high level of water scarcity within its own boundaries. In order to maintain self-sufficiency, it followed a supply-

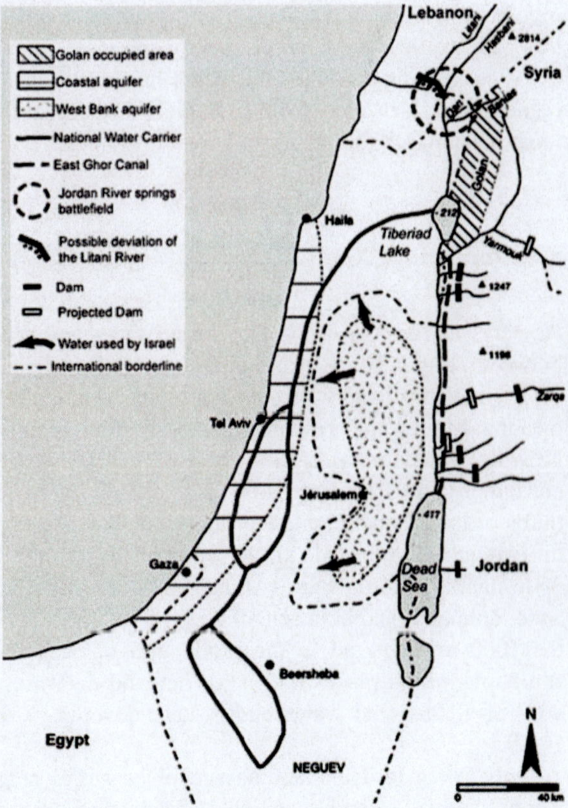


**Fig. 4** National Water Carrier (Israel) and East Ghor Canal (Jordan) are the consequences of supply-side management policy. Such infrastructures, and later dams, initiate the Dead Sea level lowering from the 1960s. (Adapted from Mutin 2000)

side management during this period, developing large and expensive water transfer schemes. An example of this can be seen in the construction of the Water Carrier in the early 1960s (Fig. 4). Nowadays, this pipeline carries 1.7 million of m$^3$ of water per day from the Tiberiad Lake, 200 m below sea level, to the coastal and desert farms in the Negev (for example Falkenmark and Lindh 1993).

In 1985, Israel's agricultural sector was consuming 80% of the water supplies yet only contributing 5% to the annual Gross Domestic Product; by 1995 it was consuming 58% and still only contributing 5% to the annual Gross Domestic Product (FAO 1997).

Industry, on the other hand, has been consuming 20% of water supplies on average, while contributing 50% to the annual Gross Domestic Product (EIU 1997). These figures reflect the reductions made to the agricultural sector during the droughts of 1986 and 1991, after which Israel was forced to take notice of its inefficient and unsustainable policies. The droughts created a situation of emergency, which in turn allowed for the subject of "water scarcity and misallocation" to come onto the political agenda and facilitated the imposition of the "reserve sector role" on agriculture (Turton 1999a). Overall, this allowed for a 60%

reduction in water allocation to agriculture between 1984 and 1991, which prior to this date would have been politically suicidal.

By modifying its approach, Israel has in effect turned its unsustainable development trajectory away from ecological disaster and conflict and instead towards a more sustainable future.

## 5.2 Jordan

As early as the 1960s, Jordan started to experience water levels of somewhere between 20 and 50 $m^3$ per capita a year, depending on the location (FAO 1997). This is not surprising when one considers Jordan's circumstances: the lowest level of annual renewable resources in the region, accompanied by one of the fastest growing populations in the world (FAO 1997). Estimates of domestic water consumption in Jordan are difficult to obtain, but, on average, water levels are two thirds of Israel's per capita consumption. Considering that a value of at least 1100 $m^3$ is required to provide all the water needs of an individual each year, the nature of Jordan's deficit is stark (Allan 1997). An individual requires 1 $m^3$ of drinking water annually, and between 50 and 100 $m^3$ for other domestic uses. However it is the 1000 $m^3$ required in the production of each individual's annual food consumption, which places the major demand on water supplies. As a result, about 90% of all national water budgets are devoted to the agricultural sector (Allan 1997).

Only 5% of Jordan's landmass receives sufficient rainfall to support cultivation, where even in a good year, wheat production covers only 20% of food requirements and 85% of the cultivable land is still totally dependent on rainfall (EIU 1997). Therefore state intervention in the agricultural sector remains essential in order to increase staple food production in response to population growth.

Already in 1958, intensive irrigation projects were implemented, the first of which diverted water from the Yarmuk River, resulting in the construction of the East Ghor Canal, now a major water artery for Jordan. Jordan has also built a dam on the Yarmuk River with Syria as its partner in 2007. However, water remains scarce and expensive, yet low-value crops and water demanding production systems have been encouraged up until the mid-1990s (Rogers and Lydon 1994). Private irrigation projects in the southern and eastern desert areas continue to fail to meet the increasing demands. The agricultural sector remains dominated by acute water shortages, where withdrawal continually exceeds renewal, and the cost of supplying irrigation outweighs any economic returns. Under the continuous pressure to mobilize water resources, Jordan now consumes 110% of its water stock annually, and has accumulated a deficit equivalent to one year's supply, through over-pumping of groundwater aquifers (Rogers and Lydon 1994).

In the 1990s, Jordanian water planners became aware of the general misallocation between agriculture and industry (Rogers and Lydon 1994). The current production of low-value crops contributes little to the Gross Domestic Product.

Agriculture requires 240 millions of $m^3$ to contribute 1% to the Gross Domestic Product whereas large industry only requires 2.5 millions of $m^3$ to contribute 1% to the Gross Domestic Product. In other words, the economic return per unit of water used by industry is one hundred times the economic return from agriculture. This, however, is politically strenuous as the intensively irrigated Jordan valley, allocated to wealthy Amman families since the 1960s, is highly subsidized by the government, and is also responsible for the 80% of the total agricultural production, which is produced there (EIU 1997). These families wield great political power, suggesting that political acceptability will continue to win over economic rationality in the Jordan valley.

Jordan's industrial sector has only been partially efficient since the 1980s. However, the service industry dominates the economy, contributing 66% of the Gross Domestic Product (tourism activities: 8–12%), and more importantly uses insignificant amounts of water (EIU 1997). Industry contributed 3,777 jobs and, at least, 7,000 indirect jobs per million of $m^3$ of water consumed in 2004. Tourism contributed 1,693 jobs per million of $m^3$. Agriculture employed just 148 workers per million of $m^3$—half of whom are foreign laborers—and contributed no income tax revenue to the treasury.

If Jordan can further develop industrial and tourism sectors, it has the potential to purchase its way out of its water problems, and at the same time avoid difficult politically suicidal decisions.

## 5.3 A Look in the Rear-View Mirror

The intensity of regional politics in the Middle East reflects the importance of each state's perception of its rights of water access. When the downward trajectory of regional surface water was first noticed in the 1960s, competition between states in the region for scarce water resources was such that it led to hostilities on a number of occasions:

- 1951: Jordan makes public its plans to irrigate the Jordan Valley by tapping the Yarmouk River; Israel responds by commencing drainage of the Huleh swamps located in the demilitarized zone between Israel and Syria; border skirmishes ensue between Israel and Syria (Wolf 1997; Samson and Charrier 1997).
- 1953: Israel begins construction of its National Water Carrier to transfer water from the north of the Tiberiad Lake out of the Jordan basin to the Negev Desert for irrigation. Syrian military actions along the border and international disapproval lead Israel to move its intake to the Tiberiad Lake (Samson and Charrier 1997).
- 1965–1966: Fire is exchanged over "all-Arab" plan to divert the Jordan River headwaters and presumably pre-empt Israeli National Water Carrier; Syria halts construction of its diversion in July 1966 (Wolf 1995, 1997).
- 1967: Israel destroys the Arab diversion works on the Jordan River headwaters. During the Arab-Israeli War Israel occupies Golan Heights, with Banias

tributary to the Jordan; Israel occupies the West Bank (Gleick 1993; Wolf 1995, 1997; Wallenstein and Swain 1997). Consequently, Jordan became a net importer of foodstuffs when it lost access to its principal growing areas on the West Bank of the Jordan River.
- 1969: Israel, suspicious that Jordan is over-diverting the Yarmouk, leads two raids to destroy the newly-built East Ghor Canal; secret negotiations, mediated by the US, lead to an agreement in 1970 (Samson and Charrier 1997).

However, the large-scale water wars, previously predicted by water pessimists, have not yet taken place. International relations over water have, if anything, become less tense since 1970 (Allan 2002; Caffrey 1993). The reason is that water has been available on the international market in the form of "virtual water." Indeed, economies that can import grain avoid having to mobilize scarce freshwater from their own resource base to produce wheat themselves. By the year 2000, the Middle East countries were importing tens of millions tons of grain annually, satisfying the largest demand for water in the region—food production. The remaining of water demand for drinking, domestic, and industrial use may soon be met through low-cost desalinated seawater via the Red Sea–Dead Sea Canal.

## 6 The Red Sea–Dead Sea Water Conveyance Concept

Jordan faces a rather grave domestic water scarcity situation, and is quite supportive of the idea of desalination by means of a sea-to-sea canal. In the 1990s its government pushed forward its version of the Red Sea–Dead Sea Canal. The plan was submitted to Israel for review, as required by the 1995 peace accords. In 1996, the Israeli Foreign Ministry convened a large team of experts to examine and respond to the Jordanian plan. The team deemed it inferior to the proposed alternatives (Mediterranean-Dead Sea options) in all respects. Nevertheless, ten years later, the governments of Israel, Jordan and the Palestinian Authority have articulated a shared vision of the Red Sea–Dead Sea Water Conveyance Concept, centered on:

- Saving the Dead Sea from environmental degradation;
- Desalinating water and generating energy at affordable prices for Jordan, Israel, and the Palestinians;
- Building a symbol of peace and cooperation in the Middle East.

Connecting the two seas is not a new idea. A possible inter-basin transfer has been studied in many forms since the mid-1800s. The nowadays 423 m difference in elevation between the Dead Sea and the Red Sea (or the Mediterranean Sea) has long been enticing because of the gravity flow advantage and the considerable potential for hydropower generation. From that time, the desalination of seawater has become the major goal replacing hydroelectric power generation. The concept also places rehabilitation of the Dead Sea as its top priority. It proposes a 180 km long alignment from the Red Sea at Aqaba to the Dead Sea that generally follows the border between Jordan and Israel and lies entirely in Jordanian territory.

A 1998 pre-feasibility study (financed by the Italian Government and managed by the World Bank) considered fourteen alternatives for alignments and conveyance elevation. The study defined one preferred alignment along Wadi Araba to be further investigated.

In 2008, the World Bank selected Coyne & Bellier a subsidiary of Tractebel Engineering (GDF SUEZ Group) to carry out the feasibility study of the project. This study encompasses technical, socio-environmental, economical and financial, as well as legal and institutional aspects. The study started in June 2008 and, according to the original schedule, was expected to last 2 years. The project should include the biggest Reverse Osmosis desalination plant in the world (seven times greater than the biggest existing one), providing Jordan with enough water for the next 40–50 years. Israel and the Palestinian West Bank would also benefit.

## *6.1 The Pre-Feasibility Study*

In the framework of the peace treaty between Israel and the Hashemite Kingdom of Jordan, the integrated development Master Plan for the Jordan Rift Valley was studied in the mid-1990s. The Red Sea–Dead Sea Canal was considered to be one of the most important potential elements for implementing this Master Plan (Harza JRV Group 1996).

The principal development objective should be to provide desalinated drinking water for the people of the area. The following two major tasks were defined:

- The Dead Sea level restoration at around 400 meters below the sea level (bsl). The Dead Sea level restoration is a consequence of the regional irrigation projects carried out about 45 years ago. Most of the running water in the Dead Sea drainage basin is used and the water balance of the lake is negative. Also the potash industry of the Arab Potash Company and the Dead Sea Works at the Dead Sea southern basin are increasing the rate of evaporation. As a result the average annual Dead Sea drop will be greater than 1.20 meter per year in the next decades. The lake level is around -423 meters in 2010, compared with -392 meters at the beginning of the 1960s. This extreme drop causes a lot of damage to the area around the Dead Sea. The only solution for such a drop in a desert environment with a growing population is to import seawater as a compensation for the water loss. The hydrogeologist Prof. Elias Salameh (University of Jordan) has already said "the Dead Sea's fate is clear: it will dry out and shrink to a very small pool, consisting of very, very salty brine. There is no hope unless man brings water into it from the Red Sea or the Mediterranean". The Red Sea - Dead Sea project is "vital and essential because it will also conserve the groundwater resources of the different riparian countries such as Jordan, Palestine and Israel"

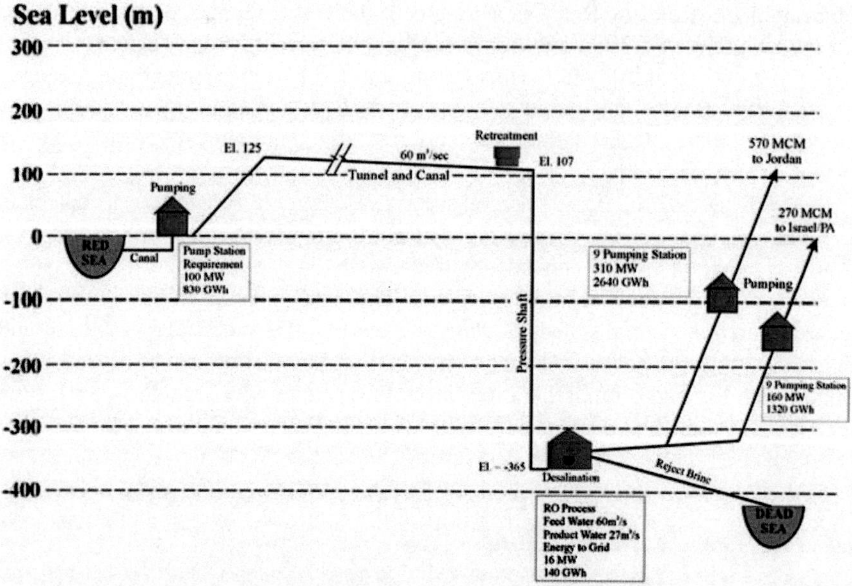


**Fig. 5** The Red Sea–Dead Sea Canal project (The Harza JRV Group 1996)

- Desalination of 800 to 850 millions m$^3$ annually. This task is suggested as a partial solution for the water deficit existing and predicted for the region. The duration of the construction should be around 7-10 years, and should cost about 7 billion USD.

There are six major components in the project (Fig. 5):

1. *Sea intake and pumping station* The seawater is pumped to +125 m above sea level at the Read Sea.
2. *Pressure pipeline* The first part of the conveyance system transmits the seawater to the planned elevation. The length is 5 km from Aqaba (3% of the whole alignment).
3. *Canal and tunnel—the major conveyance system* Seawater is transmitted to the regulating and pretreatment reservoirs with a design flow of 60 m$^3$/s. A 121 km tunnel with 7 m diameter and 39 km canal were designed.
4. *Regulating and pretreatment reservoirs* Several reservoirs were designed at +107 m at Wadi G'mal at the southeastern margin of the Dead Sea.
5. *Desalination plant* The desalination plants are designed to be operated by using the process of hydrostatically supported reverse osmosis to provide desalinated seawater. The plant will be located at Zafi at 365 m below the sea level with a water column of 475 m. With an annual flow of 2 $\times$ 10$^9$ m$^3$ and a recovery of 45%, 1,100 $\times$ 10$^9$ m$^3$ of rejected concentrates will enter the Dead Sea.
6. *Fresh water and reject brine carriers* Two-thirds of the 845 millions m$^3$ of desalinated water are for Amman, Jordan and one-third for the Palestinian Authority, Hebron and only a minor amount for Israel.

- For the transmission of the water to Amman a double pipeline of 200 km with 2.75 m diameter was designed with nine pumping stations for the uplift of 1,500 m.
- For the transmission to Hebron a double pipeline of 125 km with an elevation difference of 1,415 m was designed. 1,100 millions $m^3$ per year of brine reject water will enter the Dead Sea.
- The brine reject water would be conveyed from the desalination plant via a 7 km canal to the Dead Sea (Weiner 1996).

The Israeli study team estimated costs ranging from 1.5 to 5 billion USD, which included the cost of infrastructure and annual operational cost around 5 million USD, but not of conveyance from the desalination plant to the end users. Incorporating a desalination plant would require an additional 3 billion USD and take 10 years to finish.

The Harza study estimated that construction of the canal would take 10 years with an estimated cost of 5 billion USD. While the Harza report also did not include an extensive economic analysis of conveyance costs, Harza engineers compared the value of 1 $m^3$ of water per second diverted with the value of desalinating the same unit of water. They estimated that the cost of water diverted by the canal would range from 1.30 to 1.55 USD per $m^3$. These calculations are similar to the cost of desalinated seawater from current desalination projects. Therefore, the canal would not necessarily be the most cost efficient solution. Harza concluded that further studies about the economic benefits of a Dead Sea canal project are necessary.

## *6.2 The Feasibility Study in March 2010*

Since 2009, the World Bank is financing a two-year feasibility study for the project, which is too short period to conduct sufficient research that bridges knowledge gaps, especially with regard to the environmental implications.

For the feasibility study, the above mentioned six components have been integrated into four major elements, referred to as Sub-Studies, which need different expertise:

1. Sub-Study A: the Gulf of Aqaba
2. Sub-Study B: the water conveyance from the Red Sea to the Dead Sea
3. Sub-Sudy C: Issues related to the Dead Sea and surrounding area
4. Sub-Study D: Hydropower facilities and desalination plant.

The work is overseen by a Technical Committee consisting of representatives from the World Bank, Jordan, Israel and the Palestinian Authority. The total cost of the study is estimated at 7 million USD. The financing for it will be provided by France, Japan, the United States of America, the Netherlands, Greece, South Korea and Sweden.

Around twenty international groups had expressed their interest in this study and six of them were selected to present a tender. After a hotly contested process, the proposal from the group led by Coyne & Bellier was judged by the World Bank to be the most appropriate. Coyne & Bellier will work in association with Tractebel Development Engineering, a Belgian subsidiary of Tractebel Engineering (Suez), and Kema Nederland BV in the Netherlands.

At the time of writing this article, the feasibility study began about 6 months ago. Fifteen possible routes were analyzed and three options remains under consideration:

- a buried pipeline;
- a low-level tunnel all the way;
- a higher-level tunnel-and-canal system.

Broadly, the project established in the 1990s is not questioned. Major problems are being studied and presented below.

## 6.3 Possible Damage to the Ecological System of the Gulf of Aqaba

The impacts of the intake of water are regarded through entrainment and impingement. The Gulf is characterized by a northernmost tropical ecosystem, with natural conditions that hatch marine creatures. It has a dense marine biological population of more than 100 species of corals, 800 species of fishes and hundreds of species of crustaceans and mollusks.

During the pumping of water from the Gulf of Aqaba, the marine species suspended in the water are taken away and killed in the pumping process. Due to the high density of marine creatures in the water column of this area, this impact could be enormous. The pumping disturbs a large area of water column. The kinetic movement of water will destroy the habitats and kill the marine creatures by violent collision of water near the water inlets. Also, the noise caused by the pumps and chemical spill by the pumping facilities can disturb the marine creatures' life within a larger area as well as cause species loss and long-run ecological problems and pollutions. The environmental impacts on the fragile environmental equilibrium during the facility construction, such as pollution, noise, will also be taken into account.

## 6.4 Possible Infiltration of Sea Water into the Fresh Water Araba Aquifer

Earthquakes (Abou Karaki 1987) and flash floods are able to provoke leaks and spill of seawater. The path of the proposed Red Sea–Dead Sea conduit along the Araba Valley places a variety of design challenges due to numerous geo-environmental conditions:

- immediate proximity to shallow groundwater resources;
- flash-flood risks to open/near surface sections of the conduit;
- a cluster of tectonic related features, such as seismicity, active faults, high amplification potential and recent surface displacements.

These will have to be analyzed in the course of a substantial study. Once a final alignment for the Red Sea–Dead Sea conduit is chosen, a comprehensive study of its route will be carried out, consisting of the following issues:

- The vulnerability of the groundwater resources to leaks from the Red Sea–Dead Sea conduit is a major concern. A systematic hydrogeological survey of these resources and their vulnerability should be based on a comprehensive database of all the data available throughout the region and on additional new data in areas of information gaps.
- Provided that the conduit will follow the current eastern alignment, an array of monitoring wells will be required, mostly along the eastern margins of the Araba Valley. The array will have to be based on at least few tenths of wells, up to depths of 200–300 m, due to the heterogeneity in water quality and the possibility of sea water percolation at different horizons.
- Flash-flood hazard to the near surface/open sections of the final alignment of the conduit will be evaluated, on a basis of an accurate digital terrain model.
- A detailed study of potentially active faults along the final alignment should be carried out using all available methodologies, and summarized in a catalogue. Since almost every fault along the Araba Valley is suspected to be an active fault, this study and the relations of the faults to the alignment are highly important.
- Detailed observations along the final alignment should be carried out based on advanced space techniques (differential radar interferometry) in order to locate areas in which recent surface displacements take place.
- The seismicity and amplification potential along the conduit will be studied in detail.

A thorough study of the above mentioned parameters will allow for the best possible choice of the future site of the Red Sea–Dead Sea Conduit.

## *6.5 Mixing Dead Sea Water with Red Sea Water*

The project considers an eventual annual water transfer of 2 billion cubic meters from the Red Sea to the Dead Sea. Initially, most of this volume would flow into the Dead Sea to raise the water level and thereby compensate for the inflow reduction due to lower Jordan River diversion and industrial and natural evaporation losses. This could be combined with hydropower generation for potential sale in the region and/or energizing a potential desalination plant. Over time, part of the conveyed water could be desalinated for potable water distribution to municipalities in the region.

Through mass balance calculation it was calculated that the dissolved solids in the reject brine (from the desalinization process) is around 75 g/l. This salinity of the rejected brine is much smaller than the salinity of the Dead Sea.

The mixing is likely to produce a phenomenon of stratification due to different densities, precipitation of gypsum and halite due to mixing of the high concentrations of sulphate with high concentrations of calcium, blooming of algae and bacteria due to dilution of the surface layer. Evaporation rate will change depending on salt concentration: the higher the salinity the lower the evaporation rate and vice versa.

### 6.6 Possible Damage to the Environment of the Region

The outcome of mixing of these two water bodies over a time scale of decades is unknown and is difficult to model and predict. The Dead Sea will change its composition and characteristics as they are known today or were in the past if it receives large volumes of water from the Red Sea.

Since the Dead Sea is hydraulically connected with surrounding aquifer system, the continuing decrease in the Dead Sea water level causes an increase in the hydraulic head between the Dead Sea and the groundwater, which has the consequence that the groundwater will continuingly flow toward the Dead Sea, because the Dead Sea is the lowest point.

This flowing of groundwater from the surrounding aquifer to the hyper-saline water of the Dead Sea causes a loss for the shared countries every year estimated in millions of cubic meters of fresh water from its groundwater storage.

Rising the Dead Sea level will slow this phenomenon but will move downward the Dead Sea/fresh water interface with consequences that are difficult to predict.

### 6.7 High Cost of the Project

The magnitude of the conveyance concept is not unprecedented. Some major inter-basin water transfer projects exist elsewhere. Examples include: the Lesotho Highland Project in Lesotho and South Africa; San Francisco River Water Transfer in Brazil; Central Arizona Project in the USA. These projects transfer similar volumes of water from one basin to another as the Red Sea–Dead Sea conveyance concept, and their cost is in the same range.

## 7 Conclusion

The need to develop additional water resources in Jordan, to sustain the tourism development along the Dead Sea shore in Israel and Jordan, and to bring an image of peace have led Jordan and Israel to promote the concept of water conveyance from the Red Sea to the Dead Sea. The concept was also agreed to by the Palestinian Authority.

The global political economy of water use and trade has had important impacts on the way water is perceived in the Middle East. But at the same time, the impact of the global system has been perverse in that the availability of virtual water has slowed the pace of reforms intended to improve water efficiency.

Through the expensive storage and mobilization of their water resources, governments have created a myth of national self-sufficiency. This policy, which favors the traditional water user, the farmer, through the subsidization of water, in turn only serves to hinder the adoption of more durable and sustainable approaches, to the use of this scarce environmental capital. Therefore, governments, through the employment of viable strategies, will have to address the politically stressful issue of misallocation and equitable management of water resources.

## References


Abou Karaki N (1987) Synthèse et carte sismotectonique des pays de la bordure de la Méditerranée: sismicité du système de faille du Jourdain—Mer Morte. Université Louis Pasteur, Strasbourg, France, 430 pp

Allan JA (1996a) Water use and development in arid regions: environment, economic development and water resource politics and policy. Rev Eur Community Int Environ Law 5(2):107–115

Allan JA (1996b) The political economy of water: reasons for optimism but long term caution. In: Allan JA, Court JH (eds) Water, peace and the Middle East: negotiating resources in the Jordan Basin. I.B. Tauris Publishers, London, pp 137–168

Allan JA (1996c) Water in the Middle East: successful responses to a resource in deficit. In: Conference on water in the Middle East—a source of conflict or cooperation? University of Odense, August 26th, Denmark

Allan JA (1997) "Virtual Water": a long term solution for water short Middle Eastern economies? Columbia international affairs online working paper; The School of Oriental and African Studies (SOAS) Water Issues Study Group, Occasional Papers. http://www.ciaonet.org/wps/aln02/

Allan JA (2002) Hydro-peace in the Middle East: why no water wars? A case study of the Jordan River Basin. SAIS Rev XXII(2):255–272

Caffrey SC (1993) Water, politics and international law. In: Gleick PH (ed) Water in crisis: a guide to the world's fresh water resources. Oxford University Press, Oxford, pp 92–104

Closson D, Abou Karaki N (2009) Human-induced geological hazards along the Dead Sea coast. Env Geol 58(2):371–380

Closson D, Abou Karaki N (2010) Landslides along the Jordanian Dead Sea coast triggered by the lake level lowering. Environmental Earth Sciences 59(7):1417–1430

EIU Economist Intelligence Unit (1997) Country Profile: Jordan. London

Engelman R, Leroy P (1993) Sustaining water: population and the future of renewable water supplies. In: Population Action International, Washington DC, pp 6–47. http://www.populationaction.org/About_PAI/Index.shtml

Falkenmark M, Lindh G (1993) Water and economic development. In: Gleick PH (ed) Water in crisis: a guide to the world's fresh water resources. Oxford University Press, Oxford, pp 80–91

FAO Food and Agriculture Organization of the United Nations (1997) Trade statistics. http://www.fao.org

Gleick PH (ed) (1993) Water in crisis. a guide to the world's fresh water resources. Oxford University Press, Oxford

Jordan News Agency—Petra (2010) Jordan food imports climb to nearly JD2b in 2009. http://www.zawya.com

Karshenas M (1994) Environment, technology and employment: towards a new definition of sustainable development. Development and Change 25:723–756

Mitchell J (1998) Before the next doubling. World Watch 11(1):20–27

Mutin G (2000) L'eau dans le Monde Arabe. Ellipses, Collection Enjeux et Conflits, Paris, 156 pp

Postel S (1992) Last oasis: facing water scarcity. Norton, New York (revised 1997)

Rogers P, Lydon P (eds) (1994) Water in the Arab world: perspectives and prognoses. Harvard University Press, Cambridge, p 369

Salameh E, El-Naser H (1999) Does the actual drop in Dead Sea level reflect the development of water sources within its drainage basin? Acta Hydrochemica et Hydrobiologica 27:5–11

Salameh E, El-Naser H (2000a) Changes in the Dead Sea Level and their impacts on the surrounding groundwater bodies. Acta Hydrochemica et Hydrobiologica 28:2–33

Salameh E, El-Naser H (2000b) The interface configuration of the Fresh/Dead Sea water—theory and measurements. Acta Hydrochemica et Hydrobiologica 28(6):323–328

Samson P, Charrier B (1997) International freshwater conflict: issues and prevention strategies. Green Cross Report. http://www.gci.ch

The Harza JRV Group (1996) Red Sea–Dead Sea Canal Project, Draft Pre-feasibility Report, Main Report, Jordan Rift Valley Steering Committee of the Trilateral Economic Committee

Turton AR (1999a) Water scarcity and social adaptive capacity: towards an understanding of the social dynamics of water demand management in developing countries. MEWREW Occasional Paper No. 9, SOAS Water Issues Group, Department of Geography, School of oriental and African Studies (SOAS), University of London

Turton AR (1999b) Precipitation, people, pipelines and power: towards a "Virtual Water" based political ecology discourse. MEWREW Occasional Paper No. 11, SOAS Water Issues Group, Department of Geography, School of Oriental and African Studies (SOAS), University of London

Wallenstein P, Swain A (1997) International freshwater resources—conflict or cooperation? Comprehensive assessment of the freshwater resources of the world. Stockholm Environment Institute, Stockholm

Weiner D (1996) Red Sea–Dead Sea Canal Project, Draft Pre-feasibility report (The Harza JRV Group), in M. Beyth, Comp., Comments—RSDSC Pre-feasibility Draft Report

Wolf AT (1995) Hydropolitics along the Jordan River: scarce water and its impact on the Arab-Israeli conflict. United Nations University Press, Tokyo, Japan, 283 pp

Wolf AT (1997) "Water Wars" and water reality: conflict and cooperation along international waterways. NATO Advanced Research Workshop on Environmental Change, Adaptation, and Human Security. Budapest, October 9–12th, Hungary

# Red Sea Heliohydropower: Bab-al-Mandab Sill Macro-Project

**Roelof D. Schuiling, Viorel Badescu, Richard B. Cathcart, Jihan Seoud and Jaap C. Hanekamp**

## 1 Introduction

Macro-engineering has purposely developed, and is currently applying, proved construction and operation technologies and modern management techniques to effectively connect the various natural and anthropogenic components of certain Earth biosphere regions. "Industrial metabolism" is an apt metaphor to describe the mobilization and control of materials and energy through industrial activities. Predictive geomorphology involves the assessment of the "Anthropic Force", the combined effect on the Earth's surface, direct and indirect, of humankind's planned and unplanned industrial activities (Haff 2003). Twenty-first century technology enables us to move earth materials, and to create artificial constructions at an unprecedented scale—from the miniscule scale of Land Art to vast Macro-engineering mega-projects. In the near-term future it will become possible to divide suitable sea gulfs from the world-ocean through the building of appropriate physical barriers (dams) in order to create economically feasible power-drop macroproject sites.

R. D. Schuiling (✉)
Utrecht University, Utrecht, The Netherlands

V. Badescu
Polytechnic University of Bucharest, Bucharest, Romania

R. B. Cathcart
Geographos, Burbank, CA, USA

J. Seoud
UNDP, Beirut, Lebanon

J. C. Hanekamp
CEO, HAN, Zoetermeer, The Netherlands

V. Badescu and R. B. Cathcart (eds.), *Macro-engineering Seawater in Unique Environments*, Environmental Science and Engineering,
DOI: 10.1007/978-3-642-14779-1_6, © Springer-Verlag Berlin Heidelberg 2011

In the comedy, Birds, the Greek playwright Aristophanes tasked his dramatically contrived human character Peisthetaerus with the job of convincing the King of the Birds and his obedient subjects that they ought to help Peisthetaerus construct an ideal city, juxtaposed between eternal Olympus and the ever-changing less-than-ideal surface of the Earth. Peisthetaerus and the Kingdom's people schemed to intercept all the sacrifices of human beings rising from the Earth-surface that fed the Gods. By controlling the Gods' nourishment intake, Peisthetaerus and his pals would possess the profound coercive power to starve the resident Gods of Olympus into cooperating with them in Anthropic Force efforts of their own selection. Aristophanes' truly whimsical Utopia became, in the English language, "Cloud-Cuckoo-Land". (Aristophanes used Birds to ridicule the Philosophers in their shifting vagaries since clouds are the best metaphor for our world's caprice.) In analogy to Birds, Macro-engineering has strongly utopian undertones, but macro-engineers may soon boldly challenge the world-ocean's geographical extent at constrictive places where huge impermeable barriers can be emplaced. Already, a macro-project has been planned for the building of a dam in the Strait of Hormuz (Schuiling et al. 2005). Such barriers to seawater's flow will capitalize on the Earth's Hydrologic Cycle. Dams emplaced at Hormuz Strait and Bab-al-Mandab (BAM) will interdict and control the inflow of seawater into two strongly evaporative closed-off sea gulf regions. If macro-projects such as these are realized, they will dramatically affect the affected region's economy, political situation and ecology, and their cumulative effects may be felt far beyond the physical and political limits of the macro-projects (Schuiling et al. 2007).

## 2 Red Sea on Today's Maps

Bab-al-Mandab—BAM is Arabic for "Gate of Tears"—is an ocean strait that limits the seawater exchange with the Indian Ocean and therefore influences the seawater properties of the Red Sea. The narrow Red Sea (Fig. 1) is about 2,000 km long and has a maximum width of 354 km with a present-day watery area of ~451,000 km$^2$. Despite its millennia-long history of use by commercial shipping, the Red Sea remains one of the Earth's most unexplored underwater regions. Indeed, it was not until the 1820s that the gulf was first bathymetrically mapped (Searight 2003); accurate ship navigation charts became an absolute maritime necessity with the final seawater filling of the excavated channel on 15 August 1869 and its official celebrated opening later that year on 17 November, of the 190 km-long Suez Canal (Karabell 2003), the world's longest canal without locks. More recent bathymetric maps were published by Biton et al. (2008), Laughton (1970) and Siddall et al. (2004).

The Red Sea's freshwater deficit is ~2 m/year due to strong evaporation, very small Red Sea Basin precipitation and negligible river mostly flash-type runoffs (Sofianos et al. 2002). This figure was essentially confirmed in the recent study by Smeed (2004), who proposes 2.1 m/year. [Last century the Red Sea's annual

**Fig. 1** Satellite image of Red Sea

freshwater deficit was assumed to be ~3.5 m (Kettani and Scott 1974).] At its narrowest, BAM is ~18 km wide; its shallowest sill—not co-located with its narrowest segment—at present lies at −137 m at the Hanish Sill in the vicinity of the Hanish Islands, >100 km to the north of the BAM. During the most-recent Ice Age, when global sea level was 120 m lower than today's sea level, the Red Sea had an area of ~219,000 $km^2$ and humans were able to migrate across the narrow water body from Africa to Arabia (Bailey et al. 2007).

## 3 BAM Dam Design and Function

The concept of helio-hydroelectric (HHE) power generation utilizes the conversion of solar energy used for evaporation into hydraulic kinetic energy, and then to turbine-generated electricity. When topographical and hydrological conditions are favorable, a hydraulic head can be established and maintained by the Sun-powered evaporation of water. A system of two reservoirs is required to establish the hydraulic head: one is the reservoir, an "infinite" constant-level source of water, the other, a "sink" for evaporation. The Red Sea is completely closed, except for a narrow BAM strait, existing only for several million years, linking it to the Indian Ocean, and, of course, the man-excavated connection to the Mediterranean through Egypt's Suez Canal. If the Red Sea is selected to be the seat of such a system, construction of a big dam spanning the Bab-al-Mandab (BAM) between Arabia and Africa would transform the entire Red Sea into a

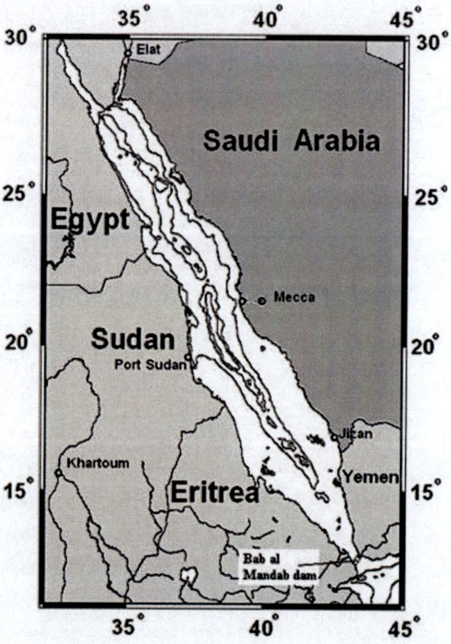


**Fig. 2** Red Sea and location of the proposed Bab-al-Mandab Dam. Map produced by using kk + w digital cartography (Koenigsweg, Germany) software

"sink" reservoir of seawater (Fig. 2). The system (Red Sea–Indian Ocean) thus formed, once the effect of induced evaporation over a number of years has reduced the Red Sea's level well below the world-ocean, may be harnessed for HHE energy conversion.

The Strait of Bab-al-Mandab is divided, at its narrowest total cross-section, into two navigable channels that are interrupted by Perim Island. The smallest channel, that between Perim Island and the coast of the Arabian Peninsula, has a width of ~2.7 km; the largest channel, lying between Perim Island and the continent of Africa is ~16.7 km wide. It appears quite feasible, in terms of Macro-engineering, to transform the entire Red Sea into a seawater evaporation reservoir with the construction of a colossal dam incorporating a hydro-electric powerhouse permanently blocking and harnessing the BAM. Such a scheme was previously proposed in 1961 (Hassan 1961). To some degree, the BAM Dam will replicate the HHE facility planned for the Strait of Hormuz (Schuiling et al. 2005), but as it is located at a tectonically moderately active site, some construction engineering adaptations may be necessary. Essentially, the Basin of the Red Sea is a rift zone where seafloor spreading occurs. The modern outline of the Red Sea was fixed about 4–5 million years ago but vertical movements of settled and unsettled Red Sea strands are still possible. Ongoing plate tectonic movements cause a ~9 mm/year widening of the Red Sea, that started ~4–5 million years ago. In addition, the rapid unloading caused by evaporation of seawater, followed by the slow loading caused by the increasing density of the residual brines and the subsequent deposition of a massive salt layer will cause disturbances from isostatic equilibrium, probably resulting in more (man-induced) tectonic movements (van den Belt and de Boer 2007; Talbot et al. 2009). Consequently, the planned BAM Dam must be capable of safely accommodating vertical movements as well as lateral expansion of its foundation and its superincumbent structural mass. It ought to be a stable macro-engineered structure that maintains its integrity during a significant historical period.

As the Red Sea is connected to the Mediterranean via the Suez Canal, another powerhouse should be located at Suez. In this respect it is worth noting that "…the Gulf of Suez… has had remarkably few earthquakes…" (Fairhead and Girdler 1970). The widest part of the Gulf of Suez does not exceed 50 km and unlike the Gulf of Aqaba the water is no more than 100 m deep (Abdel-Gawad 1970). Because the Gulf of Suez is so shallow, an interesting alternative might be to delay commencement of construction of the northern dam until the Gulf of Suez has fallen absolutely dry and then to construct the second dam at the outlet of the Gulf, on newly exposed dry land. After the construction of this dam, the Gulf of Suez can be refilled with Mediterranean seawater in order to provide a hydraulic head.

This will make construction costs lower, because the expenditure for the northern dam construction can be delayed by ~50 years, the time required to evaporate 100 m of sea water. Electric power production can commence and become widely available shortly after its completion. Moreover this will shorten very significantly the overall length of the required conduit pipes for bringing the seawater into the powerhouse turbines.

There are several location alternatives for the BAM Dam, because the shallowest part of the southern entrance to the Red Sea does not coincide with the narrowest passage. Hassan (1961) showed the cross-sections of several possible dam locations. A BAM Dam should have a piled rubble/sealed core width of ~1,000 m. Simple calculation shows that a BAM Dam ~1 km wide, ~100 km long and >150 m high from base to crest will have a maximum emplaced structural volume of $\sim 15 \times 10^9\ \mathrm{m}^3$. The dam at the southern end of the Gulf of Suez will have a significantly smaller volume, because the width that must be spanned is only ~25 km, and the required height is <100 m. After the refilling of the Gulf of Suez, it can be attempted to restore its ecology as well as possible, in order to make it again attractive for tourism (Hilderbrand et al. 2005).

## 4 Power Optimization

To evaluate the power produced at the Bab-al-Mandab Dam plant we used the procedure proposed in Kettani and Gonsalves (1972) and Kettani and Scott (1974) and very accurate bathymetry data provided by Smith and Sandwell (1997). One denotes by $Q_{\mathrm{evap}}$ the volumetric flow rate of evaporated seawater. The volumetric flow rate of fresh water entering the Red Sea by rivers and Red Sea Basin pluviosity is neglected, in agreement with Smeed (2004). At steady-state the

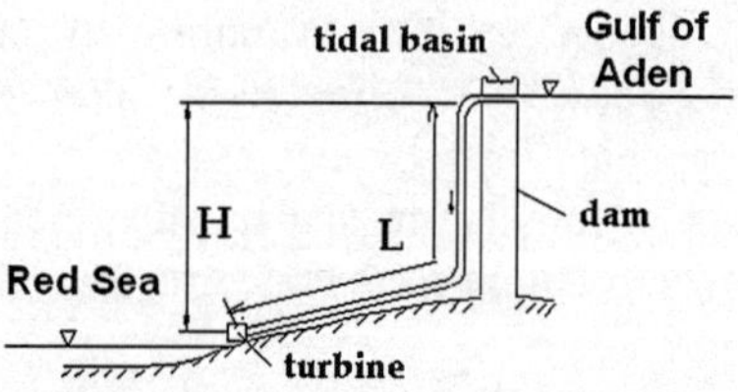


**Fig. 3** Powerhouse flow system at Strait of Bab-al-Mandab Dam. *H* seawater hydraulic head; *L* length of sweater duct

volumetric flow rate of seawater $Q_{sw}$ entering the Red Sea from the Gulf of Aden must balance the evaporation flow rate, i.e.

$$Q_{sw} = Q_{evap} \tag{1}$$

The volumetric flow rate of evaporated seawater depends, of course, on the surface area *S* of the active basin by

$$Q_{evap} = w_{evap} S \tag{2}$$

where $w_{evap}$ ($\approx$2.1 m/year) is the seawater evaporation rate (Smeed 2004). The hydraulic head *H* provided by the BAM Dam is shown in Fig. 3. The surface area *S* of the evaporative basin decreases as the head *H* increases. When the hydraulic head is zero, *S* reaches its maximum value $S_{max}$ = 451,000 km$^2$ (Smeed 2004). When *H* is a maximum (say $H_{max}$), there will be no seawater left and $S = 0$. The Red Sea maximum associated hydraulic head is $H_{max}$ = 2,789 m. The available bathymetry data was used to produce the dependence *S*(*H*), shown in Fig. 3.

The potential energy of the seawater flow transferred from the Indian Ocean through the Gulf of Aden to the Red Sea is transformed, in part, into mechanical power *P* collected at turbine shafts. The remaining part of the initial potential energy is dissipated due to friction in the seawater duct. There is a relationship between the length *L* of the seawater duct (connecting the dam and the turbine) and the hydraulic head *H* (Fig. 3). The dependence *L*(*H*) was modeled numerically by using the available bathymetry data. Twenty-nine equidistant isobaths between the present day sea-level surface (0 m) and the depth of 2,800 m were used in computation. Figure 4 shows the first 11 isobaths, between 0 and 1,000 m depth. The seawater duct is supposed to follow the maximum slope of the mapped terrain.

The following simple relationship applies for the power provided by the plant:

$$P = g\rho_{sw} Q_{sw}(H - \Delta H_f)\eta_t = g\rho_{sw} S(H) w_{evap}(H - \Delta H_f)\eta_t \tag{3}$$

Here Eqs. 1 and 2 were also used. In Eq. 3, *g* (=9.78 m/s$^2$) is gravitational acceleration, $\rho_{sw}$ (=1,030 kg/m$^3$) is the density of the falling seawater and $\eta_t$ (= 0.9) is the efficiency of the hydraulic turbine. Also, $\Delta H_f$ in Eq. 3 is the loss of head due to friction, which may be computed by using the simple relationship:

$$\Delta H_f = \Delta H_{f,lin} + \Delta H_{f'loc} \tag{4}$$

where "lin" and "loc" denote linear and local losses, respectively. The linear loss is evaluated as usual by:

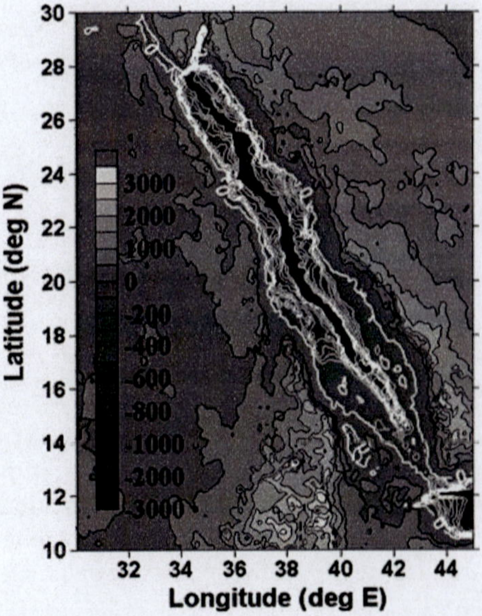


**Fig. 4** Red Sea regional topography and its contemporary known bathymetry. The *white lines* show 11 isobaths between 0 and 1,000 m (step-size 100 m)

$$\Delta H_{\mathrm{f,lin}} = \lambda \frac{L(H)w_{\mathrm{duct}}^2}{d_{\mathrm{duct}}\ 2g} \tag{5}$$

where $d_{\mathrm{duct}}$ and $w_{\mathrm{duct}}$ are duct diameter and seawater speed in the duct, respectively. The friction factor $\lambda$ may be evaluated by using the Blasius and Nikuradse relationships, respectively, as follows (Ionescu 1977):

$$\lambda = \begin{cases} \dfrac{1}{(100\mathrm{Re})^{0.25}} & 4000 < \mathrm{Re} < 10^5 \\ 0.0032 + \dfrac{0.221}{\mathrm{Re}^{0.237}} & 10^5 < \mathrm{Re} \end{cases} \tag{6}$$

Here the Reynolds number is computed by $\mathrm{Re} = w_{\mathrm{duct}} d_{\mathrm{duct}} / \nu_{\mathrm{sw}}$, where $\nu_{\mathrm{sw}}$ ($= 1.3 \times 10^{-6}$ m$^2$/s) is the kinematical viscosity of sea water. Among the local losses just that due to the change of diameter when passing from the duct to the turbine nozzle was considered:

$$\Delta H_{\mathrm{f.loc}} = \zeta \frac{w_{\mathrm{duct}}^2}{2g} \tag{7}$$

where the friction factor $\zeta$ is given by (Ionescu 1977):

$$\zeta = \left[\frac{n(n^{1.8} - 1)}{1.43n^{1.8} + 1}\right]^2 \sin\delta \tag{8}$$

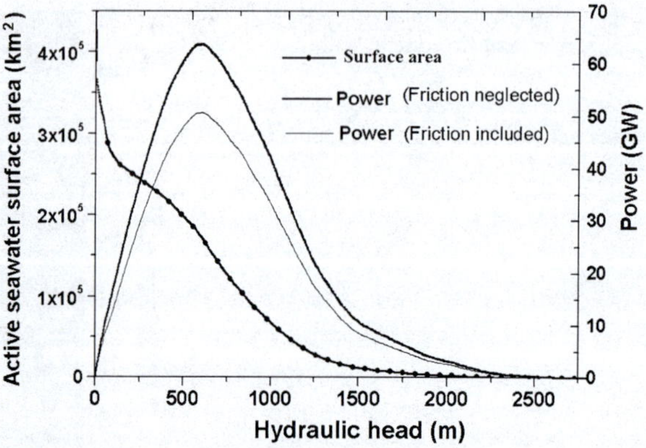


**Fig. 5** Dependence of the surface area $S$ of the evaporative basin and plant power $P$ on the hydraulic head $H$. Two cases were considered: friction neglected in the duct (i.e. $\Delta H_f = 0$) and friction included ($\Delta H_f$ given by Eq. 4). $w_{duct} = 4$ m/s, $d_{duct} = 15$ m

where $n = (d_{duct}/d_{nozzle})^2$, with $d_{nozzle}$ is the equivalent diameter of turbine nozzles and $\delta$ is the nozzle's angle. In calculations a value $\delta = 30°$ was adopted.

Mass and energy conservation require the following two relationships to be fulfilled, respectively, by the seawater speed $w_{nozzle}$ at nozzle exit:

$$\frac{w_{nozzle}}{w_{duct}} = \left(\frac{d_{duct}}{d_{nozzle}}\right)^2 \tag{9}$$

$$w_{nozzle} = \sqrt{2g(H - \Delta H_f)} \tag{10}$$

Equations 4–10 may be solved by using an iterative procedure, with values for $w_{duct}$ and $d_{duct}$ as entries. The unknowns are $d_{nozzle}$ and $w_{nozzle}$. Finally, this allows computing $\Delta H_f$ to be used in Eq. 3 giving the output power $P$.

The dependence of the power $P$ on the hydraulic head $H$ is shown in Fig. 5 for a particular HHE plant configuration. The following input values were used: $w_{duct} = 4$ m/s, $d_{duct} = 15$ m. These values are accessible to present day technology (compare for example with $w_{duct} = 4$ m/s and $d_{duct} = 6.1$ m at the penstocks of the 5,428 MW hydraulic power plant provided by eleven Francis turbines at Churchill Falls, Labrador, Canada). The computed water speed entering the turbine rotor is $w_{nozzle} = 97.5$ m/s, in good concordance with current practice.

The following scenario about the power-plant at Bab-al-Mandab Dam emerges from Fig. 5, which shows that there is an optimum hydraulic head (i.e. $H_{opt} = 611$ m), for which the power is a maximum (i.e. $P_{max} = 50{,}716$ MW). The ideal case (with friction neglected, i.e. $\Delta H_f = 0$) is also represented in Fig. 5. In this ideal case the maximum power is $P_{max,ideal} = 63{,}830$ MW. The loss of power due to friction is therefore significant. Interestingly, a major part of the losses are of local rather than linear nature (compare $H_{f,lin} = 7.9$ m to $H_{f,loc} = 117.5$ m). This gives macro-engineers some perspective for a further power increase by improving nozzle design.

The associated optimum surface of the evaporative basin is about one-third of its present value (i.e. $S_{opt} = 171{,}380$ km$^2$, see Fig. 6). Compare these results with the optimum head $H_{opt} = 500$ m associated to a maximum power plant

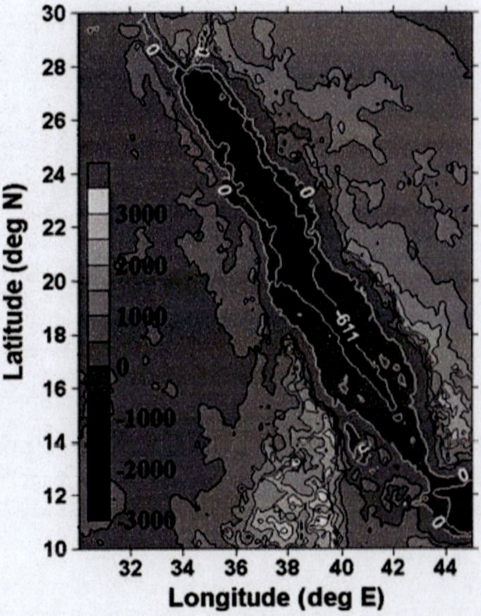


**Fig. 6** The evaporative basin of the Red Sea, present-day (isobath of 0 m) and after HHE plant construction (*white lines* of isobath of sealevel and −611 m below sealevel)

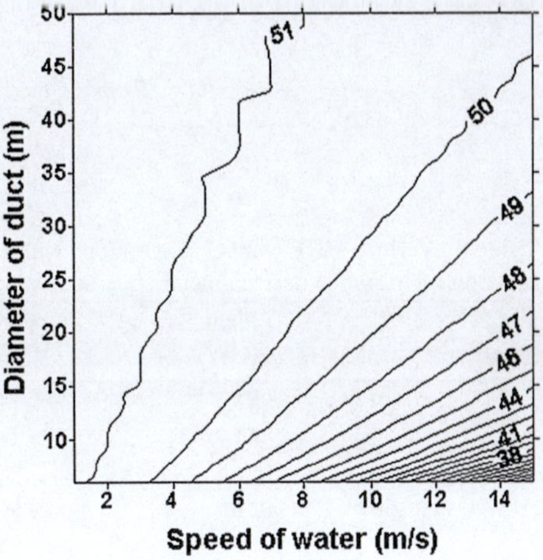


**Fig. 7** Dependence of HHE power (GW) on speed of water and diameter of the duct

$P_{max} = 67{,}000$ MW obtained in early calculations by Kettani and Scott (1974). The remaining area left for evaporation found by these authors was $S_{opt} = 133{,}000$ km$^2$. These small discrepancies are due to the fact that Kettani and Scott assumed a smaller initial total Red Sea surface (i.e. $S_{max} = 437{,}969$ km$^2$) than we have used here.

The output power $P$ of the HHE powerhouse turbines increases by increasing both $w_{duct}$ and $d_{duct}$, as expected (Fig. 7). However, in all cases the same optimum head $H_{opt} = 611$ m is obtained. A more detailed analysis is necessary to decide for the most appropriate and buildable design solution. Details of our particular site-specific macro-project computation follows. The water volumetric flow in a turbine is simply given by

$$Q_{v,turb} = w_{duct}\frac{\pi d_{duct}^2}{4} \tag{11}$$

while the power provided by a single turbine is given by

$$P_{turb} = g\rho_{sw}Q_{v,turb}(H - \Delta H_f)\eta_t \tag{12}$$

The number of turbines $N_{turb}$ in the power plant is simply computed by:

$$N_{turb} = P/P_{turb} \tag{13}$$

The speed of a turbine usually depends on the speed of a directly-coupled alternator. If this is the solution chosen at BAM power plant, one may assume a number of poles $N_{poles} = 96$ for the alternator and the electric grid frequency $f = 50$ Hz. Then, the turbine speed $n_{turb}$ is given by:

$$n_{turb} = 120\frac{f}{N_{poles}} \tag{14}$$

One finds $n_{turb} = 62.5$ rpm. Finally, the specific speed $n_s$ of the turbine is given by:

$$n_s = n_{turb}\sqrt{P_{turb}}(H - \Delta H_f)^{-5/4} \tag{15}$$

**Table 1** Various HHE plant configurations with electrical output power greater than 50 GW as a function of seawater duct diameter $d_{duct}$ and water speed $w_{duct}$ in that diameter duct

| $w_{duct}$ (m/s) | $d_{duct}$ (m) | Plant power (GW) | Turbine power (MW) | Number of turbines | Turbine type |
|---|---|---|---|---|---|
| 1 | 9 | 51.2 | 29.2 | 1,754 | Francis |
| | 15 | 51.2 | 81.1 | 632 | Francis |
| | 30 | 51.3 | 324.7 | 158 | Kaplan |
| | 50 | 51.3 | 902.0 | 57 | Kaplan |
| 2 | 9 | 51.0 | 58.1 | 877 | Francis |
| | 15 | 51.1 | 161.9 | 316 | Francis |
| | 30 | 51.2 | 648.8 | 79 | Kaplan |
| | 50 | 51.3 | 1,803.6 | 28 | Kaplan |
| 4 | 9 | 50.2 | 114.4 | 439 | Francis |
| | 15 | **50.7** | **321.2** | **158** | **Kaplan** |
| | 30 | 51.1 | 1,293.2 | 39 | Kaplan |
| | 50 | 51.2 | 3,603.6 | 14 | Kaplan |
| 6 | 15 | 50.1 | 475.6 | 105 | Kaplan |
| | 30 | 50.8 | 1,931.7 | 26 | Kaplan |
| | 50 | 51.1 | 5,396.0 | 9 | Kaplan |

The line in bold is associated to the power station treated in Fig. 5

The specific speed may be used to select the appropriate machine type among the usual Kaplan, Francis and Pelton turbines. Table 1 shows details for a few design solutions associated to electric power output larger than 50 GW. The design solution adopted in Fig. 5 (i.e. $w_{duct} = 4$ m/s and $d_{duct} = 15$ m) corresponds to a power station consisting of 321.2 MW Kaplan turbines. This fits the standards of present-day technology. [As an example, the Bajina Basta Hydropower Station on the Drina River near Perucac, Serbia, was completed in 1966–1967 and by 2013, after refurbishment and modernization, will then be capable of generating ~422 MW. Currently, it is provided with 315 MW pump turbines with 621 m head]. However, the necessary number of turbines (i.e. 158) at the BAM Dam powerhouse is unusually high. A decrease in the number of turbines requires unusually large duct diameter or seawater speed and yields very powerful turbines, sometimes far beyond the present-day figure of merit (Table 1). Note that one of the largest water turbines in operation today is a 730 MW turbine with 146 m head built by Toshiba in Japan.

The optimum energy production will not be affected if a second dam and powerhouse are constructed at the outlet of the Gulf of Suez after about 50 years of evaporation. This may have some financial advantages in the sense that the powerhouse built at the Bab-al-Mandab Dam can be smaller, as it will never have to carry the full load and the expenses for the second dam and installations can be delayed by 50 years.

The period of time necessary to reach a −611 m hydraulic head at an annual evaporation rate of 2.1 m is about 291 years. This is more than double the 143-year interval obtained in Kettani and Scott (1974) who used an overestimated value of the evaporation rate (i.e. 3.5 m/year). Here we use the more realistic value 2.1 m/year reported by recent studies (Smeed 2004). The removal of such an amount of freshwater from the Red Sea would lead to a 30 cm rise in the world-ocean after 291 years.

The Bab-al-Mandab Dam powerhouse must have a finite design and working lifespan, just like any other macro-project ever built. After many years of fresh-water evaporation caused exclusively by the region's justifiably famed prevalent sunshine, a salt deposit will eventually fill the Red Sea completely until the salt surface level reaches the outlets of the high-pressure conduits. A simple calculation shows that it would take tens of thousands of years for the Red Sea to be filled with salt to its optimum head level. Once the Red Sea's isolated seawater is saturated with salt the amount that will precipitate yearly is equal to the quantity of new salt introduced that year. A column of seawater 2.1 m in depth, with a cross-section of 1 $cm^2$ contains 210 cc of seawater with a salinity of about 35‰. Consequently, it contains about 7.35 g of salt that will be deposited on 1 $cm^2$. The density of salt being 2,240 $kg/m^3$, this means that once salt deposition begins, a new layer of about 3.15 cm thickness will form annually. Assuming a maximum Red Sea depth around 2,789 m and a hydraulic head of 611 m, the final salt layer will be 2,178 m thick, which will take about 69,000 years to form. To this span of Earth's Geological Time we must add another 5,000 years for the period it takes for the brine to reach saturation.

## 5 Post-BAM Dam Red Sea Seaport Fates

The Suez Canal will quickly lose its importance for navigation and the main seaports of the Red Sea—such as Yanbo, Jeddah (Saudi Arabia) and Port Sudan—will fall into disuse as the totally enclosed Red Sea evaporates. It will be out of the question to build low-cost ship locks at Suez that could bridge a difference in sea level of 100 m or more in a single step at Port Tawfik on the Red Sea, and present-day ships would be manifestly ill-adapted to travel in denser-than-seawater brines; it is possible that specially designed ships might be constructed to cope with the denser fluid medium of floatation and it is also possible that a ship railway, canal lift or canal incline might suffice to serve the shipping traffic (Uhlemann 2002). The voyage distance between the Saudi Arabia seaport of Jeddah and the Black Sea port of Constanta, Romania is shortened by 86% of the distance compared to the sea route that passes the Cape of Good Hope. The Suez Canal could serve to transport Mediterranean seawater to a powerhouse located at the southern end of the Gulf of Suez. The present-day Suez Canal has a cross-section of $\sim$4,600 $m^2$ so it can transport a considerable quantity of seawater (Abril and Abdel-Aal 2000). If the inflow through the canal would have a velocity of 2 m/s, the Suez Canal could just on its own supply the total annual evaporation from the residual Red Sea. The output from a Suez Canal Powerhouse will have to be coordinated with that of the BAM Dam powerhouse to balance the electrical grid. It would appear reasonable to assert speculatively that only a Suez Canal Powerhouse need be built since its geographical site is closer to present-day heavily populated urban regions, and its electrical transmission losses thereby can be saved, but if the BAM Dam site will not support economically a powerhouse, this will delay the first power production and hamper the development of the climatically dryer and economically more needy region at the Red Sea Basin's entrance/exit to the Indian Ocean. A caution needs to be stated: Port Said on the Mediterranean Sea could, anytime in the future, be impacted by severely destructive tsunamis generated in the eastern Mediterranean Sea. Alexandria and the Nile River's delta were impacted by a gigantic tsunami on 23 July 365 that was generated by a steep fault in the Hellenic Trench near the island of Crete that may have a recurrence period of approximately 800 years (Shaw et al. 2008). In order to avoid possible complications it might be wise to build a tsunami impediment or seawater surge obstruction at Port Said, functioning like the movable steel Maeslant Barrier (completed in 1997) that nowadays shields from North Sea storm surges.

The total amount of electricity generated exceeds by far the power needs of the countries currently bordering the Red Sea, considering that Egypt possesses already a large strategic hydropower capacity—approximately 2.1 GW—at the Aswan High Dam on the Nile River (Dumont 2009).

The excess should be used to establish energy-intensive industries, like aluminum and magnesium metallurgical plants, large-scale desalination units or to generate large volumes of hydrogen (Shaltout 1998). In case an aluminum processing plant is established, this could better be done near the Bab-al-Mandab Dam, because the required bauxite will mainly come from India and Australia. By having an aluminum processing plant near the BAM Dam site, these ore carriers would not have to detour around the Cape of Good Hope.

## 6 Environmental Consequences of Damming the Red Sea

The ecological consequences of closing off the Red Sea in order to construct one or more heliohydroelectric plants are irreversible and far-reaching, both in terms of regional as well as global environmental impacts.

It is evident that the construction of the BAM Dam, which will take years to complete, will be accompanied by negative but localized ecological impacts resulting from the building process. Construction will disturb the marine and coastal ecosystems in the vicinity of the dam, as well as increase air pollution and noise from the operation of dredgers and massive motorized building equipment. Depending on the location of the constructive work-sites, the severity of these impacts may vary since severity is a function of the level of disturbance to the environment as well as the natural sensitivity of the region, which in turn depends on the specific nature of the habitat and the specific plant and animal communities it currently sustains. However, these effects remain comparatively short-term and are limited to the construction site, and can be minimized by proper environmental management procedures. Long-term, drastic changes will occur upon completion of this macro-project and the initiation of powerhouse electricity generation.

### *6.1 Increased Salinity*

The most significant impact of damming the Red Sea is the increase in the salinity of the remaining seawater due to evaporation. Salinity is expressed in weight of salts per 1 kg of seawater, and in most seas and oceans its value varies between 32 and 38 g/kg, which is the range to which most marine creatures have adapted. In the Red Sea however, salinities range from about 36.5 g/kg at the BAM to up to 41 g/kg at the entrances of the Gulfs of Aqaba and Suez (Sheppard et al. 1992). Since marine organisms exist in an osmotic balance with their environment, increased concentration of salts in the marine environment results in the dehydration of cells, decrease in turgor pressure and finally death (mainly of larvae and young organisms). The sensitivity to the increase in salinity varies from species to species. Some of the planktonic algae, and in particular the siliceous ones, can tolerate high salinities; these species appear in coastal salt marshes. The sensitivity of invertebrates (mainly that of crabs) varies also and, in general, it is known that long-abdomen invertebrates are more sensitive to an increase in salinity than shortabdomen invertebrates. The larvae of crabs and of other invertebrates, which float in the water, are more sensitive than adults to changes in salinity. A significant increase in salinity, however, may not be tolerated even by organisms that are otherwise adapted to high values. Sheppard et al. (1992) described the situation when seawater salinity rises above ~42 g/kg, or where temperatures fluctuate through more than about 15°C, reef growth fails, coral growth becomes severely reduced, and macro-algae replace the corals as dominant biota. Corals require a range of physical conditions for healthy growth and reproduction. It is noteworthy that the "...human degradation of coral reefs began hundreds of years ago" (Briggs 2009). Changes in seawater quality, including raised nutrient levels and changes in salinity, temperature or turbidity, high levels of sedimentation and changes in water currents can all damage coral reefs. All of the previous mentioned changes would occur once the HHE macro-project is initiated. The Red Sea corals will completely disappear. Future coral transplantations, actually best termed "coral re-introduction", could be only employed to renew coral reefs in the Gulf of Suez or even to create new, and entirely artificial, reefs (Ammar 2009; Abuzinada et al. 2008). The timing of the coral re-introductions in the refilled Gulf of Suez would be critical to the final success, or possible ultimate failure, of the renewal macro-engineering event-process.

The northern part of the Red Sea is nutrient-deprived, and is characterized by an almost continuous fringe of contiguous coral reefs along much of its coastline, which physically protects the nearby shoreline from storm waves and tsunami, and provides food and shelter for a large and diverse underwater fauna and flora; while the southern part, which is influenced by, and replenished with, oceanic water (coming into the Red Sea from the Gulf of Aden) that is nutrient rich. The southern Red Sea is, as well, dominated by fine-grained sediments and associated mangroves, seaweeds, beds of sea grass and calcareous algae (Hoepner and Lattemann 2002).

The Red Sea's relative geographical isolation, and remarkable physical conditions, which range from near-shore shallows to depths of over 2,000 m in the central rift, have given rise to an extraordinary range of ecosystems and biological diversity, culminating in the elaborate system of coral reefs composed of approximately 200 species of stony corals, representing the highest diversity in any section of the Indian Ocean. Species endemism in the Red Sea is extremely high, particularly among some groups of reef fish and reef-associated invertebrates. Sheppard characterises the Red Sea as a remarkable and very important area, which exhibits a species and biotope diversity exceeding that of the world-ocean and the Mediterranean Sea (Sheppard 2000).

## 6.2 Unnatural Reduction of Sea-Level of the Red Sea

The proposed HHE macro-project would result in a lowering of the Red Sea by ~2.1 m/year. Consequently, coastal wetlands, mangroves and coral reefs would begin to dry and die almost as soon as the BAM Dam is closed to commence its operation. The most northerly reefs are those near the city of Suez, some at elevations of 1–3 m. Coral communities in the Red Sea are particularly rich in terms of species diversity, even for the Indian Ocean region. The most spectacular coral reefs for which the Red Sea is renowned occur in its northern and central parts. Fishes are a dominant component of reef fauna; the Red Sea has a total of approximately 1,000 species of fish (including non-reef species). Furthermore, the Red Sea is relatively diverse in sea grass species –11 have been recorded. Sea grass as well as mangroves growth patterns are similar—the greatest development of most species is higher in the south attributed principally to the wider and shallower shelf, as well as the greater prevalence of unconsolidated sediments. These ecological factors, together with the generally less extreme temperatures and seawater salinities, may have enabled sea grass beds to develop to their fullest extent in the south part of the Red Sea in the same manner as mangroves, and the reverse pattern is shown by corals. Sea grasses stabilize the shallow seabed against vigorous wave action and other common erosional forces; promote sedimentation and accumulation of organic and inorganic matter; provide a direct source of food for a few herbivorous animals such as some sea-urchins and fish, green turtles and dugongs, and more importantly, as a major input to marine organisms; provide a nursery or refuge region (also feeding zone) for resident and migratory fauna, and provide the site for important biogeochemical processes (e.g. the sulphate cycle). On the other hand, mangroves are one of several recognised critical marine habitats which serve as nurseries (Dawes 1998), and assume a number of ecological functions including land formation, primary production, sediment stabilisation and filtration of land runoff.

Overall, the many and diverse ecosystems of the Red Sea provide an ideal habitat for a large number of sea birds. Some of the important resident species include the Lesser Flamingos (*Phoenicopterus minor*) and the Yellow-vented Bulbul (*Pycnonotus xanthopygos*), while important wintering species include the Greater and Lesser Sand Plover (*Charadrius leschenault, C. mongolus*). The Red Sea is a flyway for many bird species, which seasonally migrate between Europe and Africa. The islands of the southern Red Sea, in particular the Farasan Islands, are utilized by many hundreds of thousands of birds in the Spring and Autumn migrations. Here there are internationally important populations of Saunders' Little Tern (*Sterna saundersi*), Bridled Tern (*Sterna anaethetus*) and the resident Egyptian Vulture (*Neophron percopterus*). If the coastal and island ecosystems are altered due to the lowering of the sea-level within the Red Sea Basin and a change in seawater quality caused by anthropogenic factors occurs, both the resident and migratory bird populations will be adversely affected.

An unnatural reduction in sea-level will expose new strands, which will form highly salty desert-like land regions. That, in addition to the arid to hyper-arid climate that predominates in the Red Sea Basin, may result in the scattering of millions of metric tons of salt and finely dispersed dust into the atmosphere, similar to what happened when the Aral Sea disappeared by evaporative drying (Stanev et al. 2004).

## 6.3 Rise in World-Ocean Sea-Level

It is speculated that operating the proposed HHE project over 50 years would result in an approximate 12 cm rise in world-ocean sea level, slowly increasing to a maximum of 30 cm once the hydraulic head reaches 650 m after 310 years. This, in addition to the existing possibility of an increase in global sea-level by an average of 6 cm per decade (with an uncertainty range of 3–10 cm) as a result of expected climate change, if no specific measures are taken to abate greenhouse gas emissions (according to the Intergovernmental Panel of Climate Change IPCC), would be very serious ecologically as well as economically. Higher sea-levels would threaten low-lying coastal areas and small islands, potentially putting millions of people and large areas of coastal land at risk. These negative environmental aspects are only partly offset by the fact that if 50,000 MW is never generated by fossil fuel use, an emission of about 425 million tonnes of $CO_2$ annually will be prevented, thus ensuring that the world-ocean's sea-level will rise less quickly than is nowadays anticipated by geoscientists, if there were no BAM Dam.

## 6.4 Mitigating Measures

As with most macro-engineering projects, the alterations they impose on the natural environment are immense and often irreversible. We envision some possible measures that may lessen their Earth-biosphere impact. In the case of the BAM Dam macro-project, the following measures may be considered:

- Temporary preservation of the living communities of the Gulf of Suez in huge aquaria, from which the Gulf can be repopulated after the Gulf is refilled 50 years from the start of the macro-project (Spotte 1992). Although this looks like an ideal option, it is unlikely to be viable in practice, since firstly, the whole bio-physical nature of the Gulf would be altered as a result of the macro-project; secondly, the reintroduction of the marine fauna and flora is unlikely to succeed; and thirdly, the preservation of all the living organism species originally present in the Gulf is probably extremely difficult to achieve in practice. This scale of ameliorative measure is unprecedented.
- Selection of a BAM Dam site as far to the north inside the Red Sea as possible, to preserve some typical biotopes of the southern Red Sea, as well as resting and feeding places for migratory birds. This option would mitigate the impacts to a certain extent in terms of the area of the region impacted, but will not change the damage that is inflicted on the remaining part of the Red Sea.
- Creating pay-to-view Nature Heritage Parks on the newly emerged land in selected regions, for which some of the desalination waters should be made available. Although this measure would not actually mitigate the impacts of the BAM Dam macro-project on the Red Sea's environment, it would create a new

financial source of income for coastal countries and would form a new habitat for certain land flora and fauna.

- Creation of a nature conservation fund from the income of the electrical power production, from which wildlife conservation programs in the BAM Dam region can be financed.

## 7 Geopolitical Consequences of Closing the Red Sea

Political consequences are not easy to gauge, but it is certain that the proposed macro-project will require considerable cooperation between the different countries in the affected region. The riparian countries are Egypt, Sudan, Eritrea, Yemen, Saudi Arabia and possibly Djibouti, depending on the precise location of the BAM Dam (Fig. 2), while Israel and Jordan are connected to the Red Sea via the Gulf of Aqaba. Apart from these directly involved countries, and in accordance with present-day stake-holder models, even countries like Ethiopia and Somalia, not directly adjacent to the Red Sea, yet impacted by the ecological, economic and political changes should be consulted in some fashion by the main contracting parties. A prerequisite, however, for most of these stakeholder models, is a functioning democratic environment, which cannot today be taken for granted in the region. Indeed, a stake-holder event-process (as developed in Europe and the USA) might not work well in this region. Nonetheless, different peoples will be affected both positively and negatively in terms of politics, economics and ecology, highlighting the numerous pitfalls for a macro-engineering project of the Red Sea. It will be essential, therefore, to find a modus through which their voices are effectively heard and their responses fully assessed. Evidently, the Middle East crisis is an example of historic as well as current geopolitical struggle. One needs only to think of the fact that both Israel and Jordan now have access to the Red Sea through the Gulf of Aqaba, which they will lose if the macro-project is implemented. The Gulf of Aqaba harbors a number of important submarine adventure ecotourist sites, which will also be lost. It stands to reason that Egypt, Israel and Jordan should be compensated for these significant economic losses by being allotted a fair share of the electricity produced at the far-distant BAM Dam and elsewhere closer that might exist in the Red Sea Basin.

Other macro-problems of a geopolitical nature, related to existing international law are the demarcations of the future borders between the ecosystem-countries over the newly emerged land, and the jurisdiction over the various proposed Red Sea dam sites. The Suez Canal Dam site is anchored at both ends on Egyptian soil, but the BAM Dam site has Eritrea (or Djibouti if the southernmost possible dam site is selected) on one side, and Yemen on the other. A possibility could be to establish a legal organization in which all riparian countries are represented, that has the overall supervision, but leaves the operational control and jurisdiction in the hands of Egypt for the Suez Canal Dam, and to Eritrea (or Djibouti) and

Yemen for the BAM Dam site. The BAM Dam's proposers are optimistic since it has recently been revealed in newscasts that a Yemen-Djibouti Fixed Link, anticipated to be emplaced by circa 2020, is being planned by Middle East-based macro-engineers (Sawyer 2007) (see COWI website, http://www.cowi.com).

In order to make parts of the former sea bottom fit for normal land agriculture, it will be necessary to desalinize and irrigate it. This makes it obligatory to reserve part of the electricity produced for large-scale seawater desalination. It must then be decided how this freshwater is distributed among the BAM Dam participants, because the region that can be irrigated is limited in area, even if a sizable part of the produced power is set aside for desalination. All these pending issues and all of these obvious massive changes must be steadfastly faced in what is still a fairly conservative and underdeveloped social environment. The risk of final authoritative decisions being taken favoring or disfavoring the implementation of this macro-project at high political levels without hearing the true voices of the people is very pronounced. Considering the current political situation in the countries affected by the macro-project, any consultation and/or decision taken will be at the level of governments, and may not necessarily be a reflection of the traditional populations' well informed opinion. The realization that cooperation is necessary in order to ensure the safe provision of sufficient energy may prevail over the many existing political and geopolitical differences.

## 8 The Go/No Go Decision Process in Macro-Engineering, Ethics and Economics in a Cautious Twenty-first Century

Obviously, ecological and economical impacts can only be discussed here in the broadest sense. Some of the salient environmental impacts were presented in a preceding chapter section, but a detailed impact assessment on ecological and economical is not part of the discussion here.

The first consequence that needs to sink in firmly—as put forward earlier—is that such a macro-project will massively change the region's ecology, with some potential global impacts. Regional economics will also be impacted as a considerable amount of electrical energy will become suddenly available. Conversely, economic activities like oceanic shipping, fishery or submarine adventure eco-tourism dependent on the present ecological make-up of the Red Sea will drastically change or may even disappear as a result of the Red Sea BAM Dam's closure.

Apart from the response from fully industrialized countries everywhere, the local and regional publics will have to deal with this regional change, which will be notably gradual, albeit considerably faster than most people are historically accustomed to experiencing.

The major Go/No Go macro-problem herein is the issue of uncertainty in relation to the ecological changes and the economic impacts spawned by the BAM Dam, as Science can give comprehensible or reliable answers as to the effects of newly implemented technologies only for a limited number of instances and effects.

There are several reasons for this: lack of data (ignorance), indeterminacy (non-causal random events or badly understood non-linear interactions), measurement errors, variability (observed or predicted variation of individual responses to an identical stimulus) or a combination of these (Klinke and Renn 2003).

Apart from these Go/No Go decisional uncertainties, normative valuations of such a large scale project highly influence the feasibility. In other words: major ambiguity will ensue as a result of propositions such as the BAM Dam (Douglas and Wildavsky 1982). Notably, this ambiguity is enhanced by the precautionary culture, of which the so-called Precautionary Principle is the lynchpin of environmental and public health policies. We need to discuss the Principle here as the frequently encountered international (non) governmental response to projects such as the BAM Dam is inspired by a cautious outlook.

The main gist of so-called "precaution" is best captured in the Rio definition that is considered the most authoritative among the many formulations of the Principle that can be found in the Science literature nowadays (Graham 2001):

> ...Where there are threats of serious or irreversible damage, lack of full scientific certainty shall not be used as a reason for postponing cost-effective measures to prevent environmental degradation.

Precaution and sustainability are closely related to each other. Precaution therefore is regarded as the lodestar on the road to sustainability. As the European Commission puts it (European Commission 2000):

> The dimension of the Precautionary Principle goes beyond the problems associated with a short or medium-term approach to risks. It also concerns the longer run and the well-being of future generations.

In fact, one could open-mindedly argue that they are both sides of the same medallion. As such, the Precautionary Principle impresses upon us a moral obligation to take care of the environment, of mankind, our children, and our children's children. The Precautionary Principle, therefore, carries a profound inter-generational perspective on anthropogenic activities and their potential Earth-biosphere risks.

In that sense, the macro-project described in this chapter is highly interesting in relation to the application of the principle as it is, more often than not, applied to new technological and scientific advancements (Pieterman and Hanekamp 2002). Indeed, Hanekamp et al. (2005) described the contemporary use of the Precautionary Principle as an expression of an over-arching pessimistic (if not to say misanthropic) conceptualization of Science and Technology.

When taking a closer look at the remark of the European Commission on Sustainability, in particular when 'the well-being of future generations' is addressed, the BAM Dam macro-project gains—in view of the Precautionary Principle—monumental contradictory characteristics. At first glance, the principle's definition on irreversible environmental effects, to all intents and purposes, seems to seal the fate of the macro-project straight off. Indeed, the ensuing ecological effects from such a dam, viewed on all sorts of time-scales, will be irreversible and indubitably large-scale, whereby the Precautionary Principle, more or less, bars the advancement of the BAM Dam. The Dutch public and political discussion on extracting natural gas from beneath the Wadden Sea, which was prohibited for a number of years for fear of ecological damage that was on all accounts regarded as being quite remote, is instructive in this matter (Pieterman and Hanekamp 2002).

However, without acquiring extra but fundamentally believable Science data on the actual ecological risks, this Precautionary Principle-based decision not to extract natural gas from beneath the Wadden Sea was eventually reversed, seemingly on economic and financial grounds only. This example might be indicative for the opportunistic justifications with which the Precautionary Principle is actually implemented (Hanekamp et al. 2005). In reality, when confronting the trade-offs induced by the realization of the BAM Dam, the ecological and economic cards might be shuffled in a completely different manner than currently would be the case, certainly if one emphasizes concern for the well-being of future human generations. Especially this aspect highlights the contradictory consequences of BAM Dam macro-project in view of a precautionary regulatory stance. Namely it shows that the Precautionary Principle cannot be a directing principle, as the BAM Dam mega-project, on the one hand, generates huge ecological changes generally regarded as unwanted, yet on the other hand will sustain and enhance the living conditions of innumerable people of future generations. Moreover, the ecological deficit might be off-set by the reduction of oil consumption and its concomitant waste production (e.g. carbon dioxide gas). As Sunstein (2002) observes:

> …I have argued not that the [P]recautionary [P]rinciple leads in the wrong directions, but that it leads in no direction at all. The reason is that risks of one kind or another are on all sides of regulatory choices, and it is therefore impossible, in most real-world cases, to avoid running afoul of the principle. Frequently risk regulation creates a (speculative) risk from substitute risks or from foregone risk-reduction opportunities. And because of the (speculative) mortality and morbidity effects of costly regulation, any regulation, if it is costly, threatens to run afoul of the [P]recautionary [P]rinciple. … The [P]recautionary [P]rinciple seems to offer guidance only because people blind themselves to certain aspects of the risk situation, focusing on a mere subset of the hazards that are at stake.

We live in a multi-risk world in which risks are on all sides of the decisional equation (Sunstein 2002; Wiener 2001). When regarding the inevitable demise of oil production—an argument we use here detached from its apocalyptic 'The Limits to Growth' context—the realization of macro-projects such as discussed in this chapter need to be viewed in a different revealing light, especially considering the humanity's growing energy needs. Secondly, man-made global changes to the Earth-biosphere are not limited to the highly technological segment of the world's human population today. Indeed, with the invention of agriculture, the face of the Earth was changed more drastically and more extensively than by any other kind of human technology, contemporary or historical (Grübler 1998). The onset, soon, of open-ocean seawater aquaculture, an expansion of ages-old close-to-shore mariculture by humans which was an extension to the ocean of land-based agriculture, will merely spread this anthropic impact more widely. By 2009, aquaculture, as it is currently constituted, accounted for ~50% of the total fish and shellfish consumed by all humans (Naylor et al. 2009). Viewed from that perspective, the ecological effects of the BAM Dam are much smaller than the ecological impact generated by agricultural and future aquacultural practices. One obvious difference is, of course, that agriculture and its global ecological impacts developed over millennia, and are widely dispersed geographically while the BAM Dam and its effects will be realized in the course of mere decades in one specific region.

Thirdly, macro-projects have the capability to localize ecological impacts of human intervention in terms of harnessed energy production. Energy use by means of burning petrochemical products has a global effect through its pervasive global presence as an economic activity (production, logistics, refinement and use of oil and gas products) and the emission of carbon dioxide gas to the atmosphere generally regarded as an important "Greenhouse" gas. Although discussions on carbon dioxide gas and its potential infrared forcing impact on air might alter drastically in the coming years, this does not change the necessity to explore new ways to generate energy without the large-scale use of petrochemistry.

## 9 Conclusions

It is technically possible to close off the Red Sea, and, by using the high seawater evaporation rate, provide a large hydraulic head of 650 m at the BAM Dam site that can generate more than 50,000 MW of hydropower. Our reasoned calculations show that a period of 310 years is necessary to reach this optimum hydraulic head, but power production at less than the maximum output can start well before this period has elapsed. The associated optimum surface of the evaporative basin is about one-third of its present-day value. After about 64,000 years of evaporation, salt deposition would eventually fill this residual Red Sea completely until the salt level reaches the outlets of the high-pressure conduits.

Such a BAM Dam macro-project will have large repercussions on ecology, tourism, fishery and transport in the region, and some of the effects may become globalized.

Decision makers will have to weigh the negative consequences, like the massive and irreversible destruction of valuable ecosystems, against the benefits for the development of the region and the quality of life of its native peoples.

No model or guideline for such macro-engineering decisions exists yet. And, contrary to the popular perception of many politicians, the Precautionary Principle does not provide such a guideline, as risks are on all sides of the regulatory equation. The critical choice is in the types of risks (whether socio-political, ecological or financial) to take relative to the possible benefits. If the riparian countries of the Red Sea would decide in favor of the macro-project, it is their responsibility to limit the negative consequences as much as possible.

## References

Abdel-Gawad M (1970) The Gulf of Suez: a brief review of stratigraphy and structure. In: A discussion on the structure and evolution of the Red Sea, Gulf of Aden and Ethiopia rift junction. Phil Trans Roy Soc London A 267:41–47

Abril JM, Abdel-Aal MM (2000) Marine radioactivity studies in the Suez Canal. Part I: hydrodynamics and transit times. Estuar Coast Shelf Sci 50:489–502

Abuzinada AH, Barth H-J, Krupp F, Boer B, Abdessalaam TZA (2008) Protecting the Gulf's Marine Ecosystems from pollution. Birkhauser, Basel

Ammar MSA (2009) Coral reef restoration and artificial reef management, future and economic. Open Environ Eng J 2:37–49

Bailey GN, Flemming N, King GCP, Lambeck K, Momber G, Moran L, Al-Sharekh A, Vita-Finzi C (2007) Coastlines, Submerged Landscapes, and Human Evolution: the Red Sea Basin and the Farasan Islands. J Island Coastal Archaeol 2:127–160

Biton E, Gildor H, Peltier WR (2008) Red Sea during the last glacial maximum for sea level reconstruction. Paleoceanography 23:PA1214

Briggs JC (2009) Atlantic coral reefs: the transplantation alternative. Biol Invasions 11:1845–1854

Dawes CJ (1998) Marine botany, 2nd edn. Wiley, New York

Douglas M, Wildavsky A (1982) Risk and culture. An essay on the selection of technological and environmental dangers. University of California Press, Berkeley

Dumont HJ (ed) (2009) The Nile: origin, environments, limnology and human use. Springer, Berlin

European Commission (2000) Commission of the European Communities, Communication from the Commission on the Precautionary Principle. Brussels

Fairhead JD, Girdler RW (1970) Seismicity of the Red Sea. A discussion on the structure and evolution of the Red Sea, Gulf of Aden and Ethiopia rift junction. Phil Trans Roy Soc Lond A 267:49–74

Graham JD (2001) Decision-analytic refinements of the precautionary principle. J Risk Res 4(2):127–141

Grübler A (1998) Technology and global change. Cambridge University Press, Cambridge

Haff PK (2003) Neogeomorphology, prediction, and the anthropic landscape. In: Wilcock PR, Iverson RM (eds) Prediction in geomorphology. American Geophysical Union Press, Washington DC, pp 15–26

Hanekamp JC, Vera-Navas G, Verstegen SW (2005) The historical roots of precautionary thinking: the cultural ecological critique and the limits to growth. J Risk Res 8:295–310

Hassan ESM (1961) Power from the sea. J Power Division Proc Am Soc Civil Eng 87:21–27

Hilderbrand RH, Watts AC, Randle AM (2005) The myths of restoration ecology. Ecol Soc 10:19–29

Hoepner T, Lattemann S (2002) Chemical impacts from seawater desalination plants—a case study of the northern Red Sea. Desalination 152:133–140

Ionescu GD (1977) Introducere in hidraulica (in Romanian). Tehnica, Bucharest, pp 224–225

Karabell Z (2003) Parting the desert: the creation of the Suez Canal. AE Knopf, New York

Kettani MA, Gonsalves LM (1972) Heliohydroelectric (HHE) power generation. Sol Energy 14:29–39

Kettani MA, Scott RE (1974) A Red Sea heliohydroelectric (HHE) plant. Rev Int Heliotechnique, 2nd semester, pp 19–25

Klinke A, Renn O (2003) A new approach to risk evaluation and management: risk-based, precaution-based, and discourse-based strategies. Risk Anal 22(6):1071–1094

Laughton AS (1970) A new bathymetric chart of the Red Sea. A discussion on the structure and evolution of the Red Sea, Gulf of Aden and Ethiopia rift junction. Phil Trans Roy Soc London A 267:21–22

Naylor RL, Hardy RW, Bureau DP, Elliott ChiuA, M (2009) Feeding aquaculture in an era of finite resources. Proc Natl Acad Sci 106:15103–15110

Pieterman R, Hanekamp JC (2002) The Cautious Society? An essay on the rise of the precautionary culture. Zoetermeer, Heidelberg, Appeal, Netherlands

Sawyer T (2007) Notice-to-proceed launches ambitious Red Sea Crossing. ENR.com International, http://enr.construction.com

Schuiling RD, Badescu V, Cathcart RB, Van Overveld PALC (2005) The Hormuz Strait Dam Macroproject. 21st century electricity development infrastructure node (EDIN)? Mar Georesour Geotechnol 23:25–37

Schuiling RD, Badescu V, Cathcart RB, Seoud J, Hanekamp JC (2007) Power from closing the Red Sea: economic and ecological costs and benefits following the isolation of the Red Sea. Int J Global Environ Issues 7:341–361

Searight S (2003) The charting of the Red Sea. Hist Today 53:40–46

Shaltout MAM (1998) Solar hydrogen from Lake Nasser for 21st century in Egypt. Int J Hydrog Energy 23:233–238

Shaw B, Ambraseys NN, England PC, Floyd MA, Gorman GJ, Higham TFG, Jackson JA, Nocquet J-M, Pain CC, Piggott MD (2008) Eastern Mediterranean tectonics and tsunami hazard from the AD 365 earthquake. Nat Geosci 1:268–276

Sheppard C (2000) The Red Sea. In: Sheppard C (ed) Seas at the Millennium. Pergamon, Amsterdam

Sheppard C, Price A, Roberts C (1992) Marine ecology of the Arabian region. Academic Press, London

Siddall M, Smeed DA, Hemleben C, Rohling EJ, Schmelzer I, Peltier WR (2004) Understanding the Red Sea response to sea level. Earth Planet Sci Lett 225:421–434

Smeed DA (2004) Exchange through the Bab el Mandab. Deep Sea Res II 51:455–474

Smith WHF, Sandwell DT (1997) Global sea floor topography from satellite altimetry and ship depth soundings. Science 277:1956–1962 (see also: http://topex.ucsd.edu/cgi-bin/get_data.cgi at Scripps Institution of Oceanography, University of California, San Diego, 9500 Gilman Drive, La Jolla, 92093-0225)

Sofianos SS, Johns WA, Murray SP (2002) Heat and freshwater budgets in the Red Sea from direct observations at Bab el Mandeb. Deep-Sea Res II 49:1323–1340

Spotte S (1992) Captive seawater fishes. Wiley, New York

Stanev EV, Peneva EL, Mercier F (2004) Temporal and spatial patterns of sea level in inland basins: recent events in the Aral Sea. Geoph Res Lett 31:L15505

Sunstein CS (2002) Beyond the precautionary principle. John M. Olin Law & Economics Working Paper No. 149 (2nd series), the Law School, University of Chicago

Talbot C, Aftabi P, Chemia Z (2009) Potash in a salt mushroom at Hormoz Island, Hormoz Strait, Iran. Ore Geol Rev 35:317–332

Uhlemann H-J (2002) Canal lifts and inclines of the world. Internat Limited, ISBN 0 95431 811 0

Van den Belt FJG, De Boer PL (2007) A shallow-basin model for 'saline giants' based on isostasy-driven subsidence. In: Nichols G, Williams E, Paola C (eds) Sedimentary processes, environments and basins. Special Publication Number 38 of the International Association of Sedimentologists. Blackwell, London, pp 241–252

Wiener JB (2001) Precaution in a multi-risk world. Duke Law School Public Law and Legal Theory. Working Paper Series, Working Paper No. 23. Durham, North Carolina

# The Hormuz Strait Dam Macroproject

**Roelof Schuiling, Viorel Badescu, Richard Cathcart and Piet van Overveld**

## 1 Introduction

"Noah's Ark", millennia ago, was caulked with tar mined from the ground's surface to make it a waterproof lifeboat capable of enduring "Noah's Flood". Today, petroleum is pumped from beneath the same desert region where, putatively, "Noah's Ark" was assembled. "Edin" is the Sumerian word for "plain"; the "Garden of Edin", planted by God as man's first home, is speculated to have been located somewhere on the trough-like alluvial plain situated north of the Strait of Hormuz, some of which has become seafloor while another part has become covered by river-borne sediments shifted from the Tigris-Euphrates watershed (Isaev and Mikhailova 2009; Kennett and Kennett 2006). "Edin" was Paradise, the true cradle of mankind, according to Sumerian myths and later religious mythologies (Hamblin 1987; Scafi 2006). Is there a possibility that Twenty-first Century Macro-engineering may make the arid region north of the Strait of Hormuz more livable? We assume no significant future changes in the existing wind climatology of the Persian Gulf—that is, during the summertime, the Indian monsoon-created light northwesterly winds over the Gulf, which allow the formation of thermally driven air circulations—sea breezes on land all year and land breezes on land during

R. Schuiling (✉)
Utrecht University, Utrecht, The Netherlands

V. Badescu
Polytechnic University of Bucharest, Bucharest Romania

R. Cathcart
Geographos, Burbank CA, USA

P. van Overveld
Volker Wessels, Woerden, The Netherlands

V. Badescu and R. B. Cathcart (eds.), *Macro-engineering Seawater in Unique Environments*, Environmental Science and Engineering,
DOI: 10.1007/978-3-642-14779-1_7, 

nighttime (Eager et al. 2008). Sadly, and with some trepidation bordering on alarm, we also must assume that the historically famous "Fertile Crescent" could "disappear" during the Twenty-first Century owing to climate change related to global warming, however that air warming might be caused (Kitoh et al. 2008). In other words, the ecosystem-states of the Gulf must find, during the Twenty-first Century, some new economic status of energy supply security and price stability, amongst other macro-problems in need of macro-engineering solutions.

We think that a technically and economically feasible dam, emplaced in the Strait of Hormuz, could cause significant social progress in the region (Schuiling et al. 2005). A renewable electricity-generating facility there may foster a near-term future Twenty-first Century Electricity Development Infrastructure Node, or EDIN, that, in purely geographical terms, removes seawater from a large region of the seafloor, exposing an area of new land. We propose that EDIN's realization will instigate a basin-wide habitat that can harmoniously accommodate humans and wildlife simultaneously.

## 2 Persian Gulf Recent Geologic History

Extending westward and towards the north from the Strait of Hormuz, the shallow oceanic gulf rests entirely on the Arabian Tectonic Plate. The Gulf is underlain by continental crust that is currently colliding with the Eurasian Tectonic Plate along the Zagros Thrust. Earthquakes have been a factor affecting the constructive efforts of structural engineers (Ambraseys 2009). The Strait of Hormuz has a complex tectonic and climatic history, which is not yet fully comprehended (Uchupi et al. 2002). Circa 14,000 years BC, the "…Strait of Hormuz opened up as a narrow waterway and by about 12,500 years ago the marine incursion into the Central Basin had started. The Western Basin of the Gulf flooded about 1,000 years later. Momentary sea-level still-stands may have occurred during the Gulf flooding phase at about 11,300 and 10,500 years BC. The present shoreline was reached shortly before 6,000 years ago and exceeded as relative sea level rose 1–2 m above its present level, inundating low-lying areas of lower Mesopotamia" (Lambeck 1996).

## 3 Conditions at the Strait of Hormuz

The Gulf, a semi-enclosed marginal sea connected to the Arabian Sea through the Strait of Hormuz, occupies an area of $\sim$233,100 km$^2$. Two main sources of water affect its volume ($\sim$8,630 km$^3$, with an average depth of $\sim$35 m) and its material and biotic contents: an inflow of 326 km$^3$/year from the Arabian Sea into the Gulf to replace evaporation losses and $\sim$37 km$^3$ of freshwater coming to the Gulf via the Tigris-Euphrates after their confluence known as the Shatt-al-Arab. Separating

**Fig. 1** Satellite view of Strait of Hormuz, One sees the Southern Iran (*top part*), the Qeshm island and the extremity of the Musandan peninsula (*bottom right*)

the Arabian Peninsula from Iran, the Strait of Hormuz, at its narrowest, is ~50 km wide (Subba Rao 2000) (Fig. 1).

Both inflow, occurring at depth of 0–50 m, and outflow, occurring at depth of 50–100 m, traverse a sill ~100 m below present-day sea-level. From data cited above, we calculate ~1.4 m evaporation/year; however, we have selected a net evaporation rate over the Persian Gulf of ~2 m/year (Johns et al. 2003).

Maps of the natural gas and oilfields in and around the Gulf, tanker terminals and natural gas and oil pipelines illustrate vividly a networked production, storage, processing and shipping infrastructure of great technical complexity and economic value. Fortunately for the EDIN macro-project's planners, there are no obstructive pipelines, terminals or oil and natural gas production platforms situated in the Strait of Hormuz itself that might hinder its realization. More importantly, oil and natural gas exports from the Gulf will start to decline soon as a consequence of resource depletion (ASPO 2004). Abu Dhabi is the first hydrocarbon-producing nation to have taken the ecologically significant step towards sustainable living by its construction of the new real estate development "Masdar City". By the time the Hormuz Dam nears completion, exports from the Gulf will have dwindled, making most of the existing production, storage and shipping infrastructure obsolete. These forecasts underline the need for a local alternative energy resource for the Persian Gulf's socially diverse peoples. It is worth noting that owing to the crowding of ship traffic in the Strait of Hormuz, since the early 1980s a land traversing shipping canal, excavated through the northern emirate of Ras Al-Kheima (25°46′60″N latitude by 55°56′00″E longitude) to the Indian Ocean has been under planning consideration by the Gulf Cooperation Council (organized by Bahrain, Kuwait, Oman, Qatar, Saudi Arabia and the United Arab Emirates in

1981). Since this macro-project is merely a proposal, just like ours, we have chosen to ignore its realization in this specific exposition on EDIN.

## 4 Planned/Built Pre-EDIN Macro-projects

Circa 1970, M. Ali Kettani and L. M. Gonsalves proposed a macro-project that would use a steady inflow of seawater into an artificially closed Dawhat Salwah, a small sub-gulf located between Qatar, Bahrain and Saudi Arabia. The regulated inflow was planned to spin hydroelectric turbines installed within the enclosing dam (Kettani and Gonsalves 1972). Their macro-project was never built. However, a 25 km-long causeway, existing since 1986, connects Saudi Arabia and Bahrain via Umm Nasan Island and Bayn Island (van Tongeren 1986).

A complicating factor in evaluating the economy of the Hormuz Dam project is the construction of a series of tourist attracting offshore artificial islands in Dubai, with broad beaches to be emplaced in the Gulf (Kumar 2009) which will lose considerably in monetary investment value if, in the near-term future, these real estate and infrastructure properties will no longer be luxury islands in the Gulf, but small hills in a dry country. The first island was completed in 2003 and additional islands are under construction. Two islands remain still under construction—"The Palm Jumeirah" and "The Palm, Jebel Ali"—rest on the Gulf's seafloor in depths of ~4 m, and the islands will all rise at least ~4 m above the present-day Gulf sea-level. The construction cost is estimated to be ~2009 US$ 3 billions or more. The islands increase the length of Dubai's coastline by ~120 km. A fourth artificial group of islands, named "The World" since it replicates a map of the Earth's continents and many islands, is also still being created (Bellini 2008). All of these islands would, of course, become hills when the Gulf's sea level is reduced. As Mahmoud Ali Abdelfattah pointed out in his oral presentation, *Land Degradation Indicators and Management Options in the Desert Environment of Abu Dhabi, United Arab Emirates* at the International Conference on Soil Degradation, Riga, Latvia 17–19 February 2009, dry erosive wind is the main cause of land degradation in the arid landscape Abu Dhabi, making dunes, blocking highways and reducing the depth of waterways. Dubai Creek is polluted by organics detrimental to macro-benthic fauna (Saunders et al. 2007). Masdar City, a 6 $km^2$ urban region being developed on the geographical periphery of Abu Dhabi that will service 50,000–75,000 persons, is touted globally as the world's first "zero-carbon, zero-waste and car-free city". Slated to be fully operational by 2016, Masdar City is already the world-headquarters for the International Renewable Energy Agency (IRENA), formed in 2009 with more than 131 participating members. At Pearl City Al-Khiran in Kuwait, macro-engineers have fostered new habitats by excavating marine waterways in saline sabkha and hyper-saline khors of low ecological value (Jones et al. 2007).

Dubai has begun excavation of the largest canal since the Suez Canal's digging commenced in 1859. The 75 km-long "Arabian Canal", ~6 m deep, will link two palm tree-shaped artificial islands (Palm Jebel Ali and Palm Jumeirah) and be capable of accommodating yacht-type sea-going vessels as much as ~40 m long. The estimated final cost of the "Arabian Canal" is 2009 US$11 billion. Canals do have a beneficial effect on the sale prices of adjacent real estate (Nelson et al. 2005) and this will become evident when the 12,500 ha canal-side city is built, at a 2009 cost of US$55 billion, astride the Arabian Canal. This new city will be emplaced, in part, atop an artificial topography featuring machine-mounded stabilized soil and rock hills—some as much as ~250 m high—that will be constructed using the >1 billion cubic meters of solid and granular materials excavated to make the waterway.

Louis Vadot recommended a dam be installed in the Strait of Hormuz that would be followed, subsequently, by natural evaporation of the seawater isolated in the Persian Gulf and, thereby, create a useful power-head (Vadot 1954). Vadot feared that a heliohydropower macro-project emplaced thereabout could have several significant drawbacks: (1) the increasing salinity of the tail-water (i.e., the areally smaller reducing Persian Gulf) and (2) distance from consuming electrical power markets. However, he foresaw intriguing macro-economic possibilities for electro-chemical processors and desalination for future freshwater public, private and industrial supplies. In *Barrages par Louis Vadot*, which appeared in the October 1953 issue No. 91 of *Atomes. Tous les aspects scientifique d'un nouvel age*, the macro-engineering projections of major hydroelectric dams imagined by Louis Vadot were delightfully recounted to the public in a popular style. Louis Vadot was a purveyor of a large-scale technological fix at the Strait of Hromuz; whilst Vadot did not use the term technological fix, which in its literal rendering implies an environmental and/or social fix produced entirely by applied technology, he did clearly intimate its future "presence" in Macro-engineering's lexicon, first as an English-language vogue term coined circa 1967 by Alvin M. Weinberg (1915–2006). Vadot's ideas were earlier high-lighted in Atomes Issue No. 77 (August 1952) in *A propos des marees Louis Vadot*. [For the record: Louis Vadot authored Etude de l'ecoulement de l'eau dans les diffuseurs de turbines hydrauliques (1940: Besancon, Nancy) and Cours de machines hydrauliques (1945: Ecole des ingenieurs hydrauliciens, Grenoble)].

By 1961, El Sayed Mohamed Hassan presented a similar macro-engineering plan for a dam (and associated infrastructure) suitable for isolating the Red Sea from the Arabian Sea at Bab el Mandeb (Hassan 1961). Hassan's dam was designed for emplacement at a selected place with a depth to seafloor of approximately 27 m, with the dam's crest situated at an elevation above sea-level of ~9 m. It would have a total width of ~405 m, including a crest roadway ~9 m wide. Of course, a Strait of Hormuz dam, the most important enabling feature of our proposed EDIN regional development macro-project for the Gulf, would be bigger and have a shape fitted to the tectonic and climatic conditions obtaining when it is undertaken during the mid-Twenty-first Century. In a sense, it would simply be a "linear" artificial island that is connected to the mainland at both ends, more or less imitative of the on-going macro-projects being undertaken by Dubai. If a detailed survey shows this to be possible, its trajectory would

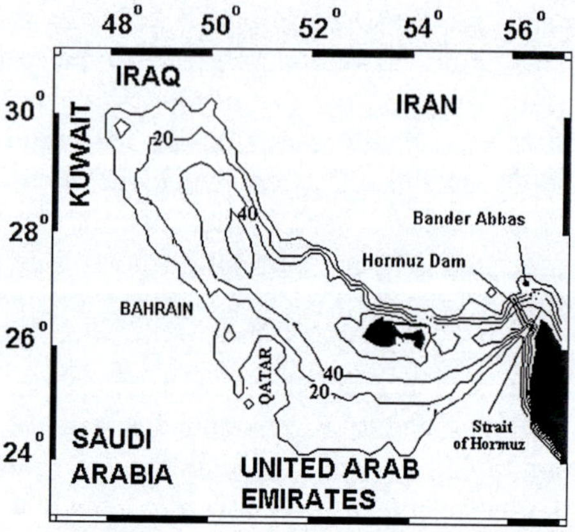


**Fig. 2** Persian Gulf bathymetry (m) and location of proposed Gulf dam

preferably run from the Musandan peninsula (Oman) in a north-easterly direction across the island of Qeshm, in order to preserve the open connection of the important Iranian harbor of Bander Abbas (27°11′N latitude by 56°17′E longitude) to the Arabian Sea (see Fig. 2).

If seawater from the Arabian Sea were continuously passed through a powerhouse ~35 m below the Arabian Sea's existing sea-level (into a reduced sea-level Gulf), ~2,050 MW may be generated, which may be increased to approximately 2,500 MW by some special measures; reduction of the Gulf's sea level by ~35 m would (at the same time) elevate the world-ocean's sea-level by less than 2 cm, a small but not completely negligible global elevation of the ocean. The Gulf's surface area would be reduced to ~47% of what it was before the emplacement and operation of the Hormuz Dam.

## 5 Particulars of Strait of Hormuz Dam

The magnitude of the annual mean deep outflow of Gulf seawater through the Strait of Hormuz is approximately 0.20 Sv, implying a net Persian Gulf evaporation rate of ~2 m/year; a Hormuz Dam powerhouse, drawing from a head-pond (i.e., the Arabian Sea) can, thus, take advantage of this difference in sea levels within a timeframe of only several years. We will not discuss any consequences for the Arabian Sea of a cut-off of the very saline seawater mass nowadays entering that segment of the Indian Ocean. Because we plan to return a saline underflow to the Arabian Sea, in order to prevent salt deposition in the Gulf, this issue will largely be solved anyway.

Built of crushed rock transported to the dam worksite, or from uplifted carbonate rocks (see Sect. 6), a Hormuz dam might total 2 km$^3$. According to the globalized cement industry, Twenty-first Century constructors pour >6 km$^3$ of concrete every year. And, furthermore, it has been alleged, "The total earth moved in the past 5,000 year [by humans] would be sufficient to build a 4 km-high mountain range, 40 km wide and 100 km long". (Hooke 2000). (Recalculated, that is $\sim 8 \times 10^3$ km$^3$, if we imagine the mountain range to have a triangular cross-section).

## 6 Prospects for "Geochemical Macro-engineering"

The sediments underlying the Strait of Hormuz, in front of the Musandan Peninsula are predominantly shallow-water carbonates. This particular geologic constellation in the Strait of Hormuz allows us to envision an unusual method of dam construction. The following is a brief summary of the principle, and the results obtained thus far. For a more extended description of the technology, the reader is referred to Schuiling (2004). Sulfuric acid reacts readily with limestone and water to form carbon dioxide gas and gypsum $CaCO_3 + H_2O + H_2SO_4 \rightarrow CaSO_4{\cdot}2H_2O + CO_2$. The molar volume of $CaCO_3$ (limestone) is 36.9 cm$^3$/mol → and it forms gypsum with a molar volume of 74.7 cm$^3$/mol.

The molar volume of the precipitated gypsum is twice that of the dissolved limestone. This means that the injection of industrial waste sulfuric acid into subsurface limestone causes a large expansion of the in situ rock. This expansion can only be accommodated by surface uplift. This "Geochemical Macro-engineering" (Schuiling 1988) method was developed, supported by a grant from The Netherlands Technology Foundation (de Graaff et al. 2000) and is considered a viable chemo-geotechnical process to discard waste sulfuric acid at all conceivable pressures, because even at moderate acid strength the equilibrium $CO_2$ pressures are extremely high. Thermodynamic data show that gypsum is stable below 50°C, down to a depth of ~1 km at normal geothermal gradient. At higher temperatures and greater depth, gypsum transforms to the less voluminous anhydrite (45.9 cm$^3$/mol). "Geochemical Macro-engineering" is a fast-developing field of professional endeavor for macro-engineers. Researchers in India, for instance, are perfecting means to shield storm-affected coasts with creating offshore mudbanks artificially by introducing ~3–5% zaherite (hydrated aluminium sulphate) into near-shore clayey seafloor sediments (Dinesh and Jayaprakash 2008). "Geochemical Macro-engineering" complements the recently inaugurated "Metabolic Engineering" (Qian et al. 2009).

The wasteful flaring of natural gas is being phased out in the countries to be directly affected by our proposed Strait of Hormuz Dam. This may result in the stockpiling of huge quantities of native sulfur, from the desulphurization of gases containing $H_2S$. If this sulfur is converted to sulfuric acid, and injected into limestone in the sub-surface at the Hormuz Strait, it should be possible to start raising some segments of the intended Hormuz Dam construction site at modest monetary cost. Less crushed rock will need to be emplaced if the foundation upon which the Hormuz Dam rests is uplifted via "Geochemical Macro-engineering".

## 7 Electric Power Generation Optimization

To evaluate the power produced at the Hormuz plant one shall use the procedure proposed in Kettani and Gonsalves (1972) and Kettani and Scott (1974) and bathymetry data for the Persian Gulf (Fig. 2). One denotes by $Q_{\text{evap}}$ the volumetric flow rate of evaporated seawater and by $Q_{\text{fw}}$ (=1,173 m$^3$/s) the volumetric flow rate of freshwater entering the Gulf by perennial and flash rivers. At steady-state the volumetric flow rate of seawater $Q_{\text{sw}}$ entering the Persian Gulf from the Gulf of Oman must balance the previous flow rates, i.e.

$$Q_{\text{sw}} = Q_{\text{evap}} - Q_{\text{fw}} \tag{1}$$

The volumetric flow rate of water coming by rain was neglected. The volumetric flow rate of evaporated seawater depends of course on the surface area $S$ of the active basin by

$$Q_{\text{evap}} = w_{\text{evap}} S \tag{2}$$

where $w_{\text{evap}}$ (=2 m/year) is the seawater evaporation rate. The hydraulic head $H$ provided by the dam is shown in Fig. 3. The surface area $S$ of the evaporative basin decreases as the head $H$ increases. When the hydraulic head is zero, $S$ reaches its maximum value $S_{\text{max}}$. When $H$ is a maximum (say $H_{\text{max}}$), one could believe that there will be no seawater left and $S = 0$. However, because $Q_{\text{fw}}$ is not a vanishing quantity (unless the Shatt-al-Arab is used for irrigation of the new coastal areas), keeping balance between the three quantities entering Eq. 1 requires a non-vanishing evaporative surface area of at least 19,000 km$^2$. The maximum associated hydraulic head is $H_{\text{max}} = 79$ m. The bathymetry of Fig. 2 was used to produce a rough estimate of the dependence $S(H)$, shown in Fig. 4.

The potential energy of the seawater flow transferred from the Gulf of Oman to the Persian Gulf is transformed into mechanical power $P$ collected at turbine shafts. The following simple relationship applies:

$$P = g\rho_{\text{sw}}(Q_{\text{sw}} - Q_{\text{fw}})H\eta_t = g\rho_{\text{sw}}\left[S(H)w_{\text{evap}} - Q_{\text{fw}}\right]H\eta_{\text{t}} \tag{3}$$

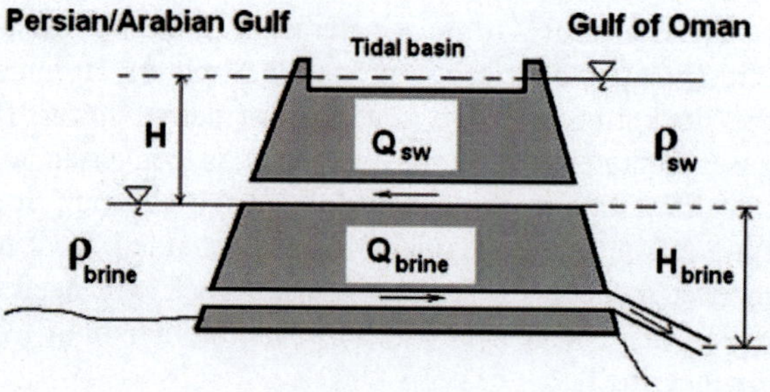


**Fig. 3** Cross-section, idealized, of the proposed dam controlling the Strait of Hormuz. $H$ seawater hydraulic head; $Q_{\text{sw}}$ volumetric seawater flow rate, $Q_{\text{brine}}$ volumetric brine flow rate, $H_{\text{brine}}$ brine hydraulic head; $\rho_{\text{sw}}$ and $\rho_{\text{brine}}$ seawater and brine density, respectively

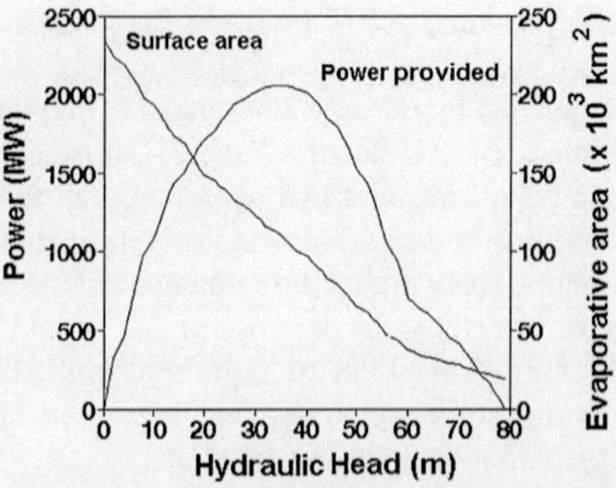


**Fig. 4** Dependence of the surface area $S$ of the evaporative basin of the Persian Gulf and plant power $P$ on the hydraulic head $H$. The maximum evaporative basin surface area is $S_{max} = 233{,}100\ km^2$ and the maximum head is $H_{max} = 79$ m

Here Eqs. 1 and 2 were also used. In Eq. 3, $g$ (=9.78 m/s$^2$) is gravitational acceleration, $\rho_{sw}$ (=1,030 kg/m$^3$) is the density of the falling seawater and $\eta_t$ (=0.9) is the efficiency of the hydraulic turbine. The dependence of the power $P$ on the hydraulic head $H$ is shown in Fig. 4.

Two different scenarios could be envisaged about the power plant at Hormuz Dam. In the first scenario, a rather small hydraulic head could be used (say, $H = 10$ m). This has the advantage that the evaporative basin is still large (its surface area is $S = 193{,}000\ km^2$) and the Persian Gulf keeps its main present-day features. The electric power provided by the plant is $P = 1{,}083$ MW in this case. Because of the low head, the hydropower plant should be provided with Kaplan turbines.

The second scenario emerges from Fig. 4, which shows that there is an optimum hydraulic head (i.e. $H_{opt} = 35$ m), for which the electric power is a maximum (i.e. $P_{max} = 2{,}053$ MW). The associated optimum surface of the evaporative basin is, however, less than half of its present value (i.e. $S_{opt} = 110{,}000\ km^2$). This maximum power-plant is our choice in the following. Due to the medium hydraulic head, Francis or even Pelton turbines should be envisaged in this case.

The period necessary to reach a 35 m hydraulic head at an annual evaporation rate of 2 m is about 18 years. It is not an unreasonable period to wait. It is of course possible to commence power plant operations as soon as the dam at the Strait of Hormuz has been closed, but at a power rate less than the maximum possible. Note, however, that the minimum head for proper operation of Francis and Pelton turbines is 20 and 30 m, respectively. The consequence is that power generation cannot start earlier than 10 years after the Hormuz Dam's construction. The critical element in the Strait of Hormuz/EDIN macro-project is the Hormuz Dam. It is larger and more ambitious than any dam which has so far been constructed anywhere on Earth, similar in approach but at a vastly larger scale than the 30 km-long "Afsluitdijk" (Barrier Dam) completed in The Netherlands in 1932 to close off the Zuiderzee created by a fierce North Sea storm circa 1,300. Once separated from the North Sea, the Zuiderzee was re-named Ijssselmeer. Is it possible the Persian Gulf might be re-named after closure of the Hormuz Dam? The Gulf will then be much smaller geographically than our proposed Red Sea

Dam-created gulf—the BAM Dam is fully detailed in another chapter of this book. The electricity produced at the powerhouse of the Hormuz Dam will probably be regulated by the consumer demand and not by the sheer availability of the energy, unless excess electrical energy is used for desalination or for the production of liquid and gaseous hydrogen fuels and feedstocks. Another possibility would be to establish a magnesium industry near the Hormuz Dam, as the two major ingredients, concentrated brine and cheap electricity are available. The overall average over several years of consumption should equal the average energy capacity of the facility. This kind of power generation affords, therefore, a unique means of storing solar energy and leveling out its diurnal and seasonal variations, as well as the variation in power demand.

The Hormuz Dam powerhouse must have a finite design and working lifespan, just like any other macro-project ever built. After many years of evaporation, salt deposition would eventually fill the Gulf completely until no seawater surface remains for solar evaporation. A simple calculation shows that it would take thousands of years for the Gulf to be filled with salt to its optimum head level. Once the Gulf's isolated water is saturated with salt—as it currently is in a region lying between Qatar and Saudi Arabia—the amount that will precipitate yearly is equal to the quantity of new salt introduced that year. A column of seawater of 2 m depth, with a cross-section of 1 $cm^2$ normally contains 200 cc of seawater with a salinity of about 3.5 ‰. Consequently, it contains about 7 g of salt, which will be deposited on 1 $cm^2$. The density of salt being 2,240 $kg/m^3$, this means that once salt deposition begins, a new layer of about 3 cm thickness will form annually, in line with the observed thickness of annual rhythms in evaporite deposits elsewhere. Assuming a maximum gulf depth around 120 m and a hydraulic head of 35 m, the final salt layer will be 85 m thick. This means the gulf filling time with salt would be about 2,830 years. We will not discuss here in detail the complications arising from the isostatic rebound related to the rapid 35 meter drop in Gulf level, nor the isostatic sinking of the residual Gulf when it is loaded with precipitated salts, but if precipitation of salt is allowed to proceed, a residual basin depth of 50 m will be converted into a 125 m thick layer of halite by the effects of isostatic compensation.

It may be possible, however, to prolong the operational life of the Persian Gulf's Hormuz Dam system indefinitely and avoid ending up with a huge Gulf salt pan by releasing very saline bottom water through a tube, installed inside the base of the dam, back into the Arabian Sea (see Fig. 3). The removal of the brine can even be carried out without any energy expenditure by extending the outlet tube out into the Arabian Sea to a depth $H + H_{\text{brine}}$ (see Fig. 2) of >215 m. As soon as the condition is fulfilled

$$H_{\text{brine}}\rho_{\text{brine}} > (H_{\text{brine}} + H)\rho_{\text{sw}} \quad (4)$$

then the brine exported from the closed Persian Gulf will spontaneously flow down the outlet tube, without requiring any additional pumping. For a brine concentration of 30% (density 1.230 $kg/m^3$) this solution holds for $H_{\text{brine}} = 180$ m, i.e. for any depth of the outlet of the tube exceeding 215 m. Such a depth is encountered in the Arabian Sea at a distance of approximately 100 km from the projected worksite of the proposed Hormuz Dam. This idea was first proposed in Schuiling et al. (2005) and was later taken up by Purnama and Al-Barwani (2006) and Thoppil and Hogan (2009) in terms of anthropogenic brine waste discharges that nowadays flow naturally into Gulf of Oman.

An even more important reason to maintain a return flow of saline water is the fact that evaporation of water becomes less effective, the more saline the water is, for two reasons:

- The vapor pressure of saline water is less than the vapor pressure over fresh water and
- Water becomes more viscous when more saline, so at equal wind speed saline waters will be less turbulent, which decreases their dynamic evaporation.

As the amount of hydropower produced is directly proportional to the amount of water that evaporates, keeping the salinity behind the dam as low as possible is very important.

If the Shatt-al-Arab freshwater outflow is used to irrigate and desalinize the newly emerged coastal areas, a positive contribution to potential energy production would come from the reduced inflow of freshwater into the Gulf, as most of the freshwater will be lost by evaporation and evapo-transpiration. A reduced inflow from the Shatt-al-Arab can be compensated by an increased inflow of water from the Arabian Sea, thereby increasing the energy production. At the conditions assumed (a 35 m drop of Persian Gulf level, and a reduced Gulf area of 110,000 km$^2$), a reduction of the annual inflow of fresh water by 30 km$^3$ would increase the potential energy production by $\sim$15%.

Another modest increase of electric power production could be obtained by the construction of a shallow basin along the crest of the dam. The bottom of this basin should be situated at low-tide level. By letting the water in from the Arabian Sea at high tide, and using this stored water at low tide for power generation, it is possible to bridge the production minimum at low tide (=lowest head). The present tidal range is around 2.5 m and is likely to increase if the entrance to the Gulf is closed. This tidal range caused some problems for Alexander the Great, during his conquest of India. Arrianus, in his book on Alexander's conquest of the Persian Empire gives the following account:

> When they had anchored there, the tide changed, as happens in the oceans; it became ebb, leaving their ships on land. Alexander's troops had never experienced tides, and were very terrified. But their fright even increased when after some delay the water came back, and lifted their ships. Ships that were caught in the mud were lifted again undamaged. Ships, however, that were left on dry land in precarious equilibrium collided with each other when the waves returned, or were smashed on land and destroyed. Alexander, as best as he could repaired the damage.

Table 1 presents several proposals of heliohydroelectric (HHE) plants. The power provided by the Hormuz Dam HHE is comparable with that of Qattara Depression power-plant.

**Table 1** Proposals of heliohydroelectric plants

| Quantity | Dawhat Salwah area (Kettani and Gonsalves 1972) | Bab Al Mandab Strait (Kettani and Scott 1974) | Qattara depression (Bassler 1972) | Hormuz Strait (this chapter) |
|---|---|---|---|---|
| Dam length (km) | Not available | 22 | Not necessary | 50 |
| Dam height (M) | Not available | 205 | Not necessary | 100 |
| Basin actual surface ($km^2$) | 28,000 | 437,969 | 19,500 | 234,000 |
| Basin active surface ($km^2$) | 12,000 | 133,000 | 12,100 | 110,000 |
| Evaporative rate (m/s) | 3.5 | 3.5[a] | 1.7 | 2 |
| Volumetric flow rate ($m^3/s$) | 255 | 14,760 | 650 | 5,803 |
| Hydraulic head (m) | 14 | 500 | 275 | 35 |
| Hydraulic Power (MW) | 34 | 67,000 | 1,000–4,000 | 2,053 |
| Time until full power (years) | 3.5 | 143 | 0 | 18 |
| Operation time (years) | Not available | 1,175 | Not available | 4,250 |

[a] More recent studies shows an evaporative rate of 2.1 m/s (see Chap. 6 in this book)

# 8 Spin-offs Consequent to Persian Gulf Lowering

A major potential source of income, as well as the largest economic uncertainty is the value of the newly emerging land. If it can be used for agriculture, the value of the ~120,000 $km^2$ of new land is tremendous. To achieve this, simultaneous with the Hormuz Dam's construction, a canal ought to be dug approximately along the present-day coastline, diverting the waters from the Shatt-al-Arab. This unnamed canal, or aqueduct, can serve to irrigate and desalinize at least part of the new coast lands emerging from the reduced Persian Gulf, making them suitable for agriculture and habitation. It stands to reason that Iraq (and Iran?) should be compensated by the Gulf States for making the freshwater from the Shatt-al-Arab available. From a consideration of the bathymetry of the Gulf, it appears that Saudi-Arabia and the United Arab Emirates acquire most of the new, arable land area. How the land should be distributed between the riparian states must be sorted out by international agreement. In order to avoid legal hassles, the Gulf eco-system states should create a supra-national body that will be responsible for all legal, financial and organizational aspects of the Hormuz Dam; an option is that this body delegates the operational aspects of running the powerhouse to Iran and Oman, the two states at both terminals of the Hormuz Dam, as these states will have the formal legal jurisdiction over the region occupied by the Dam closing the Persian Gulf.

The resulting international legal issues and the implication of emergence of new coastal regions need examination by competent international law experts trusted by the nations involved. It is worth noting that international law is only just now

Table 2 Overview of 2009 capital costs (MEuro)

| | |
|---|---|
| Dam construction | 44,200 |
| Power plants | 3,350 |
| Irrigation canal | 14,270 |
| Total capital costs | 61,820 |

beginning to seriously address the international legal problems of land submergence that might be caused by a prospective global sea-level rise.

## 9 Estimate of Costs and Benefits

The major costs of the macro-project are the capital expenditures for dam construction, power plant installations and the irrigation canal. The major economic benefits are the capital income from the reclaimed land and the annual revenues of the power-plants and the irrigation freshwater. A rough estimate (2009 price level) with an uncertainty of +30/−20% is given in Table 2.

The largest uncertainty is the real estate market price of the newly uncovered land. If it is possible, which seems increasingly unlikely as the regional effects of global climate change takes place during the Twenty-first Century, to irrigate 50,000 $km^2$ by the use of an irrigation canal derived from the Shatt-al-Arab, this land should bring about 3 eurocent/$m^2$ to cover the capital cost of the irrigation canal, assuming that the non-irrigated land has no value.

The annual revenues from the electric power production (a plant of 2,500 MW produces 22 billion kWh) will be in the order of 3.3 billion 2009-value Euro, if an electricity price is assumed of 15 eurocent/kWh, about halfway between the price for large industrial users and for households.

Although it is unclear what price the dwindling freshwater from the Shatt-al-Arab will fetch, when used for desert irrigation, it will probably be in the same order of magnitude as the revenues from electric power production.

It is clear, then, that from a strictly economical point of view, this macro-project will have a too-long return on investment, unless some of the cost estimate used above is unduly pessimistic. The macro-project will only be adopted when the countries involved are convinced that the enterprise is necessary for their national development, or even for their survival as modern states. The urgency will grow as the oil and gas reserves of the region are depleting as available natural resources.

## 10 Environmental Impact; The Eco-ethical Dilemma

Hydro-electric power generation systems are sustainable and produce no Greenhouse gases. If the power generation as foreseen for the Hormuz Dam were generated by coal-powered plants, this would lead to an emission of about

23.8 megatonnes of $CO_2$ gas annually. So, a Hormuz Dam hydro-electric macro-project will lead to a considerable reduction of greenhouse enhancing atmospheric gases. The negative ecological aspects, however, are considerable. In ways as yet unfathomable, the whole Persian Gulf region's hydrogeology, and probably its climate as well will be affected. If the large tracts of uncovered seafloor are vegetated, this may be reflected in increased precipitation. Macro-engineering will profoundly affect the ecology of the region. Sea becomes land, and vegetation and wildlife will be affected in dramatic fashion. This will cause many people to oppose such large interventions. It probably comes only as a small consolation that most of the Gulf was, in fact, land some 12,000 years ago, so in a sense the Hormuz Dam macro-project only restores the original constitution of the Persian Gulf. Jihan Seoud, an environmentalist based in Beirut, Lebanon has phrased some of her well founded ecological objections to the Hormuz Dam macro-project:

1. Changes to the marine environment due to pumping such vast quantities of seawater will be drastic, not to mention irreversible: the coastal marine eco-system, and maybe even the deep ocean ecosystems maybe altered. Potential changes may have consequences as severe as those that occurred in the Aral Sea.
2. As a consequence to the changes in the marine environment, particularly with respect to the coastal environment, dramatic changes in other inter-dependent natural ecosystems may occur, such as changes in bird migratory patterns, living and mating patterns of sea turtles or other coastal breeding animals.
3. The increase in the coastline's length and area may increase land available for agriculture, but it is highly likely that this land would require large amount of fertilizers to become arable, and so that would increase contaminated run-off to the Gulf, among other negative ecological impacts. Furthermore, the cost of increasing such landmass is likely to be too high to be cost-beneficial.
4. The ecotourism industry would be affected significantly, especially in a city like Dubai which is pumping billions of dollars into it's tourism sector, particularly along the coastal area in mega-projects like the Palm Islands, World City, et cetera. These projects would be left high and dry by reducing the Gulf's sea-level.
5. The cost, both economically and environmentally, of rerouting pipelines is expected to be very high. Who would bear these costs?
6. Lastly, and of course not least of all, changing public opinion towards the use and need of renewable energy, particularly of those people living in natural gas and oil-rich Gulf countries, is a difficult task as is, let alone trying to convince them of the need for such a drastic mega-project.

The questions about the tourist and ecotourism industries, the cost of rerouting existing pipelines and the future fate of the oil reserves (Jones 2009) have been dealt with in previous sections of this chapter, and an Aral Sea-type environmental disaster can be avoided by brine recirculation into the Arabian Sea. The ecological questions remain. Marine and most of the coastal wildlife will indeed disappear, and migratory patterns for birds are likely to be affected. These are the dilemmas that every proposed macro-engineering project faces. There are only a few, and not always satisfactory answers. The projected reduction of Greenhouse gas emissions is an important positive factor. As far as the future ecology of the region is concerned, dramatic changes will sometimes surprisingly rapidly create new and valuable ecological systems. As a trade-off for the ecologies that are destroyed by the macro-project, some of its income should be diverted to finance the creation of new, or the preservation of existing nature reserves. Other ecological comments on this project may be found in Sheppard et al. (2010).

In the end, the overriding question will be, are similar macro-projects justified for the future survival of the Persian Gulf region and its peoples?

## 11 Closing Remarks

It is likely that ~2,500 MW will be generated at one place in EDIN's realm. Cheap hydro-electricity production, combined with efficient and inexpensive electricity transmission, makes the fresh-water shortages of the Gulf ecosystem-states (Yang et al. 2003) solvable with a distributed factory-scale desalination industry (Alspach and Watson 2004). We are aware the desalination costs will rise as the Gulf's seawater becomes more hyper-saline (Smith et al. 2007), but by using the flow of water coming out right after the turbines before it gets mixed with the more saline Gulf waters this problem can be solved to a large extent. After the Gulf is blockaded by the Hormuz Dam, reducing its elevation by ~35 m and increasing its salinity, it will assume an aquarium-like status as a seawater body. Undoubtedly, nautical archeology will discover and recover many once hidden historic time-capsules (preserved shipwrecks). Some of the uncovered lowlands may even be planted with salt-tolerant crops able to grow when irrigated only with seawater (Glenn et al. 1998). Biologists think the Earth-biosphere—and its sub-regions—will be both monitored and transformed by small organisms with altered DNA (Macalady and Banfield 2003). Genetic micro-engineering that allows the transfer of genes across taxonomic boundaries enables the production of new products and permits the development of new hazards for the Earth-biosphere's natural life. It is not too far-fetched for us to imagine the aquarium-like EDIN-operated Gulf greatly changed by microscopic and macroscopic forms of life (Azam and Worden 2004). We must imagine simply because the transformed Gulf will be a totally new physical reality.

**Acknowledgments** We wish to thank Mr. J. de Nie of the Rotterdam Harbor for drawing our attention to the impending decline of oil and natural gas production. Miss Jihan Seoud is thanked for pointing out some of the severest ecological consequences of the building of the Hormuz Dam; in doing so, she has forced us to recognize the ethical dilemmas associated with Macro-engineering. Poppe de Boer and Pieter Kleingeld of the Utrecht University are thanked for the discussion on isostatic compensation and the calculation of the reduction of Greenhouse gas emissions, respectively. Marieke Schuiling is thanked for the reference to Arrianus.

## References

Alspach B, Watson I (2004) Sea change. Civil Eng 74:70–75

Ambraseys NN (2009) Earthquakes in the eastern Mediterranean and the Middle East: a multidisciplinary study of 2,000 years of seismicity. Cambridge University Press, Cambridge

ASPO (2004) The Association for the Study of Peak Oil. Uppsala Hydrocarbon Depletion Study Group, Oil and gas liquids 2004 Scenario. http://www.peakoil.net/uhdsg/

Azam F, Worden AZ (2004) Microbes, molecules, and marine ecosystems. Science 303:1622–1624

Bassler F (1972) Solar depression power plant of Qattara in Egypt. Sol Energy 14:21–28

Bellini OS (2008) New frontiers in architecture: Dubai between vision and reality. White Star Publishers, Vercelli, pp 274–275

de Graaff JWM, Speck PJHR, Zijlstra JJP (2000) Method for locally elevating the ground: some aspects of sulfuric acid injection into subsurface limestone. Final project report, Netherlands Technology Foundation, The Netherlands, pp 59

Dinesh AC, Jayaprakash C (2008) Zaherite—the key mineral in Alleppey mudbank formation and on the possibility of creating mudbanks artificially. Curr Sci 95:962–966

Eager RE, Raman S, Wooten A, Westphal DL, Reid JS, Al Mandoos A (2008) A climatological study of the sea and land breezes in the Arabian Gulf region. J Geophys Res 113:D15106

Glenn EP, Brown JJ, O'Leary JW (1998) Irrigating crops with seawater. Sci Am 279:76–81

Hamblin DJ (1987) Has the Garden of Eden been located at last? Smithsonian 18:127–135

Hassan ESM (1961) Power from the Sea. J Power Div Proc Am Soc Civil Eng 87:21–27

Hooke RL (2000) On the history of humans as geomorphic agents. Geology 28:843–846

Isaev VA, Mikhailova MV (2009) The hydrology, evolution, and hydrological regime of the mouth area of the Shatt Al-Arab River. Water Resour 36:380–395

Johns WE, Yao F, Olson DB, Josey SA, Grist JP, Smeed DA (2003) Observations of seasonal exchange through the Straits of Hormuz and the inferred heat and freshwater budgets of the Persian Gulf. J Geophys Res 108(C12):3391

Jones JC (2009) Technical note: total amounts of oil produced over the history of the industry. Int J Oil Gas Coal Technol 2:199–200

Jones DA, Ealey T, Baca B, Livesey S, Al-Jamali F (2007) Gulf desert developments encompassing a marine environment, a compensatory solution to the loss of coastal habitats by infill and reclamation: The case of the Pearl City Al-Khiran, Kuwait. Aquat Ecosyst Health Manage 10:268–276

Kennett DJ, Kennett JP (2006) Early state formation in Southern Mesopotamia: sea levels, shorelines, and climate change. J Island Coast Archaeol 1:67–99

Kettani MA, Gonsalves LM (1972) Heliohydroelectric (HHE) power generation. Sol Energy 14:29–39

Kettani MA, Scott RE (1974) A Red Sea heliohydroelectric (HHE) plant. Rev Int Heliotechnique, 2eme semestre, pp 19–25

Kitoh A, Yatagai A, Alpert P (2008) First super-high-resolution model projection that the ancient "Fertile Crescent" will disappear in this century. Hydrol Res Lett 2:1–4

Kumar A (2009) Reclaimed islands and new offshore townships in the Arabian Gulf: potential natural hazards. Curr Sci 96:480–485

Lambeck K (1996) Shoreline reconstructions for the Persian Gulf since the last glacial maximum. Earth Planet Sci Lett 142:43–57

Macalady J, Banfield JF (2003) Molecular geomicrobiology: genes and geochemical cycling. Earth Planet Sci Lett 209:1–17

Nelson G, Hansz JA, Cypher ML (2005) The influence of artificial water canals on residential sale property. Appraisal J LXXII:167–174

Purnama A, Al-Barwani HH (2006) Spreading of brine waste discharges into the Gulf of Oman. Desalination 195:26–31

Qian Z-G, Xia X-X, Lee SY (2009) Metabolic engineering of *Escherichia coli* for the production of putrescine, a four carbon diamine. Biotechnol Bioeng 104:651–662

Saunders JE, Al Zahed KM, Paterson DM (2007) The impact of organic pollution on the macrobenthic fauna of Dubai Creek (UAE). Mar Pollut Bull 54:1715–1723

Scafi A (2006) Mapping paradise: a history of heaven on Earth. University of Chicago Press, Chicago

Schuiling RD (1988) Method to artificially raise the level of the land. Utrecht University, Dutch patent application no. 8800838

Schuiling RD (2004) Palk Strait: repairing Adam's bridge with gypsum? Curr Sci 86(10):1351

Schuiling RD, Badescu V, Cathcart RB, Van Overveld PALC (2005) The Hormuz Strait Dam MAcroproject—21st century electricity development infrastructure node (EDIN)? Mar Georesour Geotechnol 23:25–37

Sheppard C, Al-Husiani M, Al-Jamali F, Al-Yamani F, Baldwin R, Bishop J, Benzoni F, Dutrieux E, Dulvy NK, Durvasula SRV, Jones DA, Loughland R, Medio D, Nithyanandan M, Pilling GM, Polikarpov I, Price ARG, Purkis S, Riegl B, Saburova M, Namin KS, Taylor O, Wilson S, Zainal K (2010) The Gulf: a young sea in decline. Mar Pollut Bull 60:13–38

Smith R, Purnama A, Al-Barwani HH (2007) Sensitivity of hypersaline Arabian Gulf to seawater desalination plants. Appl Math Model 31:2347–2354

Subba Rao DV (2000) Arabian Gulf. In: Sheppard C (ed) Seas at the Millennium, Vol II. Elsevier Science, Oxford Chap. 53

Thoppil PG, Hogan PJ (2009) On the mechanisms of episodic salinity outflow events in the Strait of Hormuz. J Phys Oceanogr 39:1340–1360

Uchupi E, Swift SA, Ross DA (2002) Tectonic geomorphology of the Gulf of Oman Basin. In: Clift PD, Kroon D, Gaedicke C, Craig J (eds) Tectonic and climatic evolution of the Arabian Sea. Geological Society of London Special Publication no. 195, pp 37–69

Vadot L (1954) Les centrales hydro-solaires (in French). La Houille Blanche 9:557–568

van Tongeren H (1986) Saudi Arabia-Bahrain Causeway. Land Water Int 57:19–28

Yang H, Reichert P, Abbaspour KC, Zehnder JB (2003) A water resources threshold and its implications for food security. Environ Sci Technol 37:3048–3054

# Construction Techniques for Deep-water Immersed Tunnel Using Real-time Sea Strait Current Forecast

**Yukinobu Oda, Kazunori Ito, Takahide Honda and Solomon Yim**

## 1 Introduction

The Marmaray Project (Lykke and Belkaya 2005) in Istanbul, which provides an upgraded urban railway system approximately 76 km long with a 13.4 km underground section (see Fig. 1), is under construction to overcome Istanbul's chronic traffic congestion. Istanbul is divided into a European side and an Asian side by the Bosphorus Strait, thus the new railway system needs to cross the strait. An immersed tunnel method was applied for the construction of the tunnel, called Bosphorus Crossing Immersed Tunnel, and it is at the present time the world's deepest immersed tube tunnel. The tunnel elements were prefabricated off-site and towed by ship to the work-site for installation and connection to the previously installed underwater elements. The remaining parts on the Asian side and the European side are being constructed using TBMs (tunnel boring machines) and are to be connected to both ends of the immersed tunnel, which consists of 11 individual submarine elements each approximately 135 m long, 15.3 m wide and 8.6 m tall. The total length of the immersed tunnel is 1,387 m and the maximum depth of the bottom of the tunnel is 60 m under the water of the Bosphorus Strait.

The seawater current in the Bosphorus Strait is of a complex nature, and it often varies dramatically and irregularly within a few hours due to ambient daily meteorological changes. It has a two-layer current flow system consisting of a southward upper flow, which flows from the Black Sea with low salinity, and a northward lower flow, which moves underneath from the Marmara Sea with high

Y. Oda (✉), K. Ito and T. Honda
Taisei Corporation, Yokohama, Japan
e-mail: od-ykn00@pub.taisei.co.jp

S. Yim
Oregon State University, Corvallis, USA

V. Badescu and R. B. Cathcart (eds.), *Macro-engineering Seawater in Unique Environments*, Environmental Science and Engineering,
DOI: 10.1007/978-3-642-14779-1_8, © Springer-Verlag Berlin Heidelberg 2011

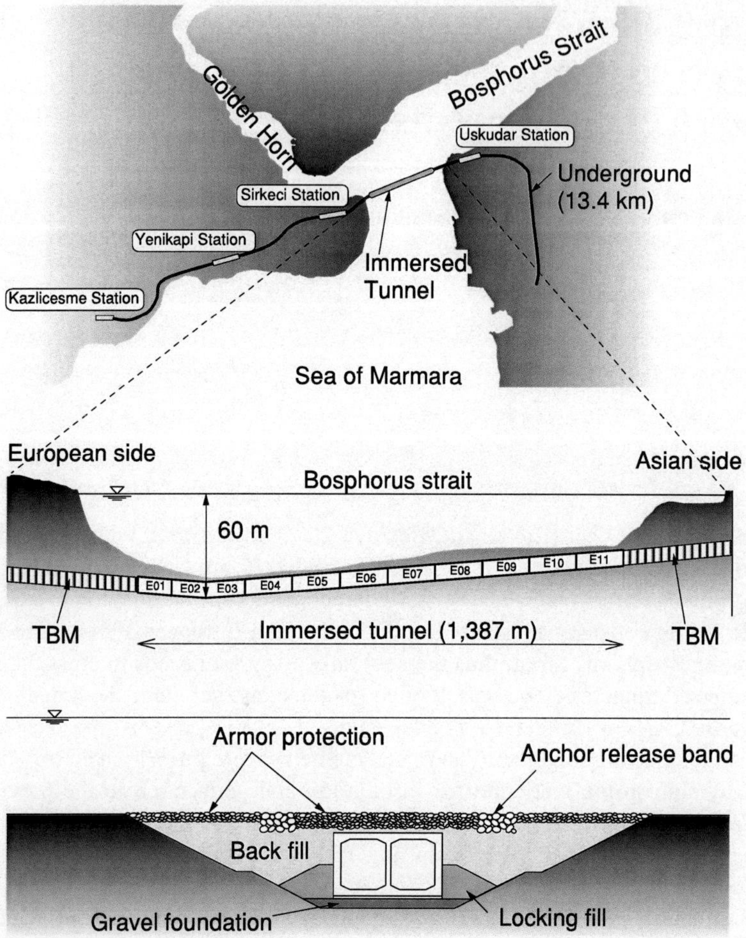


**Fig. 1** Location of the immersed tunnel and its cross section

salinity (Arisoy and Akyarli 1989). The seawater current speed of the surface layer frequently surpasses 2 m/s.

A number of new technologies were developed and applied in this Marmaray Project for the construction of the immersed tunnel under such severe environmental conditions. In this chapter, the following newly developed building and installation technologies specifically to overcome these severe environmental conditions are presented.

- **System of gravel foundation construction:** Prior to the tunnel element installaion, a gravel foundation was formed on the Bosphorus Strait's bottom (see Fig. 1). Flatness and elevation accuracy of the foundation were important factors, and the roughness surface causes severe macro-problems in the final positioning of submarine tunnel elements. To overcome the conditions of deep water, fast flow and two-layer current, gravel foundation construction system being composed of a gravel dumping simulator and a System for Underwater Grading (SUG) was developed.

- **Method of tunnel element installation:** The tunnel element immersion is generally performed by using pontoons or a placing barge in deep water (Grantz 1993). Here it was performed by using a wiring system similar to the one applied for pontoons and a placing barge. The procedure of the tunnel element installation will be described in detail.
- **Device of access shaft installation:** At the end of the Asian side of the first immersed tunnel element, an access shaft (AS) was installed through which materials, machines and laboring manpower were supplied into the tunnel elements. However, in a hydraulic laboratory experiment, it was found that the AS might oscillate during the installation when it was supported by lifting wires in the strongly flowing current. To suppress the predicted oscillation, a countermeasure using a Double-fin was developed by the mega-project's engineers.
- **Real-time Bosphorus Strait seawater current forecast system:** For construction in the severe marine environment, one of the most important issues is to estimate the current conditions accurately in advance. Especially for the tunnel element immersion construction, the go/no-go decision was governed by the current condition, namely the maximum current speed had to be less than 1.5 m/s, which was the threshold for the tunnel element immersion construction.

## 2 Seawater Current Condition of the Bosphorus Strait

Figure 2 shows the typical patterns of the cross-sectional velocity distribution along the tunnel alignment in relatively weak and strong current situations obtained by boat measurements with an acoustic Doppler current profiler (ADCP). Here negative velocity indicates southward flow and positive northward of the absolute current velocity. It is observed that the current is stratified and the velocity varies gradationally in the vertical direction. This two-layer current system is due to density effect. The salinity of the upper layer is low and that of the

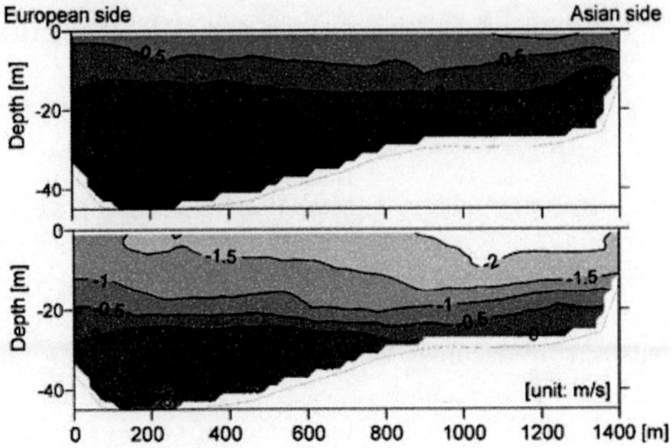


**Fig. 2** Velocity distribution at the site of the immersed Bosphorus strait tunnel

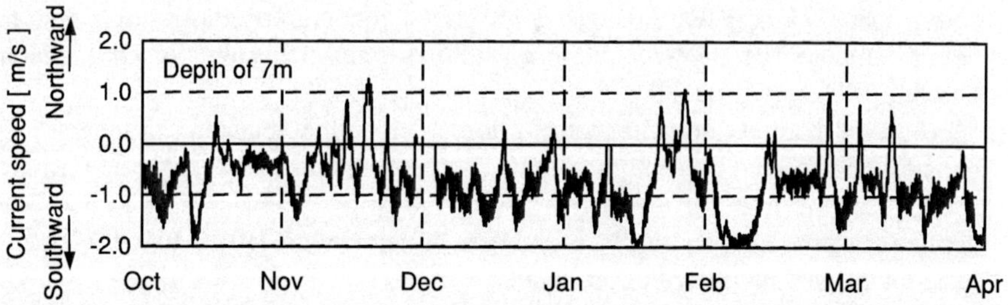


**Fig. 3** Current speed changes at the midpoint of the immerse tunnel railway (depth of 7 m)

lower layer is high. The density interfaces between the two layers stay at a relatively constant depth of approximately 15 and 25 m in the weak and the strong current situation. The maximum velocities of the surface seawater flow is almost 0 and over 2.0 m/s, respectively, but the velocities of the reverse seawater flow in the lower layer are both less than 1.0 m/s. Normally the surface velocities vary over a range of 0–2.0 m/s. Regarding the lower layer, the maximum velocities of the reverse flow are almost unchanged, but the area varies vertically.

Figure 3 shows a time series of the seawater current speed at 7 m water depth at the midpoint of the tunnel line. The current varies frequently and irregularly in the range between southward (normal direction) with a current speed of 2.0 m/s and northward with a current speed of 1.0 m/s. When the current direction becomes northward (the velocity is positive), the density interface reaches the vicinity of water surface. This is caused by a strong southerly wind and is a rare event (mentioned later).

The current is basically caused by the seawater level difference between the two ends of the strait. These variations are mainly due to regional weather conditions, air pressure overhead the mega-project and local wind, which have no regularity (like some tidal variations), thus making prediction of the seawater current conditions extremely difficult. The detailed current structure will be discussed later in this chapter.

# 3 System of Submarine Gravel Foundation Construction

## *3.1 Gravel Foundation Construction*

Prior to tunnel element installation, a seabed gravel foundation was constructed (see Fig. 1). For placement of the installed elements, accurate elevation and evenness of the foundations were important. The roughness (variations in elevation) tolerance of the foundation was 0.3 m. To ensure this high degree of bathymetric accuracy throughout the construction under the condition of strong seawater current and deep water working conditions, the mega-project's engineers developed a comprehensive system of gravel foundation formation.

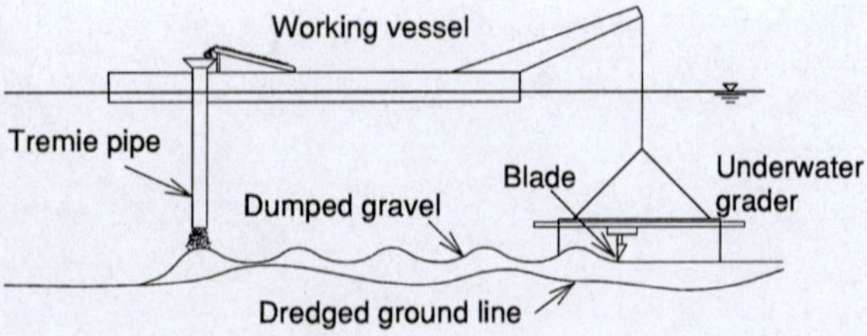


**Fig. 4** Devices to construct the gravel foundation. Gravel (and concrete) can be carefully placed via the Tremie pipe

In this system, two devices shown in Fig. 4, a tremie pipe and an underwater grader, were used. Trenches where the elements were buried had been excavated in advance by a grab bucket (see Fig. 1). Gravel for the foundation was dumped on the dredged ground through the tremie pipe, and then the dumped gravel was graded by the underwater grader. To minimize the unevenness of the dredged ground, the amount of gravel at each dumped point would be optimized in accordance with the undulation of the dredged ground. To estimate the optimal dumping point and the amount of gravel, a gravel dumping simulator (GDS) was developed (Ito et al. 2009). Since the foundation was to be formed on the Bosphorus Strait bottom, as deep as 60 m, forming by divers was practically impossible. Therefore the system for underwater grading (SUG), where the underwater grader was controlled to create an even foundation surface, was developed.

## 3.2 *Development of Gravel Dumping Simulator (GDS)*

The Bosphorus Strait has a complex current structure and the seawater depth at the construction site is demandingly deep. In the case that gravel was directly dumped from a work vessel to the sea-bottom, the dumped gravel would, because of the opposing strong seawater currents, disperse unevenly making it difficult to form the planned target foundation shape. To improve the gravel laying process, a tremie pipe was used for gravel dumping. Because the depth varied along the tunnel alignment, the tremie pipe had to have a telescopic mechanism. The gravel dumped from the lower end of the pipe at several points formed overlapped conic shapes shown in Fig. 5. Although the underwater grader would later level these dumped cones, to achieve the target foundation form, the amount contained in each gravel cone should be optimized to reduce the unevenness.

The gravel dumping simulator (GDS) predicts the dumped gravel geometry based on the dredged ground bathymetry and the amount of dumped gravel. It is assumed that the dumped gravel dispersion is basically Gaussian distributed. The height of the dumped gravel is expressed as Eq. 1 (see Fig. 6)

$$h_g(x,y) = \frac{V}{S}\iint_S \frac{1}{2\pi\sigma^2}\exp\left\{-\frac{1}{2}\frac{R^2+r^2-2Rr\cos\theta}{\sigma^2}\right\}drd\theta \tag{1}$$

where $h_g$ denotes the height of the dumped gravel and (r, $\theta$) the angular coordinate on the lower end of the tremie pipe; (x, y) the horizontal rectangular coordinate with

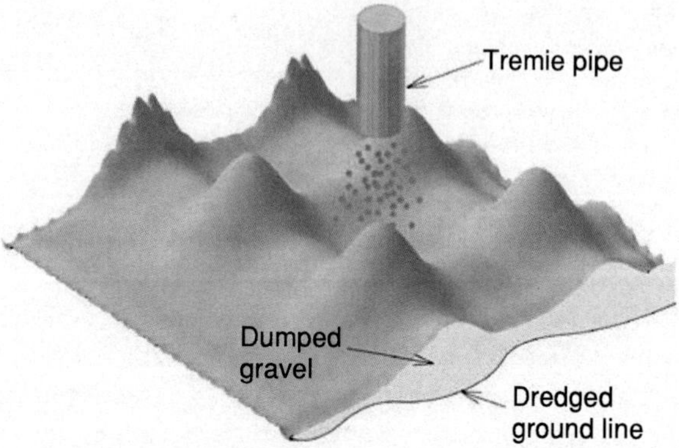


**Fig. 5** Gravel dumping with a tremie pipe

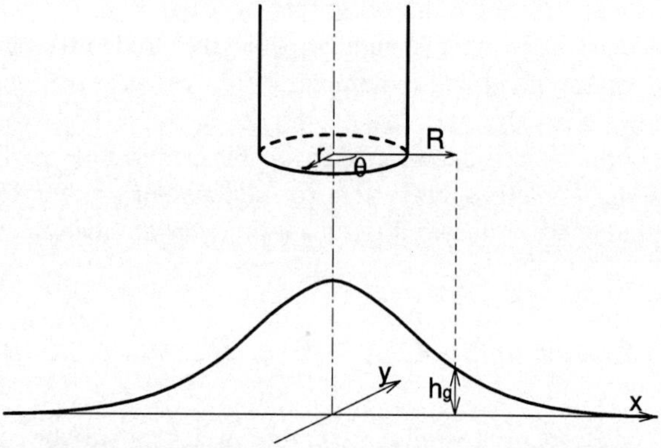


**Fig. 6** Definition of the GDS model

the origin of the center line of the tremie pipe; R(x, y) the horizontal distance of the referring point from the x–y origin; V the amount of dumped gravel; S the cross section area of the tremie pipe; $\sigma$, the variance and is function of the height of the lower end of the tremie pipe from the bottom, and is determined by in situ field test.

Equation 1 can be numerically integrated but may yield excessive slope predictions because it does not take into account the limiting angle of repose of the dumped gravel. Thus after estimating the dumped gravel geometry using Eq. 1, a small amount of the gravel is numerically dropped down to the next grid at the steepest slope. This operation is repeated until the slopes of the dumped gravel surface at all points are less than or equal to the angle of repose.

## *3.3 Underwater Grading System*

The foundations should be graded horizontally to achieve the accuracy of the elevation (flatness) error within 0.3 m at the maximum depth of 60 m. A system for underwater grading (SUG), composed of an underwater grader shown in Fig. 7 and a control system, was developed and employed.

**Fig. 7** Underwater grader resting on land

The underwater grader had a blade that could be moved arbitrarily, both horizontally and vertically, as well as four super-structure support legs with adjustable lengths to compensate for the undulation of the dumped gravel surface to level the entire machine. The blade can be also rotated to sweep and level the top of the dumped gravels. Leveling of the machine could be measured from the working vessel by echo sounding and the detail surface elevation of the foundation could be also measured by an echo sounder installed at the vicinity of the blade. All devices were remotely operated from the floating working vessel and there was no need for hard-hat or SCUBA divers.

Figure 8 shows the elevation and the discrepancy (the difference between the measurement after grading and the target elevation) of the graded foundation by SUG at the European side of the element E03, which is at the largest water depth. Note that there is a ramp at the 30 m point to fit the configuration of the foundation to the existing drainage tank below element E03. It can be observed that the differences between the surveyed elevation and the target were less than 0.3 m, which was the maximum allowable error. The foundations were formed quite exactly even at the extremely seawater depth of 60 m.

## 4 Immersed Tunnel Element Installation

There was no precedent for applying a tunnel immersion method to such deep and strong current condition. In this mega-project, a placing barge was used for immersion of tunnel elements. Commonly in a placing barge method, a tunnel element is simply suspended by only sling wires and the position of the element is indirectly adjusted by means of positioning the barge. However it was difficult to adjust the element position because of the large water depth in this case, so adjusting wires, a method used in pontoons, was applied. Adjustable wires were securely connected to the placing barge and anchors via pulleys at the corners of

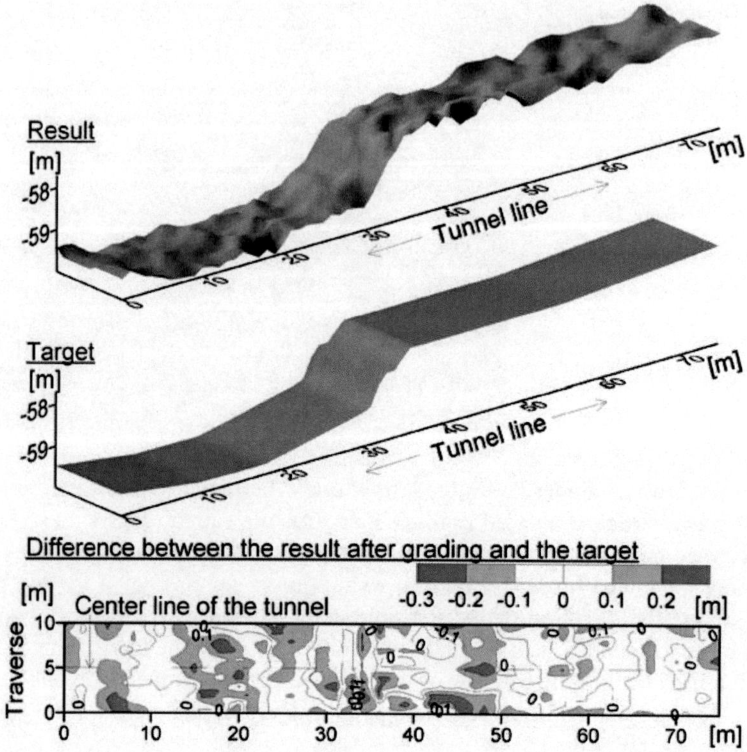


**Fig. 8** Result of the foundation grading by SUG

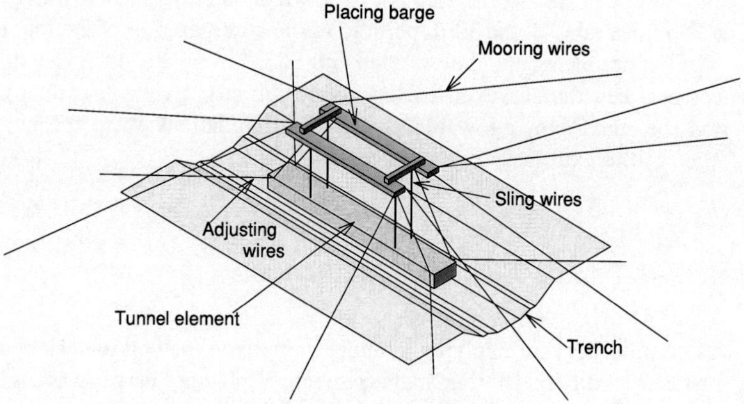


**Fig. 9** Mooring system for tunnel element installation

the tunnel element. They performed the role of not only adjusting the element position but also decreasing its oscillation (see later sections). In this mega-project, a total of 17 wires were used for the tunnel installation, as is shown in Fig. 9.

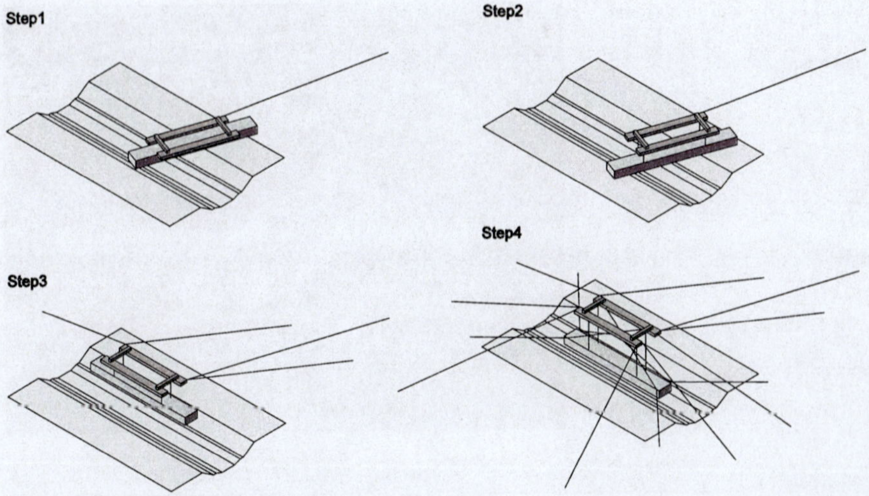


**Fig. 10** Procedure of the tunnel element installation

The procedure employed to perform the tunnel element installation is shown in Fig. 10. At the first step of the installation process, the placing barge was moored at the mega-project construction site in the streamlined position parallel to the surface seawater current (step 1). Then the tunnel element was lowered to mid-water depth at the density interface, where the flow is relatively calm (step 2). Next the element was rotated using mooring winches to an optimal operational direction (step 3) where its longitudinal axis was exposed to predominantly north–south seawater current. This is the essential reason the element was lowered to the slow current depth beforehand at step 2 to avoid extreme forces that could and likely would occur in the uppermost seawater layer. In the final step, the tunnel element was lowered to the sea-bottom and connected to the adjacent prefabricated tunnel element with adjusting wires (step 4).

To validate the procedure of the tunnel installation experimentally, and to estimate the forces on all wires in each step, a 1/120-scale physical model test, shown in Fig. 11, was carried out prior to the field operation (Lykke and Kerk 2005; Oda et al. 2005). The mooring equipment was designed based on the results of the revealing physical model test.

When long sling wires are needed at large water depth, the horizontal restoring forces of the element at the position just below the placing barge provided by the sling wires would be relatively small. Although the current forces were not large at the final step (step 4) because the element was already lowered into the trench, sway resonance was often excited and hard to suppress once the element started swaying. The adjusting wires played a part by minimizing this oscillation in the final step. Figure 12 shows the effect of the adjusting wires for suppressing the oscillation in the final step. These sub-figures show the horizontal oscillations at the corner of the submerged tunnel element. Comparing these figures we observed

**Fig. 11** Physical model test of the immersed tunnel element installation

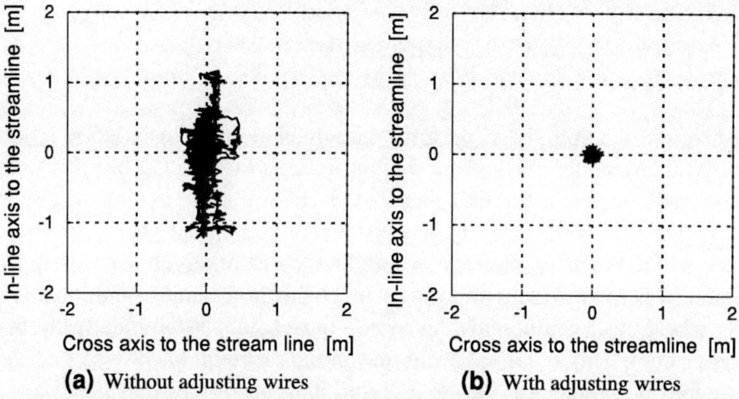


**Fig. 12** Experimental result of the horizontal oscillations on the tunnel element installation

that, by using the adjusting wires, the amplitude of the oscillation could be reduced from 1.0 m to less than 0.2 m, which was a amplitude less than the allowable excursion for setting the immersed tunnel element successfully on the sea-bottom rack equipped with a function of ultimate position adjustment.

## 5 Access Shaft (AS) Oscillation Counter-Measure

In an immersed submarine tunnel method, a pit structure, used for access of human workers and materials into the immersed element, is commonly constructed on the landward-end of the tunnel element prior to its installation. In this mega-project, to minimize the total work period and cost, unlike the conventional procedure, a cylindrical structure called "access shaft" (like a chimney) was installed on an

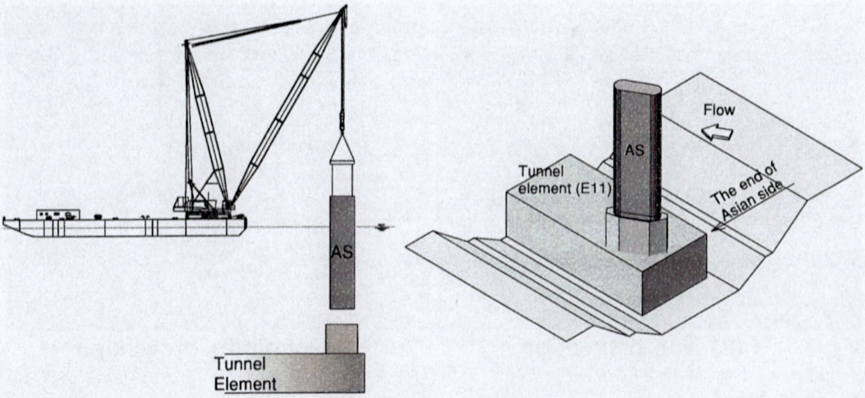


**Fig. 13** Schematic view of the access shaft (*AS*) installation

**Table 1** Size of the access shaft

| Weight in air | Height | Length | Width |
|---|---|---|---|
| 2,000 kN | 26.0 m | 8.6 m | 3.15 m |

already immersed element (see Fig. 13). The access shaft has the shape of the hollow cylinder with dimensions listed in Table 1. It was to be installed to the guide base attached to the first installed element at the Asian side-end by using a floating crane. The AS was suspended by four sling wires and connected to the floating crane with two adjusting wires to control the rotational position of the AS. It is well known that a slender structure in fluid flow may oscillate due to Karman vortex shedding. A 1/50-scale physical model test was carried out to verify the safety of the AS installation.

The results indicated that the AS could oscillate with large amplitude when the current velocity was more than 1.0 m/s. However, when the motion of the AS was constrained at the experiment, it is observed that the formation of large vortices was suppressed and the periodical lift forces were negligible. For the fixed AS, the vortices separated at the upstream shoulders did not continue down to the downstream. On the other hand, when the AS was allowed to freely oscillate, both shoulders at the down-stream side were exposed to the flow and vortices separated there. Then the vortices alternately shed behind the AS, causing oscillatory forces.

Several possible counter-measures were examined through physical model testing (Ito et al. 2009). It was found that Double-fin was the most suitable countermeasure in this instance (see Figs. 14 and 15). Additional physical model tests were performed to obtain the effective range of the Double-fin form. The results are shown in Fig. 15. D and $\delta$ are short diameter of the AS (=3.15 m) and averaged full amplitude of oscillation at the measurement point, which was measured by using a laser displacement sensor at the AS bottom. $\delta$/D denotes the normalized average amplitude and $F_x$ is the fluid force in-line to the seawater

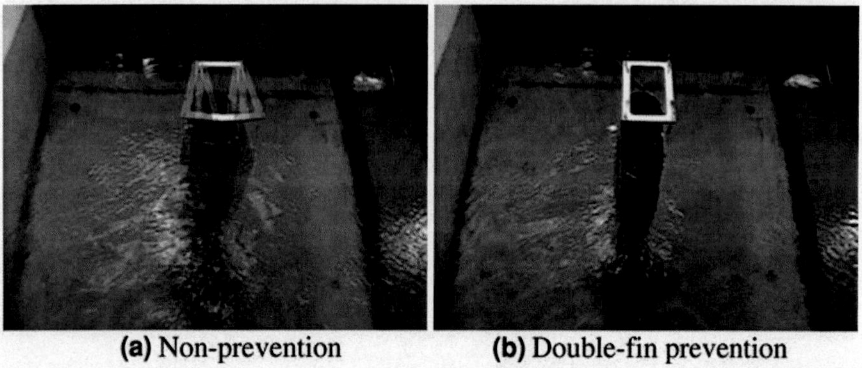


**Fig.14** Prevention against the AS oscillation by Double-fin **a** non-prevention **b** double-fin prevention

**Fig. 15** Access shaft oscillation and fluid force

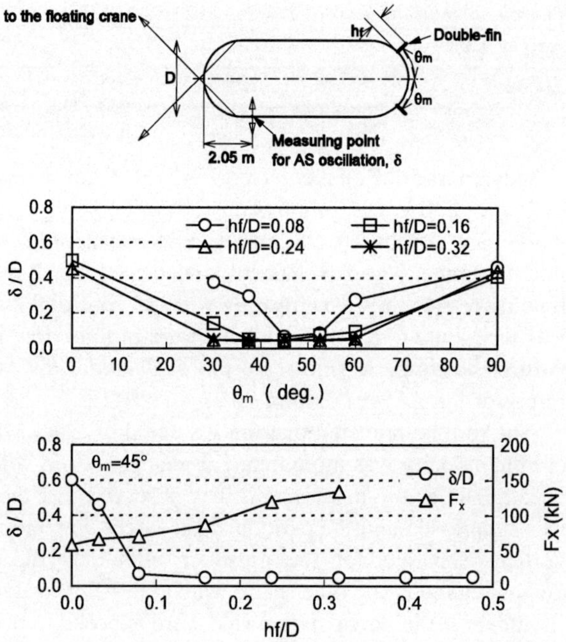


current, measured with the AS model fixed. Figure 15 shows that the AS without any counter-measures oscillated with an amplitude of approximately 60% of the AS short diameter, D. On the other hand, the amplitude of the AS oscillation with the appropriate Double-fin suppressor was successfully reduced to less than one-tenth of the non-suppressed case.

In respect to the setting angle of fins, $\theta_m$, the AS oscillation was reduced in the range of $\theta_m = 30°–60°$ to except the relatively small fin height case, $h_f/D = 0.08$, in which the effective range became narrow ($\theta_m = 45°–52.5°$). Regarding the effective fin height, in the range that $h_f/D$ was larger than 0.08, the oscillation was effectively reduced. On the other hand, the fluid force, $F_x$, became larger with increasing $h_f/D$, especially when $H/D > 0.2$. Thus, it was concluded that the height of the fin height was effective in the range from $H/D = 0.1–0.2$ and the setting angle and height of the Double-fin were designed as 45° and 0.5 m accordingly in this project.

# 6 Development of Real-Time Bosphorus Strait Current Forecast

## 6.1 Outline

The Bosphorus Strait has a very complex seawater current structure as mentioned above. It connects the Black Sea in the north to the Marmara Sea in the south, which is in turn connected to the Aegean Sea by the Dardanelles (see Fig. 16). The rainwater flowing into the Black Sea runs into the Aegean Sea through the

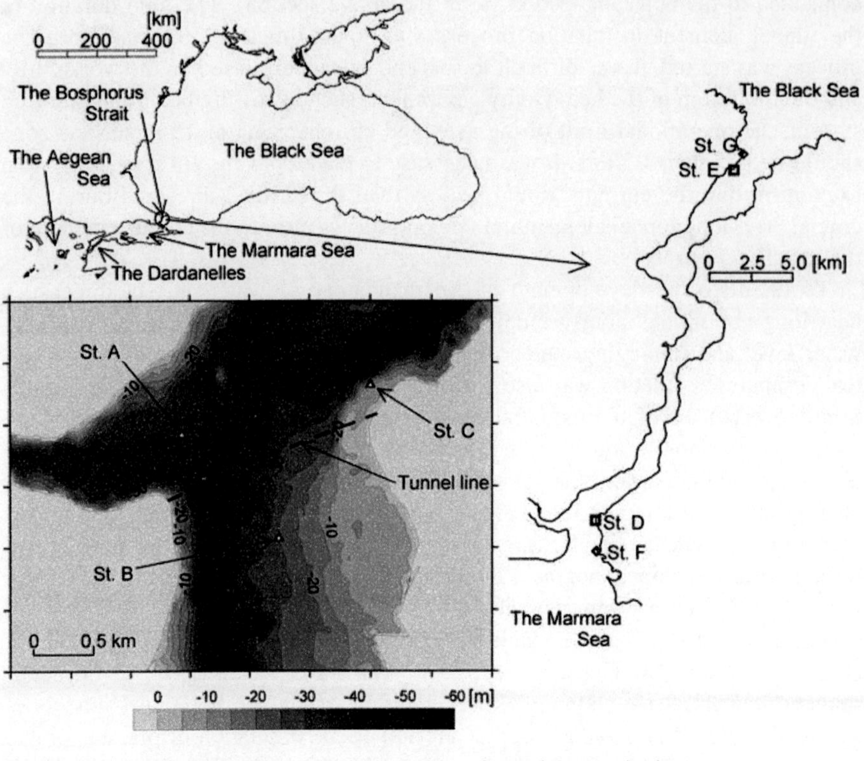


**Fig. 16** Geography of the Bosphorus strait and locations of the monitoring

Bosphorus and the Dardanelles. The current in the Bosphorus Strait is basically generated by the seawater level differences between the adjacent basins, the Black Sea and the Marmara Sea. These adjacent basins are almost closed to the open ocean, thus the influence on seawater level variations by the tidal forces is not conspicuous (Gregg et al. 1999) but the meteorological conditions, air pressure and wind, are dominant to the seawater level variations, mentioned later. Moreover, the Bosphorus Strait has a two-layer current system. The Marmara Sea water is more saline compared to the main flow coming from the Black Sea, and therefore goes underneath it and it then becomes a backward lower layer seawater flow (Arisoy and Akyarli 1989; Gregg and Özsoy 2002).

To ensure safety and quality of the submarine work, it is important to be able to accurately predict the seawater current condition before any work is undertaken on the mega-project. If the current was solely caused by tidal effect, seawater current condition prediction would not be difficult. However, the above-mentioned environmental influences make it difficult to estimate the current state in the Bosphorus Strait, i.e. information that is essential in the railway tunnel immersion work for making "go/no-go" decision. Prior to the mega-engineering installation process, the placing barge carrying a single tunnel element was temporary moored at the southern mouth of the strait. First, the placing barge was towed to the construction work-site and moored, and then the submarine tunnel element was lowered and connected to the adjacent section (refer the above section). The total duration of the tunnel element installation process was approximately 36 h and once the process was started, it was difficult to suspend or cancel. Based on the workability and the limitation of the heavy-duty equipment, such as towing boats and mooring system, the operational limit of the averaged current value of 15 m surface zone should be less than 1.5 m/s. It was necessary to develop a current forecast system to confirm that the currents would be less than this critical threshold during the crucial 36 h-long tunnel element immersion process prior to the commencement of the installation process.

To facilitate the development of the mandatory seawater current forecast, a year-long continuous hourly monitoring and data-collection of the current, seawater level and impinging meteorological conditions was conducted. 3D numerical computer simulation was also performed on the whole area of the Bosphorus Strait to catch all of the essential features of the seawater current(s) structure. Based on the monitoring data and the simulation, the relationship amongst the meteorological condition, the seawater levels and the seawater current was modeled by means of a regression analysis. Note that the current forecast system was used in daily marine operations, requiring both robustness and efficiency. Therefore, it was developed not as a numerical simulation model but a functional relation model among various factors.

There are few preceding studies on the development of forecast models of seawater current flow (Jensen et al. 2000; Tanaka et al. 2005) based on deterministic numerical simulations. However the deterministic numerical models of stratified current may contain significant numerical simulation errors, as pointed out by Komatsu et al. (1997). Jensen et al. (2000) minimized the errors of the numerical simulations, caused by uncertainty in natural phenomena and factors using an artificial neural network, and Tanaka et al. (2005) employed real-time monitoring data using an ensemble Kalman filters. However, to predict the stratified seawater current flow behavior quantitatively, highly accurate numerical schemes and/or finer grid are required using numerical simulations. For the mega-project construction operations planning, such calculations were undesirable because providing current condition prediction on demand was essential and the long computational turnaround time would be unacceptable. Therefore, instead, the current forecast model was developed based on a regression analysis with a yearly observation and numerical simulations were carried out only to confirm the qualitative current flow behavior for model development.

## 6.2 Monitoring

Monitoring of the Bosphorus Strait seawater current conditions was conducted continuously for a year, from October 2004 to September 2005. The locations of the monitoring stations are shown in Fig. 16 (Stations A through G). The current speed conditions were measured as vertical current profiles with acoustic Doppler current profilers (ADCPs) at three stations A, B and C. Water levels were measured with differential pressure gauges at Stations D and E located at the both ends of the strait. Meteorological conditions, air pressures, wind speeds and directions, were observed at the Station F and G. All data were collected hourly. In addition, the cross-sectional current measurements along the entire tunnel alignment were carried out monthly using a boat with an ADCP, as shown earlier in Fig. 2.

Figure 17 shows the time-series of sample monitoring data, where seawater levels at Station E (the Black Sea side) and Station D (the Marmara Sea side), air pressure at Station F, northerly wind at Station F, water level difference between the Black Sea side and the Marmara Sea side, and current speed of 7 m depth at Station B are indicated. Here, the current speed is indicated as an absolute value, positive for northward flow.

It can be observed that the seawater level on the Marmara Sea side responds to the air pressure; the seawater level decreases as the air pressure increases, while the seawater level on the Black Sea side does not appear to be significantly influenced by variable ambient air pressure. As for the wind effect, on the strong northerly wind situations, the seawater levels show a tendency to follow the wind speed; the seawater levels become higher on the Black Sea side and lower on the Marmara Sea side. When the northerly wind blows, the water levels at the south coasts of both seas rise up due to wind set-up and the water level at the north side of the Marmara Sea falls in compensation. Yüce (1993) confirmed this phenomenon in their observation for the Marmara Sea. However, such a trend is not clearly observed on the Marmara Sea side in Fig. 17 because the effects of both air pressure and wind effects are included.

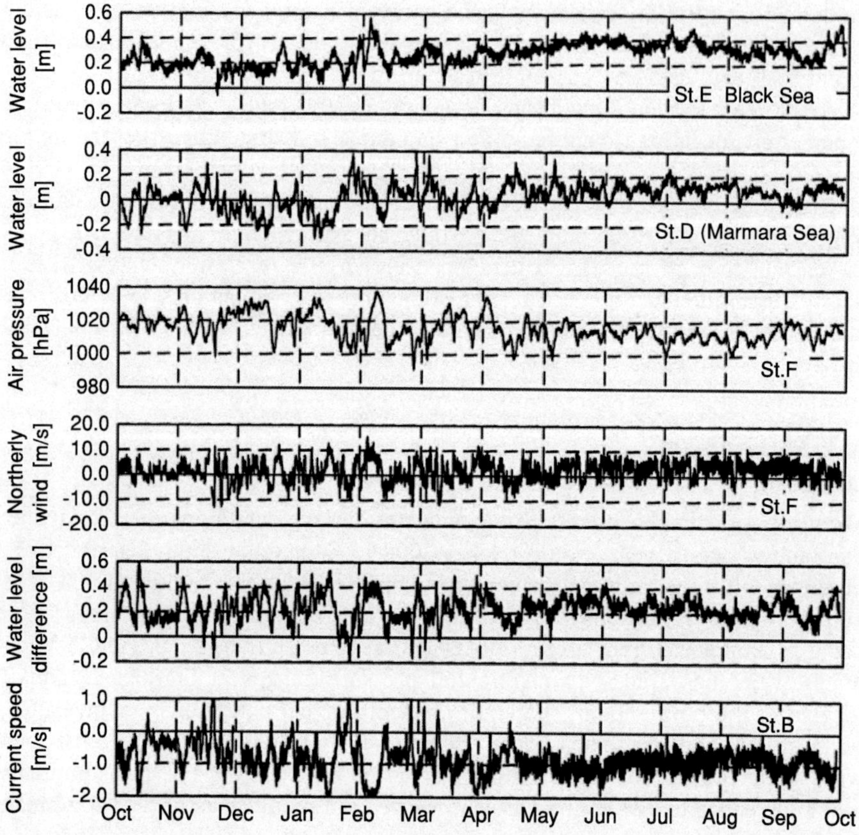


**Fig. 17** Monitoring results

A strong relation observed between the seawater level difference and the current speed and the seawater level difference is expected to dominantly drive the current in this internationally important strait.

## *6.3 Numerical Simulation*

Numerical simulations were laboriously carried out to reproduce the qualitative general seawater current flow behavior based on a parametric study, focusing on the pronounced stratification effect naturally present in the Bosphorus Strait. The numerical code used for developing the Bosphorus Strait model was Delft3D-FLOW, developed by Delft Hydraulics (Lesser et al. 2004). Turbulence effects were taken into account using a turbulence model with k-ε closure. A curvilinear coordinate system was applied in a horizontal direction with the minimum

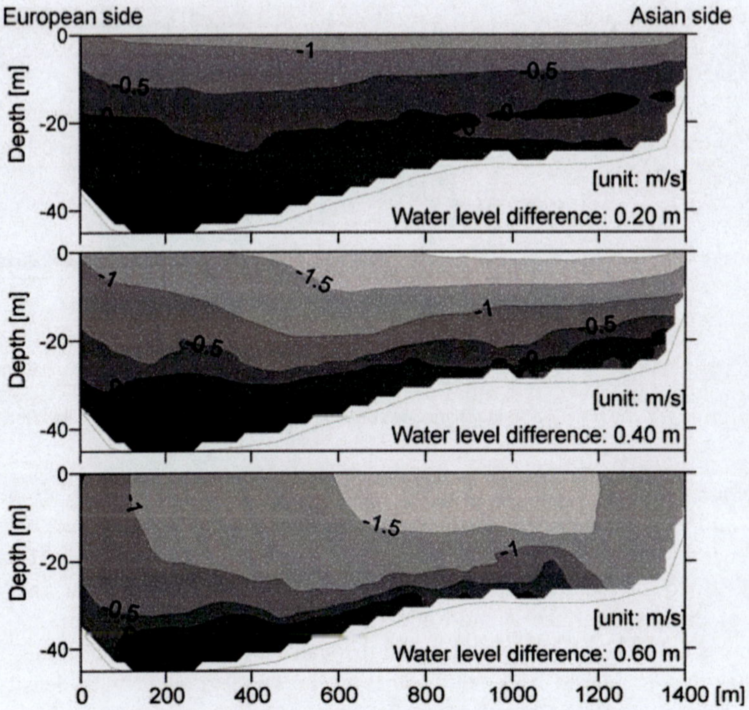


**Fig. 18** Numerical prediction of velocity distribution along the tunnel line

horizontal grid size of 80 m around the tunnel location and the maximum size of 400 m at distant places. A sigma coordinate system was applied in the vertical direction with ten layers.

Figures 18 and 19 show examples of the computer simulation results, which are assumed to be the steady-state results where the boundary conditions were given as constant values and the calculations were continued until equilibrium is achieved. Simulation results for three steady water level difference cases are shown here.

Figure 18 presents the cross-sectional current distributions along the tunnel alignment. Comparing the simulation results with those obtained from field monitoring shown in Fig. 2, it can be observed that the patterns of the cases of 0.2 and 0.4 m seawater level difference are similar. Wherein particular, fast current regions spread widely and thinly on the surface and the reverse flows with the maximum current speed of approximately 0.5 m/s stay in the lower layers. On the other hand, for the case of 0.6 m, the maximum current speed, about 2.0 m/s, does not increase compared to the case of 0.4 m but the fast current region, faster than about 1.0 m/s, extends through the depth.

Figure 19 shows the calculated longitudinal salinity distribution along the center of the strait. It is observed that the high salinity water from the Marmara Sea creeps under the low salinity water from the Black Sea. As the water level

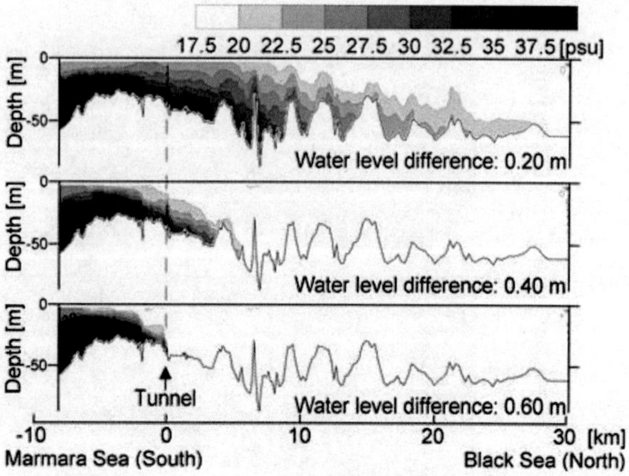


**Fig. 19** Numerical prediction of longitudinal salinity distribution

difference increases from 0.2 to 0.4 m, the front of the density interface retreats southward significantly (about 20 km right to left in Fig. 19). However, when the difference increased from 0.4 to 0.6 m, the front retreats only slightly (less than 5 km). When the front of the density interface extends northward, the (negative) slope of density interface is mild and the pressure gradient with density distribution is small. On the other hand, when the water level difference becomes larger and the front is pushed back southward, the pressure gradient with density distribution becomes large, thus the front movement becomes more insensitive to the water level difference variation. Moreover, in this situation, the stratification is relatively strong (i.e., the contour lines are closer to each other), thus the large velocity region extends to the depth without the maximum velocity increasing in Fig. 18.

## 6.4 Seawater Level Modeling

The current system in the Bosphorus Strait is dominantly caused by the seawater level difference mentioned above. Therefore the current forecast model should include the calculation procedure of the water levels at Station E at the Black Sea side and Station D at the Marmara Sea side, respectively, to predict the cross-sectional current distribution along the tunnel line. Tidal forces, air pressures and winds were taken as the dominant parameters for the water level variations in this model. Each effect will be discussed as follow.

### 6.4.1 Tidal Effects

Figure 20 shows the spectra of the yearly seawater level variations for both the Black Sea side and the Marmara Sea side. Four peaks clearly appear around 0.04

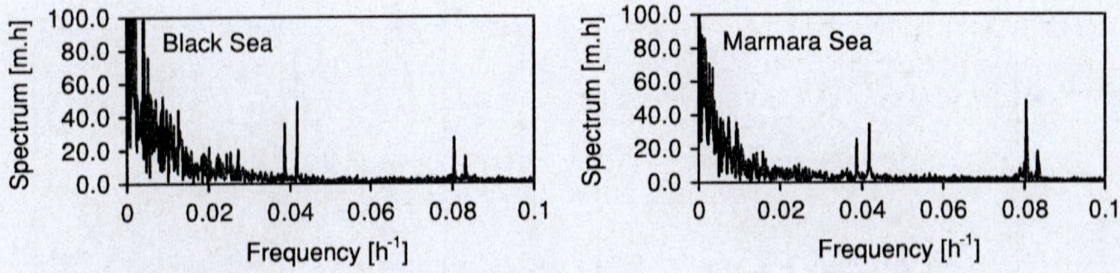


**Fig. 20** Spectrum of seawater levels for a year

**Table 2** Principal tidal components at St. D and St. F

| Component | $Q_1$ | $K_1$ | $M_2$ | $S_2$ |
|---|---|---|---|---|
| Period [h] | 25.82 | 23.93 | 12.42 | 12.00 |
| The Black sea (St. F) [m] | 0.009 | 0.010 | 0.005 | 0.004 |
| The Marmara sea (St. D) [m] | 0.007 | 0.010 | 0.015 | 0.006 |

and 0.08 /h frequencies (i.e. 25 and 12.5 h periods). They correspond to four principal tidal constituents, $Q_1$, $K_1$, $M_2$ and $S_2$, the periods of which are listed in Table 2. The amplitudes of tidal constituents can be analyzed with long-term water level measurements by means of harmonic analysis and the analyzed results are summarized in Table 2. The amplitudes are relatively small with the maximum spring tide of 0.028 m at the Marmara Sea side and 0.038 m at the Black Sea side. The analyzed results, including phases, were applied in the model as the tidal components.

### 6.4.2 Air Pressure Effects

Theoretically, the seawater level varies linearly with local air pressure and 0.00993 m/hPa is commonly taken as the proportional constant, which is derived from the unit air pressure divided by the specific gravity of the water (Gil and de Toro 2005; Spellman 2009). Figure 21 shows the relationships between air pressure and seawater level fluctuation, in which the long periodic components, considered to be as seasonal variations, are removed by means of 90-day moving average. Note that these water level fluctuations are still affected by the wind and other factors not taken into account in the model.

As for the Marmara Sea side, a linear relationship between the seawater level and the air pressure can be readily observed. The proportional constant estimated by using the least square method is 0.00821 m/hPa and slightly smaller than the commonly used value of 0.00993 m/hPa. The inflow/outflow through the Dardanelles connecting the Marmara Sea to the open ocean is essential to the seawater level variation caused by air pressure. Since the discharge rate of this flow encounters some geographical and frictional resistance in the Dardanelles, the practical constant value is expected to be slightly smaller.

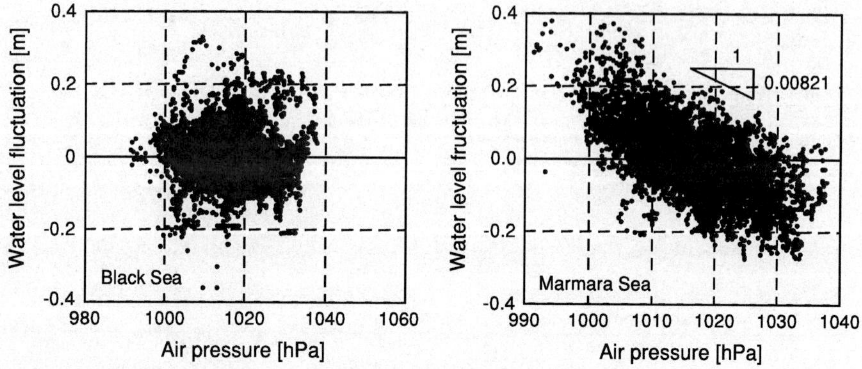


**Fig. 21** Relationship between air pressure and water levels (Stations G and F)

**Fig. 22** Relationship between air pressure deviation and water level (Station G)

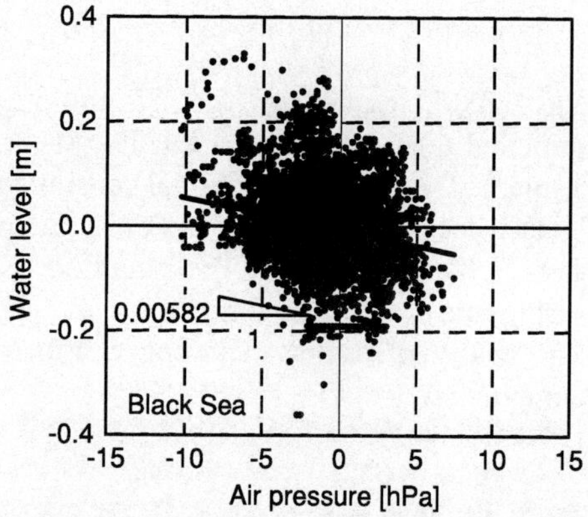


As for the Black Sea side, no obvious relationship is observed in Fig. 21 and the contribution of the air pressure to the seawater level variation is expected to be small. Since the Black Sea is practically closed to open sea, the amount of seawater in it is not affected much by the air pressure variation (Ducet et al. 1999). However, the Black Sea has a very large surface area, thus the seawater level is considered to vary locally depending on the air pressure distribution over the whole region of the Black Sea. Figure 22 shows the relationship between the local air pressure deviation (at Station G) from the mean air pressure over the entire surface of the Black Sea and the seawater level fluctuation. A weak but linear relationship can be observed compared with Fig. 21. The estimated proportional constant is 0.00582 and much smaller than in theory. Because the Black Sea is usually stratified and the stratification counters the air pressure effect, the constant is expected to be small. In this model, the above-mentioned proportional constants were used for the seawater level models on both nautical ends of the Bosphorus Strait.

### 6.4.3 Wind Effects

Figure 23 shows the example of the time series of the seawater level and the southerly wind speed at the Marmara Sea side. We can see that seawater level clearly responds to the wind speed but with a delay on the order of hours. Moreover, there is a short periodic fluctuation in the wind speed but it does not appear in the seawater level. The wind set-up in the closed seawater area is considered to be proportional to the square of the wind speed. It is assumed that the wind set-ups are independent in the frequency domain and linear superposition can be assumed. The wind set-up in an arbitrary direction can be then expressed as:

$$\eta_{\text{wind}}(f) = R(f)|W(f)|W(f) \tag{2}$$

where R denotes the response function, obtained by dividing the spectrum of the seawater level by that of the square of wind speed, and W is the wind speed.

Figure 24 depicts the spectra of the year-long time series of the seawater level and the square of the wind speed of N–S component, and the response function in the Marmara Sea side. Note that the seawater level spectrum includes the tidal components mentioned above and the wind spectrum includes a daily cycle component, which is considered due to land and sea breeze circulation.

Figure 24a shows that the seawater level responds less to the wind in the high frequency region compared with the low one. The wind set-up is developed with the wind-driven current generation, so some set-up time is necessary for the full growth thus the short periodic component becomes smaller. Hence the fetch-wind effect must be introduced as not a static but a dynamic model. To inject the periodical response difference into the wind set-up modeling, the application of response functions is proposed.

Figure 24b shows the 12 monthly response functions derived by dividing the spectra of the water level by the square of wind speed in each month. The functions appear to fluctuate abnormally. This can be explained as follows: the spectrum of the seawater level or the wind speed does not necessarily contain a domain value in the whole frequency. For the frequency where the value of the wind speed spectrum is small or zero, the response function, which is derived by the water level spectrum being divided by the square wind spectrum, becomes very large or

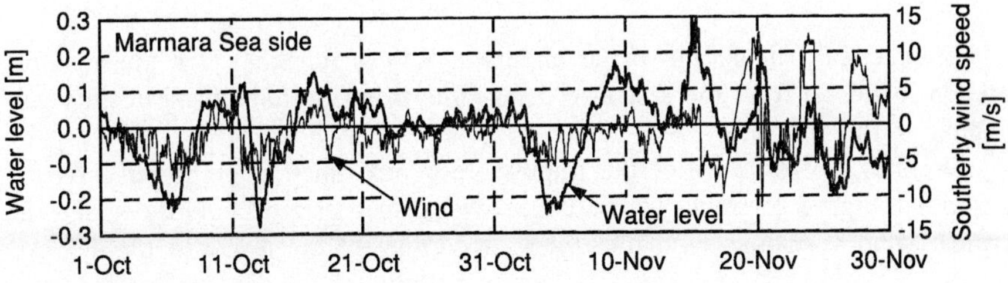


**Fig. 23** Seawater level and wind speed at the Marmara sea side

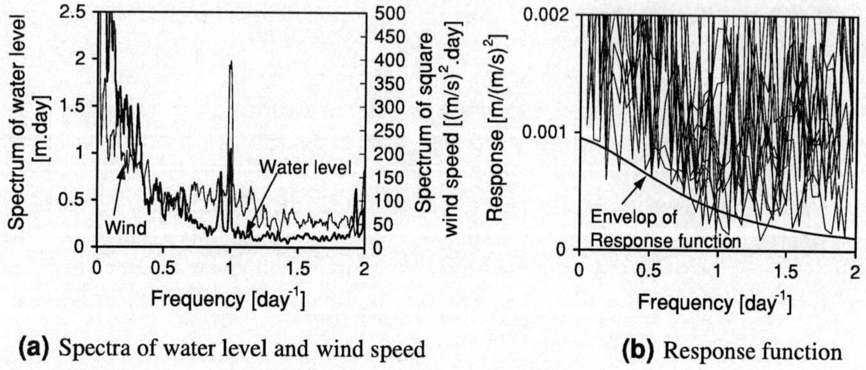


**Fig. 24** Spectra of the seawater level and the wind speed and its response function

infinite. This divergence tendency should be understood as a numerical artifact and the lower value of the response function can be considered more realistic, because only the response of the seawater level induced by (not small or zero but) large wind speed should be included in the analysis. Hence, we proposed to use the response function, which was obtained as the lower envelope curve of the monthly response functions with violet fluctuation, as shown in Fig. 24b. We would present later how this response function performed.

Although response functions are given in the frequency domain, the time domain expression is required for the wind effects modeling. Thus an impulse response function, which is derived from the frequency response function, is introduced

$$K(\tau) = 4 \int_{0}^{\infty} R(f) \cos(2\pi f \tau) df \tag{3}$$

and the time series of the seawater level can be calculated by convolving the impulse response function and the time series of the wind speed.

$$\eta_{\text{wind}}(t) = \int_{-\infty}^{t} K(t-\tau)|W(t)|W(t) d\tau \tag{4}$$

For practical calculation of the impulse response functions, the phase information, obtained from the response delay time from the monitoring results was applied in this model.

Hereafter, the efficacy of the impulse response function proposed above is described. Figure 25 shows the seawater level responses with bold lines, and the typical wind variations with thin lines. According to the response for the step functional wind variation shown in the upper subfigure, we can confirm the basic characteristic of this model reproducing the response delay, which is 32 h here and

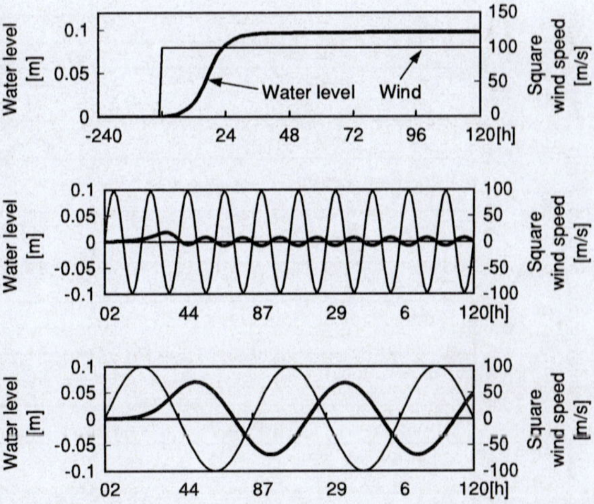


**Fig. 25** Parametric calculation for the wind component model

is corresponding to the monitoring data in the Marmara Sea side. As for the response for the cyclic wind variation of periods of 12 and 48 h, the middle and the bottom sub-figures depicts that the wind set-up does not fully develop by the 12-h periodic wind variation, while it develops substantially by the 48-h periodic wind variation. Yüce (1993) pointed out that the sea-level variation is only affected by winds sustained over 1 or 2 days in the Marmara Sea and the model proposed here can reproduce this phenomenon.

### 6.4.4 Long Periodic Variations

The contributions expressed from the above-mentioned models vary in a relatively short period of days. The seawater level variations shown in Fig. 17 include longer periodic ones as well. It is considered to be caused mainly by inflow into the Black Sea of rainwater and/or melted snow water entering the Black Sea as river runoff. Therefore it changes from year to year and cannot be predicted. It is reasonable to assume that the variation is quite slow and relatively constant over a short duration, say 48 h forecasting time period. In this model, the long periodic component was estimated at the beginning of the forecast procedure from the observed seawater level and the predicted short periodic components using corresponding models mentioned above. Note that it contains other components not considered in this model and also errors within each component model.

### 6.4.5 Validation of the Seawater Level Forecast Model

Figure 26 shows calculation results of the seawater level prediction models. In Fig. 26, the seawater models of the air pressure component, the wind component

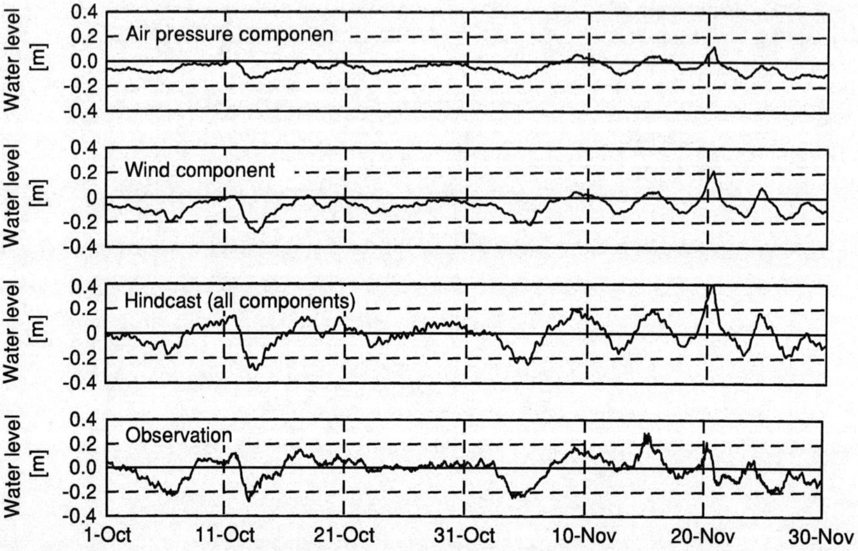


**Fig. 26** Seawater level forecast model (October to November, Station D)

and the total of all components, which includes the tidal component and the long periodic variation predicted by 30-day moving average of the water level monitoring data. The air pressure and wind components are calculated from the monitoring results using their corresponding models mentioned above. According to Fig. 26, the air pressure and wind components are comparably large in the Marmara Sea side and these two components are found to be principal dominant factors. On the other hand, the air pressure component in the Black Sea side is much smaller (as shown in Fig. 21) and only the wind component is dominant.

Validation of the seawater level model for both the Marmara Sea side and the Black Sea side are shown in Fig. 27. Hindcasts calculated using the monitoring results of the air pressure and the wind, and monitoring results are presented in the figure. The monitoring results varied irregularly with large amplitude and the hindcast reproduces such phenomenon with good accuracy. However, the model sometimes overestimates the seawater level fluctuation at the Marmara Sea side (e.g. early February and early April). In these situations, the seawater level of the Black Sea side increases significantly and the discharge from the Black Sea to the Marmara Sea, which makes the seawater level of the Marmara Sea higher, may become large. This model does not take the effect of this inflow from the Black Sea into account, thus the seawater level lowering at the Marmara Sea side may be overestimated. Note that this estimation error results in the predicted Bosphorus Strait current being stronger than the actual condition encountered, thus is on the conservative (safe) side.

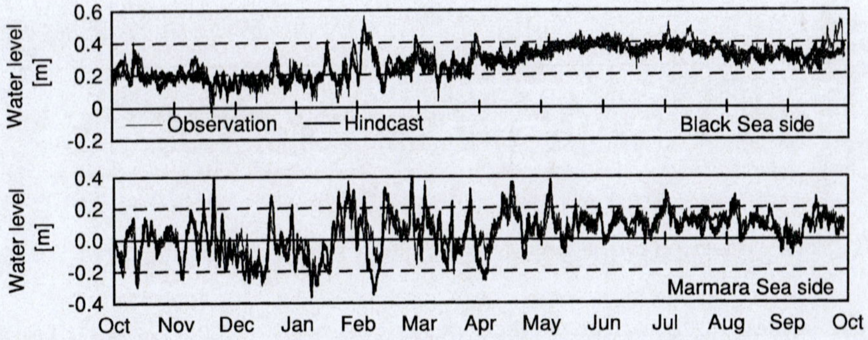


**Fig. 27** Verification of the seawater level forecast model (Stations E and D)

**Fig. 28** Relationship between seawater level difference and current speed

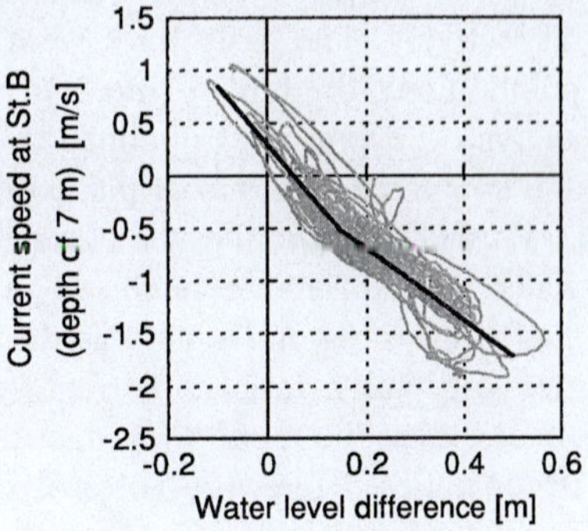


## *6.5 Seawater Current Modeling*

### 6.5.1 Current Model Based on the Basic Line

Figure 28 shows the monitoring result of the relationship between the seawater level difference and the current speed at 7 m depth of Station B. Theoretically, for a simple channel without density distribution effects, the square of speed should be proportional to the water level difference. In the Bosphorus Strait, however, the two-layer current effect causes the behavior to deviate from the simply theory. Figure 28 shows a strong correlation between the current and the water level difference though with some dispersion. The scatter shown in the figure may be primarily due to the stratification effect. As a first step of constructing the current model, the data was fitted by an optimal continuous multi-linear line (called the basic line here) shown in Fig. 28. Although the basic line may be defined as the typical relation between the seawater level difference and the current velocity, there is dispersion of the relation on the fringe of the basic line because of the

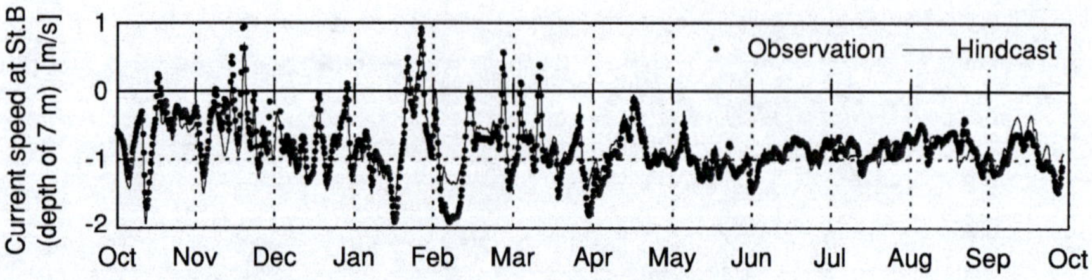


**Fig. 29** Hindcasting of current speed by means of basic function

unsteadiness of the current and the seawater density distribution. Using the basic line, the current speed can be approximated from the seawater level difference as,

$$u_{\text{estimation}} = f(\Delta\eta) \tag{5}$$

where u and $\Delta\eta$ indicate the current speed and the seawater level difference, respectively. The basic lines were predicted with respect to any depths and any points along the tunnel line, and were obtained through an iterative regression analysis. Several lines obtained from a variety of procedures had been examined and compared in terms of prediction capabilities.

Using the basic line, the current speed hindcasted from the monitoring data of water level difference is shown in Fig. 29. The hindcasted current speed shows almost the same movement as the monitoring result with some slight discrepancies, especially, in large current velocity events. The estimated current speed is smaller than the monitoring data there, hence not conservative and likely unsafe for operational forecast. To resolve this error, the current model was modified in the following section.

### 6.5.2 Current Model Modification

With the basic line shown in Fig. 28, the seawater level difference between the Marmara Sea side and the Black Sea side can be calculated from the monitoring data of current speed using the inverse relationship of Eq. 5,

$$\Delta\eta^* = f^{-1}(u_{\text{observation}}) \tag{6}$$

Here, the modification factor, $\Delta\eta_r$, to adjust the discrepancy between the actual seawater level difference, $\Delta\eta$, and the estimated one calculated by Eq. 6, $\Delta\eta^*$, is defined as,

$$\Delta\eta_r = \Delta\eta - \Delta\eta^* \tag{7}$$

The modification factor, $\Delta\eta_r$, reflects the difference between the actual relation and the typical relation expressed by the basic line, and it represents the difference between the basic line and the monitoring data on the horizontal axis in Fig. 28. Figure 30 shows the time series of $\Delta\eta$, $\Delta\eta^*$ and $\Delta\eta_r$. The figure shows

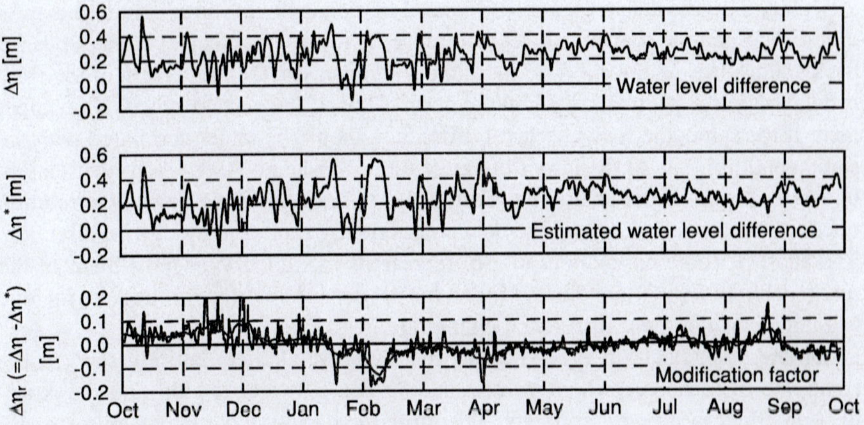


**Fig. 30** Modification of the water level difference

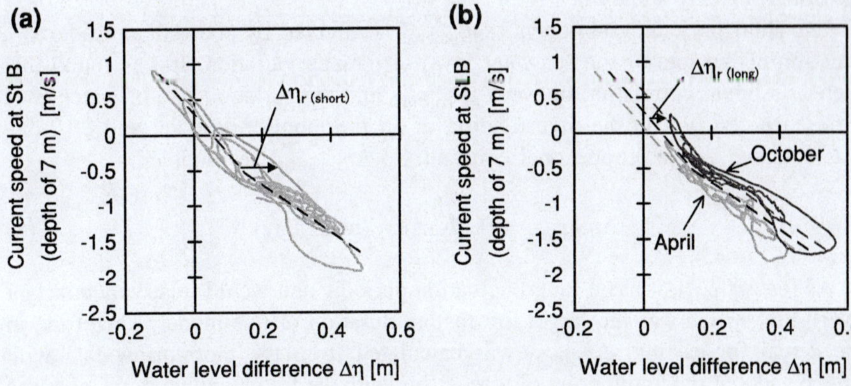


**Fig. 31** Description of the seawater level difference modification

that the modification factor, $\Delta\eta_r$, is composed of two periodic components, the short one over a day or so and the longer one over weeks or a month, which is depicted in bold line in the figure and has been obtained by means of 10-day moving average.

Here, the physical meaning of the modification factor of the seawater level difference, $\Delta\eta_r$, is explained. Figure 31 shows the relationship between the water level difference and the current speed as same as Fig. 28. Figure 31a depicts the 3 months consecutive data from December to January and Fig. 31b shows the comparison of the 2 months on October and April. The two figures delineate the short periodic nature of $\Delta\eta_r$ and the long one, respectively. The discrepancies from the basic line appeared in Fig. 31, as expected, are due to the dynamic behavior of the stratification, described as follows.

In Fig. 31a, the data traces an oval shape in the clockwise direction on the order of a day. When the seawater level difference increases, the density front is pushed back southward. As the short periodic motion increases, the movement of the front is delayed and it stays relatively in the north. Thus, if the current speed in the upper layer is the same, the seawater level difference should be larger compared with the static condition due to the larger effect of the lower layer. As a result, the relation traces right side of the basic line. Contrarily, in the decreasing phase, the relation traces along the left side, and the data consequently rotates in clockwise direction. The short periodic component of $\Delta\eta_r$ represents this horizontal movement of the relationship in Fig. 31a and is the index for the density interface front delaying due to rapid seawater level difference changing.

On the other hand, in Fig. 31b, the data curve in October entirely stays at the right side of that in April. In winter, the surface seawater is cold and the stratification effect is strongly apparent. Therefore the density front may stay relatively in the north and the current speed is expected to decrease significantly. As a result, the relation line in winter is located on the right side. The long periodic component of $\Delta\eta_r$ means this horizontal movement and is the index of the long durational (seasonal) density situation.

The short periodic component, $\Delta\eta_{r(\text{short})}$, is affected by and thus considered a function of the rate of the seawater level difference variation. It was found that there is a linear correlation between $\Delta\eta_{r(\text{short})}$ and the variation rate of water level difference $\partial(\Delta\eta)/\partial t$ in the investigation of all the monitoring data, and 0.17 day was obtained as the proportional constant. So $\Delta\eta_{r(\text{short})}$ was modeled as,

$$\Delta\eta_{r(\text{short})} = 0.17\frac{\partial}{\partial t}\Delta\eta \quad (\text{unit : day}) \tag{8}$$

As for $\Delta\eta_{r(\text{long})}$, it has a long fluctuation periodic and would relatively constant over 48 h, which was the target forecasting duration of this model. Therefore, in the actual forecasting, $\Delta\eta_{r(\text{long})}$ was calculated from the monitoring data with Eqs. 6 and 8 at the initial point of forecasting with the following equation, and was assumed to be constant in the forecasting duration.

$$\Delta\eta_{r(\text{long})} = \Delta\eta_{\text{observation}} - \Delta\eta^* - \Delta\eta_{r(\text{short})} \tag{9}$$

Consequently, the modified current model using the modified water level difference, $\Delta\eta - \Delta\eta_r$, which is the prediction of $\Delta\eta^*$, is expressed as,

$$u = f(\Delta\eta - \Delta\eta_r) = f(\Delta\eta - \Delta\eta_{r(\text{short})} - \Delta\eta_{r(\text{long})}). \tag{10}$$

### 6.5.3 Validation of the Current Forecast Model

Hindcasts of the seawater current speed shown in Fig. 29 is re-analyzed with the modified current model mentioned above. The result is shown in Fig. 8. Here, $\Delta\eta_{r(\text{short})}$ is directly calculated from the monitoring water level difference with

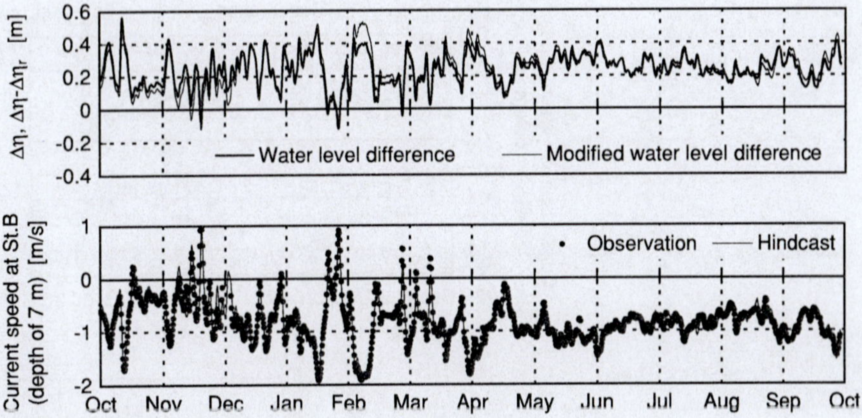


**Fig. 32** Verification of the modified current forecast model

Eq. 8. As for $\Delta\eta_{r(long)}$, in the actual forecasting, the value obtained at the beginning point of the forecasting with the monitoring data was applied as the constant value during the forecasting duration of 48 h. Hence the values of 48 h before each hindcasting time is calculated with Eq. 9 and used for the hindcasting in Fig. 32. The hindcast of the current speed shows good agreement with the observed data. The dispersions of error defined in the following equation are 0.838 for Fig. 29, and 0.118 for Fig. 32.

$$\overline{Error} = \frac{1}{n}\sum_{n} (u_{\text{observation}} - u_{\text{hindcast}})^2 \qquad (11)$$

where n is number of data. It is confirmed that the current model has been improved well by using the modified water level difference.

## *6.6 Forecasting Results*

For forecasting, monitoring data at the present/past time and weather forecast are required. In this mega-project, the monitoring data of the seawater levels and relevant weather conditions (air pressure and wind speed) at the two ends of the strait, and the current speed at a reference point were obtained at the monitoring stations presented in Fig. 16. The weather forecast was obtained as 48-h forecast from the local weather service.

First, seawater level forecasting was executed for both ends of the strait. The tidal components, the air pressure components and the wind components at the present time on both ends of the Strait were directly calculated with the corresponding models with the monitoring data. The forecast of these components were also calculated in the same manner. The long periodical variations (seasonal

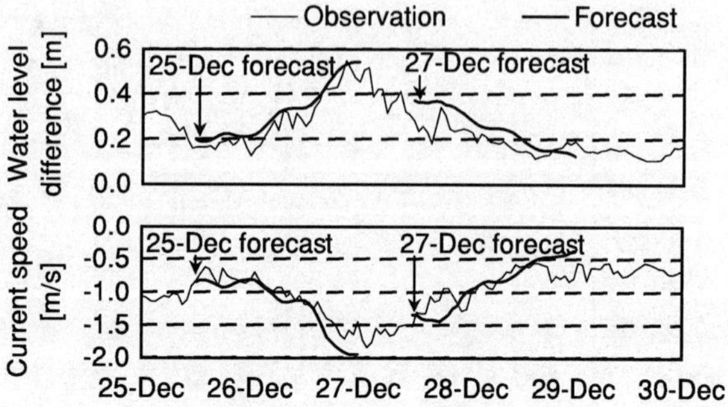


**Fig. 33** "Go/no-go" decision of the tunnel immersion by the current forecasting

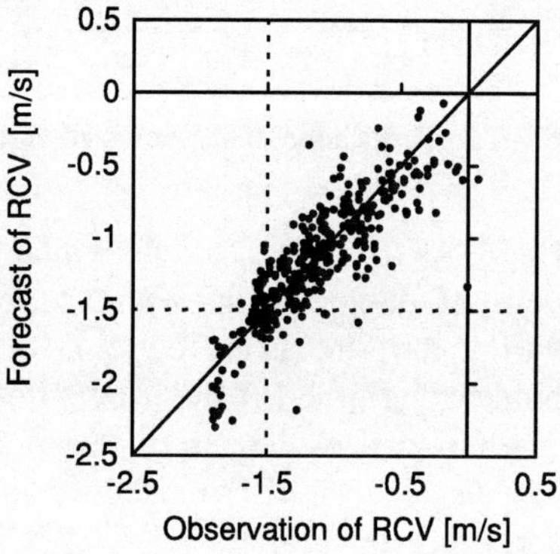


**Fig. 34** Current forecasting results

components) were calculated as the differences between the all above-mentioned components and the water level monitoring data at the present time, and were assumed to be constant for the next 48 h.

Then the seawater current speed at a selected single current monitoring station, called reference point, and the monitoring water level difference were used to calculate the long periodic component of $\Delta\eta_r$, $\Delta\eta_{r(long)}$, and the short periodic one, $\Delta\eta_{r(short)}$, was estimated for the forecasting time with Eq. 8. Finally, the modified water level difference and the current speed were forecasted with the corresponding basic lines defined at selected points along the tunnel line with Eq. 10.

An example of the application of this model is shown in Fig. 33. It presents the decision practice when the immersion training session was carried out. The forecast issued on December 25, 2006 was that current speed was increasing and became larger than 1.5 m/s, the threshold value, on the next day. Thus a "no-go" decision was made and the session was postponed. The forecast on 27th was that the current speed was decreasing and became less than 1.5 m/s, so that a "go" decision was made. In comparison with the actually observed data, it is found that the forecast was accurate and the "go/no-go" decisions were made correctly.

Figure 34 shows a comparison between the observed data and the forecasted data over the entire project period. In Fig. 34, reference current velocities (RCVs), which were an average of the 15 m surface layer current speed above the where the tunnel element was due to be immersed, were compared. The data show the maximum values of 48-h daily forecasting and the monitoring during the same duration. The data closely align with the diagonal line with a high rate of 91% of correct "go/no go" decisions. Note that in the range of the observed value of over 1.5 m/s, the forecast values are a bit larger than the observed values in Fig. 34. The water level at the Marmara Sea side was sometimes estimated to be lower than the actual condition, as shown in Fig. 27. Therefore the forecasted current speed might be overestimated. The error was however on the conservative (safe) side for marine construction operation.

## 7 Conclusion

The idea of a railway tunnel crossing the Bosporus Strait was first raised in 1860. However, it was impossible to construct the tunnel under such large depth by the mega-engineering building techniques available more than a century ago. Later, the method of the immersed submarine tunnel was developed since late in the 19th century, which gave mega-engineers the technical capability. Nevertheless, construction of an immersed tunnel has long been hindered by the deep water with strong complex current structure present in the Bosphorus Strait. It is believed that this is a condition that is the most difficult ever experienced.

In this chapter, a number of techniques developed to overcome such severe conditions were presented. Especially, the current information was the most important for the marine operations. It was cross-sectionally predicted with the numerical simulations and the monitoring data including the current velocities at the certain three points. Moreover the seawater current forecast system was developed as a prompt, stable and robust system based on the regression analysis with the monitoring result, not depending on the numerical calculation, which is the present mainstream to predict natural phenomena but the computation may be too time consuming. The seawater current forecast system played a key role for achieving safety and optimizing the submarine construction operation. A variety of techniques, including those not presented here, were examined. Immersion of the last Bosphorus Strait submarine tunnel element was successfully completed by 2008. We believe that the mega-project planning and construction techniques developed here can be applicable not only to immersed tunnels but also to other various kinds of marine and underwater constructs.

## References

Arisoy Y, Akyarli A (1989) Long term current and sea level measurements conducted at Bosphorus. In: Pratt LD (ed) The physical oceanography of sea straits, NATO/ASI series. Kluwer, Dordrecht, pp 225–236

Ducet N, Le Traon PY, Gauzelin P (1999) Response of the Black Sea mean level to atmospheric pressure and wind forcing. J Marine Syst 22:311–327

Gil E, de Toro C (2005) Improving tide-gauge data processing: a method involving tidal frequencies and inverted barometer effect. Comput Geosci 31:1048–1058

Grantz WC (1993) Catalog of immersed tunnels. Tunn Undergr Space Technol 8:175–263

Gregg MC, Özsoy E (2002) Flow, water mass changes, and hydraulics in the Bosphorus. J Geophys Res 107(C3):2.1–2.23

Gregg MC, Özsoy E, Latif MA (1999) Quasi-steady exchange flow in the Bosphorus. Geophys Res Lett 26:83–86

Ito K, Oda Y, Honda T, Koyama F (2009) Construction techniques for deep water immersed tunnel in the Bosphorus Strait. In: Chung JS, Prinsenberg S, Hong SW, Nagata S (eds) The nineteenth international offshore and polar engineering conference. International society of offshore and polar engineers, California, pp 217–224

Jensen HR, Babovic V, Rasmussen EB (2000) Real-time current forecasts for tunnel element towing in Øresund. In: Øresundsbro konsortiet (ed) Øresund link immersed tunnel conference. Øresund tunnel contractors and international tunnelling association, Copenhagen, p D8

Komatsu T, Ohgushi K, Asai K (1997) Refined numerical scheme for advective transport in diffusion simulation. J Hydraul Eng (ASCE) 123:41–50

Lesser GR, Roelvink JA, van Kester JATM, Stelling GS (2004) Development and validation of a three-dimensional morphological model. Coast Eng 51:883–915

Lykke S, Belkaya H (2005) Marmaray project: the project and its management. Tunn Undergr Space Technol 20:600–603

Lykke S, Kerk F (2005) Marmaray project: marine operations, the Bosphorus crossing. Tunn Undergr Space Technol 20:609–611

Oda Y, Ito K, Ueno S, Sakaeda H (2005) Hydraulic physical model test of mooring for the immersed tunnel construction under the Bosphorus Strait. In: Choi BH, Suh KD, Yoon SB (eds) Third international conference on Asian and Pacific coasts. Hanrimwon, Korea, pp 260–274

Spellman FR (2009) The science of air: concepts and applications. CRC Press, Boca Raton, FL, p 291

Tanaka M, Serafy GE, Gerritsen H, Adachi T (2005) Development of a new real time flow forecast model in estuary based on ensemble kalman filter (in Japanese). J Coast Eng (JSCE) 52:321–325

Yüce H (1993) Water level variations in the sea of Marmara. Oceanol Acta 16:335–340

# Ecological Energy Conversion of Oceanic and Afferent River Water Currents

**Mircea Dimitrie Cazacu and Sergiu Nicolaie**

## 1 Small and Great Power Plants Using the Hydrokinetic Energy of Flowing Seawater/Freshwater Without Damming

The special importance of these hydroelectric hydrokinetic power plants consists, such as in energy production from renewable and non-polluting sources, as by their ecological building in the absence of dams so as not to hinder the normal circuit of migratory fishes as well as the passage of ships (Cazacu 1999).

To extract the maximum mechanical power at the axial rotor-shaft we have applied our original method (Cazacu 2005a, b, 2006, 2007, 2008; Cazacu and Nicolaie 2002; Nicolaie 2009), which permits us to obtain the best profile shape and its settlement relative angle for any blade radius.

Due to the important role of the turbo-machine efficiency, concerning the energy economy and environment pollution diminishing (Cazacu 1999), we shall present here two original methods to maximise their performances, establishing the optimum blade profile and its setting angles at different blade radius.

### *1.1 Hydropower by a Run-of-River Hydraulic Turbine*

Projecting the two components of hydrodynamic resultant on the rotational peripheral direction, we shall obtain the mechanical power expression $P$ (Fig. 1):

M. D. Cazacu (✉)
Polytechnic University of Bucharest, Bucharest, Romania
e-mail: cazacumircea@yahoo.com

S. Nicolaie
National Institute for R&D in Electrical Engineering, Bucharest, Romania

V. Badescu and R. B. Cathcart (eds.), *Macro-engineering Seawater in Unique Environments*, Environmental Science and Engineering,
DOI: 10.1007/978-3-642-14779-1_9, 

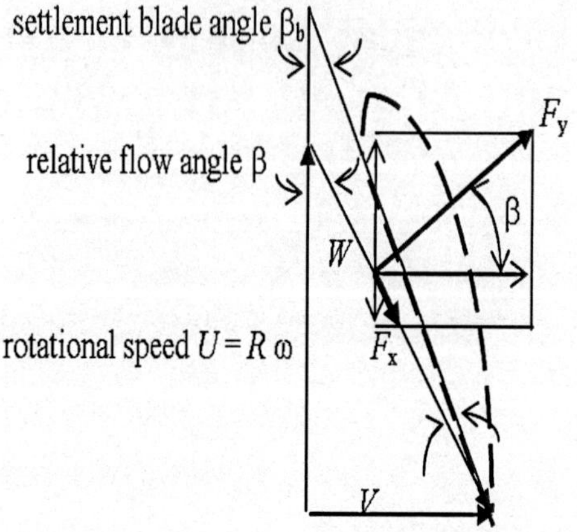


**Fig. 1** Velocity triangle and hydrodynamic resultant components

$$P = UF_u = U\big(F_y \sin\beta - F_x \cos\beta\big) = \frac{\rho}{2}V^3 bl\left[c_y(i)\frac{\cos\beta}{\sin^2\beta} - c_x(i)\frac{\cos^2\beta}{\sin^3\beta}\right] \quad (1)$$

where $U$ is the rotational speed, $F$ is the force acting on the blade, $\beta$ is the relative flow angle, $\rho$ is the mass density of blade's material, $V$ is the wind velocity, $b$ and $l$ are the width and depth of wing, while $c_x(i)$ and $c_y(i)$ are the lift and resistance coefficients, both of them as functions of the attack angle $i$. The subscripts $u$, $x$ and $y$ in Eq. (1) denote the projections on different directions.

Cancelling the partial derivative of $P$ given by Eq. (1) with respect to the relative angle $\beta$ one finds:

$$\frac{\partial P}{\partial \beta} = -c_y(i)\frac{1+\cos^2\beta}{\sin^3\beta} + c_x(i)\frac{\cos\beta(2+\cos^2\beta)}{\sin^4\beta} = 0, \quad (2)$$

Using the notation $\sin^2\beta = x$, Eq. (2) turns into the algebraic equation

$$P(x) = \left[f^2(i)+1\right]x^3 - \left[4f^2(i)+7\right]x^2 + \left[4f^2(i)+15\right]x - 9 = 0, \quad (3)$$

from which the under-unitary solution is the one which maximises the power for any chosen profile shape. Once more, introducing these values $i$ and $\beta$ in the power expression (1), the maximal power value will indicate the best profile to use. In the present case, the best solution is given by the profile Gő 450 (Hutte 1947) (Tables 1, 2).

## 1.2 The Best Incidence Angle of Blade Profile for Other Radii

Applying the relation $V = Utg\beta$ at the outskirts, we obtain the optimal angular velocity $\omega$, which being the same for all the blade radius $r$, determines the rotation velocity at any other radius $R_j$ and because the relative angle will be done $\beta$ $(r)$, the power maximization will be obtained only by the variation of the incident angle in the case of the considered profile.

**Table 1** Various quantities related to the optimum blade profile Gő 450

| $i$ (°) | $c_y$ | $c_x$ | $f = c_y/c_x$ | $x = \sin^2\beta$ | $P(x) \approx 0$ | $\beta$ (°) | Power (W) | U/V[a] |
|---|---|---|---|---|---|---|---|---|
| −3 | 0.20 | 0.023 | 8.70 | 0.02917 | $2.75 \times 10^{-5}$ | 9.83 | 2.27 | 5.77 |
| 0 | 0.41 | 0.020 | 20.50 | 0.00534 | 0.000318 | 4.19 | 25.59 | 13.65 |
| 3 | 0.63 | 0.032 | 19.69 | 0.00578 | $1.83 \times 10^{-6}$ | 4.36 | 36.28 | 13.11 |
| 6 | 0.85 | 0.055 | 15.46 | 0.00936 | 0.00012 | 5.55 | 30.22 | 10.29 |
| 9 | 1.05 | 0.081 | 12.96 | 0.01327 | $3.44 \times 10^{-6}$ | 6.62 | 26.31 | 8.62 |
| 12 | 1.15 | 0.112 | 10.27 | 0.02104 | $3.70 \times 10^{-5}$ | 8.34 | 18.15 | 6.82 |
| 15 | 1.21 | 0.147 | 8.23 | 0.03249 | $5.22 \times 10^{-6}$ | 10.4 | 10.38 | 5.46 |

[a] $V$ is the wind velocity

**Table 2** Optimum parameters of the blade profile as function of relative radius $r$

| Relative radius $r$ | Relative angle $\beta$ (°) | Optimal angle $i_{opt}$ (°) | Settlement blade angle $\beta_b$ (°) | Blade twist $\delta\beta$ (°) |
|---|---|---|---|---|
| 1 | 16.87 | 11.54 | 5.33 | 0 |
| 0.9 | 18.62 | 11.85 | 6.77 | 1.44 |
| 0.8 | 20.76 | 12.16 | 8.59 | 3.26 |
| 0.7 | 23.42 | 12.47 | 10.94 | 5.61 |
| 0.6 | 26.81 | 12.78 | 14.03 | 8.69 |
| 0.5 | 31.23 | 13.09 | 18.14 | 12.81 |
| 0.4 | 37.17 | 13.39 | 23.77 | 18.44 |
| 0.3 | 45.31 | 13.69 | 31.62 | 26.28 |
| 0.2 | 56.60 | 13.99 | 42.61 | 37.28 |

The extracted power $P_i$ is a function of the incidence angle $i$:

$$P_j = \frac{V}{tg\beta_j}\frac{\rho}{2}V^2 bl_j(R_j)\left[c_y(i)\frac{1}{\sin\beta_j} - c_x(i)\frac{\cos\beta_j}{\sin^2\beta_j}\right] = \frac{\rho}{2}V^3 bl\left[A(R_j)c_y(i) - B(R_j)c_x(i)\right] \quad (4)$$

For a flow channel, placed at radius $r_j = R_j/R_p$ (where $R_p$ is the peripheral radius), we obtain the maximization of the extracted power and the blade relative angle $\beta_b = \beta - i$ by cancelling the partial derivative of $P_i$ with respect to the incidence angle $i$

$$\frac{\partial\left[A(R)c_y(i) - B(R)c_x(i)\right]}{\partial i} = 0 = \left[A\left(c_{y1} - 4i^3 c_{y4}\right) - B\left(c_{x1} + 2ic_{x2}\right)\right] \quad (5)$$

Here we considered the usual expressions of variation with the incidence angle of the lift and drag coefficients, assumed to have the form:

$$c_y(i) = c_{y0} + ic_{y1} - i^4 c_{y4} \quad \text{and} \quad c_x(i) = c_{x0} + ic_{x1} + i^2 c_{x2} \quad (6)$$

After some algebra one obtains the relationship giving the best incidence angle for any radius

$$i^3 + i\frac{\omega_{\text{opt}}c_{x2}}{2Vc_{y4}}R + \frac{\omega_{\text{opt}}c_{x1}}{4Vc_{y4}}R - \frac{c_{y1}}{4c_{y4}} = 0 \tag{7}$$

where $\omega$ is the blade angular velocity. Note that the optimum incidence angle $i$ increases by decreasing blade radius. This allows us to obtain a larger velocity around the carefully selected profile (Cazacu 2005a, b, 2006, 2007, 2008; Nicolaie 2009).

For a three blade rotor the value of the power coefficient is $C_p = 0.42$ while for a four blade rotor the coefficient is $C_p = 0.56$ (Cazacu 2005a, b, 2006, 2007, 2008; Nicolaie 2009). These very high values validate the maximization method.

## 1.3 Lighting Buoys with Energetic Autonomy

A first realization in the field was the luminous signalling system for safe navigation on rivers. Devoted to the security of the Danube River traffic, this type of lighting buoy with energetic autonomy was tested in the Glass Window Channel (Cazacu et al. 2003a, b, c) in the Hydropower Laboratory *Professor Dorin Pavel* of the Politehnica University in Bucharest (Romania), Faculty of Power Engineering, Department of Hydraulics, Hydraulic Machines and Environment Protection (Fig. 2). It was realised in collaboration with the National Institute for R&D in Electrical Engineering ICPE-CA, in Bucharest. ICPE manufactured the axial rotor

**Fig. 2** The hydroelectric unit tested in the Glass Window Channel of the Polytechnic University of Bucharest, Romania

**Fig. 3** Launching on the Danube River of the lighting buoy with 100% energy autonomy

blades, designed after an original method, which assure the extraction of the maximal mechanical power from the inherent kinetic energy of the flowing liquid, as well as the submersible electric generator, endowed with permanent magnets executed from the rare-earths with Neodymium–Boron–Iron (Mihaiescu et al. 2006).

The system ensures autonomy of energy for the signalising lighting buoys on the Romanian part of the Danube River ($\sim$1,000 km long). Presently, one needs to change the electric accumulators (batteries) of about 300 buoys and to carry them ashore to be recharged. Our system allows elimination of these heavy accumulators, but supposed to loss by common robbery. Also, our system will stop the classical fuel consumption and personnel necessary to the servicing fleets which continuously navigate, as well as environmental pollution (Fig. 3).

This realization has been patented in Romania. It was presented at the 32nd International Exhibition of Inventions, New Techniques and Products of Geneva in 2004. There, it was awarded with a Bronze Medal. Simultaneously, it received the Praise of the Society for Promotion of Recoverable, Inexhaustible and New Energies—SPERIN, for his contribution to the sustainable development of the human societies.

## *1.4 The Ecological Small Hydropower Plants on Rivers*

Presently, we are working to implement the run-of-river hydrokinetic power plants (without dams) in collaboration with the Republic of Moldova on the Prut River

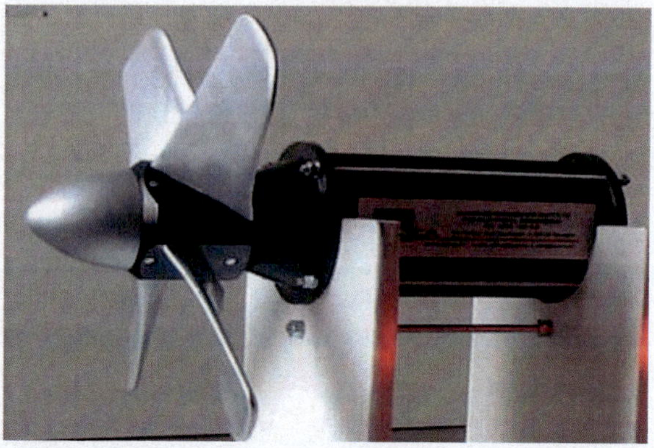

**Fig. 4** Hydraulic turbine rotor directly coupled with the submersible electric generator

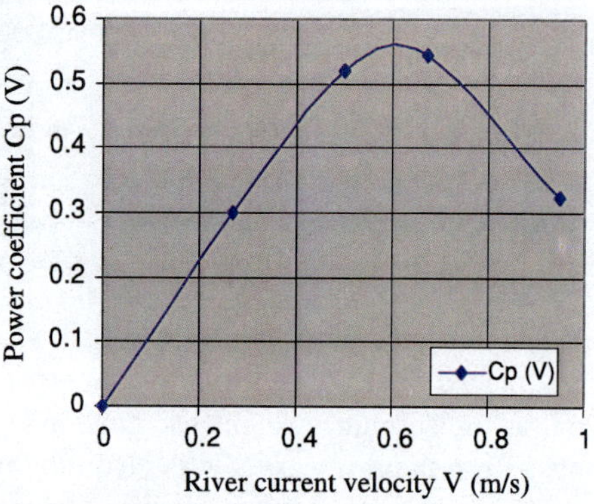


**Fig. 5** The variation of the power coefficient with the current velocity, simultaneously with the special economic advantages due to the great number of rotations and elimination of the cavitations phenomenon by direct coupling with the submersible electric generator, endowed with permanent magnets

and in Bulgaria on the Danube River. For these Micro Hydraulic Power Plants we propose two variants (Cazacu et al. 2005; Cazacu and Nicolaie 2005):

- The first version is represented in Fig. 4, where we have experimented with the new 4-bladed axial-free rotor, designed by our original method to extract the maximum mechanical power at the rotor shaft from the hydrokinetic river energy. We obtained a maximum power coefficient value $C_p = 0.56$ (Fig. 5);
- The second version has the rotor placement in a hydrodynamic housing as in the Fig. 6, with the goal to increase the fluid kinetic energy and also to extract more mechanical power at the turbine shaft, taking into account the same rotor area.

## 2 The Ecological Hydropower Station in the Bosphorus Strait

A short history of the Bosphorus Strait genesis may be consulted in the beginning of Chap. 13 of this book (also, see Ozsoy et al. 2002. There, adjacent to the

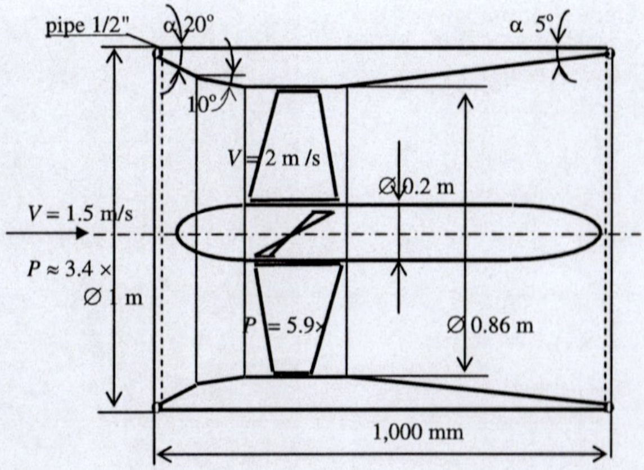


**Fig. 6** The scheme of the micro-hydraulic turbine placed in a profiling housing

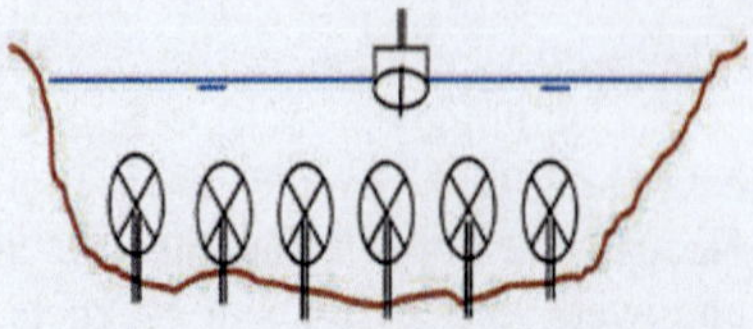

**Fig. 7** Scheme of the hydroelectric power arrangements in Bosphorus Strait

sea-bottom, a permanent deep seawater current (from the Mediterranean Sea to the Black Sea) flows. There, mega-engineers can place many hydrokinetic hydroelectricity production units (Fig. 7). This is a very favourable place for such hydroelectric units, taking into account of the large seawater velocities and the fact that there is no need of building barrages. Consequently, one allows unobstructed passage for ships and fishes.

The idea of such a mega-project came from our famous professor Dorin Pavel (1900–1979), a member of the Romanian Academy of Sciences, the greatest Romanian hydropower engineer, who realised the hydropower station on the Danube River, at Iron Gates. He lectured us on the more important long-proposed massive hydropower station such as that placed at Gibraltar Strait (see, for example, Cathcart 1995). Then, we thought on the special building of such hydraulic turbine, on its actual afferent transport, mounting and exploitation conditions.

## 3 Ecological Boats for Traffic Inside the Romanian Biosphere Reservation Danube River Delta

Another of our recent realizations is the Project: Hydro Energetic Systems for the Conversion, Storage and Distribution of Renewable Energies, devoted to the Ecologic Fluvial Transport from the protected Areas. This project has been exposed at the Romanian International Fair during 2007 and was awarded by the Romanian Research, Education and Youth Ministry with the 1st Praise.

**Fig. 8** General view of the ecological boat at the Romanian International Fair

The ecological boat is presented in Fig. 8 and consists in a two-body craft of the catamaran type, having on its roof photovoltaic panels to generate a continuous electric current for the electric accumulators and submersible electric motor, endowed with permanent magnets, which drive a specially-designed propulsion screw by our performing maximization method, mounted in the front of the boat to eliminate the rudder (Cazacu and Nicolaie 2007; Cazacu et al. 2006).

For the nighttime, when the boat is anchored at a landing pontoon, the hydroelectric unit lanced in the kinetic river energy, and composed by an efficient axial turbine rotor, directly coupled with a submerged electric generator, also with permanent magnets, can store this energy in the electric accumulators.

Due to the important role of the turbo-machines efficiency, concerning the energy economy and the environment pollution reduction (Cazacu 2007, Cazacu et al. 2007, 2008; Nicolaie 2009), we shall present the original method to maximise their performance, establishing the optimum blade profile and its setting angles at different radii.

### *3.1 Classical Expressions of Lift and Drag Components of the Hydrodynamic Resultant*

Taking into account the expressions of the lift and drag components of the aerodynamic and/or hydrodynamic resultant force (Fig. 9),

$$F_y = c_y(i)\frac{\rho}{2}W^2bl(R), \quad F_x = c_x(i)\frac{\rho}{2}W^2bl(R), \tag{8}$$

exerted on the profiled blade, laid at the incidence angle $i$ with respect to the relative angle $\beta$, corresponding to the relative velocity $W$ from the velocity triangle, we can then calculate the axial component of these forces, representing the propulsion force:

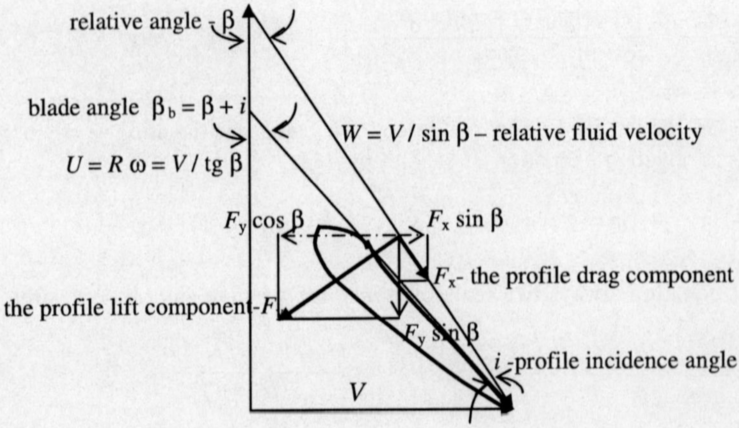


**Fig. 9** The velocity triangle and the hydrodynamic resultant components

$$F_a = F_y \cos\beta - F_x \sin\beta = \frac{\rho}{2} V^2 bl(R) \left[ c_y(i) \frac{\cos\beta}{\sin^2\beta} - c_x(i) \frac{1}{\sin\beta} \right] \tag{9}$$

We can also calculate the expression of the negative shaft driving mechanical power:

$$P_m = U\left(F_y \sin\beta + F_x \cos\beta\right) \quad \text{or} \quad p_m = \frac{2P_m}{\rho V^3 bl} = c_y(i) \frac{\cos\beta}{\sin^2\beta} + c_x(i) \frac{\cos^2\beta}{\sin^3\beta} \tag{10}$$

## 3.2 *Maximization of the Ratio Between the Axial Force and the Consumed Mechanical Power*

In this case one start from the following dimensionless ratio (the negative sign is not important in this stage):

$$\frac{f_a}{p_m} = \frac{F_a V}{P_m} = \frac{c_y(i) \frac{\cos\beta}{\sin^2\beta} - c_x(i) \frac{1}{\sin\beta}}{c_y(i) \frac{\cos\beta}{\sin^2\beta} + c_x(i) \frac{\cos^2\beta}{\sin^3\beta}} = \frac{f \cos\beta \sin\beta - \sin^2\beta}{f \cos\beta \sin\beta + \cos^2\beta} \tag{11}$$

By cancelling the partial derivative of this ratio with respect to the relative angle $0 \le \beta \le \pi/2$, and taking into account that the denominator is never equal with infinite one finds:

$$\frac{\partial(f_a/p_m)}{\partial\beta} = \frac{f\left(\cos^2\beta - \sin^2\beta\right) - 2\sin\beta\cos\beta}{f^2 \sin^2\beta \cos^2\beta + 2f \sin\beta \cos^3\beta + \cos^4\beta} = 0$$
$$\rightarrow f\left(1 - 2\sin^2\beta\right) = 2\sin\sqrt{1 - \sin^2\beta} \tag{12}$$

$$\frac{\partial(f_a/p_m)}{\partial\beta} = \frac{fctg^2\beta - 2ctg\beta - f}{\cos^2\beta(f^2 + 2fctg\beta + ctg^2\beta)} = 0 \rightarrow fctg^2\beta - 2ctg\beta - f = 0 \quad (12')$$

This is the maximisation condition. By using the notation $x = \sin^2\beta$, one obtains an algebraic equation of second degree:

$$f(i)x^2 - 2x - f(i) = 4(f^2 + 1)x^2 - 4(f^2 + 1)x + f^2 = 0 \quad (13)$$

This equation always two real solutions, one positive and other negative

$$x_{1,2} = \frac{1 \pm \sqrt{1 + f^2}}{f} = \frac{f^2 + 1 \pm \sqrt{f^2 + 1}}{2(f^2 + 1)} \quad (14)$$

## 3.3 Results for the Case of Many Profiles

By using this method we put into evidence the possibility to enlarge the operation radius of an airplane or a boat, adopting the real solution having before the root the positive sign.

Comparing the values presented in Table 3, obtained for different profiles by this constraint optimisation method for the axial force reported to the mechanical driving power at the shaft of the screw, one put into the evidence the following results:

- The optimum incidence angle $i$ of a profile is the same for both real solutions $x$ (+) and $x$ (−),

**Table 3** Results obtained for various profiles

| Profile | Angle $i$ (°) | $f_a/p_m$ (+) | $p_m$ (+) | $f_a/p_m$ (−) | $p_m$ (−) | $\beta$+ (°) | $\beta$− (°) |
|---|---|---|---|---|---|---|---|
| NACA 4412 | 6 | 0.969 | 1.262 | 0.963 | 3.654 | 45.47 | 28.22 |
| CLARK Y | 5.3 | 0.973 | 1.238 | 0.968 | 3.553 | 45.47 | 28.27 |
| SG 6042 | 2.15 | 0.980 | 1.031 | 0.980 | 1.062 | 45.30 | 44.75 |
| SG 6043 | 2.11 | 0.984 | 1.404 | 0.984 | 1.437 | 45.23 | 44.80 |
| Gö 451 | 3 | 0.892 | 0.849 | 0.874 | 2.969 | 46.64 | 27.05 |
| Gö 450 | 1 | 0.906 | 0.564 | 0.890 | 1.902 | 46.42 | 27.27 |
| Gö 449 | −3 | 0.892 | 0.534 | 0.874 | 1.867 | 46.64 | 27.05 |
| Gö 448 | 3 | 0.853 | 1.310 | 0.830 | 5.083 | 47.26 | 26.42 |
| Gö 447 | 0 | 0.885 | 0.833 | 0.865 | 2.972 | 46.76 | 26.93 |
| Gö 446 | 0 | 0.887 | 0.738 | 0.868 | 2.615 | 46.72 | 26.97 |
| PlanePlateGe | −3 | 1.244 | −0.228 | 1.251 | −0.308 | 48.15 | 41.90 |
| Ro | 5 | 0.784 | 0.499 | 0.789 | 0.729 | 48.40 | 41.64 |
| CurvedPlateGe | 5 | 0.827 | 1.390 | 0.830 | 1.868 | 47.69 | 42.35 |
| Ro | 6 | 0.821 | 1.521 | 0.824 | 2.065 | 47.78 | 42.27 |

- All the realised axial forces reported to the consumed mechanical powers at the shaft of the turbo-machine rotors have approximately the same relative values. For this reason we considered also for any incidence angle $i$ the value of the mechanical power dimensionless Eq. (11), which allows to adopt the profile with the minimal consumed power;
- Excepting the profiles SG 6042 and SG 6043, for which the relative angle $\beta(+)$ and $\beta(-)$ have approximately the same value for both real solutions, for all other profiles these two angles differ appreciably, and the optimised solution must be determined by experimental methods.

## 3.4 Determination of the Optimum Profile Setting Angle for Other Blade Radii

For different blade radii $R_j$, the relative angle $\beta$ is determined by:

$$tg\beta_j = \frac{V}{R_j\omega} = \frac{R_p}{R_j} tg\beta_p \rightarrow \beta_j = \operatorname{arc} tg\left(\frac{R_p}{R_j} tg\beta_p\right) \tag{15}$$

One shall take into consideration all relations that give the expressions of lift and drag coefficients in the case of profile Gö 450 (Hutte 1947). One shall consider the variation of profile lift coefficients as function of the attack angle $i$ (named now *incidence angle*)

$$c_y(i) = c_{y0} + c_{y1}i - c_{y2}i^2 = 0.414 + 0.085,8i - 0.002,18i^2 \tag{16}$$

Also, one shall consider the parabolic variation of the drag coefficients with the incidence angle:

$$c_x(i) = c_{x0} + c_{x1}i + c_{x2}i^2 = 0.02 + 0.004,12i + 0.000,29i^2 \tag{17}$$

Next, we shall maximise the ratio between the axial force and the mechanical consumed power, given in this case by the next Eq. (18):

$$\begin{aligned} \frac{f_a}{p_m} = \frac{F_a V}{P_m} &= \frac{c_y(i)\frac{\cos\beta}{\sin^2\beta} - c_x(i)\frac{1}{\sin\beta}}{c_y(i)\frac{\cos\beta}{\sin^2\beta} + c_x(i)\frac{\cos^2\beta}{\sin^3\beta}} \\ &= \frac{\left(c_{y0} + c_{y1}i - c_{y2}i^2\right)\cos\beta\sin\beta - \left(c_{x0} + c_{x1}i + c_{x2}i^2\right)\sin^2\beta}{\left(c_{y0} + c_{y1}i - c_{y2}i^2\right)\cos\beta\sin\beta + \left(c_{x0} + c_{x1}i + c_{x2}i^2\right)\cos^2\beta} \end{aligned} \tag{18}$$

Here we assumed the blade portions of wing-spread to be $b_j = \delta R =$ constant on the radius, as well as the wing depths $l_j =$ constant. Next, one shall cancel the derivative of Eq. (18) with respect to the incidence angle $i$. One obtains the following condition, taking into consideration that the denominator is ever different by infinite

$$\frac{\partial\left(\frac{f_{aj}}{p_{mj}}\right)}{\partial i}=\frac{\begin{bmatrix}(c_{y0}+c_{y1}i-c_{y2}i^2)\sin\beta_j\cos\beta_j+\\+(c_{x0}+c_{x1}i+c_{x2}i^2)\cos^2\beta_j\end{bmatrix}\begin{bmatrix}(-c_{y1}+2ic_{y2})\sin\beta_j\cos\beta_j+\\+(c_{x1}+2ic_{x2})\sin^2\beta_j\end{bmatrix}}{\left[c_y(i)\sin\beta_j\cos\beta_j+c_x(i)\cos^2\beta_j\right]^2=\text{Denominator}}$$
$$-\frac{\begin{bmatrix}(c_{y1}-2ic_{y2})\sin\beta_j\cos\beta_j+\\+(c_{x1}+2ic_{x2})\cos^2\beta_j\end{bmatrix}\begin{bmatrix}-(c_{y0}+c_{y1}i-c_{y2}i^2)\sin\beta_j\cos\beta_j+\\+(c_{x0}+c_{x1}i+c_{x2}i^2)\sin^2\beta_j\end{bmatrix}}{\text{Denominator}\neq\infty}=0 \tag{19}$$

The optimal values of the incidence angle for the relative radii $r_j$ are obtained by solving the following equation of third degrees

$$\begin{aligned}&\sin^3\beta_j\cos\beta_j\left[(c_{x1}c_{y0}-c_{x0}c_{y1})+2i(c_{x2}c_{y0}+c_{x0}c_{y2})+i^2(c_{x2}c_{y1}+c_{x1}c_{y2})\right]\\&\quad+\sin\beta_j\cos^3\beta_j\left[(c_{x1}c_{y0}-c_{x0}c_{y1})+2i(c_{x2}c_{y0}+c_{x0}c_{y2})+i^2(c_{x2}c_{y1}+c_{x1}c_{y2})\right]\\&=\sin\beta_j\cos\beta_j(C+2Bi+Ai^2)=0\end{aligned} \tag{20}$$

From Eq. (20) one sees that the two real solutions are independent of the peripheral angle $\beta_{\text{per}}$ and also of the relative radius $r_{\text{j}}$. They are given by:

$$\begin{aligned}C=c_{x1}c_{y0}-c_{x0}c_{y1},\quad B=c_{x2}c_{y0}+c_{x0}c_{y2}\quad\text{and}\\A=c_{x2}c_{y1}+c_{x1}c_{y2}\rightarrow i_{1,2}=\frac{-B\pm\sqrt{B^2-AC}}{A}\end{aligned} \tag{21}$$

## 4 Conclusions and Remarks

Our original method to maximise the hydro-mechanical performances of the axial turbo machine rotors, of the turbine as well as of the screw, are very valuable methods which give good resultants concerning energy consumption, as well as the possibility to protect the environment.

## References

Cathcart RB (1995) Mitigative anthropogeomorphology: a revived ‘plan’ for the Mediterranean Sea Basin and the Sahara. Terra Nova 7:636–640

Cazacu MD (1999) Tehnologii pentru o dezvoltare durabila (Technologies for a sustainable development). Academy of Romanian Scientists. The Congress “Dezvoltarea in pragul mileniului al III-lea” (Development in the third millennium brink), Section “Dezvoltarea durabila” (Sustainable development), 27–29 September 1998, Bucharest. Publishing House EUROPA NOVA, pp 533–539

Cazacu MD (2005) Cresterea eficientei tehnico-economice a hidro-agregatelor de mica putere (The technical-economic efficiency growth of the hydro units of small power). A V-a Conferinta Nationala multidisciplinara "Profesorul Dorin Pavel-fondatorul hidroenergeticii romanesti", vol 7, 3–4 June 2005, Sebes, Editura AGIR, Bucharest, pp 195–200

Cazacu MD (2005) Microagregat hidroelectric pentru asigurarea autonomiei energetice a balizelor luminoase sau a unor barci fluviale (Hydroelectric micro aggregate to ensure the energetic autonomy of the lighting buoys or river boats). Review in science, industry, technology, nr. 2, Bucharest, pp 44–45

Cazacu MD (2006) Amenajarea tehnico-economic avantajoasă a micro-potenţialului hidroenergetic (The technical-economic advantageous arrangement of the hydro energetic micro potential). A 4-a Conf. Dorin Pavel a Hidroenergeticienilor din Romania, vol II, 26–27 May 2006, University of Politehnica, Bucharest, pp 863–872

Cazacu MD (2007) Maximization methods of turbo machines performances. International conference on applied and industrial mathematics, 17–19 August 2006, Chisinau, Science Academy Bulletin of Moldova Republic, Mathematical Section, nr. 3 (55), pp 49–54

Cazacu MD (2008) Numarul optim de pale la turbomasini (The optimum number of turbo machine blades). Stiinta si Tehnica, vol 13. Publishing House AGIR, Bucharest, pp 181–186

Cazacu MD, Nicolaie S (2002) Micro-hydro turbine for run-of-river power station. The second conference of the hydroenergeticians from Romania, 24–25 May 2002, vol II. University of Politehnica, Bucharest, pp 443–448

Cazacu MD, Nicolaie S (2005) Experimental research of an axial micro turbine. The second international symposium on renewable energies and sustainable development, Session 3—Wind and waves energy, 22–24 September 2005, Danube Delta, Tulcea, Romania, CD, p 36

Cazacu MD, Nicolaie S (2007) Building optimisation of the river ecological boats. Modelling and optimisation in the machines building field, vol 2 (MOCM 13). ALMA MATER Publishing House, University of Bacau, Romania, 2007, pp 108–115

Cazacu MD, Baran G, Nicolaie S (2003a) Incercarea in laborator a microturbinei pentru asigurarea autonomiei energetice a balizelor luminoase pe Dunare (The laboratory test of the micro turbine for the energetic autonomy assurance of the lighting buoys on the Danube). International conference on energy and environment, 23–25 October 2003, University of Politehnica, Bucharest, Section 3—Hydroenergetics, Session 2—Machines and hydraulic Equipment, vol 3. Hydrogenerators, pp 119–124

Cazacu MD, Mihaiescu MG, Nicolaie S (2003b) Baliza luminoasa folosind energia cinetica a fluviilor navigabile (Lighting buoy using the kinetic energy of the navigable rivers). National conference of renewable energies, 11–14 September 2003, University of Valahia, Targoviste, CD

Cazacu MD, Mihaiescu MG, Nicolaie S, Dorian D (2003c) Instalatie de semnalizare luminoasa pe cursurile de apa (Lighting signalising installation on the waterways). Patent nr. A/00211/2003 appeared in RO 120397 B1/01.2006. Bronze Medal at the Geneva International Salon in 2004, Salon International des Inventions des Techniques et Produits nouveaux, Geneve, 31 Mars–4 Avril 2004, Geneva Palexpo, p 141

Cazacu MD, Mihaiescu MG, Nicolaie S, Hondrea M (2005) Experimental research of an axial wind turbine. The second international symposium on renewable energies and sustainable development, Session 3—wind and waves energy, 22–24 September 2005, Danube Delta, Tulcea, Romania, CD, p 36

Cazacu MD, Samoilescu G, Nicolaie S, Mihaiescu MG, Jula N, Racuciu C (2006) Sisteme cu autonomie energetica pe Dunare (Systems with energetic autonomy on the Danube). Naval Academy Publishing House "Mircea the Old", Constanta, 269 pp, SPERIN Prize in 2006

Cazacu MD, Mihaiescu GM, Nicolaie S (2007) Ecological boat using the renewable energies. The third international conference on energy and environment—CIEM, June 28–July 4, 2007, Section 1—renewable energies. University of Politehnica of Bucharest. CD, 4 p

Cazacu MD, Isbasoiu C, Nicolaie S, Mihaiescu MG, Marin D, Ilie C, Zus M, Diana B, Georgiana D (2008) Experimental tests with electric propulsion boat supplied by renewable energies.

Homage volume of the conference at the Prof. M.D. Cazacu 80 years Anniversary, 27 October 2008, University of Politehnica, Bucharest, pp 39–48

Hutte I (1947) Manualul Inginerului (Engineer's hand-book). Publishing House AGIR, Bucharest, pp 510–511

Mihaiescu GM, Nicolaie S, Cazacu MD (2006) Constructii speciale de generatoare electrice, utilizate pentru conversia in energie electrica a cursurilor de apa (Special buildings of electric generators, used for the conversion in electric energy of the river courses). A 4-a Conf. Dorin Pavel of the Hydroenergeticians from Romania, vol II. University of Politehnica, Bucharest, 26–27 May 2006, pp 779–788

Nicolaie S (2009) Miscarea fluidului vascos in rotoare de turbomasini axiale (Viscous fluid motion in axial turbo machine rotors). PhD thesis, 2009, Polytechnic University of Bucharest

Őzsoy E, Latif MA, Besiktepe S (2002) The current system of the Bosphorus Strait based on the recent measurements. The second meeting on the physical oceanography of sea straits, Villefranche

# Wave Energy Assessments and Modeling of Wave–Current Interactions in the Black Sea

**Eugen Rusu**

## 1 Introduction

The Black Sea is semi-isolated sea located between southeastern Europe and Anatolia, that can be considered as a distant arm of the Atlantic Ocean by way of the Mediterranean Sea. It is connected to the Mediterranean by the Turkish Straits System (composed of the Bosporus and the Sea of Marmara), and to the Sea of Azov by way of the Strait of Kerch.

A bathymetric map of the Black Sea, including the Sea of Azov, is presented in Fig. 1. The sea-basin is divided into two sub-basins by a convexity extending south from the Crimean peninsula. The north-west of the basin is characterized by a relatively large shelf up to 190 km wide, which has a relatively shallow apron with gradients between 1:40 and 1:1000. The southern edge around Turkey and the eastern edge around Georgia however, are typified by a shelf that rarely exceeds 20 km in width and an apron that is typically 1:40 gradient with numerous submarine canyons and channel extensions. The Euxine abyssal plain in the centre of the Black Sea reaches a maximum depth of 2,206 m just south of Yalta on the Crimean peninsula.

The Black Sea is the world's largest meromictic basin where the deep waters do not mix with the upper layers of seawater that receive oxygen from the atmosphere. As a result, over 90% of the deeper Black Sea volume is anoxic seawater. Another important feature of the Black Sea is an unusually high freshwater river discharge into the relatively small semi-enclosed Sea. The average annual river runoff into Black Sea is 350 km$^3$, whereas the Black Sea's volume is 550,000 km$^3$. This relatively important river discharges dilute Black Sea water. Its surface layer salinity is 17‰ (gram salt in 1 l of seawater), two times less than that

E. Rusu (✉)
University Dunărea de Jos of Galaţi, Galati, Romania
e-mail: erusu@ugal.ro

V. Badescu and R. B. Cathcart (eds.), *Macro-engineering Seawater in Unique Environments*, Environmental Science and Engineering,
DOI: 10.1007/978-3-642-14779-1_10, 

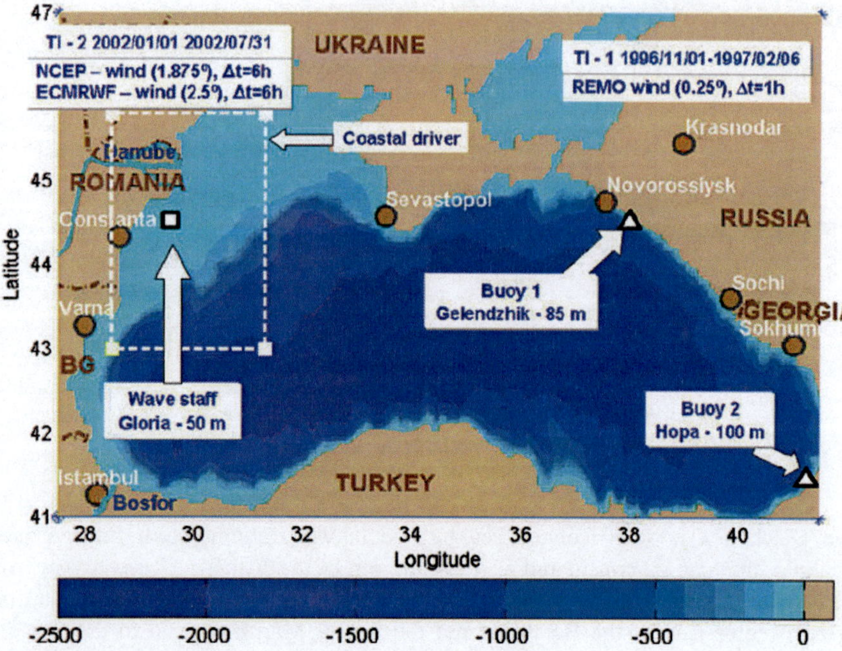


**Fig. 1** The bathymetric map of the Black Sea basin corresponding to the computational domain and the locations of the three data sources available

of the Ocean (average 35‰). This seawater head creates a current from the Black Sea, through the Straits and the Sea of Marmara, to the Mediterranean.

Circulation in the Black Sea (see, for example, Oguz et al. 1992), is mainly driven by the winds that sweep across it and the buoyancy differences between inflowing freshwater and the saltier Mediterranean inflow. The consequence is a cyclonic circulation throughout the year though not at constant speed. The high intensity storms that develop in the sea, particularly in winter, transfer energy to the circulation causing it to speed up as well as intensifying mixing patterns. A permanent feature of the upper layer circulation is the Rim Current which encircles the entire Black Sea. In places it is several tens of kilometers wide and can attain a maximum speed of 0.8–1 knot (40–50 cm/s), increasing sometimes up to 1.6–2 knots (80–100 cm/s). This has been determined by direct observations of the current velocity from surface buoys. Rim Current is directed counter-clockwise forming the two rings over basin apron in western and eastern parts of the Sea. Most of Black Sea Rim current flow is limited to the upper 100–200 m layer. Formation of anticyclonic eddies is also characteristic for the Black Sea coastal waters, and they are more pronounced and stable at the Caucasian and Anatolian coasts. Tidal oscillations of the Black Sea level do not exceed 10 cm because Mediterranean tidal waves extinguish in the Straits, and Black Sea own tidal waves development is limited by the upper water layer above the pycnocline. Most noticeable fast sea level changes in the Black Sea are the results of the strong on-shore and off-shore winds: Persistent North-East wind driven negative water setup at Caucasian coast of the Black Sea can reach 30 cm in 24 h.

Since the Black Sea is an inland sea, the fetch is considerably smaller than in Open Ocean and, as a consequence, the average wave conditions are usually less energetic. Nevertheless, very strong storms, that are sometimes characteristic of that region, can generate in the Black Sea waves comparable, in terms of wave heights, with the big ocean waves. The particularities of such environment do not stop here and many wave processes (especially in deep water) often have a dynamics somehow different than in the case of the common ocean waves. This is also due to the fact that the low frequency domain contribution in the total energetic budget of a wave train is usually not as relevant as in the ocean. This behavior might induce the idea that, although the models describing the mechanisms of wave generation, whitecapping dissipation and quadruplet interactions are valid in the same mathematical form, some parameters associated to these models should be rather different.

Presently, the most accurate estimation of the wave fields is given by the spectral phase averaging models and, among them, WAM (acronym for Wave Modeling, WAMDI Group 1988), and SWAN (acronym for Simulating Waves Nearshore, Booij et al. 1999) are considered as being the state of the art models for wave generation and coastal transformation, respectively. The basic scientific philosophy of these two models is similar, as they are both third generation wave models based on solving the energy balance equation in spectral form and they use the same formulations for the source terms, although SWAN uses the adapted code for the DIA technique (acronym for Discrete Interaction Approximation) in the parameterization of the quadruplet interactions. SWAN contains also some additional formulations primarily for shallow water referring to processes as triad non linear interactions, bottom friction and depth induced wave breaking.

The capacities of the SWAN model were substantially extended in the last years both in offshore and nearshore directions. As regards its offshore extension the main improvements are: the high order propagation scheme S&L (Stelling and Leendertse 1992), almost free of numerical diffusion, that is associated to large scale propagations in the non stationary mode, and the parameterization to counteract the Garden Sprinkler effect (Booij and Holthuijsen 1987), that may show up due to this small numerical diffusion associated with a reduced resolution in the spectral space. On the other hand, for extending the model performance in the nearshore direction the most recent improvement concerns in designing a phase decoupled approach to account for the diffraction effect (Holthuijsen et al. 2003). Together with the wave induced set up, diffraction is adequate to local scale simulations and both these two processes are associated mainly with a Cartesian coordinate system.

Hence, although probably not so efficient from computational point of view for oceanic scales than WW3 (Wave Watch 3) or WAM, for the sub oceanic scales SWAN seems to be now more appropriate. The main reason is its larger flexibility given by including various alternatives for modelling and tuning the physical processes, either in shallow or deep water. In this way, by model calibration, a more appropriate combination of the physical processes and parameters can be designed in any particular site, leading to a substantially better quality in the predictions of the main wave parameters. Another advantage is that one single model can cover the full scale of the wave modelling process and the focusing of the wave prediction system in the nearshore direction is thus straightforward.

Motivated by the above reasons, the first objective of the present work is to implement and calibrate a wave prediction system for the Black Sea basin, entirely based on the SWAN model, which can be used also for nearshore focusing. Once a reliable wave prediction system, two further applications will be also discussed. The first is related with the assessment of the wave energy resources in the Black Sea basin and the second was to model the wave–current interactions at the entrance of the Danube delta, where usually strong interactions between waves and currents are encountered.

Second generation wave models have been widely used in the Black Sea and they often provided good results some of them being still operational. However, third generation models include a more complete description of the physical processes and that is why their reliability is usually considerably higher. During the time, various implementations of the WAM model have been carried out in the Black Sea area and their calibrations have been made either with in situ or remotely sensed data. The implementation of Valchev et al. (2004) is considered here as a reference mainly because the availability of the same wind field allows a parallel analysis of the performances of the two model systems developed using WAM and SWAN, respectively.

# 2 Theoretical Formulations

## 2.1 Action Balance Equation in Spectral Form

As most phase averaging models, SWAN solves the spectral energy balance equation that describes the evolution of the wave spectrum in time, geographical and spectral spaces (Holthuijsen 2007). The spectrum that is considered is the action density spectrum ($N$), rather than the energy density spectrum ($E$), since in the presence of currents action density is conserved whereas energy density is not. The action density is equal to the energy density divided by the relative frequency ($\sigma$). Hence the spectral action balance equation has the expression:

$$\frac{\partial N}{\partial t} + \nabla\left[\left(\vec{c}_g + \vec{U}\right)N\right] + \frac{\partial}{\partial \sigma} c_\sigma N + \frac{\partial}{\partial \theta} c_\theta N = \frac{S}{\sigma}, \tag{1}$$

where: $\theta$ is the wave direction and $\vec{U}$ the velocity of the ambient current considered uniform with respect to the vertical coordinate. The propagation velocities of the wave energy are the group velocity $\vec{c}_g$ for the geographical space ($\vec{c}_g = \partial\sigma / \partial\vec{k}$, with $k$ is the wave number), and $c_\sigma = \dot{\sigma}$ and $c_\theta = \dot{\theta}$ for the spectral space.

For large scale applications this equation is related to the spherical coordinates defined by longitude $\lambda$ and latitude $\phi$:

$$\frac{\partial N}{\partial t} + \frac{\partial}{\partial \lambda} c_\lambda N + \frac{1}{\cos\varphi}\frac{\partial}{\partial \varphi} c_\varphi N \cos\varphi + \frac{\partial}{\partial \sigma} c_\sigma N + \frac{\partial}{\partial \theta} c_\theta N = \frac{S}{\sigma}. \tag{2}$$

In the right hand side of the action balance equation is the source expressed in terms of energy density. In deep water, three components are significant in the expression of the total source term. They correspond to the atmospheric input ($S_{in}$), whitecapping dissipation ($S_{dis}$) and nonlinear quadruplet interactions ($S_{nl}$), respectively. Besides these three terms, in shallow water additional source terms induced by the finite depth effects ($S_{fd}$) may play an important role and they correspond to phenomena like bottom friction, depth induced wave breaking and triad non linear wave-wave interactions. Hence the total source becomes:

$$S_{total} = S_{in} + S_{dis} + S_{nl} + S_{fd}. \tag{3}$$

## 2.2 Wave Energy

In SWAN, the energy transport components (expressed in W/m, i.e. energy transport per unit length of wave front), are computed with the relationships:

$$\begin{aligned} E_{trx} &= \rho g \iint c_x E(\sigma,\theta) d\sigma d\theta \\ E_{try} &= \rho g \iint c_y E(\sigma,\theta) d\sigma d\theta, \end{aligned} \tag{4}$$

where $x$, $y$ are the problem coordinate system (for the spherical coordinates $x$ axis corresponds to longitude and $y$ axis to latitude), and $c_x$, $c_y$ are the propagation velocities of wave energy in the geographical space defined as:

$$\frac{d\vec{x}}{dt} = (c_x, c_y) = \vec{c}_g + \vec{U}. \tag{5}$$

Hence the absolute value of the energy transport (denoted also as wave power) will be:

$$E_{tr} = \sqrt{E_{trx}^2 + E_{try}^2}. \tag{6}$$

## 2.3 Wave–Current Interactions

If waves propagate on a non stationary and non uniform current $\vec{U}(\vec{x}, t)$ the intrinsic frequency ($\sigma$) may vary in space and time. This frequency (called also relative) is related with the wave number ($k$) by the dispersion relationship:

$$\sigma^2 = gk \tanh(kd), \tag{7}$$

where ($g$) is the acceleration of gravity and ($d$) the water depth. In the presence of currents the absolute radian frequency ($\omega$) is given by the usual Doppler shift:

$$\omega = \sigma + \vec{k}\vec{U}. \tag{8}$$

It is known that the number of wave crests passing a given point per unit time is determined by the kinematical conservation equation (Massel 1996):

$$\frac{\partial \vec{k}}{\partial t} + \nabla_h \omega = 0. \tag{9}$$

For a steady state the equation above reduces to:

$$\nabla_h \omega = 0. \tag{10}$$

This implies that the absolute frequency (or observed frequency) $\omega$ is an invariant under a steady state.

The methods for the computation of regular-wave refraction over water of variable depth are well established. However, two significant new features occur when a current is present. The first is that the wave energy (or rather wave action) propagates with the sum of the current velocity and the relative group velocity, so that energy is not transmitted at right angles to the wave crests, and the second that wave energy is not conserved even in the absence of friction, since there is a transfer of energy between the waves and the current.

Mathematically, the general case of waves propagating through a current field over an arbitrary bottom is a complex initial-boundary value problem. Refraction theory (coming from geometrical optics) has the assumptions of linear and locally plane waves, which restricts the consideration to large-scale currents and wave fields. These conditions were assumed in Jonsson (1990) and so the kinematical and dynamical results found there will apply.

Taking the ratio between wave action flux and the same quantity at the initial position of the ray, yields for the wave height ($H$)

$$\frac{H}{H_{\mathrm{st}}} = K_c K_s K_{\mathrm{ra}} K_f = K_c J_R^{-1/2} K_f, \tag{11}$$

where the subscript (st) indicates the initial (start) position of the wave ray and the $K$-coefficients have the following significations: $K_c$ is the Doppler coefficient, $K_s$ is the shoaling coefficient, $K_{\mathrm{ra}}$ is the refraction coefficient and $K_f$ is the friction coefficient (see Christoffersen 1982; Jonsson and Christoffersen 1985, for the detailed expressions of these coefficients). The first three coefficients are determined by the kinematics of the wave field, whereas the fourth is found by the integration along the wave ray. Thus it appears that $J_R = (K_s K_{\mathrm{ra}})^{-2}$. Note that even in constant water depth $K_s \neq 0$, when a current is present.

Particularly simple relations emerge when steady wave transformation by currents over a horizontal bed is considered. In these conditions wave refraction due to currents is governed by the Snell's low according to which $k{\cdot}\sin\beta = \text{const.}$, where $\beta$ is the angle between the normal to the current direction and the normal to the wave front (Jonsson 1990).

If the dissipation, generation and wave-wave interactions are negligible, the conservation of the wave action density reduces to (Massel 1996):

$$\frac{\partial}{\partial t}\left(\frac{\psi}{\sigma}\right) + \nabla_h\left[\vec{C}_g\frac{\psi}{\sigma}\right] = 0, \tag{12}$$

where $\psi(\vec{k},\vec{x},t)$ is the wave number spectrum and $\left(N\left(\vec{\text{k}}\right) = \psi\left(\vec{k},\vec{x},t\right)/\sigma\right)$ the wave number action density. The equation above states that the local rate of change of the wave action is balanced by the divergence of the flux of action, a quantity that flows relative to the moving medium with the absolute group velocity $C_g$:

$$C_g = \frac{\partial\omega}{\partial k} = \frac{\partial\sigma}{\partial k} + \vec{U}\frac{\vec{k}}{\left|\vec{k}\right|} > 0. \tag{13}$$

This result was first obtained by Bretherton and Garret (1968) who demonstrated that the wave number action density is conserved in a moving medium. This implies that under a steady state and along wave rays we obtain:

$$\frac{\psi}{\sigma} = \text{const.} \tag{14}$$

In order to pass to from the wave number spectrum to the frequency spectrum $E(\omega,\theta,\vec{x},t)$ (where $\theta$ is the wave direction) the variable transformation (Massel 1996) yields to:

$$\psi\left(\vec{k},\vec{x},t\right) = JE(\omega,\theta,\vec{x},t), \tag{15}$$

where $J$ is the Jacobian determinant of the transformation defined as:

$$J = \frac{\partial(\omega,\theta)}{\partial\left(k_x,k_y\right)} = \left[\frac{\partial\left(k_x,k_y\right)}{\partial(\omega,\theta)}\right]^{-1} = \frac{C_g}{k}. \tag{16}$$

Under these assumptions results:

$$\frac{C_g}{k}\frac{E(\omega,\theta,\vec{x},t)}{\omega - \vec{U}\vec{k}} = \text{const.} \tag{17}$$

Assuming now that the group velocity, wave number and two dimensional spectral density in deep water, in absence of currents, are: $C_{g0}$, $k_0$ and $E_0(\omega,\theta)$, respectively, the above equation can be rewritten as:

$$E(\omega,\theta,\vec{x}) = \left(\frac{k}{k_0}\right)\left(\frac{C_{g0}}{C_g}\right)\frac{\omega - \vec{k}\vec{U}}{\omega}E_0(\omega,\theta)\,. \tag{18}$$

This equation expresses the variation of the frequency spectrum due to the presence of currents.

The absolute group velocity $C_g$ (as given by Eq. 13) defines the direction of the wave ray. The local group velocity $c_g = \partial\sigma/\partial k$ is associated with the normal to the wave front. In extreme conditions, when:

$$c_g = \frac{\partial\sigma}{\partial k} = -\vec{U}\frac{\vec{k}}{\left|\vec{k}\right|}, \tag{19}$$

the spectral component no longer propagates against the current in that direction. Theoretically the local spectral energy density becomes infinite. This suggests that these components will tend to attenuate by wave breaking or by lateral stretching in the crest direction, before this point is reached.

Like most phase averaging models, SWAN solves the spectral energy balance Eq. 1 that describes the evolution of the wave spectrum in time, geographical and spectral spaces (Holthuijsen 2007). SWAN is thus fully spectral (in all directions and frequencies) and computes the evolution of wind waves in coastal regions with shallow water and ambient current. The spectrum that is considered is the action density spectrum ($N$), rather than the energy density spectrum ($E$), since in the presence of currents, action density is conserved whereas energy density is not.

If ambient currents are present, an iteration process for spectral propagation (current-induced refraction and frequency shift) is carried out in SWAN. The fluxes in the spectral space are not approximated with the first order upwind scheme since this turns out to be very diffusive for frequencies near the blocking frequency (when the waves are blocked by the currents). Note that in the absence of mean current there are no shifts in the frequency. Wave-induced currents are not computed by SWAN so if such currents are relevant they should be included as a component in the input current field.

SWAN cannot handle wave propagation on super-critical current flow that is when the flow velocity is larger than the local group wave velocity. If such flow is encountered during SWAN computations, the current is locally reduced to sub-critical flow. The currents may mismatch with given water depth in the sense that the Froude number ($F_\gamma = U/(gd)^{1/2}$) might become larger than unity, situation in which the flow becomes supercritical. From this reason in SWAN a parameter *froudmax* (default value 0.8) is used. Thus, if the current velocity is relatively large, i.e. the Froude number is larger than *froudmax*, it will be reduced such that the Froude number to become equal to *froudmax*.

Another option available in SWAN that is related with the wave–current interactions refer to the possibility to switch the numerical scheme for refraction from a central scheme that has the largest accuracy (diffusion zero) but the computation may more easily generate spurious fluctuations to a first order upwind scheme that it is more diffusive and therefore preferable if (strong) gradients in depth or current are present.

# 3 Implementation of the Wave Prediction System

## 3.1 General Settings and Data Sources Considered

The computational domain considered for the present SWAN implementation is illustrated in Fig. 1. The system origin corresponds to the lower left corner point and has the coordinates (27.5°, 41.0°), whereas the lengths are 14° in *x*-direction (longitude) and 6° in *y*-direction (latitude), covering both the Black Sea and the Sea of Azov. In the geographical space the computational grid was chosen identically with the bathymetric grid and has 176 points in *x* direction and 76 points in *y* direction, the points are equally spaced with $\Delta x = \Delta y = 0.08°$. In the spectral space 24 directions and 30 frequencies were assumed, the frequency range considered being between 0.06 Hz and 1.2 Hz.

The objective at this level was to implement and validate a wave prediction system for the Black Sea basin. This system is entirely based on the SWAN model, which it is used for both wave generation and nearshore transformation.

Three check points were considered for model calibration and performing validation tests. First, data coming from two directional buoys operating in deep water, close to the east coast, were considered. As a second step, the reliability of the SWAN model results was tested on the west coast using two different wind fields. The comparisons were made this time against the measurements provided by a wave staff located at the Gloria drilling platform that is operating in that area.

An enlarged presentation of the model system implementation in the Black Sea basin and of the validation tests performed is given in Rusu et al. (2006). These studies were focused on the performances of the formulations available in SWAN to account for whitecapping dissipation. This is mainly because whitecapping is still widely considered the weak link in deep water wave modeling and it is closely related with the process of wave generation.

Four different formulations are available to account for the processes of wave generation by wind and whitecapping dissipation. These are: a) Komen's parameterization (Komen et al. 1984) to describe the transfer of wind energy to the waves coupled with the pulse-based model of Hasselmann (1974) for whitecapping dissipation, as adapted by the WAMDI group (1988) to be applicable in finite water depth; b) Janssen's model for atmospheric input (Janssen 1991) coupled with the same pulse-based model of Hasselmann (1974) for whitecapping; c) Komen's model coupled with the Cumulative Steepness Method (Van Vledder and Hurdle 2002) for whitecapping; d) Yan's model for atmospheric input (Yan 1987) coupled with the saturation based model of Alves and Banner (2003) for whitecapping (that will be denoted in this work as A&B).

The first parameterization (Komen–Hasselmann) corresponds also to WAM cycle 3 while the second one (Janssen–Hasselmann) corresponds to WAM cycle 4. CSM (Cumulative Steepness Method) is an alternative formulation for whitecapping and with this method dissipation due to whitecapping depends on the steepness of the wave spectrum at and below a particular frequency. A second alternative for the whitecapping expression uses the formulation of Alves and Banner (2003). This expression is based on experimental findings that whitecapping dissipation appears to be related to the nonlinear hydrodynamics within wave groups. This yields a dissipation term that primarily depends on quantities that are local in the frequency spectrum, opposed to ones that are distributed over the spectrum, as in the expression of Komen et al. (1984). This whitecapping expression is used together with a wind input term that is based on Yan (1987).

As indicated in Fig. 1, two time intervals were considered for model system calibration and for performing the validation tests. The first phase corresponds to the time interval between 1$^{st}$ of November 1996 and 6th of February 1997, denoted as TI-1. For this interval, model data were compared against two buoys operating on the east coast of the sea. These are Gelendzihik (37.98°E, 44.51°N), denoted as Buoy 1 and Hopa (41.38°E, 41.42°N), denoted as Buoy 2. They are both located in deep water at 85 and 100 m depth, respectively.

The wind field for this first period was provided by the project HIPOCAS "Hindcast of Dynamic Processes of the Ocean and Coastal Areas of Europe", developed in the framework of the European Program "Energy, Environment and Sustainable Development" (Guedes Soares et al. 2002; Guedes Soares 2008), which gave the reanalysis wind conditions for 44 years, between 1958 and 2001. In this first phase, for the wave simulations corresponding to the time interval TI-1, the global NCEP (acronym for US National Center for Environmental Prediction) reanalysis wind was used as a driver for the regional atmosphere model REMO (Regional Environmental Model). The spatial resolution of the wind model output was 0.25° and the time step of 1 h.

## *3.2 Calibration Process (SWAN Against Buoy 1)*

Model system simulations were carried out, and most of the possibilities for tuning the SWAN model, in relationship with the deep water wave processes, were experimented. As shown before, four options are currently available in SWAN to account for whitecapping dissipation and they are coupled with the process of wind generation. The first two consider the standard formulations as they use the pulse based model of Hasselmann for whitecapping coupled with the wind input formulations of Komen and Janssen, respectively. Two alternative formulations can be activated in the currently operating SWAN version (40.72), and they were described above. These are the cumulative steepness method (denoted as CSM), and the saturation based model (denoted in the present work as A&B). For this second alternative, another wind input formulation (Yan 1987) is used. The corresponding results are presented in Table 1 for the time interval TI-1.

The analysis of the statistical data shows that, for most of the parameters analyzed (significant wave height, mean period, peak period and mean wave direction) the best results are provided by the A&B formulation followed closely by Komen's formulation. The above results show also that the A&B formulation

**Table 1** Gelendzhik directional buoy (Buoy 1), wave statistics for the period TI-1 (1996/11/01–1997/02/06), all the four formulations available for modeling whitecapping dissipation were tested

| $n = 684$ | $X_{mes}$ | $X_{sim}$ | Bias | RMSE | SI | $r$ | |
|---|---|---|---|---|---|---|---|
| Hs (m) | 1.004 | 1.014 | −0.010 | 0.365 | 0.364 | 0.886 | K |
| Tm (s) | 3.975 | 3.411 | 0.564 | 0.893 | 0.225 | 0.797 | O |
| Tp (s) | 5.609 | 5.164 | 0.444 | 1.486 | 0.265 | 0.640 | M |
| Dir (°) | 214.9 | 204.28 | 10.66 | 53.21 | 0.248 | 0.449 | |
| Hs (m) | 1.004 | 1.043 | −0.039 | 0.469 | 0.467 | 0.808 | J |
| Tm (s) | 3.975 | 4.032 | −0.058 | 0.855 | 0.215 | 0.701 | A |
| Tp (s) | 5.609 | 5.566 | 0.043 | 1.588 | 0.283 | 0.525 | N |
| Dir (°) | 214.9 | 224.52 | −9.568 | 68.60 | 0.319 | 0.318 | |
| Hs (m) | 1.004 | 1.107 | −0.104 | 0.429 | 0.428 | 0.848 | C |
| Tm (s) | 3.975 | 3.321 | 0.653 | 0.991 | 0.249 | 0.745 | S |
| Tp (s) | 5.609 | 5.873 | −0.264 | 1.554 | 0.277 | 0.576 | M |
| Dir (°) | 214.9 | 221.92 | −6.971 | 67.11 | 0.312 | 0.395 | |
| Hs (m) | 1.004 | 1.004 | 0.000 | 0.361 | 0.360 | 0.888 | A |
| Tm (s) | 3.975 | 3.790 | 0.185 | 0.826 | 0.208 | 0.767 | & |
| Tp (s) | 5.609 | 5.563 | 0.045 | 1.485 | 0.265 | 0.618 | B |
| Dir (°) | 214.9 | 208.9 | 6.000 | 53.54 | 0.269 | 0.432 | |

not only provides the best predictions, but also requests the smallest rate of change for the tunable coefficients.

The results were also compared with those provided by the WAM model for the same time interval at the end of 1996 and the beginning of 1997 (TI-1) that used exactly the same wind data field, Valchev et al. (2004). Looking at the values of the statistical parameters it has to be noticed that, probably due to the better flexibility in tuning the parameters, the SWAN results are considerably better when comparing to those provided by WAM for any of the whitecapping parameterization considered. Thus, the biases obtained for the same time interval were for Hs: 0.27 m, for Tp: 0.43 s and for Dir: 33.1°.

Referring to the RMSE the values obtained were for Hs: 0.53 m, for Tp: 1.74 s and for Dir: 92.7°, which means considerably higher root mean square errors at least as regards the significant wave height and the mean direction. The scatter indices obtained were: 0.68 for Hs, 0.34 for Tp and 0.46 for Dir, which means also that the results of SWAN are better. Finally the computed correlation coefficients were for Hs: 0.73, for Tp: 0.55 and 0.36 for Dir. This means also better correlations with the measured data as concerns the SWAN results in comparison with WAM.

### *3.3 First Validation Test (SWAN Against Buoy 2)*

The performances of the system were also assessed in the southern part of the east coast for the same time interval (TI-1). In that area the Hopa directional buoy was operating and its location is also indicated in Fig. 1. This time only Komen and

**Table 2** Hopa directional buoy (Buoy 2), wave statistics for the period TI-1 (1996/11/01–1997/02/06)

| $n = 465$ | $X_{mes}$ | $X_{sim}$ | Bias | RMSE | SI | $r$ | |
|---|---|---|---|---|---|---|---|
| Hs (m) | 0.565 | 0.646 | −0.081 | 0.325 | 0.576 | 0.781 | K |
| Tm (s) | 4.142 | 3.722 | 0.420 | 0.998 | 0.241 | 0.699 | O |
| Dir (°) | 253.5 | 301.4 | −47.95 | 107.4 | 0.424 | 0.133 | M |
| Hs (m) | 0.565 | 0.552 | 0.013 | 0.297 | 0.526 | 0.814 | A |
| Tm (s) | 4.142 | 4.309 | −0.166 | 1.110 | 0.268 | 0.585 | & |
| Dir (°) | 253.5 | 294.3 | −40.81 | 102.7 | 0.405 | 0.241 | B |

A&B whitecapping formulations, that were found more effective in the previous case, were evaluated. The corresponding results are presented in Table 2.

As it can be seen from Table 2, in terms of significant wave height the results provided by the A&B formulation are still better than those obtained when the Komen–Hasselmann approach is used, while for the mean period, Komen formulation gives slightly better results. As regards the mean directions although the model predictions are not very accurate the A&B formulation performs better. When comparing with the results at the Buoy 1 (as presented in Table 2) it can be seen that the significant wave heights are better estimated at Buoy 2 in terms of RMSE and at Buoy 1 in terms of scatter indices. The mean periods and mean directions are better estimated in all parameters at Buoy 1.

## 3.4 Second Validation Test (Validation on the West Coast)

As a second validation test, model simulations were performed for the time interval between 1st of January 2002 and 31st of July 2002 (denoted as TI-2). Two different wind fields were used in this phase: NCEP (1.875° spatial resolution) and ECMRWF (acronym for European Center for Medium Range Weather Forecast) with 2.5° spatial resolution, the temporal resolution is 6 h for both fields. Data measured in situ by a wave staff placed at the Gloria drilling platform were used as a reference. The drilling unit operates in the western coastal environment of the Black Sea at about 50 m water depth (Fig. 1), and both wind and wave measurements were available for the period considered. Since the wave staff was located in relatively intermediate water depth, a second area was connected to the main computational domain in order to account better for the coastal wave transformation, as illustrated in Fig. 1.

A comparison of the measured wind velocities against the predictions coming from the two different model systems was first carried out. The measurements of the wind velocities have been performed at 28 m height and the wind speeds were adjusted from 28 to 10 m using the simple relationship from Hsu et al. (1994):

$$u_2 = u_1 (z_2/z_1)^P, \tag{20}$$

where $u_2$ denotes the wind speed at the reference height $z_2$ (10 m), and $u_1$ represents the wind speed measured at height $z_1$(28 m). The exponent, $P$, was set to 0.11 which is considered typical for sea conditions. Direct comparisons between measured and predicted wind velocities were also performed and they show in general reliable wind model predictions. A first conclusion is that the NCEP wind fields are in general more accurate. However, the wind velocities coming from the atmospheric models seem to be systematically smaller than the measurements (positive biases).

Using the model parameterizations defined in Sect. 2, corresponding to Komen and A&B formulations, respectively, SWAN simulations were performed considering these two different wind fields. As expected, NCEP field provided better results in terms of the wave parameters than ECMRWF. For significant wave heights and mean periods direct comparisons between the model results, using NCEP wind field, and the measurements are illustrated in Fig. 2.

The corresponding statistical results are presented in Table 3. The results are again slightly better when A&B formulation is used. Nevertheless, the results are

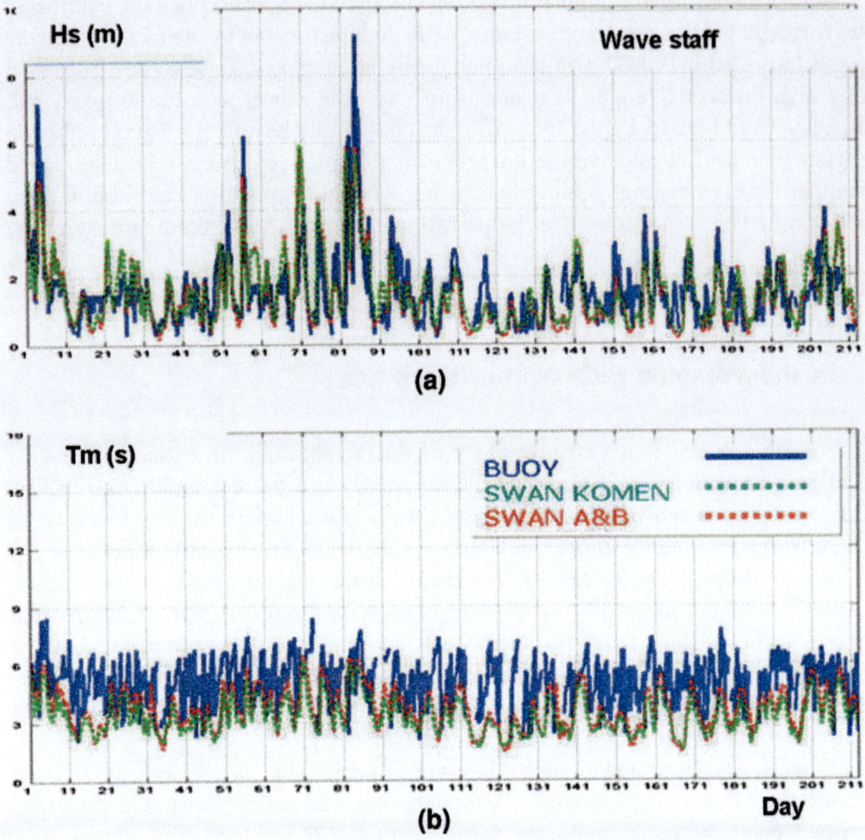


**Fig. 2** Direct comparisons on the west coast, SWAN, forced with NCEP wind field, against wave staff measurements (**a** Hs and **b** Tm), day 1-2002/01/01, day 212–2002/07/31 (TI-2)

**Table 3** Gloria drilling platform (wave staff), wave statistics for the period TI-2 (01/01/2002–31/07/2002)

| $n = 781$ | $X_{mes}$ | $X_{sim}$ | Bias | RMSE | SI | $r$ | |
|---|---|---|---|---|---|---|---|
| Hs (m) | 1,535 | 1,551 | −0,016 | 0,762 | 0,496 | 0,709 | KOM |
| Tm (s) | 5.08 | 3.339 | 1.741 | 2.191 | 0.431 | 0.221 | |
| Hs (m) | 1,535 | 1.520 | 0.015 | 0.758 | 0.494 | 0.724 | A & B |
| Tm (s) | 5.08 | 3.744 | 1.336 | 1.941 | 0.382 | 0.234 | |

in general less accurate than for the cases presented in Sects. 3.2 and 3.3 when the wind resolution was higher both in space and time.

The wave measurements at the staff were non directional. However, based on visual observations, the wave direction was indicated appreciatively by the human operator (S, SW, W, etc.). These indications were compared with the model results and it was found that 88.4% of the model predictions corresponded to the above observations.

An important aspect refers to the high energetic conditions when the significant wave height peaks seem to be systematically underestimated by the model when the A&B formulation is used, while Komen formulation appears to be more effective in this high energetic range. In relationship with this aspect, a solution taken into account would be to define two different configurations for the wave prediction system, one for low and average energetic conditions, using A&B formulation, and another for high energetic conditions, using Komen formulation, and switch them after case. This would improve the performances of the wave prediction system.

## 4 Evaluation of the Wave Energy Resources in the Western Side of the Black Sea

The Earth's ocean is a vast repository of energy that can be extracted from its motion, temperature, or chemistry. Ocean energy can be recovered from waves, tides, marine currents, thermal gradients, and differences in salinity. Among all these forms may be the most challenging is the wave energy extraction.

Wave energy technologies extract energy from the movement of the ocean's waves, which are created by wind interacting with the ocean surface, while tidal energy technologies use the powerful water currents created by the rise and fall of the tides. Although not yet quantified, tidal technologies' overall availability is much less than wave energy because it is only feasible in areas where the tidal currents are strong enough to run turbines efficiently.

Wave energy is abundant and, while not as predictable as tidal energy, is more predictable than wind or solar. Although the amount of energy that can be created using wave technologies varies from site-to-site and from day-to-day, depending on location and weather conditions, wave energy can be accurately predicted using numerical models within a window of a few days. Constancy and predictability enable a more straightforward and reliable integration into the electric utility grid. Wave energy also offers much higher energy densities, allowing devices to extract more power from a smaller volume at consequent lower costs.

Interest in extracting power from waves (as distinctly different from flowing water, such as rivers or tides) began in earnest in the 1970s. Unfortunately, none of the schemes developed at that time was able to get a full-scale trial. The field remained largely quiet until the past decade, when the success of the wind power industry spurred renewed interest in discovering what might work in the sea. Wave-energy extraction is at this moment, in a real sense, where wind power was 15 or 20 years ago, with no clearly superior engineering solutions. However, around the world, various research teams are making now progress in devising effective and efficient means for tapping wave energy and as going forward it becomes more and more obvious that wave power will play an important role in the global energy portfolio.

The highest energy ocean waves are concentrated off the western coasts in the 40–60° latitude range north and south. Waves are bigger and more powerful along the western edge of the earth's continents because of the prevailing west-to-east winds. The annual average power in the wave fronts varies in these areas between 30 and 70 kW/m, with peaks up to 100 kW/m in the Atlantic SW of Ireland, in the Southern Ocean and close to the Cape Horn. Chile, Australia, New Zealand, Ireland, UK, Portugal, and Norway have substantial wave power potential. That is due to features they all share: a location in relatively high latitude and a long stretch of ocean immediately to the west.

Although not included in the richest areas in wave energy resources, the Black Sea basin may be however relevant from this point of view. This is due to the fact that very strong wind patterns are characteristic to those areas inducing an energetic regime often comparable in terms of wave heights with the oceanic one.

In this context, the objective of the present section would be to explore the wave energy resources in the Black Sea. The study will be focused mainly on the western part, which is the most energetic from the entire sea.

After analyzing in spatial frames several cases of wave energy distributions (a few hundreds), four case studies considered more relevant will be presented and discussed below. According to Rusu (2009) they correspond to the most relevant energetic patterns for average and high energetic conditions, respectively. Two of them correspond to the entire Black Sea basin while the other two correspond to a medium resolution domain covering the Romanian coastal environment in the western side of the Black Sea. For each case study the significant wave height scalar fields were first represented together with the wind and wave vectors followed by a second representation giving the normalized wave power in background and the energy transport vectors in foreground.

The non dimensional normalized wave power was expressed as:

$$E_{\mathrm{TRn}} = \frac{E_{\mathrm{TR}}}{E_{\mathrm{TRmax}}}. \tag{21}$$

In the present work $E_{TRmax}$ was defined separately for each individual case study and it is a rounded off value selected close to the maximum wave power corresponding to the computational domain.

## 4.1 Case Study 1 (CS1): 1997/01/12/h12

This first case study presents an average energetic situation for the entire Black Sea basin and is illustrated in Fig. 3. Thus, Fig. 3a presents the significant wave height scalar fields in background and in foreground the wave (black arrows) and wind vectors (white arrows). Figure 3b illustrates in background the normalized wave power and in foreground the energy transport vectors (in kW/m of wave front). The locations in the computational domain of the maximum values for the significant wave height, wind velocity and wave power are marked with circles. For this case study the value $E_{TRmax}$ was set to 20 kW/m.

Some observations arise from the analysis of the above results. Although there is an obvious relation between the significant wave height and the wave power, the energetic peak in a computational domain is not necessarily located in the same point (or even close) with the maximum significant wave height.

This is very well illustrated by the results presented in Figs. 3a and b where the overall energetic peak (20 kW/m) is located in the southern side of the sea while the maximum significant wave height (3.1 m) in the western side.

## 4.2 Case Study 2 (CS2): 1997/01/21/h18

This second case study presents some storm conditions in the Black Sea and is illustrated in Fig. 4. Nevertheless, it has to be mentioned that this is not an extreme event, but a regular storm (moreover it is a typical storm when the western part of the Black Sea is more energetic due to the dominant wind patterns). Figure 4a presents in background the significant wave height scalar fields, and in foreground the wave (black arrows) and wind vectors (white arrows). Figure 4b illustrates in background the normalized wave power and in foreground the energy transport vectors (in kW/m of wave front). For this case study the value $E_{TRmax}$ was set to 150 kW/m. This time there is a correspondence between the locations of the maximum significant wave height (6.9 m) and of the maximum wave power (142.8 kW/m).

## 4.3 Case Study 3 (CS3): 2002/02/24/h03

This case study, illustrated in Fig. 5, presents average to high energetic conditions for the western medium resolution computational domain. The geographical extent

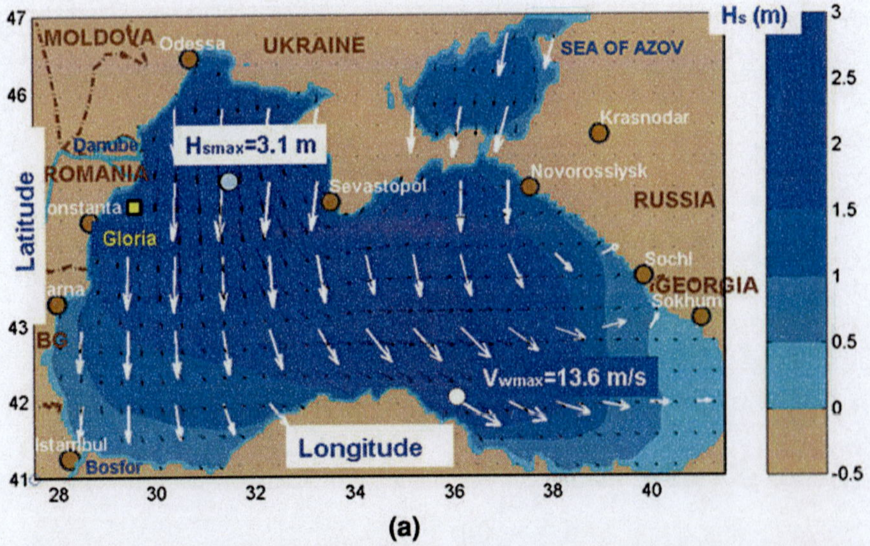


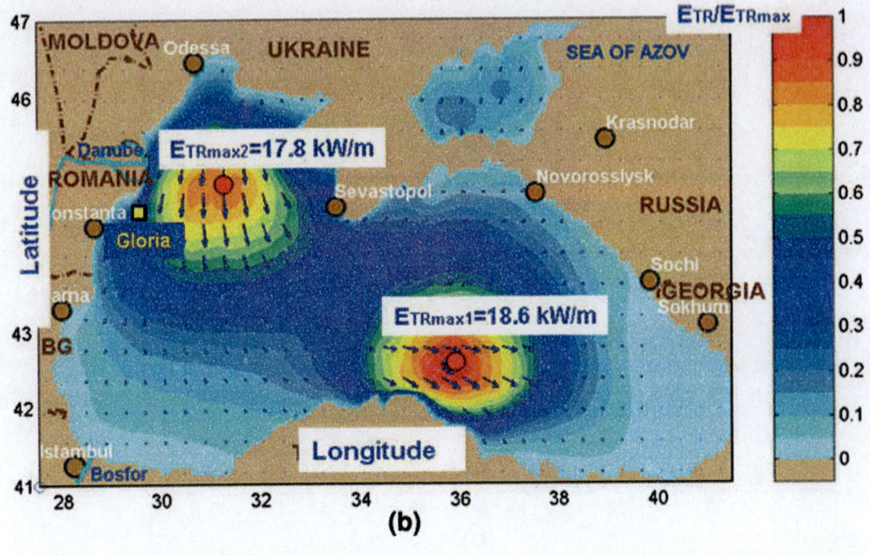


**Fig. 3** CS1-1997/01/12/h12, average energetic situation, representation for the entire Black Sea basin. **a** In background significant wave height scalar fields, in foreground wave vectors (*black arrows*) and wind vectors (*white arrows*); **b** In background normalized wave power, in foreground energy transport vectors (in kW/m of wave front)

of this area is illustrated in Fig. 1 and the boundary conditions are generated using the nesting procedure from model simulations performed in the entire basin of the sea. Figure 5a presents in background the significant wave height scalar fields, and in foreground the wave (black arrows) and wind vectors (white arrows).

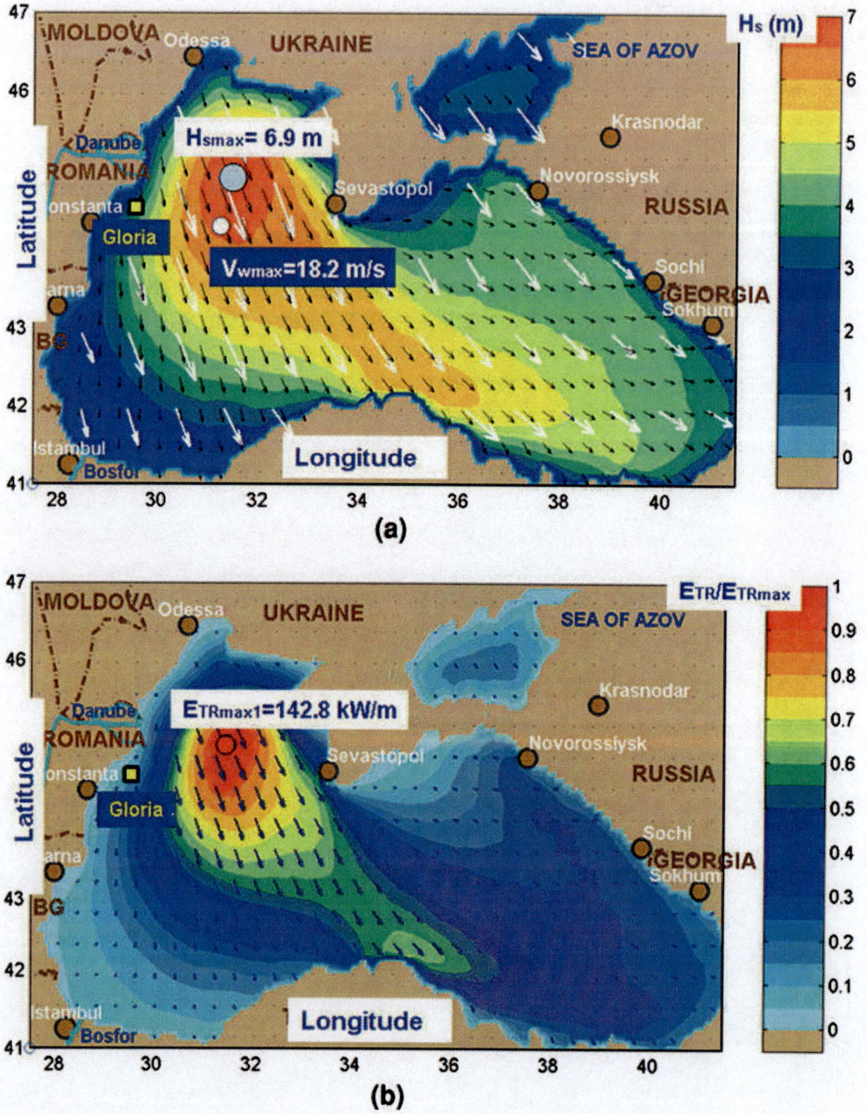


**Fig. 4** As Fig. 3 for CS2—1997/01/21/h18, storm pattern

Figure 5b illustrates in background the normalized wave power and in foreground the energy transport vectors (in kW/m of wave front). A reference line crossing the most energetic area, which is offshore to the Sulina channel, was also represented. For this case study the value $E_{TRmax}$ was set to 50 kW/m. Figure 5c presents the variations along the reference line of some relevant wave parameters ($H_s$, $C_g$—group velocity, water depth and wave energy), and in this way the energetic peaks can be better observed. The locations of the two

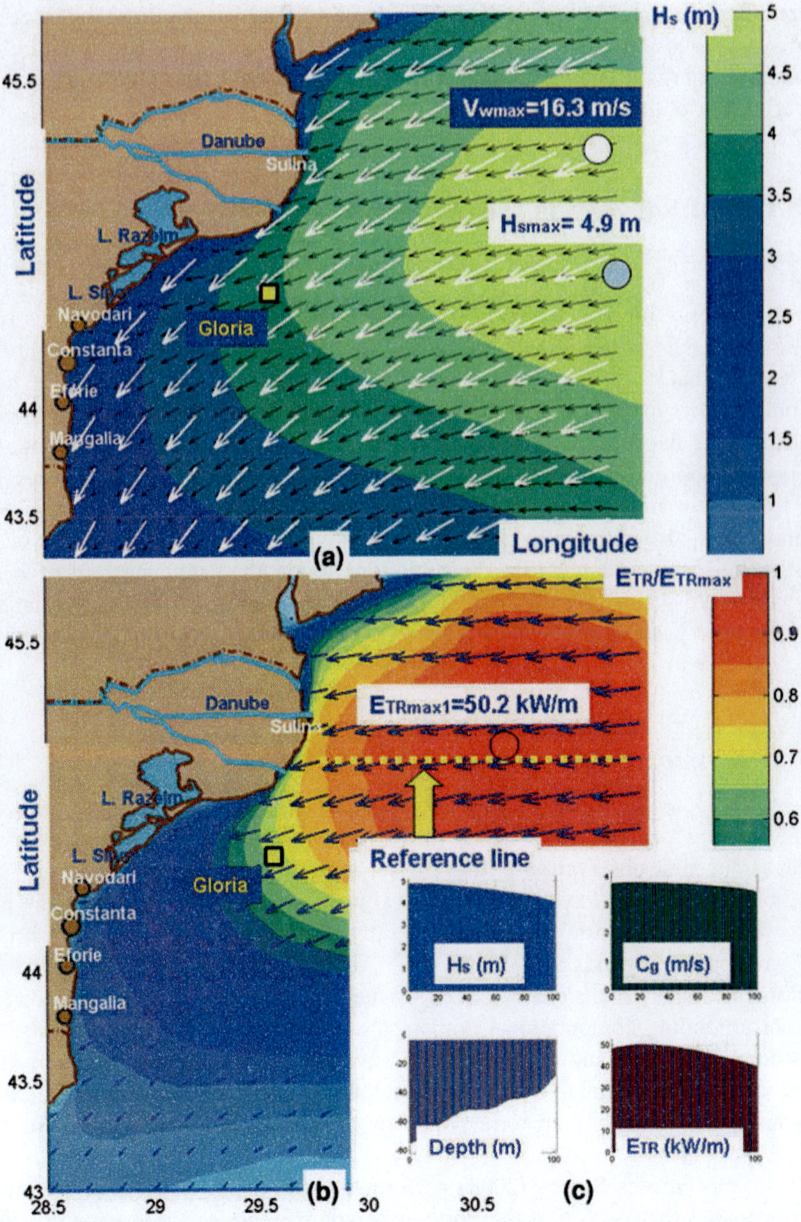


**Fig. 5** CS3—2002/02/24/h03, average to high energetic pattern, representation for the western coastal computational domain. **a** In background significant wave height scalar fields, in foreground wave vectors (*black arrows*) and wind vectors (*white arrows*); **b** In background normalized wave power, in foreground energy transport vectors (in kW/m of wave front), a reference line in the field is also represented; **c** Variation along the reference line of wave parameters ($H_s$, $C_g$—water depth and wave energy, respectively)

maximums ($H_{smax} = 4.9$ m and $E_{TRmax1} = 50.2$ kW/m) are close but not the same. A second important observation would be that the local effects may induce sometime energetic peaks that exceed the global peak from the coarse area covering the entire Black Sea basin.

## 4.4 Case Study 4 (CS4): 2002/03/11/h18

This last case study considered in the present section, which is illustrated in Fig. 6, presents the coastal impact of an extreme storm as resulting from simulations in the western medium resolution computational domain. Figure 6a presents in background the significant wave height scalar fields, and in foreground the wave (black arrows) and wind vectors (white arrows). Figure 6b illustrates in background the normalized wave power and in foreground the energy transport vectors (in kW/m of wave front). A reference line crossing the most energetic area, which is this time in the southern part of the computational domain, was also represented. For this case study the value $E_{TRmax}$ was set to 300 kW/m. Figure 6c presents the variations along the reference line of some relevant wave parameters ($H_s$, $C_g$—depth and wave energy). In these variations the energetic peak is very well evidenced. The locations of the two maximums ($H_{smax} = 8.9$ m and $E_{TRmax1} = 318.9$ kW/m) are very close.

## 4.5 Discussion of the Results

Some general observations can be made from the analysis of the above case studies. The first observation is that the locations of the energetic peak and the maximum significant wave height do not always coincide. This is due to the fact that, while the significant wave height is only a function of the energy spectrum, the wave power depends also on the group velocity as seen in Eq. 5.

Some nearshore areas characterized by their high energy relative to the rest of the corresponding computational domain also were detected in this analysis. As illustrated by Figs. 9 and 10, the richest wave energy resource locations are offshore the Sulina channel and in the southern part of the medium resolution computational domain that corresponds to the coastal area from north of the Bulgarian nearshore.

Rusu and Guedes Soares (2008) performed an in-depth analysis of the wave energy spatial distribution in the Portuguese continental nearshore which is considered to be among the coastal areas richest in wave energy resources. This analysis was based on in situ measured data for the period 1994–2003 and on simulations with a model system based on WAM for wave generation and SWAN for nearshore transformation in the coastal environment.

Some insight can be gained at this point by comparing the results of that study of the west Iberian coast with those from the present investigation of the Black Sea.

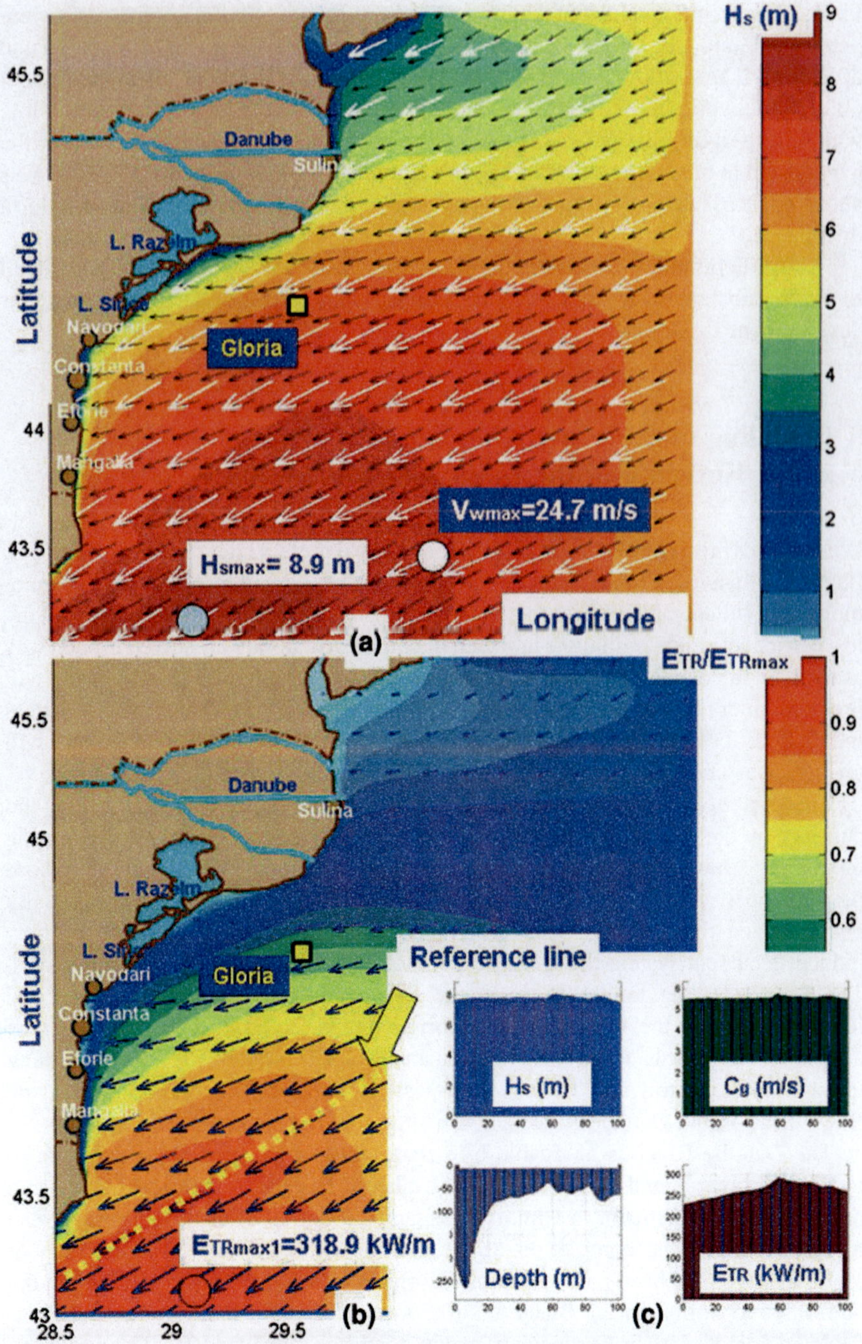


**Fig. 6** As Fig. 5 for CS4—2002/03/11/h18, extreme wave conditions

First, the maximum values of wave power in the Iberian coastal environment were between 50 and 100 kW/m, while in the Black Sea these values lie between 20 and 50 kW/m. The highest energetic conditions in the Iberian nearshore were about 600 kW/m, while in the Black Sea they are around 320 kW/m. At the Sines buoy, which is operating in the central part of the Portuguese continental nearshore, the average value of significant wave height recorded during the period 1994–2003 was about 1.7 m. This is very close to the average value in the Black Sea at the Gloria drilling unit.

In consideration of the above, the Black Sea, which was not generally believed to have substantial wave energy resources, may actually have potential for development especially on its western side.

## 5 Modeling of the Wave–Current Interactions at the Entrance of the Danube Delta

When waves propagate on a variable current several mutual interactions might occur. Physically, wave-current interaction is a problem of wave propagation in an inhomogeneous, dispersive, dissipative and moving medium. Since surface waves belong to a particular group of motions in which the fluid particles are organized in such way as to show the oscillatory surface elevation, any concurrent motion in the fluid will produce interactions; for example, wave-wave interactions and wave–current interactions. From these interactions, certain changes in the wave characteristics will arise such as changes in wave amplitude, in wave direction and in phase speed.

The aim of this section is to evaluate the interactions between waves and currents in the Black Sea at the entrance of the Danube Delta by performing numerical simulations with the wave prediction system SWAN based whose performances were discussed and evaluated Sect. 3. The targeted region is usually characterized through high wave and wind conditions and the Danube outflow induces a current system that generates sometimes strong interactions with waves. The Danube is the longest river in the European Union and Europe's second longest river after the Volga. The river originates in the Black Forest in Germany and flows eastwards for a distance of about 2,850 km (1,771 miles), passing through four Central and Eastern European capitals, before emptying into the Black Sea via the Danube River Delta.

The Danube River is navigable by ocean-going ships from the Black Sea to Brăila in Romania and by river ships to Kelheim, Bavaria; smaller watercraft can navigate further upstream to Ulm, in Germany. About 60 of its tributaries are also navigable. Since the completion of the German Rhine–Main–Danube Canal in 1992, the river has been part of a trans-European waterway from Rotterdam on the North Sea to Sulina on the Black Sea (3,500 km). In 1994 the Danube River was declared one of ten Pan-European transport corridors (corridor VII). The waterway is designed for large scale inland vessels (110 × 11.45 m) but it can carry much larger vessels on most of its watercourse. The Danube branches into three main

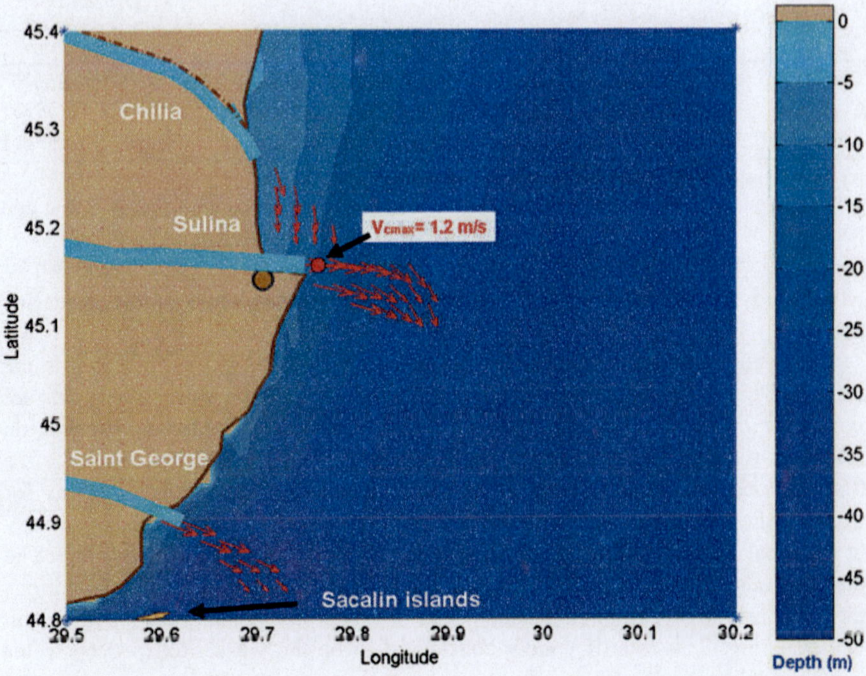


**Fig. 7** Map of the target area. In background the bathymetry is represented while in foreground the current induced by the river outflow in normal conditions

arms at its delta, Chilia, Sulina, and Sfântu Gheorghe as illustrated in Fig. 7. The last two branches form the Tulcea arm, which continues as a single body for several kilometers after the separation with Chilia arm. At the mouths of each arm alluvium processes take place, as a sign that the delta is continuing its expansion.

Chilia, in the north, is the longest, youngest, and most vigorous, with two secondary internal deltas and one in full process of formation at its mouth (to Ukraine). Sulina is the central and thus the shortest arm, which consequently led to its extensive use for traffic and severe transformation. At its mouth is located the main port (Sulina) and the single settlement with urban characteristics of the Romanian part of the delta. Because of the alluvium deposited at its mouth, a channel gradually advancing into the sea (presently it has 10 km), was built in order to protect the navigation. Sfântu Gheorghe (Saint George), in the south, is the oldest and more sparsely populated. Its alluvium has led to the creation, beginning with 1897, of the Sacalin islands, which nowadays measure 19 km in length.

## *5.1 System Focusing Towards the Target Area*

The wave prediction system, SWAN based, that was implemented in the entire Black Sea basin was focused now on the west side of the sea. This methodology

**Table 4** Computational grids used in the SWAN simulations

| Grids | $\Delta x \times \Delta y$ | $\Delta t$ (min) | nf | n$\theta$ | ngx × ngy = np |
|---|---|---|---|---|---|
| Global | 0.08° × 0.08° | 20 non-stat | 35 | 24 | 176 × 76 = 13376 |
| Coastal | 0.02° × 0.02° | 20 non-stat | 35 | 36 | 141 × 141 = 19881 |
| Local | 0.005° × 0.005° | 20 non-stat | 35 | 36 | 141 × 121 = 17061 |

$\Delta x$ and $\Delta y$ represent the resolution in the geographical space
$\Delta t$ time resolution, *nf* number of frequencies, $n\theta$ number of directions, *ngx*, *ngy* number of grid points in *x* and *y* direction, respectively, *np* total number of grid points

has the advantage that one single model covers the full scale of the modeling process.

Three computational levels were defined in the system focusing towards the Danube Delta and the characteristics of the corresponding computational grids are presented in Table 4. They correspond to the generation scale, the transformation scale and the local computational domain, respectively. The bathymetries for the first and second computational levels were obtained from the internet (as for example the web site of the US Naval Oceanographic Office) while the bathymetry of the local level was digitalized from a high resolution hydrographical map. The physics considered for each level is rather different. For the large scale simulations the emphasis was put on the deep water wave processes and among them on whitecapping. This is still widely considered to be the weak link in deep water wave modeling.

For the high resolution domain some processes specific to shallow water as triad wave-wave interactions and diffraction were also activated. Moreover, in order to account for the wave–current interactions, the current field was introduced in the model simulations. In this situation the currents accounted for are those generated by the river outflow combined with the currents induced by the local wind and waves. This is based on the relationship:

$$\vec{v}_{\mathrm{c}} = \vec{v}_{\mathrm{Rc}} + \vec{v}_{\mathrm{wc}}, \tag{22}$$

where $\vec{v}_{\mathrm{Rc}}$ is the river current and $\vec{v}_{\mathrm{wc}}$ are the wind and wave induced currents. The currents induced in the sea by the river in normal to high flowing conditions usually varies from 1.2 m/s at river mouths to 0.3 m/s about seven kilometers offshore (Almazov et al. 1963). Further descriptions of the current systems at the mouths of the river Danube including the results of some measurements and field experiments are presented in Stănica and Panin (2009). Using the above results an average to high current field corresponding to the river outflow was constructed. The wind and wave induced currents were computed as:

$$\vec{v}_{\mathrm{wc}} = \vec{U}_{\mathrm{C_{wind}}} + \vec{U}_{\mathrm{Stokes}}, \tag{23}$$

Where $\vec{U}_{\mathrm{C_{wind}}}$ is the wind induced current that has the value $U_{\mathrm{Cwind}} = 0.035U_{\mathrm{wind}}$, with $U_{\mathrm{wind}}$ the wind speed measured at 10 m height above the sea. The direction of the current motion induced by the wind is at a non-zero angle to the direction of the wind as a result of the Ekman effect. The formula proposed by

Samuels et al. (1982) is considered to compute the deflection angle. According to this the wind-driven current is deflected with an angle $\theta$ to the right (clockwise) from the wind direction (for the Northern Hemisphere):

$$\theta = 25^\circ \exp\left(-10^{-8}\frac{U_{\text{wind}}^3}{\upsilon g}\right), \tag{24}$$

where $g$ is the gravitational acceleration and $\upsilon$ the kinematic viscosity of the seawater. The vertical variation of wind induced surface current can be modeled by the inclusion of a logarithmic velocity profile. In analogy with flow structure within a bottom boundary layer it may be assumed that the surface layer moves at a velocity which decays with the depth as presented by Elliott (1986). Nevertheless, since SWAN model deals only with surface currents this effect was not considered in the present work. $\vec{U}_{\text{Stokes}}$ is the Stokes drift that was estimated using the mass transport relationship from the second order theory (Lakshmi and Clayson 2000):

$$\bar{u}(z) = \left(\frac{\pi H}{L}\right)^2 \cdot \frac{C}{2} \cdot \frac{\cosh[4\pi(z+d)/L]}{\sinh^2(2\pi d/L)}. \tag{25}$$

where $C$ is the phase velocity defined using the standard definitions (CERC 1984) and $z$ the water particle position relatively to the still water level. Since the goal was to estimate an average value for the Stokes drift at the surface it was considered $z = 0$ and the root mean square wave height ($H_{\text{RMS}}$) was used instead of $H$ in the above relationship. $H_{\text{RMS}}$ was deduced from the significant wave height considering the standard Rayleigh distribution ($H_s = 1.416H_{\text{RMS}}$).

For the simulations performed in the high resolution computational domain, the parameter *froudmax* (which represent the maximum value of the Froude number) was set to 0.9 (instead of 0.8) and the parameter CDD (which is the coefficient that controls the diffusion in the refraction scheme) was set to 1 this switching from a low diffusive central scheme that has larger accuracy to a first order upwind scheme that it is more diffusive and therefore preferable to avoid numerical oscillations if (strong) gradients in depth or current are present.

## 5.2 *Additional Validation Tests*

Validations against measured data were carried out for each computational level. As regards the generation scale, the performances of the SWAN based predictions in the Black Sea basin using data sources located in both the eastern and western sides of the sea were discussed in Sect. 3. The emphasis was put on the processes of whitecapping and wave generation by wind, but some studies on the results provided by various parameterizations for quadruplet wave-wave interactions were also carried out.

Model simulations were performed in the non stationary mode, with a time step of 20 min. The time interval considered (denoted as TI-2) covers the first 7 months of 2002 (2002/01/01–2002/07/31). As check point, the Gloria drilling platform was considerd where registrations of the wave parameters were systematicaly performed using various devices. The statistical results corresponding to TI-2 in terms of significant wave height and wave period were presented in Table 3.

A high energetic situation from this time interval is illustrated in Fig. 8 where the coastal focusing corresponding to the time frame 2002/03/25/h14 is presented.

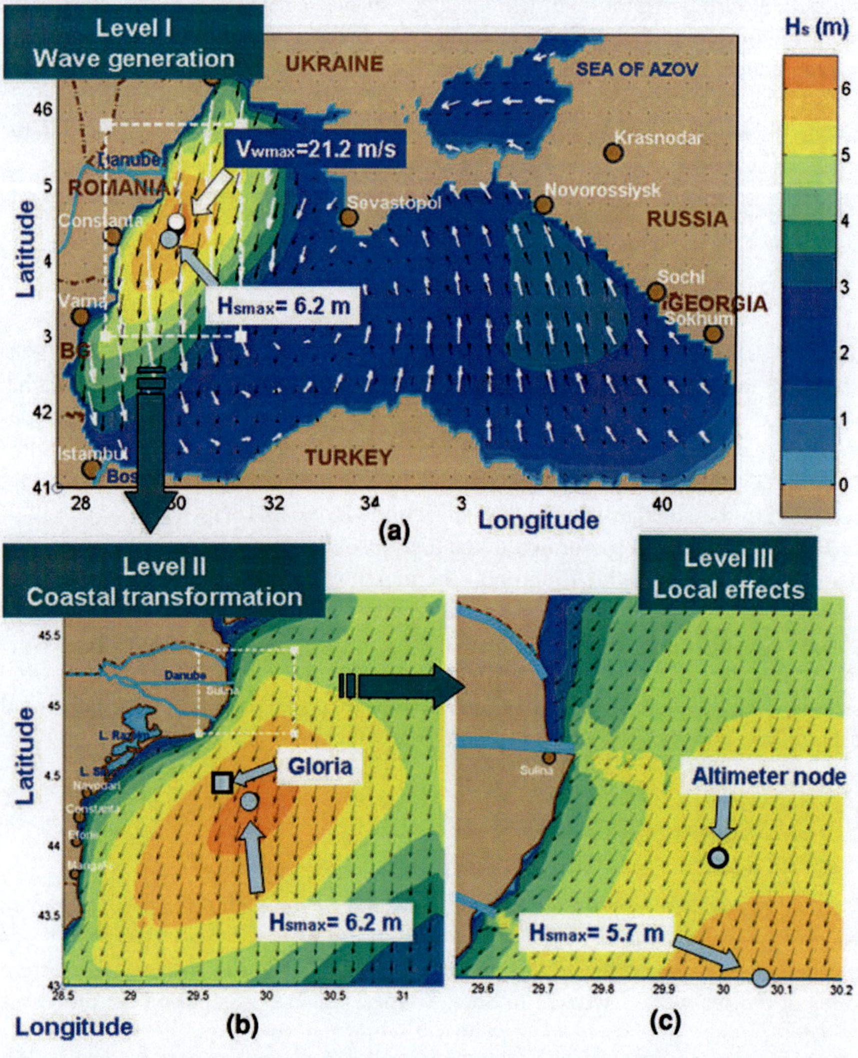


Fig. 8 SWAN model zooming towards the target area (Danube delta), time frame 2002/03/25/h14, average to high energetic situation in the west side of the Black Sea. In background significant wave height scalar fields, in foreground wave vectors (*black arrows*) and wind vectors (*white arrows*)

**Table 5** Wave statistics in the medium and high resolution domains

| | *X* | *Y* | Bias | RMSE | SI | *r* | Device | Time interval |
|---|---|---|---|---|---|---|---|---|
| Hs (m) | 1.54 | 1.52 | 0.02 | 0.76 | 0.49 | 0.72 | Wave staff | TI-2 ($n = 684$) |
| Tm (s) | 5.08 | 3.74 | 1.34 | 1.94 | 0.38 | 0.23 | | |
| Hs (m) | 2.36 | 2.45 | −0.09 | 1.02 | 0.43 | 0.80 | Wave staff | TI-3 ($n = 247$) |
| Tm (s) | 5.12 | 4.48 | 0.64 | 1.38 | 0.27 | 0.41 | | |
| Dir (°) | 263 | 270 | −6.6 | 51.8 | 0.20 | 0.89 | | |
| Hs (m) | 1.19 | 1.16 | 0.03 | 0.37 | 0.31 | 0.81 | Altim. | TI-4 ($n = 91$) currents |
| Hs (m) | 1.19 | 1.13 | 0.05 | 0.40 | 0.34 | 0.78 | Altim. | TI-4 ($n = 91$) Without |

*X* represents the average value of the measured wave parameters and *Y* the corresponding average value for the period considered of the wave parameters resulting from model simulations. *n* represents the number of data points

Thus, Fig. 8a illustrates the entire basin of the sea, in background being presented the significant wave height scalar fields, and in foreground the wave vectors (black arrows) and the wind vectors (white arrows). Figures 8b and c, respectively, illustrate the coastal and local levels presenting in background the significant wave height scalar fields and in foreground the wave vectors (black arrows). The locations of the two data sources considered are also represented in Fig. 8. From the simulations carried out in the 7 month period above considered, five case studies were selected as most relevant patterns and they will be discussed in Sect. 5.3.

Since in the period December 2003–January 2004 directional wave data were available at the Gloria drilling platform, this time window was considered as a second interval for validations in the medium resolution domain (TI-3, 2003/12/01–2004/01/31). The corresponding statistical results of the main wave parameters for this time interval are presented also in Table 5 while the direct comparisons between the measurements and the SWAN models simulations are illustrated in Fig. 9a. For the significant wave height a scatter plot is also presented in Fig. 10a. By comparing now the results of the model from the two time intervals considered it can be observed that they are in general coherent. In terms of significant wave height although in TI-2 the results are better as regards RMSE, in relationship with the scatter index and the correlation coefficient they are better for TI-3. In terms of mean period the results are balanced in the two time intervals considered although they are slightly better in TI-3. As regards the wave direction the results show a very good correlation.

In the last years satelite data became available on various internet sites (as for example http://las.aviso.oceanobs.com maintained by NOAA, US National Oceanic and Atmospheric Administration). An altimeter node that gives near real time multi-mission merged non interpolated values of the significant wave height is located in the target area (30°E, 45°N) and its position is illustrated in Fig. 8c.

The check-point considered is in intermediate to deep water (about 40 m water depth) where no relevant local effects are encountered and hence these data can be

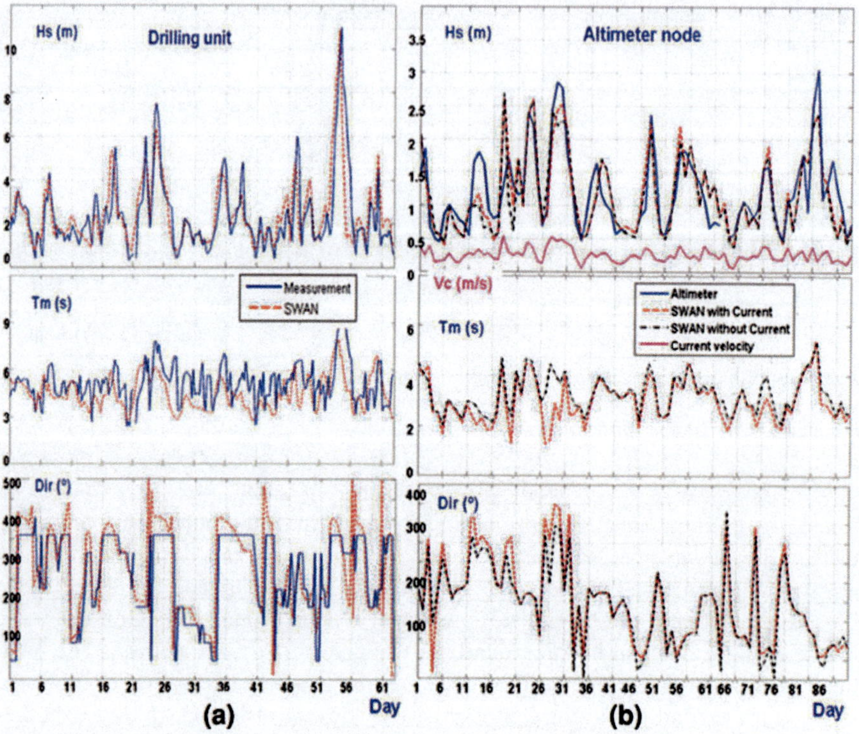


**Fig. 9** Direct comparisons SWAN against measurements. **a** Simulations in the second computational level against measurements performed at Gloria drilling platform (TI-3). **b** Simulations in the target area both with and without currents against satellite data (TI-4)

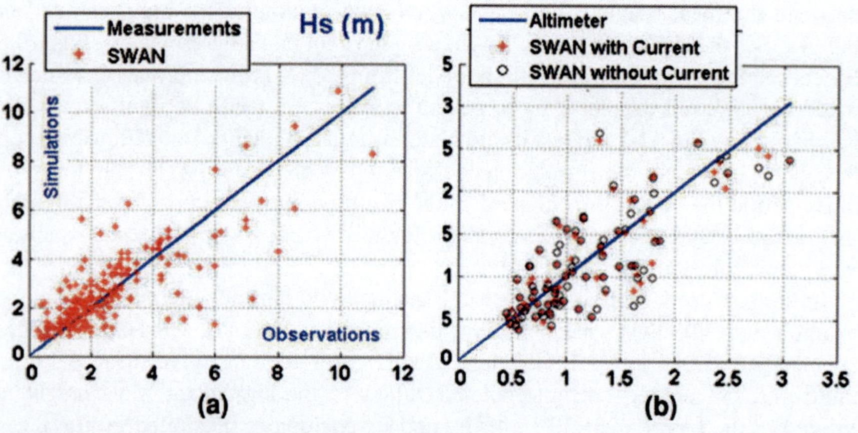


**Fig. 10** Hs scatter plots, SWAN against measurements. **a** simulations in the second computational level; **b** Simulations in the target area

considered as reasonable accurate for validating the results of the simulations performed in the high resolution computational domain. In order to assess the influence of the currents, simulations with and without including the current fields were performed in parallel. This third time interval considered for validations in the high resolution domain was denoted as TI-4 and is the 3-month period at the beginning of 2007 (2007/01/01–2007/03/31). The direct comparisons are illustrated in Fig. 9b. Although only the significant wave height was actually compared at this level with the measurements, for the other parameters considered (mean period and wave direction) a comparison between the SWAN simulations with and without currents was also performed. The scatter plot for the significant wave height is illustrated in Fig. 10b and the statistical results for both simulations are presented in Table 5.

The analysis of the statistical parameters makes clear that, although slight, the improvement of the results is visible when the current field is included. Moreover, since the check point is located practically outside the influence of the river outflow, it is supposed that the influence in some areas strongly affected by the river current should be considerably higher.

The results of the validation tests carried out at each computational level highlighted that, under certain limitations, the system developed provides reliable information on the wave conditions in the target area where the local effects induced by the Danube River's outflow will be next studied.

## *5.3 Analysis of Five Case Studies*

Since the average wind velocity and the periods of calm have a quite pronounced cyclic evolution in the Black Sea region, some relevant patterns can be defined. Such cases will be forward analyzed in more detail. Hence, in order to assess the interactions between waves and currents at the entrance of the Danube Delta, five case studies were considered. They belong to the simulations performed in TI-2 and correspond to some environmental patterns that can be considered the most significant for average to high energetic conditions in the target area.

### 5.3.1 Case Study 5 (CS5): Weak Wind Conditions

This case study refers to the situation encountered in 2002/04/16/h22, characterized by average wave conditions and weak wind. Although the coastal sector considered is usually characterized by strong wind conditions the weak wind may still represent a significant case. In this situation the currents accounted for are mainly those generated by the river outflow.

The results of the wave model simulations performed in the high resolution area are illustrated in Fig. 11. Thus, Fig. 11a presents the significant wave height scalar fields and the wave vectors corresponding to the SWAN model simulation without

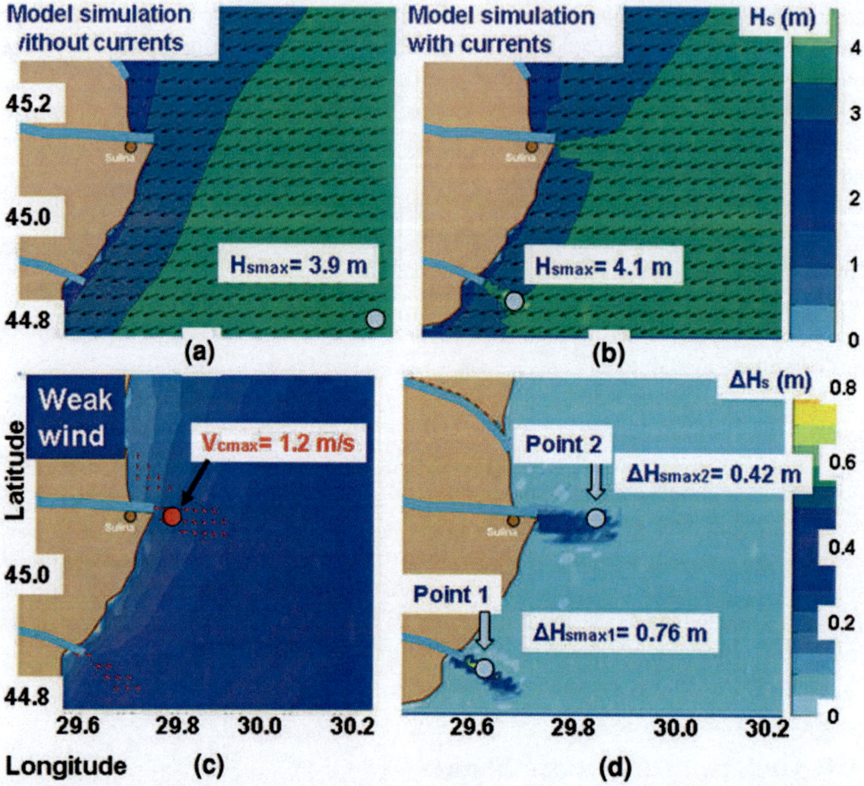


**Fig. 11** CS5 (time frame 2002/04/16/h22), average to high energetic conditions and zero local wind; **a** wave model simulation without currents; **b** wave model simulation with currents. In background significant wave height fields are represented while in foreground the wave vectors (*black arrows*). **c** Representation of the current field (*red arrows*) and information on the average wind conditions in the target area, in background the bathymetric map is illustrated. **d** Variation of the significant wave height due to the presence of the currents

accounting for the currents, while Fig. 11b presents the results corresponding to the same conditions but accounting for the current field. Figure 11c illustrates in foreground the current field (with red arrows) and in background the bathymetry is presented. Figure 11d presets the significant wave height bias fields as resulting from the SWAN model simulations due to the presence of currents. The significant wave height bias was computed as:

$$\Delta H_s(i,j) = H_{sc}(i,j) - H_s(i,j), \tag{26}$$

where $H_{sc}$ is the significant wave height computed considering the current field and $H_s$ the significant wave height computed without considering the current field while $i$ and $j$ are integer numbers defining the position of the grid point (the row and the column respectively in the grid matrix).

As evident from Fig. 11d the currents induce an increase of the significant wave height of 0.76 m at Saint George arm and of 0.42 m at the entrance of Sulina channel, corresponding to relative increases of about 23 and 12%, respectively.

The 1D spectral shapes corresponding to these two reference points for the two situations considered (with and without currents) are presented in Fig. 12

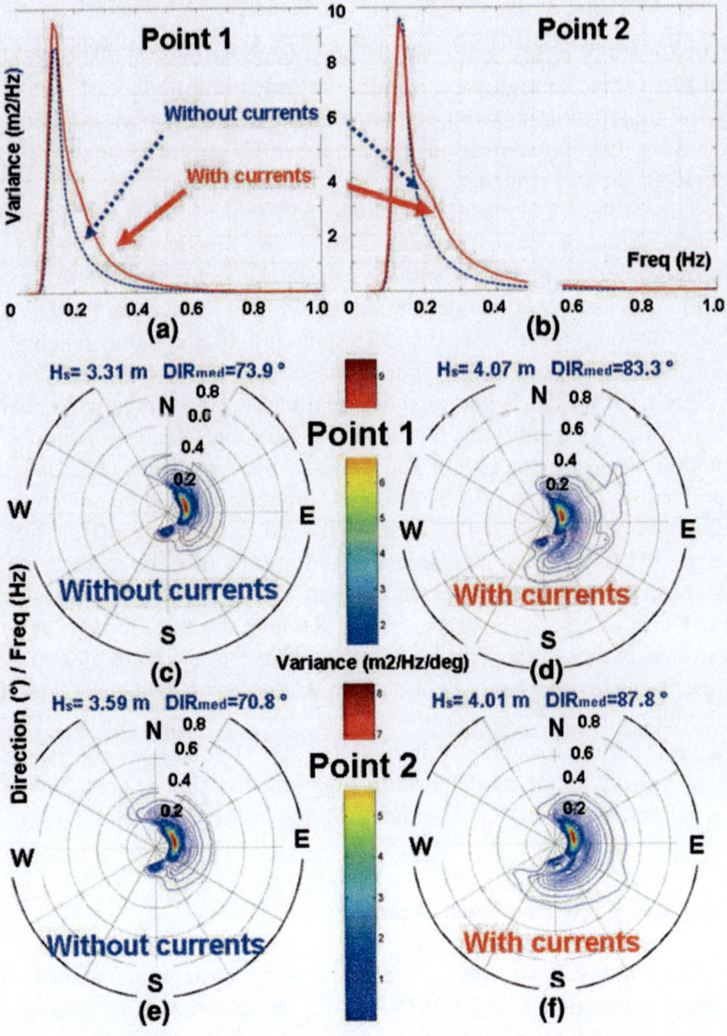


**Fig. 12** CS5, spectral analysis. Reference points considered: P1 at the mouth of Sf. Gheorghe arm and P2 at the mouth of Sulina channel. **a** P1—comparison of the 1D spectrum (as resulting without and with currents, respectively). **b** P2—comparison of the 1D spectrum; **c** P1—representation of the 2D spectrum, without currents. **d** P1—representation of the 2D spectrum, with currents. **e** P2—representation of the 2D spectrum, without currents. **f** P2—representation of the 2D spectrum, with currents

(a and b). In Figs. 12c and e the 2D spectra at the reference points 1 and 2, respectively, corresponding to the simulation without currents are represented while Figs. 12d and f illustrate the 2D spectra at the same points but corresponding to the simulation with currents.

### 5.3.2 Case Study 6 (CS6): Strong North East Wind

The next case study refers to the situation encountered in 2002/02/04/h09, characterized by average to high wave conditions and strong north east wind. This is actually the most common wind pattern for that coastal region and corresponds also to a very common situation of average to high energetic waves that is encountered in the target area.

The results of the wave model simulations performed for this second case study in the high resolution computational domain are illustrated in Fig. 13. Thus, Fig. 13a presents the significant wave height scalar fields and the wave vectors corresponding to the SWAN model simulation without accounting for the currents, while Fig. 13b presents the results corresponding to the same conditions but accounting for the current field. Figure 13c illustrates in foreground the current field (with red arrows) and gives also information on the average wind conditions in the target area by specifying the mean velocity and mean direction is represented with a white arrow. Figure 13d presents the significant wave height bias fields as resulting from the SWAN model simulations due to the presence of currents. As Fig. 13d shows, the currents induce an increase of the significant wave height of 1.06 m at Saint George arm and of 0.66 m at the entrance of Sulina channel, corresponding to relative increases of about 27 and 18%, respectively.

A rather similar case is illustrated in Fig. 8 where the system focusing towards the target area is presented. This corresponds to the time frame 2002/03/25/h14 and represents a high energetic situation with wind coming from North- Northeast. Although more energetic than the present case study, the case illustrated in Fig. 8 shows a very similar spatial distribution for the significant wave height. The analysis of various other model results suggests that this is actually the most probable pattern for the high energetic situations in the target area.

### 5.3.3 Case Study 7 (CS7): West Wind

The third case study related with the process of wave–current interactions refers to the situation encountered in 2002/04/18/h15, characterized by average wave conditions and wind blowing from west. Although usually such situation is not the most probable, according to the statistics it may be however encountered about 20% of the time. In this situation the currents generated by the river outflow are increased by the wind induced currents.

The results of the wave model simulations performed for this third case study in the high resolution computational domain are illustrated in Fig. 14. Figure 14a

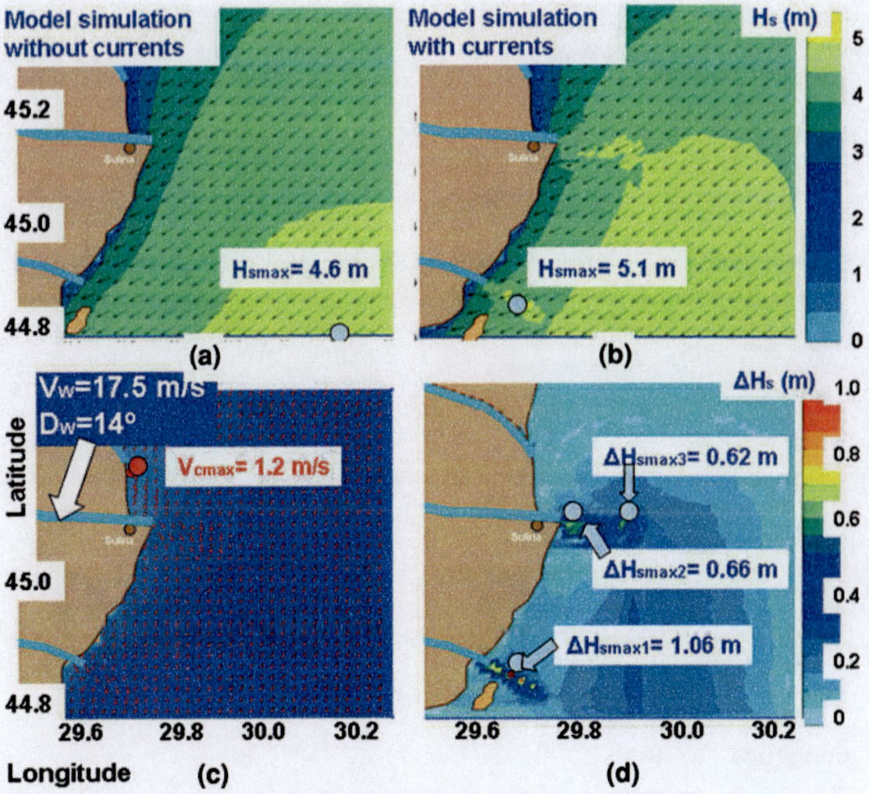


**Fig. 13** CS6—(time frame 2002/02/04/h09; **a** wave model simulation without currents; **b** wave model simulation with currents. In background significant wave height fields and in foreground the wave vectors (*black arrows*). **c** Representation of the current field (*red arrows*) and information on the average wind conditions in the target area (mean velocity specified and direction represented with a *white arrow*). In background the bathymetric map is illustrated. **d** Variation of the significant wave height due to the presence of the currents

presents the significant wave height scalar fields and the wave vectors corresponding to the SWAN model simulation without accounting for the currents while Fig. 14b presents the results corresponding to the same conditions but accounting for the current field. Figure 14c illustrates in foreground the current field (with red arrows) and gives also information on the average wind conditions in the target area by specifying the mean velocity and mean direction is represented with a white arrow. Figure 14d presents the significant wave height bias fields as resulting from the SWAN model simulations due to the presence of currents.

As resulting from Fig. 14d the currents induce an increase of the significant wave height of 0.4 m at Saint George arm and of 0.29 m at the entrance of Sulina channel, corresponding to relative increases of about 37 and 18%, respectively.

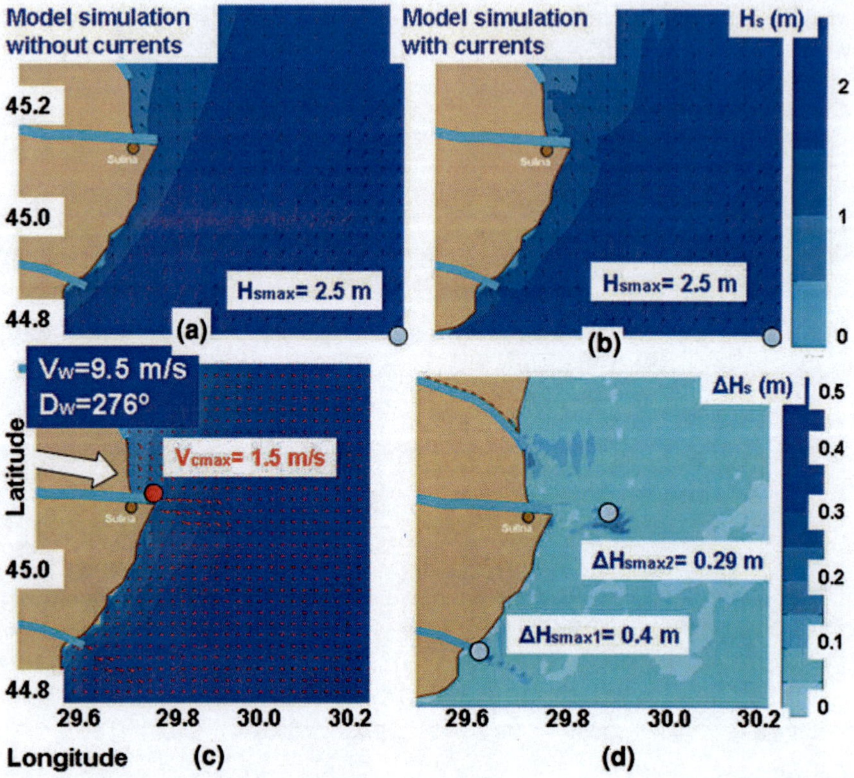


**Fig. 14** As Fig. 13 for CS7 (time frame 2002/04/18/h15)

### 5.3.4 Case Study 8 (CS8): East Wind

The next case study refers to the situation encountered in 2002/05/27/h12, characterized by average wave conditions and wind blowing from east.

The results of the wave model simulations performed for this case study in the high resolution computational domain are illustrated in Fig. 15. Figure 15a presents the significant wave height scalar fields and the wave vectors corresponding to the SWAN model simulation without accounting for the currents while Fig. 15b presents the results corresponding to the same conditions but accounting for the current field. Figure 15c illustrates in foreground the current field (with red arrows) and gives also information on the average wind conditions in the target area by specifying the mean velocity and mean direction is represented with a white arrow. Figure 15d presents the significant wave height bias fields as resulting from the SWAN model simulations due to the presence of currents.

As resulting from Fig. 15d the currents induce an increase of the significant wave height of 0.65 m at Saint George arm and of 0.45 m at the entrance of Sulina channel, corresponding to relative increases of about 37 and 22%, respectively.

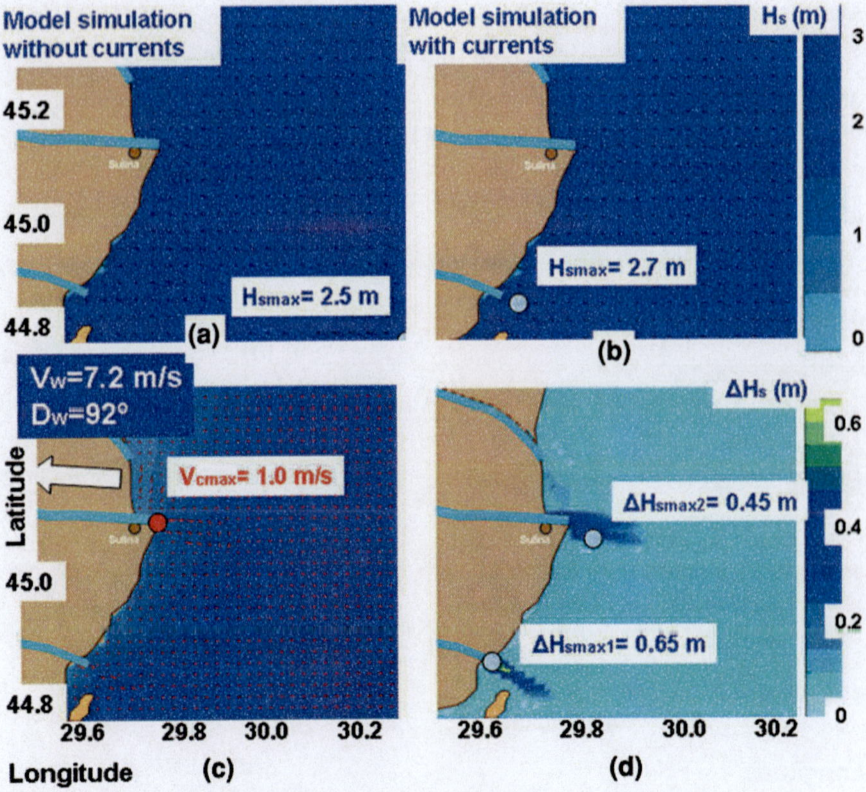


**Fig. 15** As Fig. 13 for CS8 (time frame 2002/05/27/h12)

### 5.3.5 Case Study 9 (CS9): Wind Coming from South

Finally, the fifth case study refers to the situation encountered in 2002/06/03/h18, characterized by average wave conditions and wind blowing from south.

The results of the wave model simulations performed for this last case study in the high resolution computational domain are illustrated in Fig. 16. Figure 16a presents the significant wave height scalar fields and the wave vectors corresponding to the SWAN model simulation without accounting for the currents while Fig. 16b presents the results corresponding to the same conditions but accounting for the current field. Figure 16c illustrates in foreground the current field (with red arrows) and gives also information on the average wind conditions in the target area by specifying the mean velocity and mean direction is represented with a white arrow. Figure 16d presents the significant wave height bias fields as resulting from the SWAN model simulations due to the presence of currents. As resulting from Fig. 16d the currents induce an increase of the significant wave height of 0.65 m at Saint George arm and of 0.32 m at the entrance of Sulina channel, corresponding to relative increases of about 55 and 21%, respectively.

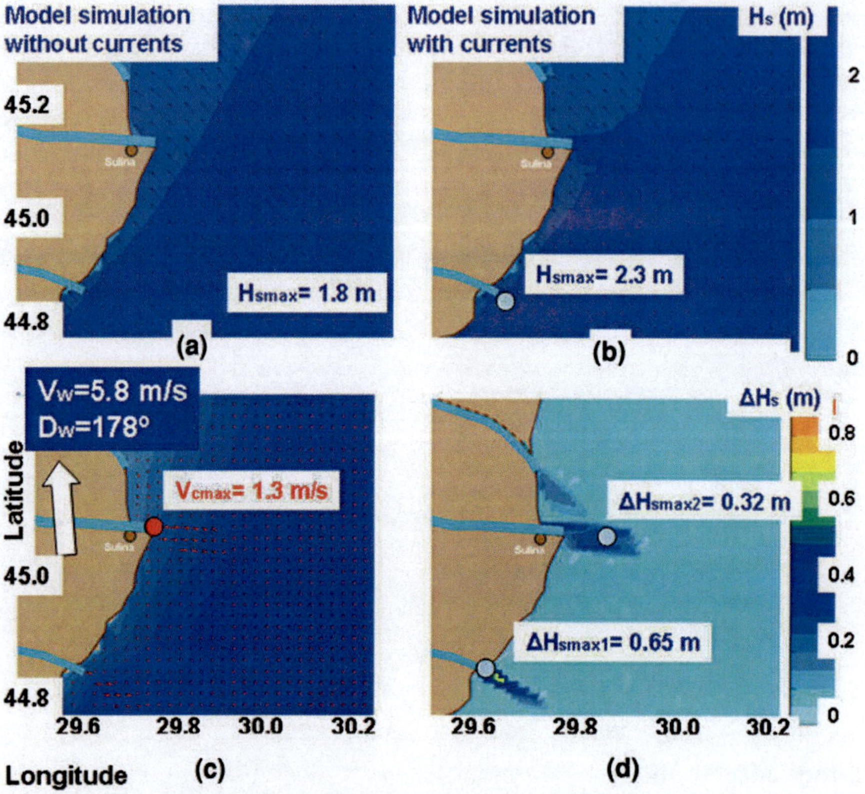


**Fig. 16** As Fig. 13 for CS9 (time frame 2002/06/03/h18)

## *5.4 Discussion of the Results*

From the analysis of the five case studies above presented some discussions can be made in relationship with the effects of the interactions between waves and currents. These effects actively drive an alluvium processes which is very dynamic in the coastal sector under analysis.

The Danube branches into three main distributary arms into its delta: Chilia channel, Sulina channel, and Sfântu Gheorghe channel. As regards the Chilia arm, in the north, especially due to the angle of the flow (see for example Fig. 11) the interactions between waves and currents do not usually induce notable transformations in the wave field. The strongest wave–current interactions occur at the arm Sfântu Gheorghe (Saint George), in the south. Its alluvium has led to the creation, beginning with 1897, of the Sacalin islands, which are already 19 km in length and still enlarging.

On the other hand, of particular importance are the local effects at the entrance of the Sulina channel which is actually the main gate for the seventh Pan-European transport corridor.

Five case studies (CS5-9) corresponding to real situations that cover in some sense the most relevant features of the environmental matrix in that particular area were considered and analyzed in this section. CS5 (2002/04/16/h22) refers to average wave conditions coming from north east and weak wind conditions.

Figure 12a and b illustrates the variations due to the presence of the currents of the 1D energy spectra in the two reference points considered for CS5. A general observation would be that although not necessarily increasing the value of the spectral energetic peak, the presence of currents induces globally more energy to the waves.

A qualitatively better image of the wave transformations due to the current effect is given in Fig. 12c and d (for Point 1) and Fig. 12e and f (for Point 2) where comparisons of the 2D spectra are presented in parallel for the situations without currents and with currents corresponding to the first case study considered. Thus the effects of the opposite currents on waves that have been discussed in Sect. 2, as refraction and increasing of the wave heights in terms of directional energy distribution can be well noticed from these figures.

CS6 (2002/02/04/h09) consider average to high wave conditions coming from north east and high wind (17.5 m/s in the target area) blowing from north east. This case study represents in fact the most relevant pattern for the sea traffic because the probability that such case may occur in the 6-month highest energetic period (from October to March) is usually very high. In fact, a rather similar situation is illustrated also in Fig. 8 where the coastal focusing of the system on the Danube Delta is presented for a different time frame (2002/03/25/h14).

CS7 (2002/04/18/h15) illustrates a more unusual pattern that may however occur. In this case the wind is blowing from west, north-west (with an average intensity of 9.5 m/s) and the waves are propagating from south to north. Such cases might be important because the wind generated currents are acting in the same direction with the currents induced by the river outflow increasing in this way the total current. Theoretically such case could lead also some time to supercritical flows that are not accounted for in the wave model. However, it has to be mentioned that for the particular situation presented in CS7 the supercritical flow was not reached although this was really close.

In order to give a better perspective of the possibilities concerning the combinations of the elements of the environmental matrix that might be encountered in the target area, two more case studies are also discussed. These are: CS8 (2002/05/27/h12) and CS9 (2002/06/03/h18) corresponding to wind coming from east and south, respectively (with average intensities of 7.2 and 5.8 m/s, respectively). An analysis of the local data in the regions most affected by the wave–current interactions indicated by reference Point 1 in front of Saint George arm and by reference Point 2 at the entrance of Sulina channel is supported by Table 6 and Fig. 17. Thus the values of the most relevant wave parameters corresponding to the model simulations performed with and without currents are presented in parallel. These parameters are the significant wave height ($H_s$), mean period (Tm01), wavelength, and mean wave direction.

**Table 6** The values of the main wave parameters corresponding to the model simulations without and with currents for the five case studies considered

| Case study | Coordinates (long-lat °) | Current speed $V_c$ (m/s) | Hs (m) | Tm (s) | Wl (m) | Dir (°) |
|---|---|---|---|---|---|---|
| CS5 | (P1) $X_1 = 29.6594$ $Y_1 = 44.8801$ | 0.0 | 3.34 | 5.05 | 26.2 | 73.6 |
| | | 0.74 | 4.1 | 4.76 | 20.2 | 82.4 |
| | (P2) $X_2 = 29.8652$ $Y_2 = 45.1547$ | 0.0 | 3.5 | 5.15 | 27.6 | 70.8 |
| | | 0.58 | 3.92 | 4.65 | 19.2 | 83.5 |
| CS6 | (P1) $X_1 = 29.6594$ $Y_1 = 44.8801$ | 0.0 | 4.0 | 5.37 | 28.7 | 67.4 |
| | | 0.88 | 5.06 | 4.8 | 21.2 | 81.4 |
| | (P2) $X_2 = 29.7648$ $Y_2 = 45.1967$ | 0.0 | 3.68 | 5.24 | 26.6 | 65.9 |
| | | 0.95 | 4.34 | 5.08 | 24.1 | 70.6 |
| CS7 | (P1) $X_1 = 29.6097$ $Y_1 = 44.9064$ | 0.0 | 1.07 | 4.34 | 16.7 | 156.1 |
| | | 1.17 | 1.47 | 4.28 | 14.0 | 129.0 |
| | (P2) $X_2 = 29.8652$ $Y_2 = 45.1567$ | 0.0 | 1.64 | 4.32 | 19.8 | 188.2 |
| | | 0.87 | 1.93 | 3.97 | 11.7 | 165.1 |
| CS8 | (P1) $X_1 = 29.6097$ $Y_1 = 44.9084$ | 0.0 | 1.77 | 5.71 | 31.5 | 113.7 |
| | | 0.60 | 2.42 | 5.18 | 25.3 | 112.0 |
| | (P2) $X_2 = 29.7902$ $Y_2 = 45.1304$ | 0.0 | 2.07 | 5.63 | 35.3 | 107.6 |
| | | 0.56 | 2.52 | 5.64 | 33.7 | 112.5 |
| CS9 | (P1) $X_1 = 29.6087$ $Y_1 = 44.9084$ | 0.0 | 1.18 | 4.4 | 20.9 | 119.1 |
| | | 0.78 | 1.83 | 3.27 | 8.1 | 96.0 |
| | (P2) $X_2 = 29.8176$ $Y_2 = 45.1304$ | 0.0 | 1.51 | 4.89 | 25 | 142.0 |
| | | 0.51 | 1.83 | 4.71 | 20.5 | 138.6 |

The wave parameters were estimated for the branches Saint George and Sulina in reference points (P1—for Saint George arm and P2—for Sulina channel) where the maximum variation of the significant wave heights were encountered due to the presence of currents for each case study

Table 6 presents also the position of the reference points considered for each case study, these points indicating the location where the maximum variation of the significant wave height is encountered. A first important observation would be that the positions of these points is usually very close (some time even identical) from one case study to another. Hence this observation allows the identification of the most dangerous locations for the coastal navigation.

As resulting also from Table 6 the currents induce in the two reference points variations of the main wave parameters. The variations in magnitude for the significant wave height were already discussed in Sect. 5.3. As regards the mean periods, the currents usually induce decreases less than one-second with the exception of CS9 where in P1 a decrease of 1.13 s was encountered, this representing a relative decrease of about 26%.

In relationship with the wavelengths these are decreased due to the currents usually with less than 10 m (about 40% relative decrease). However for the CS9 in P1 a decrease of 12 m was encountered, this representing a relative decrease of about 60%.

Finally, as regards the wave direction the currents induced in the target area variations of about 20°, the maximum variations being encountered in the CS7 when a variation of 27° was encountered for P1 and 23° for P2.

Up to this point the discussion was concentrated on the five case studies that although represent relevant patterns refer however to some punctual situations. Hence in order to give a better image on the overall effect of the wave–current interactions in the target area, the transformation of the waves due to currents will be next assessed for a longer time window. Figure 18 illustrates direct comparisons between the results of the SWAN simulations in the target area carried out without and with currents for the 3-month interval denoted as TI-4 when model validations against satellite data were performed. The two points considered for these comparisons are the reference points 1 and 2 as resulting from the previous analysis as being between the most affected and they correspond to CS6. The same parameters were compared as those from Fig. 17. The current velocities were also represented. Figure 19 illustrates the $H_s$ scatter plots corresponding to the simulations without currents against simulations with currents. Since the variation of the main wave parameters due to the currents for the point corresponding to the altimeter node is illustrated in Fig. 9b, in order to give an overall perspective on the influence of the currents Table 7 presents in statistical terms the changes induced by the currents on the waves in the three locations under consideration (Altimeter node, Point 1 and Point 2).

The analysis of the above data show that in the coastal environment considered even in the points where the average current is relatively weak, as the location of the altimeter node is, the current produces some changes in the wave field and for the case considered these mean especially changes of the wave periods (average decrease of 0.26 s) and directions (average variation of 7.5°).

As discussed in Sect. 5.3, the area in front of the Saint George arm is strongly affected by the currents induced by the river outflow and this is fully illustrated by the results presented in Figs. 18a and 19a and from Table 7. Here the average changes of the wave parameters are substantial. This implies an increase of about 0.2 m for the significant wave height, decreases of 1.85 s and 7.7 m for the mean period and wavelength, respectively and an average variation of the mean wave direction with about 15°.

Somehow unexpected weak are the average changes resulted at the reference Point 2 in front of Sulina channel.

Thus an increase of 0.07 m occurs for the significant wave height while the variations of the other wave parameters considered are not so high (decreases of 0.3 s and 1.9 m for the mean period and wavelength, respectively and an average variation of only 5° for the mean wave direction).

A possible explanation for this behavior of the main wave parameters is related with the energetic conditions. While in CS6 relatively strong energetic conditions were considered, the period TI-4 is mainly characterized by average to weak energetic conditions.

Finally, in order to give a more complete picture on the wave–current interactions impact, the BFI (Benjamin-Feir Index) variations were also evaluated for all the three points considered and they are illustrated in Fig. 20.

Benjamin-Feir Index, or the steepness-over-randomness ratio, has been introduced formally by Janssen (2003) and is defined as:

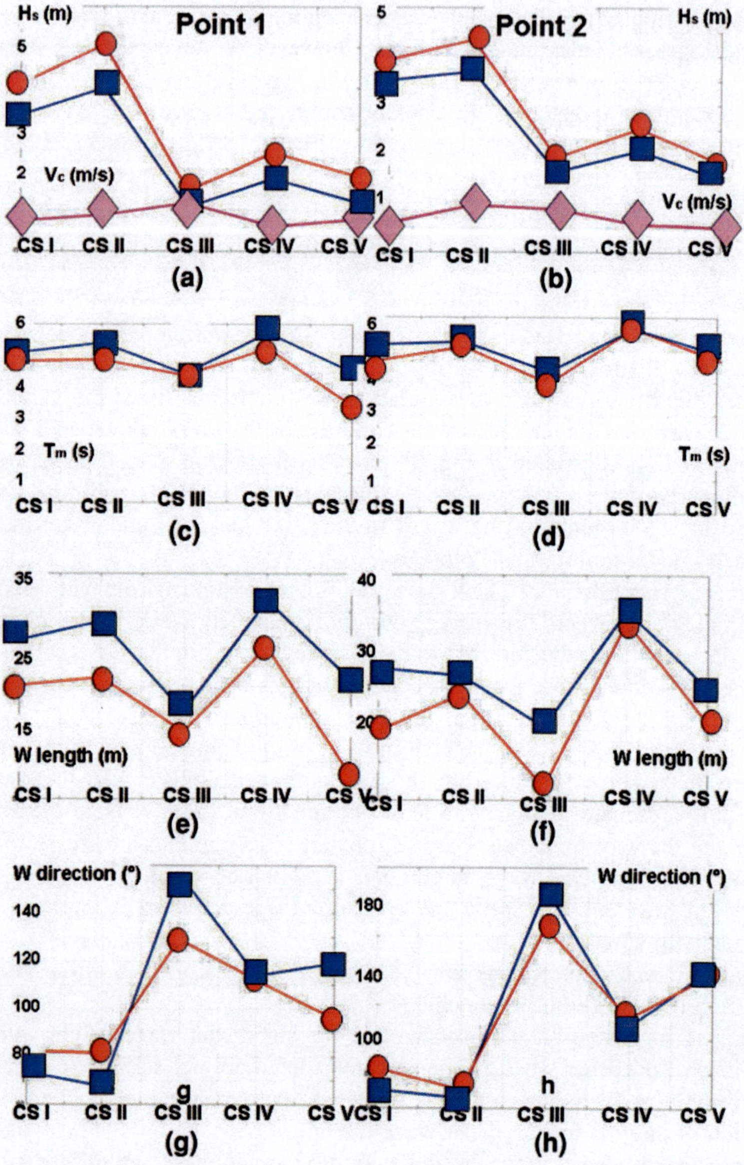


**Fig. 17** Variation of the main wave parameters corresponding to the model simulations without (represented with *blue squares*) and with currents (represented with *red circles*) for the five case studies considered. **a** P1—Hs and Vc; **b** P2—Hs and Vc; **c** P1—Tm; **d** P2—Tm; **e** P1—wave length; **f** P2—wave length; **g** P1—mean wave direction; **f** P2—mean wave direction

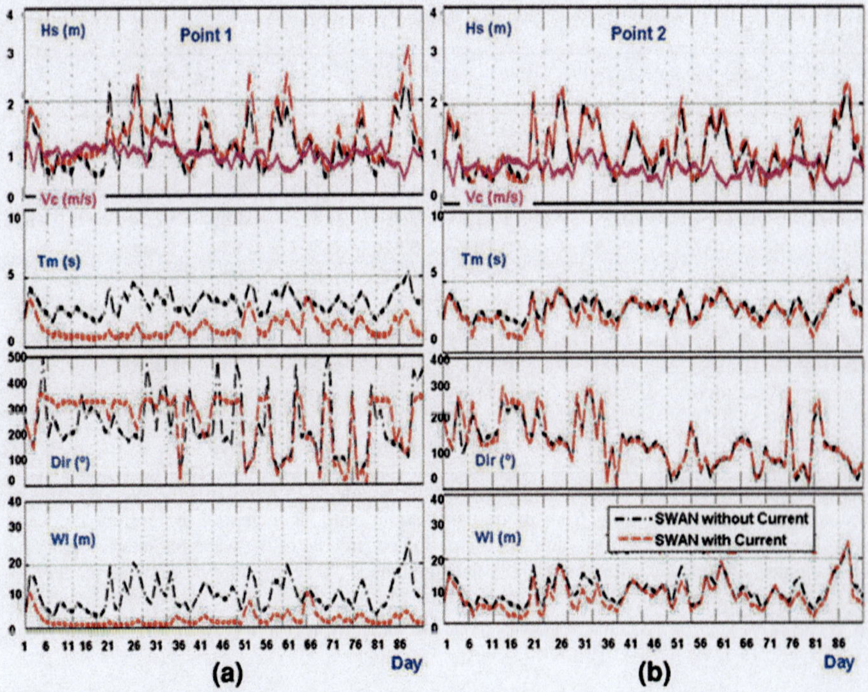


**Fig. 18** Comparisons SWAN simulations without and with currents in the target area (TI-3, 2007/01/01–2007/03/31). **a** Reference Point 1, in front of Saint George arm, **b** Reference Point 2, in front of Sulina channel

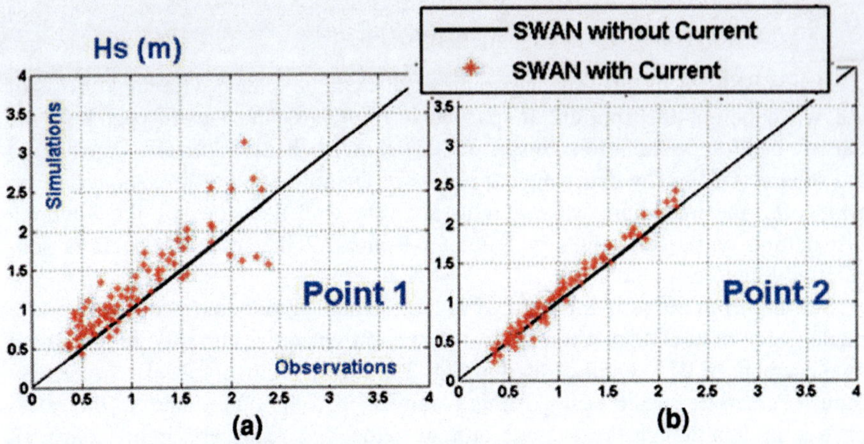


**Fig. 19** Hs scatter plots SWAN, simulations with currents against SWAN simulations without currents (TI-3, 2007/01/01–2007/03/31). **a** Reference Point 1, in front of Saint George arm, **b** Reference Point 2, in front of Sulina channel

**Table 7** Evaluation of the overall influence of currents on the wave fields in the three reference points considered

| $n = 91$ data | $X_{sim1}$ (*X*) | $X_{sim2}$ (*Y*) | Bias | RMSE | SI | *r* | Point |
|---|---|---|---|---|---|---|---|
| Hs (m) | 1.128 | 1.130 | −0.002 | 0.197 | 0.174 | 0.941 | Altimeter |
| Tm (s) | 3.249 | 2.987 | 0.262 | 0.469 | 0.144 | 0.879 | Node |
| Dir (°) | 167.1 | 174.5 | −7.4 | 21.82 | 0.131 | 0.976 | (weak current) |
| Wl (m) | 11.16 | 10.04 | 1.118 | 2.704 | 0.242 | 0.885 | Vcmed = 0.25 m/s |
| Hs (m) | 1.010 | 1.203 | −0.193 | 0.320 | 0.317 | 0.884 | Point 1 |
| Tm (s) | 3.070 | 1.215 | 1.855 | 1.97 | 0.642 | 0.586 | Saint George |
| Dir (°) | 238.4 | 253.2 | −14.8 | 89.6 | 0.376 | 0.737 | (strong current) |
| Wl (m) | 9.84 | 2.159 | 7.681 | 8.748 | 0.889 | 0.459 | Vcmed = 0.83 m/s |
| Hs (m) | 0.996 | 1.066 | −0.07 | 0.118 | 0.118 | 0.988 | Point 2 |
| Tm (s) | 3.065 | 2.753 | 0.312 | 0.497 | 0.162 | 0.924 | Sulina |
| Dir (°) | 162.0 | 167.0 | −5.0 | 27.89 | 0.172 | 0.959 | (strong current) |
| Wl (m) | 9.811 | 7.881 | 1.93 | 2.548 | 0.260 | 0.931 | Vcmed = 0.68 m/s |

Wave statistics in the high resolution computational domain. $X_{sim1}$ represents the average value of the wave parameters resulting from model simulations without including the current field and $X_{sim2}$ the corresponding average value for the period considered of the wave parameters resulting from model simulations with the current field included. The time interval considered is TI-3 (2007/01/01–2007/03/31)

$$\mathrm{BFI} = \sqrt{2\pi}\mathrm{St} \cdot Q_\mathrm{p}, \tag{27}$$

where St represents the integral wave steepness and is computed as the ration between the significant wave height and the wave length and $Q_\mathrm{p}$ represents the peakedness of the wave spectrum and it is defined as:

$$Q_\mathrm{p} = 2\frac{\int \sigma E^2(\sigma,\theta)d\sigma d\theta}{\left(\int \sigma E(\sigma,\theta)d\sigma d\theta\right)^2}. \tag{28}$$

Hence BFI is a spectral shape parameter that can be related with the kurtosis of the wave height distribution. In particular for Gaussian-shaped spectra in the narrow band approximation Janssen (2003) showed that the kurtosis depends on the square of BFI. The experimental results of Onorato et al. (2009) show that for BFI = 0.2 the maximum wave heights are very well described by the Rayleigh distribution while for values of BFI of 0.9 and 1.2 the ratio Hmax/Hs is substantially underestimated.

Starting from the results presented in Fig. 20 some other comments can be also made. The first is that almost always the presence of currents produces an enhancement of BFI. As illustrated in Fig. 20a even a relatively weak current may induce sometime strong peaks of this parameter. On the other hand, in the reference point 2 although the average current velocity is relatively high (0.68 m/s), Fig. 20c illustrates smaller enhancements for BFI and relatively smooth variations. Finally, Fig. 20b illustrates very clear the fact that the sector in front of Saint George arm presents a high risk of occurrence of strong waves not only from the

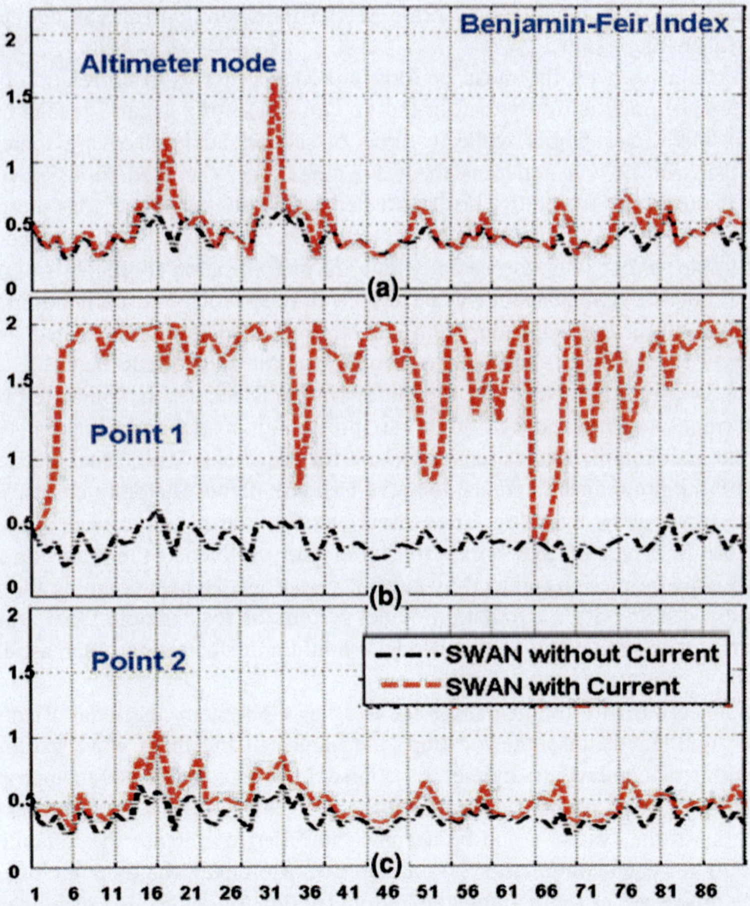


**Fig. 20** Variation of the Benjamin-Feir Index (TI-3, 2007/01/01–2007/03/31). **a** Altimeter node (30°E, 45°N); **b** Reference Point 1, in front of Saint George arm, **c** Reference Point 2, in front of Sulina channel

point of view of their significant wave height but also with ratios between the maximum wave height and the significant wave height often greater than standard value of 1.86 as resulting from the Rayleigh distribution.

## 6 Concluding Remarks

A first conclusion of the present work would be that, under certain limitations, the model system implemented and tested herewith can provide reliable information concerning the wave climate in the Black Sea basin, especially as regards predictions of the significant wave heights where all the statistical parameters analyzed (bias, root mean square error, scatter index and correlation coefficient) show reasonable results.

In relationship with the mean periods, although in terms of correlation coefficients the accuracy is lower than in the previous case, the general results can be still considered acceptable while in terms of scatter indices they are sometimes better than for $H_s$. As concerns the wave directions, the predictions obviously follow the reality in the field. This is reflected by the direct comparisons presented. Nevertheless, the model estimations for wave directions can be still subjected to considerable improvements especially in terms of correlation coefficients. It is also true that both directional buoys were located in deep water and no depth induced refraction affects the wave direction so it is expected that in intermediate and shallow water the direction estimated would become statistically better.

As shown in introduction, the circulation in the Black Sea is dominated by the Rim current characterized by a high variability both spatial and temporal. Moreover, the anticyclonic eddies, characteristic for the Black Sea coastal waters, are usually more pronounced and stable at the Caucasian and Anatolian coasts, where the two buoys were operating. These flow patterns that have not been accounted for in the first part of the work can significantly influence the wave direction through wave refraction induced by current variations. Hence, coupling the wave prediction system with a circulation model system (as for example POM which is the acronym for Princeton Ocean Model) should contribute to an increase in the accuracy of the wave model results.

In connection with the nearshore focusing of the system, the local effects can change significantly in some locations the values of the main wave parameters predicted from one level to another. Of course, in order to carry out an appropriate calibration, measured data are required for each level. In this way various configurations of the system could be designed in order to increase the reliability of the results for each computational level defined. Moreover, the dynamic behavior of some processes, as for example the bottom friction, might be also accounted for.

A first application of the wave prediction system developed is related to the evaluations of the wave energy resources in the Black Sea basin. The potential for the wave energy extraction can be obtained from the analysis of the wave climate. Buoy data can give a general idea of the existing conditions as well as valuable information concerning some tendencies. Nevertheless, this approach has some limitations especially due to the facts that the time period of the measurement is in general limited and they usually operate in deep water. On the other hand, wave energy can be accurately predicted within a window of a few days using numerical models.

It is thus of major interest to develop a system that is able to predict the waves characteristics in various coastal locations, not necessarily considered as deep water. That is why it is essential to predict the wave climate with numerical models. These models not only allow predictions of the wave fields, but also can be forced with atmospheric models and produce long term hindcasts for periods of about 40–50 years, offering in this way the possibility of creating a solid statistical base for consistent analysis and wave predictions in both offshore and nearshore.

The present work is aiming to demonstrate the effectiveness of using such a complex wave prediction system, based on spectral phase averaging numerical models, for assessing the spatial distribution patterns of the wave energy in the Black Sea basin. From an energetic point of view, the most resourceful areas in the coastal environment can be thus better identified.

The target area of the second application considered is the coastal sector at the entrance of the Danube Delta in the Black Sea. In this place, high energetic wave conditions are usual and the currents induced by the Danube river outflow make this coastal environment representative from the point of view of the relevance of the wave-current interaction process. Moreover, since Sulina channel is the main entrance in the seventh pan European transport corridor the target area is subjected to high navigation traffic that might be affected by the interactions between waves and currents.

Besides the shifting of the wave frequency due to the Doppler Effect the presence of currents leads to a redistribution of the wave energy over the spectrum that may produce significant changes in the wave direction but also as regards the magnitude of the wave height and of the wavelength. As regards the system focusing towards the Danube Delta validations against measured data were performed for each computational level. In the medium resolution domain data registered at the Gloria drilling unit were used while in the high resolution area satellite data. In this high resolution computational domain the effect of currents was also accounted for.

The results of these comparisons show in general a good concordance for the main wave parameters between the model and the measurements (significant wave height, mean period and mean direction) and also that the presence of the current field improves the model results. Hence an important conclusion of the present work would be that the wave prediction system developed has in general the capacity to provide reliable information concerning the wave generation and coastal transformation towards the coastal areas and also that by including the current field in the coastal simulations the accuracy of the results increases, although in the sectors where the currents intensities are not so high the relevance of this process is also not high.

At this point another discussion that should be employed would be: in the case when strong current conditions are encountered what is the reliability of the wave predictions. Experimental studies as those performed by Guedes Soares and Pablo (2006), but also various field experiments as those presented in Rusu and Guedes Soares (2007) show that under certain limitations the model can perform predictions with reasonable accuracy also in such situations.

After an analysis of the main patterns concerning the components of the environmental matrix five relevant case studies were presented and discussed in more detail. From these analyses resulted the most affected areas by the process of the interactions between waves and currents, but also the absolute and relative changes induced by this process in the magnitude of the main wave parameters (significant wave height, mean period, mean wave direction and wave length).

As a further step, after the identification of the most affected locations a study of the overall changes induced in these locations by the presence of currents in the wave field was carried out for a larger time window (the 3-month period at the beginning of 2007 denoted as TI-4).

Finally, an important conclusion resulting from this work would be that the presence of strong opposite currents from the coastal sector at the Danube's mouth produces considerable changes in the significant wave heights, wavelengths and in the mean wave directions. Another second conclusion resulting also from this work would be that in the most affected locations the currents produce even deeper changes into the wave fields and as a result the wave heights are no longer Rayleigh distributed and the maximum wave height becomes sometimes greater than twice of the magnitude of the significant wave height. Such waves are usually called abnormal waves or freak waves. The occurrence of such abnormal waves that affect the entrance in the Saint George channel as predicted by the model simulations was confirmed by the practice and this problem is in the attention of the local navigation authorities.

## References

Almazov A, Bondar C, Diaconu C, Ghederim V, Mihailov V, Miţă P, Nichiforov I, Rai I, Rodionov N, Stănescu S, Stănescu V, Vaghin N (1963) Hydrological monography of the Danube's mouths, Romanian State Department of the Water, Bucharest

Alves JHGM, Banner ML (2003) Performance of a saturation-based dissipation-rate source term in modelling the fetch-limited evolution of wind waves. J Phys Oceanogr 33:1274–1298

Booij N, Holthuijsen LH (1987) Propagation of ocean waves in discrete spectral wave models. J Comput Phys 68:307–326

Booij N, Ris RC, Holthuijsen LH (1999) A third generation wave model for coastal regions. Part 1: model description and validation. J Geophys Res 104(C4):7649–7666

Bretherton FP, Garrett CJR (1968) Wavetrains in inhomogeneous moving media. Proc Roy Soc (Lond) Ser A 302(1471):529–554

CERC (1984) Shore protection manual. Coastal Engineer Research Center, US Army Corps of Eng, Washington DC: US Govt Printing Office

Christoffersen JB (1982) Current depth refraction of dissipative water waves. Series Paper 30, Inst Hydrodyn and Hydraul Eng, Tech Univ Denmark, Lyngby

Elliott AJ (1986) Share diffusion and spread of oil in the surface layers of the North sea. Dt Hydrogr Z 39:113–137

Guedes Soares C (2008) Hindcast of dynamic processes of the ocean and coastal areas of Europe. Coast Eng 55(11):825–826

Guedes Soares C, de Pablo H (2006) Experimental study of the transformation of wave spectra by a uniform current. Ocean Eng 33:293–310

Guedes Soares C, Weisse R, Carretero JC, Alvarez E (2002) A 40 years hindcast of wind, sea level and waves in European waters. OMAE2002–SR28604 (Offshore Mechanics and Arctic Engineering Conference held in Oslo, Norway, 23–28 June 2002)

Hasselmann K (1974) On the spectral dissipation of ocean waves due to white-cap. Bound-layer Meteor 6(1–2):107–127

Holthuijsen H (2007) Waves in oceanic and coastal waters. Cambridge University Press, New York

Holthuijsen LH, Herman A, Booij N (2003) Phase-decoupled refraction-diffraction for spectral wave models. Coast Eng 49:291–305

Hsu SA, Meindl EA, Gilhousen D (1994) Determining the power-law wind-profile exponent under near-neutral stability conditions at sea. J Appl Met 33(6):757–765

Janssen PAEM (1991) Quasi-linear theory of wind-wave generation applied to wave forecasting. J Phys Oceanogr 21:1631–1642

Janssen PAEM (2003) Nonlinear four-wave interactions and freak waves. J Phys Oceanogr 33:863–883

Jonsson IG (1990) Wave–current interactions. In: Le Mehaute B, Hanes DM (eds) The sea, vol. 9, part A, Chapter 3. Wiley, New York, pp 65–120

Jonsson IG, Christoffersen JB (1985) Current depth refraction of regular waves. In: Nineteenth coastal engineering conference held in Houston, Texas USA, 3–7 September 1984, pp 1103–1117

Komen GJ, Hasselmann S, Hasselmann K (1984) On the existence of a fully developed wind sea spectrum. J Phys Oceanogr 14:1271–1285

Lakshmi HK, Clayson CA (2000) Small scale processes in geophysical fluid flows. In: AP – International geophysics series, vol 67, Academic Press, San Diego, p 888

Massel SR (1996) Ocean surface waves: their physics and prediction. In: Advanced series on ocean engineering, vol 11. World Scientific, Elsevier, Amsterdam

Oguz T, La Violette PE, Unluata U (1992) The upper layer circulation of the Black Sea: its variability as inferred from hydrographic and satellite observations. J Geophys Res 97(C8):12569–12584

Onorato M, Waseda T, Toffoli A, Cavaleri L, Gramstad O, Janssen PA, Kinoshita T, Monbaliu J, Mori N, Osborne AR, Serio M, Stansberg CT, Tamura H, Trulsen K (2009) Statistical properties of directional ocean waves. Phys Rev Lett 102:114502

Rusu E (2009) Wave energy assessments in the Black Sea. J Mar Sci Tech 14(3):359–372

Rusu L, Guedes C (2007) Modelling of the wave–current interactions in the Tagus Estuary. In: Guedes Soares C, Kolev P (eds) Maritime industry, ocean engineering and coastal resources, vol II. Francis & Taylor, London, pp 801–810

Rusu E, Guedes Soares C (2008) Wave energy assessments in the coastal environment of Portugal Continental. OMAE08-57820 (offshore mechanics and arctic engineering conference – OMAE2008 held in Estoril, Portugal, 15–20 June 2008)

Rusu E, Rusu L, Guedes Soares C (2006) Assessing of extreme wave conditions in the Black Sea with numerical models. In: Ninth international workshop on wave hindcasting and forecasting held in Victoria, Canada, 25–29 September 2006

Samuels WB, Huang NE, Amstutz DE (1982) An oil spill trajectory analysis model with a variable wind deflection angle. Ocean Eng 9:347–360

Stănica A, Panin N (2009) Present evolution and future predictions for the deltaic coastal zone between the Sulina and Sf. Gheorghe Danube river mouths. Geomorphology 107:41–46

Stelling GS, Leendertse JJ (1992) Approximation of convective processes by cyclic AOI methods. In: Estuarine and coastal modeling conference held in Tampa, Florida USA, 13–15 November 1992, pp 771–782

Valchev N, Pilar P, Cherneva Z, Guedes Soares C (2004). Set-up and validation of a third-generation wave model for the Black Sea. In: Seventh conference on marine science and technology BLACK SEA2004 held in Varna, Bulgaria, 7–9 October 2004, pp 273–279

Van Vledder GPh, Hurdle DP (2002) Performance of formulations for whitecapping in wave prediction models. In: OMAE2002-28146 (offshore mechanics and arctic engineering conference held in oslo, Norway, 23–28 June 2002)

WAMDI Group (1988) The WAM model—a third generation ocean wave prediction model. J Phys Oceanogr 18:1775–1810

Yan L (1987) An improved wind input source term for third generation ocean wave modelling. Scientific report WR-No 87-8, De Bilt, The Netherlands

# The Sea Highway Macro-Project: A Multiple-Advantage Infrastructural Installation for Necessary Twenty-First Century Stabilization of Romanian Coastline Erosion

Mircea Dimitrie Cazacu and Dan Aurel Machita

## 1 Present-Day Romanian Coastal Erosion

Romania's segment of the existing early twenty-first century Black Sea shoreline, due to the disequilibrium between the alluvia quantity delivered by the Danube River via its distributaries at the coast, and their transport effect south of the littoral under the sea-wave action directed by the energetic resultant of the predominant winds from northeast, was disturbed in the twentieth century by the observable diminishment and shoreline regression of the alluvia volume, due to: the inland flood-control and Danube River navigational struggle of large-scale works against the soil erosion by numerous torrent-control arrangements (dams) and the riverbed's regularization against the annual flood dangers (canalization), the longitudinal dams built by the dredging of the navigable Sulina Ship Channel, as well as by the hydropower structures realized on the Danube River, so on its tributaries and its distributaries (Bondar et al. 1973; Bondar 1975, 1984; Spataru 1980, 1981).

Some macro-project solutions, with very important complex human uses, in these kinds of infrastructure, indicated by our renowned professor Dorin Pavel (1900–1979), were offered in our previous papers (Cazacu 1999, 2000, 2002a, b, 2009; Cazacu and Machita 2003a, b, 2005; Machita 2009).

Our earlier, previous work (Cazacu and Machita 2005) has been based on the traveling wave potential $\Phi$ of Franz Joseph Gerstner (1756–1832). The theory involves two unknown constants. One of these constants is obtained by using the constraint that the vertical velocity component vanishes on the foundation mat:

M. D. Cazacu (✉)
Polytechnic University of Bucharest, Bucharest, Romania
e-mail: cazacumircea@yahoo.com

D. A. Machita
Ministry of Transports and Infrastructure, Bucharest, Romania

V. Badescu and R. B. Cathcart (eds.), *Macro-engineering Seawater in Unique Environments*, Environmental Science and Engineering,
DOI: 10.1007/978-3-642-14779-1_11, © Springer-Verlag Berlin Heidelberg 2011

$$\Phi(X, Y, T) = C\mathrm{ch}k(H - Y)\sin(kX - \omega T) \tag{1}$$

Here X, Y and T stand for the horizontal coordinate, vertical coordinate and time, respectively. Also, C is a constant, H is the height to the water's free surface, k is a wave number and $\omega$ is the frequency. Then, we calculated the maximums of the erosion horizontal velocities on the foundation mat, determining the second constant of the sea-wave potential as function of the allowed wave height to the free seawater surface, by neglecting the square of the small velocities and applying Daniel Bernoulli's relation:

$$\frac{1}{g}\Phi'_{\mathrm{t}} + \frac{U^2 + V^2}{2g} + \frac{P}{\gamma} + \eta = c(t) \tag{2}$$

Here g and $\gamma$ is the gravitational acceleration and the specific weight, while U, V and P denote the velocity components on the X and Y directions and the pressure, respectively. The derivative of the potential in Eq. (2) is taken in respect to time. Also, $\eta$ is a position variable (to be defined below) while c(t) is a (space) constant, depending on time.

## 1.1 The Determination of the C Constant

To obtain the value of the constant C in Eq. (1) we shall use Bernoulli's Eq. (2). The same (space) constant value c(t) will be used in the whole domain occupied by the ideal fluid. This constant is evaluated at the same moment of time (for example, at the initial time $t = 0$), independent of space. Also, we shall consider the constant pressure $P_0$ on the watery wave free surface and, taking into account that $\eta = -Y$, the wave height will have consequently the relationship:

$$\begin{aligned} h &= \eta(0,0) - \eta\left(\frac{\lambda}{2}, 0\right) = -Y(0,0) + Y\left(\frac{\lambda}{2}, 0\right) \\ &= \frac{U^2\left(\frac{\lambda}{2}, \frac{h}{2}, 0\right) - U^2\left(0, \frac{-h}{2}, 0\right)}{2g} + \frac{1}{g}\left[\Phi'_{\mathrm{t}}\left(\frac{\lambda}{2}, \frac{h}{2}, 0\right) - \Phi'_{\mathrm{t}}\left(0, \frac{-h}{2}, 0\right)\right], \end{aligned} \tag{3}$$

where $\lambda$ is an appropriate wavelength. In Eq. (3) we also have to take into consideration the relation $V(0, -h/2, 0) = V(\lambda/2, h/2, 0) = 0$ for the vertical velocity deduced from Eq. (1), and where by calculation of the maximum and minimum values of the horizontal velocity at the water wave's free surface

$$U\left(0, \frac{-h}{2}, 0\right) \cong kC\mathrm{ch}k\left(H + \frac{h}{2}\right), U\left(\frac{\lambda}{2}, \frac{h}{2}, 0\right) \cong -kC\mathrm{ch}k\left(H - \frac{h}{2}\right) \tag{4}$$

and by the usage of the velocity potential derivatives with respect to the time

$$\Phi'_t\left(0,\frac{-h}{2},0\right) \cong -\omega C\mathrm{ch}k\left(H+\frac{h}{2}\right),\ \Phi'_t\left(\frac{\lambda}{2},\frac{h}{2},0\right) \cong \omega C\mathrm{ch}k\left(H-\frac{h}{2}\right) \tag{5}$$

the Eq. (3) will lead to a second degree equation. From that equation we can obtain the relationship for the constant $C$

$$\begin{aligned} &k^2\left[\mathrm{ch}^2k\left(H+{}^h\!/\!_2\right)-\mathrm{ch}^2k\left(H-{}^h\!/\!_2\right)\right]C^2 \\ &-2\omega\left[\mathrm{ch}k\left(H+{}^h\!/\!_2\right)+\mathrm{ch}k\left(H-{}^h\!/\!_2\right)\right]C+2gh=0, \end{aligned} \tag{6}$$

In the general case the solution of Eq. (6) is:

$$C(\lambda,H,h)=\frac{\omega\begin{bmatrix}\mathrm{ch}k(H+h/2)\\+\mathrm{ch}k(H-h/2)\end{bmatrix}+\sqrt{\begin{matrix}\omega^2\left[\mathrm{ch}k(H+h/2)+\mathrm{ch}k(H-h/2)\right]^2\\-2ghk^2\left[\mathrm{ch}^2k(H+h/2)-\mathrm{ch}^2k(H-h/2)\right]\end{matrix}}}{k^2\left[\mathrm{ch}^2k\left(H+{}^h\!/\!_2\right)-\mathrm{ch}^2k\left(H-{}^h\!/\!_2\right)\right]} \tag{7}$$

Consequently, the motion potential Eq. (1) has the implicit expression

$$\Phi(X,Y,t,\lambda,H,h)=C(\lambda,H,h)\mathrm{ch}k(H-Y)\sin(kX-\omega t) \tag{8}$$

### 1.1.1 The Condition to Obtain the Real Solutions of the Constant C

With a view to obtain non-imaginary solutions for the constant $C$, it will be a must that in the Eq. (7) the quantity under the root sign is positive. One denote by $\aleph=H/\lambda$ and $\chi=h/\lambda$, the relative heights of the wave and of the static water level in the river channel, respectively. After simple algebra we obtain the condition

$$\mathrm{ch}2\pi\left(\aleph+\frac{\chi}{2}\right)+\mathrm{ch}2\pi\left(\aleph-\frac{\chi}{2}\right)\geq\frac{2gh}{c^2}\left[\mathrm{ch}2\pi\left(\aleph+\frac{\chi}{2}\right)-\mathrm{ch}2\pi\left(\aleph-\frac{\chi}{2}\right)\right] \tag{9}$$

Now, we take into account the over-unitary value of the ratio calculated from Eq. (10). This ratio is developed is series. We obtain:

$$1\leq\frac{ch2\pi(\aleph+\chi/2)}{ch2\pi(\aleph-\chi/2)}\leq\frac{2gh/c^2+1}{2gh/c^2-1}=\frac{x+1}{x-1}\cong1+2\left(\frac{1}{x}+\frac{1}{x^2}+\frac{1}{x^3}+\cdots\right)>1 \tag{10}$$

In the second ratio of Eq. (10) we used the notation $x=2gh/2ghc^2$. $c^2\succ1$. This quantity is over-unitary in practice. Next, the propagation velocity expression of the traveler wave, obtained from Gerstner's theory, $c^2=\frac{g\lambda}{2\pi}\mathrm{th}\,2\pi\aleph$, is used. One finds the necessary condition to be fulfilled in the channel by the relative values of

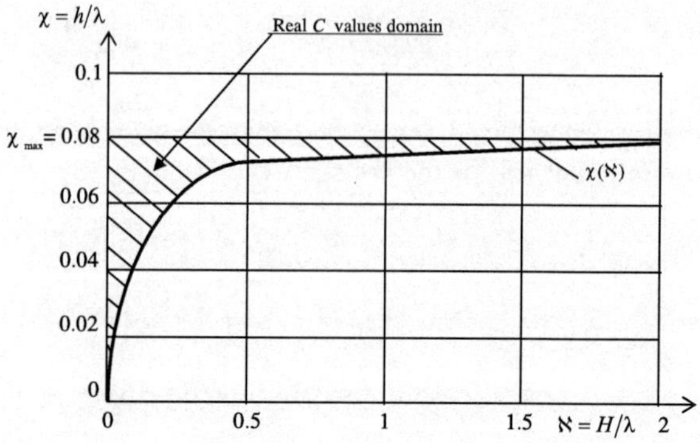


**Fig. 1** The dependence of the relative wave height by the water relative depth in the channel

the wave height and the water depth, respectively (reported to the wave length) (Fig. 1). This relation ensures obtaining real values for the constant *C*:

$$\chi_{\max} = 0.08 \geq \chi \succ \frac{1}{4\pi}\mathrm{th}2\pi\aleph \tag{11}$$

## 1.2 Effect of the Prevailing Northeast Black Sea Surface Winds

The interaction of the northeast winds with the seawater surface at the base of the air's boundary layer determines a circular seawater current in the Black Sea (Muntean 2004). Due to the presence of the Crimean Peninsula this current divides into an eastern and a western branch (Fig. 2).

Through the Bosphorus Strait, a seawater surface current circulates from the Black Sea towards the Mediterranean Sea. It consists of not-very-salty seawater. Another, much saltier and deeper seawater current circulates from the Mediterranean Sea towards the Black Sea (Fig. 3).

## 1.3 The Tidal Erosion Effect?

The tidal effect is caused by the revolving Moon and the Sun gravitational attractions. In the Black Sea, due to its small latitude difference between the bordering meridians, the tidal effect has a very small intensity (of maximum ~28 cm), when the Moon and Sun are above the same part of the Earth's rotating surface.

The time moment of the maximum tidal is when the strongest winds blow onshore, towards the Romanian and Bulgarian coasts. The tidal erosion effect is due to the fact that at the tidal retreat the seawater velocity near the sea-bottom is larger than at ordinary tidal coming.

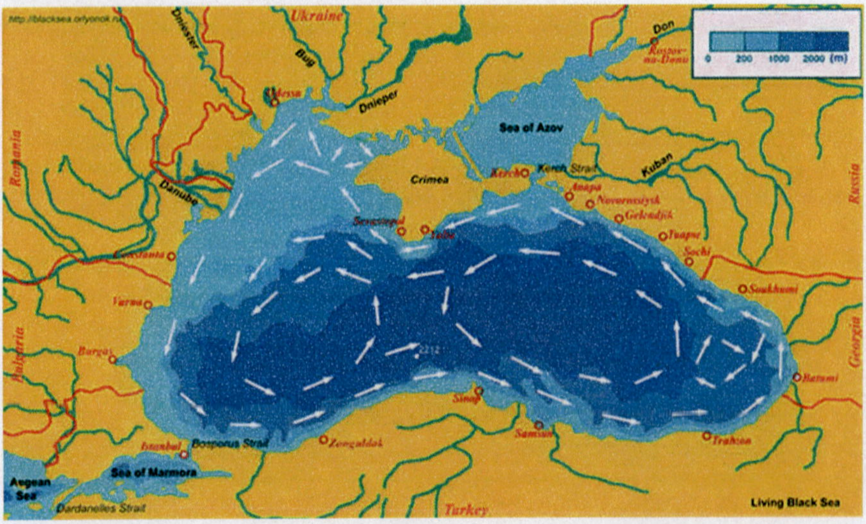


**Fig. 2** The seawater surface currents, the different major depth zones and the most important river runoffs that enter the almost enclosed Black Sea

## *1.4 Solid Particle Motion Under the Traveler Wave-Action*

In this paragraph a more accurate determination of the second constant $C$ of Gerstner's potential will be performed (Cazacu and Machita 2005). Next we shall present the results obtained concerning a solid particle motion on the near-shore sea-bottom (Cazacu and Machita 2003a, b, 2005, 2006, 2007; Machita 2009).

We shall consider the following forces acting on a solid particle in its relative motion with respect to the unsteady traveler wave motion: inertia, heavy, Archimedean lift and hydrodynamic resistance (Cazacu and Machita 2003a, b, 2005). These forces determine the trajectories of solid particles placed in different positions into the traveler wave interior. Of a special interest for the full appreciation of the region's erosion macro-management problem is the motion of the particles placed on the near-shore bottom.

A spherical solid particle is considered. Its drag coefficient is a function of O. Reynolds' hydrodynamics/hydrostatics number $Re = d_j\, w_{ij}/\upsilon$ (Fig. 4) (Hütte 1947):

$$c_{ij} = \frac{24}{\mathrm{Re}}\left(1 + \frac{3\mathrm{Re}}{16}\right) = \frac{24\upsilon}{d_j w_{ij}}\left(1 + \frac{3 d_j w_{ij}}{16\upsilon}\right) = \frac{9}{2} + \frac{24}{\mathrm{Re}} = 4,5 + \frac{24\upsilon}{d_j w_{ij}} \qquad (12)$$

Here $d_j$ is particle diameter, $w_{ij}$ is particle velocity while $\upsilon$ is water's kinematic viscosity. The drag coefficient may be found by numerical solution of the motion equation on the wave propagation direction:

$$\rho_i \frac{\pi}{6} d_j^3 \ddot{Y}_{ij} = (\rho - \rho_i)\frac{\pi}{6} d_j^3 g + \mathrm{sign}\{\dot{Y} - \dot{Y}_{ij}\} c_{ij}\left(\frac{d_j w_{ij}}{\upsilon}\right)\frac{\rho}{2} w_{ij}^2 \frac{\pi}{4} d_j^2 \sin\alpha \qquad (13)$$

**Fig. 3** Surface current of “sweeter water” from the Black Sea towards the Mediterranean Sea

**Fig. 4** Spherical particle drag coefficient variation with O. Reynolds' number

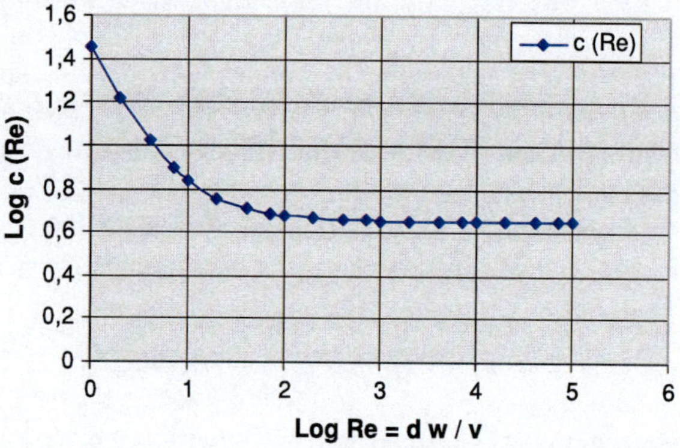


Here $\rho_i$ and $\rho$ is particle and water mass density, respectively, $Y_{ij}$ and $Y$ is the vertical displacement of particle and water, respectively, while $\alpha$ is the angle between the wave and the solid particle trajectory.

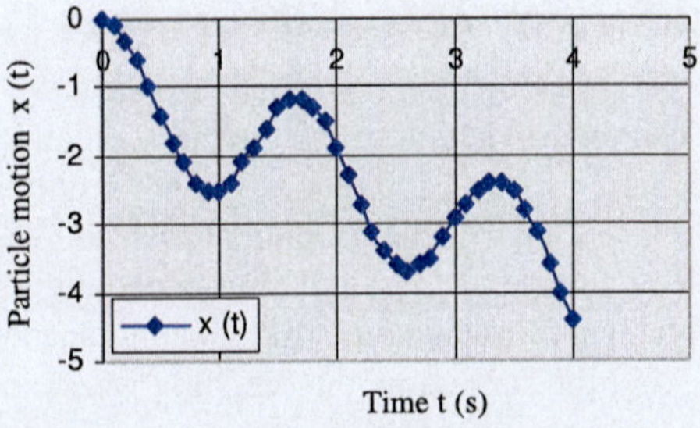


**Fig. 5** The regressive motion of a solid particle on the sea-bottom mat

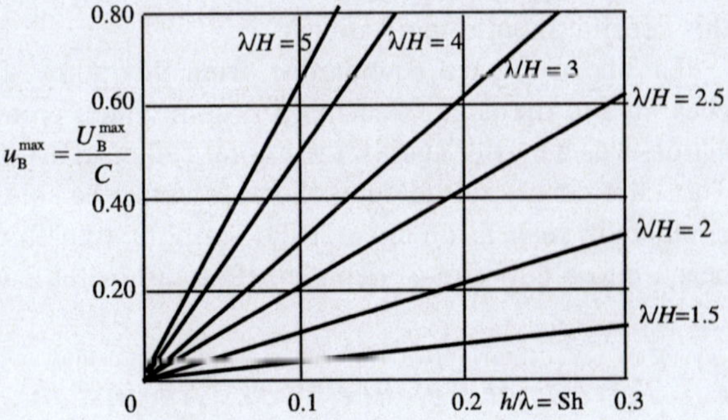


**Fig. 6** Maximal relative velocity $u$ on the wave erosion bottom as function of wave high $h$ for different seawater depth $H$, $h/\lambda$ being the Strouhal's number for a specific wave geometry

By solving Eq. (14) we obtained the regressive trajectory with respect to the wave propagation direction of $h = 0.4$ m height, for a sand particle, having a diameter of $d_j = 1$ mm, situated on the channel bottom at the depth $H = 2$ m (Fig. 5). This elucidates the transport mechanism in/out shore of the alluvial from the littoral under the incident wave action.

$$h = Y_{\max} - Y_{\min} = \frac{2A\omega}{g}(e^{kH} + e^{-kH}), \quad A = \frac{gh}{4k\mathrm{ch}kH} \tag{14}$$

the new expression of the Gerstner's potential

$$\Phi(X, Y, t) = \frac{\omega h(e^{\mathrm{kY}} + e^{-\mathrm{kY}})}{2k\,(e^{\mathrm{kH}} + e^{-\mathrm{kH}})}\sin(kX - \omega t) \tag{15}$$

allows us to calculate the value of erosion velocity on the sea-bottom

$$U_{\mathrm{B}} = \Phi'_X = \frac{\omega h(e^{\mathrm{kY}} + e^{-\mathrm{kY}})}{2(e^{\mathrm{kH}} + e^{-\mathrm{kH}})}\cos(kX - \omega t) \tag{16}$$

having the maximal values (see Fig. 6):

$$U_{\mathrm{B}}^{\max} = \frac{\omega h}{e^{\mathrm{kH}} + e^{-\mathrm{kH}}} = \frac{2\omega h}{\mathrm{ch}kH} = \frac{2\omega h}{\mathrm{ch}2\pi H/\lambda} = \frac{4\pi h}{T\mathrm{ch}2\pi H/\lambda} \tag{17}$$

### 1.5 Effect of Hydropower Dam Construction on the Danube River

The hydroenergetic arrangements realized for a historically long time on the Danube River have resulted in an important diminishing of the quantity of alluvia transported by the Danube River to the Black Sea and later on deposited on Romania's beaches, and afterwards in the southern direction, to Bulgarian beaches under the influence of the predominant winds from the northeast.

At the same time the dredging works in the Sulina Ship Channel and Black Sea harbor facilities are interposed on the natural sea-current trajectories, which have been determined practically by the induced unnatural sediment deposit on the shore, causing a strong erosion of the Mamaia beaches especially in the south of this specific monitoring station.

The most affected strand zone, from this point of view, are the Mamaia beaches, due to the dam extension till 5 km which protect Midia Harbor and whose action is as a barrier against the littoral cell currents on the North–South direction. This dam pushes to the outer shore, towards the southeast, the sediment flux that circulate in suspension towards the seashore, transforming practically the Mamaia beach into a gulf almost deprived (in total) of the natural sediments influx.

### 1.6 Terrestrial Defence Works Against the Rain-caused Torrent Erosions

The ground arrangements defending against landslide and erosive action of the rain-caused river torrents inland have an analogous diminishing effect on the Danube River runoff-transported alluvia.

## 2 The Sea Highway with Multiple Technical-economical Advantages

Developing the *multiple utilities idea* for renewable energies as the hydropower arrangements, introduced in Romania by the famous scientist, our professor and renowned macro-engineer Dorin Pavel (Cazacu 2000, 2002a, b), we present in this chapter a special hydroenergetic building (Cazacu 2009), placed in front of the Black Sea shore existing today at the isobaths between 5 and 10 m, which achieves the following utilities (see Fig. 7):

- There is no horizontal force on this Sea Highway, obtained by the alternative disposition of the flap plate 1, oscillating with respect to some fixed on the sea-bottom stage 2 and transmitting the slow mechanical motion to the pistons 3, having an upstream cofferdam 5 against the damages and a downstream cofferdam 6 to reflect the wave created by the plate 1. It does not have the effect of an impermeable dike.

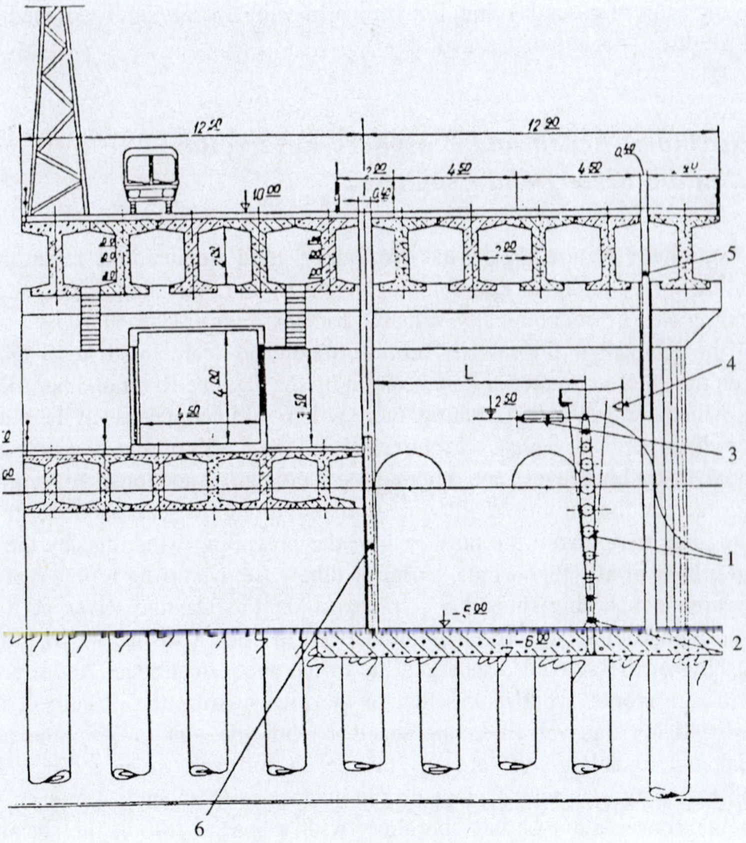


**Fig. 7** Cross-section in the sea highway with multiple utilities

- Captation of nonpolluted and inexhaustible saltwater wave energy, which on the Romanian littoral segment of ~200 km has the same size order of that of all interior river potential energy (Bondar et al. 1973; Bondar 1975, 1984; Spataru 1980, 1981; Cazacu 2000),
- Shoreline protection against the sea-wave destructive action in the conditions of diminishing of Danube River alluvial transport, as fact of the already built hydroenergetic dams and of the general dispositions to struggle against the continual shoreline erosion,
- Captation of nonpolluted and inexhaustible energy of wind, by laying a row of wind turbine along the Sea Highway (Bondar et al. 1973; Bondar 1984; Cazacu 2002a, b),
- The high-speed sea links by means of slider ships (hovercraft) riding on sustained air-cushion in this zone without large waves, between the littoral and the energy collectors captation devices (Cazacu 1999),
- Terrestrial high-speed transport on this Sea Highway is parallel to Romania's litorral (Cazacu 2000) and will affect inland landuse beneficially,

- The transport of gaseous or liquid hydrocarbons by the pipelines mounted on the Sea Highway construct (Cazacu 2000).

## 2.1 Seashore, Beach and Biosphere Reservation, Danube River Delta Protection

Conservation of Nature's gifts has extremely good implications regarding the balneal therapy, ecological tourism, human sports, and the aquatic fauna and flora developing will be economically valuable for this region (Cazacu 1999).

All the Romanian littoral has been formed, and maintained, as a resultant between the alluvium material flow brought by the Danube River and her tributary rivers (which are, generally speaking, more wild or quieter because of the presence of hydroelectric power dams), corroborated with the works of torrent arrangements such as river-training levees and other anti-soil erosion protections, cliff crumbling or landslide and the activity of alluvium transport along the seacoast under the sea-wave regime driven from the northeast by the prevailing wind, and by the energetic resultant of all other winds, storm or otherwise, occurring in this region.

Research concerning the alluvia transport on the Danube River at its first entrance onto Romanian territory, commenced in 1840 (Bondar 1975), shows a diminishing of the alluvial discharge, due to the above mentioned anthropogenic infrastructure works, what makes that in the present time the placement of the expensive dams transversal on the seacoast with the aim to keep the alluvia material and to enlarge the area of the beach surfaces, to be inefficient, the shoreline deterioration taking place on the average with about 1–2 m each year.

For the conservation of this shoreline, with a special role in the sustainable twenty-first century development of Romania, we propose not a mad struggle against the sea-wave's destructive energy, but their capture and subsequent mechanical hydro-kinetic conversion into electricity, realizing in this way the shoreline stabilization as well as beach enlargement and protection, with advantageous implications concerning the balneal therapy, tourism, sports, aquatic fauna and flora.

At the same time, the Sea Highway is realized as a potential high-speed terrestrial and naval North–South offshore link with a reduced consumption of fuel and diminished pollution, as well as a favorable transport of loose and packaged materials by pipeline—that is, by fluid conveyance, slurries and freight capsules—can be materialized.

## 2.2 Sea-Wave Energy Conversion and Energy of Wind Turbines

Sea-waves and wind are two nonpolluting and renewable energy sources, which can assure the future sustainable twenty-first century development of Romania (Cazacu 1999, 2002a, b).

In Romania, and in the frame of the Society for Promotion of Renewable, Inexhaustible and New Energies —SPERIN, I have made a series of theoretical researches (Cazacu and Stavarache 2002, 2004) as well as experimental researches (Cazacu and Machita 2003a, b, 2005, 2008; Machita 2009) concerning the sea-wave and wind characteristics at the Romanian shoreline of the Black Sea and the devices which can assure a conversion (with good efficiency) of these freely-available energies of Nature.

Also, by a lot of technical and scientific research contracts we have designed the equipments for a Black Sea pilot-plant station to convert, with a reasonably good efficiency, the static and dynamic sea-wave energy (Cazacu 1999).

A very important twin problem, which we are still studying intensively, both theoretically and experimentally, is: the interaction between the sea-wave and the oscillating flap valve as catcher, and the optimum blade profile for a wind turbine sited atop the offered offshore Sea Highway.

## 2.3 The Sea Highway for Speedy Terrestrial Circulation

The building of a Sea Highway along the Romanian Black Sea littoral should achieve, in our meridian zone, a multiple link between the North and South of Europe, favorable to the sustainable development of human civilization, and with a reduced impact on the region's physical and social environment (Marco Polo European Program).

The high-speed transport of persons, goods and materials would solve the following problems by elimination of terrestrial highway road transport difficulties, as:

- Energy consumption by crossing the mountains and hills,
- Direction deviations, consumers of time by velocity diminishing,
- Atmosphere pollution with toxic gases, aerosols and noises,
- Risk increase of terrestrial road accidents
- The fact that a terrestrial highway represents an ongoing barrier in the growth and survival of the regional fauna and flora.

## 2.4 Marine High-Speed Circulation with Air-Cushion Vehicles

Its advantage is due to the elimination of commercial and naval high-speed transport difficulties, such as:

- The immersed keels of ships are very important consumers of time, due to its reduction of ship speed,
- Especially great propulsion motor energy consumption of immersed keel ships at high speeds.

### *2.5 Freight-Pipeline Transport of Containers, Oil and Natural Gas*

A number of advantages are due to the elimination of the pipeline transport difficulties, for hydrocarbons, ground materials or containers, such as:

- The level or direction differences may lead to clogging of the pipes and their untimely wear, making repair and replacement necessary long before obsolence
- The non-corrosive coating of pipes or their burial in the ground,
- The possibilities for theft of their content.

## 3 Conclusions

Building of the Sea Highway macro-project is possible, taking into account the knowledge level (human experience) and reliability of present-day geographical information and engineering data. It is also feasible from a technical point of view, because the standard Benotto's installation for the pillar plantation is very efficient.

The linear physical character of the Sea Highway investment presents also another important advantage; we can build it when we have the financial resources, as they become available for expenditure, and we can obtain the new theoretical improvements and macro-engineering practical experiences.

For the littoral conservation, having a special role in Romania's future sustainable development, we propose not only a valiant and necessary fight against the destructive sea-wave energy, which could well be amplified by future (unforecasted) regional climate change, but its capture and the transformation into electricity and, in virtually the same time, the capture and utilization of wind energy, realizing in this manner the shoreline and beach protection with multiple advantages concerning the: balneal therapy, tourism, sport, aquatic flora and fauna (Cazacu 1999).

Once the Sea Highway is realized, a rapid terrestrial commercial and naval link will exist on the direction North–South in Europe (which is crossed by mountainous terrain), with an important reduction of motor fuel consumption and reduction of general environmental pollution, as well as an advantageous transport by pipelines of the gaseous and liquid hydrocarbons, of grainy materials and of the container-packages of freight.

The Society for Promotion of Recoverable, Inexhaustible and New Energies (SPERIN) directed a series of theoretical and experimental researches concerning the wind and sea-wave characteristics on the Black Sea shore, as well as the macro-projects, which can assure a conversion with good results of these energies. Thus, by several technical and scientific contracts was macro-projected the building and preparation of the mechanical and electrical equipment of a prototypical pilot-plant station to convert with good efficiency the static and dynamic sea-wave energy and also a wind turbine with a power coefficient $C_p = 0.56$.

In this way, by construction of the Sea Highway along just a small part of Romania's extant littoral, one can provide renewable and unpolluting energy; the sea-wave energy equals the energy of all dam-harnessed Romanian interior rivers (Bondar et al. 1973) while the wind energy equals the energy of the Nuclear Power Plant at Cernavoda. Also, the region Dobrogea and the Biosphere Reservation Danube River Delta will be protected against an anticipated 2 m erosion of the shorelines and beaches annually.

## References

Bondar C (1975) Consideratii asupra potentialului energetic al valurilor Marii Negre pe litoralul romanesc (Considerations about the energetic potential of Black Sea waves on the Romanian littoral). 2nd part—hydrology vol XLIV, studies and researches. Meteorology and Hydrology Institute, Bucharest

Bondar C (1984) Stadiul cunoasterii si perspectivele cercetarii potentialului energetic al valurilor Marii Negre pe litoralul romanesc (The knowledge stage and the research expectations of the Black Sea waves energetic potential on the Romanian littoral) studies and researches. Meteorology and Hydrology Institute. Meteorological and Hydrological Foundation of the no conventional energetic sources, Bucharest

Bondar C, Roventa V, State I (1973) Marea Neagra in zona litoralului romanesc (The Black Sea in the Romanian littoral zone). Hydrologic Monograph, Meteorology and Hydrology Institute of Bucharest

Cazacu MD (1999) Tehnologii pentru o dezvoltare durabila (Technologies for a sustainable development). Acad. of Romanian Scientists. The Congress "Dezvoltarea in pragul mileniului al III-lea" (Development in the 3rd millennium brink), Section "Dezvoltarea durabila" (Sustainable development), 27–29 September 1998, Bucharest. Publishing House EUROPA NOVA, pp 533–539

Cazacu MD (2000) Dorin Pavel—the Founder of Romanian Hydroenergetics. In: The 1st conference of Romanian Hydroenergeticians, 26–27 May 2000, University of POLITEH-NICA in Bucharest, pp 21–32

Cazacu MD (2002a) Sea highway with multiples utilities. Scientific Symposium 25 years of Superior Education. 18–20 April 2002, Ovidius University Annals, Constanta, vol 1, pp 367–370

Cazacu MD (2002b) Autostrada marina, caz interesant de folosinte complexe. In: The 2nd technical—scientifically conference "Professor Dorin Pavel—the founder of Romanian Hydropower", 31 May–2 June 2002, Sebes, vol I—science and engineering, publish. House of Romanian Engineers General Association—AGIR, pp 27–34

Cazacu MD (2009) Metoda si instalatie de captare si conversie in electricitate a energiei valurilor (Method and installation for catching and convertion in electricity of the wave energy). RoPatent A/00736/17.09.2009

Cazacu MD, Machita DA (2003a) Theoretical and experimental modeling of the wave erosion action and an advantageous possibility of shore protection. In: International symposium coastal erosion: problems and solutions, 26–28 June 2003, Mangalia–Romania. Publishing House S.C. AQUAPROIECT S.A, Bucharest, pp 171–180

Cazacu MD, Machita DA (2003b) Wave erosion action and the possibilities to protect the shore, using their energy. In: International conference on energy and environment, 23–25 October 2003, University of Politehnica, Bucharest, CD, 6 pp

Cazacu MD, Machita DA (2005) Determination of the constant C for the Gerstner's traveller wave potential. Ovidius University Annals Series—Civil Engineering, vol 1, Nr 7, pp 203–206

Cazacu MD, Machita DA (2006) Conditiile specifice valului calator si erodarea tarmului (Traveller wave specific conditions and the coast erosion). In: The 4th conference on Dorin Pavel of the Romanian hydroenergeticians, 26–27 May 2006, vol I. Politehnica Univ, Bucharest, pp 37–42

Cazacu MD, Machita DA (2007) On the kinetic of the traveller wave. In: The 3rd international conference on energy and environment—CIEM, June 28–July 4, Section 1—renewable energies. Politehnica University of Bucharest, CD, 9 pp

Cazacu MD, Machita DA (2008) Metoda si dispozitiv de captare a energiei cinetice orizontale a valurilor (Method and device to catch the horizontal wave kinetic energy). Patent nr. A/00172/03.03.2008

Cazacu MD, Stavarache SC (2002) Wave generation by oscillating flap valve. In: The 2nd conference of the Romanian hydropower engineers, 24–25 May 2002, vol I. Politehnica Univ, Bucharest, pp 31–36

Cazacu MD, Stavarache SC (2004) Energy catchment's experimental study of traveling waves by inclined plane device. A III-a Conf.Dorin Pavel a Hidroenergeticienilor din Romania, 28–29 May 2004, vol I. University of Politehnica, Bucharest, pp 53–56

Hütte I (1947) Manualul inginerului (The Engineer Handbook), vol I. Publishing House of Romanian Engineer's General Association-AGIR, Bucharest, p 493

Machita DA (2009) Contributii privind actiunea valurilor asupra erodarii plajelor (Contributions concerning the wave action on the beach erosion) PhD doctoral Thesis, Politehnica University of Bucharest, Romania

Muntean A (2004) Contributii la studiul curentilor marini, cu aplicatie la zona litoralului romanesc al Marii Negre (Contribution to the marine current study, with applications to the Romanian littoral zone of Black Sea) PhD doctoral Thesis, Ovidius University of Constanta, Romania

Spataru A (1980) Considerente privind potentialul energetic al valurilor marilor. Particularitati la litoralul romanesc (Considerations regarding the energetic potential of the sea waves. Peculiarities at the Romanian littoral). The future of seas and oceans. Romanian Academy Publishing House, Bucharest

Spataru A (1981) Potentialul energetic al valurilor (Wave energy potential). Rev. Hydrotechnics, vol 26, nr 10, Bucharest

# The Black Sea: A Georeactor to Immobilize Metal Wastes

**R. D. Schuiling**

## 1 Introduction

Industries in many countries produce a range of metal wastes (metal sludges, residues from etching baths or galvanization processes, metal-rich fly ashes, tailings from ore dressing operations, metallurgical slags etc.). If the metals cannot be recycled, these wastes are disposed in specially adapted isolated landfills, often after conversion into (hydro-)oxide sludges. Such metal-rich waste deposits will require "eternal" monitoring, and even under the best conditions offer no guarantee that the heavy metals will not contaminate the surrounding soil and ground water at some unknown time in the future. Each of these deposits is, therefore, like a time bomb, which is going to go off at some unknown moment in the (near) future. There is an urgent need to solve the problem of metal waste disposal in a safer and more sustainable way. A number of mines use STD (Submarine Tailing Disposal), so they use the bottom of the sea instead of a land-based disposal site behind dams, that run the danger of bursting, causing gigantic pollution downstream. STD is currently being practiced in the following places (Coumans 2008):

- In Chile at the Huasco Iron Pelletising Plant operated by Compania Minera del Pacifico;
- In Indonesia at Minahasa Raya and Batu Hijau mines both operated by Newmont Corporation;
- *In Turkey at the Cayeli Bakir Mine operated by Inmet Mining*;
- In Papua New Guinea at the Lihir Mine operated by Lihir Management Company and Rio Tinto;
- In Papua New Guinea at the Misima Mine operated by Placer Dome;

R. D. Schuiling (✉)
Utrecht University, Utrecht, The Netherlands
e-mail: schuiling@geo.uu.nl

V. Badescu and R. B. Cathcart (eds.), *Macro-engineering Seawater in Unique Environments*, Environmental Science and Engineering,
DOI: 10.1007/978-3-642-14779-1_12, © Springer-Verlag Berlin Heidelberg 2011

- In England at the Boulby Potash Mine operated by Cleveland Potash;
- In the Philippines at the Atlas Mine operated by Atlas Consolidated Mining and Development Corporation.

Although the immediate danger to humans may be less in the case of sea bed disposal than for disposal on land, such a solution is still equivalent to dumping of the wastes, often causing damage to marine ecosystems. There is one exception, the Cayeli zinc mine in Turkey, where the metal-rich wastes are disposed in the sea bed of the Black Sea. In the following we will explain the difference between natural immobilization of metal wastes in the Black Sea, as opposed to dumping in other marine waters.

## 2 Behavior of Metals under Anoxic Conditions

It is well known that most metals form metal sulfides, which are very insoluble under anoxic conditions, but are not stable under oxidizing conditions. If reducing, oxygen-free conditions can be guaranteed for a particular disposal site over a long period of time, this would offer a sustainable and safe solution for the disposal of metal wastes. Such conditions have existed a number of times in geologic history, where they have led to the formation of black shales. Black shales commonly contain trace metals in concentrations more than a hundred times their average crustal abundance (Wilkin and Arthur 2000).

Sulfate is reduced by organic molecules, which means that we can write the precipitation of metal sulfides, e.g. mercury, in the following way

$$SO_4^{2-} + Hg^{2+} + CH_4 \rightarrow HgS + CO_2 + 2H_2O \quad (1)$$

Arsenic can be removed as realgar (AsS) under similar conditions (O'Day et al. 2004).

## 3 The Black Sea

There are a number of anoxic basins in the world, of which the Black Sea is by far the largest. The Black Sea, with a depth up to 2,200 m, is essentially a two-layered system (Murray et al. 1991). A rather thin layer of oxygenated less saline surface water of no more than 150 m thick floats on a deeper stagnant more saline water body, that is anoxic, has a high concentration of $H_2S$, and is completely lifeless, except for some chemo-autotrophic cold water Archaeans (Fig. 1).

Methane emissions from the sea bottom are common in the Black Sea. The sea bottom is partly covered with a blackish organic-rich mud, which is full of metal sulfides, and partly by widespread carbonate-rich muds that are distinctly laminated, with greater than 3% organic matter (Ross and Degens 1974). The sulfides have formed because dissolved metals that are carried into the Black Sea by the

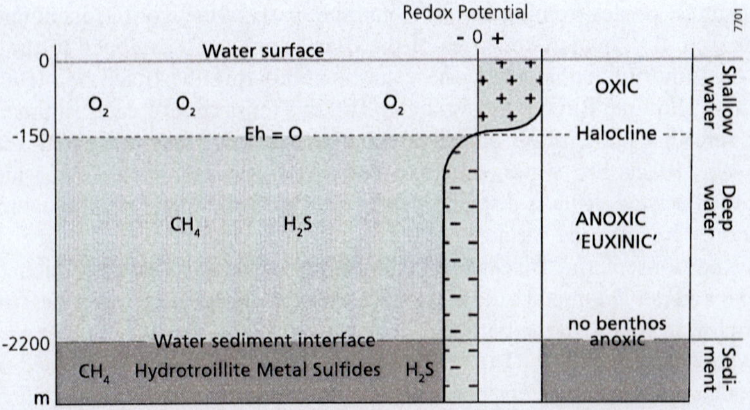


**Fig. 1** Schematic representation of the Black Sea redox system (Müller 2005)

Danube and the Dnjepr rivers react with the $H_2S$ in the Black Sea. The products of this reaction are metal sulfide particles, which sink to the bottom, and are incorporated in the bottom sediments. Similar black sediments, called black shales, rich in metal sulfides have formed a number of times in the geological past in anoxic basins. The sedimentary conditions under which these black shales have formed are known as "euxinic" facies, after the name of the Black Sea in Antiquity. Several of these metal-enriched black shales are mined for their metal content, like the famous "Kupferschiefer" (copper shales) in Poland and Germany.

## 4 Use of the Black Sea as a Georeactor

The bottom sediments of the Black Sea are black shales "in the making". We propose to investigate the possibility of using the Black Sea as a georeactor, in which metal waste streams are converted into insoluble metal sulfides (Schuiling 1996; Mazur et al. 1997; Schuiling et al. 2007). This environmental option is known as "immobilization", and is part of geochemical engineering, the discipline that seeks to apply natural processes and natural materials to solve (mainly environmental) engineering problems (Schuiling 1998). It is interesting to note that the late G. Müller arrived independently at the same solution at the same time (Müller 1995). The technology proposed here should not be confused with dumping, because the wastes are in fact "treated" by nature, converted into harmless substances, and incorporated in a secure and stable environment. We consider this solution to be the only safe way to store metal wastes, because even well isolated waste deposits under oxic conditions will start to leak and will require permanent monitoring. Experience teaches us that such a monitoring set-up will be stopped or lapse after some time, and the waste disposal site will be forgotten until the moment that it will leak and poison the surrounding soil and groundwater.

The metal wastes from industries from other parts of the world (including metal wastes from the riparian countries of the Black Sea) can be shipped to the Black Sea. An authority appointed by the countries bordering the Black Sea (Bulgaria, Rumania, Ukraine, Russia, Georgia and Turkey) must control each shipment and make sure that the harmful constituents are effectively immobilized by reaction with deep Black Sea water, and that the cargo does not include non-metallic dangerous wastes. After a disposal permit has been obtained from this authority, the wastes are released through a 200 m long free dipping tube into the deeper anoxic portions of the Black Sea. This will prevent any contamination of the surface waters. The metal sulfide particles that are formed by reaction with the $H_2S$-rich waters sink to the bottom, where they will join the sulfides that were put there already by nature. The impermeable nature of the fine-grained muds, approaching a closed system behavior, guarantees that these sulfides will not be oxidized and remobilized until their host rock sediments are exposed again to air after the next orogenic period, maybe 200 million years from now.

A minor environmental advantage is that a small part of the potentially dangerous huge amount of the poisonous $H_2S$ in the Black Sea (Schuiling et al. 2007) is removed by the precipitation reactions of the metal wastes. For metal immobilization, however, $H_2S$ is a valuable asset. It is self-evident that the riparian countries should impose a tax for the use of their natural resource—the $H_2S$ of the Black Sea.

## 5 Elaboration

Before such a disposal scheme can be put into practice, a number of investigations must be carried out and aspects of public acceptance and political support must be addressed. These include among others:

- Inventory of major metal waste streams (types and volumes);
- Experimental study of the behavior of typical metal waste streams under Black Sea conditions (quality and completeness of immobilization, rate studies);
- Testing procedure and disposal protocol;
- Legal aspects (how must a regulatory representative authority that issues permits be organized; what happens when certain bordering states are opposed, and others are in favor);
- How can the public be convinced that this is not only a sustainable way, but that it is in fact the safest and most sustainable way to deal with metal wastes?

The most important issue will be:

- *How can the public and the politicians be convinced that over longer times nature is far more trustworthy than any man-made management and monitoring system that does not take into account the nature of the material to be disposed?*

The experimental verification of the proposed immobilization method will first be done under laboratory conditions. The efficiency of the immobilization and the required time of exposure must be verified under natural conditions, by transporting different metal waste samples in closed systems to a depth of 200 m in the Black Sea, and then exposing them for a pre-determined length of time to Black Sea deep water. Thereafter the system is closed again and brought to the surface for inspection. In this way it can be investigated which types of metal waste streams are amenable to immobilization under Black Sea conditions.

## 6 Conclusions

- The safest way to store metal wastes is to store them in the form of metal sulfides in an anoxic environment;
- The world's largest anoxic basin is the Black Sea, and the bottom sediments of the Black Sea provide the safest place for sustainable storage of metal wastes;
- The Black Sea can serve as a giant georeactor that converts metal wastes into stable solid metal sulfides;
- No ecological damage is done, no biotopes are threatened, because the deep waters of the Black Sea are lifeless;
- Dangerous disposal sites of metal wastes, that are a constant threat to their environment can be eliminated;
- The countries bordering the Black Sea should control and monitor metal waste disposal, and should levy a tax on the use of their "treatment and storage facility".

**Acknowledgments** I wish to thank Prof. Panin from Geoecomar/Romania, Dr. Reynolds, coordinator, GEF/UNDP Black Sea Ecosystem Recovery Project, Istanbul, as well as Prof. Zhovinsky, Kiev and Prof. Mazur, Odessa for their interest and support. Basem Shomar/Heidelberg helped me to trace Müller's papers containing the same idea of disposal in the Black Sea.

## References

Coumans C (2008) Submarine tailings disposal. MiningWatch Canada, November

Mazur VA, Tsykalo AL, Schuiling RD (1997) Biogeochemical processes in the depths of the Black Sea as a model of wastes transforming technology (in Russian). Emerg Situat Civil Def 2:49–50

Müller G (1995) Das Schwarze Meer. Ein sicheres Endlager für schwermetallkontaminierte Feststoffe? (The Black Sea. Safe storage for metal-contaminated wastes?). Naturwissenschaften 13:202–206

Müller G (2005) The stagnant anaerobic deep water body of the Black Sea: a safe disposal site for solid wastes contaminated with heavy metals. In: Third international conference on remediation of contaminated sediments. Poster Session, New Orleans, pp 24–27

Murray JW, Top Z, Ozsoy E (1991) Hydrographic properties and ventilation of the Black Sea. Deep Sea Res 38(suppl.2):663–689

O'Day PA, Vlassopoulos D, Root R, Rivera N (2004) The influence of sulfur and iron on dissolved arsenic concentrations in the shallow subsurface with changing redox conditions. Proc Natl Acad Sci USA 101(38):13704–13708

Ross DA, Degens ET (1974) Recent sediments of Black Sea, in The Black Sea—geology, chemistry and biology. Am Ass Petr Geol pp 183–199

Schuiling RD (1998) Geochemical engineering: taking stock. J Geochem Exp 62:1–28

Schuiling RD (1996) Geochemical engineering; principles and case studies. In: Reuther R (ed) Geochemical approaches to environmental engineering of metals. Springer, Berlin, pp 3–12

Schuiling RD, Cathcart RB, Badescu V, Isvoranu D, Pelinovsky E (2007) Asteroid impact in the Black Sea. Death by drowning or asphyxiation? Nat Hazards 40(2):327–338

Wilkin RT, Arthur MA (2000) Partitioning of trace metals in anoxic Black Sea sediments. In: Proceedings of 9th Annual Goldschmidt conference

# Advantageous Techno-Naturalization of the Seawater in the Black Sea

Mircea Dimitrie Cazacu and Raducu Viorel Iancu

## 1 A Short History of the Black Sea

At the beginning we shall present a short vision concerning the recounting of the Black Sea's history, constituted by some remembrance writings from Antiquity, as well as on some more recent pertinent historical writings (Ballard 1999, 2000) based on the research made in situ, with the aim to elucidate the present day chemical evolution of the degradation process of the seawater resident in the Black Sea.

At the same time, we have our duty to present some methods and proceedings to technically naturalize the deep waters of this important body of seawater in the topical imperious tendency to assure the sustainable development of our Earth-biosphere.

### *1.1 The Ancient Period of Black Sea Sweet Water*

Looking backwards in time thousands years, the Black Sea, like the Caspian Sea, Aral Sea and other lakes, held sweet water deposited by many rivers, supplied with rain water runoffs, yet its freshwater level remained inferior to the nearby world-ocean (Mediterranean Sea level), which discharged a small runoff of its salted waters coming from the Atlantic Ocean through a small saline river existing in the present place of the Bosphorus Strait, but which could not be greatly influenced in

M. D. Cazacu (✉)
Polytechnic University of Bucharest, Bucharest, Romania

R. V. Iancu
FujiServLtd, Bucharest, Romania

V. Badescu and R. B. Cathcart (eds.), *Macro-engineering Seawater in Unique Environments*, Environmental Science and Engineering,
DOI: 10.1007/978-3-642-14779-1_13, 

an important mode by the life which was, more or less, permanently resident on the Black Sea's lake-like shoreline (Georgievski and Stanev 2006).

But, it is popularly told, a catastrophic phenomenon intervened at a given moment that resulted in the creation of the Black Sea as we know it today (Giosan et al. 2009). The Ancients even had a story of a human eyewitness to the sudden creation-event in the land region between the Mediterranean Sea and the Black Sea—the Greek story of Dardanus, alleged founder of Troy, after he had survived an overwhelming flood when an earthen barrier in the Bosphorus–Dardanelles crumbled, permitting the rapid influx of seawater from the Mediterranean Sea in memorably magnificent great volume.

## 1.2 Creation of the Bosphorus Strait

With the occasion of this geomorphologic transformative cataclysm, possibly produced by a great earthquake or series of earthquakes for which even the present-day region is infamous, that had the effect of creating a crevice for the saline river's bed between the two seawater basins—the Mediterranean Sea and the lake we today name the Black Sea—and the making of the Bosphorus Strait (Ballard 1999), followed by the salty water overflow of the Mediterranean Sea towards the sweeter water of the Black Sea, a process which continued for a prolonged time period, permitting to the resident peoples and mobile freshwater fishes to retreat gradually to higher places dry land and more inland parts of the many river basins, respectively, in the rivers of sweet water (Turney and Brown 2007). But, the fixed flora and less mobile fauna which could not adapt to the rapid salinity modification perished and fell, gravitationally, to the sea-bottom where they rotted, producing hydrogen sulphide emissions, according to general scientific popular opinion (Piper and Calvert 2009; Poort et al. 2005).

However, we cannot allow this process, still ongoing nowadays, although we have in the cave "Movile" (beside Mangalia, a small town on the Romanian seashore) a fauna and flora adapted to hydrogen sulphide, whose high concentration level growth in present in the Black Sea with about 2 m each year, probably also due to the organic matter brought by the rivers from tributary human settlements and industries, which over saturate the rivers with wastes, the pollutant waters insufficiently treated, the seawater fishes living at present only to a depth of ~50 m under the sometimes very stormy Black Sea's surface, a marine top layer in which the wind and the waves created by this natural process can still refresh the oxygen content of the seawater in the Black Sea.

Faced with this disagreeable twenty-first century pollution/contamination situation, we have propose a patented macro-engineering method and a specified macro-project installation (Cazacu and Iancu 2008) to assure not only the maintenance of the present-day situation in the Black Sea Basin, but even its improvement for the socially sustainable development assurance, however "sustainable" may be defined, obtaining also an advantageous winning of useful energy.

On the other hand, we must mention the thermal freshwater springs, rich in hydrogen sulphide, which on the Romanian territory can be found in the Mangalia region, and which made us to suppose logically the existence of many other similar submarine freshwater springs offshore and in the Black Sea's unknown depths. It is known that in many submarine coastal regions of the world, continental aquifers discharge a continuous flow of freshwater to the world-ocean (Ranjan et al. 2009).

## 1.3 The Present-Day Situation in the Black Sea

After the recent data (Sorokin 1983), the Black Sea salinity is comprised between the 1.7% at the free seawater surface and 2–3% in the deeper seawater.

The high-level hydrogen sulfide concentration uppermost layer, ascending by 2 m each year, rose in the most recent years from 180 to ~150 m depth. The fishes are living in a fluid strata only to the 50 m seawater depth aerated zone because of the action of stirring sea waves and wind. The maximal concentration of the hydrogen sulfide in the deepest seawaters of 2,000 m, presented also in the Fig. 1, is up to 1.15%.

The discovery in the 1980s years of the cave "Movile" in the Mangalia region, made it possible to simulate the difficult research in this field of environmental amelioration using technology on a macro-engineering scale.

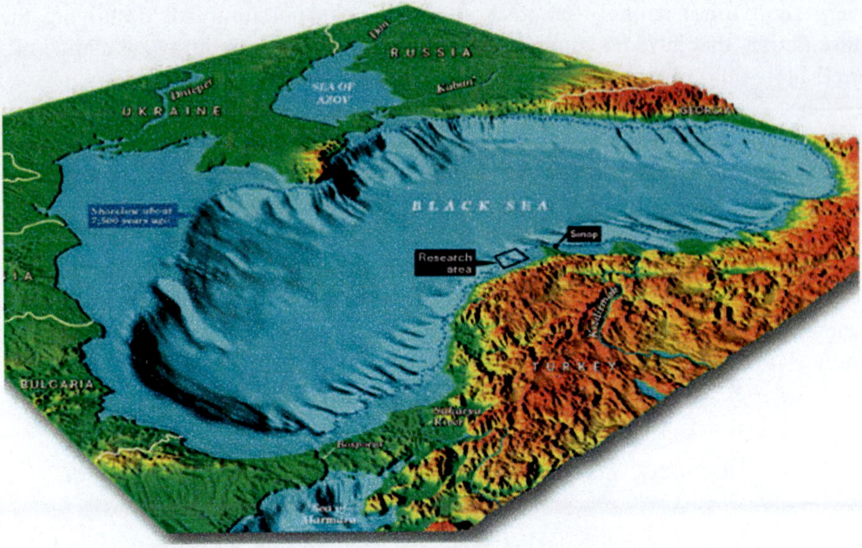


**Fig. 1** Robert D. Ballard's research sites on the Black Sea shelf

## 2 The Possibilities to Rehabilitate the Aquatic Ecosystem

Due to the undeniable importance of Black Sea waters techno-naturalization, which is menaced to become in a few decades a hideous dead sea, we are proposing now some macro-projects (Cazacu 1999, 2004, 2009).

### 2.1 Previous Proposals

In the aim of Black Sea deep seawater techno-naturalization is known the patented invention nr. 66972/30.11.1979, *Proceeding to obtain the hydrogen sulfide from Black Sea water*, authors: Eng. Nicolae Barlea and his collaborators Alexandru Bacalu and Ioan Epure from the Institute of Research and Design for Water Treatment in Romania—ICPEAR of Bucharest, in which to extract the deep seawater one uses a dept pump and also an installation to burn the hydrogen sulphide for the purpose of energy production (Barlea et al. 1979), having two important disadvantages.

The primary disadvantage of this particular macro-project is the utilization of an axial turbo pump, which cannot give motion to the seawater from the vertical pipeline of 2 km due to its big inertia, the pump working intermittently because of the seawater column breaking due to the cavitations phenomenon, destructive even for the turbo pump, outside of the mechanical shocks exerted on the rotor blades due to its working with a heterogeneous unfiltered liquid. Except that, if the seawater column is put into motion, by emitting of the hydrogen sulphide ascendant bubbles in the vertical water column, the phenomenon becomes (in a very short time) strongly eruptive, presenting the possibility of destroying the installation, that have no automatic closing valve like a very sturdy submarine oil-well blow-out valve.

The secondary disadvantage is due to the burning of hydrogen sulphide to collect the chemical reaction products, which will be conducive to the formation of damaging acid rainfall downwind, with polluting particle depositional action over the ground and vegetation receiving the descending acid rainfall.

Recently, another group of Romanian collaborators have proposed a special macro-engineered installation to produce electricity directly from a fuel cell (Pacala et al. 2005), the proceeding consists of the hydrogen sulfide electro-chemical oxidation using a novel fuel cell, presented in Fig. 2, in which hydrogen sulfide is dissolved in the liquid electrolyte.

### 2.2 The Proposed Useful Black Sea Clean-Up Macro-Project

In 2008, we proposed a patented industrial method and macro-project installation to techno-naturalize polluted water from the depth of lakes, seas or the

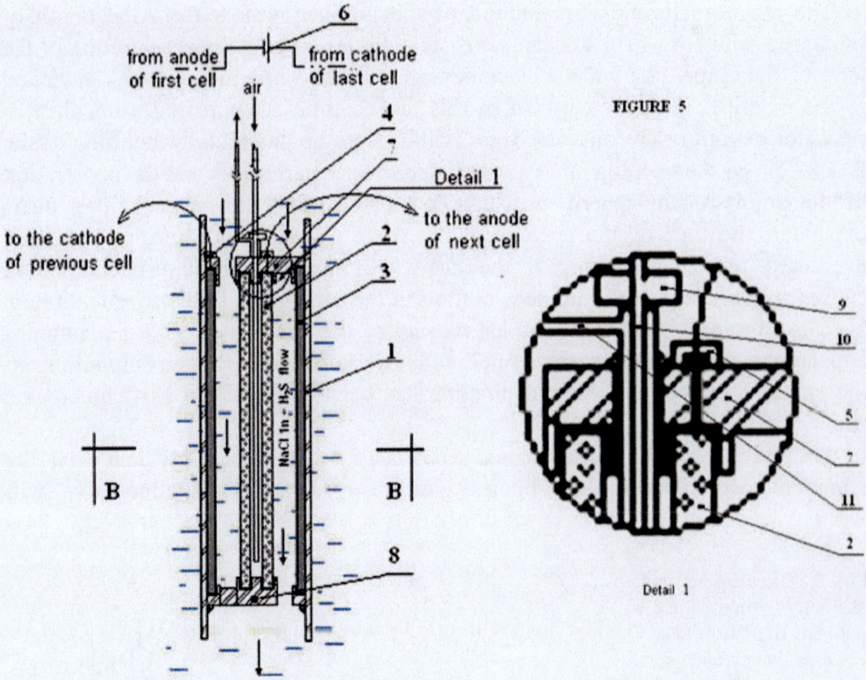


**Fig. 2** Cylindrical cell, according to the patent—an open tubular insulated waterproof and corrosive resistant material (1) that has inside coaxially affixed a central tubular carbon porous cathode (2) catalyst coated preferably with platinum and a tubular anode (3) preferably made of smooth carbon black, maintained at spaced by a top binding element (7) and a bottom binding element (8) each made of an insulated, waterproof and corrosive resistant material such as Teflon or glass-fiber reinforced polyester, a pressure regulator (9) that controls the air injected into said cathode 2 through a tube (10) made of an insulated, waterproof and corrosive resistant material such are Technoflon (FKM) or IIR, that has a coaxial larger but shorter tube (11) around made of the same material as said tube (10) and an anodic connector (4) and cathodic connector (5) both waterproof and corrosion resistant and mounted on said top binding element (7)

global ocean (Cazacu and Iancu 2000, 2003, 2008), which permits the elimination of the dangerous and potentially destructive naturally-eruptive gas phenomenon from the ascendant seawater column, due to the expansion of gas bubbles as they move upwards, by periodical and progressive kicking of the big diameter, long vertical pipe and by its correct positioning in a superior recipient attached place of the unpolluted seawater with a generous view to its ultimate elimination into the Black Sea.

Simultaneously, one uses the bubble gas energy by their induced expansion in a turbo expander, as well as the energy obtained by the burning of the hydrogen sulphide, with the aim to transform the pollutant into their useful elements, as: energy production, sulphuric acid commercial product and, eventually, of deuterium harvesting (Petersen 1994).

The techno-naturalization method of the polluted waters from the depth of lakes, seas and our world's ocean, consists in liquid progressive acceleration of the long vertical pipeline by the achievement of a vacuum in a superior tank, attached to the pipeline. In the lower part of this tank can be accumulated a quantity of seawater devoid of any dissolved gas, which may be periodically returned to the Black Sea, by the opening of a valve located at its superior part and the penetration of the ambient atmospheric pressure. We can recover this energy in a turbo expander, and then by the opening of the valve placed at its inferior part, using eventually moreover a pump; in the aim to eliminate the eruptive phenomenon, caused by the ascendant fluid acceleration in the pipeline with hydrogen sulphide bubbles in expansion, by the sudden closing of the valve at the first gas bubbles appearance, until that moment no polluted seawater with hydrogen sulphide from the superior part of the vertical pipeline has been accumulated as liquid in the superior tank.

The hydrogen sulphide, extracted from deep water (>50 m), which after the complete accumulation in the superior part under the vertical pipeline valve, will

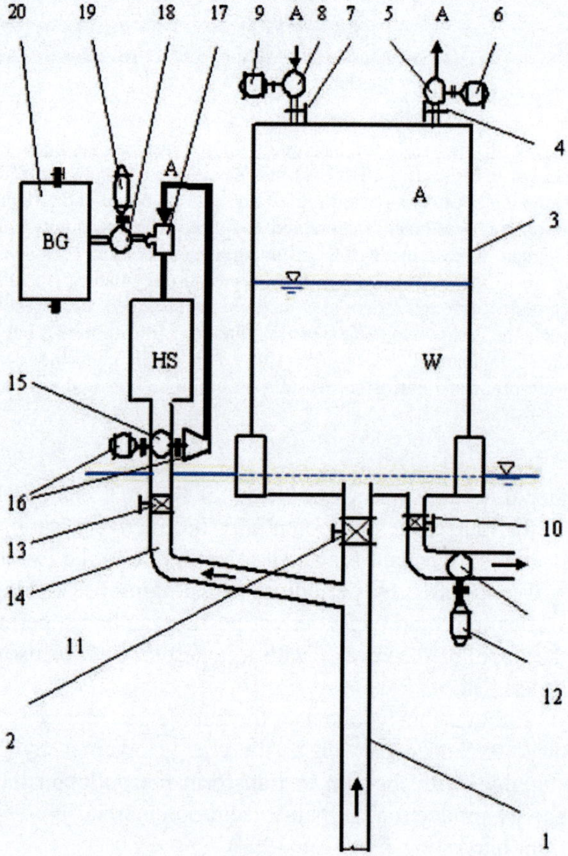


**Fig. 3** The proposed macro-engineering method and a sample installation to advance the advantageous techno-naturalization of the Black Sea's deepest water layer (>50 m)

be at high pressure depending of the weight difference of both columns: the first with seawater and the second with gas, in that case we can use the energy produced by the gas expansion conducted through a lateral connection to a turbo expander directly coupled with an electric generator, or with a turbo blow, used for the air necessary to burn the hydrogen sulphide and to obtaining other utile substances included in these, as: the sulphuric acid, as well as the expansion of these burned gases in a turbine, for the utilization of the transferred energy, it being coupled with another electric generator.

The macro-project installation for the techno-naturalization of polluted water from the depth of lakes, seas or Earth's ocean (Fig. 3), consists in a pipe 1, preferably to be vertical for the shortest length, equipped towards its superior end with a valve 2, whose closure will prevent the eruptive gas phenomenon until the bubbles will come to the upper part of the pipe 1, that is a common body with a tank 3 of calculated size to store the unpolluted water *L*, which will lift through the vertical pipe 1 in the starting period of the liquid column by the progressively vacuum-producing in this tank, opening a vane 4 and using a vacuum pump 5 driven by a motor 6, the water *L* elimination producing then periodically by closing of the valve 4 and opening another valve 7 and the introduction of the air *A* from atmospheric pressure through a turbo expander 8 acting an electric generator 9, using the liquid *L* gravitational flow by the opening of a valve 10 placed in the inferior part of the tank 3 and eventually by using a pump 11 coupled with an electric motor 12.

## 3 Conclusions and Recommendations

The advantages of the proposed macro-engineering method and installation for the techno-naturalization of the deep water of lakes, nearly closed seas or parts of our planet's ocean are the following:

- The easy, progressive starting of the accelerated unsteady flow of the biphasic fluid from the vertical pipe of very great length,
- The elimination of the eruptive dangerous phenomenon, created by the accelerated lifting of the gas bubbles by their expansion within the vertical pipe,
- The deposition of the water without solvent gases in a superior reservoir, from where these may afterwards be restored periodically into the Black Sea,
- The utilization of the compressed gases detention energy by their elimination from the deep water,
- The use of energy given by the hydrogen sulfide burning,
- The utilization of the eliminated gases from the deep water to the sulfuric acid production,
- The eventual separation of the deuterium contained in the deep waters and the gases from the depth, necessary for the present heavy water production and for the future fusion nuclear reaction activity.

## References

Ballard RD (1999) Black Sea expedition. National Geographic Society. http://www.nationalgeographic.com/ngnews/blacksea.html

Ballard RD (2000) Further evidence of abrupt Holocene drowning of the Black Sea shelf. Mar Geol 170:253–261

Barlea N, Bacalu A, Epure I (1979) Procedeu de obţinere a hidrogenului sulfurat din apa Marii Negre (Proceeding to obtain the hydrogen sulphite from the Black Sea water). Patent of invention nr. 66972. OSIM, Bucharest

Cazacu MD (1999) A survey of the technics which might preserve the biosphere reservation Danube Delta. The 8th symposium "Technologies, installations and equipments for improvement of environment quality", Polytechnic University, Bucharest, vol I, pp 85–92

Cazacu MD (2004) Black Sea deep water restoration. In: The 3rd Congress of the Romanian Scientists Academy "The water a miracle", Ovidius University, Constanta

Cazacu MD (2009) Sustainable development problems. Academy of Romanian Scientists Annals, Series on Engineering Sciences. Publishing House of Romanian Scientists Academy, Bucharest, pp 33–44

Cazacu MD, Iancu VR (2000) Water inertia time for a long pipeline start. In: The 5th international conference on hydraulic machinery and hydrodynamics, Scientific Bulletin of the Polytechnic University, Timisoara, vol 45(59), pp 55–58

Cazacu MD, Iancu VR (2003) Advantageous technologies for Black Sea water restoration by elimination of hydrogen sulfide. In: International symposium coastal erosion: problems and solutions, Mangalia, Romania, CD

Cazacu MD, Iancu RV (2008) Method and installation to naturalize the lake, sea and ocean waters. Patented invention nr. A/01011, OSIM, Bucharest

Georgievski G, Stanev EV (2006) Paleo-evolution of the Black Sea watershed: sea level and water transport through the Bosphorus Straits as an indicator of the Late glacial-Holocene transition. Clim Dyn 26:631–644

Giosan L, Filip F, Constatinescu S (2009) Was the Black Sea catastrophically flooded in the early Holocene? Quat Sci Rev 28:1–6

Pacala O, Pacala M, Ciocanea A (2005) Process and Fuel Cell for Generating Electric Power. RO 120736 B1 and European Patent Office PCT/RO 2005/000010, International Publication Number WO 2006/025758 A2

Petersen U (1994) Mining the hydrosphere. Geochim Cosmochim Acta 58:2387–2403

Piper DZ, Calvert SE (2009) A marine biogeochemical perspective on black shale deposition. Earth Sci Rev 95:63–96

Poort J, Vassilev A, Dimitrov L (2005) Did the postglacial catastrophic flooding trigger massive changes in the Black Sea gas hydrate reservoir? Terra Nova 17:135–140

Ranjan P, Kazama S, Sawamoto M, Sana S (2009) Global scale evaluation of coastal fresh groundwater resources. Ocean Coast Manag 52:197–206

Sorokin YI (1983) The Black Sea. Estuaries and Enclosed Seas. Ecosystem of the World. B.N. Ketchum, New York

Turney CSM, Brown H (2007) Catastrophic early Holocene sea level rise, human migration and the Neolithic transition in Europe. Quat Sci Rev 26:2036–2041

# Proto-Type of Replicable Industrial Black Sea $H_2S$ Gas Extraction Plant

Salah A. Naman, I. Engin Ture and T. Nejat Veziroglu

## 1 Introduction

Hydrogen sulfide gas ($H_2S$) is flammable and poisonous. It is soluble in water, and it can also corrode pump and pipe metals such as iron, steel, copper and brass. The equilibrium concentration of $H_2S$ gas in the Black Sea is 10 ppm at 1,000 m depth. For mega-engineering purposes, $H_2S$ should be extracted without conversion to other molecules and it should be concentrated to 10,000 ppm or more, in order to bring it to concentrations similar to natural gas (methane), for which the technology has already been well developed for a long period of time, in order to produce useful hydrogen gas fuel. The solution of $H_2S$ gas in seawater is non-ideal and the extraction of this gas from seawater should be through Henry's law and it depends on the physical and chemical variables (such as: concentration, pH, salinity, pressure of stripping pump, temperature, height of the stripping tower, etc.). These variables can be studied through Le Chatelier's principle to find the equilibrium concentrations of $H_2S$ in the Black Sea.

Black Sea has layers of seawater that are oxic, suboxic and anoxic. The daily production of $H_2S$ by sulfur reducing bacteria (SRB) is about 10,000 metric tons [tonnes] in the Black Sea and the reservoir of $H_2S$ is estimated to be 4.587 billion tonnes (Naman et al. 2008). The mixture of $H_2S$ in pure water is considered as a non-ideal (gas–liquid) solution (Duan et al. 2007). Both molecules can produce hydrogen, but $H_2S$ is much easier to dissociate than water due to the chemical bonding structure of the molecules.

S. A. Naman (✉)
University of Duhok, Duhok, Iraq
e-mail: salah.naman@yahoo.com

I. Engin Ture and T. Nejat Veziroglu
UNIDO-ICHET, Istanbul, Turkey

V. Badescu and R. B. Cathcart (eds.), *Macro-engineering Seawater in Unique Environments*, Environmental Science and Engineering,
DOI: 10.1007/978-3-642-14779-1_14, © Springer-Verlag Berlin Heidelberg 2011

## 1.1 Concentration of $H_2S$ Gas at Different Black Sea Depths

Literature survey for the distribution of $H_2S$ and oxygen in the Black Sea shows different concentrations of this gas at different depths of the sea, ranging from 7–14 ppm (Fig. 1) has been treated statistically and shows that at the Black Sea's floor, at a depth of 2,200 m, $H_2S$ concentration is about 14 ppm, which is due to the formation of this gas from sulfides and $SO_4^{2-}$ ions by anaerobic bacteria (sulfur reducing bacteria, SOB), while at the surface the concentration of $H_2S$ is zero down to the depth of about 100 m due to the action of SOB (Fig. 1).

Some of these reports have been using different units for the concentration of $H_2S$, which give conflicting results. In this chapter only SI units will be adopted for all the physical and chemical variables. The concentration of oxygen is about 8–9 ppm at the surface of the Black Sea and it declines to zero at about 100 m depth due to the activities of sulfur oxidizing bacteria on the seawater surface which oxidize $H_2S$ to sulfate. Therefore, the concentration of $H_2S$ is zero at the surface, and it starts increasing at the depth of 100 m. There is a natural equilibrium between these two types of bacteria at the bottom and surface of Black Sea. Figure 1 shows that Le Chatelier's principle is applicable to all these reactions at the depth of 1,000 m. Bio-activities of both SRB at the seafloor and SOB at the sub-aerial seawater surface, and at different depths in the Black Sea, as well as the activities of photosynthesis bacteria in the Black Sea as shown in Fig. 1, fully explains the dissociation constants $K_1$ and $K_2$ of $H_2S$ with its ions $HS^-$ and $S^{2-}$ at salinity and pH of Black Sea seawater. In the natural equilibrium zone in Fig. 2 the concentration of $H_2S$ is 10 ppm at 1,000 m depth, when the temperature is 8°C and the salinity is 20,000 ppm.

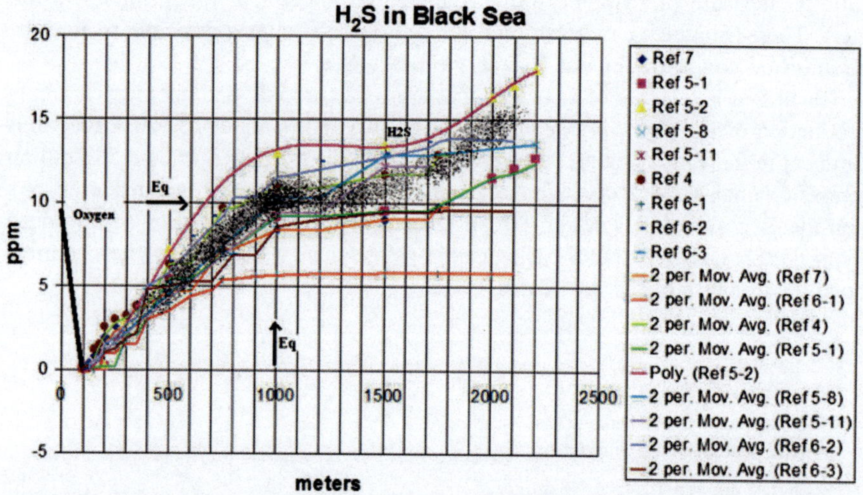


**Fig. 1** Concentration of $H_2S$ and $O_2$ in Black Sea water. *Mov* moving average. Number of references cf. Naman et al. (2008)

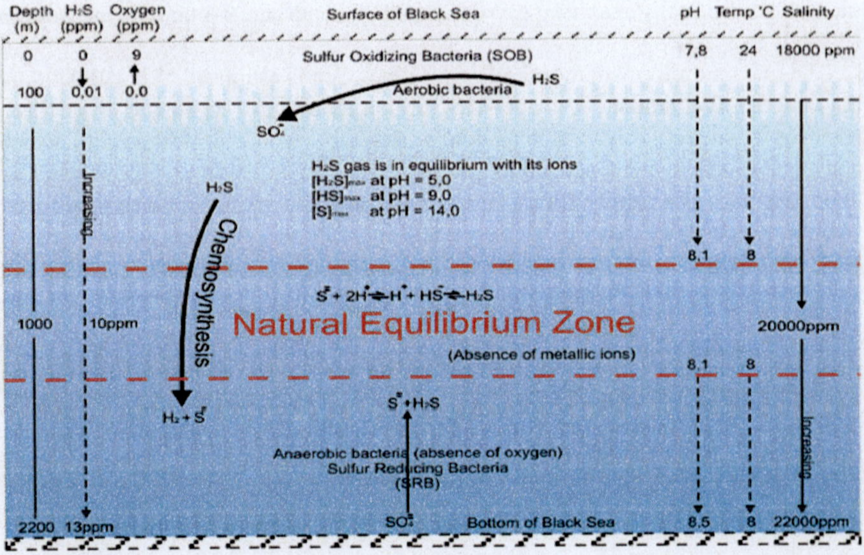


**Fig. 2** Natural equilibrium of Black Sea

Figure 2 shows the concentrations of $H_2S$ and oxygen as a function of depth of water from the results of different research (Carroll 2010; Blabev 2010). Le Chatelier's principle can be summarized, for our purposes here, as:

If a chemical system at equilibrium experiences a change in concentration, temperature, or total pressure, the equilibrium will shift in order to minimize that change (Le Chatelier's principle 2010).

According to Le Chatelier's principle for chemical equilibrium the characteristic of $H_2S$ in the Black Sea at equilibrium depth will not be affected when this gas is pumped to the surface of the sea, and that was the main reason why this seawater extraction depth was selected for our proto-type of a future mega-scale extraction system.

### *1.2 The Black Sea's Natural Equilibrium*

The Black Sea is well-known to be rich in $H_2S$ gas. Large amounts of $H_2S$ are formed by the activity of SRB bacteria in the sulfur deposits at the sea-bottom, or from organic matter accumulation deposited by the larger rivers pouring their runoffs into the Black Sea, and it may be due to the fractures and mud volcanoes (Kopf 2003), as well as the destroyed gas-hydrate deposits, which are transformed by this SR Bacteria into $H_2S$ gas. Figure 2 shows this picture. Surveys on the concentration of $H_2S$ gas in the Black Sea have been treated statistically and show that the concentration is zero at the surface, and it increases gradually after 100 m then it will normally reach 10 ppm at 1,000 m depth and 14 ppm at the bottom (Figs. 1, 2).

The maximum formation of the $H_2S$ in Black Sea is estimated to be 10,000 ton/day and the total amount is about 4.587 billion ton. And ~25% of $H_2S$ has been removed from the Black Sea by photosynthesis bacteria (purple bacteria) which convert $H_2S$ gas to elemental sulfur and hydrogen gas to freshwater (Baykara et al. 2004, 2005). At the surface of the Black Sea there is a species of sulfur oxidizing bacteria (aerobic bacteria) which convert $H_2S$ into sulfate and, that is the chief reason there is no $H_2S$ above a depth of 100 m in the sunlit zone, as illustrated in Fig. 2.

$H_2S$ gas is dissolved in the seawater of Black Sea where salinity varies between 17,000 ppm at surface to 22,000 ppm at the seabed; it exists as a natural non-ideal solution gas–liquid. The solubility of $H_2S$, according to Henry's Law is dependent on temperature, salinity, pH, pressure, and also the nature of $H_2S$ and water molecules. According to the following dissociation reaction constants Eqs. 1–3 (Chidgopkar 2006; Carroll 2010):

$$H_2S \leftrightarrow H^+ + HS^- K_1 \tag{1}$$

$$HS \leftrightarrow H^+ + S^- K_2 \tag{2}$$

$$H_2O \leftrightarrow H^+ + OH^- K_w \tag{3}$$

The concentrations of these ions in the Black Sea seawater are a function of pH, temperature and salinity. Considering the biological reactions for formation and consumption of $H_2S$ at different of depths of the Black Sea the reactions Eqs. 4–6 take place:

$$SO_4^- \xrightarrow[]{\substack{\text{anaerobic bacteria} \\ \text{sulfur reducing bacteria}}} H_2S + S^-(\text{SRB}) \tag{4}$$

$$H_2S \xrightarrow{\text{photosynthesis (purple bacteria)}} S + H_2 \tag{5}$$

$$H_2S \xrightarrow[]{\substack{\text{aerobic bacteria} \\ \text{thiobacillus}}} SO_4^-(\text{SOB}) \tag{6}$$

All these reactions will lead to certain concentration of $H_2S$ at the equilibrium zone at a certain depth in the Black Sea. According to Le Chatelier's principle, the concentration gradient of $H_2S$ with the depth of the seawater, as reported by several researchers, has a plateau region (equilibrium zone) which has a concentration of 10 ppm at a depth of ~800–1,200 m (Fig. 1). We suggest that this concentration of $H_2S$ at this equilibrium depth is a result of the conditions in Black Sea (pH, temperature, salinity) and extraction (stripping) of $H_2S$ from Black Sea seawater (10 ppm and depth of 1,000 m) can be possible. Figure 2 shows all these bio and inorganic reactions with their dissociation constant at the Black Sea as a function of the depth, assuming the proper application of Le Chatelier's principle.

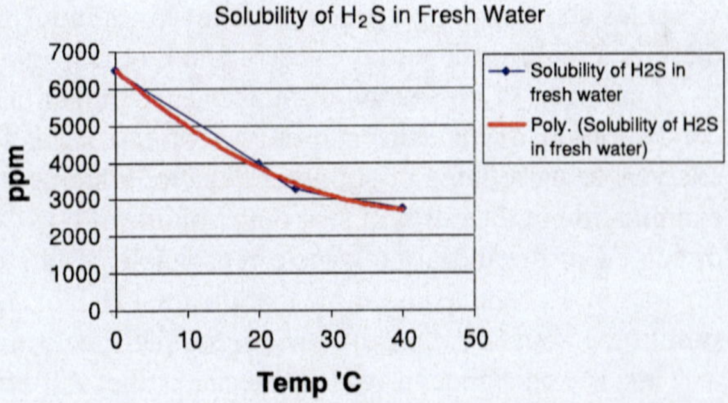


**Fig. 3** Solubility of $H_2S$ gas in freshwater at different temperatures

# 2 Extraction of $H_2S$, from Gas–Liquid Solution

Extraction of hydrogen sulfide molecules is very important in the process of hydrogen production from $H_2S$. Environmental regulations, prohibits the evaporation of unpleasant rotten-egg odors of $H_2S$ inside the household living area. Therefore, in the present $H_2S$ removing, units of conversion to metal sulfide/iron removable filters containing manganese green and or carbon surface filters or, for higher concentrations chlorination are most common treatment methods. As it is well known $H_2S$ gas can easily produce hydrogen (Blabev 2010; USEPA 2010). The solubility of $H_2S$ in fresh water is very high and it is a function of temperature and Fig. 3 shows this effect, comparing this solubility with the concentration of this gas in the Black Sea, as there are differences in salinity, pH, temperature and the chemical and biochemical reactions at different depths of the sea.

A method for massive extraction of $H_2S$ from seawater is by spraying it at the top of an enclosed tower in the atmosphere. This type of extraction of $H_2S$ will be very useful for separation without conversion of the molecules of $H_2S$, and that can be done through Henry's Law. Air strippers are widely utilized equipment for mass transfer, where air and water are contacted, and the contaminants are transferred from water into the air phase. But, unfortunately, some $H_2S$ will be oxidized by air, being converted to sulfur or sulfate. Therefore, exact theoretical and practical extraction of $H_2S$ from the Black Sea can be explained by the extensive application of Henry's Law (Le Chatelier's principle 2010; Chidgopkar 2006; Carroll 2010).

## 2.1 *Application of Henry's Law for Seawater Extraction of $H_2S$ Gas*

Extraction of $H_2S$ from seawater depends on several variables and changing these variables in our scalable laboratory pilot-plant, resulting in big changes in the efficiency of separation. The values of Henry's Law constantly change with all variables and this dangerously toxic gas gives immediate warning indication when there is a leak in the aerial experimental system (Naman 2006).

Monitoring of $H_2S$ came through measurement of flow rate or measurement of concentration with a sensor through septums of gas followed by chromatographic analysis using a flame photometer detector which gave conflicting results. But the examination of the ratios of stripping volume of $H_2S$ to the initial volume of this gas dissolved in the initial tank gave reasonable results corresponding to 80–95% of separation. Recent results show that the total $H_2S$ in aqueous solution is equal to the sum of the concentrations of various sulfide species in the solution (Carroll 2010).

This is independent of pH, because the salt-effect would have significant influence on the solubility of $H_2S$, as follows: at pH = 7, concentrations of $H_2S$ and $S^{2-}$ are equal, and at pH = 8, the concentration of $S^{2-}$ is 10 times greater than of $H_2S$ (Carroll 2010).

## 3 Survey of our R&D Activities Focused on the Black Sea

Hydrogen production from $H_2S$ gas includes four technological steps as listed below:

1. Concentration of $H_2S$ gas in the Black Sea.
2. Pumping of $H_2S$ gas from correct seawater depth.
3. Extraction of $H_2S$ gas in an efficient, economic industrial process without conversion.
4. Decomposition of $H_2S$ gas to derive a pure hydrogen fuel product.

Concentration profile of $H_2S$ in Black Sea is given by various researchers (Baykara et al. 2004, 2005; Dimitrov et al. 2003; Neretin et al. 2001; Midilli et al. 2005). Daily production of $H_2S$ is 10,000 tonnes. The new plot of this concentration distribution with the depth of the Black Sea (Fig. 1) after statistical treatment of the results (shadow region) shows that the concentration is 10 ppm at the depth of 1,000 m which is the equilibrium concentration value for $H_2S$ and this concentration is due to the SRB bacteria to form $H_2S$ at the bottom and the SOB bacteria at the seawater surface, as well as the first and second dissociation constants of $H_2S$ in the mid-section of the Black Sea's water column. According to Le Chatelier's principle, artificial pumping of $H_2S$ gas from the Black Sea would cause a shift in favor of the forward reaction (Fig. 2). We came to the conclusion that the important technological steps are 1, 2 and 3 in above because the concentration of 10 ppm of $H_2S$ gas in seawater is very low, and it should be industrially concentrated to above 10,000 ppm, similar to the concentration in natural gas. The decomposition of $H_2S$ to produce $H_2$ and S is a well established technology due to extensive research, and now is at the investment stage (T-Raissi 2001; T-Raissi and Huang 2004; Zaman and Chakma 1995). In order to develop steps 1, 2 and 3, extraction (stripping) of $H_2S$ gas at high concentration in ethanolamine tanks is very important. Table 1 gives us important information.

**Table 1** Physical constants for $H_2S$ in Black Sea

| Variables | Ranging at different depth during 12 months | At 1,000 m value equilibrium |
|---|---|---|
| Hydrogen sulfide concentration (ppm) | 6–14 | 8–12 |
| Temperature (°C) | 24–28°C | 8.3°C |
| pH | 7.9–8.45 | 8 |
| Salinity (ppm) | 17,000–21,000 | 18 |
| Geographical Location of $H_2S$ maximum concentration | 42.4 N. Lat. 30.9 E. Long. | North 42.9 East 39.0 |
| Daily formation of $H_2S$ (metric tons, tonnes) | $10^4$ tonnes | $6.75 \times 10^3$ tonnes (net) |
| Total sulfur (metric tons) | 4.587 billion tonnes | |

**Table 2** Comparison between extraction outside and inside Black Sea water

| | Outside water (on beach) (Fig. 6) | Inside water at 1,000 m depth (Fig. 7) |
|---|---|---|
| 1 | There is no need for pipes to transfer seawater from near seafloor to the surface pumping station | No need to pipeline to pump seawater, only pipe for transporting $H_2S$ gas to the surface |
| 2 | Same quantity of seawater should be pumped to sprayer | Same quantity of seawater should be pumped to sprayer |
| 3 | Pump with certain specification in order to spray seawater. | Immersion pump, with same power, to spray seawater under seawater surface |
| 4 | There is a need for a large, closed tank on the beach for temporary seawater storage | There is no need for tanks for seawater storage |
| 5 | Spraying unit on the beach surface operates at 1 atm | Spraying unit, at depth of 1,000 m, at pressure of 100 atm (floating unit) |
| 6 | There are pumping stations for pumping seawater to the surface | No need for pumping station to pump seawater to the surface. Only a small pump for pumping $H_2S$ gas from sprayers lower to the surface |
| 7 | There is the possibility of pollution by $H_2S$ gas escaping to the atmosphere. | There is very weak possibility for escaping $H_2S$ gas to the atmosphere |

## 4 Decomposition Unit for the Mega-Scale Extraction of $H_2S$ Gas from Seawater

Table 2 gives the difference in extraction of $H_2S$ from outside and inside Black Sea water. It is economic to pump $H_2S$ gas from the bottom of water instead of pumping water containing $H_2S$. Figure 4 shows our conceptual pilot-plant for extraction of the $H_2S$ gas from seawater (Black Sea). It is based on the principle of Henry's Law, but it is huge, and it can extract large amounts of seawater containing the $H_2S$ gas, the extracted $H_2S$ will be gradually collected into a large tank of ethanolamine. Afterwards, it will be moved to the decomposition unit to produce hydrogen and sulfur.

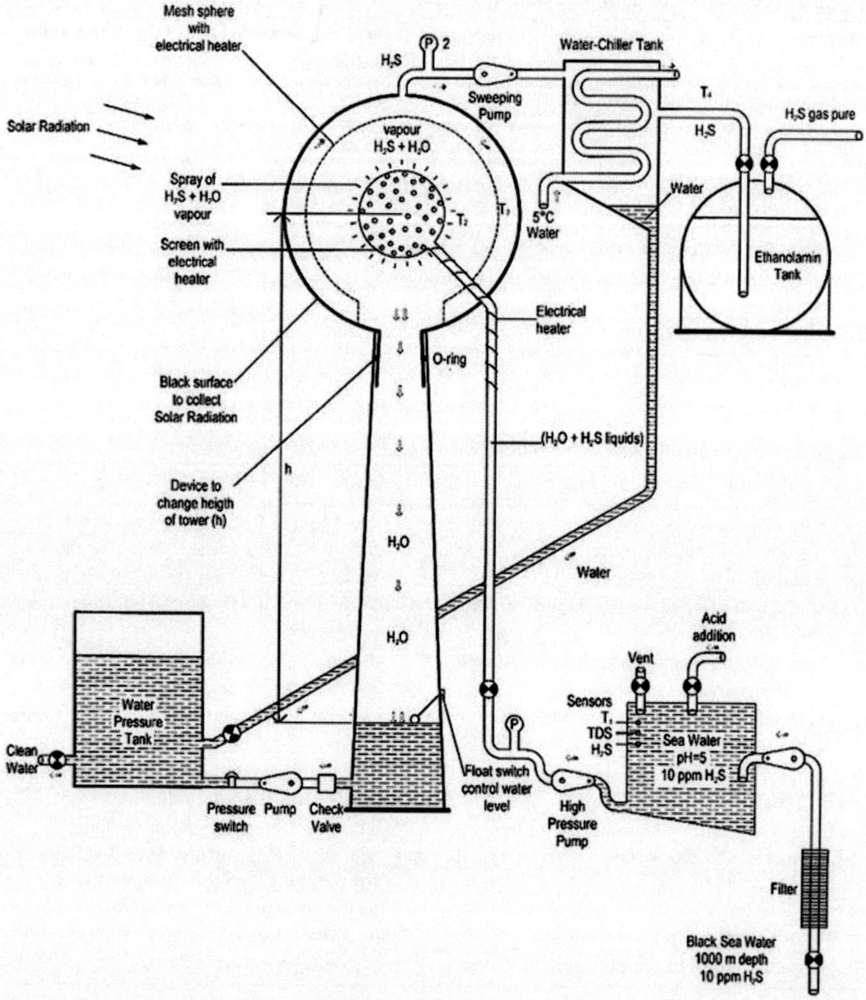


**Fig. 4** Pilot-plant industrial extraction of $H_2S$ gas

This pilot-plant is, in its working principle, similar to our laboratory-scale pilot-plant, but it can pump seawater directly from Black sea. It contains an electrical heater to stabilize the temperature, a water chiller at the top to separate any freshwater vapor. The spraying tower is evacuated by a sweeping pump located at the tower's top. In order to measure most of the variables, there are special sensors at sited at different locations of the pilot-plant. Also, it is possible to change (extend or reduce) the height of the processing tower. This special extraction tower has an option to change height (h) between 5 and 20 m in order to separate the $H_2S$ gas from the seawater droplets by gravity force (Another design, which might be made suitable to operate inside the Black Sea, is described in Fig. 8).

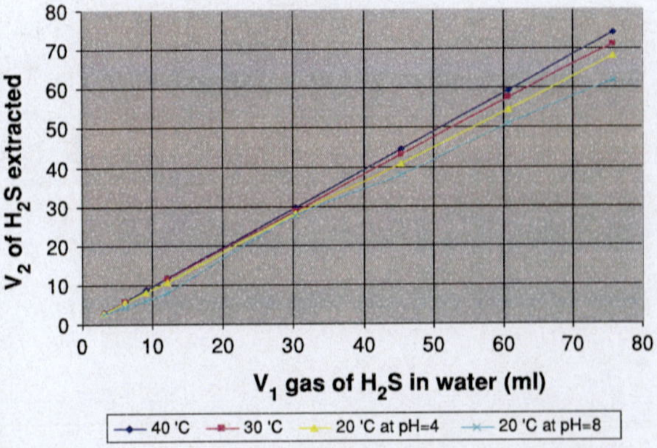


**Fig. 5** Efficiency of separation of $H_2S$ gas from water

Preliminary results from laboratory extraction decomposition unit (see Fig. 7), shows that the value of $V_1$ and $V_2$ measurement is not easy due to the fact that it depends on several variables such as temperature, salinity, pH, tower height, pumping pressure and so forth. Also this extractive unit should be situated within a closed laboratory fume-hood. For fixed pumping rate at pH 4, and salinity of 18,500 ppm, with height of tower of 100 cm, a simple calculation has been done to reduce the higher concentration of $H_2S$ gas 4,000 ppm to 10 ppm, as follows: 242 mL of 4,000 ppm = 734 mL of $H_2S$ gas at STP considering that the density of $H_2S$ vapor is 1,363 g/L.The tank contains 10 ppm of $H_2S$ gas. Each 400 mL of 10 ppm contains 3,0033 mL of $H_2S$ at STP.

Figure 5 shows that the efficiency of the stripping of $H_2S$ from our $H_2S$ solution of 10 ppm concentration which it is depends on temperature, pH, and also the salinity of water and the height of stripping tower.

The results show that the value of $V_2/V_1$ efficiency of stripping process is between 80 and 98%, depending on the physical variables (temperature and pH).

This pilot-plant gives conflicting results when we change pH, temperature, and the height of the tower is changed or if there is any leak from the unit. The temperature measurements through different parts of this unit were carried out by thermocouples and the final volume was measured by displacement of $H_2S$ to the water vessel at pH 14 or by sensors detectors and by gas-chromatographic (GC) analysis using flame photometer detectors. These sets of experiments have been done employing special sensors for measuring $H_2S$ gas (as ppt), but unfortunately the results are not reproducible or even reasonable values, especially when there is the noticeable trace odor of $H_2S$ gas in the laboratory that is due to instrumental leaks, or we have to recalibrate all of these industrial sensors (caused by destructive corrosion of sensors by $H_2S$ gas). Obviously, this laboratory extraction decomposition unit should be improved to safer material standards; especially some parts that are plastic glass and held together with glue.

Our experiments have been done using pure $H_2S$ gas from Fluka AG cylinder equipped with a special gas pressure regulator. Some other experiments have been

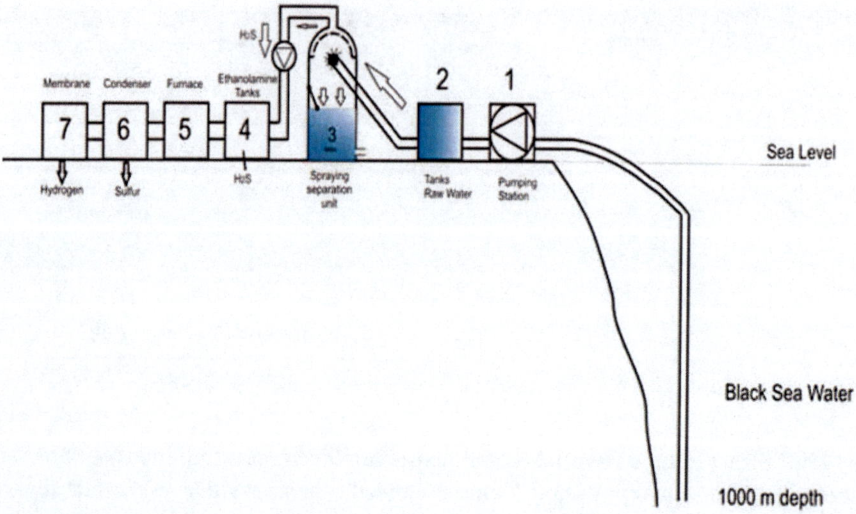


**Fig. 6** Extraction outside water (on beach)

done using Kipp's apparatus using FeS and $H_2SO_4$ to produce $H_2S$ gas without any contact to the ambient laboratory air. Any gas leak in this pilot-plant extractor unit will change some $H_2S$ gas to sulfur of sulfate by oxidation with air-obtained oxygen.

The resulting experimental data will guide us, eventually, to the optimum condition for maximum efficiency of separation of $H_2S$ gas from seawater. Also, the results from laboratory scale pilot-plant will be more helpful to operate the industrial extraction unit.

## 5 Technology for Production of Hydrogen from $H_2S$ in Black Sea

The stages of production of hydrogen gas from Black Sea $H_2S$ gas in Black consists of:

1. Pumping seawater containing $H_2S$ gas (10 ppm at 1,000 Black Sea depth).
2. Extraction of $H_2S$ from seawater and concentration of the gas up to 10,000 ppm in the ethanolamine tank.
3. Decomposition of $H_2S$ gas to produce hydrogen and sulfur

The Decomposition Unit of extraction of this economically valuable gas may be done outside or inside the Black Sea, according to the feasibility of each method given in Table 2, which compares the technical requirements for proposed units 1, 2, 3 (pumping, extraction), at the seawater's surface, with these units inside the

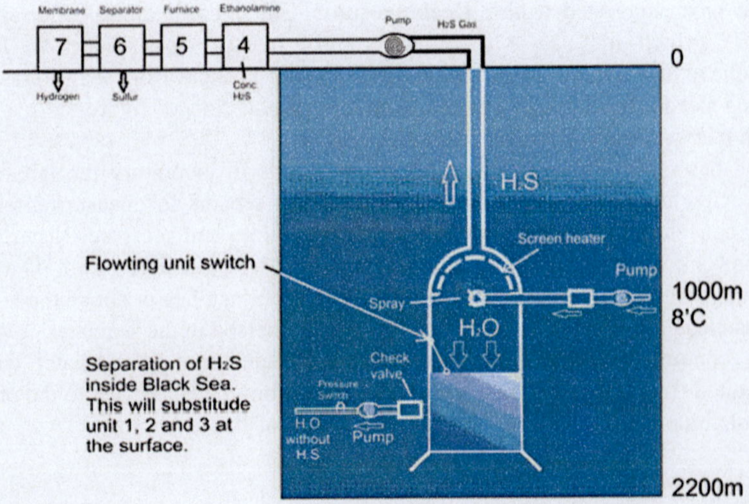


**Fig. 7** Extraction inside water at 1,000 m depth; Extraction of $H_2S$ gas by submerged unit

**Fig. 8** Novel industrial extraction unit of $H_2S$ gas

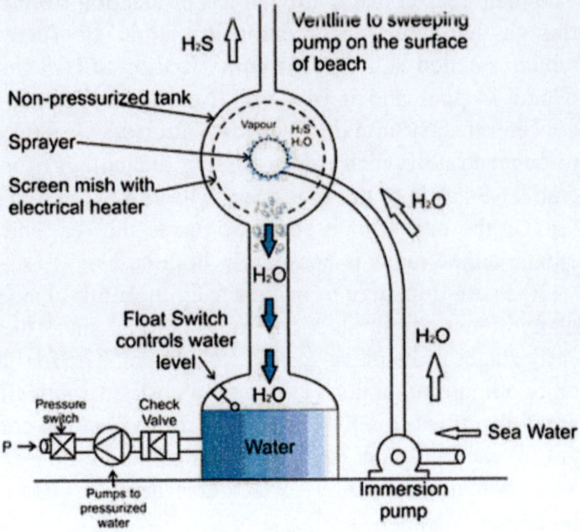


seawater at a depth of 1,000 m assuming that each processor giving $10^9$ $m^3$ seawater will give 0.833 tonnes of gaseous hydrogen.

The feasibility (economical and mega-engineering) of flow sheet (Fig. 7 or 8) can be calculated according to the specifications given in Table 2. If the monetary cost of tanks, pipes for pumping of $10^9$ $m^3$ water from Black Sea depth of 1,000 m will be higher than the price of floating spraying unit at depth of 1,000 m, the second choice flow sheet (Fig. 7) will be more feasible.

We have suggested a new Decomposition Unit located entirely in seawater (Fig. 7), which may be a modified type of our previous model in Fig. 4. We believe that a viable extraction unit—removing $H_2S$ gas from seawater—is the missing needed technology for the economic production of hydrogen gas from the anoxic seawater of the Black Sea.

The best extraction unit depends on our results in laboratory the laboratory-scale extraction unit, especially the positioning of sensors for measuring temperature, concentration, salinity, flow rate, pumping rate and pressure. Materials of this unit are glass and plastic, and the unit should be gas-leak proof. The other essential point is the location of this unit within the Black Sea or above the level of the sea close to the shoreline. We believe that, immersed in the seawater, it will be most economical, Table 2. Because piping and pumping of seawater will be substituted (by pumping of hydrogen gas from within the Black Sea to the outside ethanolamine tanks) that is also an improvement of the technology.

## 6 Conclusion

The main reason that hydrogen gas production from $H_2S$ gas is feasible is because this gas has high concentration in natural gas (between 35,000 ppm and higher which is called SONG). The concentration of $H_2S$ gas in the Black Sea is between 6 and 14 ppm and it is a very toxic gas which is difficult to deal with or to concentrate. The aim of our investigation was to extract this gas from seawater and to concentrate (which it is a missing technology) for production of hydrogen gas from $H_2S$ gas in of the Black Sea. The unit extraction for $H_2S$ can be outside of the sea, on the beach, or it could be inside the sea, and pumping of $H_2S$ gas to the ethanolamine tanks is possible in both cases.

It is still premature to make a truly meaningful assessment for the financial cost of hydrogen gas production from $H_2S$ gas since the industrial process is in the early stages of its mega-scale development. The feasibility of this process depends on two important units: (1) pumping cost, energetically and financially, of shifting great quantities of seawater and (2) the extraction, concentration processes of $H_2S$ gas. These two processes could be operated on the seacoast or beneath the Black Sea's sometimes stormy surface at a depth of 100 m. Comparison of these two processes is shown in Figs. 6 or 7.

The application of Henry's Law in this process will lead us to find the optimum condition with respect to physical properties and the costs of each method of operation. The industrial extraction units, multiples of our perfected Decomposition Unit, would best be situated at the seawater water surface or floating below the surface of the Black Sea at a depth of 1,000 m.

Figures 4, 7 and 8 show that, this process should be modified with respect to units 1, 2 and 3 (pumping, extraction and concentration), and $H_2S$ gas concentration should be increased from 10 ppm to above 10,000 ppm.

**Acknowledgments** We express our appreciation to the UNIDO-ICHET, Istanbul, Turkey for financial support under the contract (TF/INT/03/002/11-51) and University of Duhok, College of Education in Iraq for co-funding this mega-project R&D.

## References

Baykara SZ, Figen EH, Kale A, Veziroğlu TN (2004) Hydrogen from $H_2S$ in the Black Sea. Proc. 15th World Hydrogen Energy Conference, Yokohama, Japan, June 27–July 2 (published on CD-ROM)

Baykara SZ, Kale A, Veziroğlu TN (2005) Possibilities for hydrogen production from H2S in Black Sea. Proc. International Hydrogen Energy Conference, Istanbul, Turkey, July 13–15 (published on CD-ROM)

Blabev M (2010) General chemistry. http://wine1.sb.fsu.edu/chm1046/notes/chmEquil/Lechat/LeChat.Htm

Carroll JJ (2010) Software for phase equilibric in natural groundwater system. AQV Alibrium. http://www3.telus.net/public/jcarroll/ion.htm

Chidgopkar VR (2006) Applying Henry's law to groundwater treatment. http://dbw.eni.edu/Teacher/SNT/log. Accessed 26 Feb 2006

Dimitrov D, Gemov I, Kozhuhavov E (2003) Alternative resources and energy sources from Black Sea bottom, http://www.io-bas-bg/products/volume4.html

Duan Z, Sun R, Liu R, Zhu C (2007) Accurate thermodynamic model for the calculation of $H_2S$ solubility in pure water and brines. Energy Fuels 21:2056–2065

Kopf AJ (2003) Global methane emission through mud volcanoes and its past and present impact on the Earth's climate. Int J Earth Sci (Geol Rundsch) 92:806–816

Midilli A, Ay M, Kale A, Veziroğlu TN (2005) Hydrogen energy potential of the Black Sea Deep water based on $H_2S$ and importance for the region. Proc. Inter. Hydrogen Energy Congress Exhibition IHEC 2005, Istanbul, Turkey 13–15 July

Naman SA (2006) Hydrogen production from $H_2S$ of Black Sea pilot project. Turkey, 7th October

Naman SA, Ture IE, Veziroglu TN (2008) Industrial extraction pilot plant for stripping $H_2S$ gas from Black Sea water. Int J Hydrogen Energy 33:6577–6585

Neretin LN, Volkov II, Bottcher ME, Grineuko VA (2001) A sulfur budget for the Black Sea anoxic zone. Deep Sea Res 8:2569–2593

T-Raissi A (2001) Proceeding & US. DOE Hydrogen program annual review, Baltimore (April 18)

T-Raissi A, Huang C (2004) Innovation technology award 15th WHEC, Yokohama, Japan, June 27–July 2

USEPA (2010) US Environmental Protection Agency. http://www.epa.gov/eogapti/module4/absorbtion/absorbtion.htm (appendix 2)

Wikipedia (2010) Le Chatelier's principle. http://en.wikipedia.org/wiki/Le_Chatelier's_principle

Zaman J, Chakma A (1995) Production of hydrogen and sulfur from hydrogen sulfide. Fuel Process Technol 41:159

# Oxygenation of Large Volumes of Natural Waters by Geo-Engineering: with Particular Reference to a Pilot Experiment in Byfjorden

**Anders Stigebrandt and Bengt Liljebladh**

## 1 Introduction

Deeper parts of enclosed natural water bodies may at times be depleted in oxygen because consumption is higher than supply. When the oxygen concentration becomes less than 2 ml $O_2$ $l^{-1}$ the water is said to be hypoxic, and many animal species will have problems with the oxygen supply. If the water becomes completely depleted in oxygen, i.e. anoxic, no higher forms of life are possible. Since the 1950s, the Baltic proper is usually hypoxic in the semi-permanent halocline and below this the deepwater is often anoxic for long periods. There is then a lack of biomass as compared to a hypothetic case when anoxia and hypoxia do not occur (Diaz and Rosenberg 2008), a problem of increasing global concern. Hypoxia and anoxia influence the biogeochemistry. For instance, hypoxia stimulates the loss of nitrate and nitrite to nitrogen gas by denitrification and anoxia inhibits the binding of phosphorus to iron and manganese. Conley et al. (2002) showed that there is a correlation between decreased phosphorus content in the water and decreased area of anoxic bottoms in the Baltic proper. Stigebrandt and Gustafsson (2007) estimated that earlier anoxic bottoms in the Baltic proper that become overlain by oxic water may more or less instantaneously bind 4 tons P per square kilometer. The process is reversible why phosphorus will dissolve again if the bottom water becomes anoxic.

It has been shown that the lowering of the P content in the Baltic in the mid 1990s was due to a period with changed climate which led to less and fresher inflow of new deepwater and increased wind driven vertical mixing (Stigebrandt and Gustafsson 2007), a large-scale experiment by Nature itself. The combined effect of these changes led to a lowering of the top of the semi-permanent

A. Stigebrandt (✉) and B. Liljebladh
University of Gothenburg, Gothenburg, Sweden
e-mail: anst@gvc.gu.se

V. Badescu and R. B. Cathcart (eds.), *Macro-engineering Seawater in Unique Environments*, Environmental Science and Engineering,
DOI: 10.1007/978-3-642-14779-1_15, © Springer-Verlag Berlin Heidelberg 2011

halocline from about 60 to about 100 m in the Baltic proper. About 30,000 $km^2$ of often anoxic bottoms above about 120 m were thereby kept oxygenated for a couple of years which coincided with a lowering of the phosphorus content. At the end of the 1990s the inflow and salinity of new deepwater increased again and the top of the halocline returned to its usual depth. After this, the area of anoxic bottoms gradually increased and the total amount of dissolved phosphorus returned to about the same amount as before the short natural experiment.

Stigebrandt and Gustafsson (2007) suggested that it would be possible to lower the phosphorus content of the Baltic proper permanently by adding oxygen to the deepwater down to about 125 m. The idea was to mimic the large-scale experiment performed by Nature in the 1990s. It was suggested that the needed supply of oxygen, estimated to 100 kg $s^{-1}$, which corresponds to a specific load of 46 g $O_2$ $m^{-2}$ $year^{-1}$ through the 80 m depth level, should be supplied by pumping down so-called winter water from ca 50 m depth, which is above the semi-permanent halocline. This water is always well oxygenated with concentration >10 mg $O_2$ $l^{-1}$ (1 mg $O_2$ $l^{-1}$ $\approx$ 0,7 ml $O_2$ $l^{-1}$). Thus, about 10,000 $m^3$ $s^{-1}$ of winter water should be added to the deepwater. For comparison, this flow is approximately equally large as the long term mean inflow of new deepwater through the Belts and the Sound (e.g. Stigebrandt 2001). However, before a full-scale oxygenation system can be recommended, a number of crucial biogeochemical, ecological and technological questions have to be answered. Some of these will be answered by the pilot experiment BOX ("a pilot study to evaluate possible effects of Baltic deep water Oxygenation", see the homepage of BOX at http://www.marsys.se) in which coastal basins will be oxygenated. Here we describe the principles of oxygenation of isolated basins with particular reference to the oxygenation experiment that will be done by BOX in Byfjorden, a small fjord on the Swedish west coast.

Byfjorden has almost permanent anoxia in the basin water below the halocline, which is centered at about 14 m depth, c.f. Fig. 2. By studying biogeochemical processes in the basin water of Byfjorden, described in Sect. 2 below, before and after oxygenation it should be possible to draw conclusions about phosphorus dynamics valid also for oxygenation of the Baltic Sea deep water. It should be mentioned that investigations will be done also in a much fresher coastal basin in the Baltic proper in order to study if there is a salinity effect on the biogeochemical processes, mainly due to a decreasing amount of available sulfate with decreasing salinity which may influence the sink of iron and thereby the sink of phosphorus.

The main objective of the experiment in Byfjorden is to study phosphorus exchange between previously anoxic sediments and overlying water which is kept oxygenated artificially. The intention is therefore to keep the deep waters of Byfjorden oxygenated for a longer period. Oxygenation can be done in two ways; either by stimulating an increase of the frequency of inflow of new oxygen-rich deepwater or by supply of oxygen to the deepwater. The frequency of inflow of new deepwater can be increased by increasing the rate of density reduction of the deepwater by pumping down less dense surface water. Adding oxygen rich surface water will also add oxygen to the deep water.

In this chapter we present estimates of the rate of change of density and the rate of change of oxygen concentration achieved by pumping. The main purpose of this chapter is to develop a simple method to estimate the flow of surface water needed to keep the oxygen concentration above a certain level sustaining preferred biogeochemical and ecological states of the deep water.

## 2 Topography, Hydrography and Circulation of Byfjorden

Byfjorden is a typical sill fjord with the sill in Sunninge Sund (Fig. 1). Owing to a series of dredgings in Sunninge Sund, the last in 1996, the strait has been both broadened and deepened and has now a sill depth of 13.5 m and a vertical cross-sectional area at mean sea level of about 2,730 m$^2$. Before 1958, the sill depth was 9 m and the cross-sectional area 720 m$^2$ (Göransson and Stigebrandt 1998). Maximum depth in Byfjorden is about 50 m. Fjord area at sea level is around 6 km$^2$, at 16 m around 3.4 km$^2$ and at 30 m depth approx. 2.3 km$^2$. The total volume is 0.1377 km$^3$, of which 0.0641 km$^3$ is below 16 m and 0.0252 km$^3$ below 30 m. River Bäveån with a mean flow of about 4 m$^3$ s$^{-1}$, accounts for about 85% of the freshwater supply to Byfjorden (Göransson and Svensson 1975). The tide is essentially semi-diurnal with mean amplitude of about 0.15 m.

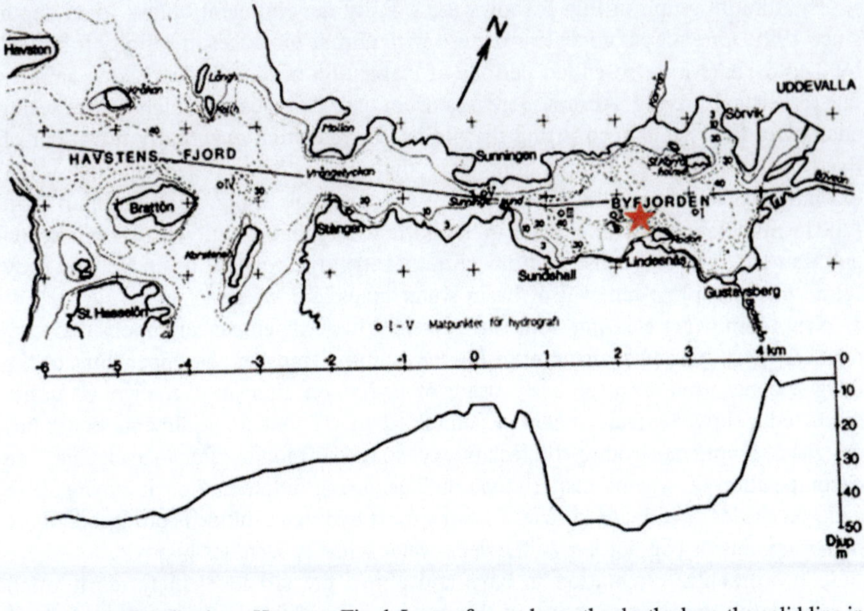


**Fig. 1** Map of Byfjorden—Havstens Fjord. Lower figure shows the depth along the solid line in the map. (From Göransson and Svensson 1975). The *star* shows the position of the pump

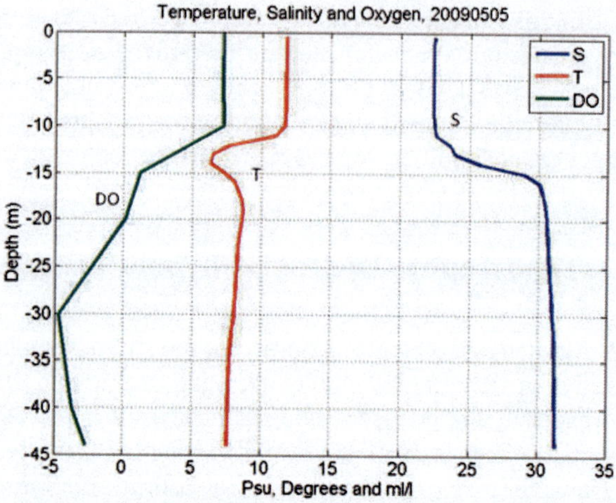


**Fig. 2** Vertical profiles of salinity (S, PSU), temperature (T, °C) and oxygen (ml $O_2$ $l^{-1}$) taken 5 May, 2009. Negative oxygen concentration is the amount of oxygen required to oxidize the measured concentration of hydrogen sulfide

An example of vertical profiles of salinity, temperature and oxygen are shown in Fig. 2. Above the sill level, all variables vary a lot with season. Here, the water residence time is only about 1 week why the conditions are quite similar to those in the Havstensfjord, e.g. Viktorsson (2007). In periods with large local freshwater supply there may be a few meters thick surface layer of low salinity. The conditions below sill level are rather stable as discussed below. The transition zone between the surface water and the basin water is usually characterized by strong vertical gradients in density (the pycnocline) and other quantities.

The middle graph in Fig. 3 shows the salinity development below 15 m depth since 1990. One may note a clear pattern with abrupt increases in salinity followed by 2-to-6-years-long so-called periods of stagnation of slowly decreasing salinity due to vertical mixing. Abrupt increases occur at exchanges on which denser water enters from Havstensfjord lifting the old basin water that eventually flows out of the fjord. Deep-water temperature is in the range 4°–8°, reflecting that water exchange occurs during the winter months (upper graph in Fig. 3). Liungman et al. (2001) modeled water exchange in Byfjorden in great detail with emphasis on entrainment into the dense bottom current carrying the new basin water. They found that complete renewal of basin water takes 2–3 weeks.

New basin water entering from Havstensfjord has oxygen concentrations that vary over the year, see lower diagram in Figs. 3 and 4b. This sets the upper limit of the oxygen concentration in the basin water of Byfjorden after an exchange as further discussed below. Organic matter accumulated in the bottom sediments consumes oxygen at a rapid rate leading to a lack of oxygen in 3–6 months after an exchange. The decomposition of organic matter then continues using nitrate and nitrite, manganese and iron oxides, and finally sulfate, whereby toxic hydrogen sulfide is produced. Many higher organisms can not live in the deep water when it becomes hypoxic.

The horizontal circulation in bays and above the sill level in fjords is often driven mainly by the variability of the density field of the coastal water, c.f. Aure

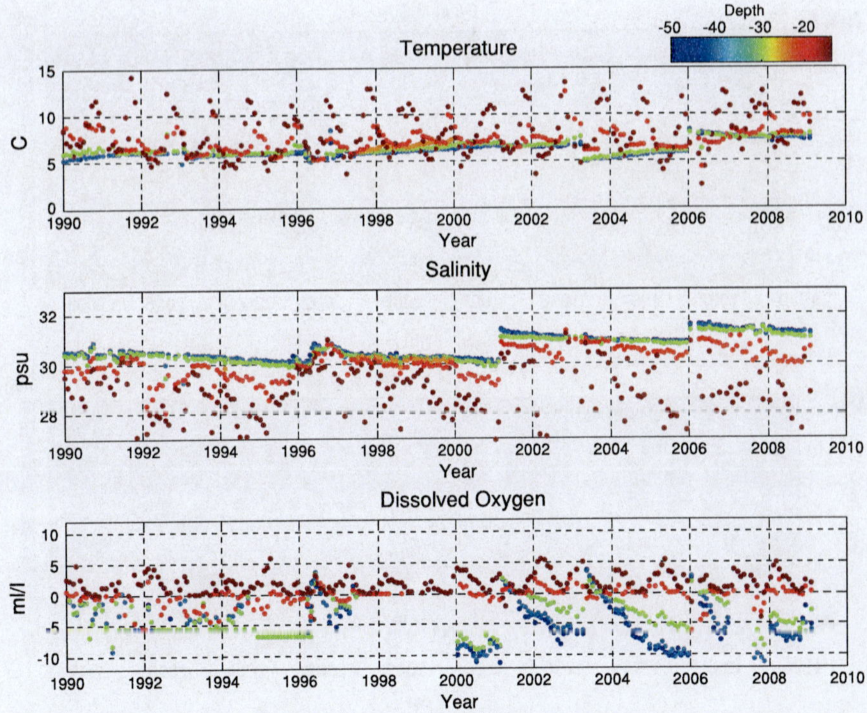


**Fig. 3** Temperature (°C), Salinity (PSU) and Oxygen concentration (ml $O_2$ $l^{-1}$) at depths below 15 m in Byfjorden since 1990. Negative oxygen concentration is the amount of oxygen required to oxidize the measured concentration of hydrogen sulfide

et al. (1996) who called this mode of circulation the intermediary circulation. However, also local freshwater supply and local wind mixing contribute through estuarine circulation. Even tides can contribute. The various modes of water exchange were estimated by Göransson and Stigebrandt (1998) and later by e.g. Viktorsson (2007).

Local winds and internal tides, generated at the entrance strait (Sunninge Sund), drive the very weak vertical mixing below sill level, c.f. Stigebrandt and Aure (1989) for analysis and discussion of mixing of basin waters of fjords. The fact that the deep water is renewed very seldom is due to the combination of very weak vertical mixing and large variability of the density in the coastal water on time-scales of several years as explained by Stigebrandt and Erlandsson (2006).

## 3 Oxygen Conditions in the Deep Water

In the BOX project we want to keep the basin water oxygenated to study the phosphorus chemistry of sediments that earlier have been essentially oxygen-free.

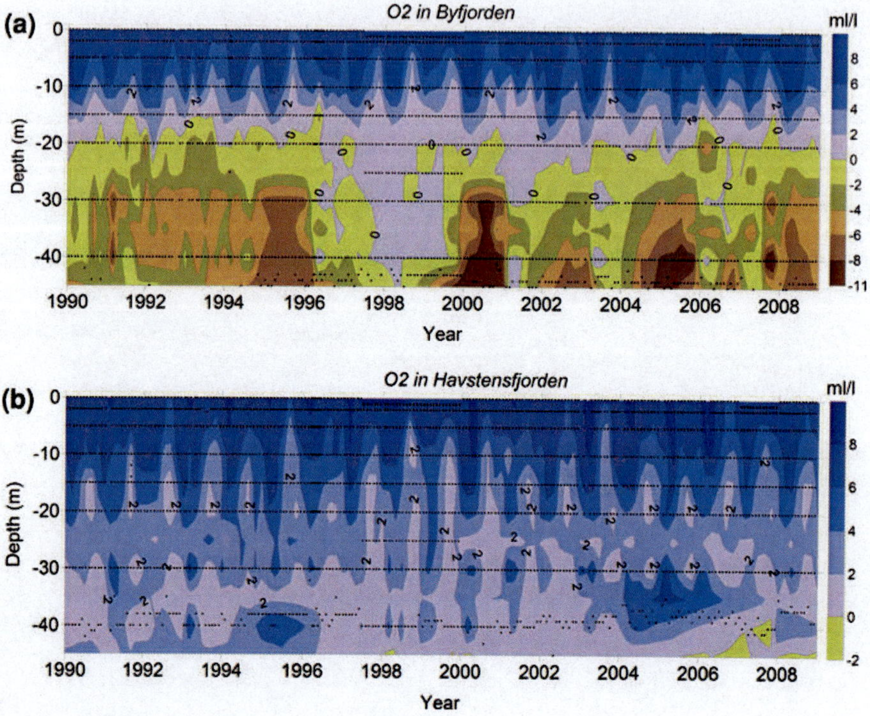


**Fig. 4** **a** Dissolved oxygen in Byfjorden between 1990 and 2009. Hypoxic zone is *dotted*. **b** Dissolved oxygen in Havstensfjorden between 1990 and 2009. Hypoxic zone is *dotted*

We also want to keep the oxygen concentration high enough for animals to live in the fjord basin because it is believed that inhabited bottom sediment may be ventilated by burrowing animals and this might increase the volume of oxygenated sediments, e.g. Karlsson et al. (2007). This should also increase the capacity of the sediments to bind phosphorus. To estimate how much oxygen that should be supplied by pumping, the oxygen consumption must be known. Neglecting the diffusion of oxygen down into the basin, the following simple oxygen budget should be valid during periods of stagnation:

$$\frac{\mathrm{dO}_{22}}{\mathrm{d}t} = -\frac{\mathrm{Cons}}{V} \tag{1}$$

Here $\mathrm{dO}_{22}/\mathrm{d}t$ is the rate of change of the mean oxygen concentration in the basin water, $\mathrm{O}_{22}$, which we estimate from measurements. Cons is the total apparent consumption and $V$ the volume of the basin water. From the lower graph in Fig. 3 (blue dots) we can estimate that $\mathrm{dO}_{22}/\mathrm{d}t$ in the deep water equals about $-1$ g $\mathrm{O}_2$ $\mathrm{m}^{-3}$ $\mathrm{month}^{-1}$. With $V = 60 \times 10^6$ $\mathrm{m}^3$ this gives a total consumption of 60 tons $\mathrm{O}_2$ $\mathrm{month}^{-1}$. Actually, the total consumption may be slightly higher because we have neglected some transport of oxygen into the basin water by vertical diffusion. The apparent consumption of 60 tons $\mathrm{O}_2$ $\mathrm{month}^{-1}$, or 720 tons

$O_2$ year$^{-1}$ corresponds to oxidation of approximately 270 tons C year$^{-1}$ in the deep water, or approximately 80 g C m$^{-2}$ year$^{-1}$ based on the fjord area at 15 m depth. This corresponds to the net production in the fjords outside Byfjorden as estimated by Kajrup (1997).

If the water at 1 m depth in Byfjorden has an average concentration of 10 g $O_2$ m$^{-3}$ the pumping of 2 m$^3$ s$^{-1}$ would supply about 52 tons $O_2$ month$^{-1}$ which is almost sufficient to cover the apparent oxygen consumption. However, pumping surface water into the basin water will also reduce the density of the basin water and thereby trigger inflows of new basin water and these will also bring oxygen to the basin water. This is discussed in Sect. 4 below where we derive a model that then in Sect. 5 is used to compute the evolution of the oxygen concentration in the basin water when pumping down surface water.

## 4 Model of Basin Water with Addition of Surface Water by Pumping

Surface water is usually saturated by oxygen and has lower density than the basin water. By pumping down surface water into the basin water one expects that both

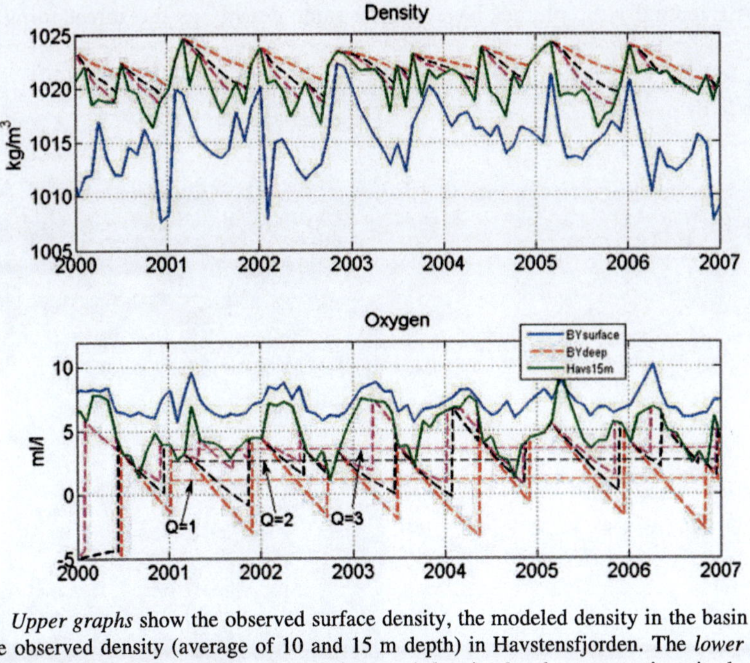


**Fig. 5** *Upper graphs* show the observed surface density, the modeled density in the basin water and the observed density (average of 10 and 15 m depth) in Havstensfjorden. The *lower graph* shows the corresponding oxygen concentrations and the simulated concentrations in the basin water with pump flow equal to 1, 2 and 3 m$^3$ s$^{-1}$, respectively. *Horizontal dashed lines* show mean oxygen concentrations

the density and the rate of oxygen decrease should diminish. Below we describe how pumping increases the rate by which the salinity (density) decreases in the basin and increases the frequency of exchange of basin water. Equations for salinity, temperature and oxygen concentration in the basin water are derived which makes it possible to quantify the effects of pumping.

When applying the basin water model one should know what oxygen concentration that should be aimed at in the basin water. If the goal of the pumping is to achieve oxic basin water, the lowest acceptable oxygen concentration should be above zero. If the goal is to have a healthy benthic ecosystem, the lower limit should be higher because many species of bottom living animals require that the oxygen concentration is at least 2 ml $O_2$ $m^{-3}$.

## 5 Circulation and Mixing in the Deep Water

The pumping of surface water introduces three different dynamical regimes in the deepwater; the jet regime, the buoyant plume regime and the horizontal spreading regime, c.f. Fig. 6. Surface water is ejected from the pump through orifices as jets in order to initially mix as much as possible. The jets entrain surrounding water while they gradually are slowed down. After a certain distance water rises as buoyant plumes towards the pycnocline with additional entrainment of ambient water. Eventually the plumes have got the same density as the surrounding water

**Fig. 6** Sketch of the pump station in Byfjorden

and the plume water will interleave in the lower part of the pycnocline, c.f. Fig. 2. The third dynamical regime then takes over with horizontal spreading of the plume water. Owing to the entrainment, a vertical downward flux of deepwater is induced to replace the water that has been entrained into the plumes. The actual speed will depend on the rate of entrainment and the fjord area, respectively, as functions of depth as further discussed at the end of the present section.

To compute the rate of changes of density and oxygen in the basin water in the fjord basin during periods of stagnation we use the following model. The volume of the basin water is $V$ ($m^3$) and the volume flow of surface water that is pumped down into the deepwater is $Q$ ($m^3\ s^{-1}$). The salinity, temperature and oxygen concentration of the surface water pumped down into the deepwater are $S_1$, $T_1$, and $O_{21}$, respectively. The mean salinity, temperature and oxygen in the deepwater equal $S_2$, $T_2$, and $O_{22}$, respectively. The volume $V$ of the deepwater is considered constant which means that there is a flow $Q$ out of the top of the deepwater and into the intermediate layer. The speed $w(z)$ of this flow equals $Q/A(z)$ where $A(z)$ is the horizontal surface area at the top of the deepwater at the level z. From continuity we may formulate the following equations for salinity, temperature and oxygen in the deepwater of the fjord during periods of stagnation:

$$\frac{dS_2}{dt} = -\frac{Q}{V}(S_2 - S_1) \tag{2}$$

$$\frac{dT_2}{dt} = -\frac{Q}{V}(T_2 - T_1) \tag{3}$$

$$\frac{dO_{22}}{dt} = \frac{Q}{V}(O_{21} - O_{22}) - \frac{\text{Cons}}{V} \tag{4}$$

Here Cons is the apparent rate of oxygen consumption that naturally occurs in the fjord basin. In the equations above we have for simplicity neglected the effect of vertical mixing across the pycnocline that also will contribute to changes of temperature and salinity in the deepwater. This should be a good approximation as long as the pumping induces much greater changes than natural diffusion. Please note that since we use the apparent oxygen consumption, the effect of oxygen diffusion is already included in Eq. (4).

If time series for $S_1$, $T_1$, and $O_{21}$ are available, one may integrate the equations above to compute the response in time of the properties of the deepwater due to pumping. The equations are valid during stagnation conditions which require that the water flowing in from the neighboring sea area is less dense than the basin water. If not, the inflowing water will descend into the basin water, and thereby contribute to renewal of the basin water. In our model we assume that the basin water is completely renewed when such inflows occur.

The model expressed by Eqs. (2), (3) and (4) assumes that the top of the pycnocline is continuously eroded, by the vertical speed $w$, so that the volume of the deepwater $V$ is constant. As long as the top of the pycnocline corresponds to the sill depth, intermediary circulation will erode the top and secure that the pycnocline remains close to the depth of the sill.

# 6 Model Simulations of Pumping

We have run the simple basin water model derived in Sect. 5 for the period 2000–2007. Properties at 1 m depth in Byfjorden and at 12.5 m depth in Havstensfjorden (potential new deepwater) are calculated from monthly observations (Fig. 5). The model was run with pump flow rates $Q$ equal to 1, 2 and 3 $m^3s^{-1}$. In Fig. 5 (lower graph) it can be seen that pumping causes renewals to take place much more often than once in 2–6 years which is the natural interval. The dashed black line shows the result for the pump capacity, 2 $m^3$ $s^{-1}$. The average oxygen concentration is then almost 3 ml $l^{-1}$. Increasing the capacity of the pumps gives less and less changes in the average oxygen concentration since this is set by the concentrations in Havstensfjord. One may see that the lowest pumping rate cannot prevent the water from becoming anoxic in early 2004. However with the pumping rate 2 $m^3$ $s^{-1}$ the lowest oxygen concentration during the period is about 0 ml $l^{-1}$ and with the pumping rate 3 $m^3$ $s^{-1}$ the lowest concentration is slightly less than 2 ml $l^{-1}$. As discussed below we believe that the model underestimates the water exchange so that oxygen conditions might be better than estimated by the model.

## *6.1 Discussion of Model Simulations*

It has previously been argued that sulfidic water can spread to Havstensfjord associated with natural water exchanges (Göransson and Stigebrandt 1998). Partial exchanges of the upper part of the basin water seem to occur almost every winter, c.f. Fig. 4a. When the rare complete exchanges have occurred they might have caused ecological problems in Havstensfjord but this has never been studied.

When pumping triggers the first exchange of water it has already reduced the amount of hydrogen sulfide by adding oxygen-rich surface water to the deep water. Thus, the amount of hydrogen sulfide at the forced exchange of water will be less than it would have been at the next natural water exchange. The potential damage to Havstensfjord will therefore be less than it would have been at a natural water exchange with significantly more sulfide. The improved situation in Byfjorden will eliminate leakage of hydrogen sulfide over the sill to Havstensfjord as long as pumping continues. This should imply that the oxygen concentration of new basin water from Havstensfjord should increase as pumping proceeds (positive feedback). This is one reason why we believe that our computations underestimate the oxygen conditions in the fjord. Another reason is that we have not included the natural vertical diffusion occurring in the fjord that will help to reduce the density a little faster than our computations show and thereby tend to increase the frequency of water exchange. However, the diffusion of oxygen is already accounted for in our estimate since we use the apparent consumption Cons, see Sect. 3.

The observations used to force the model are obtained once a month. Variability on shorter periods is then contained in the observations only through aliasing. In this part of the world, passing weather systems typically induces a lot of variability on biweekly, weekly and shorter periods. It is not clear how this influences our calculations. We believe that our computations may lack shorter partial renewals of the basin water and this should be an additional reason why the model likely underestimates the oxygen concentration in the basin water.

After having taken the uncertainties discussed above into account we decided to chose $Q = 2\ \mathrm{m}^3\ \mathrm{s}^{-1}$ for our experiment in Byfjorden. To use unnecessarily large pumping rate costs a lot of money, c.f. the following section, and should therefore be avoided.

## 7 The Pump's Location and Design

The pump is positioned between the islands Ringburen and Rödön at the southern side of the fjord, see Fig. 1, where the bottom topography allows deep enough inflow of water to the entraining jets from the pumps and the pump is far enough from the harbor area. The pump station consists of a 15 m long, 4 m wide and 0.8 m deep concrete pontoon bridge. Two pumps hang loosely under the gables. This allows the pontoon to swing without risking damage to the pumps. The pumps are suspended from spring boxes to minimize transmission of vertical motion of the pontoon. The pumps themselves are quiet, relatively slowly rotating mixers with a capacity of $1\ \mathrm{m}^3\ \mathrm{s}^{-1}$ each. A stainless steel reinforced rubber pipe leads the water from each pump to 4 spray nozzles at an arbitrary depth. In Byfjorden we use two pumps with outlets at 30 and 40 m depth, respectively, for experimental reasons

The water from the pumps is ejected as horizontal jets with relatively high speed (typically $2\ \mathrm{ms}^{-1}$) to ensure efficient initial mixing with ambient water. After this jet-phase, the released water loses speed and rises slowly as buoyant plumes due to its lower density while mixing with the surrounding water. The power needed in this setting can be calculated using an energy equation:

$$P = Q\rho_1\left[\frac{U_0^2}{2} + \lambda\frac{U_p^2}{2} + \frac{g}{\rho_2}(\rho_2 - \rho_1)h\right]\Big/\ \mathrm{Eff} \qquad (5)$$

Here $Q$ is the pumped flow of surface water $\rho_1$ down a length, $h$, to the lower layer of density $\rho_2$. $U_\mathrm{p}$ is the speed in the pipe and $U_0$ is the speed at the nozzle. The friction factor $\lambda$ depends on properties of the wall of the pipe. The first term on the right hand side is the power used to accelerate the water through the nozzles, the second term the power used to overcome wall friction in the tube and the third term the power used to overcome the higher density in the water column between the inlet and the outlet. With the present setup the pump has an efficiency factor, Eff, of about 0.6. For a smooth pipe of length 40 m and diameter 0.8 m, $Q = 1\ \mathrm{m}^3\ \mathrm{s}^{-1}$ and $U_0 = U_\mathrm{p} = 2\ \mathrm{m\ s}^{-1}$, the electrical power needed equals 12.5 kW. This is supplied by a cable from land.

Using formulas in Fischer et al. (1979), we calculate that the jets ejected from each nozzle will have a horizontal length of about 4 m before being transformed to buoyant plumes. In the jet regime the volume flow increases about three times. The volume flow increases about 3–4 times in the plume regime. Thus, the flow increases by about a factor of 10 before it is interleaved below the halocline. During our experiments in Byfjorden we intend to take the opportunity to study the dynamics of jets and buoyant plumes and compare our findings with various models..

If the volume flow due to entrainment into the pumped plume equals XQ, the turnover time $T_T$ of the deep water due to pumping equals:

$$T_T = \frac{V}{XQ} \tag{6}$$

With $Q = 2\ \mathrm{m^3\ s^{-1}}$, $V = 60 \times 10^6\ \mathrm{m^3}$ and $X = 10$, $T_T \approx 1.2$ months. A water particle will thus descend through the deep water volume in about 1.2 months, or put another way, a deep-water particle will on average be entrained into the jet/plume every 1.2 month. The pumping significantly reinforces the mixing in the deep water which will result in a smooth vertical gradient of oxygen in the deep water and in a steep vertical gradient at pycnocline depth. The (upward) vertical velocity of the pycnocline, $w = Q/A(z)$ equals about 2/3,400,000 m $\mathrm{s^{-1}}$ (at the depth of 16 m) or about 1.5 m $\mathrm{month^{-1}}$.

## 8 Conclusions

In future it might be interesting to oxygenate large natural basins for, e.g. nutrient management or promotion of preferred ecological states. But before such enterprises in geo-engineering should be undertaken we must learn about the biogeochemical and ecological response upon oxygenation. The basin water of Byfjorden on the Swedish west coast will be oxygenated as a pilot investigation aiming at nutrient management through large-scale oxygenation of the Baltic proper. In the present chapter we have described how one may estimate the required pump capacity and the electrical power needed for this, starting from knowledge of the natural oxygen consumption in the basin. Our analysis is quite general and should be applicable to many other natural basins. In order to reduce the errors and get reliable results it is very important to have good field data. Our application of the model to Byfjorden suffers from the sparseness of field data. Instead of the observation interval 1 month one would have preferred 1 week. The pump designed in this chapter was installed in Byfjorden in the beginning of June, 2010.

## References

Aure J, Molvaer J, Stigebrandt A (1996) Observations of inshore water exchange forced by a fluctuating offshore density field. Mar Pollut Bull 33:112–119

Conley DJ, Humborg C, Rahm L, Savchuk O, Wulff F (2002) Hypoxia in the Baltic Sea and basin scale changes in phosphorus biogeochemistry. Environ Sci Technol 36:5315–5320

Diaz R, Rosenberg R (2008) Spreading dead zones and consequences for marine ecosystems. Science 321:926–929

Fischer HB, List JE, Koh CR, Imberger J, Brooks NH (1979) Mixing in inland and coastal waters. Academic Press, 483 pp

Göransson CG, Stigebrandt A (1998) PM angående recipienten. Göteborg 1998-06-12,VBB VIAK (In Swedish)

Göransson CG, Svensson T (1975) Byfjorden: Vattenomsättning. Swedish environmental protection agency, SNV PM 594: p. 152 (In Swedish)

Kajrup N (1997) Net production, oxygen consumption rate and flux of organic carbon in some Swedish Fjords. M.Sc. thesis, Oceanography, Gothenburg University, Dept of Earth Sciences. Report B36

Karlsson K, Bonsdorff E, Rosenberg R (2007) The impact of benthic macrofauna for nutrient fluxes from Baltic Sea sediments. Ambio 36:161–167

Liungman O, Rydberg L, Göransson C-G (2001) Modeling and observations of deep water. Renewal and entrainment in a Swedish Sill Fjord. J Phys Oceanogr 31:3401–3420

Stigebrandt A (2001) Physical oceanography of the Baltic Sea, Chap. 2. In: Wulff F, Rahm L, Larsson P (eds) Ecological studies 148, a systems analysis of the Baltic Sea. Springer, New York, pp 19–74

Stigebrandt A, Aure J (1989) Vertical mixing in the basin waters of fjords. J Phys Oceanogr 19:917–926

Stigebrandt A, Erlandsson C (2006) Dynamics of water exchange in fjord basins. In: Erlandsson C (ed) Influence of physical oceanographic processes on water quality in coastal and inshore waters. Univ of Gothenburg, Dept of Earth Sciences, PhD thesis A 110

Stigebrandt A, Gustafsson B (2007) Improvement of Baltic proper water quality using large-scale ecological engineering. Ambio 36:280–286

Viktorsson L (2007) Water quality response to reduced phosphorus and nitrogen loads to Byfjorden. Univ of Gothenburg, Dept of Earth Sciences, Report B 507

# Aral Sea Partial Refill with Imported Caspian Sea Water

**Richard B. Cathcart and Viorel Badescu**

## 1 Introduction

In 1960, the Aral Sea's volume was slightly more than 1,000 km$^3$ with a salinity of $\sim$10 g/L. Its level stood at $\sim$53 m above the world-ocean's mean sea-level—almost what it was ca. 200 AD—but, by 2007, its level had dropped to $\sim$30 m above the world's prevailing ocean level (Glantz 2007; Micklin 2007) (Fig. 1.)

And, by July 2009, it was officially predicted the Aral Sea's southernmost "Large Aral" section, a reducing sub-division of the former Aral Sea shared by Kazakhstan and Uzbekistan that first became present in 1989, will dry out completely by AD 2020. Environmental change is so rapid in this economically stressed independent region that an issue of Journal of Marine Systems (Volume 76, Issue 3, pages 251–366, of 10 March 2009) wholly dedicated to "Aral Sea Basin Hydrological, Chemical and Biological Dynamics Today, Compared with Trends of the Past 50 Years" cautioned its readers that all geographical data presented therein were extremely transitory in true use-value, becoming merely historical rather quickly. Of course, the same advice is applicable to The Aral Sea Encyclopedia (Zonn et al. 2009) since it is likely none of these contributory cautioners wishes to become known as prognosticative provocateurs. The former seabed, exposed by the regressing body of increasingly saline water, became a wind-modified, salt-strewn arid landscape, the Aral Karakum Desert. Salt particles and dust blown off the exposed seabed nowadays constitutes a huge agricultural and public health menace to the remaining peoples trying to live and work around the

R. B. Cathcart (✉)
Geographos, Burbank, CA, USA
e-mail: rbcathcart@gmail.com

V. Badescu
Polytechnic University of Bucharest, Bucharest, Romania

V. Badescu and R. B. Cathcart (eds.), *Macro-engineering Seawater in Unique Environments*, Environmental Science and Engineering,
DOI: 10.1007/978-3-642-14779-1_16, 

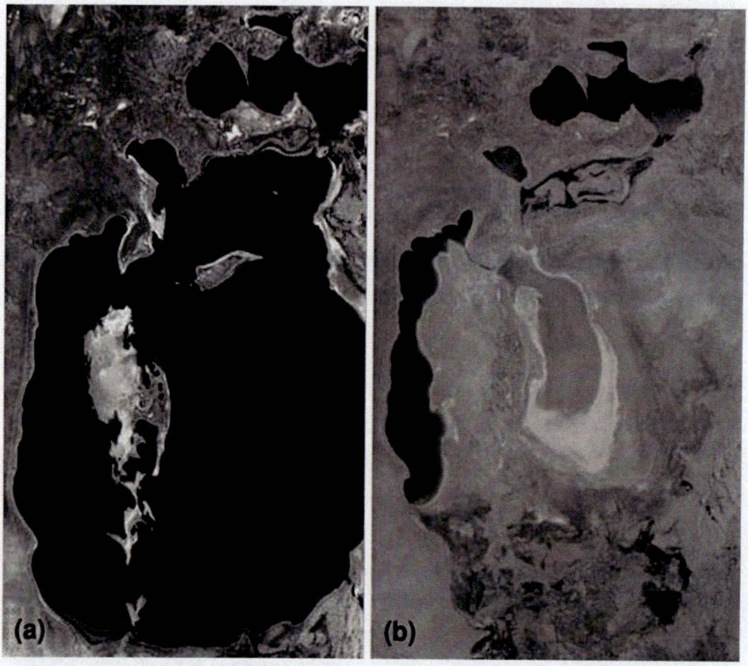


**Fig. 1** Satellite view of Aral Sea. *Left* **a** July–September 1989; *Right* **b** October 5, 2008. The Western, Eastern (almost dry "Large Aral") and Northern basins are clearly visible in **b** (Permission: PD–USGOV–NASA)

drastically desiccated Aral Sea. The "Large Aral" section finally commenced its disappearance in 2006 and, probably, will be a desert landscape about eleven years hence (Roget et al. 2009). The Aral Sea's once-sustainable fisheries, where halieutics was a prized human skill-set, fostered as much as technically possible by uncoordinated Kazakhstan and Uzbekistan management, finally collapsed during 1983 (Fig. 2).

The estimated economic damage is between 1.25 and 2.5 GUSD (1 GUSD = $10^9$ USD), essentially of 2009 value, annually just for human health, tourism and agriculture. Lacking the skilled perspective of restorationist eco-aesthetics, we cannot estimate usefully the economic costs of repairing the Aral Sea to its recent past glory. Neither can we predict future Central Asian river flows insofar as they are extra-regionally dependent on global climate factors related to atmospheric teleconnections (Barlow and Tippett 2008).

Here, we propose a comprehensive latitudinal control strategy to partially recreate the endorheic Aral Sea. This involves regulation of several hydrological factors: (1) stabilization of the Aral Sea's water level by real-time hydrological management of the freshwater inputs from the Syr Darya river as well as regulated evaporation in the Western Basin of the Aral Sea; (2) overland pipeline conveyance of water extracted from the Caspian Sea and deposited into the Aral Basin; (3) overland pipeline conveyance of salty water extracted from the Aral Sea and

**Fig. 2** Stranded fishing vessel, abandoned in place on an ever-enlarging "beach" because of the Aral Sea's remarkable constantly shrinking, and near-absent, areal presence. Photograph by courtesy of Michael Shamshidov, Orex CA.com

deposited into the Caspian Sea; (4) overland pipeline importation of seawater from the Black Sea to the Caspian Sea, to induce constancy on its free water level. Appropriate outline economic evaluations are provided for all proposed macro-engineering technical solutions (Badescu and Cathcart 2009a, b).

Our goal is a partial restoration of the ecosystem services present before 1960 in the Aral Sea region (Benayas et al. 2009). Our macro-project proposal, in some of its most significant aspects, essentially mirrors our still contemplated water-body flushing mega-project proposal to pump 30,575,000 $m^3$/year of saline water—now moderately more saline than common world-ocean seawater—out of California's Salton Sea for final deposition in the upper Gulf of California or to the Pacific Ocean (Leslie et al. 2009) and to import to the Salton Sea an equal volume of seawater from the global ocean. The importation and exportation pipes would each have diameters of ~284.48 cm with equivalent flow rates of ~152.4 cm/s. Regulatory government mandates require the State of California, by 2018, to devise and implement a long-term restoration plan before 2025 AD for the Imperial Valley lest the evaporating Salton Sea lose ~60% of its fluid volume, triple its water salinity and expose ~300 $km^2$ of contaminated former lakebed. Geoscientists, not sales representatives from "Clairvoyant Reality", have determined that this anthropogenic beachfront property will be contaminated and publicly undesirable new land (Dillon et al. 2009).

## 2 Central Asia's Previously Proposed Freshwater Management Plans

### *2.1 Aral Sea Basin*

The modern-day, and volumetrically diminished Aral Sea, consists of three or four reducing lakes (Fig. 3b), ugly sub-divisions of the former fecund Aral Sea.

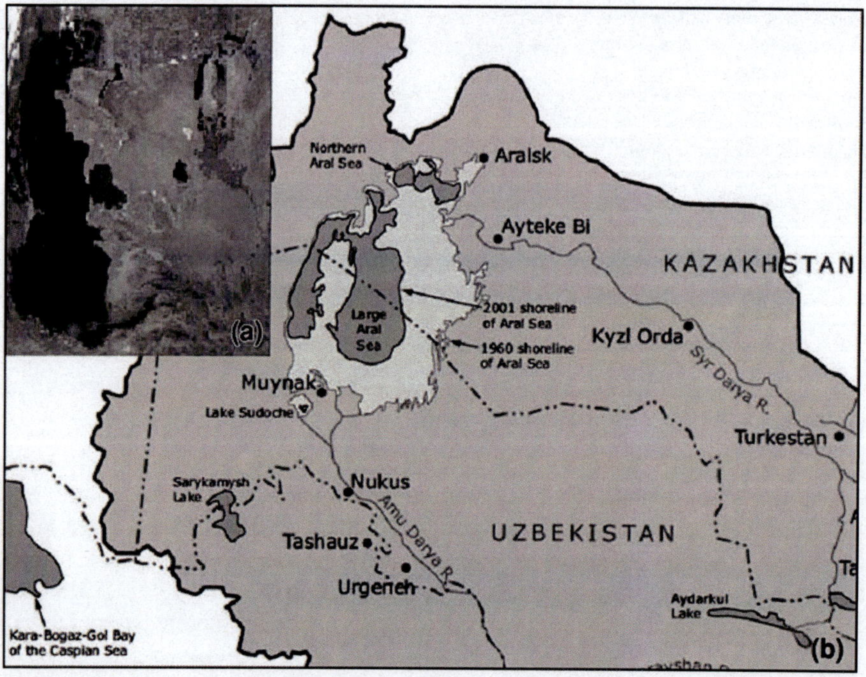


**Fig. 3** **a** The Aral Sea's (*top right*) proximity and relative size to the Caspian Sea (*left*). (MODIS/Terra image from June 10, 2006, courtesy of NASA). **b** The northwest portion of the Aral Sea's watershed and toponomy (Alles 2007; Roll et al. 2006). The "Large Aral Sea" may become nonexistent by 2020 AD under present-day environmental circumstances

The presence of the Aral Karakum Desert has caused remarkable short- and long-term meteorological changes in the various bordering downwind climate regimes of Central Asia (Elguindi and Giorgi 2007; Shibuo et al. 2007). If it were possible to provide these relict lakes with the Central Asian river runoffs extant during the pre-1960 AD period, at least 200 years would be necessary for the Aral Sea to be recreated (Salokhiddinnov and Khakimov 2004).

However, it is presently feasible through macro-engineering to bring some water to the Aral Sea from outside the boundaries of Central Asia (Micklin 2007). Micklin continues to contemplate a 27–30 km$^3$/year input of freshwater, such as from the long-discussed Siberian River Diversions (Davies et al. 2006; Duke 2006). Such postulated freshwater transfers will certainly be impacted by near-term future global climate change (Dobrovolski 2007; White et al. 2007). Even in the post-USSR twenty-first century, some Russian geopoliticians seek to revive the Siberian River Diversion macro-projects (Pearce 2002) to import freshwater from the Ob River and its tributary, the Irtysh River, to Kazakhstan via a proposed "SibAral Canal Project", a 2,500 km-long, 200 m-wide, 16 m-deep concrete-lined canal conveying about 6–7% of the Ob River's yearly runoff to Central Asia, overcoming a 126 m-high topographic pass in the Turgai Depression. Although this still unrealized 1949 "Mitrofan Mikhailovich Davydov Plan" sought to convey Siberia's excess freshwater to Central Asia (Davydov 1949), the estimated 2009 GUSD 100 concept of diverting massive freshwater flows to Central Asia originated with N. Demchenko in 1871. In Chap. 17 of this book, a different macro-engineering version of the Irtysh River freshwater diversion is proposed.

## 2.2 *Caspian Sea and Black Sea Basins*

Whilst world sea-level rose ~13 cm during the twentieth century, the Caspian Sea rose 13 cm during the period 1977–1995. Coastal erosion, infrastructure damage and other macro-problems stimulated by such fluid volume changes are chronic drags on the economies of the bordering ecosystem-countries (Demin 2007; Leroy et al. 2007).

Dam-building Russian macro-engineers tried to utilize the Garabogaz Bay's high overall yearly evaporation as an effective hydrological regulator of the Caspian Sea's ordinarily fluctuating level; Garabogaz Bay comprises only ~3% of the Caspian Sea's surface but, under natural circumstances, it evaporates ~5% of the Caspian Sea's annual hydrologic loss (Krapivin and Phillips 2001; Kosarev and Kostianoy 2005; Clauer et al. 2009). A dike enclosed the Garabogaz Bay from March 1980 until June 1992. The former USSR's macro-engineers also proposed, ca. 1960, a very expensive macro-project dividing the fresher northern Caspian Sea from the saltier southern Caspian Sea with a nearly impermeable >200 km-long dike, sited atop the Apsheron Sill (Kuprin 2002), for the sole purpose of supposedly preserving a stabilized body of seawater (Bobrov 1961; Kosarev et al. 2009).

On 10 June 2007, Kazakhstan's President (Nursultan Nazarbayev) proposed construction of a 650 km-long "Eurasia Canal" between the Black Sea and the Caspian Sea. The Eurasia Canal could require 3–5 years to build and cost more than six 2009 GUSD. As proposed in 2007, the Eurasia Canal would be 80 m-wide and have a standard vessel navigational depth of 16.5 m; theoretically it should be capable of carrying ships of 3,500–10,000 metric tons, allowing cargo delivery schedules of 9–12 days with a cargo traffic capability more than twice that of the Volga-Don Canal.

In April 2007, the Russian Federation President (Vladimir Putin) had proposed an upgrading of the over-used Volga-Don Canal and, on 15 June 2007, RF First Deputy Prime Minister (Sergei Ivanov) announced that Kazakhstan, Azerbaijan and Turkmenistan might be disposed to participate in a renewal and modernization of the Volga-Don Canal. The President of Kazakhstan, however, foresees a separate Eurasia Canal utilizing the present USSR-era navigable freshwater reservoirs in the Kuma-Manych Depression of southern RF which would shorten the shipping route by ~1,000 km, transforming landlocked Kazakhstan and all Central Asian ecosystem-states into semi-maritime nations. This structural economic and geopolitical transformation would accompany, maybe augment, the further development of the "New Silk Road", wherein Central Asia becomes more than merely an ancient cultural and commercial link-up zone for Europe and Asia reinvigorated, but nowadays is increasingly becoming a major production zone for oil and natural gas sought by outsiders (Carls 2009).

## 3 A New Macro-Project Proposal

Solving the reduction and exposure problems experienced by the Aral Sea requires new macro-project solutions (Mainguet and Letolle 1997; Dukhovny 2007). Indeed, full restoration of the ca. 1960 Aral Sea through vigorous and intensified freshwater conservation alone seems unrealistic since such a restrictive consumptive use program would impose significant economic hardships on the increasing human populations of Central Asia's post-1991 independent ecosystem-states (Kazakhstan, Kyrgyz Republic, Tajikistan, Turkmenistan and Uzbekistan). (Only Kazakhstan and Uzbekistan are riparian.) A reasonable estimate of the freshwater inflow to the vicinity of the Aral Karakum Desert during the twenty-first century is ~12 km$^3$/year; to restore the Aral Sea to the water body's ca. 1960 size would require a total annual fluid inflow of at least 56 km$^3$/year.

The best nearby source for clean, or technically cleanable at reasonable economic cost, seawater is the Caspian Sea, lying presently ~27 m below the world-ocean's mean sea level and, approximately 600 km west of the Aral Karakum Desert (Fig. 3). One may conceive that people in the riparian states, including non-riparian Russia through its freshwater reservoir operations, organize to control seawater levels in both the Caspian Sea and the partly renewed Aral Sea, using low-cost pipelines.

A simple form of the time-dependent water volume balance equation for the Aral Sea is:

$$\frac{dV}{dt} = Q_{river} + Q_{prec} + Q_{g-w} + Q_{other(+)} - Q_{\inf} - Q_{evap} - Q_{other(-)} \quad (1)$$

where $V$ is the volume of water while the five terms in the right-hand-side (r.h.s.) member of Eq. 1 refers to the incoming volume flow rates due to rivers, precipitation, groundwater and from other sources, respectively, and to the lost volume flow rates due to infiltration, evaporation and other causes, respectively. In writing Eq. 1 we assumed all water flows have the same mass density. Thus, the mass balance equation turned into a volume balance equation. Presently, the term $Q_{evap}$ dominates the r.h.s. member of Eq. 1.

The time-dependent salinity balance equation for the Aral Sea is:

$$\frac{d(sV)}{dt} = s_{river}Q_{river} + s_{g-w}Q_{g-w} + s_{other(+)}Q_{other(+)} - s_{other(-)}Q_{other(-)} \quad (2)$$

where $s$, $s_{river}$, $s_{g-w}$, $s_{other(+)}$ and $s_{other(-)}$ are the average salinities (in g/L) of the Aral Sea, of the water incoming from Amu Darya and Syr Darya rivers (Ismaiylov et al. 2007), of the groundwater, incoming from additional sources and leaving the

Aral Sea due to "other" causes, respectively. The salinity content associated to precipitation and evaporation in the r.h.s. member of Eq. 1 is zero and these terms do not contribute to Eq. 2.

There are three main means to artificially change the Aral Sea's hydrographic regime: (1) by better managing the existing resources of natural freshwater, (2) by significantly changing the magnitude of the additional sources and sinks of water and (3) by changing the evaporation rate. All these ways are studied in the following.

### 3.1 Managing the Existing Freshwater Resources

Savoskul et al. (2003) focused on the future of the Syr Darya to 2099. The results given in their Tables 8.4 and 8.5 shows the amount of freshwater expected to be added to the Aral Sea for various envisioned macro-engineering scenarios. Associated costs estimations would be very useful.

A study of the Amu Darya by Froebrich and Kayumov (2004) demonstrated that 100% of the inflow to the Aral Sea from the Amu Darya is "return flow water"—that is, water that is entirely composed of liquid runoff from upstream irrigated farm fields. Such return water is laden with silt, salt and pollutants such as pesticide residues, human and animal wastes. Its reuse causes widespread soil salinization, which is adverse to good agricultural practices, as well as further contamination of groundwater in the Amu Darya's catchment.

### 3.2 Using Additional Sources and Sinks of Water

The volume water flow rates $Q_{other(+)}$ and $Q_{other(-)}$ in Eq. 1 are presently negligible and the Aral Sea volume decreases in time. For a steady-state Aral Sea one needs from Eq. 1 that

$$Q_{river} + Q_{prec} + Q_{g-w} + Q_{other(+)} - Q_{\inf} - Q_{other(-)} = Q_{evap} = r_{evap}S(V) \qquad (3)$$

where $r_{evap}$ is the evaporation rate (in m/s) and $S(V)$ is the free water surface area. The evaporation rate $r_{evap}$ in Aral Sea is evaluated now at 1.3 m/year (this has to be compared with the rate 1 m/year accepted in previous studies; for discussions see (Letolle et al. 2005, p. 54).

First, we have to evaluate the evaporated water flow rate necessary to keep the Aral Sea in steady-state, for a given water level height $h$ in Fig. 4.

To this purpose we use Table 1b and Fig. 2 from Letolle et al. (2005). Figure 5 shows the free water surface area and the water volume $V$ in the North Basin ("Small Aral Sea") and in the Western and Eastern Basins.

For given value of $Q_{evap}$, Eq. 3 provides the associated free water surface area $S$ that, in turn, allows us to estimate by interpolation in Fig. 5 the free water surface

**Fig. 4** Aral Sea bathymetry (m)

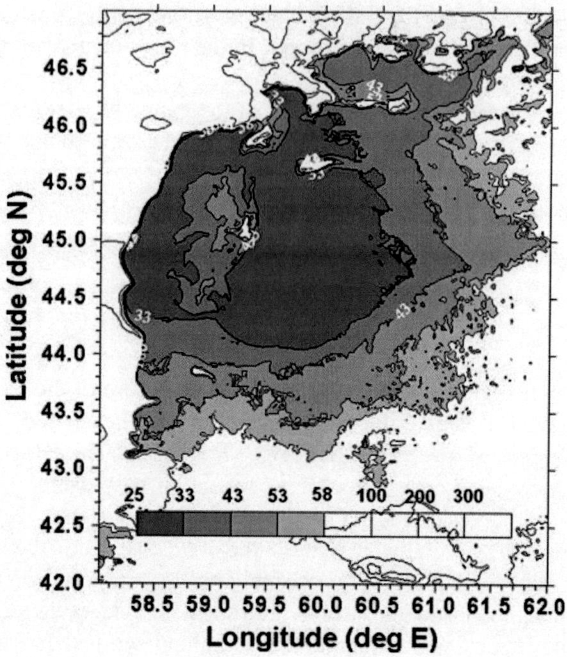


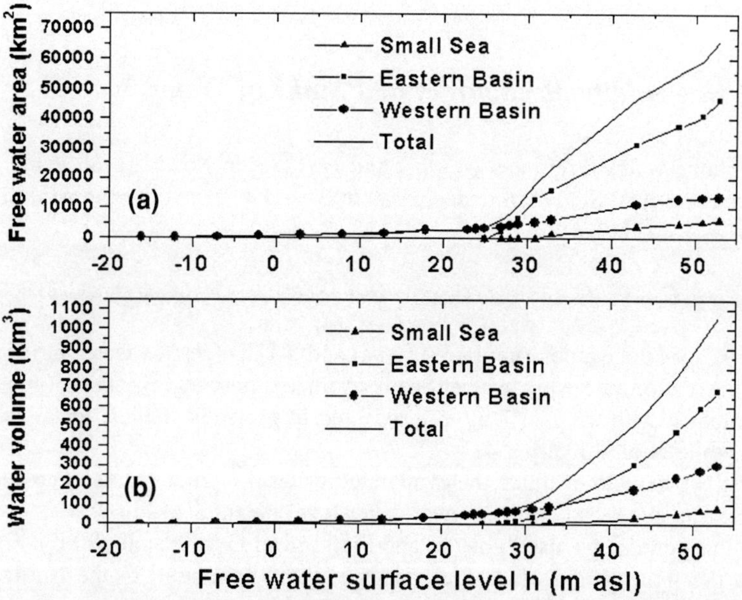


**Fig. 5** **a** The free water surface area and **b** the water volume in the North Basin ("Small Aral Sea") and in the Western and Eastern basins of the Aral Sea, respectively, as a function of the free water surface level (Letolle et al. 2005)

**Table 1** Water flow rate $Q$ necessary to keep the Aral Sea in steady-state, for a given water level height $h$ and evaporation rate $r_{vap}$

| Free water surface level $h$ (m asl) | Inflow rate $Q$ (km$^3$/year) | |
|---|---|---|
| | $r_{vap} = 1$ m/year | $r_{vap} = 1.3$ m/year |
| 1 | 0.8 | 1 |
| 12 | 1.5 | 2 |
| 16 | 2.3 | 3 |
| 25 | 3.8 | 5 |
| 27 | 7.7 | 10 |
| 30 | 15.4 | 20 |
| 33 | 22.3 | 29 |
| 36 | 30 | 39 |
| 39 | 37 | 48 |
| 43 | 47 | 61 |
| 47 | *57* | *75* |
| 50 | *61* | *80* |
| 53 | *65* | *85* |
| 58 | *72* | *90* |

Values in italics are taken from Fig. 2 of (Letolle et al. 2005). Other values are computed by using the procedure described in the text

level $h$. The procedure may be used for values of $h$ smaller than 33 m above global sea-level (asl). In case $h$ exceeds 47 m asl, the associated value of the flow rate $Q_{evap}$ may be obtained by direct interpolation from Fig. 2 of Letolle et al. (2005). For $h$ values between 33 m and 47 m asl the following procedure has been adopted. Knowing the values of $Q_{evap}$ associated to $h = 33$ m and $h = 47$ m, respectively, linear interpolation has been used to associate $Q_{evap}$ values to $h$ values in this range. Results are given in Table 1 which shows the value of $Q_{evap}$ estimated by using Eq. 3 for several values of $h$ and two values of $r_{evap}$.

### 3.2.1 Future Infrastructure

The Aral Sea, the Caspian Sea and the Black Sea (Fig. 6) are the modern-day remainders of a formerly existing single Pleistocene sea. Presently, the Black Sea and the Caspian Sea do not merely differ in salinity (17.5 vs. 10–13 g/L) but also in chemistry.

The Black Sea, connected to the world's ocean, is diluted seawater, whereas the Caspian is concentrated Volga River runoff, with very different ion ratios. The Aral Sea presently has a very unique ionic composition and dissolved gases content in its remaining water (Zavialov et al. 2009). Recent super-computer laboratory modeling of the Caspian Sea Basin's climate "…implies the possibility of several meters decrease in the (Caspian Sea's extant level) for the twenty-first century" (Elguindi and Giorgi 2007).

A complex scheme for inter-basin water transfers between the three separated water bodies may be conceived. First, water is imported from the Caspian Sea to

**Fig. 6** Reading *right to left*: Aral Sea, Caspian Sea and Black Sea. Relative geographical positions

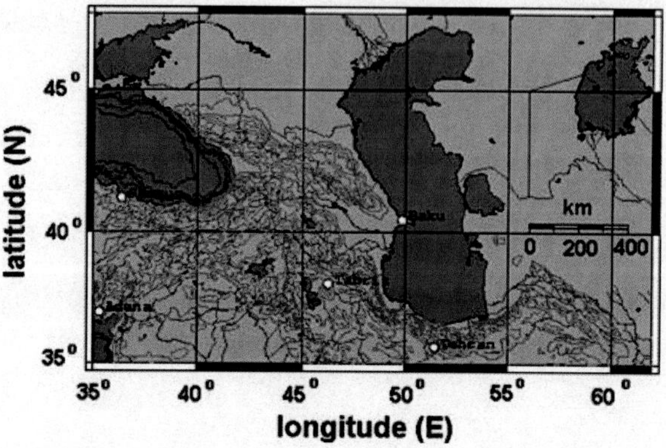


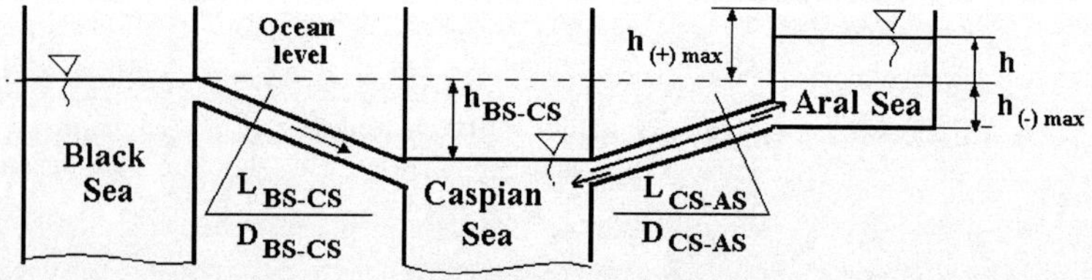


**Fig. 7** Aral Sea, Caspian Sea and Black Sea. The geometry of the problem studied here. $h$ is the water level in Aral Sea in respect to the Black Sea water level (which might be used as a rough measure for the average world-ocean sea-level). $h_{BC-CS} = 27$ m is the present-day level of the Caspian Sea's water surface below the average planetary ocean level, $h_{(+)max} = 58$ m is the maximum possible height of the Aral Sea above the average planetary ocean level, $h_{(-)max} = 16$ m—the minimum height of Aral Sea Level below the average planetary ocean level (it corresponds to the Western Basin)

the Aral Sea. Second, a smaller flow rate of saltier water from Aral Sea may be discharged into the Caspian Sea. In this way the salinity of the Aral Sea is kept within selected limits. Finally, the deficit of water in the Caspian Sea may be compensated from importing true seawater from the Black Sea. Figure 7 shows the essentials of this inter-basin saltwater transfer macro-project.

Note that the highest possible water surface level of the Aral Sea in Fig. 7 is $h_{(+)max} = 58$ m, due to the low altitude and gentle gradient of the Amu Darya's delta. This corresponds to the largest possible water surface of the Aral Sea (Fig. 4). Before 1960, the Aral Sea's free mean water level oscillated ~53 m asl, induced by variable rates of yearly aerial evaporation (see Fig. 4).

### 3.2.2 Technical and Economical Model

Inexpensive Seawater Pipeline Solution

Conventional pipelines are usually made of steel. A less expensive solution is a reinforced tensioned textile pipe which can tolerate pressure to about 100 bar.

Purely textile pipes of very large diameter are possible because of the advancement of materials science (Beylerian and Dent 2007).

Our macro-project invokes a seawater pipeline-integrated photovoltaic flexible power module or membrane that won't ever—or, at least, for a period of tens of years—debond under very harsh conditions of summertime sunshine, wintertime ice, seasonal cold air temperatures and windblown particle abrasion. The pliable thin-film PV coating will need to cover part of the upper hemisphere of the (steel, plastic, concrete or other material) pipe, which is the basic support structure for the proposed PV electricity-generation system to power water pumps. By the start of 2009, the average price of a 1-W PV cell-module was <4 USD, and prices are estimated nowadays to reduce to ~2 USD/watt by 2010, with thin-film PV cell production costs to reach ~1 USD/watt in 2010.

The pipeline as a whole includes the water pumps and other monitoring devices—devices which will verify the design, record all incident physical events and monitor the overall structural health of the pipeline (Peeters et al. 2009)—and regulatory equipment such as valves. Any tensioned textile pipeline will, in fact, have many of the characteristics of a very large diameter flexible hose. Therefore, it is possible to imagine multiple hoses instead of merely one, two or more "pipelines" transferring water to the Aral Sea (or to the Caspian Sea) in a special zoned infrastructure security corridor. Such secured hoses would be aggregated, but not bundled.

A wide-diameter hydraulic tube is more expensive but such a pipe, hose or tube has the advantage of decreasing greatly the pressure loss and increases significantly the efficiency and decreases the friction loss and pump power. Tubes near the pump stations have high pressure up to 30 bars. They must be made from something even stronger—perhaps from some composite fiber material. Such composed material has higher maximum stress (up to 600 kg/mm$^2$, steel has only about 120 kg/mm$^2$) and low specific-density (1,800 kg/m$^3$, steel has about 7,900 kg/m$^3$). It may be, or soon become, markedly cheaper quite soon. The tensioned textile tube must be supported along its entire length by a special chute, a cradle-like trough of discrete air-inflated pillows.

## Water Exchange Between Caspian Sea and Aral Sea

The construction of the pipelines connecting Caspian Sea and Aral Sea will be treated in more detail here. This involves some macro-engineering and macro-economic calculations. However, these evaluations are very rough, taking into account that there are so many unknown and missing parameters involved.

As a starting point, a single duct connecting the Caspian Sea with Aral Sea is considered. Taking into account the land's topography, route A in Fig. 8 is envisaged. This pipeline will draw lower salinity water from the northern region of Caspian Sea (near the Volga River's mouth) to the Aral Sea.

The pipeline B in Fig. 8 may be used to export saltier water from the Aral Sea to the closable Garabogaz Bay (Fig. 3b), a regulator of the Caspian's Sea's salinity due to evaporation. Pipeline B follows a natural route from the present-day Aral

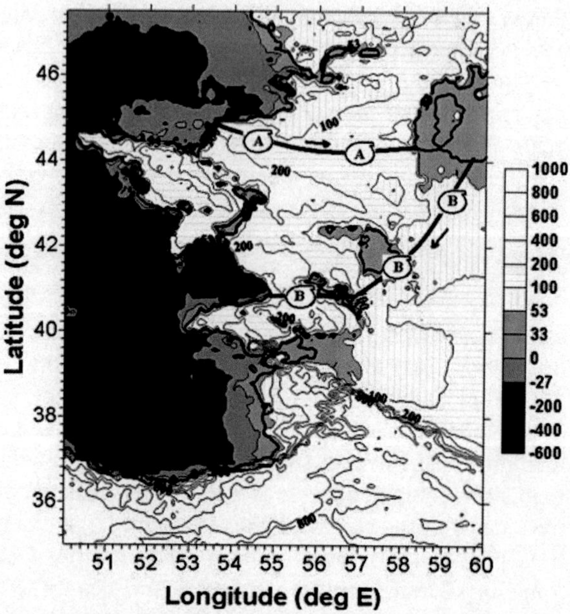


**Fig. 8** Routes for the pipes connecting Caspian Sea to Aral Sea, with altimetry shown (m)

Karakum Desert to Caspian Sea, i.e. the abandoned natural Uzboi Channel which, as recently as 1500 BC, connected the Aral Sea with the Caspian as an Aral Sea peak flood freshwater discharge overflow route (Letolle et al. 2007). Uzboi Channel had a maximum discharge capacity of ~2,000 $m^3/s$ owing to channel topography limitations during the Holocene. The maximum topographic elevation of the Uzboi Channel's ancient waterway is about 57 m asl. Therefore, the fluid discharge by suggested pipeline B from Aral Sea to Caspian Sea is greatly assisted by gravity with important reduction in pumping power consumption.

Both pipelines A and B are approximately 600 km long.

The next computations refer to the pipeline on route A. The duct length and diameter is denoted $L_{duct}$ and $D_{duct}$, respectively, while the height difference between its two heads is $H_{duct}$.The seawater volume flow rate through the duct is $Q_{sw}$. If a single pipe is considered, then $Q_{other(+)} = Q_{sw}$. The speed $w_{sw}$ of seawater in the duct of diameter $D_{duct}$ is given by:

$$w_{sw} = \frac{4Q_{sw}}{\pi D_{duct}^2} \tag{4}$$

The duct includes multiple electricity-powered pumping stations along the route. The power $P_{pump}$ required to transport seawater in the flexible or non-flexible conveying duct is obtained from:

$$P_{pump} = g\rho_{sw} q_{sw} H / \eta_p \tag{5}$$

In Eq. 5, $g$ (=9.78 m/s$^2$) is gravitational acceleration, $\rho_{sw}$ (=1,030 kg/m$^3$) is the density of the seawater, $H$ is the hydraulic head and $\eta_p$ ($\approx$0.75) is the efficiency of the pump. The hydraulic head is obtained by summing the mean height of the erg with the lost pressure height $\Delta H$ due to friction:

$$H = H_{duct} + \Delta H \tag{6}$$

Only linear pressure losses are considered next and:

$$\Delta H = \lambda \frac{L_{duct}}{D_{duct}} \frac{w_{sw}^2}{2g} \tag{7}$$

where $\lambda$ is the linear pressure loss coefficient given by:

$$\lambda = \begin{cases} \frac{1}{\sqrt{4100 Re}} & \text{for } Re < 10^5 \\ 0.0032 + \frac{0.211}{Re^{0.237}} & \text{for } Re > 10^5 \end{cases} \tag{8}$$

where the Reynolds number is defined by

$$Re = \frac{w_{sw} D_{duct}}{\nu_{sw}} \tag{9}$$

where $\nu_{sw}$ is the kinematic viscosity of seawater. The constant value $\nu_{sw} = 13 \times 10^{-4}/\rho_{sw}$ is adopted in this study.

The energy consumed per year with efficient pumping $E_{pump,year}$ is obtained (in J/year) from:

$$E_{pump,year} = 365 \times 24 \times 3600 \times P_{pump} \tag{10}$$

Part of the necessary fluid-pumping electric power is provided by PV cell modules. The literature is poor on detailed solar radiation data in the vicinity of Aral Sea. In Table 2 there are some data that could be a starting point for computations (Solar Irradiation Database 2008). Noteworthy is that the average annual temperature varies from 8.3°C in the region of pipeline A to 12.6°C in the region of pipeline B, and the average maximum temperatures are 28.3 and 32.3°C, respectively, in July, and −8.6 and −4.6°C in January. Data in Table 2 allows us to usefully evaluate the following yearly averaged daily solar global irradiation on a horizontal plane on the route A: $G_{day}$ = 13.64 MJ/(m$^2$ day). On the route B the corresponding value is $G_{day}$ = 15.05 MJ/(m$^2$ day). The yearly solar global irradiation $G_{year}$ is:

$$G_{year} = 365\, G_{day} \tag{11}$$

The energy provided per unit surface by PV cells during a year, $E_{PV,year,1}$ is given by:

$$E_{PV,year,1} = G_{year} \eta_{PV} \tag{12}$$

**Table 2** Monthly mean of daily global solar irradiation on a horizontal surface ($G$) and air temperature at 10 m above local ground surface ($T$)

| Site (Lat/Long) Month | Route A | | Route B | | Route P | |
|---|---|---|---|---|---|---|
| | 46°N, 57°E | | 41°N, 57°E | | 46°N, 42°E | |
| | $G$ [kWh/(m$^2$ day)] | $T$ (°C) | $G$ [kWh/(m$^2$ day)] | $T$ (°C) | $G$ [kWh/(m$^2$ day)] | $T$ (°C) |
| Jan | 1.40 | −8.61 | 1.79 | −4.36 | 1.39 | −7.33 |
| Feb | 2.25 | −7.18 | 2.59 | −4.41 | 2.23 | −8.11 |
| Mar | 3.34 | −0.43 | 3.77 | 4.15 | 3.13 | −0.76 |
| Apr | 4.80 | 9.93 | 4.99 | 13.3 | 4.15 | 9.49 |
| May | 5.81 | 17.6 | 6.08 | 20.0 | 5.57 | 15.3 |
| Jun | 6.59 | 25.3 | 6.78 | 28.0 | 5.82 | 19.8 |
| July | 6.27 | 28.3 | 6.68 | 32.3 | 5.86 | 23.0 |
| Aug | 5.47 | 24.9 | 5.86 | 28.7 | 5.25 | 22.1 |
| Sep | 4.20 | 17.4 | 4.71 | 21.1 | 3.89 | 16.1 |
| Oct | 2.74 | 7.38 | 3.30 | 11.5 | 2.62 | 7.80 |
| Nov | 1.53 | −1.53 | 2.11 | 4.03 | 1.33 | −0.14 |
| Dec | 1.12 | −7.98 | 1.54 | −2.35 | 1.09 | −6.52 |
| Yearly Mean | 3.79 | 8.78 | 4.18 | 12.6 | 3.53 | 7.56 |

where $\eta_{PV}$ is PV cell efficiency (yearly average). In computations we have used a rather high (optimized) value $\eta_{PV} = 0.15$ (Badescu 2006). The energy provided by the whole PV cells surface during a year, $E_{PV,year}$ is obtained from:

$$E_{PV,year} = E_{PV,year,1} D_{duct} L_{duct} \tag{13}$$

The energy consumed per year with pumping, $E_{pump,year}$, is provided in part by the PV cells. The remaining part, $E_{classic,year}$, should be provided from classical energy sources. One has:

$$E_{classic,year} = E_{pump,year} - E_{PV,year} \tag{14}$$

For these macro-engineered components (i.e. the tensioned textile duct and the solar-powered pumping system) there are two associated costs:

- Costs of construction (monetary investments). They are proportional with the main extensive quantity of each component. These costs refer to building the duct Caspian Sea–Aral Sea, including the pumps and the PV cell modules. These costs are functions of $L_{duct}$, $H_{duct}$, $D_{duct}$, $Q_{sw}$ and of the duct material;
- Costs of subsequent operation and maintenance. They are proportional with the main extensive quantity of each component. They are related to seawater pumping, maintenance and operation.

First, we must estimate the cost of the wide-diameter flexible duct, of the PV cells and of the pumps enabling seawater movement. The cost $c_{duct}$ of the selected tubular duct is given by:

$$c_{duct} = c_{duct,1} L_{duct} \tag{15}$$

where $c_{duct,1}$ is the cost of a unit length of duct. Similarly, the cost of installing the duct, $c_{inst,duct}$ is given by:

$$c_{inst,duct} = c_{inst,duct,1} L_{duct} \tag{16}$$

where $c_{inst,duct,1}$ is the cost of installing a unit length of duct. The unitary costs depend, of course, on various factors, such as the duct diameter $D_{duct}$ as well as the material of the duct. Table 25.2 shows the input values used in this work.

The cost of the pumping installation, $c_{pump}$, is obtained from:

$$c_{pump} = P_{pump} c_{pump,1} \tag{17}$$

where $c_{pump,1}$ is the cost a pump of unit power.

Table 25.3 shows statistically average values for $c_{pump,1}$, as a function of pumping device quality. The gold-plated cost in Table 25.3 is for a custom built. The inexpensive cost is for an off-the-shelf, mass-produced pump. The maintenance cost per year for the pumping system, $c_{pump,maint}$ is

$$c_{pump,\mathrm{maint}} = f_{pump,\mathrm{maint}} c_{pump} \tag{18}$$

where $f_{pump,maint}$ is a given fraction. Here $f_{pump,maint} = 0.01$ has been adopted.

The monetary cost $c_{PV}$ of the PV cells is obtained from the product of the cost of a unit surface PV cell, $c_{PV,1}$ and the total surface covered by PV cells:

$$c_{PV} = c_{PV,1} L_{duct} D_{duct} \tag{19}$$

Using the estimate for future PV Cells technology, the cost per square meter of flexible PV cells sheeting is $c_{PV,1} \approx 100$ USD/m$^2$ (Liu et al. 2007). The maintenance cost per year for PV cells, $c_{PV,maint}$ is

$$c_{PV,\mathrm{maint}} = f_{PV,\mathrm{maint}} c_{PV} \tag{20}$$

where $f_{PV,maint}$ is a given fraction. Here $f_{PV,maint} = 0.01$ has been adopted. The cost of the classical energy consumed during a year, $c_{classic,year}$, is given by:

$$c_{classic,year} = E_{classic,year} c_{classic,year,1} \tag{21}$$

where $c_{classic,year,1}$ is the cost of a classical energy unit. Here electricity is considered and $c_{classic,year,1} \approx 0.1$ USD/kWh (PelletHeat 2007).

The initial investment cost $c_{invest,Aral}$ for the duct, PV cells and pumping installation, is given by

$$c_{invest,Aral} = c_{pump} + c_{PV} + c_{duct} + c_{inst,duct} \tag{22}$$

The total yearly maintenance and operation cost $c_{tot,year}$ of the PV cells and pumps is

$$c_{tot,year} = c_{pump,\mathrm{maint}} + c_{PV,\mathrm{maint}} + c_{classic,year} \tag{23}$$

Previous computations apply to pipeline B, too, except for the pumping power evaluation, which is obviously smaller in this case rather than in case of pipeline A.

## Seawater Importation to Caspian Sea from Black Sea

Route P from the Black Sea to the Caspian Sea in Fig. 9 appears to be ~700 km long and to pose a topographic pumping height sill of between 50 and 100 m. It follows the classic lowland route: the Kuma-Manych Depression, a river valley. The Manych Pass's highest point is located at ~26 m asl. Calculations for the Black Sea–Caspian Sea duct were performed by using the same procedure described in the case of the Caspian Sea–Aral Sea duct.

## Simulation

A single pipe on the route A in Fig. 8 is considered first. Then, for given flow rate $Q_{other(+)} = Q_{sw}$, the water speed in the duct decreases by increasing the duct's diameter $D_{duct}$ (Fig. 10a). This is associated to a decrease in the value of the power $P_{pump}$ needed to move the water inside the pipe (Fig. 10b). The flow rate $Q_{other(+)} = 22$ km$^3$/year needs a pumping system exceeding the present day technological scale. A more reasonable pumping system is needed for $Q_{other(+)} = 5$ km$^3$/year, especially at lower fluid speed values. For given flow rate

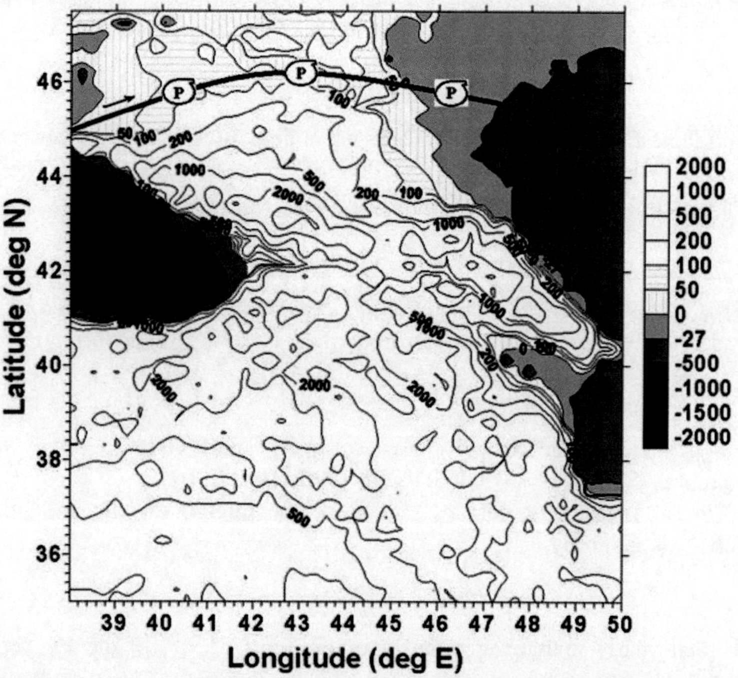


**Fig. 9** Route for the pipeline proposed to connect the Black Sea to the Caspian Sea, with altimetry shown (m)

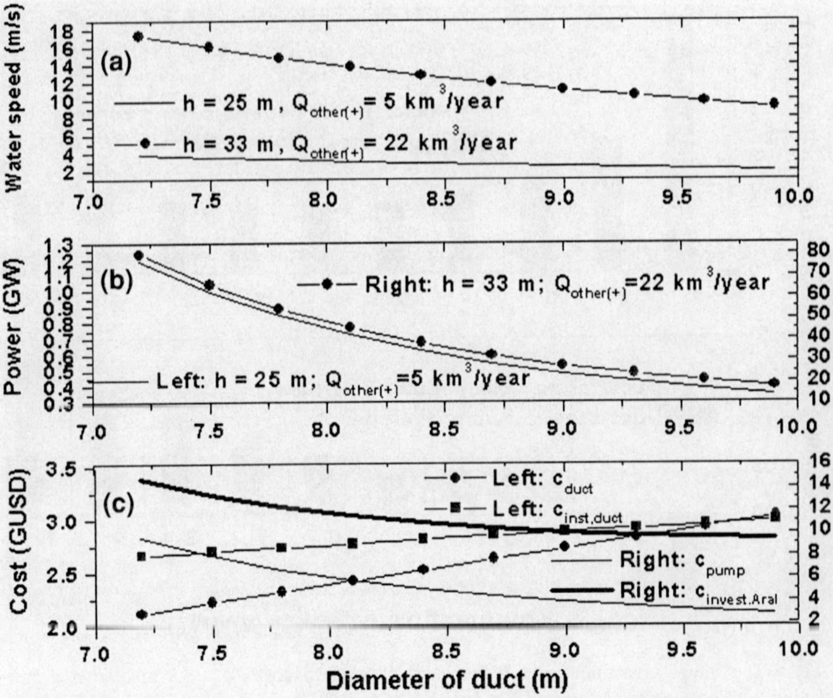


**Fig. 10** Dependence of various technical and economical parameters on the duct diameter. **a** Water speed in the duct; **b** Power necessary to move the water in the pipe; **c** Cost of the duct and of installing the duct, cost of installing the pumping system and the total investment cost. The route A in Fig. 7 connecting the Caspian Sea to the Aral Sea has been considered

$Q_{other(+)}$, by increasing the duct diameter $D_{duct}$ both the cost of the duct ($c_{duct}$) and of duct installation ($c_{inst,duct}$) increases (Fig. 10c). The cost of the associated pumping system ($c_{pump}$) then remarkably decreases. Per total, the initial investment $c_{invest,Aral}$ decreases by increasing the duct diameter, $D_{duct}$. However, too-large diameters are not recommended because of difficulties in construction and operational maintenance.

Figure 11 associates to a duct diameter $D_{duct} = 9.9$ m. Two design solutions are considered. The first case (inexpensive) corresponds to a reinforced tensioned textile fabric duct and a cutting corners pumping system while the second case (average cost) corresponds to a steel duct and an average cost pumping system. In both cases the free water surface level in the Aral Sea is $h = 33$ m. The investment is about three times larger in the average cost case than in the inexpensive cost case. In both cases, the initial investments increase by increasing the flow rate $Q_{other(+)}$. However, an optimum value exists for the ratio $c_{invest,Aral}/Q_{other(+)}$ (Fig. 11a). This optimum value corresponds to $Q_{other(+)} = 7$ and 6 $km^3$/year in the inexpensive and average cost cases, respectively. The initial ca. 2009 investments are of the order of 3.85 GUSD and 11 GUSD in the two optimum cases, respectively.

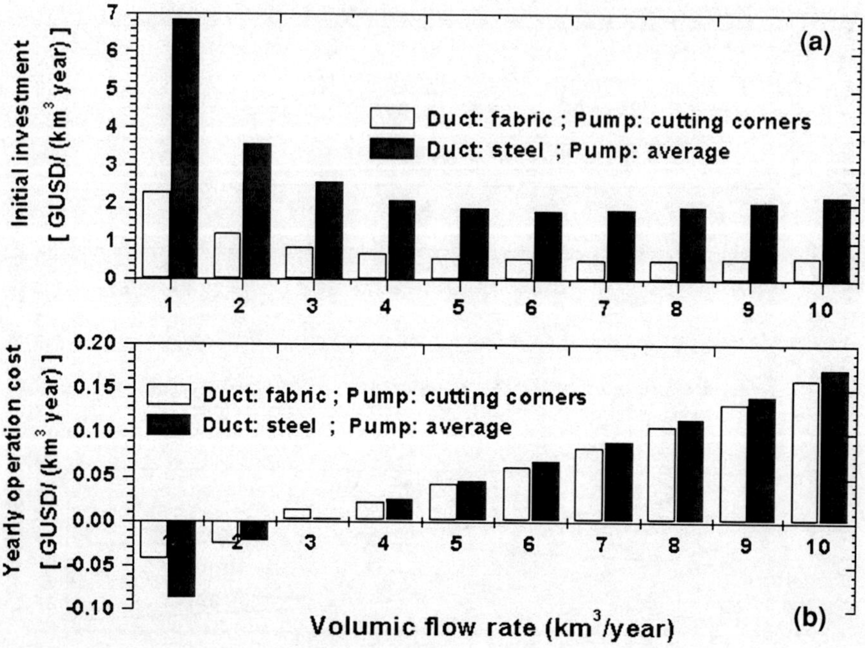


**Fig. 11** **a** The initial investment and **b** the yearly operation cost for a 9.9 m diameter pipe connecting the Caspian Sea to the Aral Sea (route A in Fig. 7), as a function of the yearly volume flow rate

When higher volume flow rates are necessary, it is recommended to use two or more ducts, each of them transporting fluid at a flow rate as close as possible to near the optimum.

Both the yearly operation and maintenance cost $c_{tot,year}$ and the relative operation cost $c_{tot,year}/Q_{other(+)}$ increase by increasing the volume flow rate $Q_{other(+)}$ (Fig. 11b). At small values of $Q_{other(+)}$ (<2 km$^3$/year) the total yearly cost $c_{tot,year}$ and the relative operation cost $c_{tot,year}/Q_{other(+)}$ are negative. This means that the PV cell modules provide enough electrical power to drive the pumping system and any excess power is directed to the regional electrical grid. This way, for a volume flow rate $Q_{other(+)} = 1$ km$^3$/year for example, the money saved will pay back the initial investment in about 40 years, for the inexpensive case, and in about 75 years, for the average cost case.

The cost of the PV cell array, $c_{PV}$, as well as the money saved yearly by using the PV modules increases by increasing the duct diameter (Fig. 12). These monies saved were computed by the difference between the cost of the saved classical energy and the maintenance cost of the PV array.

Note that for given flow rate $Q_{other(+)}$ the cost of the PV arrays is the same for both route A and B in Fig. 12. However, the investment money saved is slightly larger in case of route B, where the useful incident solar radiation is greater. The payback

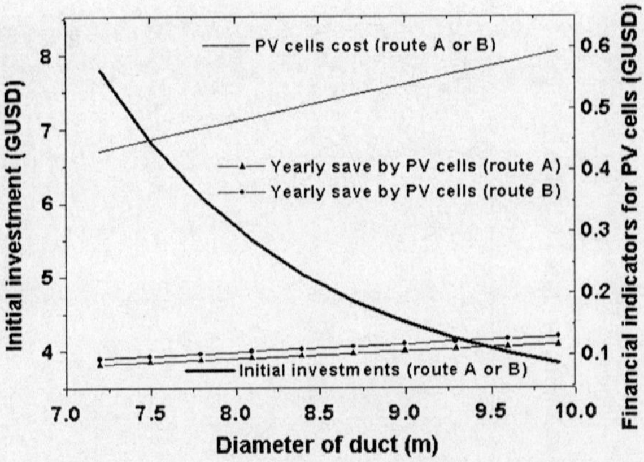


**Fig. 12** Initial investments and various financial indicators related to PV cells system as a function of duct diameter. Both route A and B in Fig. 7 for the pipe connecting the Caspian Sea to the Aral Sea were considered

period for the investment in the PV cell array is slightly increasing by increasing the duct diameter. For a diameter of ~10 m the payback period is about 6 years.

The relative financial importance of the PV array within the initial investment increases strongly by increasing the duct diameter, from approximately 5% when the diameter is 7.2 m to about 16% at a 10 m duct diameter. Using PV cells to provide part of the energy necessary to drive the pumps might be a solution to decrease the operation costs. However, the economic impact of the PV cells is larger at higher duct diameters, when the investment in the thin-film PV array is also important.

Now, we will consider the route P for the duct between the Black Sea and the Caspian Sea. Again, two design solutions are considered. The first case (inexpensive) corresponds to a tensioned fabric duct and a cutting corners pumping system while the second case (average cost) corresponds to a common steel duct and an average cost pumping system. In both cases, the free water surface level between the Black Sea and Caspian Sea is −27 m. This means that when seawater flows from the Black Sea to the Caspian Sea part of the head lost by friction is made up by the difference of level between the two water bodies. In both cases, the initial investments increase by increasing the flow rate $Q_{other(+)}$. However, an optimum value exists for the ratio $c_{invest,Aral}/Q_{other(+)}$ (Fig. 13). For a given duct diameter of ~10 m, this optimum value corresponds to about $Q_{other(+)} = 7$ and 6 km$^3$/year in the inexpensive and average cost cases, respectively (see Fig. 13a, b). The minimum relative investment is about three times larger in the average cost case than in the inexpensive cost. These results are similar to those obtained in studying the pipeline A connecting the Caspian Sea and the Aral Sea (see Fig. 11a).

For a volume flow rate of 7 km$^3$/year the initial ca. 2009 investments are of the order of 4 GUSD and 12.5 GUSD in the two nearly optimum cases, respectively. Again, when higher volumic flow rates are necessary, it is recommended to use two or more ducts, each of them transporting water at a flow rate as close as possible to near the optimum.

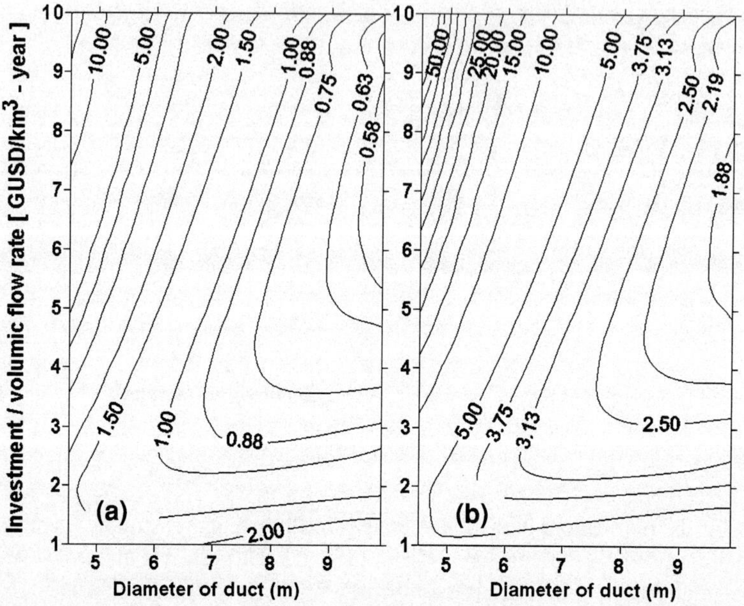


**Fig. 13** Ratio between investments and liquid volumic flow rate as a function of duct diameter. **a** the inexpensive technical solution; **b** the average cost technical solution. The route B in Fig. 7 connecting the Caspian Sea to the Aral Sea has been considered

## *3.3 Changing the Evaporation Rate*

The Aral Sea's surface may be partially covered, to diminish evaporation. We have researched various floating anti-evaporation lake covers. The most expensive method involved plastic sheeting. A floating cover made from thin polyethylene sheeting—rather like a heavy "bubble-wrap" used widely in shipping industry packaging—with 100% water imperviousness would cost 10 USD/m$^2$. But, there are other means to cover the open-air body of liquid that are just about as useful, namely, floating white-colored hollow plastic balls. The material costs will be about the same as the sheeting but deployment (spreading) will be much easier, quicker. Balls will be more affected by wind than complexly anchored sheeting—they can be piled on the shoreline. However, they can be cheaply returned to the water surface by caretakers of the present macro-project using mobile suction vehicles. Since, one way or the other, we will be using a significantly large quantity of plastic, the bulk manufacturing costs should be low since widespread bidding will offer the prospect of low-cost bulk suppliers from everywhere in the world. "Balls"—this term is one of purely constructive convenience since three-dimensional central symmetric plastic bodies different from spheres that can float in all orientations are possible (Gilbert 1991)—seem especially easy to out-source, especially considering the world economy's fragile condition in 2009.

While the sheeting is impermeable, causing 100% retention of water (zero evaporation), the floating balls will cause a retention rate of ~70%. The floating balls will cost about 5 USD/m$^2$, maybe less if worldwide, competitive production is revved up by our macro-project. Some manufacturers will be retained on contract to provide replacements for lost balls and a clean-up crew will constantly be working downwind of a partly refilled Aral Sea to remove cheaply collectible litter from the countryside caused by our macro-project.

Using a covering surface for the "Large Aral Sea" and for the Eastern Basin is not recommended. These two shrinking water bodies are shallow in depth and if they are emptied the covering material will be deposited on the dried lake's bed, losing its potential of reducing any evaporation. Only the Western Basin should be considered for covering by investment budget-limited macro-engineers. The covered water surface $S_{cover}$ is given by:

$$S_{\text{cover}} = xS_{WB*} \tag{24}$$

where $S_{WB*}$ is the free water surface area in the Western Basin for a given (fixed) water level $h*$ and $0 \leq x \leq 1$ shows how much of the surface area $S_{WB*}$ is covered.

To evaluate the financial magnitude of the macro-project, we have to estimate the cost $c_{cover}$ of partially covering the Aral Sea:

$$c_{\text{cover}} = xS_{lake}c_{\text{cover},1} \tag{25}$$

where $c_{cover,1}$ is the cost of covering the unit surface area. Here we adopt $c_{cover,1} = 5$ USD/m$^2$. The maintenance cost per year for the covering system, $c_{cover,maint}$ is

$$c_{\text{cover,maint}} = f_{\text{cover,maint}}c_{\text{cover}} \tag{26}$$

where $f_{cover,maint}$ is a given fraction. Here $f_{cover,maint} = 0.01$ has been adopted.

Let us assume $h* = 33$ m. Then, $S_{WB*} = 6203$ km$^2$. The investment cost of the covering will increase linearly between 0 and about 31 GUSD for $x$ ranging from 0 to 1 and the yearly maximum maintenance cost (for $x = 1$) will be about 0.3 GUSD.

## 4 Restoration Time-Dependent Processes in Aral Sea

Describing quantitatively the Aral Sea evolution process requires solving the time-dependent Eqs. 1 and 2. The following hypotheses are adopted. The precipitation volume rate is proportional to the Aral Sea surface area, i.e.:

$$Q_{prec} = r_{prec}S(V) \tag{27}$$

where $r_{prec}$ is the precipitation rate (in m/year). Previous studies (Letolle et al. 2005) assume the infiltration volume rate is proportional to the Aral Sea volume, i.e.:

$$Q_{\inf} = p_{\inf} V \tag{28}$$

where $p_{inf}$ is the infiltration factor (in year$^{-1}$). The evaporation volume rate is given by:

$$Q_{evap} = \begin{cases} r_{evap}[S(V) - xS_{WB*}] & \text{for} \quad S(V) - xS_{WB*} > 0 \\ 0 & \text{for} \quad S(V) - xS_{WB*} \leq 0 \end{cases} \tag{29}$$

For larger values of the free level surface, the volume river water flow income comes mainly from the Amu Darya and the Syr Darya (each river adds merely ~10 km$^3$/year). When the free water level decreases below a given value $h*$, most of the income from the Syr Darya river remains in the "Small Aral Sea" (which evolutes as a separate water body) and only a small part will go to the Eastern Basin, much larger and less deep than the Western Basin. Consequently, the volume rate income by rivers is given by

$$Q_{river} = \begin{cases} Q_{river,1} & \text{for } h > h* \\ Q_{river,2} & \text{for } h \leq h* \end{cases} \tag{30}$$

For free water surface mean levels below $h*$, the Eqs. 1 and 2 refer to the Eastern and Western Basins.

A discrete solving procedure is adopted here with the time measured in years. Then, the ordinary differential equations (1) and (2) are transformed into the following finite-difference equations, respectively:

$$\begin{aligned} V(n+1) - V(n) = Q_{river}(n) + Q_{q-w}(n) + Q_{other(+)}(n) + Q_{prec}[n, S(V(n))] \\ - Q_{\inf}[n, V(n)] - Q_{evap}[n, S(V(n))] - Q_{other(-)}(n) \end{aligned} \tag{31}$$

$$\begin{aligned} V(n+1)s(n+1) - V(n)s(n) = Q_{river}(n)s_{river}(n) + Q_{g-w}(n)s_{g-w}(n) \\ + Q_{other(+)}(n)s_{other(+)}(n) - Q_{other(-)}(n)s(n) \end{aligned} \tag{32}$$

Here $n$ refers to the number of years from the beginning of the refilling process.

Equations 31 and 32 may be solved numerically by using the Fig. 5 showing the dependence of the total Aral Sea free water surface area $S$ and the total Aral Sea water volume $V$ on the free water surface level $h$. Data used to produce Fig. 5 is used to compute by interpolation, or any given value of $S$, or $V$, or $h$ the associated value of any of the other two quantities. Knowing all information associated to the year $n$, Eq. 31 allows computing $V(n + 1)$ while Eq. 32 allows computing $s(n + 1)$.

In computation we adopt constant (in time) values for the flow rates $Q_{river}$, $Q_{g-w}$, $Q_{other(+)}$ and $Q_{other(-)}$. Note that the evaporation rate $r_{evap}$ decreases by increasing the water salinity (Letolle et al. 2005) while the precipitation rate $r_{prec}$ is slightly dependent on the free water surface area $S$. However, in our rough calculations we use constant (in time) values for the parameters $r_{prec}$, $r_{evap}$ and $p_{inf}$, as well as for the salinity contents $s_{river}$, $s_{g-w}$ and $s_{other(+)}$.

## 4.1 The Natural Water Level Stabilization Process

Presently, the total annual recorded output of the Syr Darya is ~10 km$^3$/year, but only about 2–3 km$^3$/year actually enters the Aral Sea's South Basin (Letolle et al. 2005, p. 54). Note that some predictive scenarios for the period 2010-2099 that estimate the inflow from Syr Darya range from 2.1 to 7.1 km$^3$/year, depending on the forecasting scenario adopted (Savoskul et al. 2003, their Tables 8.4 and 8.5, pp. 53–54).

First, the natural stabilization of the fluid level in the Aral Sea is roughly described. The following input values are adopted: $r_{prec} = 0.1$ m/year, $p_{inf} = 0.05$ year$^{-1}$, $r_{evap} = 1.3$ m/year. These values are close to known present-day values. A groundwater inflow originating in the Pamir and Tian Shan Mountains and ranging into the Aral Sea of ~4 km$^3$/year is assumed (Shibuo et al. 2006). Thus, a constant value $Q_{g-w} = 4$ km$^3$/year will be used in the following.

No macro-engineering is assumed: $Q_{other(+)} = Q_{other(-)}$ and $x = 0$.

Three scenarios concerning the inflow fluid rate by rivers are considered. The first scenario ($Q_{river,1} = 13$ km$^3$/year) is close to the present-day situation. The scenario ($Q_{river,1} = 4$ km$^3$/year) corresponds to the Amu Darya and Syr Darya contributing 2 km$^3$/year. This case might be appropriate if the Aral Sea's level drops excessively and part of the river water is irretrievably lost to the Aral Karakum Desert. The second scenario ($Q_{river,1} = 7$ km$^3$/year) is a middle value between the first two cases.

The Syr Darya, in its upper reaches, has a salt content between 0.3 and 0.5 g/L, but in the lower reaches, where there is a fairly brackish return flow from the post-irrigation waters, the water's salt content is between 1.3 and 1.7 g/L at the Kizil Ordu reporting station. For the Amu Darya, the measured saltiness in the lower/middle reaches was around 1.05–1.15 g/L in the period 1980–1990, but it may have come down slightly to ~1 g/L. In calculations we shall adopt an average value for the river water salinity $s_{river} = 1.1$ g/L as a compromise between the Syr Darya's higher salinity and Amu Darya's larger water flow rate. Groundwater salinity is neglected ($s_{gw} = 0$).

The characteristics of Aral Sea in the year 2005 are adopted as initial values: salinity $s_0 = 48$ g/L and water surface level $h_0 = 33$ m.

The water level stabilizes to 28.9 m (in 17 years), 27.2 m (in 20 years) and 25.7 m (in 20 years) for the three scenarios, respectively (Fig. 14a). In the last case, the Eastern Basin is almost completely dried, becoming an unmistakable part of the Aral Karakum Desert. The water surface reduces to a few thousand square

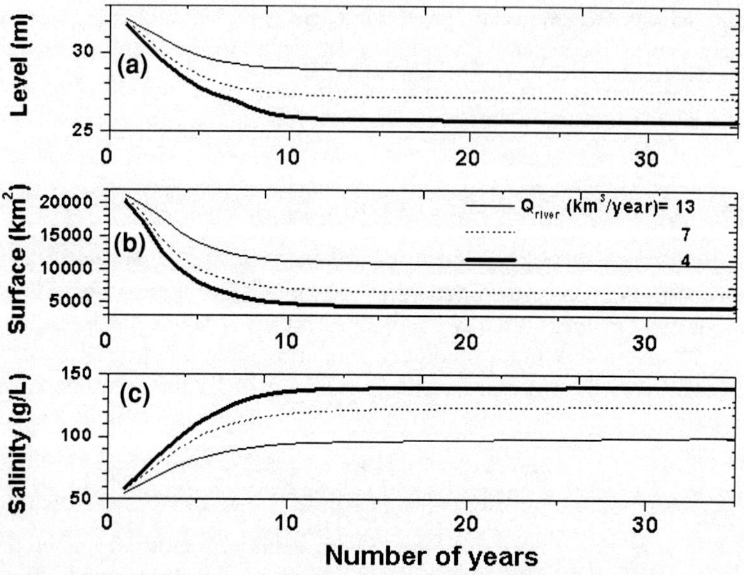


**Fig. 14** Time-dependence of various Aral Sea parameters in the next 40 years; **a** Free water surface level; **b** Water surface area; **c** salinity. Three different values of the inflow rate $Q_{river}$ were considered

kilometers (Fig. 14b) while the salinity increase about two times for first scenario and up to three times in the final scenario (when salt precipitation may occur; this phenomenon is not considered here) (Fig. 14c).

## 4.2 Simulation of the macro-engineering restoration process

In this section the macro-engineering of a naturally stabilized Aral Sea is considered. The first scenario in Sect. 4.1 in Chap. 4 is considered in the following (stabilized level and salinity content is $h_0 = 28.9$ m and $s_0 = 100.4$ g/L, respectively. The initial water body consists of the Western and Eastern basins only. These values are reached in ~20 years if the river inflow is $Q_{river,2} = 13$ km$^3$/year. First, we shall accept two possible scenario values for the improvement in managing the river inflow rate, i.e. (1) $Q_{river,1} = 27$ km$^3$/year and $Q_{river,2} = 20$ km$^3$/year and (2) $Q_{river,1} = 30$ km$^3$/year and $Q_{river,2} = 23$ km$^3$/year. No other manipulative macro-engineering measures are considered, i.e. there is no additional water source or sink ($Q_{other(+)} = Q_{other(-)} = 0$) and the evaporation process is not controlled ($x = 0$). In all cases the results are derived by simulation with runs for a period of 75 years after the natural water level stabilization happens. The end of this simulation time interval corresponds roughly

Table 3 Influence of changing $Q_{river,1}$ and $Q_{river,2}$

| $Q_{river,1}$ (km$^3$/year) | $Q_{river,2}$ (km$^3$/year) | $r_{evap}$ (m/year) | $h$ (m asl) | $\Delta h$ (m) | $S$ (km$^2$) | $s$ (g/L) |
|---|---|---|---|---|---|---|
| 27 | 20 | 1 | 31.6 | 2.7 | 19,764 | 78 |
| 27 | 20 | 1.3 | 30.3 | 1.4 | 15,800 | 96 |
| 30 | 23 | 1 | 32.4 | 3.5 | 22,019 | 69 |
| 30 | 23 | 1.3 | 30.9 | 2.0 | 17,923 | 91 |

The level difference $\Delta h = h - h_O$ is also shown, where $h$ represents the initial free mean water surface level. The other conditions are not macro-engineered

Table 4 Influence of changing the covering factor $x$

| $x$ | $r_{evap}$ (m/year) | $h$ (m asl) | $\Delta h$ (m) | $S$ (km$^2$) | $s$ (g/L) | $c_{cover}$ (GUSD) |
|---|---|---|---|---|---|---|
| 1 | 1 | 30.5 | 1.6 | 16,557 | 88 | 21.5 |
| 1 | 1.3 | 29.7 | 0.8 | 13,700 | 99 | 20.9 |
| 0 | 1 | 29.7 | 0.8 | 13,770 | 99 | 20.6 |
| 0 | 1.3 | 28.9 | 0.0 | 10,842 | 114 | 19.8 |

The difference of level $\Delta h = h - h_O$ is also shown, where $h_O = 28.9$ m asl is the initial free water surface level. The other conditions are not macro-engineered

to 2099, nearly a century hence, which is the end of the terminal prognosis offered in Savoskul et al. (2003).

The influence of changing the sweet water inflow rate by rivers is shown in Table 3. The free water surface level $h$ increases, as expected, but the North Basin does not merge the Western and Eastern Basins (i.e. $h$ is always lower than $h^* = 33$ m). The maximum excursion of $h$ ranges between about 1.4–3.5 m, respectively, in the two extreme cases (i.e., a large evaporation rate and river water reasonably managed, on one hand, and low evaporation rate and the most optimistic estimation of sweet water management in the upper basin of Syr Darya and Amu Darya, including a significant quantity of the Amu Darya's anthropogenic degraded low-quality groundwater, on the other hand). The fluid surface stabilizes to about 60–80% of the 2005 Aral Sea surface, depending on the case considered. The water salinity decreases by an improved sweet water management but it is always higher than its 2005 value, whatever the evaporation rate. Generally, the fluid level stabilizes occurs in about 20 years, whatever the evaporation rate $r_{evap}$. The salinity decreases to a minimum in about a decade and afterwards increases to the values reported in Table 3. It is obvious that only an improved management of the sweet river water can contribute significantly to the restoration of the Aral Sea.

The influence of managing the amount of evaporated water is shown in Table 4 where estimations are reported for $x = 0$ (no evaporation rate control) and $x = 1$ (maximum evaporation rate control considered in this study). An average freshwater inflow rate ($Q_{river,2} = 13$ km$^3$/year) and no additional water sources or sinks have been considered ($Q_{other(+)} = Q_{other(-)} = 0$). Using an artificial cover for the whole Western Basin of Aral Sea may lead to an increase in the free water surface mean level of about 0.8–1 m. The water surface slightly increases and the salinity slightly decreases in case of both low and high evaporation rate values. The new

**Table 5** Influence of changing $Q_{other(+)}$ and $Q_{other(-)}$

| $Q_{other(+)}$ ($km^3$/year) | $Q_{other(-)}$ ($km^3$/year) | $h$ (m asl) | $\Delta h$ (m) | $S$ ($km^2$) | $s$ (g/L) | $c_{CS\text{–}A}$ (GUSD) | $c_{BS\text{–}CS}$ (GUSD) | Total cost (GUSD) 2009 |
|---|---|---|---|---|---|---|---|---|
| 7 | 0 | 30.3 | 1.4 | 15800 | 148 | 3.85 | 4 | 7.85 |
| 14 | | 31.8 | 2.9 | 20421 | 159 | 7.7 | 8 | 15.7 |
| 21 | | 34.9 | 6.0 | 28065 | 119 | 11.55 | 12 | 23.55 |
| 28 | | 36.5 | 7.6 | 31625 | 117 | 15.4 | 16 | 31.4 |
| 37 | | 38.0 | 9.1 | 35185 | 115 | 19.25 | 20 | 39.25 |
| 42 | | 39.6 | 10.7 | 38745 | 114 | 23.1 | 24 | 43.1 |
| 7 | $0.5Q_{other(+)}$ | 29.6 | 0.7 | 13323 | 29 | 7.7 | 2 | 9.7 |
| 14 | | 30.3 | 1.4 | 15800 | 24 | 15.4 | 4 | 19.4 |
| 21 | | 31.0 | 2.1 | 18274 | 23 | 23.1 | 8 | 31.2 |
| 28 | | 31.8 | 2.9 | 20421 | 23 | 30.8 | 12 | 42.8 |
| 37 | | 32.6 | 3.7 | 22567 | 23 | 38.5 | 16 | 54.5 |
| 42 | | 35.0 | 6.1 | 28065 | 23 | 46.2 | 20 | 66.2 |

The difference of level $\Delta h = h - h_O$ is also shown, where $h_O = 28.9$ m asl is the initial free-water surface level. Other conditions are not macro-engineered. The evaporation rate is $r_{evap} = 1.3$ m/s

Aral Sea steady-state regime occurs in about 20 years, whatever the evaporation rate $r_{evap}$ is. The direct costs associated to the lake covering macro-project are quite high, ranging from 19.8 to 21.5 GUSD of ca. 2009 value.

The influence of importing seawater from the Caspian Sea is shown in Table 5 where estimations are reported for $x = 0$ (no evaporation rate control). The fresh-water inflow rate by rivers is $Q_{river,2} = 13$ $km^3$/year and $Q_{river,1} = 20$ $km^3$/year. Two cases were analyzed. In the first case only the pipeline A in Fig. 7 exists. Therefore, $Q_{other(-)} = 0$. In the second case the pipeline B in Fig. 7, is assumed and a non-vanishing saltwater flow rate $Q_{other(-)}$ is returned to the Caspian Sea. In all computations we assumed $Q_{other(-)} = 0.5Q_{other(+)}$. The fluid level $h$ in the Aral Sea increases by increasing the inflow rate $Q_{other(+)}$, as expected. The values of $Q_{other(+)}$ in Table 5 are multiples of the optimum value 7 $km^3$/year (see Fig. 11a). Thus, a value $Q_{other(+)}$=14 $km^3$/year is achieved by using two optimum tensioned textile pipes of about 10 m in diameter, and so on. This applies to the pipeline A connecting the Caspian Sea to the Aral Sea, as well as to the pipeline P connecting the Black Sea to the Caspian Sea. The pipeline B connecting the Aral Sea to the Caspian Sea transports half of the flow rate transported by pipeline A and it is smaller in the same proportion. Its cost is also half of the cost of pipeline A.

In case $Q_{other(-)} = 0$, the final salinity $s$ is always higher than the initial salinity, and decreases at high inflow rate $Q_{other(+)}$. The sudden decrease of $s$ between $Q_{other(+)} = 14$ and 21 $km^3$/year corresponds to the incorporation of the North Basin in the Aral Sea. The costs in Table 5 were evaluated by adopting the "inexpensive" case considered above (i.e., tensioned textile pipe and cutting corners with low-cost pumping system). In the case of the "average cost" case (i.e., ordinary steel pipe and average-cost pumping system) the financial costs in Table 5 should be, roughly speaking, multiplied by at least three. Our preliminary results show that systems of importing seawater, at flow rates >14 $km^3$/year, become very prohibitive financially and energetically. This is associated to a stabilized mean free water surface level in the Aral Sea of the order of ~32 m, which is about 1 m below the level prevailing during 2005. The costs may be reduced by 50%, however, if only the Caspian Sea–Aral Sea water conveyance ducts are considered. This solution may be taken into account for the short-term, in case of future higher-than-average freshwater input from rivers into, or precipitation directly onto, the ever-fluctuating Caspian Sea.

In the case $Q_{other(-)} = 0.5Q_{other(+)}$, the final water level decreases with 2–3 m in respect to the case when the pipeline B does not exists. However, the existence of this pipeline yields a significant decrease in the final salinity of the Aral Sea, which is obviously lower than the average ocean salinity and is comparable with the salinity of the Black Sea. Thus, building pipeline B certainly makes the macro-project more attractive ecologically. Also, note that the transport of fluid by this pipeline is assisted by gravity and the costs associated to pumping operations are obviously smaller than in case of pipeline A. However, this cost diminishing was not considered in Table 5. There is also a possibility that pipeline B be built only on a part of the route presented in Fig. 7, with water flowing freely on the existing natural channel towards Garabogaz Bay. This will further reduce the associated costs of the proposed macro-engineering project.

## 5 Ecological, Cultural and Social Consequences

This section is focused mainly on the technical and economical aspects of the proposed macro-project. Ecological, cultural and social consequences of the macro-project (both positive and negative) have not yet been fully considered. To give a perspective, a few considerations are presented next.

Any endorheic lake will become saltier over time, because liquid freshwater evaporates into vapor, but the solid salts remain. Temporary exceptions occur when such end-lakes in closed hydrological basins have a "trick" to dump part of their salt load in another separate place, and thus keep a fairly low salinity (such as the Caspian Sea gulf, the Garabogaz Bay). Indeed, the Caspian Sea does not owe its salinity as a remnant of its oceanic past. Considering that the Volga River replaces the total water of the Caspian Sea every 300 years it is evident that the fluid composition is completely dominated by the solute composition of the Volga River. If the Aral Sea is maintained by pumping in Caspian Sea liquid, its salinity would increase until it reaches the saturation point of mirabilite ($Na_2SO_4 \cdot 10H_2O$) in wintertime, which is the same as precipitates in the Garabogaz Bay (Cretaux et al. 2009). So, with Caspian Sea waters, it will take some hundred years before any mirabilite starts to crystallize during wintertime in the Aral Sea region, then re-dissolving during summertime, but it will take probably more than a century before all the fish die from extant hyper-saline conditions. However, if an artificial Garabogaz Bay situation is created for the Aral Sea, it is possible to keep the Aral

Sea itself at a fairly constant salinity, mainly as a mixed Na-bicarbonate-sulfate water. These aspects will have to be studied in significantly more detail by geochemical macro-engineering experts.

Alternatives exist for discharging a small flow rate of saltier water from Aral Sea into the Caspian Sea. For example, the sinks may be Turkmenistan's −81 m below global sea level Akdzhakaya Depression located at 41°02′N. Lat. by 58°18′E. Long., ~473 km distant from Aral Sea, or Kazakhstan's −132 m below world sea level Karagiye Depression at 43°27′N. Lat. by 51°45′E. Long., 752 km distant from Aral Sea. These alternatives should be carefully investigated by participating macro-engineers and other interested parties studying the region's most basic macro-problem. A macro-project precedent for land depression filling with extra-depression imported water is the example of the "Altyn Asyr" in Turkmenistan, a so-called freshwater reservoir being filled, at a cost of GUSD ~4.5, which might be completed by 2025. On 15 July 2009, the President of Turkmenistan (G.M. Berdymukhamedov) opened a dedicated water channel feeding fluid into northern Turkmenistan's Karashor Depression. As currently planned by that ecosystem-nation's indigenous macro-engineers, the finished "Altyn Asyr" will contain ultimately >130 billion $m^3$ of low-quality freshwater, present a sub-aerial lake surface of ~1,987 $km^2$ and have a maximum depth of −81 m. In other words, the Karashor Depression Reservoir will be filled to present-day world-ocean sea level (Stone 2008).

A possible catastrophic flooding looms for the Aral Sea Basin: the 1911 Usoi landslide dam on a headwater tributary to the Amu Darya Basin in the Pamir Mountains could fail structurally at any time, suddenly releasing a flashflood of ~17 $km^3$ from Lake Sarez in Tajikistan (Risley et al. 2006). If a controlled release of this freshwater reservoir were technically arranged and safely performed, then such a carefully planned freshwater release could jump-start refreshment/dilution of our proposed "Aral Sea Restoration" seawater importation/dilution macro-project.

Aside from P.P. Micklin's geographical-historical projection, however, imported seawater diluted with all available locally obtained freshwater would be useful to the region's human inhabitants. Seawater, transported from afar, must be filtered to remove all harmful biota (Dumont et al. 2004); the authors have no desire to bring alien and often invasive aquatic and terrestrial plants and animals into the Basin of the Aral Sea but since the Aral Sea is all but absent today, we do view our macro-project as somewhat akin to restocking a cleaned aquarium. It is worth noting that ~44% of all reported cases of non-indigenous plant and animal invasions of the Caspian Sea Basin were traced to international shipping activities (Grigorovich et al. 2003). Unwanted biotic invasion of the Caspian Sea and renewed Aral Sea should be prevented by thorough filtration of all pumped fluids. Inexpensive membrane technologies exist to do so efficiently; such porous membrane technologies are far easier to implement than still finer and more costly porous membrane technologies used to desalinate seawater for only human consumption.

There is a plausible and potentially significant downside to any major effort to extract and export Black Sea water: the volume of anaerobic seawater in the Black Sea is enormous and it contains a huge volume of toxic hydrogen sulfide gas

(Titov 2007), making it a deadly air pollution danger menacing all people living at the Black Sea's shore (Badescu 2007). In light of this oceanographical fact, we intend only to draw off surface water at a rate and place that won't cause a toxic gaseous eruption from the Black Sea's depths. Most emphatically, we do not desire that our "Aral Sea Restoration" macro-project inevitably led to some kind of aerial chemical fallout episode equivalent to the ca. 1628 BC volcanic eruption of Santorini/Thera, when $\sim 1.8\text{–}2.4 \times 10^{10}$ kg of chlorine and $1.8\text{–}2.7 \times 10^{11}$ kg of sulfur were released into the air during the eruption and, eventually, rained upon plants, people and animals downwind of the island's gas and dust eruption plume (Michaud et al. 2000). A remarkable toxic ecologic incident in Brittany, France during August 2009 is slightly illustrative of the anticipated macro-problem: the seaweed, *Ulva lactuca* Linnaeus, washed up on the world-famed beaches. Decaying on the shore, the rotting seaweed produced hydrogen sulfide gas (up to $\sim$1,000 ppm) in the air that caused human death subsequently as well as a worldwide anti-tourism public health scare.

Central Asia is an energy development and exporting region, with a large number of natural gas and oil pipelines. Measures should be taken to ensure that the flora, fauna and marine facets of our macro-project won't adversely impact the existing ecological values greatly, that the landscape's characteristics (topography, geology and soils) won't endure massive contamination except in the direst emergency situations, that the visual impact will be tolerable as the reinforced tensioned textile hoses/pipelines can be camouflaged, that the region's groundwater and surface waters won't be polluted, that air quality may improve (smaller mass of wind-blown salt and dust), that noise from enclosed pumps won't provide a nuisance to anyone, that construction wastes will be removed and/or discretely entombed, and that no significant cultural heritage damage will be permitted.

Long-term refilling of the Aral Sea may have some important consequences for the nearby human population. First, a filled-with-water Aral Sea may change the climate, making it more favorable for human resettlement. Actually, it may somewhat restore the climate as it was prior to the Aral Sea's technogenically induced disappearance. Second, salty water from a regenerated Aral Sea may be used for seawater agriculture on suitable, immediately adjacent, farmable land (Ozturk et al. 2005). Indeed, arable land adjacent to the present-day briny lakes forming what remains of the post-1960 Aral Sea is already contaminated with salt.

Before the proposed restoration macro-project is undertaken, it is recommended to (1) remove all rusting shipwrecks from the exposed seabed; (2) fully map the restoration work site for future hydrographic and navigational charts; (3) fully assess bathymetrically the new saline lake's bottom, just as if it were a commercial "real-estate" development lake and (4) resolve the border disputed between Kazakhstan and Uzbekistan related to the undemarcated line through the lake region since the island of Vozrozdeniy became a peninsula in 2001. ["Real estate lakes" are created to charm house and commercial property buyers. Eutrophication usually affects all small artificial freshwater lakes since such lakes evolve through a natural degradation progression (from clean water to murky-water swamp) resulting, finally, in land after refilling with sediment introduced by overland inflows of runoff. However, our macro-project will result in a really big seawater "real estate lake" that, probably, will avoid the usual physical changes involved with tiny unmanaged commercial real estate lakes.]

It is expected that a successful outcome of the proposed macro-engineering control strategy, "Aral Sea Restoration", will require a United Nations Organization-observed international treaty-codified unity of the affected region's participating geopolitical and society decision-makers (Wouters and Dukhovny 2008). Advice from the thousands of expert members of the Society for Ecological Restoration International (established 1988) must, obviously, be solicited, especially in regards to the still-forming action concepts of restorationist eco-aesthetics.

## 6 Conclusions

A comprehensive control strategy to partially recreate the endorheic Aral Sea is offered which involves several actions: (1) management of the freshwater inputs from the Syr Darya and Amu Darya rivers; (2) regulated evaporation in the Western Basin of the Aral Sea; (3) water imported from the Caspian Sea and (4) salty water extracted from the Aral Sea and transported into the Caspian Sea and/or elsewhere. A simple technical and economical model is proposed. No full ecological Aral Sea restoration to ca. 1960 status is anticipated or even possible (Jackson and Hobbs 2009).

Taking into account that covering the whole Western Basin surface to reduce evaporation is very expensive and yields a gain of about one meter in free water surface level (see Table 4), one may conclude that action (2) is not recommended. A combination of the remaining three actions may, however, yield a steady-state Aral Sea larger than its 2005 status, depending on the funding involved (see Tables 3, 5).

Finally, note that the computations reported here are very rough and in actual practice the economic costs of the macro-project may be higher by a significantly large amount.

**Acknowledgments** The authors thank Prof. R.D. Schuiling (Utrecht University, The Netherlands) and Dr. A.A. Bolonkin (C & R, New York City, USA) for useful comments and discussions.

## References

Alles DL (2007) The Aral Sea. Western Washington University. http://fire.biol.wwu.edu/trent/alles/AralSea.pdf

Badescu V (2006) Simple optimization procedure for silicon-based solar cell interconnection in a series-parallel PV module. Energy Convers Manag 47:1146–1158

Badescu V (2007) Release of hydrogen sulfide by asteroid impacts in Black Sea and risks for inland human population. Environ Toxicol 22:510–524

Badescu V, Cathcart RB (2009a) Aral Sea partial restoration. I. A Caspian water importation macroproject. Int J Environ Waste Manag (accepted)

Badescu V, Cathcart RB (2009b) Aral Sea partial restoration. II. Simulation of time-dependent processes. Int J Environ Waste Manag (accepted)

Barlow MA, Tippett MK (2008) Variability and predictability of Central Asia river flows: antecedent winter precipitation and large-scale teleconnections. J Hydrometeorol 9:1334–1349

Benayas JMR, Newton AC, Diaz A, Bullock JM (2009) Enhancement of biodiversity and ecosystem services by ecological restoration: a meta-analysis. Science. doi:10.1126/science.1172460.

Beylerian GM, Dent A (2007) Ultra materials: how materials innovation is changing the world. Thames & Hudson, London

Bobrov SN (1961) The transformation of the Caspian Sea. Soviet Geography: Review and Translation 7:56

Carls A-C (2009) Central Asia: the new Silk Road's Gordian knot? World History Connect 6:1 18

Clauer N, Pierret M-C, Chaudhuri S (2009) Role of subsurface brines in salt balance: the case study of the Caspian Sea and Kara Bogaz Bay. Aquat Geochem 15:237–261

Cretaux JF, Letolle R, Calmant S (2009) Investigations on Aral Sea regressions from mirabilite deposits and remote sensing. Aquat Geochem 15:277–291

Davies BR, Thoms M, Meador M (2006) An assessment of the ecological impacts of inter-basin water transfers, and their threats to river basin integrity and conservation. Aquat Conserv Mar Freshw Ecosyst 2:325–349

Davydov MM (1949) The Ob'–Aral–Caspian water connection. Gidrotekhnicheskoye Stroitelstvo 3:6–11

Demin AP (2007) Present-day changes in water consumption in the Caspian Sea basin. Water Resour 34:237–253

Dillon JG, McMath LM, Trout AL (2009) Seasonal changes in bacterial diversity in the Salton Sea. Hydrobiologia 632:49–64

Dobrovolski SG (2007) The issue of global warming and changes in the runoff of Russian rivers. Water Resour 34:607–618

Duke DF (2006) Seizing favours from Nature: the rise and fall of Siberian River Diversion. In: Tvedt T, Jakobsson E (eds) A history of water: water control and river biographies. I. B. Tauris, London, pp 3–34

Dukhovny VA (2007) Water and globalization: case study of Central Asia. Irrig Drain 56:489–507

Dumont HJ, Shiganova TA, Niermann U (2004) Aquatic invasions in the Black, Caspian, and Mediterranean Seas. Springer, Heidelberg

Elguindi N, Giorgi F (2007) Simulating future Caspian Sea level changes using regional climate model outputs. Clim Dyn 28:376

Froebrich J, Kayumov O (2004) Water management aspects of Amu Darya. In: Nihoul JCJ, Zavialov PO, Micklin PP (eds) Dying and dead seas: climatic versus anthropic causes. Kluwer, Dordrecht, pp 49–76

Gilbert EN (1991) How things float. Am Math Mon 98:201–216

Glantz MH (2007) Aral Sea basin: a sea dies, a sea also rises. Ambio J Hum Environ 36:323–327

Grigorovich IA, Therriault TW, MacIsaac HJ (2003) History of aquatic invertebrate invasions in the Caspian Sea. Biol Invasions 5:103–115

Ismaiylov GK, Fedorov VM, Sadati Nezhad SD (2007) Assessment of possible anthropogenic changes in the runoff of the Syr Darya river on the basis of a mathematical model. Water Resour 34:359–371

Jackson ST, Hobbs RJ (2009) Ecological Restoration in the light of ecological history. Science 325:567–569

Kosarev AN, Kostianoy AG (2005) Kara-Bogaz-Gol Bay. In: Kosarev AN, Kostianoy AG (eds) The Caspian Sea environment. Springer, Heidelberg, pp 211–221

Kosarev AN, Kostianoy AG, Zonn IS (2009) Kara-Bogaz-Gol Bay: physical and chemical evolution. Aquat Geochem 15:223–236

Krapivin VF, Phillips GW (2001) A remote sensing-based expert system to study the Aral-Caspian aquageosystem water regime. Remote Sens Environ 75:201–215

Kuprin PN (2002) Apsheron threshold and its role in the processes of sedimentation and formation of hydrological regimes in the southern and middle Caspian basins. Water Resour 29:473–484

Leroy SAG et al (2007) River inflow and salinity changes in the Caspian Sea during the last 5500 years. Quat Sci Rev 26:3359–3383

Leslie HM, Schluter M, Cudney-Bueno R, Levin SA (2009) Modeling responses of coupled social–ecological systems of the Gulf of California to anthropogenic and natural perturbations. Ecol Res 24:505–519

Letolle R, Aladin N, Filipov I, Boroffka N (2005) The future chemical evolution of the Aral Sea from 2000 to the year 2050. Mitig Adapt Strateg Glob Change 10:51–70

Letolle R, Micklin P, Aladin N, Plotnikov I (2007) Uzboy and the Aral regressions: a hydrological approach. Quat Int 173–174:125–136

Liu J, Namboothiry MAG, Carroll DL (2007) Fiber-based architectures for organic photovoltaics. Appl Phys Lett 90:063501

Mainguet M, Letolle R (1997) The ecological crisis of the Aral Sea basin in the frame of a new time scale: the anthropo-geological scale. Natur Wissenschaften 84:331–339

Michaud V, Clocchiatti R, Sbrana S (2000) The Minoan and post-Minoan eruptions, Santorini (Greece) in the light of melt inclusions: chlorine and sulphur behavior. J Volcanol Geotherm Res 99:213

Micklin PP (2007) The Aral Sea disaster. Annu Rev Earth Planet Sci 35:47–72

Ozturk M, Waisel Y, Khan MA, Gork G (eds) (2005) Biosaline agriculture and salinity tolerance plants. Birkhäuser, Basel

Pearce F (2002) Russian revives epic river plan. New Sci 181(2433):8–9

Peeters B, Couvreur G, Razinkov O, Kundig C, Auweraer H, Roeck GD (2009) Continuous monitoring of the Oresund Bridge: system and data analysis. Struct Infrastruct Eng 5:395–405

PelletHeat (2007). http://www.pelletheat.org/3/residential/compareFuel.cfm. Accessed 10 October 2007

Risley JC, Walder JS, Denlinger RP (2006) Usoi Dam wave overtopping and flood routing in the Bartang and Panj Rivers, Tajikistan. Nat Hazards 38:375–390

Roget E, Zavialov P, Khan V, Muniz MA (2009) Geodynamical processes in the channel connecting the two lobes of the Large Aral Sea. Hydrol Earth Syst Sci Discuss 6:5279–5301

Roll G, Alexeeva N, Aladin N, Plotnikov I, Sokolov V, Sarsembekov T, Micklin P (2006) Aral Sea: experience and lessons learned brief. Lake Basin Manag Initiat. http://www.worldlakes.org/uploads/01_Aral_Sea_27February2006.pdf. Accessed 20 Nov 2006

Salokhiddinnov AT, Khakimov ZM (2004) Ways the Aral Sea behaves. J Mar Syst 47:133

Savoskul OS, Chevnina EV, Perziger FI, Vasilina LY, Baburin VL, Danshin AI, Matyakubov B, Murakaev RR (2003) Water, climate, food, and environment in the Syr Darya basin. In: Savoskul OS (ed) Adaptation strategies to changing environments. Contribution to the project ADAPT funded by the Dutch Ministry of Foreign Affairs (July 2003)

Shibuo Y, Jarsjo J, Destouni G (2006) Bathymetry-topography effects on saltwater-fresh groundwater interactions around the shrinking Aral Sea. Water Resour Res 42:W11410

Shibuo Y, Jarsjo J, Destouni G (2007) Hydrological responses to climate change and irrigation in the Aral Sea drainage basin. Geophys Res Lett 34:L21406

Solar Irradiation Database (2008). http://energy.caeds.eng.uml.edu/fpdb/Irrdata.asp

Stone R (2008) A new Great Lake—or Dead Sea? Science 320:1002–1005

Titov VB (2007) Estimation of volumes of aerobic and anaerobic waters in the Black Sea and of oxygen and hydrogen sulfide contained in them. Russ Meteorol Hydrol 32:60–64

White D, Hinzman L, Alessa L, Cassano J, Chambers M, Falkner K, Francis J, Gutowski B, Holland D, Holmes M, Huntington H, Kane D, Kliskey A, Lee C, McClelland J, Peterson B,

Straneo F, Steele M, Woodgate R, Yang D, Yoshikawa K, Zhang T (2007) The Arctic freshwater system: changes and impacts. J Geophys Res 112:G04S54
Wouters P, Dukhovny V (2008) International legal and institutional issues and the Aral Sea: challenges for cooperation. International Water Association Publishing, London
Zavialov PO, Ni AA, Kudyshkin TV, Ishniyazov DP, Tomashevskaya IG, Mukhamedzhanova D (2009) Ongoing changes of ionic composition and dissolved gases in the Aral Sea. Aquat Geochem 15:263–275
Zonn IS, Glantz MH, Kostiancy AG, Kosarev AN (2009) The Aral Sea encyclopedia. Springer, Heidelberg

# Aral Sea Rehabilitation with Irtysh Imports

**Roelof D. Schuiling and Viorel Badescu**

## 1 Introduction

Many years ago a major macro-project was proposed by Russian scientists to divert water from major Siberian rivers and use this water to irrigate the steppes of Uzbekistan and Turkmenistan. At that time the plan was dismissed as megalomaniac, and it was proclaimed that it would have dire implications for the climate. Since then the runoffs from the two rivers that feed into the Aral Sea, the Syr Darya and the Amu Darya have been used to irrigate the steppes. This has made Uzbekistan (and to a lesser degree Turkmenistan and Kazakhstan) one of the leading cotton producers in the world, at the cost, however, of causing the Aral Sea (which is an endorheic lake) to evaporate and transform into a dusty salt plain. Ironically, it has led indeed to a climate change for the worse, as the winter rains have become scarcer in the region.

The Aral Sea desiccation is one of the world's major ecological disasters, and a serious attempt should be made to redress the current ecological and economic situation (Micklin and Aladin 2008). It should be noticed, however, that the decrease of the Aral Sea size can also take place without human intervention. The Aral Sea has undergone desiccation and salt deposition in prehistoric times, apparently in response to periods with lower precipitation (Boomer et al. 2000).

In this chapter we will re-examine the Russian proposal to see if waters from the Ob River and its major tributary the Irtysh River could be diverted in times of high water after the snow melt to the Syr Darya, which in turn would refill the Aral Sea.

R. D. Schuiling (✉)
Utrecht University, Utrecht, The Netherlands
e-mail: schuiling@geo.uu.nl

V. Badescu
Polytechnic University of Bucharest, Bucharest, Romania
e-mail: badescu@theta.termo.pub.ro

V. Badescu and R. B. Cathcart (eds.), *Macro-engineering Seawater in Unique Environments*, Environmental Science and Engineering,
DOI: 10.1007/978-3-642-14779-1_17, 

This should be done in conjunction with a better management of the drainage waters from the cotton fields, in order to minimize losses by evaporation. Macro-engineering projects like this have become more acceptable, and new technologies have brought their implementation closer. Acceptance will also be easier after the successful restoration of the (smaller) North Aral lake, where the local climate has measurably improved, and fishing has restarted.

The chapter summarizes the main options of restoring the Aral Sea. The solution based on water exchange between the Ob River basin and Aral Sea is briefly analyzed and a new route for the inter-basin water exchange canal is proposed. A simple water balance model is used to estimate the reasonable magnitude of the restoration process and its duration.

## 2 Options for Restoration of the Aral Sea

It has become accepted practice to move vast volumes of water by pumping or gravity from places where it is abundant to places where it is scarce, and many examples of existing or planned macro-projects, like the South-North water transfer in China are presented in (Vijayan and Schultz 2007; Liu and Zheng 2002). If the Aral Sea is to be restored, one can think of the following solutions:

1. Back to 1950. Restoring the original situation by stopping the cotton production, and let the waters of the Amu Darya and the Syr Darya flow again into the Aral Sea.
2. The water goes, the salts stay behind. Pumping waters from the Black Sea (at ~0 m asl) and/or the Caspian Sea (at −26.5 m asl) to the Aral Sea (at +53 m asl in restored state). This has been described in Aral Sea Partial Refill with Imported Caspian Sea Water of this book.
3. Competing water demands. Diverting waters from major Siberian rivers to the Aral Sea (Blagov 2002).

Each of these options has not only severe economical consequences, but geo-political and ecological consequences as well. The loss of biogeographical unity, the loss of endemic biotas, the consequent introduction of alien and sometimes invasive aquatic and terrestrial plants and animals, the genetic intermixing of formerly separated populations, the changes in water quality, the alteration of hydrological regimes of long standing, possible climate effects, spread of disease vectors are obvious ecological macro-problems. As to geopolitical macro-problems, the artificial interbasin waterway would affect the post-construction behavior of the transited countries. Once the waterway is built, the threat of operational disruption, or even closure, becomes viable as a threat pose. The existence of the new waterway would shift the bargaining power for transiting costs to the upstream countries and certainly changes the value of the freshwater moved to the downstream freshwater recipients at the Aral Sea. As in all macro-engineering projects, decisions must be based on weighing incomparable benefits and losses against each other (Schuiling et al. 2007). In case the decision goes in favor of a particular macro-project, care must be taken to limit the negative consequences as much as possible by taking appropriate mitigating measures.

In the following sections we will briefly describe each of these options, and will treat the last one in more technical detail.

## 2.1 Back to 1950

The goal of irrigating the dry steppes of Uzbekistan, Turkmenistan, Kazakhstan, Kirghizia and Tadjikistan has been achieved. Most of their irrigated fields are planted with cotton and rice, and together they produce more than 10% of the world's cotton. If irrigation is stopped, the economical losses for the countries involved, Uzbekistan and Turkmenistan in the first place, and the drop in the standard of living would be colossal and unacceptable. Revenues from fishing in a restored Aral Sea, and the expected improvement of the climate around the Aral Sea would provide some compensation for the loss of income. In its heyday, fishing from the Aral Sea landed about 50,000 metric ton of fish yearly. Fishing always played a major role in the region, as can be seen from the many geographical names in which the word *balik* or *beluga* appears (Turkish and Russian for *fish* or *sturgeon*).

One can contemplate, however, a number of measures that could reduce the negative impact of irrigation on the water balance of the Aral Sea. It should be investigated if irrigation canals can be protected from evaporation, by covering them with a low-cost reflective water-retaining material with small holes to permit the occasional rainwater to be added to the canal volume, or by replacing open canals by closed pipelines wherever possible, and making the canal bottoms impermeable. In dry countries, wherever farmland is irrigated, irrigation must always be accompanied by drainage, as otherwise salts would accumulate in the soil, making it unfit for cultivation. Care should be taken for a maximum return of these drainage waters to the rivers by not letting them evaporate along the way. One can also promote the cultivation of cash crops that are less water demanding, or more suitable than cotton or rice for drip irrigation, while maintaining the same level of income. The negative consequences of monocultures are becoming apparent worldwide, so crop rotation would be a sensible policy, not only from the point of view of saving irrigation waters, but also to maintain a sustainable agriculture as well. Diversification of crops might include the planting of orchards, more grain and corn production, and maybe potatoes as well.

Each affected growing region should also start a system of periodically leaving arable fields fallow for a year, which will allow the soil to recover. The year that fields are left fallow can be used for repair and maintenance of the local irrigation and drainage systems, among others by making the floors of the canals impermeable and thereby cut infiltration losses.

If an acceptable solution for the restoration of the Aral Sea can be found, the participating countries should set an agreed legally enforceable international limit to their planted cotton acreage, similar to oil production limits by OPEC. This way they can avoid interstate competition, leading to lower margins on their product, and it may even help them to secure a reasonable and stable price for their cotton, perhaps marketed by the cartel.

## 2.2 The Water Goes, the Salts Stay Behind

New challenging macro-project ideas have been brought forward for making up the water deficit of the Aral Sea by pumping seawater from the Black Sea into the Caspian Sea, and thence pumping saltwater more than 80 m upwards and into the (fully restored) Aral Sea (Cathcart 2008; see also Aral Sea Partial Refill with Imported Caspian Sea Water). All pumps were to be solar powered. In essence, this macro-project simply reverses what occurred naturally during the late Pleistocene (Svitoch 2008). To maintain a constant level of the Aral Sea, once restored, an annual volume of water of $\sim$40 billion $m^3$ is required. For topographical reasons, Black Sea water cannot be channeled directly to the Aral Sea, unless one would construct very long photovoltaic-powered textile pipelines passing through Georgia and Iran among others, creating more trans-boundary problems.

Limiting the problem to pumping only Caspian Sea waters raises a difficult problem, because this would lead to a significant lowering of the Caspian Sea level, although in the short-run it would alleviate the problems caused by the ongoing rapid sea level rise in the Caspian Sea. The problems one can expect in such a macro-engineering style solution are manifold, however.

The Black Sea is essentially a two-layered system, with a thin oxygenated layer of fairly low salinity, in the order of 17.5 per mil, i.e. about half the salinity of ocean water, on top of an almost stagnant anoxic body of deep water. The large outflow through the Bosporus is mainly from the surface layer, whereas a smaller undercurrent of denser saline Mediterranean waters represents the return flow. The anoxic waters start at a depth of about 50–100 m, and the anoxic boundary is sometimes even shallower, leading to massive fish kills occasionally. Roughly speaking, one can say that the salt composition represents a dilute marine water, which is also evident from the normal marine salinity/chloride ratio of just over 1.8.

The situation of the Caspian Sea is very different. Like the Aral Sea, the Caspian is also an endorheic lake. It has a salinity ranging from 10 to 13 per mil, but the ion ratios of its salts are very different from ocean water. Its water is refreshed every 300 years by the inflow of the Volga River, and this has been going on for several million years. It is clear, therefore, that its composition is essentially governed by the dissolved load of that river, and is not a relic of its marine past, many millions of years ago. The composition of the Volga River, as far as major components are concerned, is dominated by $Na^+$ and $SO_4^{2-}$ and it is low in chloride. When a water with a composition like that of the Volga evaporates, first a modest amount of calcite precipitates, followed by $Na_2SO_4$ on further concentration by evaporation, or at low temperatures by its hydrated equivalent, $Na_2SO_4{\cdot}10H_2O$.

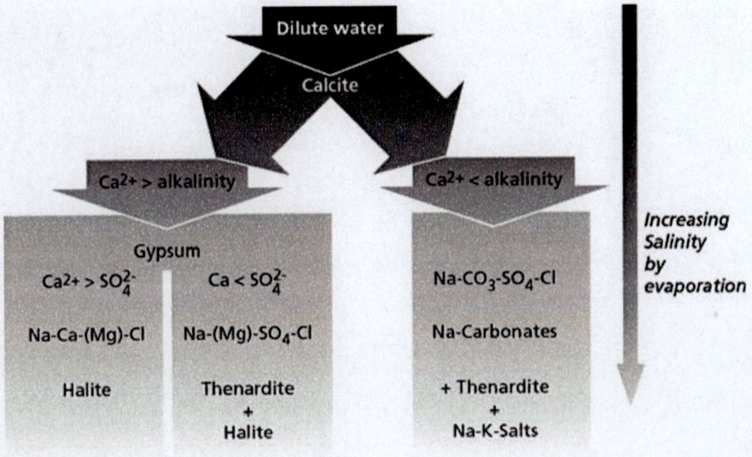


**Fig. 1** The brine tree

How saline waters develop when they evaporate, and which salts will precipitate can be seen from the "brine tree" (Fig. 1) (Schuiling et al. 1994). Evaporation of almost all natural waters will first lead to the formation of calcite. Thereafter two branches of brine evolution develop, depending on the ratio of $Ca^{2+}$ to alkalinity. If $Ca^{2+}$ > alkalinity the next mineral to precipitate on further evaporation will be gypsum. Depending on the ratio of $Ca^{2+}$ to $SO_4^{2-}$, the brines will develop into NaCl brines (the normal marine sequence), or into $Na_2SO_4$ brines, like in the Caspian Sea. In case alkalinity > $Ca^{2+}$, the brines will become alkaline, and end up as soda lakes, with the formation of minerals like trona, natron or thermonatrite.

The Caspian Sea itself does not develop into a brine, because a small outflow over a shallow threshold into the almost closed Karabogaz keeps the salinity of the Caspian Sea more or less constant. Qualitatively one can say that the salt contained in the large volume of fairly fresh Volga water flowing into the Caspian Sea is compensated by the much smaller outflow of a saline solution from the Caspian Sea into the Karabogaz. It is evident that the composition of the Caspian Sea differs drastically from that of a somewhat dilute marine water. Its salinity/chloride ratio is 2.39 compared to 1.8 for sea water, and after evaporation in the Karabogaz it precipitates mirabilite ($Na_2SO_4 \cdot 10H_2O$) in wintertime instead of halite (Fig. 2), which one would expect from evaporation of sea water (for a description of this, see Paustovsky (1932)).

If we start to pump each year 40 billion $m^3$ of Caspian Sea water, or mixed Caspian Sea/Black Sea waters into the Aral Sea, evaporation will quickly transform the Aral Sea into a salt pan, because the evaporation and infiltration during 27.5 years equals the total volume of the (restored) Aral Sea. This means that the salinity of the Aral Sea will increase by the same salinity percentage as found in the incoming waters (10 per mil for Caspian Sea water) for each period of 27.5 years. Within a century or so, the salinity will already have risen to such

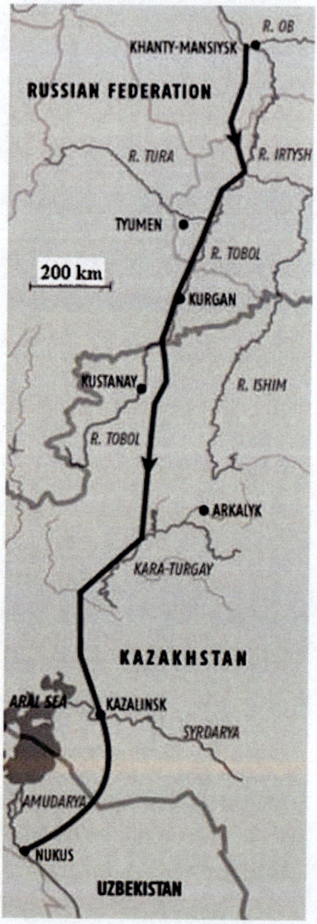


**Fig. 2** Siberia–Aral canal. Route proposed in the past

concentrations that all fish will die. So, instead of bringing a solution, in the middle long run this approach will make the situation worse than at the present. However, a solution to avoid this problem could be envisaged (see Aral Sea Partial Refill with Imported Caspian Sea Water). But they imply, roughly speaking, doubling the cost of the project.

Another disadvantage of this macro-engineering solution is the large amount of energy required to pump a volume of water, of the order of the annual discharge of the river Rhine, and lift it up by more than 80 m. Even so, this would be acceptable if that water is absolutely necessary to solve a huge environmental problem, but in this case it won't take long before the solution becomes itself the problem.

If, nevertheless, one would attempt to pump Black Sea water to the Aral Sea, the utmost care should be taken at the intake point of the Black Sea water, in order to avoid sucking up an admixture of deep, anoxic and $H_2S$ rich waters, which would cause another type of environmental disaster. For other details see Aral Sea Partial Refill with Imported Caspian Sea Water.

## 2.3 Competing water demands

In view of the problems encountered with the first two macro-project options, let us see if the third realization option, diverting part of the water of one or more of the Siberian rivers offers a useful macro-engineering solution. The combined Ob-Irtysh catchment is 2,972,497 km$^2$ with an annual runoff of $\sim$396 km$^3$. The long-discussed Siberian River Diversions project allows a 27–30 km$^3$/year input of freshwater to the Aral Sea (Davies et al. 1992; Duke 2006; Micklin 2007). Although this still unrealized 1949 "Mitrofan Mikhailovich Davydov Plan" sought to convey Siberia's excess freshwater to Central Asia (Davydov 1949), the estimated 100 GUSD concept of diverting massive freshwater flows to Central Asia originated with N. Demchenko in 1871. This is in line with the ambitious plans for the southward diversion of the Siberian rivers, as formulated in the mid 1970s. The plan was shelved in 1986, but even in 1988 Salay alleged (Salay 1988):

> If the water management problems in Central Asia remain in the future, and if the Soviets wish to manage the critical situation of the Aral Sea, it is not unthinkable that the idea of a larger water transfer from the north, at least to the Aral Sea, will be raised again. Perhaps in 10–20 years, and if then, perhaps in a reduced, less costly form.

In the post-USSR twenty-first century, some Russian scientists seek to revive the Siberian River Diversions macro-projects (Pearce 2004) to import freshwater to Kazakhstan via a proposed "SibAral Canal Project", a 2,500 km-long, 200 m-wide, 16 m-deep concrete-lined canal conveying about 6–7% of the Ob River's yearly runoff—to Central Asia, overcoming a 126 m-high topographic pass in the Turgai Depression. The planned macro-projects essentially boil down to diverting waters from the Ob and the Irtysh. The Ural River offers no, or at most a partial solution, because its average discharge is only 400 m$^3$/s, less than one third of what is required for a full recovery of the Aral Sea. Moreover, the Ural is polluted with municipal waste water from several major cities, and there may be some pollutants leftover from past aerial nuclear testing. If it would be necessary, however, to involve also water from the Ural River, an intake place should be selected upstream of the city of Uralsk, because its altitude is only 36 m asl, i.e. lower than the level of a fully restored future Aral Sea.

Figure 2 shows one of the routes proposed in the past for the Ob–Aral Canal (Blagov 2002). In Fig. 3, with tentative traces of canals to feed the Aral Sea, we have selected for the Ural an intake point at Orsk, which would avoid the induction of polluted municipal waste waters. One can also look at a possible positive side-effect of involving water from the Ural River. It might offer an elegant way to modulate the level of the Caspian Sea; if it rises, one can let more water of the Ural River into the Aral Sea Basin, and less should the Caspian Sea level drop.

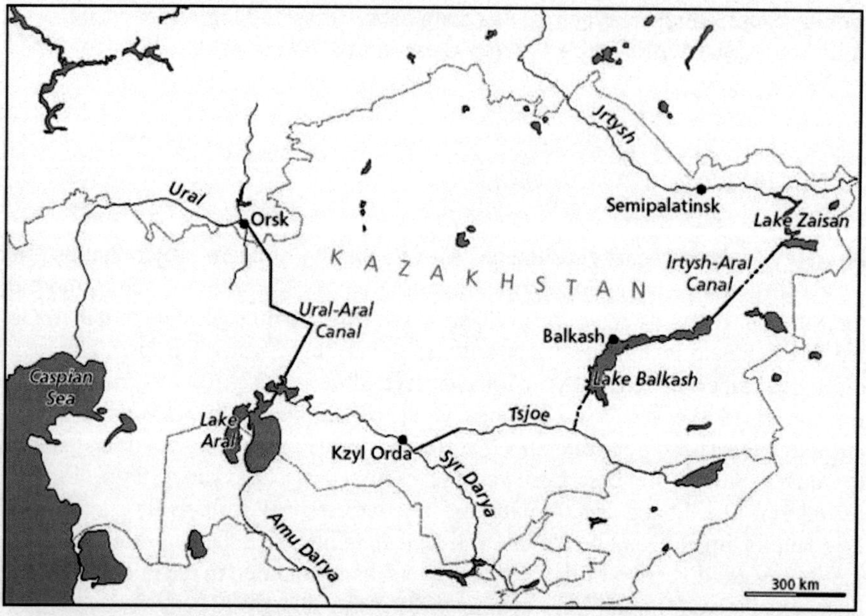


**Fig. 3** Siberia–Aral canal. Route proposed in this chapter

The discharges of both the Ob and the Irtysh rivers (respectively ~13,000 and ~3,000 $m^3/s$) would, in theory, be surplus to the Aral Sea's restoration to its 1960 form. There are other serious demands on these waters, however. Next to navigation or fishing, which do not affect the total water volume, we can distinguish three major claims on these rivers, namely:

- Hydropower generation
- Irrigation
- Human consumption

A number of hydropower stations have been installed along the Ob and the Irtysh rivers and any subtraction of water from a point higher up than the reservoirs means a diminished electric power production. A low-intake point, however, means that the required water must first be pumped up to a higher level to let it flow by gravity to the Aral Basin. It is evidently wasteful to first produce hydropower, and afterwards pump the water up to the same level again. This means that if construction costs for the canal or pipeline that will carry the river water are roughly similar, one should choose the intake point at the highest suitable elevation. We have selected, therefore, an intake point from the Zaisan Lake (alt. 386 m asl, surface area 5,500 $km^2$) along the Irtysh (Fig. 3), which would permit the water transport by gravity flow to Lake Balkash (alt. 329 m asl, surface area 18,000 $km^2$). This would require the construction of a very long tunnel for water transport through the Khrebet Tarbagataj mountain chain.

**Table 1** Dissolved load (major ions) of the waters under consideration

| | Na + K | Ca | Mg | $SO_4$ | Cl | $HCO_3$ | Total (mg/L) |
|---|---|---|---|---|---|---|---|
| Volga Astrakhan | 310 | 160 | 70 | 700 | 200 | 430 | 1,870 |
| Ob | 5.2 | 21.2 | 2.6 | 11.1 | 0.8 | 75.1 | 116 |
| Irtysh | 8.6 | 23.2 | 4.4 | 8.6 | 7.5 | 4.1 | 61 |
| Caspian | 3,260 | 330 | 740 | 3,040 | 5,350 | 100 | 12,820 |

Caspian Sea data after Blinov (1962)

Irrigation is a major claim on the waters in the middle reaches of these rivers, and the water spent for irrigation is essentially lost for any long-haul transport. Apart from measures to use the irrigation waters in the best and most economical way, by switching to modern, water-conserving methods of crop irrigation, and promote the cultivation of less water-demanding cash plantings, not much can be done about this situation. The construction of greenhouses might help to reduce water demands to some degree, while at the same time extending the growing season and increase rentability.

Human consumption is less of a problem in terms of water volumes involved. Highest demand is for potable water quality for safe consumption, but the volumes are small compared to the demands for irrigation.

Even if waters of the Ob and the Irtysh rivers could be diverted and channeled to the Aral Sea, the construction costs of the mega-canal shown in Fig. 2, capable of transporting in the order of 5–7% of their combined discharge are estimated between 12 and 20 billions US$ (Blagov 2002), or 40 billions US$ according to Pearce (2004). We do not dispose of similar calculations for the selected trace of the canal starting at Lake Zaisan (Fig. 3), but we surmise that the costs will be significantly smaller. The tunnel costs will be large. For a somewhat smaller, but comparable tunnel project of 75 km for water transport, the costs were estimated to be in the order of 1.5–2 billion US$ (Ezekiel Water Project 2008), so the costs of a Khrebet Karbagataj tunnel, which would be the largest engineering effort along the proposed Irtysh-Aral canal, probably will not exceed 3–4 billion US$, if constructed by modern TBM (Tunnel Boring Machine). No power will be needed for water transport, as the selected route allows the fresh water to be moved by gravity. Close to six hundred kilometers of the trace pass through Lake Balkash, and the final point of discharge is not in the Aral Sea itself, but at a point upstream of Kizil Ordu along the Syr Darya. These two measures will save seven to eight hundred kilometers of canal construction. Moreover, letting the water not flow directly into the Aral Sea, but into a point upstream along the Syr Darya will make it possible to restore as much as possible the valuable ecology of the Syr Darya's delta wetlands.

An advantage of this solution is that, contrary to the situation when saline solutions of Black Sea or Caspian Sea are used, the low salinities of the Ob River and the Irtysh River (see Table 1) mean that it will take about 2,000 years before the salinity of the restored Aral Sea reaches values approaching those of the Caspian Sea.

### 2.4 Water Accounting

Any freshwater redistribution scheme agreed amongst the existing Central Siberian republics will require the set-up of a trans-boundary water allotment and monitoring board. By treaty the shares of water allotted to each of the participating countries must be determined and delineated clearly in an international agreement. Freshwater is not a free commodity, but each of the users has to pay a price. This price should be kept low in order not to harm the fragile developing economies of the region, but at the same time it should be sufficiently high to encourage the users to utilize all available waters wisely and without wastage. For example, the amount charged to each country for their water intake should be reduced by the amount of their drainage waters that they release back into the eco-system, unless these waters exceed a certain limit of pollution. This will make it financially more attractive for the participants to repair and maintain their systems and to minimize water losses by infiltration or evaporation, or pollution by excessive use of pesticides or fertilizers. Every directly involved riparian country should be a full member of the governing water board, but it would be advisable to invite China and Afghanistan to become permanent observers and/or advisors.

## 3 The Restoration Process: How Long and How Much?

The model described in Aral Sea Partial Refill with Imported Caspian Sea Water will be used here to estimate the magnitude of the restoration process.

A naturally stabilized Aral Sea is considered as a starting point. Previous results (see Sect. 4.1 in Aral Sea Partial Refill with Imported Caspian Sea Water) show that the water body consists of the Western and Eastern basins only. The stabilized level and salinity content is $h_0 = 27.2$ m asl and $s_0 = 125$ g/L, respectively. These values are reached in about 20 years if the river inflow is $Q_{river,2} = 7$ km$^3$/year. We shall accept $Q_{river,1} = 17$ km$^3$/year. The two values act as initial inputs for all next simulations. In all cases the results were derived by simulation with runs for a period of 75 years after the level stabilization happened. The end of this simulation time interval corresponds roughly to 2099, which is the end of the prognosis in (Savoskul et al. 2003).

Note that the highest possible level of the Aral Sea is $h_{max} = 58$ m asl, based on the highest altitude reached by the Amu Darya sediments south of former lake Aibugir, south-west of Kungrad City (Létolle et al. 2005). This corresponds to the largest possible surface of Aral Sea. Before 1960 the Aral Sea level oscillated around 53 m asl, damped by evaporation (see Fig. 4 in Aral Sea Partial Refill with Imported Caspian Sea Water).

One possible strategy would be to refill the Aral Sea to its original (ca. 1960) size, but to do it fairly slowly. If one would go for a rapid refilling, one would need a much larger imported water volume from the Ob River Basin, and have to construct a costly over-sized canal, which would soon after lose its over-capacity

**Table 2** Influence of importing the water flow $Q_{add}$ from Ob basin on Aral Sea free water surface level ($h$), surface ($S$) and salinity ($s$) after 75 years of refill

| $Q_{add}$ (km$^3$/year) | $r_{evap} = 1.3$ m/year | | | $r_{evap} = 1$ m/year | | |
|---|---|---|---|---|---|---|
| | $h$ (m asl) | $S$ (km$^2$) | $s$ (g/L) | $h$ (m asl) | $S$ (km$^2$) | $s$ (g/L) |
| 30 | 36.3 | 31,100 | 34 | 38.7 | 36,700 | 26 |
| 40 | 38.5 | 36,200 | 27 | 41.3 | 42,700 | 21 |
| 50 | 40.7 | 41,300 | 22 | 43.8 | 48,100 | 17 |
| 60 | 42.9 | 46,300 | 19 | 46.2 | 52,300 | 14 |
| 90 | 49.3 | 57,200 | 12 | 52.7 | 65,100 | 10 |

function, once the Aral Sea stabilization is achieved. It is also possible that a compromise would be reached by not completely refilling the Aral Sea to its former level. This, of course, would reduce the evaporation, and reduce the inter-basin imported water demand for the Aral Sea.

Five Aral Sea macro-project filling scenarios will be studied here. The first scenario is $Q_{add} = 30$ km$^3$/year and corresponds to the previously proposed water income from Ob River of 27–30 km$^3$/year. This will not upset the water balance in the Ob and Irtysh river basins too much (i.e. about 6–7%) but in the beginning it would require more time to refill. However, in years of exceptionally high discharge during the snow melt in May/June the inflow can be increased to speed up the process of refilling. In the second scenario, the inflow "expected" from the Ob and Irtysh rivers would amount to $Q_{add} = 40$ km$^3$/year at steady-state conditions. The third, fourth and fifth macro-engineering scenarios correspond to $Q_{add} = 50$, 60 and 90 km$^3$/year, respectively. They are associated to too-large macro-projects to be used in practice but may be useful to give illustrative perspective for the accounting results. The salinity of Ob River (Table 1) is $s_{add} = 0.116$ g/L (in fact the salinity of Irtysh River is only 0.061 g/L).

The influence of importing water from the Ob River Basin is shown in Table 2. The water level $h$ in the Aral Sea increases and the salinity $s$ decreases by increasing the inflow rate $Q_{add}$, as expected. Note that when $h$ exceeds 33 m, the North basin of the Aral Sea becomes reconnected to the Western basin. The level reaches 99% of the stabilized value in Table 2 in about 30–40 years. The results prove to be strongly dependent on the evaporation rate $r_{evap}$. In case the evaporation rate continues to keep its present-day large value $r_{evap} = 1.3$ m/year, there is no reasonable macro-engineering solution to refill the Aral Sea to its ante-1960 volume (see Fig. 14 in Aral Sea Partial Refill with Imported Caspian Sea Water). Even if the evaporation rate drops to its old value of $r_{evap} = 1$ m/year, only part of the old 1960 Aral Sea may be re-filled by "reasonable" large macro-projects.

## 4 Other Considerations

There are probably insufficient economical arguments to justify the major diversion of Siberian rivers to restore the Aral Sea, so it must be defended in part from a moral obligation by the world not to let persist such a long-term major ecological global eco-disaster.

Its execution may become necessary for a very different reason. It is observed that the discharge of the Siberian rivers is recently on the increase, possibly as a consequence of post-Ice Age global warming. It is claimed that the additional inflow of large volumes of freshwater into the Arctic Sea may threaten the persistence of the great world ocean conveyor belt (Peterson et al. 2002). If this ever happened, it could cause a drastic deterioration of the climate regimes in north-western Europe. In that case the decision would no longer depend on the rehabilitation of an ecological disaster region, but on the dire necessity to prevent a climatic catastrophe affecting hundreds of millions of people. The authors cannot judge the seriousness of this threat, as this is outside their field of competence.

A minor argument for the restoration of Aral Sea might be the fact that the drying up of the 1960 timeframe Aral Sea and the transfer of its water to the ocean has led to an additional global sea level rise of 3 mm, equivalent to almost 2 years of sea level rise as experienced during the twentieth century. A restoration of the Aral Sea, associated with an expected increase of rainfall in the region, would, therefore, compensate 2 years of sea level rise.

## 5 Conclusions

The main options of restoring the Aral Sea are: (1) halting the cotton production, and let the waters of the Amu Darya and the Syr Darya rivers flow again into the Aral Sea; (2) pumping seawaters from the Black Sea and/or the Caspian Sea to the Aral Sea and (3) diverting waters from major Siberian rivers to the Aral Sea. The first two options are critically examined and arguments for the solution based on water exchange between the Ob Basin and Aral Sea are provided. Even if this third solution is adopted, only part of the old 1960 Aral Sea may be re-filled by "reasonable" large macro-projects.

## References

Blagov S (2002) Russian water on troubled soils. Asian Times, December 18

Blinov LK (1962) The physico-chemical properties of Caspian waters and their comparable characteristics. Trudi Gos Okeanograf Inst (GOIN) 68:7–28

Boomer I, Aladin N, Plotnikov I, Whatley R (2000) The palaeolimnology of the Aral Sea: a review. Quat Sci Rev 19:1259–1278

Cathcart RB (2008) Aral Sea refill: seawater importation macroproject. http://www.daviddarling.info/encyclopedia/A/Aral_Sea_refill.html. Accessed 15 December 2008

Davies BR, Thoms M, Meador M (1992) An assessment of the ecological impacts of inter-basin water transfers, and their threats to river basin integrity and conservation. Aquat Conserv Mar Freshw Ecosyst 2:325–349

Davydov MM (1949) The Ob'-Aral-Caspian water connection. Gidrotekhnicheskoye Stroitelstvo 3:6–11

Duke DF (2006) Seizing favours from nature: the rise and fall of Siberian River diversion, pp. 3–34. In: Tvedt T, Jakobsson E (eds) A history of water: water control and river biographies. I.B. Tauris, p 320

Ezekiel Water Project (2008) Tunnelling design and construction Methodologies. http://www.ezekielproject.org/estimate.shtml. Accessed 15 December 2008

Létolle R, Aladin N, Filipov I, Boroffka NGO (2005) The future chemical evolution of the Aral Sea from 2000 to the years 2050. Mitig Adapt Strateg Glob Change 10:51–70

Liu C, Zheng H (2002) South-to-north water transfer schemes for China. J Arid Environ 18:453–471

Micklin PP (2007) The Aral Sea disaster. Annu Rev Earth Planet Sci 35:47–72

Micklin P, Aladin NV (2008) Reclaiming the Aral Sea. Sci Am 298:64–71

Paustovsky K (1932) Kara-Bugaz (Black Gulf, English translation in Hyperion Press, Westport, Conn. 1977)

Pearce F (2004) Russia reviving massive river diversion plan. New Scientist, February 2004

Peterson BJ, Holmes RM, McClelland JW, Vorosmarty CJ, Lammers RB, Shiklomanov AI, Shiklomanov IA, Rahmstorf S (2002) Increasing river discharge to the Arctic Ocean. Science 298:2171–2173

Salay J (1988) The Soviet Union river diversion project. From plan to cancellation, 1976–1986. Research Report no 17, Uppsala papers in economic history, 30 pp

Savoskul OS, Chevnina EV, Perziger FI, Vasilina LYu, Baburin VL, Danshin AI, Matyakubov B, Murakaev RR (2003) Water, climate, food, and environment in the Syr Darya Basin, contribution to the project ADAPT, Adaptation strategies to changing environments, July 2003. In: Savoskul OS (ed) Project funded by the Dutch Ministry of Foreign Affairs

Schuiling RD, Andriessen PAM, Frapporti G, Kreulen R, de Leeuw JW, Poorter RPE, de Smeth JB, Vergouwen AA, Vriend SP, Zuurdeeg BW (1994) Introduction to geochemistry, 6th edn. Delft University Press, Delft, p 350

Schuiling RD, Badescu V, Cathcart RB, Seoud J, Hanekamp J (2007) Power from closing the Red Sea: economic and ecological costs and benefits following the isolation of the Red Sea. Int J Glob Environ Issues 7(4):341–361

Svitoch AA (2008) The Khvalynian transgression of the Caspian Sea and the New-Euxinian basin of the Black Sea. Water Resour 35:165–170

Vijayan J, Schultz G (2007) Experiences with interbasin water transfers for irrigation drainage and flood management. International Commission on Irrigation and Drainage (ICID), India

# Urumia Lake: Hydro-Ecological Stabilization and Permanence

**Hossein Golabian**

## 1 Introduction

Urumia Lake (Fig. 1), an internationally important wetland, home of a unique brine shrimp species (*Artemia Urmiana*) and seasonal settlement for thousands of migrating birds is in great danger.

Urumia Lake has been facing a grave crisis over the past 10 years. Prolonged drought is threatening the lake's biodiversity and ecology. Reduction of water depth by 6 m, increasing water salinity to saturation level (much higher than tolerance range of *Artemia* and migrating birds), appearance of huge salt fields around the lake, and huge reduction in *Artemia* population, is alarming indications of gradual total desiccation of the beautiful and unique ecosystem, the Urumia Lake. The salinity ranges from 330 to 400 ppt at different areas of the lake and one can easily see salt crystallization (Fig. 1, right picture) on the surface of the lake with naked eye. It is interesting to know that *Artemia* is still struggling and fighting against this extreme salinity and one can observe them alive swimming in the lake. The lake has reduced in depth by over 6 m and huge areas around the lake have dried up forming vast salty desert. It seems the lake is in the process of drying up and therefore *Artemia Urmiana* and the water birds depending on the lake are being threatened to extinction. Based on research done by a number of international experts, the Urumia Lake has been the only home of *Artemia Urmiana*, one of the oldest populations of *Artemia* who has inhabited the lake since millions of years ago. The lake is losing its international importance as a unique region for thousands/millions of migratory birds which used to spend their winter or lay eggs and feed their offspring with nutritious *Artemia*. As the lake is a major hub for migrating birds throughout Western Asia and Eastern Africa, the evolution in

H. Golabian (✉)
Center of Tehran University Architects, Tehran, Iran
e-mail: hossein@golabian.com

V. Badescu and R. B. Cathcart (eds.), *Macro-engineering Seawater in Unique Environments*, Environmental Science and Engineering,
DOI: 10.1007/978-3-642-14779-1_18, 

**Fig. 1** Urumia Lake in deep crisis: *left side*: oversaturated water; *right side*: salt crystals

Urumia Lake is an issue transcending Iranian borders and is truly of regional/ global importance (Agh 2008).

To tackle the grave environmental/human problem arising from the Uremia Lake desiccation has been and still is a big issue and has got hotter by the aggravation of the problem presently. In similar cases all over the world, the human intervention, as an act of *destructive engineering* in the natural elements and processes are accused. In these accusations, a return to the original circumstances and situations is implicitly intended which is in most cases nearly impossible to fulfill because many newly concluded socio-economical and political interests and dependencies will resentfully defy and resist. In the most successful cases a *complementary* and/or *amendment engineered* intervention is the only solution.

Our planet Earth has not remained the same since the human being race was forced to change her livelihood from hunting and collection to agriculture and industry. Throughout the thousands of years of recorded history, she has changed or in better words *engineered* many aspects and parts of the natural environment. In fact her survival and growth has not been possible without these *engineered changes*.

Now, in the case of Urumia Lake, we are facing a similar and typical scenario. In this chapter, on the basis of a structural and inherent analysis of the Urumia Lake phenomena, a proposal of modification engineering i.e. transfer of a regulatory quantity of water from Caspian Sea to Urumia Lake for volumetric and hydro-ecological stabilization and permanence will be described. For a better comprehension of the proposed solution of Urumia Lake, some illustrative and schematic drawings are prepared and presented in the due sections. Throughout the chapter, wherever necessary, some illustrative figures also are included to emphasis the concepts and issues in question.

## 2 Urumia Lake: A Brief Review

Most of this section is based on Eimanifar and Mohebbi (2007).

## 2.1 Urumia Lake as a Living System

Urumia Lake (or Orumiyeh) is one of the largest hyper saline lakes in the world (150–180 g $L^{-1}$ in the period 1994–1996) and the habitat of a unique bisexual *Artemia* species (*A. Urmiana*). Urumia Lake, located in northwestern Iran, is an oligotrophic lake of thalassohaline origin with a total surface area changing between 4,750 and 6,100 $km^2$ and a maximum depth of 16 m at an altitude of 1,262 m. The lake is divided into north and south parts separated by a causeway (Fig. 2, left picture) in which a 1,500-m gap provides little exchange of water between the two parts (Fig. 2, right picture). Due to drought and increasing demand for agricultural water in the lake's basin, the salinity of the lake has risen to more than 350 g/L during recent years, and large areas of the lake bed have been desiccated.

Urumia Lake is one of the largest permanent hyper saline lakes in the world which resembles the Great Salt Lake of Utah in many respects such as morphology, chemistry and sediments. These two lakes are like two twin sisters and even have similar shape and size.

The hydrology of wetland areas creates the unique physicochemical conditions that make such an ecosystem different from both well-drained terrestrial systems and deepwater aquatic systems. Hydrologic conditions are extremely important for the maintenance of a given water body's structure and function and affect many abiotic factors which, in turn, may impact the biota that develop in it. Since saline lakes occur primarily in endorheic basins, they may be particularly sensitive to environmental changes because their size, salinity and annual mixing regimes vary with alterations in their hydrologic budgets.

The total catchment area of the lake is about 51,876 $km^2$ which is 3.15% of that of the entire country, and includes 7% of the total surface water in Iran. There are thirteen main rivers in the lake basin, among them Zarrineh Rood, the largest with a total annual discharge value of about $2 \times 10^9$ $m^3$. Climate in the Urumia Lake basin is harsh and continental, affected mainly by the mountains surrounding the lake. Considerable seasonal fluctuations in air temperature occur

**Fig. 2** *Shahid Kalantari* causeway across Urumia Lake; *left side*: construction in early stages; *right side*: causeway divides the lake into two separate parts (tebyan.net 2009)

in this semi arid climate with an annual average precipitation of between 200 and 300 mm.

The air temperature usually ranges between −20 and 0°C in winter and up to 40°C in summer. From this point of view, Urumia Lake is a critical asset for the region, because it acts to moderate these extremes; the lake is a crucial climate moderator between day and night and warm and cold periods of the year. Annual draining in wet years into the lake is $6.9 \times 10^9$ m$^3$, of which $4.9 \times 10^9$ m$^3$ is from rivers, $0.5 \times 10^9$ m$^3$ from flood water (through rainfall) and $1.5 \times 10^9$ m$^3$ from precipitation. Underground springs are also a source of water, but the volume is not known.

Morphometric characteristics of Urumia Lake are among those features that have been reported variously by different authors. In general, Urumia Lake has been shrinking for a long time, so its depth has decreased significantly during recent years. Due to 10–12 years of progressive dry climate in the area, the water level is now more than 6 m <20 years ago, a dramatic change. Little is known about historical lake-level variations because direct measurements have been sparse. Urumia Lake had an increased level (2 m) in the winter 1968/1969 but the variations reached to a record height of 1,279.5 m above sea level (a.s.l.) in 1994–1996, with annual fluctuations of about 60–100 cm. Figure 6 summarizes all available data on the surface level fluctuations during the last 104 years and clearly proves that the current low level has never been experienced during this period.

A project to construct a causeway across Urumia Lake was initiated in 1979 (Fig. 2 left picture) to facilitate communication between Tabriz in the east and Urumia City in the west. Shahid Kalantari highway which was built in the narrowest part of the lake in the middle, 15 km long, divides the lake into south and north arms. To facilitate water flow between the north and south of the lake, a 1,400 m wide opening covered by a bridge is provided. Although most of the rivers runoff into the lake from the southern half and there is a continuous water flow from south to north, there is no evidence that the causeway has affected the circulating regime or salinity of the lake. In this connection it is relevant to indicate that the major ions were distributed homogenously in the lake due to the strong currents, despite the presence of the causeway across the lake.

## 2.2 Hydrochemistry

Generally, Urumia Lake is classified as oceanic, being of the sodium- chloride-sulfate type. The main cations in the lake water include $Na^{2+}$, $K^{2+}$, $Ca^{2+}$, $Li^{2+}$ and $Mg^{2+}$, while $Cl^-$, $SO_4^-$, $HCO_3^-$ are the main anions. The $Na^{2+}$ and $Cl^-$ concentration is roughly four times the concentration of natural seawater. Sodium ions are at slightly higher concentration in the south compared to the north of the lake, which could result from the shallower depth in the south, and a higher net evaporation rate.

**Fig. 3** *Artemia Urmiana*: *left side* and *center*: female and male Artemia in early stages; *right side*: a ca. 8 mm long full-grown Artemia

## 2.3 Artemia *Biology*

Hyper saline organisms adapt to high salinities by means of various physiologic mechanisms, including osmoregulation and the synthesis and accumulation of various compatible solutes. The brine shrimp, *Artemia* (Fig. 3) is the dominant macro zooplankton present in many hyper saline environments.

## *2.4 Ecology*

Compared to other aspects of brine shrimp biology, ecological studies on *Artemia* are relatively few in number and that is especially the case for Urumia Lake. The brine shrimp population in Urumia Lake exhibits seasonal fluctuations that are similar to those reported for the Great Salt Lake of Utah but differ in significant ways. Thus, hatching of cysts in the Great Salt Lake may occur as early as February when grazing pressure prevents the phytoplankton from reaching actual blooming densities in late spring as observed in other saline habitats. Adult *Artemia* densities in Urumia Lake, (3 adult $L^{-1}$) in July 1994–January 1996 can be compared to the values reported for Great Salt Lake (4 adults $L^{-1}$ in July) and 10 $L^{-1}$ for June, respectively, and 10–20 $L^{-1}$ in late spring, respectively. These low densities correlate well with low algal biomass, and agree with the findings of that in the Great Salt Lake the highest *Artemia* densities coincide with the highest food concentrations, clearly illustrating the impact of food availability on the growth and reproduction of *Artemia*.

A big difference between the two lakes is that the Great Salt Lake compared to Urumia Lake has much less fluctuation in volume, level and salinity range and therefore *Artemia* population and the food chain it is dependent on, is much less exposed and vulnerable to natural alternations. While having many similarities, the Great Salt Lake has a 60–70% share in international market of Cyst business, the Urumia Lake is not even known for its huge capacity. The natural habitat of

**Fig. 4** Coastal layout of the *Artemia Urmiana* habitat

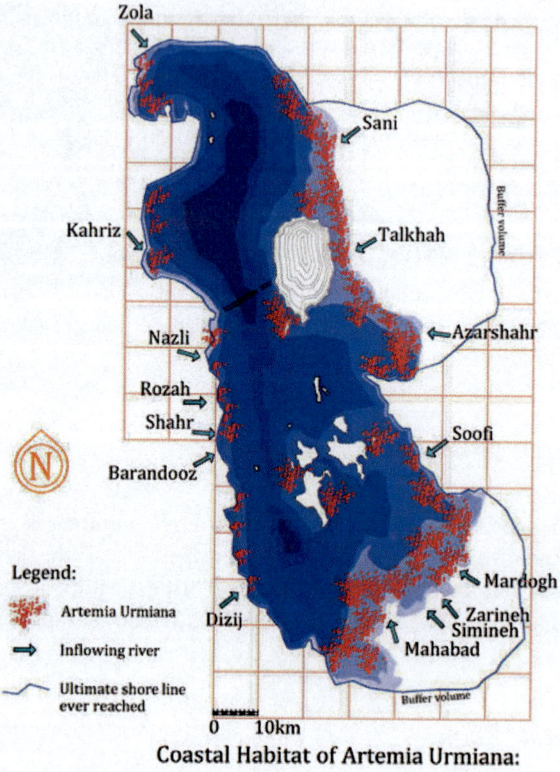


*Artemia Urmiana*, one of the largest in the world is illustrated in Fig. 4 (Reveshty 2000). The shallow coastal stripe in south, east and west of the lake if and when covered with water, becomes habitat for enormous *Artemia Urmiana* population.

Urumia Lake with its endemic *Artemia* species and its large area has been proposed by UNESCO to become a national park owing to its unique features and the recognition that it is an important natural asset, with considerable cultural, economic, aesthetic, recreational, scientific, conservation and ecological value. The lake is extremely important for breeding *Pelecanus onocrotalus*, *Egretta garzetta*, *Plegadis falcinellus*, *Platalea leucorodia*, *Phoenicopterus ruber*, *Tadorna ferruginea*, *Tadorna tadorna*, *Himantopus himantopus*, *Recurvirostra avosetta*, *Tringa totanus*, *Larus cachinnans armenicus* and *Larus genei*. Other breeding birds include several pairs of *Anser anser*, *Marmaronetta angustirostris* and *Aythya nyroca*. *Charadrius leschenaultii* has been recorded during the summer months and may breed on the saline flats around the lake (Fig. 5). Flamingos are known to breed in large numbers at Urumia Lake, and numbers still appear to be increasing slightly, with perhaps as many as 25,000 breeding pairs in recent years. Towards the end of the breeding season, the adults congregate in huge rafts to moult. The vast mudflats surrounding the lake are the most important autumn

**Fig. 5** Wild life in the Urumia Lake coastline and islands: three species of migratory birds among hundreds

staging area for migratory shorebirds and garganey Anas querquedula in Iran. The lake appears to be an important moulting area for common shelduck Tadorna.

## 2.5 Bacteriology

A review of Urumia Lake referred to bacteriological tests conducted on lake sediments, conclude that there are no bacteria in the lake that is pathogenic to humans. In previous studies *Clostridium perfringens* and *Streptococcus faecalis* were detected in Urumia Lake. It seems that these bacteria exist mainly in those areas at which rivers enter into the lake and appear to have origins from agricultural runoff and possibly sewage systems.

# 3 Urumia Lake: Nature of the Crisis

According to the available data and information, the volume of Urumia Lake's water body and its surface level has been constantly changing phenomenon. Figure 6 shows the recorded fluctuation of its surface level, from 1913 to 2009. The water body size in any period depends on two main factors: the rate of surface solar evaporation (which is almost a constant factor) and the amount of water entering the lake system from its basin of about 52,000 km$^2$. The "hydrological budget" is very useful and key concept describing the situation in the lake. The lake's hydrological yearly budget inherently is involved in either *deficits* or *surpluses*. The surplus or deficit of the hydrological budget depends on certain regional and global factors and systems rather than local conditions. Based on nearly 100 years of recorded data mentioned in Fig. 6, the fluctuative nature of the lake is clearly observed. The pattern shows that it takes about one decade or more to reach a maximum in volume and covered area as a result of gradual buildup of yearly budget surpluses. When a maximum has reached, gradual downwards inclination caused by accumulative pattern of yearly budget deficits begins to reach a bottom level, volume and surface area again. At the present time we are

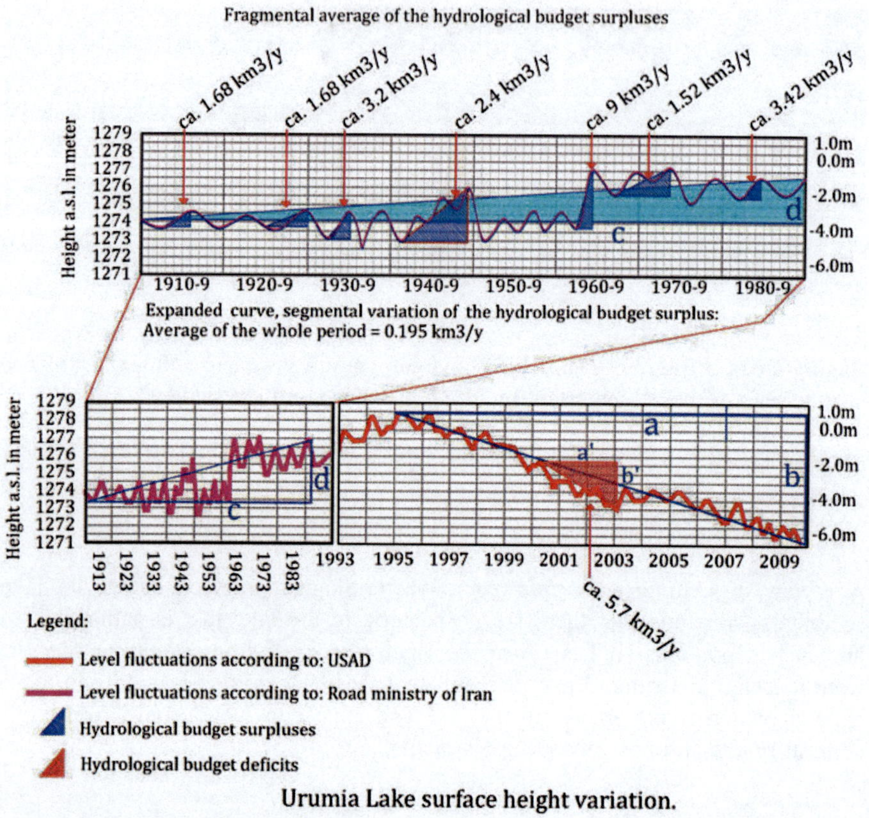


**Fig. 6** The historical pattern of the water level fluctuations in Urumia Lake

facing another bottom situation level again, but unfortunately it is the most serious one ever happened in the recorded history.

The average quantity of *annual hydrological budget deficit* according to Fig. 6 is calculated as below: The total surface level decreased during the 15 years in question (a) was about 7.5 m (b), during this period the surface area has also shrunken from 5,900 to about 4,500 km$^2$ which means an average surface area of ~5,200 km$^2$, thus the average yearly *budget deficit* can be calculated as: 5,200 km$^2$ × (7.5 m/1,000)/15 years = 2.6 km$^3$/year.

A similar calculation for *average hydrological budget surplus* for period from 1913 to 1983, nearly 70 years, leads to a very low quantity: 0.195 km$^3$/year. In fact the actual budget surpluses seen from a fragmental perspective is very different, while the value during 1940s is 2.4 km$^3$/year, the amount increased to a very high quantity of 9 km$^3$/year during 1963–1964. We will return to these values and the fundamental and key factors of hydrological budget *deficits* and *surpluses* later in the chapter.

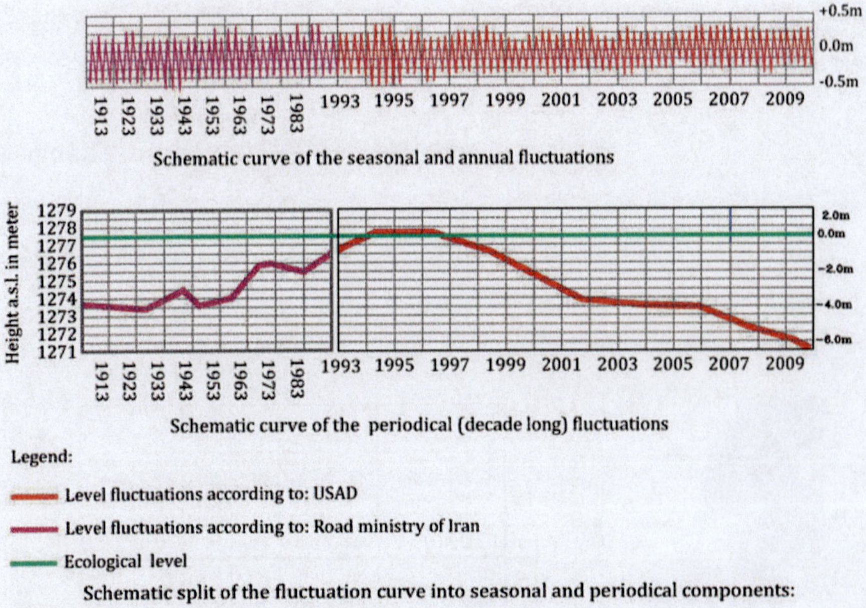


**Fig. 7** Surface water level fluctuation in seasonal and periodical perspectives

As shown in the schematic Fig. 7, the *actual fluctuation* of the lake's water level (and consequently its volume) can be splited into two integrated categories: first, *annual fluctuations*, which are approx. 60–100 cm, and second: *periodical fluctuations*, which happens during decade(s) long periods with a height of ca. 3–7.5 m. The shallow depth of the lake causes a relatively small change in surface level and volume lead to serious and fundamental changes in chemistry and salinity range as main restricting factor to the bio-ecology and living space of *Artemia Urmiana* and dependent fauna and flora population. This reality is illustrated in a schematic equation, Fig. 8. Decreasing water volume ($V_{Q\ldots}$), as the sum of discharged quantities of water into the lake, while the total salt content is constant, increases the salinity range and changes its chemistry. In a similar mechanism but at a reversed interrelation, lowering of the salinity range depends on the increase of water volume.

In other words, the *fluctuative nature* of the Urumia Lake arises from the fact that 3 of 4 main factors and inputs interacting in the lake, as a live system, are constants: topographical/geographical conditions, salt content (estimated to be ca. 8 billions MT) and total absorbed heat energy from the sun radiation, while the fourth factor i.e. precipitation and amount of water discharged from its natural basin, is strongly changing and is an uneven and variable phenomenon. Therefore, the lake could not experience a stable and fixed volume, surface level, salinity range, coastal shape and shore line during its millions years long life. The lake has always been subject to unpredictable and uncontrollable fluctuations in every aspect of its life. From a bio-chemical point of view, When water volume of the

**Fig. 8** Reverse equation between hydrological budget and salinity range

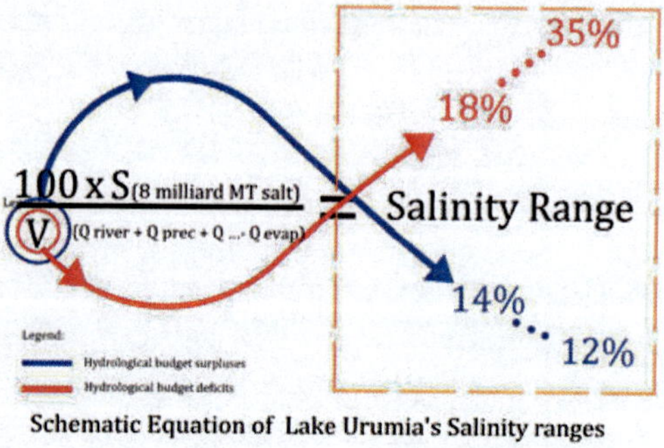


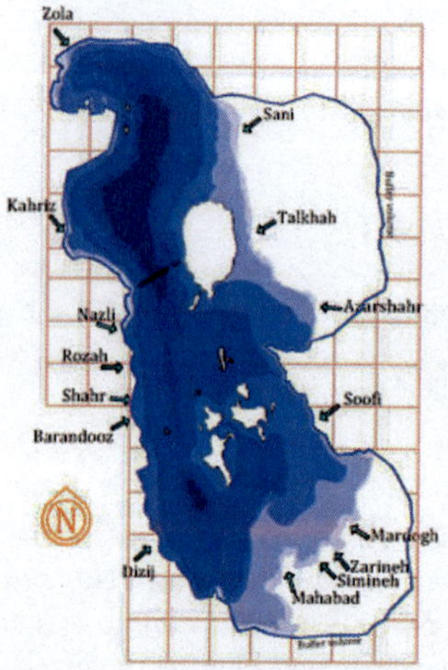


| Slice names: | 1 | 2 | 3 | 4 | 5 | 4 | 5 |
|---|---|---|---|---|---|---|---|
| | Levels in slices: thickness in m. | surface area: Km2 | % of max. surface | Slice's Volume: Km3: | Actual Volumes of Lake: Km3: | * Salinity range: % | Above free sea level:m. |
| I | 0 : +1 | 8400 | 140 | 8.4 | 58.5 | 13.6% | 1279.5 |
| H | 0 : -1 | 5957 | 100 | 5.95 | 50.04 | 16% | 1278.5 |
| G | -1 : -3 | 5669 | 84.3 | 11.34 | 46-38 | 17-20% | 1276.5 |
| F | -3 : -6 | 4534 | 76.1 | 13.6 | 34-25 | 23-32% | 1273.5 |
| E | -6 : -10 | 3025 | 50 | 9.07 | 19.15 | 42% | 1269.5 |
| D | -10 : -12 | 2190 | 36.7 | 4.38 | 10.08 | - | 1267.5 |
| C | -12 : -13 | 1560 | 26 | 3.12 | 5.7 | - | 1266.5 |
| B | -13 : -15 | 910 | 15 | 1.82 | 2.58 | - | 1263.5 |
| A | -15 : -16 | 380 | 6 | 0.76 | 0.76 | - | 1262.5 |
| | Total | - | - | 50.04 | - | - | - |

Ecological level in 1995; Stressed level: in 2005; Critical level: in 2010

*The total quantity of minerals (salt) dissolved in the lake system is estimated at 8 milliard metric tones.

Legend:
- Ecologically most desired situation (1995)
- Ecologically stressed and frigile senario(2005)
- Critical level (2010)

**The morphology of Urumia Lake : physical datas**

**Fig. 9** Structural analysis of the Urumia Lake's water body

Lake changes we do not have just a smaller or bigger lake but a quite different one. Therefore, we can not imagine any different and more stable condition in the future, especially when a big portion of the water volume naturally discharged into the lake is now being stopped and stored behind numerous reservoir dams (ca. 40) built for irrigation and domestic use in its basin territory. When the water volume shrinks, salinity range increases and directly causes destruction to the eco-system and *food chain* in the lake which consequently reduces the *Artemia* population proportionally. On the other hand, with increasing salinity, the dissolved oxygen and carbon dioxide content of the lake water decreases. Oxygen fatigue disturbs *artemia* breathing capacity and low carbon dioxide content decreases algae and

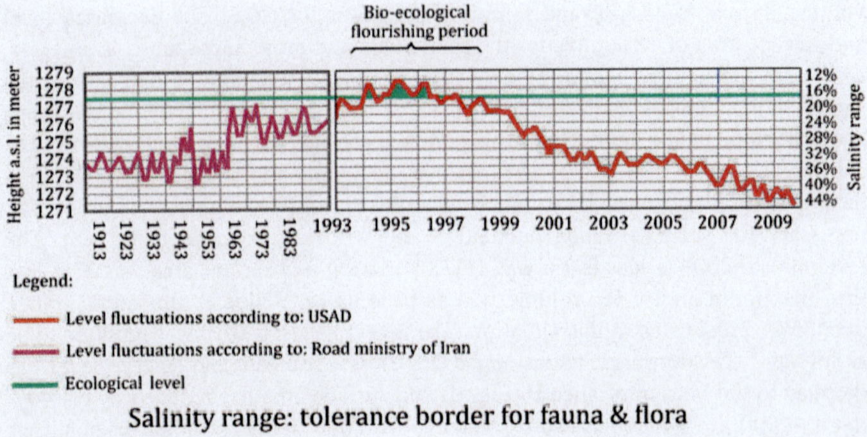


**Fig. 10** Ecological tolerance border

phytoplankton photosynthesis activities as the base of the food chain (Stappen and Coutteau 1996). Most of the oxygen and carbon dioxide is naturally supplied by the oxygen/$CO^2$ -rich water of inflowing rivers therefore when the water quantity of these rivers decreases it causes oxygen/$CO_2$ fatigue of the lake's water body.

Depending on the actual surface area of the lake and the precipitation magnitude of normal wet years, the level can increase by 60–100 cm; however in few exceptional cases more than 100 cm increase has also been recorded. While the increase of lake's surface level happens in rainy and cold seasons and months of the year (0 to −20°C), solar evaporation and subsequent surface level decrease take place in the summer times when temperature can reach up to 40°C in long, sunny, windy and hot days.

All these changes in the surface area and level, the magnitude of water body, salinity range and other time dependent specifications are summarized in Fig. 9 (Alesheikh et al. 2006). For simplification of the analysis, a sliced (layer) structure for the lake's water body is assumed. The most suitable level from ecological point of view is 1,278 $\pm$ 0.5 m a.s.l. At this level the total volume of the water increases to more than ca. 50 $km^3$ with a salinity range of 14–16% and a surface area of ca. 6,000 $km^3$.

From hydro-ecological point of view, this is a dream level which was reached once after nearly 81 years in 1994 and lasted only 3–4 years: Fig. 10. This period was the most flourishing and fruitful years for harvesting cyst and biomass of *Artemia Urmiana* which created great hopes and visions to have an *Artemia* industry in huge economical scale with intentions to enter the world market. The optimism lasted only a very short time and soon the lake level began to drop and new drought and desiccation period started which has lasted up till now. From bio-ecological point of view, the year 2009 was the lowest level recorded ever.

One to two meters decrease in the surface level (slices H and G, Fig. 9) dramatically changes everything: surface area of the lake decrease by 18% and volume shrinks to 44 $km^3$ and salinity increases to 18–20%. The decreased level and surface area of 18% includes the shallowest and most important coastal areas, which are the natural habitat of *Artemia Urmiana.* By continued decrease in the surface level, coastal line retreats further and salinity increases beyond the tolerance of biota in the lake. The long period exposure of peripheral shallow, but extremely vital parts of the lake, to sun heat and wind destroys and deeply damages the precious mudflat which contains the crucial base for "food chain" necessary for *Artemia*, fauna population and wild life dependent on them. The level in year 2009 (slices E–F), was 1,273.5 m a.s.l. and surface area was 32% less than maximum and water volume was as little as ca. 32 $km^3$ with salinity range over 35%; a real catastrophic situation. The level in March 2010 was reported to be as low as 1,371.6 m a.s.l. (Abas-Nejad 2010) which means the surface level has dropped to the bottom of slice F (Fig. 9) and surface area has reduced by ca. 40% to ca. 4,000 $km^3$. In other words, some ~200,000 hectares of the lakebed has already desiccated and converted to salty land.

The global warming and periodical dry periods and droughts are not the only causes for the shrinking of Urumia Lake, there are other reasons too. The huge and rapidly increasing demand for the very limited and inadequate fresh water sources in the basin area and its vicinity is another factor to be considered seriously. The population depending on these water resources exceeds 5–6 million. Rapid expansion of the deeply irrigation dependent agriculture in the recent decades is another important reason. Expansion of cities and urban areas with big demand for fresh water is so huge that should be included in the future calculations. The total water volume discharged to the lake is much less than severely needed of ca. 6 $km^3$ per year; therefore the hydrological budget deficit is huge and might get bigger by the year. While the water body continues shrinking, due to a constant quantity of salt and minerals, the salinity of the lake has increased to 30–40%. Salt crystals grow everywhere and turquoise water views are replaced by salt and dead birds buried in the salt.

Figure 11 is a schematic 3D model of Urumia Lake which illustrates typical and possible hydro-ecological scenarios: 1, 2, 3 and 4. Please notice that, for the sake of a discrete visualization, the dimensions in Z axis (depth) is strongly exaggerated, while the same in *X* and *Y* axis are correct and proportional. (Alesheikh et al. 2006).

*Scenario 1*: The most desirable situation from all point of views, according to the memory of the current generation and recorded information, was reached during the years 1994–1997. The satellite pictures in Fig. 12 show the lake in the highest and most favorable level. (It should be pointed out that the main goal of the present proposal is to repeat this favorable condition and try to make it sustainable and permanent).

*Scenario 2*: This is the most common and mostly repeated situation in which the ecology is very fragile and is under constant stresses. *Artemia* and wild life survive in this situation but in a very limited size and scale. The following picture was taken in 2003 (Fig. 13). The coast line has retreated and *Islami* Island, the biggest one in the lake, has partially been attached to the land and become a peninsula.

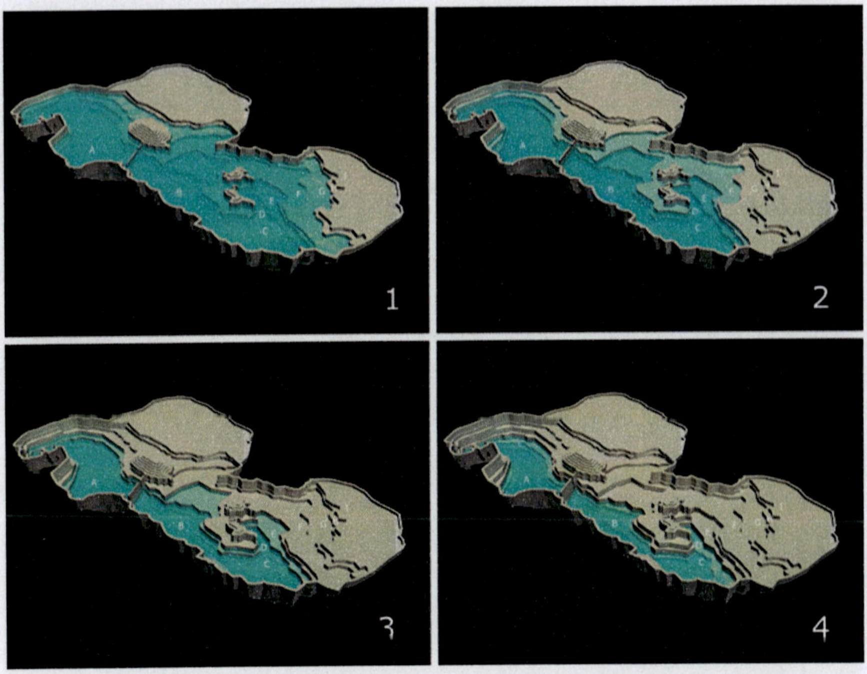


**Fig. 11** Volumetric visualization of Urumia Lake and its water body in the four hydro-ecological scenarios described in the text

**Fig. 12** Satellite pictures of the lake during its best days: *left side*: *Islami Island*; *right side*: Urumia Lake water body in its maximum volume (Earth Snapshot 2010)

**Fig. 13** Satellite pictures, the beginning of the recent crisis: shore line has retreated and *Islami Island* has converted to a peninsula (Earth Snapshot 2010)

*Scenario 3*: In this case, the ecological tragedy is a factum. In 2003, a usual low bottom level was met and it was hoped that following the previous trends, the curve will turn upwards, but surprisingly it continued its downward trend and in the years 2009–2010 reached a historical bottom level. *Artemia* life came to a standstill; migrating wild life left the coastal areas and has not returned back trying to find alternative places in other continents. Figure 14, left side, clearly shows the miserable situation of Urumia Lake by December 13th, 2009 (www.eosnap.com). The *Islami* peninsula has completely attached to the mainland and southern coast has nearly dried up.

*Scenario 4*: This is a real nightmare situation in which the Urumia Lake has become a new Aral Sea and its 8 billion MT of salt bomb has been activated and the surrounding communities in Iran and neighboring countries have come under the threat of salt dust storms and its disastrous consequences. Similar disastrous storms are already a factum in Central Asia. The salt storm over desiccated lakebed of Aral Sea on September 1, 2006 is a good example of what could happen to Urumia Lake: Fig. 14, middle and right side (Earthobservatory 2006).

As mentioned earlier, the water body of Urumia Lake has a very important function as climate moderator between day and night, summer and winter.

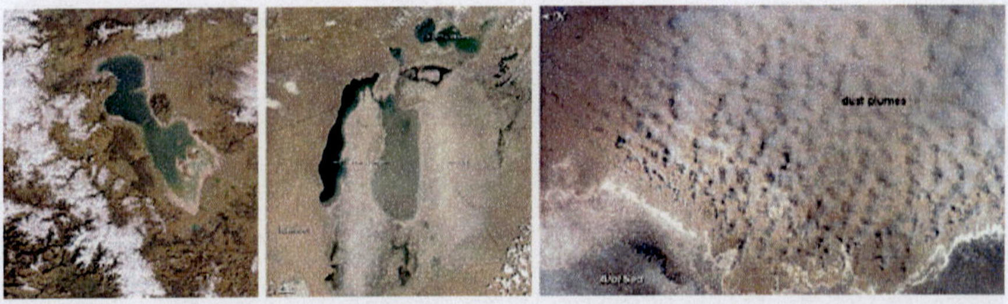

**Fig. 14** Urumia Lake crisis compared to the Aral Sea disaster. *Left side*: Urumia Lake in early 2010, southern parts have dried up; *middle*: mainly dried Aral Sea in a miserable condition; *right side*: poisonous storms from the dried lake-bed of Aral Sea

Sun heat gradually warms up the salty water which leads to the deposition of huge quantity of solar energy as heat which later is released to the adjoining atmosphere. The heat capacity of the water body is directly in proportion to the magnitude of lake's water body therefore by reduction of its volume, the heat capacity which the immediate nature is dependent on, decreases proportionately with catastrophic impact on regional climate and environment, on agriculture and fruit gardening, etc.

However Urumia Lake is some 10 times smaller in surface area and 20 times smaller in volume compared to the original Aral Sea, but while Aral Sea is situated in the middle of very huge central Asian desert and semi desert region, Urumia Lake is in the middle of a very densely populated area including north-west Iran, southern Caucasia, northern Iraq and eastern Turkey.

Salt storm as a low-lying cloud of airborne salt and minerals can hover over large areas around the lake including several big cities like Tabriz, Urumia, and even far beyond them. Salt storms in Urumia, similar to the case of Aral Sea, can pose serious health hazards to surrounding areas, as the minerals blown about by these storms are highly toxic, leading to adverse health effects such as throat and lung cancer, anemia and esophagus cancer alongside with epidemic levels of tuberculoses, kidney, liver and respiratory diseases, immunological/neurological problems, both genetic and birth defects and decreased life expectancy. Dust storm are not only hazardous to people but can even destroy their agriculture and economy as well (Whish-Wilson 2002).

## 4 Sustainable Solution for the Crisis of Urumia Lake

Generally speaking, in order to break this vain circle once and for all, the attitude toward the lake must change fundamentally, from a so-so policy to a deep-seated approach. In many years, discussion has been concentrated on the solutions which have proved unrealistic and unfeasible: from many points of view. The Urumia Lake is so unique and precious asset that deserves to be handled in a long term

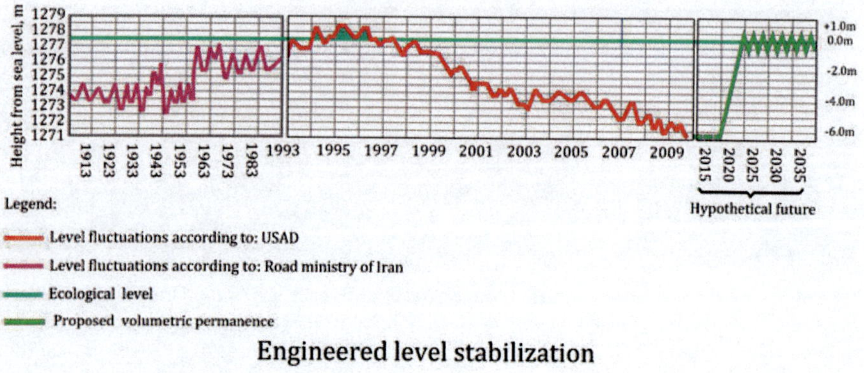


**Fig. 15** Engineered rescue and stabilization strategy

constructive and fundamental manner instead of reflective and short term actions. The attempts must be concentrated on bringing the lake to an ideal and hydro-ecologically sustainable state i.e. a permanently stabilized state. This is possible only by elimination of periodical fluctuations and careful administration and regulation of ordinarily seasonal and annual fluctuations. In other words, we have to try to find a way to realize the desire for "*volumetric and hydro-ecological stabilization and permanence of Urumia Lake*". Figure 15 tries to illustrate this idea, where the red curve, which represents the current instability and vacillative nature of the lake, gradually develops (through emergency rescuing step and following regulation/stabilization) into the green curve showing a stable and controllable surface level, water body volume, constant *bio-eco-friendly salinity* range and coastal line. According to a rough calculations, there is a need for some 15–20 $km^3$ of water for restoration of the water volume to an ecologically minimum level (scenario 2, in Fig. 11), and then for following stabilization/permanence step, in the most suitable hydro-ecological level (scenario 1) an additional amount of 10–15 $km^3$ of water. Permanence of the scenario 1 is possible only when the *yearly hydrological budget deficit* is administrated by *the macro-engineering methods*.

Where from all these relatively huge quantities of water can be provided? The available information and evidences prove that the volume of water needed for the stabilization of Urumia Lake will not be available through the rivers of its basin or even in the rivers of the neighboring basins. Usually two kinds of solutions are proposed: first one is to increase the share of the lake from total amount of precipitation magnitude and the second one is to import water from rivers of neighboring basins. The first solution is very difficult and most probably impossible to fulfill because the population depending on the available quantity of water in the region for their economy is increasing, both in number and needs and there will not be any excess of water to be devoted to the lake and even if such a source existed, it could not be reliable in the long term. The second solution is also nearly impossible and unrealistic because the same drought which has decreased precipitation in the Urumia Lake basin also has done the same in the neighboring basins which support the rivers in question. There is a general shortage of fresh water for human use and agriculture in the whole region and therefore it would not be a long term and sustainable solution to be dependent on these resources of water for which huge potential demand is growing following population increase in the whole region. Yet there is another good idea that is discussed among the experts i.e. cold cloud seeding for precipitation enhancement. This method is conditioned to the availability of the enough amounts of moisture saturated clouds which in a dry period is very rare and unreliable too.

The long term and sustainable solution is nothing less than the administration and elimination of yearly *hydrologic budget deficit*s and guarantee of *a regulative and complementary budget surplus* whenever necessary. From economical, technical and practical point of view, there are four alternatives illustrated in Fig. 16 (C0, C1, C2 and C3):

- $C_0$ is a prognosis of time and pattern of Urumia Lake's restoration in a natural manner repeated in the past. If a regime of yearly budget surpluses similar to the situation prevailed between years 1913 and 1993 repeats itself, then it will

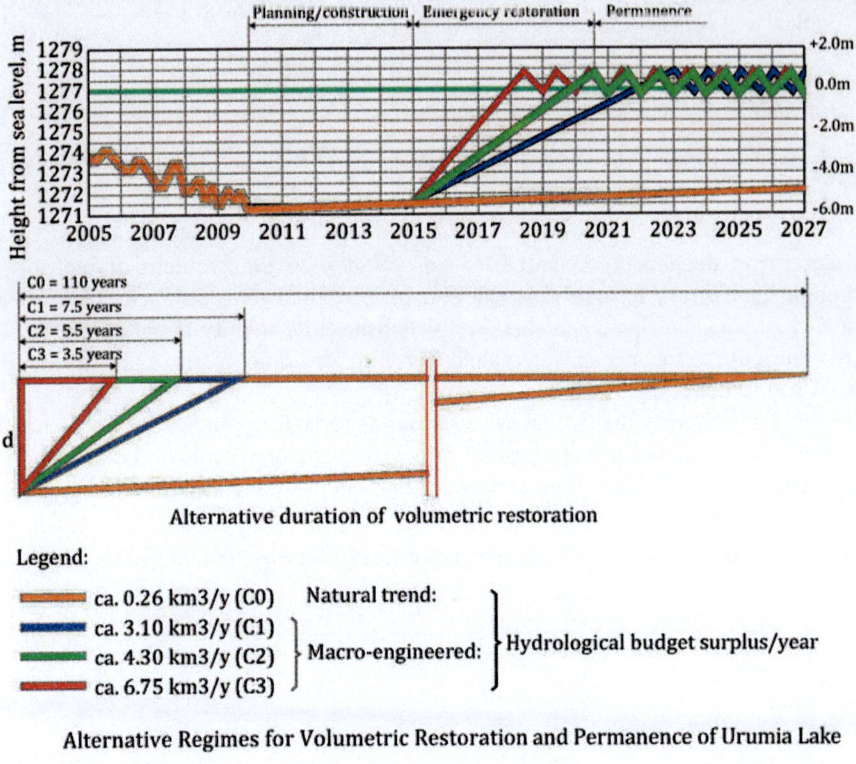


**Fig. 16** Time pattern for the restoration and permanence of Urumia Lake

probably take 70 ± 10–15 years before Urumia Lake can be restored to its ecological level again; and if restored, then how long it might last before a new era of drought returns back? This is the most hopeless and passive solution that should be abandoned.

- $C_1$ is the most low cost macro-engineered solution for the containment of the crisis which will restore the lake to its ecological level after 7.5 years. In this solution the low capacity of 3.1 $km^3$/year can not cope with a *hydrological budget deficit* of magnitude of 5.7 $km^3$/year happened during 2000–2003.
- $C_2$ is an alternative with a capacity of 4.3 $km^3$/year and will take 5.5 years for restoration and containment of the crisis which can compensate most part of the maximum yearly budget deficits of 5.7 $km^3$/year recorded ever. In this case the gap with the worst scenario of 2000–2003 is small and might be administrable with a minimum damage to the ecology of the lake. According to this alternative, after restoration of the ecological level which will take place in the first 5.5 years, there will be an excess capacity of (4.3–2.6 =) 1.7 $km^3$/year to make it possible to decrease the operation period from 12 to 10 months of the year leaving 2 months free for technical maintenance and overhaul of the whole water transferring system- after 5–6 years of nonstop operation.
- C3 is the most costly solution and can restore the ecological level in only 3.5 years time but later there will be a large excess of operational capacity which will remain idle.

## 5 Caspian Sea: Sustainable Source of Water

The only *available/accessible/reliable/sustainable* source of clean or cleanable water to be dependent on as permanent solution to the problem of repeating drought in Urumia Lake is *Caspian Sea* in a 320 km long distance. The water quality of Caspian Sea is suitable for this purpose: its salinity range is just 1.2% (one-third of the free ocean waters) and therefore would not increase the salinity of the Urumia Lake.

Table 1 compares the two lakes in various aspects. Caspian Sea is like an *elder sister* to the Urumia Lake. Ratios in surface area, water volume and depth prove this fact. 15 $km^3$ of water necessary for emergency rescuing of Urumia Lake from

**Table 1** Comparison of Urumia Lake and Caspian Sea (Tolkatchev 2004; Mansimov and Aliye 1994)

| | Urumia Lake | Caspian Sea | Ratios |
|---|---|---|---|
| Surface area ($km^2$) | 370,000 | 5,200 | 71:1 |
| Volume ($km^3$) | 78,200 | 30–50 | 1,500:1 to 2,600:1 |
| Depth (m) | 8–1,000 | 5–16 | – |
| % of minerals | 1.2% | 16–35% | 1:16 to 1:24 |
| Basin area ($km^2$) | 3,626,000 | 51,876 | 70:1 |

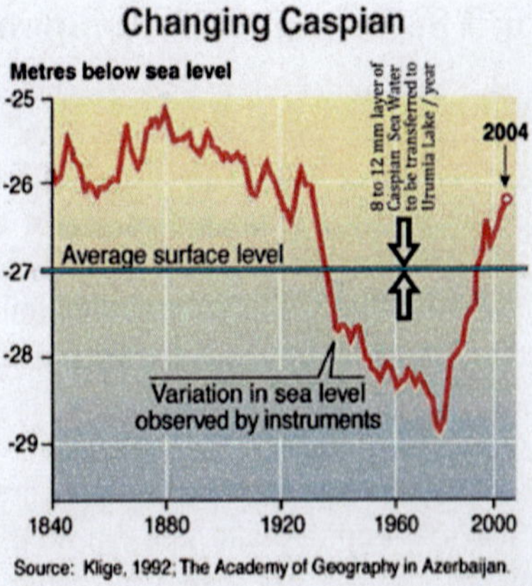


**Fig. 17** An 8–12 mm slice of Caspian Sea water per year can stabilize Urumia Lake

current crisis is just 0.015% of Caspian Sea water volume. Transfer of this quantity of water will decrease the Caspian Sea level by only 4 cm. Transfer of another 15 $km^3$ of Caspian Sea water, which can restore the ecological level in Urumia Lake, decreases its surface level by another extra 4 cm.

For the permanence of Urumia Lake's volume and level, 2.6–4.2 $km^3$/year of Caspian Sea water equal to a layer thickness of 0.8–1.2 cm per year will be necessary and enough (Fig. 17). This withdrawal will not harm the Caspian Sea, but in contrary will help it. The water level of Caspian Sea has been increasing in recent years to a dangerous level, treating numerous coastal cities in Russia, Azerbaijan and Iran. Many coastal towns like Sari in Mazandaran province of Iran, is under treat of sinking into the Caspian Sea, therefore any attempt to lower the water level is welcomed by countries around the sea (Panin 2005). It is predicted that the rise of water level following global warming will continue even in the future. Comparative analysis of surface level fluctuations between Urumia Lake and Caspian Sea during 1986–1995 showed a significant correlation (Vaziri 1998). This is probably because the basins of the two lakes geographically overlap. But current situation disproves this finding while the Caspian Sea level has stayed high, the Urumia lake level still decreases (Dating Caspian Sea 2003–2005).

Caspian Sea water should be lifted up in two steps: In the first step it must be lifted up by pump(s) near Lavandvil town from sea level at −28 m a.s.l. to +1,600 m a.s.l. for passing over the Alborz mountain chains. When the water is lifted up to this level near Neyaraq village, it will be let run by gravity downward to a level of +1,400 m a.s.l. In the second step, once again it should be lifted up by pump(s) from +1,400 a.s.l. near Ardabil to a level of +2,000 m a.s.l. to overcome the next mountainous obstacle near Nir to be then released again downward through an approx. 300–320 km long channel/river bed flowing by gravity force in the south of Sarab city and north of Tabriz and through it until its final destination, Urumia Lake.

# 6 The Layout and Components of the Proposed Project

The greatest obstacle for the transfer of water from Caspian Sea to Urumia Lake is the 1,316 m height difference between the two water bodies. The uneven geography and mountainous nature of the ca. 320 km long distance between the two lakes is another difficulty which must be overcome. In fact the height difference between the origin and destination is bigger than the nominal 1,316 m, because there is a mountainous hindrance on the path between Ardabil and Nir, which also should be overcome. As per Fig. 18b, the total height for pumping is ca. (1,600 m + 600 m =) 2,200 m.

The average amount of water proposed to be transferred annually from Caspian Sea to Urumia Lake is minimum 3.6 $km^3$ (during 10 months of the year) and maximum 4.3 $km^3$ (all year round). Two months of the year coincide with the rainiest period of the year and natural discharge of water from rivers to the lake. Having data from this wet period, will help to compute the exact amount of hydrological budget excess or surplus and administration of the water quantity to be transferred from Caspian Sea.

The whole proposed water transfer system from Caspian Sea to Urumia Lake as illustrated in Fig. 18a, b will be described briefly in the terms of axis and distances in between them (covered areas).

## *6.1 Axis 1*

Here is the beginning of the water transfer system in the shore of Caspian Sea, near Lavandvil, where the water is sucked by pump(s) after passing through a filtering system (intensive membrane technology) to remove the eventual harmful biota of Caspian Sea (Badescu and Cathcart 2010); however survival of Caspian Sea biota in high saline water of Urumia Lake is nearly impossible and they will immediately be converted in high saline environment to lifeless and harmless organic materials.

In this geographical location, Alborz mountain chain between mainland Iran and Caspian Sea is narrowest and topographically most suitable for the construction of pipe(s) line system and the related installations and constructions.

## *6.2 Distance Between Axis 1 and 2 (Area: I)*

The distance between these two axes is ca. 12 km with a height difference of ca. 1,650 m that should be overcome by pumping technology. The quantity of water to be pumped is 110–140 $m^3$ /s which can be done by one big pump or several smaller pumps arranged in parallel position. However manufacturing of pumps to work by

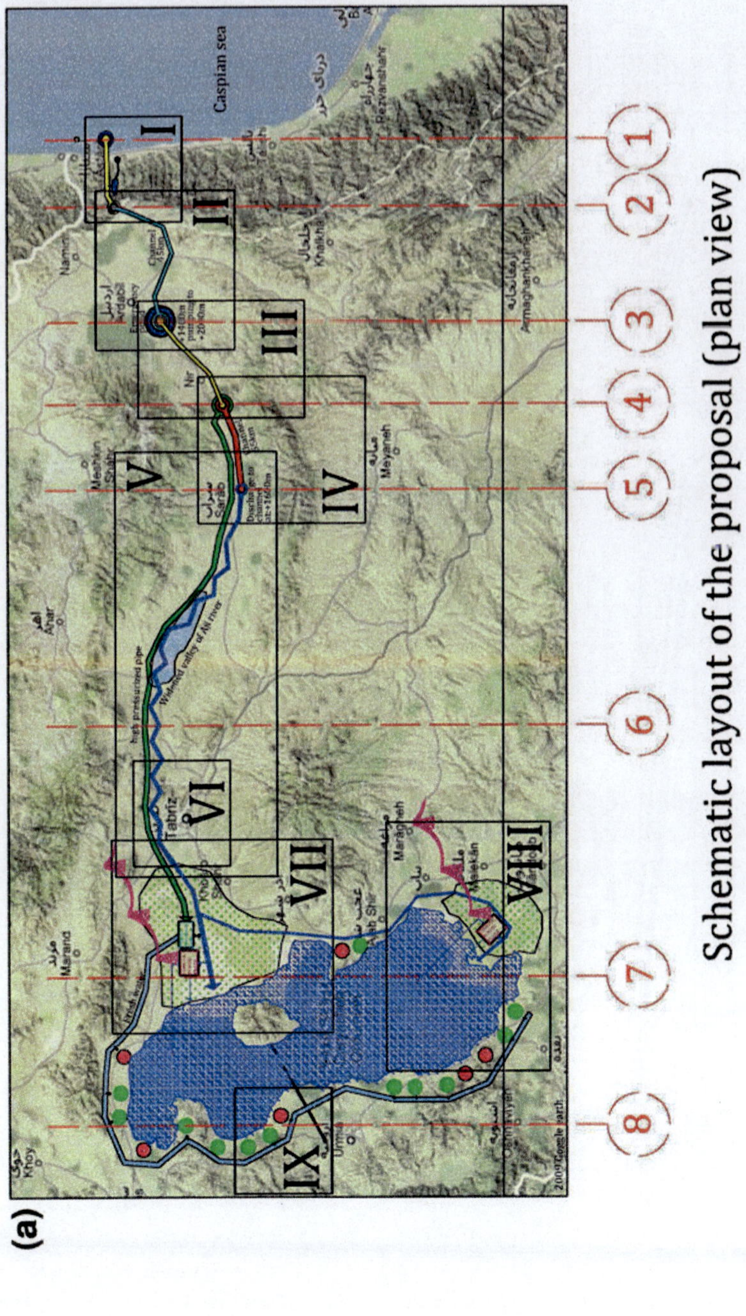


Fig. 18 Geographical layout of the proposal: **a** plan view; **b** section view

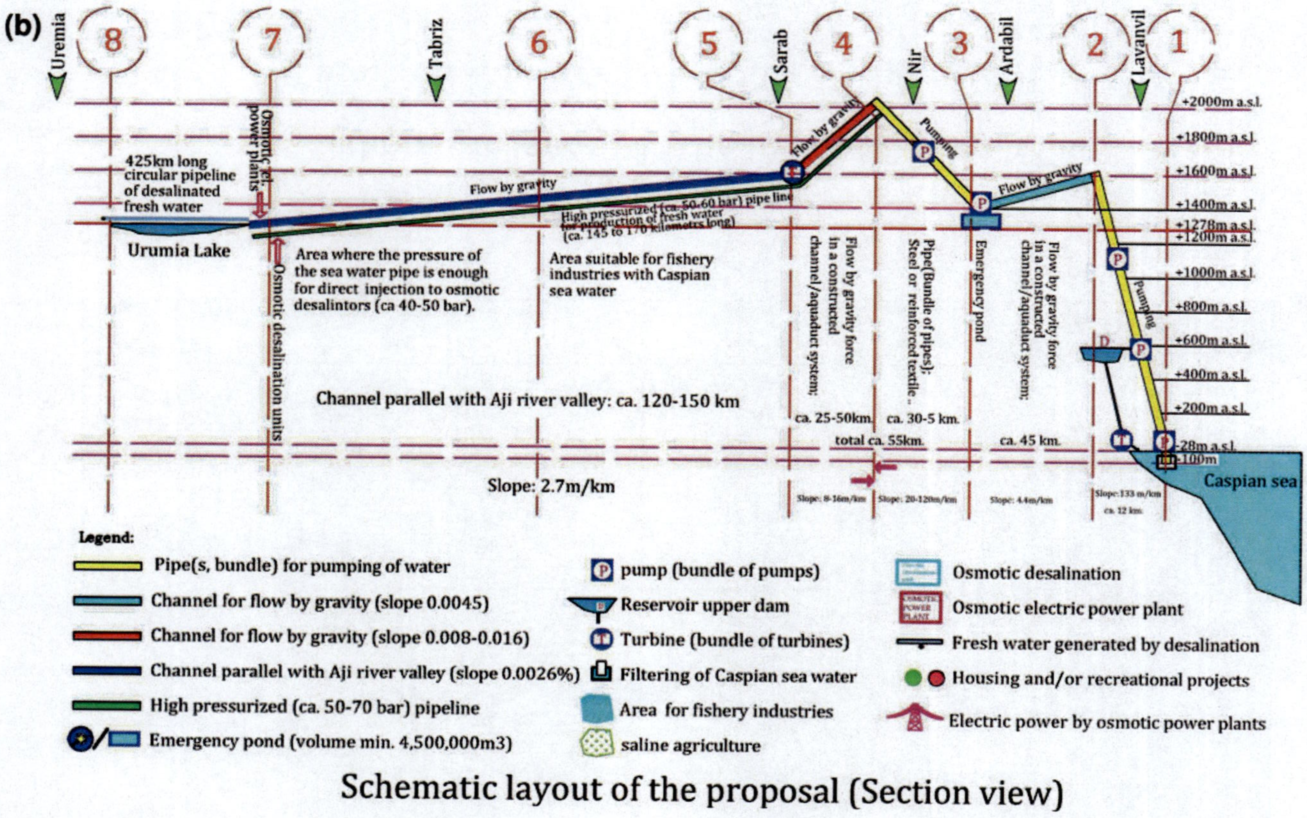


**Fig. 18** Continued

**Fig. 19** Schematic layout of transferring sea water over Alborz coastal mountains

a head of 1,650 m and capacity of the above mentioned quantity is not impossible but several pumps with lower heads arranged in serial/parallel order is more practical and economically feasible. With reference to a newly built pumped storage electrical plant in Siah- Bishe in the north of Iran and other similar projects, a stepwise arrangement of 3 series of smaller pump(s), each overcoming 550 m height, seems to be the most practical.

As smaller pumps and a bundle of steel pipes are more practical, with corrosion resistant inner lining, arranged parallel over the ground and along the perpendicular valley to the coastal line instead of graving underground, seems to be economical and easier to construct. The suitable topography of the path makes the costly tunnel(s) unnecessary.

The over-capacity of pumping system, after containment of the crisis and restoration of Urumia Lake, can be used as a parallel pumped/storage hydroelectricity generating unit similar to that of Siah-Bishe plant in the north of Iran (Siah Bishe 2010). In this case the parallel valley next to the pipe line seems to be suitable for construction of the upper dam with proportionate volume (Fig. 19).

## *6.3 Distance Between Axis 2 and 3 (Area: II)*

In this part of the water transferring system which begins near Neyaraq, the water volume pumped up to the level of 1,600 m a.s.l. will continue its path toward Urumia Lake, flowing downwardly by gravity force through a 45 km long open channel (as per Fig. 20; calculated approximately according to the manning formula) to a point near south of Ardabil City.

The area II is suitable for the establishment of fishery industries for the Caspian Sea species.

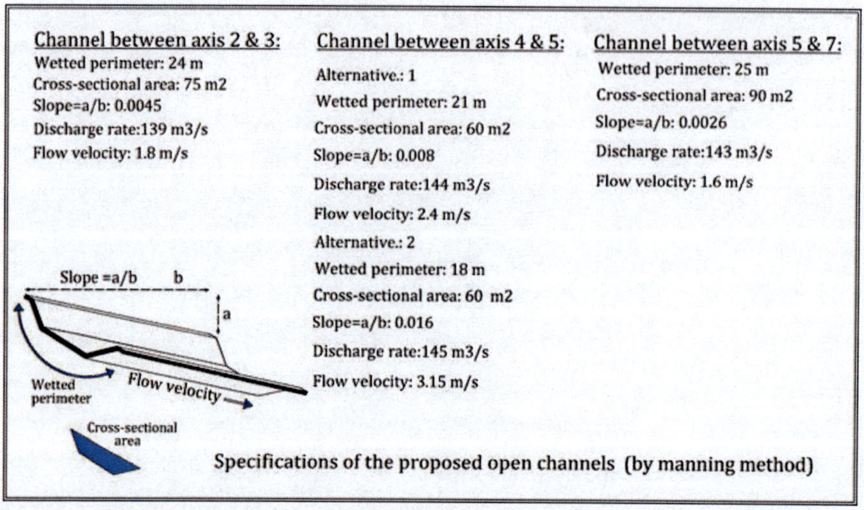


**Fig. 20** Shape and dimensions of the open channels for Caspian Sea water transfer

## 6.4 Axis 3

In this location, where the first channel terminates and second pumping system is located, an *emergency pond(s)* with a total capacity of circa 4.5–6 million m$^3$ must be provided. The function of this emergency pond capacity is temporary deposition of the flowing volume of water in the first piece of channel (between axis 2 and 3) in the case of emergency shot down of the pumping system and halting in forwarding the inflowing volume of water. Another function of this pond is emergency discharge of the pipe(s) water contents between levels +1,400 m a.s.l. and +2,000 m a.s.l (between axis 3 and 4). In such an emergency situation, the whole pumping system will be automatically stopped first and no more volume of water will enter the channel and the existing amount of water in the system will be provisionally discharged into the emergency pond(s) until the system starts its function again and water transfer system restores to its normal regime.

## 6.5 Distance Between Axis 3 and 4 (Area: III)

In this section, once again, arriving sea water should be lifted up from level 1,400 m a.s.l. to the level 2,000 m a.s.l. to overcome the 600 m high mountainous hindrance near Nir, which is similar to the first pumping system in capacity but less in vertical head. The horizontal distance can vary from 5 to 30 km depending on the most feasible, economic and engineering alternative. The capacity, number of pumps and the diameter of pipe(s) in parallel and/or serial arrangements might follow similar pattern as the first pumping section (between axis 1 and 2) mentioned earlier.

## 6.6 Distance Between Axis 4 and 5 (Area: IV)

Water pumped to the elevation 2,000 m a.s.l. now will run 400 m downwardly by the gravity force to a level of 1,600 m a.s.l. In this section, according to the specification of the previous one, there are two alternatives: long or short path as per Fig. 18a and b. According to the alternative 2, there is a possibility to utilize the water velocity (2.4–3.1 m/s) to generate electricity by installing proper turbine(s) for recovery of portion of the electrical energy consumed by the pump(s). For this purpose, instead of an open channel, a pipe line might be preferred.

## 6.7 Axis 5

As mentioned above, in this location an electric generating turbine station may be installed to recover part of the consumed electric energy by the pumps in the pumping stations between axis 3 and 4.

## 6.8 Distance Between Axis 5 and 7 (Area: V, VI and VII)

The main river of the region, Aji Chay, begins in this area and creates a very good opportunity for a natural path for Caspian Sea water to flow downwardly by gravity force in a parallel channel with the river until Urumia Lake (Fig. 21). The valley of Aji Chay, with a length of ca. 120 km, passes through a mountainous area before entering the Urumia Lake. The specification of the open channel proposed for this section is shown in the Figs. 18a and b. The area between axis 5 and 6 is suitable for the establishment of fishery aqua industries and related installations for the production of Caspian Sea fish species: enough land, plenty of Caspian Sea water, large population which is historically familiar with the seasonal fishing in Caspian Sea and experienced in the aquaculture industries and etc. are among the favorable conditions. Water borrowed for these activities from the bypassing sea water transferring channel, will be returned back after being used.

Aji River valley widen and its water gets saltier by passing through the salt mines near Tabriz city; therefore it can also be used directly for further transferring of Caspian Sea water in the final part before joining the lake. In this case there will be a need for some modifications in the valley to remove eventual physical hindrances to facilitate the Caspian Sea water flow. This opportunity can reduce the total length of the built channel between axis 6 and 7, by approximately

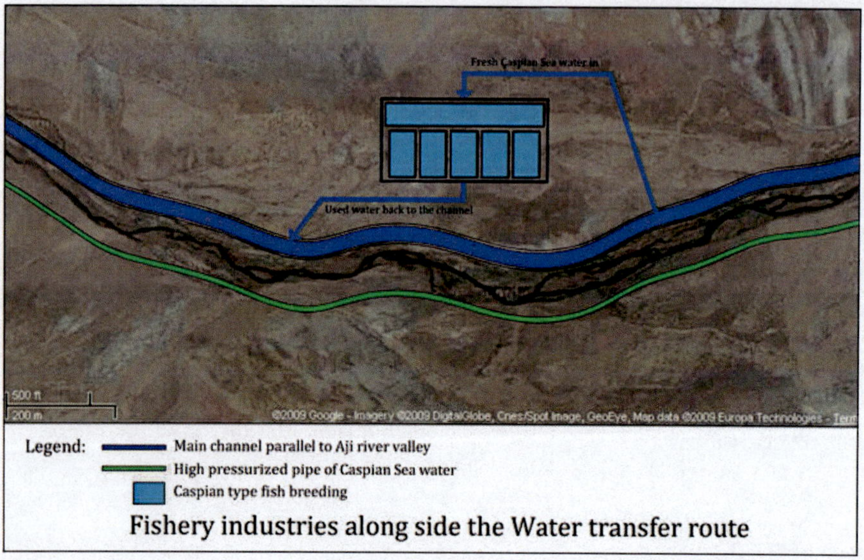


**Fig. 21** Fishery industries using Caspian Sea water

**Fig. 22** Aji river near Tabriz city in a rainy period

70–100 km. Figure 22 shows Aji River in a wet and rainy period of the year near Tabriz.

## *6.9 Axis 7 (Area: VII and VIII)*

There are two large coastal areas in the east (area VII: ~170,000 ha) and the south (area VIII: ~65,000 ha) of the lake which will function as an emergency space/area (ca. 8,400 km$^2$, including the lake, in area, and 8–16 km$^3$ in total volume) to room an eventual over-flooding of the lake (similar to the situation in years 1963–1964 as per Fig. 6), an extremely exceptional case when the surface level increases by 1–2 m over the lake's ecological level (1,279.5–81.5 m a.s.l.).

Industrial production of *Artemia Urmiana* in saline ponds arranged in the southern and eastern coastal areas: an industrial business which includes harvesting, processing and packing of cyst and matured *Artemia* biomass with tremendous opportunities for employment and incomes for the neighboring population centers (towns and villages).

In these two vast areas, beside halophyte agriculture and *artemia* related industries and activities, there is a unique opportunity to exploit the newly emerging clean/green technological possibility for the generation of electrical energy i.e. osmotic power technique. The Caspian Sea water alone or in the mixture with the fresh water from local rivers can be used as the fresh water input. The idea will be described later in Sect. 7.2.

### *6.10 Distance Between Axis 7 and 8 (Area: IX)*

The northern and western coast of the lake is very suitable for the building of recreational and/or housing agglomeration and complexes. The only limitation can be the fresh water shortages. The osmotic declinators can ease this constraint. The high pressurized pipe line providing pressurized Caspian Sea water around the lake can support the individual units of osmotic declinators. This might be done centrally too, by a separate pipe line for the carrying and distributing of fresh water from a main and central osmotic desalination plant in the east coast of the lake. The saline waste water of these desalination units would be discharged directly into the lake. Even the human sewage from these housing and recreational agglomerations can be discharged into the lake as nutritional additives for the Lake's biota stock and food chain after suitable and proper processing and disinfection.

The sea water transference from Caspian Sea to Urumia Lake with an average speed of 2 m/s will take ca. 80–100 h to reach the destination; mostly through open channels. By different means, the severely needed oxygen and carbon dioxide for the bio-ecology of the lake can be added to the Caspian Sea water, for example, by aeration through artificial waterfalls provided on the channel passageway wherever topographically possible.

## 7 A Simple Cost and Benefit Analysis

### *7.1 Costs*

The proposed project for the transfer of Caspian Sea water to Urumia Lake from its functional nature consists of two different components:

**Table 2** Siah Bishe pumped storage plant

| |
|---|
| Upper dam and reservoir: Concrete faced Rockfill Dam, Height 85 m, Crest Elevation 2,410.14 m a.s.l. |
| Dam volume 1.4 million $m^3$, Reservoir live volume: 3.5 million $m^3$ |
| Headrace: two tunnels: L = 2,015 m and 1,973 m, Ø = 5.70 m, concrete lined |
| Surge tanks: one for each tunnel, surge range = 87 m; Tanks: 223 m deep, 20 m wide, circular concrete tank; |
| Shafts: 2 × 65 m deep, Ø 6.5 m vertical shafts, concrete lined; Pressure Shafts: Two Shafts L = 2 × 760 m, Ø = 5 m, steel lined, Inclination 67° |
| Design flow: 2 × 130 $m^3$/s in turbine operation; 2 × 100 $m^3$/s in pump operation |
| Gross head: Max. 520.8 m, Min. 470.8 m; |
| Powerhouse Cavern: $L = 130.0$ m, $W = 22.0$ m, $H = 42.9$ m; |
| 4 vertical Francis Turbines; Turbine rated output: 4 × 260 MW = 1,040 MW |
| Transformer cavern: $L = 198$ m, $W = 14$ m, 4 connecting bus-bar galleries and vertical shafts; |
| Tailrace tunnels: 2 tunnels Ø 6.0 m, concrete lined; |
| Lower Dam and Reservoir: Concrete Faced Rockfill Dam, Height 106.5 m, Crest Elevation 1,911.50 m a.s.l.; |
| Dam Volume 4.93 million $m^3$, Reservoir live volume: 3.6 millions $m^3$; |
| Initial estimation for construction period: 2003–2009 (May 2010: not finished) |
| Initial estimated costs: 380 Million USD |

Project data (Siah Bishe 2010)

- Component *"a"*: pump lifting of the water to a total height of 2,200 m and a quantity of ca. 145 $m^3$/s.
- Component *"b"*: downwards flowing of this quantity of Caspian Sea water in a built open channel and/or pipe(s) with an approximate length of ca. 240 km.

In order to have a very simple estimation of the costs, a comparison with a project currently under construction in Iran with several typical similarities i.e. Siah Bishe pumped storage plant in Mazandaran province (see Table 2), can be very helpful.

We can assume that half of the cost in the Siah Bishe project belongs to the pumping and construction of the shaft/tunnel systems with a head of 520 m and a water flow of 2 × 100 $m^3$/s, i.e. 180 million USD. The other half belongs to the upper and lower dams/reservoirs, turbines and related systems. To be on the safe side, we can ignore the fact that the quantity of water to be pumped in Siah Bishe is bigger than in the case of the Urumia Lake project (200 $m^3$/s vs. 145 $m^3$/s). Now we can estimate the costs for the component *"a"* in our project as follows:

$$2,200\,\text{m} \times 180\,\text{million USD} \div 520\,\text{m} = 760\,\text{million USD}$$

For the cost of the component *"b"*, we can assume that the costs for 1 meter length of the channel as per Fig. 18a and b, is 10,000–15,000 USD (ca. 1/4 cost of 1 meter length of the deep subway tunnel construction in Tehran), thus the total cost will be:

$$240,000\,\text{m} \times 10,000 \text{ or } 15,000\,\text{USD} = 2.4 \text{ to } 3.6\,\text{billion USD}.$$

The total cost of the two components will be: $a + b = 3.16$ to 4.36 billion USD. If the cost for the feasibility studies, planning, technical documentation, calculations, design… is assumed to be 25% of the total, then the final figure for the whole water transfer project can be estimated:

$$3.16 \text{ to } 4.36 \times 1.25 = \text{ca. } 4 \text{ to } 5.5 \text{ billion USD(Siah 2006)}.$$

## 7.2 Benefits

The transfer of water from Caspian Sea to Urumia Lake alongside the fulfillment of its main goal i.e. the regeneration and stabilization of Urumia Lake's hydro-ecological systems as a necessary and precautionary step to avoid occurrence of the feared disastrous salt storms and environmental catastrophe, can create tremendous number of positive and favorable chances and vast number of opportunities for an overall development and socio-economically sustainable growth in the north-western regions and provinces of Iran. Following items are just some few examples:

- Investment in the large scale and commercial production of *Artemia Urmiana* (Fig. 23): the cyst of *Artemia* (1,600–2,000 MT/year; 50 to100 USD/kg) and *Artemia* biomass (200,000–400,000 MT/year; 5 USD/kg) which has large industrial use and can reach up to a several billion dollars/year business. Iran could become one of the main producers and exporters of these items; (Reveshty 2000; Stappen 2006).
- Restoration of the natural habitat of the wild life, including hundreds of different species;
- Huge investment in the aquaculture industries such as fisheries and etc.;
- Investment in the halophyte agriculture and Biomass using low saline (processed or unprocessed) Caspian Sea water in vast areas in the region, mainly in the south and east parts of the lake (Asri and Ghorbanli 1997);
- Increase of salt production (pure and calk free), using both traditional and industrial methods (Fig. 24);

**Fig. 23** *Artemia* nauplii as food for baby fishes (Stappen 2006)

**Fig. 24** Coastal salt production

**Fig. 25** Recreational and historical tourist attractions of the Urumia Lake region: *Top row*: *left side* and *middle*: chlorine mud therapy, *right side*: Christian heritage, historical church; *bottom left side*: *Moghadas Ardabili* mausoleum; *middle* and *right side*: historical Bazar and Mosque of Tabriz

- Production of the severely needed fresh water from 1.2% saline Caspian Sea water by R.O. technology in relatively easy and low cost method by using gravity force for creation of the necessary pressure for the input water and avoiding highly expensive electrical power consumption (pressure up to 50 Bar may be created);
- Establishment of the industries which need huge quantity of water;
- Investment in the large tourism opportunities of the region, especially recreational and health tourism (*saline/chlorine mud therapy*) in the coastal areas and using attractive touristic opportunities of the Lake and neighboring historical sites and cities (Tabriz, Urumia, Mahabad, Maragheh) (Fig. 25).
- Investment in the coastal urban and housing projects as an alternative and substitution for the natural expansion of regional big cities and metropolitans such as Tabriz, Urumia, Maragheh, Mahabad and etc.;
- Expansion of forests and fruit gardens in the area by the help of desalinated Caspian Sea water;

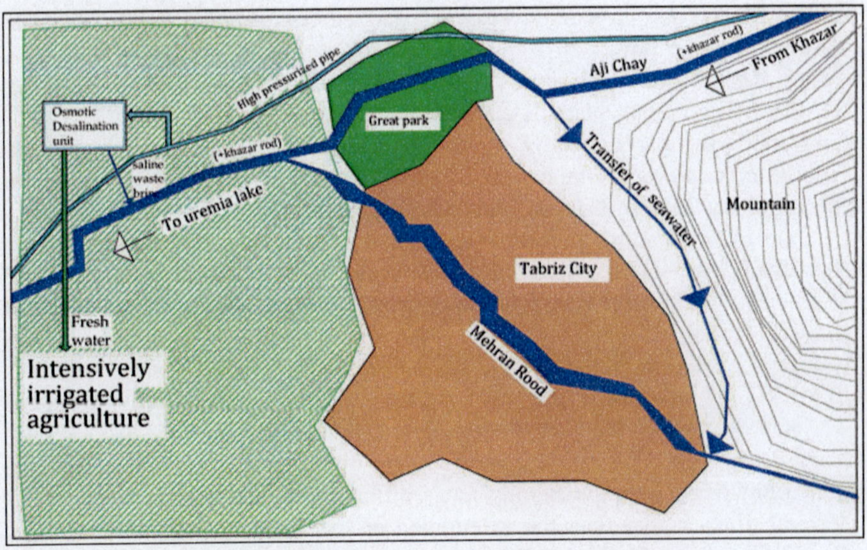


**Fig. 26** The benefits of the project for Tabriz City, a regional metropolitan

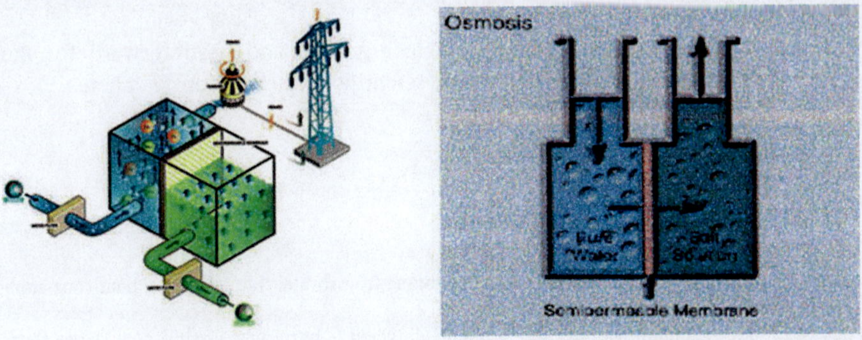


**Fig. 27** The core idea of the osmotic power plant: *left side*: schematic osmotic power plant; *right side*: main concept of the osmotic power

- Significant improvement of the city landscape in Tabriz, the metropolitan of the region, by letting Caspian Sea water to flow permanently through its two seasonal rivers: Aji and Mehran Rivers (Fig. 26).
- Using osmotic desalination units will help an intensively irrigated agriculture in the suburb areas of Tabriz and the peripherals regions of the lake;
- Freshwater from osmotic desalination units will help the industrial sector of this intensive industrialized region of Iran.
- Installation of osmotic electrical power plants (Fig. 27) in the coastal areas of Urumia Lake: The realization of this proposal will open an enormous and unique opportunity to use osmotic power technology, a new renewable source of

clean energy, for production of green and environmental friendly electrical power (Osmotic power 2010). When ca 4.2 $km^3$ of Caspian Sea with salinity range of 1.2% is annually introduced to the hyper saline water of Urumia Lake (15–20%), enormous quantities of mechanical energy are released which can be converted to electricity by osmotic technology. (http://en.wikipedia.org/wiki/Osmoticpower) The first commercial osmotic plant will be ready for operation by 2015 in Norway, Statkraft co (KEMA).

- A simple computation proves that the annual amount of water evaporated from the lake is ca. 5–6 $km^3$. Solar irradiative heat energy for this evaporation is ca. $3.15 \times 10^{11}$ $Kcal_h$ which is equal to 366$GW_h$ electrical energy. How much of these potential is refundable by the osmotic technology. According to KEMA osmotic…"Mixing fresh (river) water to sea (salinity = 3%) water generates 0.5 $KW_h$ per $m^3$ fresh water". On the basis of this assumption, by 1 $km^3$ of Caspian Sea water (1.2% salinity) imported to Urumia Lake (15–20% salinity) some $1.65 \times 10^8$ $kW_h$ can be generated? The efficiency of the reaction might be much higher because the salinity range difference in the Urumia Lake case is several times higher than the assumption of the KEMA's calculations in the Norwegian case; (Statkraft 2009);
- Creation of vast number of job opportunities in Gilan, Ardabil, west and east Azerbaijan and Kurdistan provinces in many fields and in all levels from simple to highly technological ones;
- And, a huge scientific and technological growth and push forward for the country as a whole and it's academic, scientific and engineering spheres.

## References

Abas-Nejad M (2010) East Azerbaijan Environment Institute. http://www.akhbar-rooz.com/article.jsp?essayId=28194. Accessed Feb 2010

Agh N (2008) Artemia Institute and Urumia University. http://www.isslr.org/news/newsone.asp?qnewsid=315. Accessed 10 Jan 2010

Alesheikh AA, Ghorbanali A, Nuri N (2006) Coastline change detection using remote sensing. http://www.ceers.org/ijest/issues/full/v4/n1/401008.pdf. Accessed Feb 2010

Asri Y, Ghorbanli M (1997) The halophilous vegetation of the Orumieh lake salt marshes, NW, Iran. http://www.springerlink.com/content/m64116112904q882/. Accessed Feb 2010

Badescu V, Cathcart BR (2010) Aral sea partial restoration: a Caspian water importation macroproject. Int J Environ Waste Manage. Parts I and II (to be published)

Dating Caspian Sea (2003–2005) Dating Caspian Sea level change. http://www.caspage.citg.tudelft.nl/project.html. Accessed Feb 2010

Earth Snapshot (2010) Source of the picture. http://www.eosnap.com/?s=Urmia+lake. Accessed Feb 2010

Earthobservatory (2006) http://earthobservatory.nasa.gov/NaturalHazards/view.php?id=16817. Accessed Feb 2010

Eimanifar A, Mohebbi F (2007) Urumia Lake (Northwest Iran): a brief review, saline systems 2007 3:5. doi:10.1186/1746-1448-3-5. http://www.salinesystems.org/content/3/1/5

Mansimov MA, Aliye A (1994) Fluctuating levels of the Caspian Sea. http://azer.com/aiweb/categories/magazine/23_folder/23_articles/23_caspiansea.html. Accessed Jan 2010

Osmotic Power (2010) http://en.wikipedia.org/wiki/Osmotic_power. Accessed Feb 2010

Panin NG (2005) The Caspian Sea level fluctuations as an example of local/global climate change. http://www.jstor.org/pss/4298072. Accessed Feb 2010

Reveshty AM (2000) Detecting the spatial distribution of *Artemia* at Urumia Lake. University of Zanjan, Department of Geography. http://www.gisdevelopmentnet/appliction/nrm/water/overview/ma04027c.htm. Accessed 10 Feb 2010

Siah Bishe (2010) Siah Bishe pumped storage project Iran; AF-Colenco Ltd. Täfernstrasse 26, CH-5405 Baden Dätwil, Switzerland. http://www.colenco.ch/pdf/wk/AFC_Siah-Bishe_en.pdf

Stappen GV (2006) International study of *Artemia* LXIII. Field study of the *Artemia Urmiana* population in the Urumia Lake, Iran. http://wwwartemiaworld.com/home/. Accessed Feb 2010

Stappen GV, Coutteau P (1996) FAO fisheries technical paper 361, Chap. 1. University of Gent, Belgium

Statkraft (2009) http://www.statkraft.no/energikilder/saltkraft/. Accessed Feb 2010

tebyan.net (2009) Urumia Lake, photo gallery. http://www.tebyan.net/Iranology/Proinces/East_Azarbayjan/2009/2/14/85767.html. Accessed 10 Jan 2010

Tolkatchev A (2004) Caspian Sea-level rise: an environmental emergency. http://www.pol.ac.uk/psmsl/gb2/tolkatchev.html. Accessed Feb 2010

Vaziri M (1998) Comparative analysis of the Urumia Lake and the Caspian Sea surface water lever fluctuations. http://kfki.baw.de/conferences/ICHE/1998-Cottbus/109.pdf. Accessed Feb 2010

Whish-Wilson P (2002) The Aral Sea environmental health crisis. J Rural Remote Environ Heath 1(2):29–34. http://www.jcu.edu.au/jrtph/vol/v01whish.pdf. Accessed Feb 2010

# Sediment Transport by Wind and Water: The Pioneering Work of Ralph Bagnold

**Michael Welland**

## 1 Introduction

In 1978, at the age of 82, Brigadier Ralph Alger Bagnold was awarded the honorary degree of Doctor of Science by Aarhus University in Denmark. For the occasion, he gave a lecture, later published in *Nordic Hydrology* (Bagnold 1979), titled "Sediment Transport by Wind and Water." In his introduction, Bagnold remarked, in his typically blunt style, that "it is doubtful whether modern textbooks of river and canal engineering convey any clearer understanding of the natural processes involved than was probably possessed by the great engineer Pharaohs of the 12th Dynasty 4,000 years ago, one at least of whose vast canals appears to have remained self-clearing for 1,500 years."

Bagnold as a scientist is, as discussed below, difficult to categorise. He described himself as an amateur who had never held an academic position "or had any professional status." But he felt that this gave him "the rather unusual advantage of considering problems with an open mind, unbiased by traditional textbook ideas that had remained untested against facts" (Bagnold 1990). He was, at heart, an engineer, but one with an acute and perceptive capacity to apply other disciplines—physics, mathematics, geology—to his identification and analysis of any problem. In his Aarhus lecture, he went on to attribute the limitations of textbooks, this "backwardness," to "the fact that the training of neither the earth scientist nor the hydraulic engineer has included any general grounding in the principles of physics." It was Bagnold's intuitive and radical interdisciplinary approach to sediment transport, over the course of more than 50 years and 50 scientific papers, that enabled today's earth scientists and engineers to plan and pursue projects equipped with a deep, if still imperfect, understanding of these

M. Welland (✉)
Orogen Limited, London, UK
e-mail: mw@throughthesandglass.com

V. Badescu and R. B. Cathcart (eds.), *Macro-engineering Seawater in Unique Environments*, Environmental Science and Engineering,
DOI: 10.1007/978-3-642-14779-1_19, 

critical natural processes. And textbooks today, at least in the earth sciences, will contain wording such as "In the absence of any modern readable text solely dedicated to the principles of sediment transport, a return to Bagnold's 1954 classic (for wind) and to his 1966 (for water) should provide the necessary inspiration" (Leeder 1999).

Sediment transport represents mass transfer of material on a truly gigantic and dynamic scale. If it were not for the inhibiting role of dams, it is estimated that approximately 8 $km^3$ ($2 \times 10^{10}$ tons) of sediment are transferred every year from the continental interiors to the shelves and the oceans (Leeder 1999). This does not include internal circulation of material that fails to be delivered to the coasts, such as windblown sand of the planet's deserts. Deposits of aeolian sand cover approximately 6% of the global land surface area (Pye and Tsoar 1990) and further large tracts are covered by currently stable vegetated dune systems vulnerable to reactivation. From the Sahara alone, it is estimated that $2.6 \times 10^8$ tons of dust are removed each year and distributed around the planet by atmospheric circulation (Leeder 1999). It is in the physical context of this massive and far-from-steady-state system that environmental and land-use policy must be constructed, that major engineering and resource projects must be planned, executed, and maintained, and that human quality of life must be sustained and improved.

The large-scale engineering schemes of Egypt's Pharaohs have served as examples for the country's continuing mega-projects of today that themselves serve as examples of the challenges of such endeavours in an environment of constantly dynamic sediment movement. Egypt is a country with rapidly growing population pressures (almost doubling in the last 25 years) and an urgent need for agricultural expansion. The Egyptian Environmental Affairs Agency reports that less than 6% of the country's area is populated, and less than 3.5% is under agriculture; close to 17% of the country is covered by active sand dunes and sand encroachment is a constant threat to agricultural productivity. Several "mega-projects" of large-scale irrigation are following in the footsteps of the Pharaohs and are at various stages of completion. In northern Sinai, the Al-Salam Canal project brings water from the Damietta Branch of the Nile under the (constantly dredged) Suez Canal to provide water for several hundred thousand hectares of new agricultural land; further expansion phases of the project are underway. The ways in which the canal and associated land use changes will interact with natural sediment movement and soil quality remain uncertain.

On the western side of the Nile, another mega-project is underway. The Southern Valley Development Project is a major, multiphase infrastructure initiative intended to develop not only irrigation and agriculture in the region surrounding the southern Nile, but also roads, population centres, employment, and tourism. Within this overall plan, the highly controversial Toshka Project has been underway for some years (Ministry of Water Resources and Irrigation). The Toshka Canal, which, together with its four branches, will total 250 km in length, will carry excess water from the Aswan High Dam Reservoir (Lake Nasser) into the arid heart of the Western Desert. The canal system will discharge 25 million $m^3$ of irrigation water into new agricultural areas totalling over 200,000 ha. The project is controversial for a number of reasons that arise from uncertainties over its long-term impacts. Quite apart from its effect on the water budget for the rest of the Nile system, and the associated ecological, economic, and social implications, the likely behaviour of sediment drawn from the reservoir is unknown. It is estimated that, since 1964, approximately 5 billion $m^3$ of sediments have accumulated in the reservoir (Abulnaga and El-Sammany 2004). Of the annual accumulation of 130 million $m^3$, a mere 4 million pass through the dam and enter the Nile system. The resulting large-scale disruption of the natural sediment budget has caused shrinkage of the Nile Delta and substantial ecological changes. Abulnaga and El-Sammany (2004) have suggested a project to de-silt the reservoir using slurry pipelines and distribute the sediment on a large scale to create new agricultural land. It is quite likely that discharge into the Toshka Canal will, to some extent, have the same effect, but the consequences for both canal facilities maintenance and for soil quality are unclear. Management of the canal system with respect to sediment transport will have to address two further aspects: disposal and distribution of the massive volumes of excavated material and the influx of windblown sand. The engineering knowledge of the Pharaohs—and of Ralph Bagnold—may need to be sought.

It is in the nature of societies and individuals to devise and execute mega-projects, and such ambition and creativity become increasingly necessary in the context of attempting to manage regional and global issues. The diversity of the contents of this volume illustrates the range of such projects, past and present. The environments in which these projects are set represent essentially every setting in which actively moving sediments play a key role, hyper-arid, arid, fluvial, coastal, and marine. In any of these environments any project will, inevitably, interact with the natural sediment budget, and the design and impact assessment must take this into account. Yet sediment transport is a highly complex process that operates over a range of several orders of magnitude of scale and continues to defy fully integrated attempts at modelling and prediction. In the coastal and marine environments, the Sethu Samudram Canal mega-project (Subba Rao, this volume) illustrates clearly the scale of this challenge. This shipping canal is planned for the Palk Strait region between the south-eastern coast of India and Sri Lanka. Palk Strait is a mere 40 km wide but connects the Bay of Bengal and the Arabian Sea and is therefore a highly dynamic marine environment; the regime of sediment transport, erosion, and deposition is thus equally dynamic and complex. It is a region of sand banks, shoals, sand spits, and islands, and sediment supply from three major rivers and littoral drift varies significantly on a seasonal basis reflecting monsoon activity (Chandramohan et al. 2001). Complex patterns of significant coastal accretion and erosion show variations on all timescales under normal circumstances; the effects of the 2004 Indian Ocean tsunami illustrate the profound impacts of periodic catastrophic events and emphasise the need for rigorous environmental planning and regulation in these complex environments (Shaw 2008; Cathcart 2006).

The design, planning, and execution of any mega-project are, of necessity, an inter-disciplinary process, and geological analysis, particularly of all aspects of the impacts on and of sediment transport, is a critical component. Current understanding and modelling capabilities of sediment transport are sophisticated but, inevitably, incomplete. The "state of the art" of this arena rests firmly on the foundations established over a lifetime's work by Ralph Bagnold. This chapter does not purport to be a comprehensive review of his work, or of "the state of the art" understanding of sediment transport by wind and water, but rather attempts to put his groundbreaking contributions in their historical and current contexts, to highlight their critical importance to any macro-engineering project, and to celebrate this extraordinary individual.

## 2 The Man

In one of the many phases of his eventful life, Ralph Bagnold entered into a highly productive personal and professional collaboration with Luna B. Leopold, whose wide-ranging work in fluvial geomorphology and sediment transport is, justly, renowned as groundbreaking. Leopold, together with Paul Komar and Vance Haynes, provided the introduction to Bagnold's autobiography, *Sand, Wind, and War: Memoirs of a Desert Explorer*, published in 1990 (Bagnold's preference for the title would have added the word "water" after "war"). The introduction begins as follows (all quotations in this section are from Bagnold's autobiography (1990) unless otherwise specified):

> It is unusual to find a professional soldier who is known primarily for his contributions to intellectual and scientific thought. This is true, however, of Brigadier Ralph A. Bagnold, whose research spanned several of the physical sciences, including hydrology, geophysics, oceanography, and geography. His earliest scientific publications appeared in the 1930s and dealt with the movement of desert sand; his last two papers were in 1983 and 1986, respectively, examining the random distributions of word lengths and sediment transport in water. This productivity and originality seems astonishing when one considers that Bagnold was not formally trained as a scientist and had never been a professional academician.

Ralph Bagnold (Fig. 1) was born in 1896 into a military and engineering family. His father was a Colonel in the Royal Engineers who had taken part in the unsuccessful 1885 expedition up the Nile that attempted to relieve General Gordon at Khartoum. During his time in Egypt, Arthur Henry Bagnold helped devise the means to raise the colossal statue of Rameses II into a "prone, but decently housed position" in Memphis. In 1899 Bagnold's family moved to Jamaica and it was there that he showed an early interest in fluid dynamics and apparatus-building that would never desert him. His father allowed him to "rummage around in the Royal Engineers' supply yard," from which he retrieved a variety of pipes and fittings. He was "obsessed by pipes" and, helped by his father's instructions on making concrete, constructed a wide range of hydraulic devices, together with complete working models of household drainage systems made from modelling clay.

**Fig. 1** Ralph Alger Bagnold (courtesy of Stephen Bagnold)

Back in England, the young Bagnold continued to develop his fascination with apparatus and its construction, spurred by his father's skills and equipment and, in particular, stimulated by the activities of a neighbour. Charles Phillips, who would later become Secretary of The Royal Institution, kept a home laboratory filled with scientific instruments and experimental apparatus, much of which was built by Phillips himself. Phillips clearly had a profound influence on Bagnold, who wrote that he "excited me, as a boy, with the basic idea of science." That idea very firmly included the principle of a scientist building his own experimental apparatus designed to address a specific line of enquiry; this would remain a hallmark of Bagnold's work throughout his life. From boarding school, at the age of thirteen, he wrote to his father requesting his own bench lathe, cleverly describing this as being in his father's best interests: "I am hindered in everything I do in my workshop, and as you are away all day I cannot bring the things to you to do. Consequently everything waits for years." He subsequently wrote to his mother, assuring her that he would pay for the lathe, but asking her to influence his father to "get the lathe as soon as possible" (Churchill Archives Centre, Cambridge, Bagnold papers, by kind permission of Stephen Bagnold).

In 1914, Bagnold joined the Royal Military Academy, and, the following year, the Royal Engineers; after brief training, he was posted to the Western Front. He saw service at Ypres, the Somme, and Passchendaele, but, unlike many of his friends, survived.

After the war, Bagnold completed an engineering degree from Cambridge (although he found it "too theoretical to be of much use to me"). He rejoined the army and, in 1926, was posted to Egypt as "part of that anomalous army establishment known as the 'British Troops in Egypt'," charged with protecting the Suez Canal and British interests in general. He remarked that "Peacetime soldiering leaves a lot of leisure time. Egypt fascinated me from the start...." and, together with colleagues, embarked on a series of extraordinary desert journeys. Eschewing the traditional mode of transport, the camel, and challenged by the conventional wisdom that motorised vehicles could not cross the great tracts of sand dunes, he bought a Model-T Ford "that seemed able to go anywhere."

Following a series of trial expeditions, the first major excursions took him across Sinai to Aqaba, Amman, and Petra, during which he and his companions "learned a great deal about what cars could and could not do." Bagnold and his team devised a cooling water condensing and recirculation system for the cars that enabled them, in 1927, to make a 1,800-km roundtrip from Cairo to the oasis of Siwa to the west without the need to carry extra water. But it was the unknown expanse of the Great Sand Sea that caught Bagnold's imagination, the area where the map "faded away into a blankness stippled vaguely to indicate sand, and ended with the final stimulating remark, 'limit of sand dunes unknown'." This, combined with the "irresistible challenge" of proving that motorised travel across such terrain was feasible, led Bagnold to begin laying out ambitious plans. A transfer to India was no deterrent: he took leave in 1929 and, with two friends, drove a Ford truck back from India to Cairo. Now equipped with the new (but stripped-down and re-designed) Model-A Fords, he set out on a pioneering journey that demonstrated, using the clear ground of the windswept "streets" between the chains of dunes and full-frontal assaults on the dunes themselves, that the Great Sand Sea was not the insuperable obstacle to vehicles that it had been thought. In addition to the radiator system, Bagnold developed two further keys to desert travel: the sun compass and, after the initial use of rope ladders, lengths of steel channels (originally found in a Cairo junk shop) for extricating a vehicle from soft sand; such devices remain in use today.

Over the course of the following 3 years, regardless of where he was actually based, Bagnold made a series of extraordinary and historic expeditions into the Western Desert (Bagnold 1931). These culminated, in 1932, in the longest journey into the eastern Sahara at that time: close to 10,000 km travelled in total, through Sudan and Chad (Bagnold 1933). Large areas of the map were no longer blank. Bagnold's compelling account of these travels was published in his 1935 book, *Libyan Sands* (Bagnold 1935a).

In his autobiography, Bagnold describes how boyhood holidays on Dartmoor "gave me an exciting sense of mystery, of wild open expanses with the danger of getting lost, where somewhere there might still be things to be discovered. It gave me a pervading curiosity which has never left me." This element of his character clearly led not only to his pioneering exploration of the Western Desert, but also, in the course of them, a fascination with sand and dunes that his "pervading curiosity" would continue to explore for the rest of his life. His description for the Royal Geographical Society of the 1932 expedition (Bagnold 1933), after which the Society awarded him their gold medal, includes detailed observations and illustrations of different dune types, and in his autobiography he wrote:

> In 1929 and 1930, during my weeks of travel over the lifeless sand sea in North Africa, I became fascinated by the vast scale of organization of the dunes and how a strong wind could cause the whole dune surface to flow, scouring sand from under one's feet. Here, where there existed no animals, vegetation or rain to interfere with sand movements, the dunes seemed to behave like living things. How was it that they kept their precise shape

> while marching interminably downwind? How was it they insisted on repairing any damage done to their individual shapes? How, in other regions of the same desert, were they able to breed "babies", just like themselves that proceeded to run on ahead of their parents? Why did they absorb nourishment and continue to grow instead of allowing the sand to spread out evenly over the desert as finer dust grains do? More basically, what kind of upward physical force must be exerted on the mineral grains to make them rise against the force of gravity, lifting them to such a height that they can strike one's face like little hammers? No satisfactory answers to these questions existed. Indeed, no-one had investigated the physics of blown sand. So here was a new field, I thought, one that could be explored at home in England under laboratory-controlled conditions.

Thus did a body of scientific work, now regarded as classic and as radical and pioneering as Bagnold's desert expeditions, begin. He embarked on his laboratory exploration after retirement from the army in 1935, using space and facilities provided in the Hydraulics Laboratory at Imperial College, London. Following the family tradition, and using the skills first developed in his boyhood, Bagnold designed and built a wind-tunnel himself; the results of his experiments on blown sand are discussed below and were reported in two papers in the *Proceedings of the Royal Society* (Bagnold 1936, 1937). With his typical thoroughness, he resolved to test his initial experimental results against the realities of windblown sand in the desert, and organised another Western Desert expedition in 1938. With the assistance of the Royal Geographical Society, he assembled a diverse team, including archaeologists whose goal was to investigate key sites recording ancient migration routes and to document the abundant rock art of south-western Egypt and neighbouring Sudan and Libya. The group established a base camp at the foot of the cliffs of the Gilf Kebir, a plateau the size of Switzerland lying at the junction of the three countries; Bagnold set up his instrumentation and waited for a sandstorm. As he described, his wait was not long:

> A heavy one blew up within a few days. I was well prepared for it except, alas, I had lost my sand goggles. I spent some very uncomfortable hours sitting in the open, directly exposed to a violent sand blast, trying to keep my eyes open while taking readings from an array of gauges and sand traps. The purpose of eyelashes was very evident. Fortunately, I managed to get some reliable measurements which nicely confirmed my wind-tunnel measurements made in London.

His seminal book, *The Physics of Blown Sand and Desert Dunes*, discussed below, was completed in 1939 and published in 1941.

Bagnold was recalled to active duty at the outbreak of war, and found himself on a troopship bound for a posting in East Africa. In the Mediterranean, a fluke accident would change the course of the war: Bagnold's ship suffered a collision with another vessel in the convoy, and he was forced to disembark at Port Said. His arrival in Cairo came to the attention of General Wavell, whose presence there at the time was unofficial. Wavell recognised the potential value of Bagnold's Egyptian experience and desert skills, and immediately arranged for his re-assignment to the signal unit of the army's armoured division on the coast. Bagnold was astonished at the level of military ignorance of desert travel, and "wrote a short note suggesting that we might at least buy a small assortment of desertworthy American vehicles and train a nucleus of officers and men in the art of cross-country driving…" It would take several attempts and some length of time, but, with the new threat of Italian troops in Libya, eventually Bagnold was commissioned to establish such a force, using any means necessary. Thus was established the Long Range Desert Group (LRDG) whose intention was "piracy on the high desert." Using Bagnold's inventiveness and unparalleled knowledge of the desert, the LRDG fulfilled this intention, wreaking havoc behind enemy lines across vast distances of the Sahara. Its exploits and achievements have been described by Bagnold himself (1945) and his colleague, W.B. Kennedy Shaw (1945); a sense of the impact of the LRDG as early as 1941 can be derived from General Wavell's official despatch from October of that year:

> I would like to bring to notice a small body of men who have for a year past done inconspicuous but invaluable service, the Long Range Desert Group. It was formed under Major (now Colonel) R.A. Bagnold in July 1940 to reconnoitre the Great Libyan Desert…Operating in small independent columns the Group has penetrated into nearly every part of desert Libya, an area comparable in size with that of India. Not only have the patrols brought back much information, but they have attacked enemy forts, captured personnel [and] transport and grounded aircraft as far as 800 miles inside hostile territory. They have protected Egypt and the Sudan from any possibility of raids, and have caused the enemy, in lively apprehension of the activities, to tie up considerable forces in the defence of distant outposts.

In 1944, Bagnold returned to England and was "astonished" to find that he had been elected a Fellow of the Royal Society. Before the war, his interest had moved on from the transport of sediment by wind to that by water, and he continued with this in the context of addressing questions of beach dynamics in the preparation for the Normandy landings.

The chapter in Bagnold's autobiography that covers the years immediately after the war is titled "Lyminge—Marriage—Shell," a diversity of topics. "Lyminge" refers to the house near Folkestone, in Kent, that he purchased, sufficient in size for the necessary workshop, but ultimately too small for a family. Bagnold spent time with his sister, Enid, the well-known writer: "Enid and I had two things in common—fond memories of our early childhood days in Jamaica and a disregard for convention which in our different ways amounted to a practical inventiveness. If something seemed useful to be done, we did it, even though it had never been done or thought of before." Through Enid, Bagnold met Dorothy Plank, a friend of the family, and they were married in 1946.

In 1947 Bagnold was offered the position of director of research for Shell Refining and Marketing, an offer which, with some reservations, he accepted. Official photographs of the time show an impeccably-dressed, serious-looking man, seated behind a desk in a stark office: he does not look comfortable and, by his own admission, he was not. The desert explorer and founder of the LRDG explained, with characteristic modesty, that, while he worked with congenial staff:

> I sensed, though, that my background was inadequate to cope with the great variety of industrial research projects in which the oilmen were involved, or to co-ordinate the host of different projects going on at three big centres of different nationalities. In 1949, with

some reluctance, I resigned from Shell. Some years later, Lord Rothschild took on the job. I expect he did it much better.

Another reason for my leaving Shell was that at heart I am an explorer. I longed to get back to doing my own thing, to discovering more about the natural processes that have been going on for ages, unconcerned with man's needs and convenience. I wanted to find out the basic dynamic processes by which rivers transport sand, stones and boulders along their beds, and what weight of solids a given stream can transport in a given time.

Sediment transport by water would consume much of his attention for the remainder of his life. He did for water what he had done for wind: examine the physics from scratch and design and build his own experimental apparatus (again at Imperial College) to test the implications—his "try-it-and-see" approach. He wrote that "The movement of solids along the beds of rivers and canals had become an important economic nuisance. The rates of grain movement needed to be predicted." Nevertheless, his reputation as an expert on windblown sand continued to lead to his being called in to advise on threats to desert facilities: he found himself in some demand "as a sage on the subject."

In 1956 his prolific professional collaboration with Luna Leopold began. Leopold was then Head of the Water Resources Division of the United States Geological Survey (USGS) and together they agreed that, given the state of the science, the problems in fluvial sediment transport would benefit if they would jointly "stir the pool of complacent tradition with the stick of reality." This goal they succeeded in, completing a series of innovative and radical experiments and publishing a number of landmark papers over a period of more than 10 years: "On the whole, I believe that Luna Leopold and I contributed a good deal of new knowledge to the world. We had found a sound stick and were beginning to stir the pool."

Bagnold's work and personal relationship with Leopold allowed him to continue to delight in exploration of both the scientific and the adventurous kind. In 1966, under the guise of obtaining data on water depths, Leopold organised a rafting excursion down the Colorado River through Cataract Canyon, above the Grand Canyon. The trip had only been done rarely before, and only when the river was in full flood. Aged over 70 years, Bagnold clearly relished the experience; they followed the Green River until it came to:

the deep-canyon junction with the Colorado, flowing past from left to right, swollen with snowmelt from the Rockies. High on the rock wall was painted, in large letters, "ALL CRAFT TURN LEFT" and a great arrow that pointed upstream. On the right, grim warnings forbade going downstream, further emphasised by a very large skull and crossbones. In addition, an ominous dull roar rose up from the water and echoed between the canyon walls. We ignored the warnings and turned downstream.

Bagnold begins the last chapter of his autobiography, simply titled "Later," with the comment that "With increasing age and inactivity life's memorable events occur less often." Less often, perhaps, but for someone of his character this is only a relative description. In addition to travelling extensively with his wife, both for pleasure and for award ceremonies, he worked with the U.S. National Aeronautics and Space Administration (NASA) on the early missions to Mars and published a paper with Carl Sagan on fluid transport on Earth and the red planet (Sagan and Bagnold 1975).

He also returned to some of his early work on desert sands that had required the sieving of vast numbers of samples in order to determine and analyze the size distribution of grains in each sample. His methodology of graphical display was innovative, using logarithmic scales (discussed further below), and he discovered that the distributions took the form of a hyperbola. "As this was contrary to traditional ideas about random distributions, very little notice was taken of the discovery." Notice *was* taken, some 40 years later, by Ole Barndorff-Nielsen, Professor of Theoretical Statistics at Aarhus University in Denmark, whose work had revealed hyperbolic distributions in a wide variety of fields. He invited Bagnold to visit, and thus began another productive collaboration. Together, they published on the pattern of natural size distributions (Bagnold and Barndorff-Nielsen 1980) and Bagnold continued the theme with a paper in the *Proceedings of the Royal Society* in 1983 in which he noted the hyperbolic distribution emerging in the frequency of words of a given number of letters in a variety of languages from English to Sanskrit (Bagnold 1983). Barndorff-Nielsen et al. (1983) published an internal research report titled "The Fascination of Sand" which was in many ways a tribute to Bagnold's work and included several previously unpublished pieces by him. In one of these he described the "sand organ" demonstrated to him in his boyhood by his inspirational neighbour, Charles Phillips.

Ralph Bagnold died in 1990. His final scientific paper was published by the Royal Society in 1986: he was 90 years old. Two further quotations from the final chapter of his autobiography perhaps best summarise the character of this extraordinary man:

> My main urge, from boyhood onward, was curiosity. This was often strong enough to stimulate active steps to satisfy it. Along with this, there has been a vague longing to discover something new, something previously unknown.
>
> Being an amateur, a free lance who never held any academic post or had any professional status, I had the rather unusual advantage of considering problems with an open mind, unbiased by traditional textbook ideas that had remained untested against facts. I had the further advantage that the sort of problems that interested me did not involve expensive or elaborate apparatus. I could design and make what I needed.

## 3 The Work

The Thanksgiving Service for the life of Brigadier Bagnold was held at the Church of St Martin in the Fields in London's Trafalgar Square in January 1991. His military career was celebrated through an address by Major-General David Lloyd Owen, and his scientific accomplishments by Bagnold's long-time friend and colleague, Luna Leopold. Leopold began his address by commenting that "To try to say something about the scientific accomplishments of a true genius is itself a difficult job and especially difficult if one wants to be brief about it, because there is so much to be said."

That challenge applies equally here. The following does not purport to be a comprehensive summary of Bagnold's scientific work, but rather an attempt to place it in context and review the hallmarks of his methodology and results.

## *3.1 Historical Context*

When Bagnold was born, in the last decade of the nineteenth century, the principles of sedimentology had been firmly established through the classic work of Henry Clifton Sorby, who continued to contribute until his death in 1908. Yet the understanding of desert processes and dune formation was limited. As Goudie (1999) describes in his review of the history of desert dune studies, work on deserts was influenced by the geography of the imperial times: the vast majority of the Sahara was essentially French territory, and French work dominated the literature. An important contribution was made in 1890 by Rolland, a mining engineer, whose key conclusions are summarised by Goudie. Among these was the statement that the accumulations of sand and dunes owe their origin entirely to the wind. "His need to assert that dunes were due entirely to the work of the wind is interesting in that some geologists thought they had a rock core while others invented somewhat implausible mechanisms involving ground vibrations by earthquakes…" (Goudie 1999).

In the first decades of the twentieth century, notable contributions to the descriptions of dunes and the character of sand continued by French workers, notably Chudeau in the southern Sahara (for example, Chudeau 1909, 1920). Over the same period the British, although essentially constrained by geopolitical boundaries to the Egyptian deserts, also published significant studies. The work of three men stands out: Cornish, Beadnell, and Harding King. Vaughan Cornish, although strictly an amateur, was an outstanding geographer; he was fascinated by pattern formation in the natural world, in particular wave forms (the study of which he defined as "kumatology"). His attention in this context was naturally drawn to ripples and dunes, and he published a number of important papers reporting his measurements of wave forms in sand on all scales, dune spacing and height, and the dynamics of the processes of their formation (for example, Cornish 1897, 1900, 1914). When, in 1935, Ralph Bagnold presented his paper on "The Movement of Desert Sand" to the Royal Geographical Society, Cornish was in the audience and contributed to the discussion that followed (Bagnold 1935b):

> Major Bagnold's ingenuity and industry in the conduct of experiments suggested by his observations in the field deserve our highest admiration. It is clear that the bouncing of the sand-grains, the amount of which Major Bagnold has measured, is largely responsible for the ripples which are sometimes produced when no dunes are formed and which always pattern the surface of sand-dunes. This is a definite and important step in advance which I am sure that further observations in the field will not invalidate.
>
> I admire enormously the work of Major Bagnold and I am sure when he returns to the desert he will find out a great deal more; but I am quite certain that his mechanics of

> the bounce will last and that they are a definite contribution. I congratulate him on the vividness with which he has described these collisions because when I read his paper an enormous number of episodes of the football field recurred to my mind from the days long ago when one was still a hard body capable of moving at a high speed.

Considerable debate surrounded the questions of the origins of ripples versus dunes—Cornish argued that the latter were simply overgrown versions of the former and that eddies were responsible for both—and the role of wind directions in determining dune morphology and movement. The geologist Hugh John Llewellyn Beadnell, who conducted far-ranging surveys in the eastern Sahara and Sinai over the course of a long and productive life, disputed Cornish's interpretation and added significant observations and data on wind and dune behaviour (Beadnell 1910). Another of this generation of indefatigable scientists and explorers, W.J. Harding King, also disagreed with Cornish—a lively interchange between the two men is recorded following King's (1918) paper at the Royal Geographical Society in 1918, "Study of a Dune Belt." King carefully measured and documented dune profiles, correlated them with wind directions and velocity, and, importantly, conducted simple field experiments. All of his evidence argued against Cornish's wave and eddy hypotheses. In "The Nature and Formation of Sand Ripples and Dunes," King (1916) describes two field experiments, the creation and dynamics of ripples in a wooden trough oriented into the wind during a sandstorm, and the patterns formed in smoke blowing over an obstacle. As Goudie (1999) remarks, the latter "was possibly the earliest example of the use of flow visualisation techniques."

Harding King's work, and in particular his simple experimental approach, was a significant influence on Bagnold as he contemplated addressing the physics of sand movement. He writes in his introduction to *The Physics of Blown Sand and Desert Dunes* (Bagnold 1941):

> After much desert travel, extending over many years, during which sand storms of varying intensity were frequently encountered, I became convinced that the movement of sand (as opposed to that of dust) is a purely surface effect, taking place only within a metre of the ground; and that large-scale eddy currents within the air stream play no appreciable part in maintaining the grains aloft. If this is true it is clearly possible to reproduce in the laboratory the complete phenomena of sand-driving, together with all the small-scale surface effects, such as ripples and the like, associated with it. The possibility of such an investigation under controlled conditions was first suggested to me by the experiments of Harding King with a wooden trough laid on the sand in the open. Here, though of course the wind strength was not under control, it was found possible to produce various ripple forms by placing sands of different size-grading over the floor of the trough.

At the time of Bagnold's epic journeys through the eastern Sahara, the body of literature on deserts and dunes was significant, but almost entirely descriptive in nature. Bagnold's intimate encounters with sand on all scales, from individual grains to sand seas, led to his resolve to address the processes involved in sand transport and accumulation through the combined approach of engineering, physics, and mathematics, tested through careful laboratory experiments. This work resulted in numerous papers and a book that continues to be routinely cited today and has been variously described, to quote but a few examples, as "monumental" (Warren 1986), "a classic in every sense of the word" (Greeley and Iversen 1985), and "a landmark" (Tsoar 1994). Luna Leopold, who read the book on a train between Chicago and San Francisco, described it as "one of the most exciting things I ever read in science." As Phillip Ball (2009) wrote recently in his retrospective review for *Nature*, "Bagnold [wrote] with such perception and insight that his 1941 book *The Physics of Blown Sand and Desert Dunes* became the standard work on dune formation for many decades."

## 3.2 Wind

The introduction to the original edition of the book (all quotations in this section are from *The Physics of Blown Sand and Desert Dunes* unless otherwise noted) includes a statement of the background and motivation:

> This book results from an attempt to explain on a basis of experimental physics some of the many strange phenomena produced by the natural movement of sand over the dry land of the Earth.
>
> The subject is but one aspect of a far wider problem which is still very imperfectly grasped – the transport of solid particles by fluids in general. Here the difficulty has been, and still is, that no one branch of science has attempted to deal with the problem as a whole, or to co-ordinate the vast amount of piecemeal work by students of different outlook in many unrelated fields.
>
> In the industrial world the sorting and grading of crushed material by fluid elutriation is yearly becoming of greater importance, and air-blast is used extensively for the carriage of grain, coal dust, &c. But here again our knowledge of the fundamental relations between particle size and movement on the one hand and the fluid velocity on the other is still very scanty and rests on no theoretical basis.
>
> Stimulated by the spectacular and intriguing effects of wind on dry sand in the desert – sandstorms, ripples, dunes, and sand seas – many travellers and geomorphologists have attempted to describe and classify these effects. But till recently no effort has been made to investigate the process of wind action on sand as a problem of aerodynamics, amenable to direct and precise measurement. In fact it is only within the last few years that a tardy realization of the great economic losses caused by wind erosion has directed any serious attention to the movement of silt and sand by wind.
>
> Nor can the geomorphologist rest content, in his study for its own sake of the shape and movement of sand accumulations, until he knows *why* sand collects into dunes at all, instead of scattering evenly over the land as do fine grains of dust, and *how* the dunes assume and maintain their own especial shapes.

Bagnold was the first to attack the process of windblown sand transport as a problem of fluid dynamics, quantifiable through simple basic physics, and testable in the laboratory. This approach laid the groundwork for the subsequent development and refinement of the fundamental principles, continuing today with the aid of computing power and modelling capabilities. Bagnold's work took place contemporaneously with significant progress in fluid mechanics and aerodynamics, notably through the work of Ludwig Prandtl, which Bagnold draws attention to in the book.

Desert processes take place over a vast range of scales, and are the conspiracy of a large number of variables, few of which act independently. Interaction and feedback are typical: wind entrains sand grains that, in turn, affect the wind; the morphology and roughness of the desert floor is a result of, and an influence on, sand transport and wind velocity; and Bagnold's observation of "a simplicity of form, an exactitude of repetition and a geometric order" heralded today's interpretation of aeolian dune-field patterns as symptomatic of self-organising phenomena. To attempt to dismantle this complexity, Bagnold took an essentially reductionist approach, starting with the behaviour of single grains. He described himself as having the advantage of being an "amateur" with an "uncluttered" mind, and these attributes allowed him to build his innovative mathematical analysis from scratch; in doing so, however, he found it necessary to develop his own nomenclature and notations (as he did for all his subsequent work). While he recognised "the need to make the language reasonably intelligible to specialists in a large number of widely different fields of knowledge—geophysics, meteorology, aerodynamics, hydraulics, geology, geomorphology, &c.," this requires effort on the part of the reader; his qualitative explanation of the concepts is, nevertheless, lucid and accessible.

The mathematical and fluid dynamical treatment is but one stage of Bagnold's approach. Having set out an idea, a behavioural prediction from the basic physics, he then defines an experimental approach to test this. Luna Leopold said that Bagnold's "ability to think about how experiments could be devised to test some particular idea" distinguished him from any other scientist with whom Leopold had ever worked. Many of Bagnold's experimental tests have all the hallmarks of arising from "thought experiments" based on the predictions of fluid dynamics. For example, Bagnold calculated the theoretical results of the ideal elastic impacts between sand grains and a surface pebble, and then envisioned this as a situation of a stationary observer, still air, and moving ground. "The moving ground under still air can by this artifice be represented experimentally by the edge of a horizontally rotating disc, and pebbles and other objects gummed to its surface will imitate various types of ground." Having constructed the apparatus, Bagnold proceeded to drop grains on to the rotating disc and capture them on strips of sticky paper hung at different distances from the point of impact; alternatively, they were allowed to fall to the ground and collect on oiled glass.

But Bagnold's brilliance lay not simply in devising the experiments, but in building the sophisticated apparatus himself, together with innovative measurement equipment of unprecedented sensitivity. These skills were clearly a legacy of his boyhood days spent in his father's workshop and with his own lathe. The wind tunnel that he built at Imperial College is a characteristic example, enabling him, uniquely, to make direct measurements of bed shear stress.

The experimental results he then compared to his theoretical analysis and the process, as appropriate, iterated. The third stage would then be comparison of the theoretical and experimental results with field measurements:

> The application of the laboratory results to real conditions in the open would of course depend on the truth of the above underlying assumption. But if, subsequently, a series of

quantitative predictions concerning the wind velocity and its distribution, and the rate of sand movement caused by it, based on the laboratory work, were to be corroborated by careful measurements during sand storms in the desert on a far larger scale, then there would be very strong grounds indeed for regarding both the assumption and the results based on it as being true. Such corroboration has already been obtained.

Field corroboration was the goal of Bagnold's 1938 expedition to the Gilf Kebir of south-western Egypt. Using a home-made (and home-built) manometer and sand traps, the designs of which are described in detail in the book, he collected data on wind velocity variations, velocity gradients, and sand flux (Bagnold 1938). As he reported:

Although the season was so unusually calm that the number of real sandstorms available for measurement was not as great as I could have wished, the figures obtained do corroborate very satisfactorily the predictions based on the wind tunnel experiments. The latter may therefore be accepted with some confidence as providing a foundation on which further work can be built.

*The Physics of Blown Sand and Desert Dunes* is a compact work of 250 pages, and is divided into three parts that more or less follow an increasing scale of the processes described. Part 1, "The Physics of Grain Movement," describes the basis of the fluid dynamics approach, the quantification of grain behaviour and interactions, the experimental work, and the field measurements. This is the core and the truly pioneering content of the book. Part II, "Small-scale Effects. Grain Size Distribution. Surface Ripples and Ridges," addresses the grading and size distributions of natural desert sands and the formation of surface bedforms. Finally, Part III, "Large-scale Effects. Sand Accumulation. Dunes. Internal Structure, etc.," applies the mechanisms described earlier to the major landforms of the desert.

Bagnold begins the book by developing a working definition of sand size appropriate for his fluid dynamics approach. Wentworth (1922) had set out the basis of the size scale that would be later modified by Krumbein (1934), who developed the widely used system of categories based on $\log_2$. These were, however, purely physical definitions, and Bagnold needed something that would address grain behaviour. His lower limit for a material defined as sand, without reference to shape or composition, would be "that at which the terminal velocity of fall becomes less than the upward eddy currents within the average surface wind." Particles of a smaller size (silt and dust) can be carried indefinitely by suspension. The upper limit of sand size is "that at which a grain resting on the surface ceases to be moveable either by the direct pressure of the wind or by the impact of other moving grains." He recognised the shape complexity of natural sand grains on the definition of their size by means of conversion to an "equivalent" quartz sphere through a factor that, for average windblown sand, is 0.75. He thus set out a framework of size definition and behaviour within which he could develop his quantitative approach. Bagnold stressed, essentially throughout his work, the power of plotting data such as grain sizes on logarithmic scales, preferring to use the base ten; he illustrated this with his first figure, a graph of grain size versus terminal velocity. As he wrote:

> The linear scale, since it was first cut on the wall of an Egyptian temple, has come to be accepted by man almost as if it were the one unique scale with which Nature works and builds. Whereas it is nothing of the sort. Its sole value lies in giving due prominence to the differences and sums of quantities, when these are what we want to display. But Nature, if she has any preference, probably takes more interest in the ratios between quantities; she is rarely concerned with size for the sake of size...This point is stressed, because throughout the subject with which we are dealing the relations between one quantity and another are very often to be found of the logarithmic type.

Beginning with the observation that fluid resistance and gravity are the two forces acting on a single particle, and that the ratio of these two forces is fundamental in terms of grain velocity, acceleration, and trajectory, he defined this ratio as the "susceptibility," which, for grains of the same shape, varies with size and velocity. While this term was never taken up in common usage, it provided the basis for the analysis of windblown grain paths. On a double logarithmic plot of the velocity of a sphere through the air versus susceptibility, each grain size shows an elegant, essentially linear, relationship. For the range of grain sizes typical of windblown sand (0.3–1.0 mm), and for relative speeds of a few metres per second, the susceptibility varies between 1 (terminal velocity, where drag equals gravity) and 10; fluid resistance therefore generally exceeds the particle weight, although not dramatically, and is a fundamental influence on particle trajectories.

That those trajectories in a body of moving sand are dominated by the process of saltation, and that this process can be described mathematically and measured both in the laboratory and in the field, was one of Bagnold's fundamental and radical contributions. Saltation, the movement of sand grains through leaping motions (from the Latin, *saltare*, to jump), had been observed in transport by water, and Bagnold ascribes the term to the epic work of A.K. Gilbert (1914), *The Transport of Debris by Running Water*; Gilbert, however, refers the origin of the "expressive" term to McGee (1908).

In his desert journeys, Bagnold observed that, during a sandstorm, "The bulk of the grains flowed as a dense fog, rising no higher than five feet from the ground. Over it we could see each other quite clearly, head and shoulders only, as in a swimming-bath" (Bagnold 1990). Through meticulous measurement and photography in the wind tunnel, and through his mathematical analysis, Bagnold described the characteristic saltating trajectories of grains in this "dense fog." Saltation in air is very distinct from that in water, given that the density ratio for a quartz sand grain in air is 1/2,000, in water 1/1.65. Furthermore, the viscous retardation of grain movement in water prevents grains from attaining significant heights once they have begun to rise from the bed surface, whereas in air they can create the "dense fog" and travel significant distances downwind. In addition, the relative inertia of a sand grain in air is very large and the effects of turbulence therefore negligible on the grain's trajectory. At the end of each grain's trajectory, the impact on the surface ejects further grains and the process becomes essentially self-sustaining.

The implications of the dynamics of a saltating cloud of sand grains are of fundamental importance, and Bagnold was the first to describe and quantify them. Two aspects of this work are particularly outstanding: the effect that saltation exerts on the wind and the thresholds of movement. The feedback effect of saltation on the wind velocity profile is significant; as Bagnold describes:

> In the case of sand in air the momentum extracted by each grain during one bound is equal to that given up in completely stopping a volume of air 2,000 times the volume of the grain. The saltation drag in air is therefore very large, and renders negligible the residual surface drag due to fluid friction against the stationary surface below. As a result the velocity of the air in the layer below the top of the mean grain path is entirely controlled by the intensity of the saltation of sand grains below. The greater the number of grains in saltation the greater is the drag; and experiment shows that under steady conditions a balance is maintained, such that, however large may be the velocity gradient of the wind, the velocity $V_t$ at a certain fixed height $k'$ above the surface remains constant. The velocity closer to the surface is actually reduced as the wind above it is made stronger.

For the initiation of movement, Bagnold recognised that there are two distinct thresholds of wind velocity. In the wind tunnel, at wind speeds lower than that necessary to start grain movement by shear, grain motion can be initiated by dropping grains on to the bed of sand. The grains thus ejected into the air stream continue the process of saltation as long as the supply of falling grains is sustained. The velocity at which this can occur Bagnold defined as the *impact threshold.* If the wind speed is increased, the point is reached, the *fluid threshold*, at which grain movement is initiated by shear. The effect of "saltation drag" (described above) on reducing the wind velocity close to the surface is such that "at the level of the top of the stationary surface grains, the wind velocity is less than the fluid threshold, and the wind is prevented from picking up any more grains from the surface by its own direct action." Bagnold examined the relationships of the fluid threshold velocity and velocity gradient to grain size, grain size distribution, and surface roughness, the results of which had immediate application in understanding the development of the surface bedforms of the desert.

Saltation is, of course, not the only mechanism by which sand is transported by the wind. By definition, suspension plays only a minor role. However, the movement of grains *along* the surface, whether by impact or shear, is also important, a process referred to as surface creep or, more recently, reptation. In his calculations of total sand movement, Bagnold calculated that 75% took place by saltation, 25% by creep, but that this sand flux is dependent on mean grain size, the degree of sorting or variation in size, and the velocity gradient of the wind. The equation that he derived for sand flux proved to be a classic, the basis for all subsequent work. While it has undergone further refinement and testing, it has stood the test of time and is worth reproducing here. The mass transport rate of sand, $q$, in grams per second across a centimetre width, is given by:

$$q = C\sqrt{\frac{d}{D}}\frac{\rho}{g}V_*'^3 \quad (1)$$

where $C$ is a constant (an "empirical coefficient" in Bagnold's terms), $d$ is the grain diameter of the sand in question, and $D$ that of a standard 0.25 mm sand; $V$ is the velocity gradient of the wind, specifically, and importantly, as Bagnold emphasises, that under conditions of moving sand. Sand transport is thus dependent on the square root of grain size and the cube of shear stress. Bagnold documents the value of $C$ as varying from 1.5 for a "nearly uniform sand" to 2.8 "for a sand with a very wide range of grain size." This equation, and its refinements, effectively underlies all desert processes.

The mechanisms of the *initiation* of movement of a single sand grain as the fluid threshold is reached remain the subject of some debate today. In his wind tunnel experiments, as the fluid threshold velocity was reached, Bagnold observed the initial rolling of grains beginning at the downwind end of the tunnel; the initial point of motion moved upwind as the velocity increased, and this he attributed to the effects of turbulence. In his analysis and discussion, he devotes attention to turbulence and Reynolds numbers not only in examining the distinction between transport in wind and water, but also as important considerations in understanding the formation and behaviour of ripples and dunes.

This leads into the second part of the book, which begins with a detailed discussion of the grain size distributions of desert sands, and the mechanisms of sorting and grading. The raw data for this analysis derive from Bagnold's painstaking and laborious sieving of countless samples, each set of results recorded on log–log frequency graphs. This approach was original and ground-breaking, allowing the character of an individual sand sample to be properly revealed. Bagnold writes:

> It is now for the first time apparent that outside a definite central zone the grades to the right and left of the peak fall off each at its own constant rate; this means that we are again confronted with the same logarithmic law of distribution which runs through the whole subject.

This work, together with supporting wind tunnel experimental data, allows Bagnold to define the grading of sand in desert conditions in terms of the work of suspension, saltation, and surface creep, together with the larger-scale processes operating across the profile of a dune. It would also lead to Bagnold's continuing work on the occurrence and importance of hyperbolic versus Gaussian distributions in nature (Bagnold 1983) and his collaborative work with faculty at the University of Aarhus (Bagnold and Barndorff-Nielsen 1980).

As noted above, the origins of ripples in sand (and other materials such as snow) had been the subject of considerable debate; the "wave" model of Cornish (1914) and the associated "lee eddy hypothesis" were disputed, but alternative mechanisms were elusive. While the complexities of ripples in sand remain a topic of ongoing research today (Rubin 2006), Bagnold, combining his analysis of saltation, wind velocity gradients, sand grading, surface relief, and the state of sand movement, was able to fill this gap. Saltation distances versus ripple wavelength and the influences of small-scale surface topography variations on the velocity gradient, and therefore the transporting power, of the wind, together with wind tunnel evidence, provided a working model of ripples and ridges. Key components were Bagnold's demonstrations that upstream of an obstacle, flow converges, the fluid is accelerated, and drag velocity increases, and that, if the grain size texture of a flat surface changes from larger to smaller, then the critical height $k'$ drops and the layers above are accelerated. This feedback between the nature of the surface and the capacity of the wind to transport the materials on the surface is a critical factor in the understanding of all landforms sculpted by windblown sand.

For the final section of the book, Bagnold addresses the large scale, the formation and behaviour of dunes. As a result of his extensive travels in the desert and his observations of the morphology and nature of dunes, he had asked the series of questions (see 2, above) that led to his theoretical and laboratory approach. Amongst these, he asked, "Why did they absorb nourishment and continue to grow instead of allowing the sand to spread out evenly over the desert as finer dust grains do?" This key question on the essentially self-accumulating behaviour of dunes, their distinct separation from the areas of pebble- and rock-strewn "streets" between them, he had answered through his experimental work. Sand transport, saltation, over a sand-covered surface is very different from that over pebbles, the distinction being "bouncing versus splashing." Over a pebble surface, the elastic impacts of sand grains are far more exaggerated and the rate of sand flux much greater than over a surface of soft sand. On encountering the surface of a dune, however "embryonic," the capacity of the wind and the volume of sand transported are reduced and net deposition occurs. Sand is effectively removed from the pebble surfaces and deposited on the sand; thus the dune accumulates. "A pebble surface can thus be regarded as a reservoir in which sand is stored during periods of gentle wind, and from which it is removed by a sudden storm." He demonstrated experimentally that "a strong wind causes an accretion of sand on an existing sand patch, together with an extension up-wind of the border; but this action lasts only as long as there is a plentiful supply of sand stored on a pebbly up-wind surface." It should be noted that this mechanism seems to have been qualitatively recognised some time before Bagnold's work. Infrastructure such as roads and railways threatened by dune encroachment was sometimes protected by spreading pebbles and rocks over the dune; strong winds would winnow the sand from between the pebbles in the same way that a "desert pavement" is formed naturally by the process of deflation, and any sand blowing over the pebbles will, through the effects of "bouncing," potentially be blown across the threatened infrastructure.

Bagnold went on to explore the nature of the boundary between erosion and deposition, the effects of grain size, and the influence of wind strength, duration, and direction, emphasising that total sand flow is proportional to the cube of the wind velocity and that therefore "a single day's storm may move more sand than several weeks or months of calmer weather."

Bagnold faced an inherent problem of scale in his treatment of dunes: the impossibility of reproducing them in the laboratory. Furthermore, the boundary conditions in a wind tunnel mean that "this method of investigation would break down altogether; for in order to avoid interference by the tunnel walls with the complicated air flow over and around the now curved sand surface, the tunnel would have to be made very large indeed." He was thus "forced to fall back largely on description and inference." Nevertheless, Bagnold's "description and inference" were meticulous, and contributed fundamentally to the understanding of dune and dune field processes; this work would only be significantly augmented by the advent of computer modelling approaches several decades later. While a large-scale quantitative approach was not possible, nevertheless Bagnold could apply his fluid dynamics work, wind tunnel results, and field data to a number of problems relating to dune formation, architecture (external and internal), instability, and movement. He discussed, quantitatively wherever possible, the relationships between dune shape and the rate of sand flow over it, the structure and rate of movement of barchan dunes, and the complexities of longitudinal or seif dunes and their foundations of "whalebacks." He set in place the dependence of dune forms and movement on wind direction and its variation, and suggested ways in which one form might evolve into another.

The book concludes with a chapter on "singing sands," a phenomenon that intrigued Bagnold after experiences of both the natural spontaneous sounds periodically emitted by dunes, and by his own success in artificially stimulating them. While recognising that the wide variety of documented sounds emitted by dunes originates from avalanching, a rigorous explanation eluded Bagnold; the phenomenon remains the source of vigorous scientific debate today.

*The Physics of Blown Sand and Desert Dunes* was published in 1941 with a very limited print run. It was republished in 1954 and remains in print. In the preface to the new edition, Bagnold commented that "time has not altered or appreciably extended the experimental facts concerning the movement of and by wind." And then:

> As regards the second half of the existing book which deals with full-scale dunes and other sand forms, few if any new facts have emerged. Further progress here must, I think, be in the nature of surmise only, until local wind, and possibly temperature, records taken over fairly long periods are available for the very near vicinity of dunes of various types, and until our knowledge of the meteorology of deserts is more thorough than at present. The sandy regions of the world are economically unimportant; and were it not for the recent oil developments in the Middle Eastern deserts and for the nuisance caused to these vast operations by blown sand and dunes there would be little hope of the required data being collected. Under present conditions, however, we may look forward with some confidence to a start being made on sound and well-organized lines.

By the time Bagnold had completed his field-based work in the expedition of 1938, the lack of long-term wind data that would enable a more rigorous study of dune processes led him to conclude that there was little more work to be done until such data became available; this frustration is apparent in the final chapters of the book. However, he was not deterred, and turned his analytical mind to problems of sediment transport by a different fluid, water.

### 3.3 Water

As he had noted in his introduction to *The Physics of Blown Sand and Desert Dunes*, Bagnold's fascination was with a "far wider problem which is still very imperfectly grasped—the transport of solid particles of any kind by fluids in general." He went on to say:

> The carriage of silt by rivers has received a great deal of attention from engineers. But owing to the difficulties of direct measurement, to the expense and labour of conducting full-scale experiments, and to a failure to find agreement as to the basic quantities on which a theoretical edifice may be built, the published results are far from satisfactory. Little has emerged except empirical formulae; and these are rarely capable of reliable application to conditions other than those under which they were evolved.

Having contributed everything he could to the understanding of sediment transport by wind, Bagnold set out to rectify this "far from satisfactory" situation with respect to water. He had already alluded to the dramatic differences in the fluid dynamics of water versus air, and the consequences for sediment transport; saltation as a descriptive term for grain trajectories may have originated from observation in water, but the details are profoundly different between the two different fluids. As he noted, fluid velocity reduction by saltation in air is a dominant mechanism, whereas in water the opposite is true: the effect of grains in motion on the fluid is negligible. Furthermore, while grain motion in a sandstorm can be directly observed and measured, the same is not true for a river in flood.

Bagnold's first publication (Bagnold 1939) in water research was an "attempt to throw light on the nature of the shock pressures exerted on the face of a vertical sea-wall when a wave breaks against it… The problem is made more difficult by the entire absence of any theory or mathematical background for the mechanism of the breaking wave." He approached this complex problem with exactly the same combination of laboratory (wave tank) experiment and basic physics that he had used for his work on wind. This was followed by results from the wave tank on waves and beach formation (Bagnold 1940). This new line of enquiry was interrupted by the war and Bagnold's innovative contributions to the North African campaign (Bagnold 1945), but his knowledge of beach dynamics would be sought in the preparation for the Normandy landings.

After the war, Bagnold commenced a characteristically analytical approach to sediment transport by waves. In his autobiography (Bagnold 1990), he writes:

> On the strictly practical side, engineers had long striven to build coastal defenses against the sea, while on the theoretical side, mathematicians had devised satisfactory models of wave motions under ideal conditions. But these excluded the reality of waves dissipating their energy against a beach. The real physical processes whereby shallow-water waves transport sand and shingle from place to place were still unexplored and considered

> beyond the scope of reliable theory. I found it possible to discover a lot about the general physical processes involved by undertaking well-designed experiments. Some years later, working with Douglas Inman at the Scripps Institution of Oceanography in La Jolla, California, I was able to follow up the laboratory experiments by directly seeing the effects of wave motions on the sea bed. Using a snorkel, I could lie close to the bottom and drift to and fro under the long Pacific swells, watching plumes of sand being tossed upward by water vortices created at the end of each wave oscillation.

His work was documented in a series of publications in which he developed the first physical models for the forces associated with breaking waves, beach profiles, and littoral sediment transport (for example, Bagnold 1947, 1955). But he recognised that to understand and quantify grain behaviour in water, an approach was needed that began with the basics. In his autobiography (Bagnold 1990) he writes, in a typically provocative style, that "The engineer's working knowledge of water flow rests largely on physical principles evolved in the eighteenth century." Inherent in the conventional wisdom was the assumption that any given fluid is the same density everywhere, and that therefore density differences could be ignored and a purely kinematic approach used. "This kinematic way of thinking was so useful, and had become so general, that by the start of the twentieth century it had become confused with reality…This assumption being strictly tacit, and probably often unconscious, it was some time before I realized its existence and implications…It became clear to me that I would have to start afresh, returning to the reality of dynamics."

Using as an analogy his experience as a snooker player, Bagnold "expected that the random collisions that must occur when a dispersion of solids is sheared would amount to a dispersive pressure. This pressure would be exerted on a boundary which would resist further dispersion." This insight was radical and powerful. Typically, Bagnold devised and "rigged up" the experimental apparatus necessary to measure directly, for the first time, this shear stress and demonstrated that the sediment transport rate depended on the product of stream velocity and bed shear stress. "This, I felt, began to put the subject of the transport of solids by water on a sound footing." Furthermore, it can be argued that the groundbreaking understanding of grain interactions and dispersive pressure would be the foundation for research into the physics of granular materials today. Bagnold liked the analogy of the impossibility of mixing flour and grains of whole wheat in bread-making, precisely the phenomenon of the still-challenging behaviour demonstrated in what has become known as "The Brazil Nut Effect" (Jaeger and Nagel 1997).

In 1956 he published a summary of grain transport by wind and water. In his address at the memorial service for Bagnold, Luna Leopold described "The Flow of Cohesionless Grains in Fluids" (Bagnold 1956) as "probably the most important paper he ever wrote." The paper, published in *Philosophical Transactions of the Royal Society of London*, is more than sixty pages in length, is complex, and introduces Bagnold's own nomenclature; Leopold reported that he had spent an entire year reading it. In his conclusion to the paper, Bagnold writes simply that "An attempt has been made in the foregoing paper to correlate for the first time a number of widely scattered phenomena connected with grain flow. The results appear to be mutually consistent and to afford simple explanations of much that has hitherto been somewhat mysterious."

Bagnold's association with Leopold formed an important phase of his life and work. Having addressed coastal processes and density currents (Bagnold 1962), he turned his attention to fluvial processes in which Leopold was a leading researcher. Their collaboration to "stir the pool of complacent tradition with the stick of reality" led to a radical new approach to bed load transport through theory and direct measurement in the field:

> I spent much time visiting hydraulic laboratories in various parts of the United States, but I could find no work going on other than ad hoc experiments in aid of particular commercial projects. I could find no institutions interested in studying sediment transport as a process of nature, or even in trying to discover the reason why existing experimental results remained grossly inconsistent. Discussions with Luna led to his decision that the U.S. Geological Survey would have to embark on its own experiments, designed on scientific rather than engineering lines. I outlined the series of experiments I had in mind.

A new hydraulics laboratory was constructed by the two men, and, once again, Bagnold's genius for devising and building apparatus was key, together with his insights into the mathematics of fluid dynamics. Two publications from this period, both of them U.S. Geological Survey Professional Papers, remain classics: "Flow Resistance in Irregular or Sinuous Channels" (Leopold et al. 1960), and "An Approach to the Sediment Transport Problem from General Physics" (Bagnold 1966). Bagnold regarded a river as an engine and transport rate as a function of the power extended by that engine. The power of this insight itself continues to be applied today.

Bagnold returned to publishing in the *Proceedings of the Royal Society* with "An empirical correlation of bed load transport rates in flumes and natural rivers" (Bagnold 1980) and "Transport of solids by natural water flow: Evidence for a worldwide correlation" (Bagnold 1986). The second would be his last formal publication; it was written at the age of ninety, and in the typescript (Churchill Archives Centre, Cambridge, Bagnold papers, by kind permission of Stephen Bagnold) his original wording was "Evidence for an unsuspected natural law?"

## *3.4 The "sage"*

Bagnold remarked in his autobiography that, in spite of moving on from his work on aeolian processes, "Nevertheless, I found myself in some demand as a sage on the subject." These were typically modest words that ignore the recognition he received during his lifetime for his sagacity on many subjects. He was "astonished" to find that he had been elected a Fellow of the Royal Society in 1944. International recognition followed: he was awarded the G.K. Warren Prize by the U.S. Academy of Sciences in 1969, the Geological Society of America's Penrose Medal in 1970, the Geological Society of London's Wollaston Medal in 1971, and the Sorby Medal from the International Association of Sedimentologists in 1978.

In addition, he received several awards from the Institution of Civil Engineers and two honorary doctoral degrees. In his acceptance of the prestigious Sorby Medal (presented by Luna Leopold), Bagnold said "I am deeply appreciative, the more so because I have never been a professional scientist, but merely an amateur" (Leopold 1979). In 1988, the American Society of Civil Engineers published "The Physics of Sediment Transport by Wind and Water: A Collection of Hallmark Papers by R.A. Bagnold" (Thorne et al. 1988).

As a man with a deep understanding of desert processes, and as an engineer, he was often called out to evaluate threats of sand encroachment on desert facilities, most commonly the oil and gas industry installations that had developed into major desert infrastructure during his lifetime. In 1963, for example, British Petroleum sought his "specialised experience" in advising on the construction of a pipeline across Libya's Calensio Sand Sea. On the BP letter (Churchill Archives Centre, Cambridge, Bagnold papers, by kind permission of Stephen Bagnold), Bagnold wrote "see over" and typed this note:

> This letter from British Petroleum relates to the initial development of the Calansho [*sic*] oilfield in Libya, which has subsequently made the growth of modern Libya possible. The oil lies under a great sea of sanddunes whose existence was discovered by Pat Clayton in the course of the first LRDG sortee [*sic*] into enemy territory in September 1940. Little did we dream then of what might lie underneath.

In 1970 he was summoned to Iran to evaluate oil installations at Ahwaz at the head of the Persian Gulf in temperatures of fifty degrees centigrade. His report (Churchill Archives Centre) contains several pages setting out the basics of sand transport physics, together with carefully calculated advice on the design of protective sand fences, concluding with the comment that:

> A light fence, as described, has the great advantage over a solid engineering structure that it can be quickly raised or repaired from an on-site stock of materials, whereas an engineering structure may require a 'project' to repair it.

Essentially Bagnold never stopped. Late in his life he embarked on work with Barndorff-Nielsen at Aarhus University on the statistics of natural distributions, returning to his early work on the hyperbolic nature of sand grain size distributions (Bagnold and Barndorff-Nielsen 1980, Bagnold 1983). His last public appearance was at a meeting at Queen Mary College, London, in 1988. At the age of 92 years, his enthusiasm and analytical mind remained unchanged. He discussed (Colin Thorne, personal communication) the continuing limitations of instrumentation and measurement, emphasising the need to measure the velocity of actual solids in a fluid as opposed to that of the whole "combined mix-up," which he referred to as a "conflux." He was still focussed on designing experiments, suggesting that one grain in a hundred could be coated so as to be visible in ultraviolet light and then "one could do an awful lot with multi-image photography." Researchers at the University of Chicago's Materials Research Science and Engineering Center who recently published extraordinary results from the study of streams of grains imaged by a state-of-the-art camera (Royer et al. 2009) would only agree.

## 4 The Legacy

After the publication of *The Physics of Blown Sand and Desert Dunes*, it would be more than four decades before another text on aeolian processes would appear. *Wind as a geological process on Earth, Mars, Venus and Titan* (Greeley and Iversen 1985) was published at the time when planetary exploration was adding dramatic new views of the solar system, and an understanding of aeolian processes on Mars was the ultimate purpose of the book. The revolution in remote sensing imagery of Earth and the planning for, and results of, the Mars missions provided a significant stimulus to aeolian research, and, inevitably, Bagnold played a role in this work. Together with Carl Sagan, he had published *Fluid transport on Earth and aeolian transport on Mars* (Sagan and Bagnold 1975) and his views had been sought by NASA on Mars imagery interpretation and lander preparations. As he wrote in his autobiography (Bagnold 1990):

> I spent one evening at a McDonald's with a small group of young scientists from NASA's Jet Propulsion Laboratory in Pasadena. It was fascinating for an old man of eighty-one to listen to their casual talk of navigating a spacecraft two hundred million miles away as easily as an aeroplane. Man had not begun to fly at all when I was born.

In 1978, the Geological Survey of Egypt mounted an expedition to the region of the Gilf Kebir following that of Bagnold 40 years previously. The associated research was diverse, but one of the primary objectives was the field-checking of the geological interpretations of astronaut and satellite images, and the comparison of Mars imagery with desert processes and landforms on Earth. Bagnold wrote the foreword to the report of the expedition's results (Issawy 1981).

In the preface to their book, Greeley and Iversen refer to "Bagnold's classic book" as "a classic in every sense of the word." They comment that "The fact that nearly every subsequent paper dealing with aeolian processes refers to the Bagnold book bears testimony that the basic principles described by him are essentially correct and have withstood the test of time." This remains as true today as it did a quarter of a century ago. In 2005, Doyle and Julian published their analysis of the most-cited works in the journal *Geomorphology* over the period 1995–2004. They recorded 31,696 unique works cited in the entire field of geomorphology, of which 92% were referenced only once or twice (Doyle and Julian 2005). *The Physics of Blown Sand and Desert Dunes* was cited 42 times, ranking second in the total list of both volumes and individual papers. It is of interest to note that the work that appeared in third position was the book by Luna Leopold et al. (1964), *Fluvial processes in geomorphology*. Doyle and Julian also report that ISI Citation Index (Web of Science) records 1,242 citations of Bagnold's book over the same period. These figures are remarkable and illustrate the profound influence that the work had, and continues to have, on aeolian process research. In their analysis of the bibliography of specifically aeolian research, Stout et al. (2009) comment that, in spite of being one of only 358 aeolian-research publications to appear during the war years, the book "has proven to be one of the most influential publications in the history of aeolian research."

Five years after Greeley and Iversen's book, the prominent workers in the field of desert process, Kenneth Pye and Haim Tsoar, published *Aeolian sand and sand dunes* (Pye and Tsoar 1990). In the book, they describe how Bagnold "provided an important theoretical basis which has influenced all subsequent studies of aeolian sand transport and dune formation." Bagnold's book "continues to be essential reading for any serious student of aeolian processes," but they refer to an "explosion in the scale of research" which continues today: Bagnold's work is the foundation for modern research that, employing the power of radical technology undreamed of 70 years ago, makes dramatic progress in our understanding of sand transport and how the engine of the desert works. The following discussion does not attempt to be a comprehensive review of the "state of the art" of current research, but simply a sampling to point to some of the significant paths that the work is taking.

The range of scale involved in aeolian processes and systems continues to be a substantial challenge, and the problems of accurate small-scale measurement that faced Bagnold have yet to be fully resolved today. In a wide-ranging and valuable review of the challenges of scale, process, and temporal integration in understanding aeolian system dynamics, Thomas and Wiggs (2008) describe "research routes post-Bagnold." They discuss the "two seemingly divergent routes" of small-scale analysis of wind shear and sediment entrainment versus work on the landscape and atmospheric scales, "with little linkage between them." Bagnold moved on from his work on windblown sand because of the lack of long-term wind data; that deficiency has long since been rectified, allowing a concerted sedimentological approach to desert bedforms on all scales that, as Thomas and Wiggs suggest, offers the possibility of an integrated approach linking the reductionist methodology introduced by Bagnold with analysis on the landscape scale.

On the smallest scale, the dynamics of saltation remain not fully understood, in large part because of the problems of accurately measuring shear stress close to the boundary, particularly as it fluctuates with time, and in part reflecting the complexity of sand flux in the field (see, for example, the summary in Bullard 2005). The importance of low-energy grain impacts and the consequent movement by the process of reptation was first emphasised in the 1980s (for example, Anderson and Hallet 1986); this process is also key to unravelling the mysteries of ripple formation and morphology, work that continues today (Rubin 2006; Yizhaq 2008). The challenges of accurately measuring saltation and reptation in the field are amplified by rapid temporal and directional changes in sand flux, "bursts" or flurries of sand across the surface; progress has been made (see, for example, Stout 2004 and the useful research summaries of Bullard 2005, 2006). Entrainment of sand is more complex if the effects of variations in humidity and temperature on threshold shear velocity are taken into account (work originated by McKenna Neuman 2003; Cornelis and Gabriels 2003).

The advent of significant computer modelling capabilities, particularly in computational fluid dynamics, permitted a powerful integrated approach to bridging the scale gap between sand flux and dune formation. This methodology further sheds light on the role of turbulence, a factor that Bagnold had not taken into account (Parsons et al. 2004). Nevertheless, Bagnold's principles continued to form explicitly the foundation of this work. An early example of computer modelling that took the directly opposite approach from the small scale and reductionist one was the work of Werner (1995), who treated dune dynamics as a complex system subject to simple rules (originally established by Bagnold); Werner successfully simulated many aspects of dune forms by modelling "slabs" of sand moving in a specified direction across a lattice.

The scale problems of reproducing dune dynamics in a wind tunnel have been circumvented by physical modelling using glass beads in a narrow water flow channel; Groh et al. 2009 have used this approach, combined with a particle tracking method, to suggest that the dune erosion rate from creep or reptation is comparable to that from saltation, the migration speed of the model dune being an order of magnitude smaller than that of the individual grains. For reviews of research in dune processes and dynamics, the reader is referred to Wiggs 2001, Livingstone et al. 2007, and Thomas and Wiggs 2008.

On the large scale, Bagnold's observations of spontaneous pattern formation in sand seas and "families" of dunes anticipated the science of self-organising systems and its application to understanding dune field development (Ball 2009). As Kocurek (2008) summarises, "Bedform patterns are self-organized by the interactions and behaviour of the bedforms themselves. Interactions occur between bedforms, their faster migrating defects, and remotely…Patterns emergent from these interactions alone, however, do not begin to approach the diversity seen in nature. The boundary conditions within which these complex systems evolve provide for the uniqueness of patterns in nature." Boundary conditions such as wind regime, sand supply, source area character, and area all influence significantly the development of dune fields (Ewing and Kocurek 2010). Furthermore, Andreotti et al. (2009) have shown that the scale of dunes and dune patterns is dependent on the average depth of the atmospheric convective boundary layer beneath the thermal inversion cap. All of these approaches promise significant progress in understanding landscape and regional scale evolution of deserts over time. Today's focus on issues of climate change on the global scale has stimulated an examination of arid geomorphology and processes in the context of the "earth system" (Thomas and Wiggs 2008; Tooth 2008, 2009). Critical to such studies, and particularly to an accurate understanding of "desertification," is the effect of vegetation on arid processes. Bagnold explicitly recognised this issue, but it is only recently that progress has been made with work that continues to refine his anticipation of self-organising, complex, and non-linear systems (e.g., Baas 2007).

This summary has highlighted Bagnold's legacy in aeolian research, his work in coastal and fluvial processes, while no less groundbreaking, being less broadly appreciated today. However, one significant development in the study of arid lands brings these two strands together. There has been an increasing recognition, stimulated in part through the need to better understand the geological record of aridity, that fluvial systems have a profound influence on the development of arid landscapes (Tooth 2009; Field et al. 2009; Nash et al. 2007; Bullard et al. 2007). It seems ironic that the productive exploration of the common ground between two communities of geomorphologists should be taking place so long after Bagnold's publications on sediment transport by wind and water. For dunes in coastal settings, Bagnold recognised the fundamental differences between their form and behaviours and those of the desert, together with the critical role of vegetation on dune dynamics. Bagnold did not devote a great deal of attention to these topics, but the fact remains that his analytical approach to sediment transport provides the foundation for today's research into these areas critical for coastal management and desertification issues.

Ralph Bagnold had little regard for the artificial boundaries between disciplines and, as noted above, saw the audience for his work as representing a diversity of fields of endeavour. He would no doubt be delighted by the formation of the International Society for Aeolian Research and the launch of its peer-reviewed journal in 2009. He would also be intrigued by, and in agreement with, the contents of the initial editorial (Lee and Zobeck 2009). The editorial contains the following:

> Aerosols from wind erosion affect the global atmospheric energy balance, carry biota and nutrients between continents, and may impair human health and agricultural productivity. Coastal dunes serve as crucial barricades reducing storm damage to inland homes and businesses. Increased erosion and deposition of sand from the de-stabilization of vegetated dunes is a potential threat in many regions of the world. On a more basic level, nearly 70 years after the publication of Bagnold's 1941 classic *The Physics of Blown Sand and Desert Dunes*, we still do not truly understand the controls on sand dune form or even the formation of wind ripples. Much progress has been made in the modelling of sand transport by wind and in regional and global dust transport on Earth and other planets, but our ability to accurately predict such phenomena remains far from perfect. Models of wind erosion in agricultural fields pose similar challenges. There is a strong need for additional research on all aeolian topics.
>
> Aeolian studies have always been an interdisciplinary affair, with researchers coming from such fields as agronomy, biology, soil science, geology, geography, meteorology, physics, chemistry, economics, and civil, mechanical and agricultural engineering. *Aeolian Research* serves as an intellectual melting pot for the ideas of these and other disciplines to help us understand wind effects on Earth and other planets.

This "melting pot" arises from the pioneering work of Ralph Bagnold and will no doubt endeavour to continue the enthusiasm, the intellectual alacrity, and the joy of exploration of any kind that were so characteristic of him.

While levels of awareness of his legacy may vary, the fact remains that the success of any dryland or desert macro-engineering project relies on the informed application of the principles of sediment transport first set out by Ralph Bagnold.

**Acknowledgments** The writer is indebted to Stephen Bagnold for his help and permission to use quotations from his father's works and the photograph.

## References

Abulnaga BE, El-Sammany MS (2004) De-Silting Lake Nasser with slurry pipelines. In: Critical Transitions in Water and Environmental Resources Management. Proceedings of World Water and Environmental Resources Congress 2004. Proceedings of ASCE Conference, vol 138, pp 158–171. doi:10.1061/40737(2004)158

Anderson RS, Hallet B (1986) Sediment transport by wind: toward a general model. Geol Soc Am Bull 97(5):523–535
Andreotti B, Fourrière A, Ould-Kaddourm F, Murray B, Claudin P (2009) Giant aeolian dune size determined by the average depth of the atmospheric boundary layer. Nature 457:1120–1123
Baas ACW (2007) Complex systems in Aeolian geomorphology. Geomorphology 91:311–331
Bagnold RA (1931) Journeys in the Libyan Desert, 1929 and 1930. Geogr J 78(1):13–39; (6):524–533
Bagnold RA (1933) A further journey through the Libyan Desert. Geogr J 82(2):103–129; (3):211–213, 226–235
Bagnold RA (1935a) Libyan sands: travel in a dead world. Hodder and Stoughton, London, reprinted Eland Publishing Ltd, London (2010)
Bagnold RA (1935b) The movement of desert sand. Geogr J 85(4):342–365
Bagnold RA (1936) The movement of desert sand. Proc R Soc Lond A 157(892):594–620
Bagnold RA (1937) The size-grading of sand by wind. Proc R Soc Lond A 163(913):250–264
Bagnold RA (1938) The measurement of sand storms. Proc R Soc Lond A 167(929):282–290
Bagnold RA (1939) Committee on wave pressures: interim report on wave-pressure research. J Inst Civil Eng 12:201–226
Bagnold RA (1940) Beach formation by waves: some model experiments in a wave tank. J Inst Civil Eng 5237:27–53
Bagnold RA (1941) The physics of blown sand and desert dunes. Methuen, London (reprinted 1954, 1960; 2005 Dover, Mineola, NY)
Bagnold RA (1945) Early days of the Long Range Desert Group. Geogr J 105:30–46
Bagnold RA (1947) Sand movement by waves. J Inst Civil Eng 27:447–469
Bagnold RA (1955) Some flume experiments on large grains but a little denser than the transporting fluid, and their implications. Proc Inst Civil Eng 4(2):174–205
Bagnold RA (1956) The flow of cohesionless grains in fluids. Philos Trans R Soc Lond A 249(964):235–297
Bagnold RA (1962) Auto-Suspension of Transported Sediment; Turbidity Currents. Proc R Soc Lond A Math Phys Sci 265(1322):315–319
Bagnold RA (1966) An approach to the sediment transport problem from general physics. U.S. Geological Survey Professional Paper 422-I
Bagnold RA (1979) Sediment transport by wind and water. Nordic Hydrol 10:309–322
Bagnold RA (1980) An empirical correlation of bed load transport rates in flumes and natural rivers. Proc R Soc Lond A 372:453–473
Bagnold RA (1983) The nature and correlation of random distributions. Proc R Soc Lond A 388(1795):273–291
Bagnold RA (1986) Transport of solids by natural water flow: evidence for a worldwide correlation. Proc R Soc Lond A Math Phys Sci 405(1829):369–374
Bagnold RA (1990) Sand, wind, and war: memoirs of a desert explorer. University of Arizona Press, Tucson
Bagnold RA, Barndorff-Nielsen OE (1980) The pattern of natural size distributions. Sedimentology 27(2):199–207
Ball P (2009) In retrospect: the physics of sand dunes. Nature 457:1084–1085
Barndorff-Nielsen OE, Blaesild P, Jensen JL, Sørensen M (1983) The fascination of sand. Research Report 93, Department of Theoretical Statistics, University of Aarhus
Beadnell HJL (1910) The sand-dunes of the Libyan Desert: their origin, form, and rate of movement, considered in relation to the geological and meteorological conditions of the region. Geogr J 35(4):379–395
Bullard JE (2005) Arid geomorphology. Prog Phys Geogr 29(1):93–103
Bullard JE (2006) Arid geomorphology. Prog Phys Geogr 30(4):542–552
Bullard JE, Nash DJ, North CP (2007) Drylands: linking landscape processes to sedimentary environments. Geomorphology 85:1–128
Cathcart RB (2006) Sethusamudram ship channel macroproject: anti-tsunami and storm surge textile arrestors protecting Palk Bay (India and Sri Lanka). Curr Sci 1474(91):11

Chandramohan P, Jena BK, Sanil Kumar V (2001) Littoral drift sources and sinks along the Indian coast. Curr Sci 81(3):292–297
Chudeau R (1909) Notes géologiques sur la Mauritanie (Geological notes on Mauritania). La Géographie 20(1):9–24 (in French)
Chudeau R (1920) Étude sur les dunes sahariennes (Study on the Saharan dunes). Ann Geogr 29(161):334–351 (in French)
Cornelis WM, Gabriels D (2003) The effect of surface moisture on the entrainment of dune sand by wind: an evaluation of selected models. Sedimentology 50:771–790
Cornish V (1897) On the formation of sand-dunes. Geogr J 9(3):278–309
Cornish V (1900) On desert sand-dunes bordering the Nile Delta. Geogr J 15(1):1–32
Cornish V (1914) Waves of sand and snow and the eddies which make them. T Fisher Unwin, London
Doyle MW, Julian JP (2005) The most-cited works in geomorphology. Geomorphology 72:238–249
Ewing R, Kocurek G (2010) Aeolian dune-field pattern boundary conditions. Geomorphology 114:175–187
Field JP, Breshears DD, Whicker JJ (2009) Toward a more holistic perspective of soil erosion: why aeolian research needs to explicitly consider fluvial processes and interactions. Aeolian Res 1(1–2):9–17
Gilbert GK (1914) The transport of debris by running water. U.S. Geological Survey Professional Paper 86
Goudie AS (1999) The history of desert dune studies over the last 100 years. In: Goudie AS, Livingstone I, Stokes S (eds) Aeolian environments, sediments and landforms. Wiley, Chichester, pp 1–13
Greeley R, Iversen JD (1985) Wind as a geological process on Earth, Mars, Venus and Titan. Cambridge University Press, Cambridge
Groh C, Rehberg I, Kruelle CA (2009) Particle dynamics of a cartoon dune. Cornell University Library e-print arXiv:0911.0757v1
Issawy B (ed) (1981) Annals of The Geological Survey of Egypt vol XI
Jaeger HM, Nagel SR (1997) Dynamics of granular material. Am Sci 85:540–545
Kennedy Shaw WB (1945) Long range desert group. Collins, London
King WJH (1916) The nature and formation of sand ripples and dunes. Geogr J 47(3):189–209
King WJH (1918) Study of a dune belt. Geogr J 51(1):16–33
Kocurek G (2008) Boundary-condition controls on pattern development in aeolian and fluvial dune fields. In: Parsons D, Garland T, Best J (eds) Marine and river dune dynamics. Third MARID Workshop, pp 193–196
Krumbein WC (1934) Size frequency distribution of sediments. J Sediment Petrol 4:65–77
Lee JA, Zobeck TM (2009) Aeolian research. Aeolian Res 1:1–2
Leeder MR (1999) Sedimentology and sedimentary basins. Blackwell, Oxford
Leopold LB (1979) Citation for the Sorby Medallist, Brigadier Ralph A. Bagnold. Sedimentology 26:157–160
Leopold LB, Bagnold RA, Wolman MG, Brush LM (1960) Flow resistance in sinuous or irregular channels. U.S. Geological Survey Miscellaneous Professional Paper 282-D, pp 111–134
Leopold LB, Wolman MG, Miller JP (1964) Fluvial processes in geomorphology. WH Freeman and Co, San Francisco
Livingstone I, Wiggs GFS, Weaver CM (2007) Geomorphology of desert sand dunes: a review of recent progress. Earth Sci Rev 80(3–4):239–257
McGee WJ (1908) Outlines of hydrology. Geol Soc Am Bull 19:199
McKenna Neuman C (2003) Effects of temperature and humidity upon the entrainment of sedimentary particles by wind. Bound Layer Meteorol 108:61–89
Nash DJ, Bullard JE, North CP (2007) Drylands: linking landscape processes to sedimentary environments. Sediment Geology 195:1–100

Parsons R, Walker IJ, Wiggs GFS (2004) Numerical modelling of flow structures over idealized transverse aeolian dunes of varying geometry. Geomorphology 59:149–164
Pye K, Tsoar H (1990) Aeolian sand and sand dunes. Unwin Hyman, London
Royer JR, Evans DJ, Oyarte L, Guo Q, Kapit E, Möbius ME, Waitukaitis SR, Jaeger HM (2009) High-speed tracking of rupture and clustering in freely falling granular streams. Nature 459:1110–1113
Rubin DM (2006) Ripple effect: unforeseen applications of sand studies. Eos 87(30):293–297
Sagan C, Bagnold RA (1975) Fluid transport on Earth and aeolian transport on Mars. Icarus 26(2):209–218
Shaw R (2008) Environmental aspects of the Indian Ocean tsunami recovery. J Environ Manage 89(1):24–34
Stout JE (2004) A method for establishing the critical threshold for aeolian transport in the field. Earth Surf Proc Land 29:1195–1207
Stout JE, Warren A, Gill TE (2009) Publication trends in aeolian research: an analysis of the Bibliography of Aeolian Research. Geomorphology 105(1–2):6–17
Thomas DSG, Wiggs GFS (2008) Aeolian system response to global change: challenges of scale, process and temporal integration. Earth Surf Proc Land 33(9):1396–1418
Thorne CR, MacArthur RC, Bradley JB (eds) (1988) The physics of sediment transport by wind and water: a collection of the hallmark papers by R.A. Bagnold. Book number 665, Hydraulics Division, American Society of Civil Engineers, New York
Tooth S (2008) Arid geomorphology: recent progress from an Earth system science perspective. Prog Phys Geogr 32(1):81–101
Tooth S (2009) Arid geomorphology: emerging research themes and new frontiers. Prog Phys Geogr 33(2):251–287
Tsoar H (1994) Classics of physical geography revisited: Bagnold, R.A. 1941: the physics of blown sand and desert dunes. Prog Phys Geogr 18(1):91–96
Warren A (1986) Review: Greeley R and Iversen JD: Wind as a geological process on Earth, Mars, Venus and Titan. Prog Phys Geogr 10:312–313
Wentworth CK (1922) A scale of grade and class terms for clastic sediments. J Geol 30:377–392
Werner BT (1995) Eolian dunes: computer simulations and attractor interpretation. Geology 23(12):1107–1110
Wiggs GFS (2001) Desert dune processes and dynamics. Prog Phys Geogr 25(1):53–79
Yizhaq H (2008) Aeolian megaripples: mathematical model and numerical simulations. J Coastal Res 24(6):1369–1378

# Dune: Arenaceous Anti-Desertification Architecture

**Magnus Larsson**

## 1 Introduction

Argentinian writer Jorge Luis Borges once observed that "Nothing is built on stone; all is built on sand, but we must build as if the sand were stone" (Borges 1974).

A single grain of sand is almost nothing: a splinter of rock, a miniscule fragment of a geological formation, the residue of a microcosmic event. Myriad grains together, however, become almost everything: mesmerising landscapes, vast deserts, a fluid material capable of being transformed into solid structures, and, ultimately, flourishing cities. In aggregates of sand, interlocking angular quartz grains, we find fascinating forms and emergent patterns; possibilities, potentials, substance. In short, we find a constant unfolding of interactive opportunities (Balmond 2002).

Architects work in the mineral world, in which all design is fundamentally about aggregation and erosion. Even the most austerely minimalist structures present aggregations of elements and densifications of matter that were not there before. Through accretive processes, materials become buildings that become cities. At its most fundamental level, architecture is about the manipulation of landscape.

The project presented here, *Dune*, is an architectural speculation aimed at creating a network of solidified sand dunes in the desert—a proposition that suggests precisely such an elemental ground manipulation. It also advocates a radical shift in structural thinking, away from pre-fabricated or in situ construction, towards the localised cementation of granular materials.

M. Larsson (✉)
Architectural Association, London, UK
e-mail: magnus@magnuslarsson.com

V. Badescu and R. B. Cathcart (eds.), *Macro-engineering Seawater in Unique Environments*, Environmental Science and Engineering,
DOI: 10.1007/978-3-642-14779-1_20, © Springer-Verlag Berlin Heidelberg 2011

Localised cementation of the desert sand is achieved through microbially induced carbonate precipitation (MICP) using the microorganism *Bacillus pasteurii*, an aerobic bacterium pervasive in natural soil deposits (Le Metayer-Levrel et al. 1999; Nemati and Voordouw 2003; DeJong et al. 2006; Whiffin et al. 2007). In the right circumstances, the bacterium's enzymatic urease catalyst hydrolyses urea, which—when the process occurs in a calcium-rich environment—generates calcite (the most stable polymorph of calcium carbonate), which binds the individual grains of sand together. The solidification of the sand is organised into an array of specific spatial structures to create a very narrow and roughly 6,000 km long pan-African city with the capacity to mitigate against the Sahara's shifting sands. The spatial pockets would help retain scarce water and mineral resources necessary to turn *Dune* into a micro-environmental support structure capable of assisting the formation of the Great Green Wall for the Sahara and Sahel Initiative (GGWSSI). The various spatial pockets within *Dune* would also serve as habitable and programmable space for a nodal network converging with the planned Sahara Railway.

*Dune* is an exercise aimed at investigating a climate-conscious architecture that primarily points toward adaptive responses to the potential threats of future extreme environments. The desire is to create a framework for an innovative architecture that applies a controlled biocementation process as a strategy to mitigate the continual migration of sand dunes in the circum-Saharan drylands of Africa. The final outcome is a habitable anti-desertification structure made from the desert itself, a sand-stopping device made out of sand.

While the scale and scope of the scheme inevitably positions it within the speculative realm, and while many details are left to explore before a bacterial cementation of habitable sand dune barriers can be carried out in the real world, it is important to remember that the concept outlined here is a buildable proposition. Humankind has created larger structures in the past: a recent 2-year government mapping study of the Great Wall of China (carried out by the Chinese State Administration of Cultural Heritage and the State Bureau of Surveying and Mapping), the most comprehensive archaeological survey to date of this largest of man-made structures, stretching from Lop Nur in the west to Shanhaiguan in the east, concluded that the entire wall with all of its branches stretches for some 8,851.8 km. This example is particularly pertinent since in its western section, which is mostly located in the desert, the wall was typically built with sand and mud.

Fraught with challenges and difficulties as this stratagem might be, the *Dune* project is a beginning, a blueprint, a vision. To return to the beautiful image borrowed from Borges, it is the sand; what follows is yet another move towards turning it into stone.

## 2 The Threat: Deserts and Desertification

At the heart of the world's drylands are five major zones of natural desert: the Afro-Asian Desert (a great belt stretching from the Atlantic Ocean to China, including the Sahara, Arabian, Iranian, Touranian, Thar, Takla Makan, and Gobi deserts), the North American Desert (comprising the Great Basin, Mojave, Sonoran, and Chihuahuan deserts), the Atacama and Patagonian Deserts in South America, the Namib and Kalahari Deserts in south western Africa, and the Great Sandy, Great Victoria, and Simpson Deserts in Australia (Grainger 1990; Martin 2004).

Deserts emerge from the interacting processes of atmospheric circulation and oceanic circulation (Weinstock 2010). The word 'desert' usually connotes a landscape devoid of people, but except in the most extremely hyperarid regions, people have developed agricultures and built communities either around the desert margins or in the desert itself: today, upward of a billion people live in arid or semiarid environments, coexisting with the shifting sands, sometimes struggling to get by in the wake of increasingly harsh conditions.

The apparent spread of the desert is not a new phenomenon, but one that has occurred throughout human history: In the fourth century BC, Chinese philosopher Mencius (Mèng Zǐ) wrote about desertification and its human causes, including the cutting of trees and overgrazing. Plato, in the same century, commented upon the early deforestation of Attica the historical region of Greece that contained Athens by saying that "Our land, compared with what it was, is like a skeleton of a body wasted by disease". Athens was forced to trade wine and olive oil for wheat and other items of food that its eroded soils could not supply (Burns 1995).

Dry areas cover more than one-third of the earth's land surface. Some are deserts, others are being continuously degraded by the erosion brought about by grains of sand carried by aeolian forces. Though disputed by some, desertification ("the diminution or destruction of the biological potential of the land" that is "an aspect of the widespread deterioration of ecosystems under the combined pressure of adverse and fluctuating climate and excessive exploitation" according to UN-COD's 1977 definition) is considered by many specialists in the field to be one of the most serious environmental problems facing our world (United Nations 1978). The validity of this claim is beyond the scope of this chapter: the scientific understanding of desertification is still quite limited, and the role of the climate, in particular, has until recently been thought of as mainly catalytic.

Desertification has been on the international agenda for about half a century, but we still do not know precisely how fast, or indeed whether, our deserts are growing, much less how best to address the issue should that be the case. It is estimated that firewood collection, excessive grazing, and overcultivation accounts for nearly 90% of what is perceived as desertification (Grainger 1990; Welland 2009a). *The World Atlas of Desertification* summarises the current state of scientific knowledge on the drylands of the globe, and indicates that desertification is one of the world's most pressing environmental problems, and an accelerating global issue (Middleton and Thomas 1997). In the following, that view will be presumed to be correct.

It is safe to say that the majority of those working with desertification accept that a wide variety of pressures have led to the adoption of unsustainable land management practices in the circum-Saharan drylands, including continuous overstocking and overgrazing of rangelands; continuous cropping, with reductions in fallow and rotations, repetitive tillage and soil nutrient mining; rangeland burning; and the over-exploitation and clearance of savannas (for cultivation). The impacts of these practices include loss of natural resources, changes in natural habitats and ecosystems, loss of agrobiodiversity and wild biodiversity, degradation of ecosystem services, decreases in productivity (of both arable land and rangeland) leading to poor harvests and food shortages, which in turn result in poor living conditions and poverty. Climate change might already be exacerbating these problems, with increasing weather variability (droughts and storms), and is predicted to bring further challenges in the coming decades, with rising temperatures and changes in rainfall patterns (IPCC 2007a, b).

## 3 The Site and Sahelian Droughts

Desertification is a major threat on all continents, affecting more than 100 of the world's 190+ countries. Some estimates suggest that 35% of the Earth's land surface is at risk, and that the livelihoods of 850 million people are directly affected. 75% of the world's drier lands—45,000,000 km$^2$—are affected by desertification, and every year 6,000,000 ha of agricultural land are lost and become virtual desert (Dollo and Sen 2007). More than 80% of Africa's drylands are moderately or severely desertified, a figure that equals more than one-third of all desertified land in the world. The Sudano-Sahelian region is the most affected part of the continent, mainly due to overgrazing and overcultivation, though the other two major direct causes—poor irrigation and deforestation—also play a part (Grainger 1990).

The largest volumes of shifting sands are to be found neither in Africa nor in the Middle East, but in China. The total area of China's desert is growing at around 200 km$^2$ every *month,* and innovative solutions are desperately needed to combat the thousands of tons of sand and dust that are blown into Beijing every year (Welland 2009a).

That scenario might indeed form the basis for future projects. As the site for *Dune,* however, the circum-Saharan drylands were chosen because of the frequent occurrence of droughts in this region during the past century. The name Sahel itself means "edge of the desert" (Grove 1977) and this is probably the one region in the world most closely associated with desert encroachment. In the Sahel, the availability of water (or in more precise terms, the possibility of collecting and exploiting rainwater, fog, and dew) represents the fundamental limiting factor for human survival, habitation, and production. The seasonal distribution of precipitation and its annual variation is more important than the annual, global amount of rainwater (Koechlin 1997).

In such a situation, a particularly dry year can easily spark conflict in the Sahel. Climate change, drought, increased desertification, crashing food supplies, water scarcity, famine, forced migration, political instability, warfare, crisis—that's a potential (albeit admittedly worst-case) scenario if we fail to take this seriously.

African countries share an important condition with other developing countries in that they are "especially vulnerable to climate change because of their geographic exposure, low incomes, and greater reliance on climate sensitive sectors such as agriculture" (Stern 2007). The historical climate record for Africa shows warming of approximately 0.7°C over most of the continent during the twentieth century, a decrease in precipitation over large portions of the Sahel, and an increase in precipitation in east central Africa (Desanker 2002). Droughts and floods have increased in frequency and severity across Africa over the past 30 years. Throughout the twenty-first century, the warming trend and changes in precipitation patterns of the twentieth century are expected to continue, increase in rapidity, and be accompanied by an increase in the frequency of extreme weather events—droughts, floods and storms (Stern 2007). Predictions of the magnitude of changes in temperature and precipitation are subject to considerable uncertainties, but climate change scenarios for Africa indicate future warming across the continent ranging from 0.2°C per decade (low scenario) to more than 0.5°C per decade (high scenario) (Hulme 2001; Desanker and Magadza 2001).

When drought exhausted the small harvest of 1970, an estimated three million people in the Sahel needed emergency food aid, with between 100,000 and 250,000 people dying as a result of the drought, according to a UNCOD report. The FAO estimated that 3.5 million head of cattle, 25% of the region's total, died in the Sahel in 1972–1973 alone. The drought eventually prompted the United Nations 1977 Conference on Desertification, and had a massive impact on the African drylands and the people who live there (Grainger 1990). It was a catastrophe arising from a phenomenon that gets very little attention: in our accelerated media culture, desertification is too slow to make the headlines—there are simply too few crying children and smashed-up houses to hold the attention of the general public; nothing like a tsunami or a Katrina. Despite that, this is a tragedy waiting to happen again.

Six years after the UN conference, Kenneth Hare prepared an assessment for UNEP in which he reported that "there is yet no way in which climatologists can decide whether this desiccation will continue" but that "(t)he possibility of a permanent desiccation of the drybelt climates of Africa cannot … be ruled out" (Hare 1983, 1984). If such a development were to take place during a period of unprecedented population growth and increased food demands, the situation could once again become disastrous. The population of Africa is increasing, and is set to double every 23 years (according to the UN, the fastest growing nation on the planet is Liberia, with a doubling time of about 15.5 years), while the amount of food being produced to feed each person is decreasing. Estimates show that if current trends of soil degradation and population growth continue, the continent might be able to feed just 25% of its population by 2025 (United Nations 2007). In 2007, another UN study showed that desertification currently affects 100–200 million people, and that it has become a threat to global stability, putting about 50 million people at risk of being driven away from their homes in the next decade (Adeel et al. 2007).

Such a course of events is perhaps more threatening in Nigeria than anywhere else. With a constantly growing population of more than 140 million people, it is

**Fig. 1** Helicopter perspective would become a new, narrow, pan-African, urban development built straight into the dunescape

the most populous country in Africa, experiencing serious desert encroachment issues throughout its northern states. While the largest city, Lagos, is expanding northward at an unprecedented rate, the Sahara desert is pushing the border of the fertile Sahel land southward at an alarming pace. At some point in (the distant) future, these two movements of sand and people in opposite directions may meet, creating a serious desertification refugee crisis. Every year, Nigeria loses about 600 m of its arable land mass to desert encroachment (Ezigbo 2009). During a study trip undertaken by the author in 2008, several foresters in the villages outside of Sokoto confirmed this rapid migration.

Allowing *Dune,* the 6,000 km long habitable wall proposed here (the Sahara is roughly 5,150 km across at its widest point) to pass through this part of the world could be one way of creating a unified front against the encroaching desert: a new kind of city, a narrow, pan-African, urban development built straight into the dunescape, supporting a vast shelterbelt of trees, and effectuating a permacultural evolution while connecting countries and people (Fig. 1).

## 4 Traditional Anti-Desertification Methods

Such a support structure for a shelterbelt of trees would constitute a continuation of traditional anti-desertification methods. These include the planting of trees and other plants (cactii are popular, as are different types of acacia, eucalyptus, and bushes), the cultivation of grasses and shrubs, and the construction of sand-catching fences and walls. More ambitious projects have ventured into the development of agriculture and livestock, water conservation, soil management, forestry, sustainable energy, improved land use, wildlife protection, and poverty alleviation.

Grains of sand move in three different ways in the desert: through suspension, creep, and saltation (Bagnold 1941). The way to stop a desert in its tracks is to come to terms with saltation, the mini avalanches that break the continuity of the dune surface at the top of the shear plane, which creates the slope at the leeward side of the dune. Shelterbelt plantations are usually made up of one or more rows of trees or shrubs planted so as to provide shelter from the wind while protecting the soil from erosion. The *Dune* scheme seeks to support such a shelterbelt.

Anti-desertification and afforestation projects involving large-scale planting of shelterbelts have historically been proposed by different governments to reduce soil erosion and improve the microclimate in otherwise treeless agricultural areas. These include US president Franklin Roosevelt's Great Plains Shelterbelt project, ambitiously launched in 1934 as a way of modifying the weather and preventing the Great Plains states from eroding (by 1942, this scheme had resulted in the planting of 30,233 shelterbelts containing 220 million trees that stretched for 30,000 km). In the USSR, one section of General Secretary Joseph Stalin's Great Plan for Transformation of Nature, launched in October 1948, provided planting of a gigantic network of shelterbelts across the southern steppes of the country, while the Green Wall of China is a project intended to provide 4,800 km of shelterbelts across the northern parts of that nation by 2074. Other schemes worth mentioning include Wendy Campbell-Prudie's Tree of Life trust in the 1960's, and Richard St. Barbe Baker's Men of Trees group in the 1920's (Hurt 1995; Krech et al. 2004; Campbell-Purdie 1967; St. Barbe Baker 1944).

All of the above procedures were attempts to mitigate the advancement of the desert. They were mitigatory measures. The International Panel on Climate Change (IPCC) defines mitigation as "An anthropogenic intervention to reduce the sources or enhance the sinks of greenhouse gases." The ability of a system to adjust to climate change with moderate potential damage is referred to as that system's climate adaptation. These two terms—mitigation and adaptation—are fundamental in the climate change debate. The IPCC has defined adaptation as adjustments in natural or human systems in response to actual or expected climatic stimuli or their effects, with the aim to either moderate harm or exploit beneficial opportunities. Historically, mitigation has been viewed as the primary way to proactively curb human causes of climate change—in the present case desertification—whereas adaptation has been seen as a secondary measure to cope successfully with changing conditions after the fact. However, in the case of desertification, the changing nature of the sand dunes constitute the challenge (we are not so much concerned with the theory that the Earth's deserts might be expanding as with the fact that the sand within them is moving), which turns adaptation into less of a passive measure and more of an active adjustment in response to new stimuli. Simplified, we could talk about three conditions: (1) cause mitigation (such as carbon mitigation), (2) system mitigation (for instance shelterbelts), and (3) adaptation (in this case a habitable, permacultural structure).

While the IPCC generally defines the term mitigation as cause mitigation, we are here concerned primarily with the second and third conditions.

While *Dune's* sandstone barrier mitigates against the encroaching desert through its supporting the shelterbelt, it also seeks to adapt to its environment through the creation of habitable spaces inside of this barrier. Inside the dunes, we can take care of our plants and animals, find water and shade, help the soil remain fertile, care for the shelterbelt trees, and so begin to green the desert from within (Fig. 2).

The exact strategies for this sustainable land use design lie outside the scope of the present text. A brief clarification, however, might be in order. Once the sand has been solidified into solid sandstone structures, permacultural tactics would be used to sustain their inhabitants. *Permaculture* is a portmanteau of *permanent agriculture* and *permanent culture*—a school of thought based on ecological and biological principles that often translate natural patterns into a man-made environment in order to maximise effect and minimise the amount of work needed to reach that effect. The general idea, from the writings of American agricultural scientist Franklin Hiram King and onwards (King 1911) is to educate a range of individuals in a core set of design principles with which they can then design their own environments and self-sufficient settlements, creating stable, productive systems that provide for human needs. While any local settlement along the proposed 6,000-km habitable wall would have to be meticulously analysed before any proper conclusions could be drawn, some of those permacultural design principles are likely to be fog harvesting, condensation strategies, swales (a form of rain-harvesting trenches), natural ventilation principles, edible landscaping, and waste management tactics. Some prominent permaculturalists include David Holmgren, Bill Mollison, Geoff Lawton, Patrick Whitefield, and Toby Hemenway.

**Fig. 2** Interior perspectives: Inside the solidified dunes, we can find water and shade, care for the shelterbelt trees, and begin to green the desert from within

## 5 The Great Green Wall for the Sahara and Sahel Initiative

Sand dunes cover only about one-fifth of our deserts (Welland 2009a), but those extreme areas are good places to introduce a barrier of greenery to check desert encroachment and counter soil degradation, halt the shifting sands and stop the dunes from migrating.

*Dune's* focus on adaptation seems to be in line with Jauffret and Woodfine's 2009 feasibility study for the Great Green Wall for the Sahara and Sahel Initiative (GGWSSI). As of late December 2009, 525 km out of the 7,000-km-long shelterbelt have been planted, all within Senegal (Dell'Amore 2009).

The idea of a 'Green Wall for the Sahara' was first proposed by former Nigerian president Olusegun Obasanjo in 2005, and presented first to the Community of Sahel-Saharan States (CEN-SAD) and then to the African Union (AU) that same year. The idea was further discussed in Lisbon in December 2007, and in Brussels in January 2008, after which it was agreed that a feasibility study would be carried out.

The initiative originally called for 23 African countries to come together in order to plant trees across a 7,000-by-15-km stretch south of the Sahara—a total area of 105,000 km$^2$, or 10.5 million hectares—so as to cover parts of the land that lie within the 100–400 mm annual rainfall band. The aim was to catalyse "sustainable development and poverty reduction in the desert margins north and south of the Sahara" and "to strengthen the implementation of existing continental frameworks and plans addressing the menaces of land degradation and desertification in the margin of the Sahara desert" (AU/CEN-SAD 2009).

An early press release from the African Union proposed that under the programme "300 million trees, covering three million hectares of land will be planted" and that "the goals and objectives of the green wall ... are to slow the advance of the Sahara desert, enhance environmental sustainability, control land degradation, promote integrated natural resources management, conserve biological diversity, contribute to poverty reduction, and create jobs" (African Union 2006).

Planting a shelterbelt, as we have seen, is a traditional method on a grand scale, using the trees to stop the grains from avalanching over the dunes' crests. The vegetation belt was planned to run across the entire African continent, offering mitigation against desertification and some adaptive effects through the harvesting of fruit from the trees. However, assuming 460 trees/ha and 100% survival, this plan would equal at least 4,838 million tree seedlings; unquestionably an unrealistic figure.

At the onset of the project, the solidified dunes of the *Dune* scheme were intended to support the existing GGWSSI. Following the feasibility study, a much wider range of sustainable land management practices have been proposed as "a more ecologically appropriate and holistic approach to directly benefit local land users (farmers, agro-pastoralists, and mobile pastoralists)" (Jauffret and Woodfine 2009). This move away from the original shelterbelt idea could mean that the sandstone structure no longer has any real-life trees to support. However, over and above how the structure performs in terms of tree support/desertification mitigation, its adaptive qualities—the creation of oases for people in the desert—still makes, at the very least, the nodes of this structure justifiable.

One aspect of climate change pointed out by the feasibility study is of particular concern: not so much the shift in long-term average climate, but rather the increased frequency and magnitude of climatic extremes. Climate change is eroding the social network mechanisms that have been used in the past to cope with drought, by causing climatic extremes that leave the affected people without enough time to recover. Recurrent droughts have led to the degradation of the resource base and forced millions of farmers to sell their assets, in some countries forcing them into absolute destitution. Using the *Dune* structure to physically bring people together could be one way of restoring these mechanisms and start building up the adaptive capacity of the local communities. Staying put and finding ways of adapting to the situation would be an alternative to trying to run away from the encroaching dunes. The *Dune* scheme is designed as a network around a set of nodal points at which conditions are optimal for living in these harsh environments (sheltered points with good access to water, a topography that lends itself to the sculpting of shaded areas, proximity to existing infrastructures, and so on). Just as a node in physics is a point along a standing wave where the wave has minimal amplitude, so the nodes in *Dune* would be points where the encroaching desert has a minimal impact on the way of life, and thus would be ideal places for human habitation.

## 6 Arenaceous Architecture

Designing with aggregates may commence either from the design of the elements that will make up the aggregate itself, or from the utilisation of natural grains used together with designed and/or natural constraints to guide the design and/or construction process. This project is entirely positioned within the latter realm. The systems underlying naturally occurring sand formations are interesting, as dune and ripple formations suggest a mutual modulation—airflow and water modulate dunes that in turn modulate those initial forces, which brings about formations that follow clearly discernible rules (Takahashi 2006).

When those rules and modulations are first observed and analysed, and then strategically controlled through the local solidification of sand into sandstone, turning the natural phenomena into architectural structures, we move towards a novel technology with staggering potentials. As mentioned above, the *Dune* scheme advocates a radical shift in structural thinking, away from pre-fabricated or in situ construction, towards the localised cementation of granular materials: if we can control the solidification of sand into sandstone, then we can begin to investigate a completely monolithic architecture based not on components that are attached to each other in order to create a structural system, but on the binary densification of aggregate matter: the sand is either turned into stone, or it's not. Introducing this notion of an architecture of solidification calls for a brief discussion on the nature of technology itself.

Alan Kay is a brilliant polymath with a CV that reads like a *Who's Who* of creative technology companies: Apple, Atari, Disney, Xerox. Kay's definition of technology is "anything that was invented after you were born". While neither transistors nor steam cars nor nylon carry technological connotations to me, at the time of writing this, the iPad does. If we were to ask a kindergarten teacher, however, we would probably learn that today's children feel the same way about the iPad that I do about nylon. The things that were there when we were born are simply part of the fabric of everyday life (or museum pieces). Technology is what happens next.

The US inventor, entrepreneur, and author Danny Hillis used to be Alan Kay's colleague. Hillis refined Kay's definition by concluding that technology is "everything that doesn't work yet". This points to the fact that successful inventions disappear from our awareness. We do not really consider the toothbrush a technology, not only because it was already invented when we were born, but also because it has proved so successful as to render itself essential. Essential things are not technologies. They are just indispensable parts of our existence (Kelly 2007).

The processes described here possibly constitute a technology following Kay's definition, and definitely fall within Hillis's interpretation of the term: whereas microbially induced carbonate precipitation (MICP) of grains of sand is a naturally occurring phenomenon, the exploitation of these procedures as a design and construction method is, to the best of my knowledge, an architectural speculation that has never before been proposed, and thus has yet to be implemented into a functioning system. However, just because the idea has yet to be tested does not mean that such a system could not be established—most of our tools and building materials have been subjected to hundreds if not thousands of years of research and development, but they all started out as speculations—and the success of some recently developed architectural materials (from Corian to LiTraCon) adds weight to this argument.

The dream scenario for the current proposal, then, would be to first turn this process from observable phenomenon to a technology (the controlled 'growing' of sandstone structures) that then becomes so ubiquitous as to disappear from our awareness. A common and ancient occurrence in nature becomes a technology when carried out and controlled by humans.

The word 'arenaceous' is a geological term meaning "consisting of sand or sandlike particles". Used from the mid-seventeenth century, its roots lie in the Latin *arenaceus,* from *arena* (*harena*): 'sand'. In biology, it is also used to describe animals or plants that live or grow in sand. Since the present scheme resembles, is derived from, and contains sand, while (literally) growing in highly sand-filled areas and utilising organisms that live and grow in sand, it seems natural to call it arenaceous. This novel arenaceous architecture is characterised by new material methods, new construction methods, new methods to mitigate against desertification, and new spatial, programmatic, environmental, performative, and affectual concerns.

## 7 Aggrerosion: Sand and Sandstone

I love sand. From the famous black lava grain beaches of Hawaii to the stunning stretch of white sand in Turkey known as Cleopatra's beach, from the star sands in Japan to the active dune ridges in Egypt's Great Sand Sea, this naturally occurring granular matter composed of finely divided rock and mineral particles—one of the most ubiquitous materials on the planet—is a beautifully paradoxical, seemingly magical building element.

Grains of sand are at the same time granular and massive, simple and complex, peaceful and violent, static and dynamic, too heavy to be held in suspension in the air yet light enough to be moved by the wind, individually evanescent to the point of invisibility yet capable of aggregating into highly pronounced (and huge) forms. Sand is the quintessential granular material.

It is also ever-present in architecture: its history is part of the material history of the discipline. Without sand no brick, no concrete, no glass—even wooden structures are sanded down to smoothen out their edges. How many cathedrals and mosques, how many temples and churches are made of sand? How many of our greatest cities rest on sand?

It is thrilling, in an age in which silica-based technologies control an ever-increasing part of our social, economical, and political interactions, to reassess the possibilities inherent in sand as a highly underused, exceptionally renewable building material.

One billion grains of sand come into existence around the world every *second.* It's a cyclic process: as rocks and mountains die, grains of sand are born. These grains may then become naturally glued together, or lithified (from *lithos*, greek for 'stone' or 'rock'), into a *clastic* sedimentary rock, a sandstone. When that sandstone is weathered, new grains break free. In a way, the static stone mountain becomes a moving mountain of sand. The majority of quartz sand grains are derived from the disintegration of older sandstones; perhaps half of all grains of sand have been through six cycles. Typically, the landscape of a mountain range will be lowered by a few millimetres every year (Welland 2009a).

The rule in nature, then, is that erosion follows aggregation follows erosion. Hacking into this perpetual cycle is a novel way of making exceptionally sensitive physical interventions in the landscape: exploring the poetic qualities of infinitesimal building blocks, and how they can be utilised in innovative ways to establish an understanding of (near-)microscopic stacking, packing, heaping, massing, mounding, piling, sheafing, and so on, could lead to a new building typology, one that utilises aeolian forces to define its exterior while harnessing the power of biocementation processes to sculpt its interiors. The morphological variations of dune types lie beyond the scope of the present study, but it's worth mentioning that their peculiar physics—examined first by Ralph Bagnold (1941) and later by Per Bak (1996)—yield volumes that could easily fit any existing architectural typology, from skyscrapers to railway tracks: dust sediment deposits can be more than 300 m deep; linear dunes that form parallel to the direction of

**Fig. 3** Exterior perspective: The peculiar physics of dune type variations yield volumes that could easily fit any existing architectural typology, from skyscrapers to railway tracks

strong and persistent winds may be 100 or more kilometres long; the centre of a radially symmetrical star dune can be positioned several hundred metres above ground (Weinstock 2010) (Fig. 3).

Putting aggregation and erosion together was the starting point for *Dune's* use of *aggrerosion,* a design strategy based on the accumulation and reduction of granular materials. As a building material, grains of sand can be employed across a gradient of conditions: granular mass, solid stone, dynamic medium carried by aeolian forces, compressive membrane, and so on. One of the more fascinating facts about sand dunes is that they are both stable and fluid at the same time. Individual dunes have ripples on their surface, and these are fluidly active, as is the dune itself. Their geometry is consistent over time, but since they move—travelling across a vast sand sheet that perhaps covers thousands of square kilometres while largely maintaining their form—they are also fluid (Weinstock 2010).

Thus the granularity allows for both structural and constructional inventions: the sand can for instance be used both as primary ingredient of the finished structure and as its formwork during construction. If we solidify parts of a dune to hold it in place, the non-solid parts will continue moving, making it possible to plan for the aeolian forces to excavate the structure for us.

Bagnold defined this as the power of self-accumulation, a dune's skill in using the energy of the wind to collect all its scattered grains and build them into heaps, separated by areas free of sand. "In many areas of the desert," expands Michael Welland (2009a), "the sand is piled into dunes with bare rock in between, as if some giant combination of a vacuum cleaner and a bulldozer is constantly at work". It's a construction site waiting to be deployed: aeolian forces become our extended, invisible hand in the design process—the wind carries the material to the site, and then carries the excess material away from it again.

As soon as we begin to view them in this way, the sand dunes turn into readymade buildings. All we need to do is solidify the sand wherever we need solid surfaces, and then excavate the sand we do not need—or have the wind excavate it for us. Using the existing sand as our base material, we can sculpt arches, gaps, caves, and other patterns straight into the dune. As long as we take care to design *with* the aggregation rather than *against* it, we can simply allow saltation to propel the grains in place, and then, once the sand has aggregated into a beneficial form, use an intelligent strategy for how to solidify it, petrify it, freeze it into a solid state that speaks of that one moment in time. This is the novel idea at the heart of *Dune—Arenaceous Anti-Desertification Architecture.*

The solidification process turns the sand into sandstone, a sedimentary rock composed mainly of rock grains and sand-sized minerals—usually feldspar and/or quartz, as these are the most common minerals in the Earth's crust. Resistant to weathering, yet easy to work, some sandstones have become common building and paving materials. Over and above the obvious advantages arising from the abundance of sand in the Sahel, another fact that speaks in favour of creating sandstone structures in this part of the world is that such edifices usually allow percolation of water while being porous enough to store large quantities, turning a sizeable part of the final building into a potential aquifer. Sandstone aquifers are also fine grained, and therefore better at filtering out pollutants from the surface than are, for instance, limestones (Pettijohn et al. 1987).

## 8 Thinking Big: Radical Optimism and Anti-Walls

All acts of design are inevitably and inherently radically optimistic, and architecture is all about opportunities. Old buildings give way to new, structures are razed and new ones take their place, one city turns into the foundations of the next. The moment we stop believing in the possibility of making the world a better place is the moment we give up on architecture.

So what is an architect to do when moving mountains of sand threaten to push people away from their homes and induce catastrophic scenarios? Why, adopt the position of the radical optimist. Come up with an architectural response to the predicament. Find a way of using sand dunes to stop the desert from moving. It's true that architects are trained to solve problems, but I do not believe in architectural problems. I believe in opportunities. To borrow a line from a description of the late Tibor Kalman, I hold that we need to become perversely optimistic about the opportunities around us (Hall et al. 1998).

The opportunity at the outset was the challenge of turning *Dune* into a scheme that would support the pan-African green wall, but other paths soon opened up. It became clear that an existing technology for microbially induced carbonate precipitation (MICP) could be used to suggest, in effect, an addition to our existing architectural material palette. This technology is based on the bacterium *B. pasteurii*, which has a urease enzyme that enables it to hydrolyse urea, which in a calcium-rich environment generates calcite, which in turn acts as a binding agent to cement the individual grains together. This insight opened up discussions about the first two tiers of the issue at hand: how to make it possible for people at risk of becoming desertification refugees to keep on living in their existing home area, and how to take advantage of the underused local materials in order to offer better spaces for social interaction while providing sound housing and thermal comfort in these extreme environments.

But there was a third tier: how to tackle the challenge that comes with the history of the region, a challenge that truly has to do with matters of life and death. The Sahel Belt is not very well defined, but at least covers parts of the countries of Mauritania, Senegal, Mali, Burkina Faso, Niger, Nigeria, Chad, Sudan, Somalia, Ethiopia, and Eritrea. The region includes more than 100 million of the poorest people on Earth, and is one of the most violence-prone areas on the planet. The entire Sahel is already racked by political instability and warfare—climate change and increased desertification can only make matters worse. The capriciousness of weather patterns could easily mean the collapse into violence, as in this region fluctuations in rainfall have critical implications. Without water, crops can not be grown nor animals raised. As people become increasingly desperate, many try to move to neighbouring states, and a few repatriations and droughts later, it becomes easier to understand the rationale for an armed rebellion (Fisman and Miguel 2008).

Another recent study found strong historical linkages between civil war and temperature in Africa. Warmer years lead to significant increases in the likelihood of war. When the researchers combined this fact with climate model projections of future temperature trends, the results suggests an increase in armed conflict incidence by 54% by 2030. That would mean an additional 393,000 battle deaths if future wars remain as deadly as recent wars (Burke et al. 2009).

It may be asking too much of architecture to try and come up with responses to armed conflicts, but learning more about the situation gave birth to the idea of turning *Dune* into an inverse separation barrier—a sheltered safe haven, a bridge between areas or even countries sharing the desert condition and the threat of desertification. Not so much a habitable wall as a habitable anti-wall; the wall as a cross-border connection rather than a means of national exclusion.

The Great Wall of China, the Berlin Wall, the Israeli West Bank barrier: history has taught us to think about monumental walls as physical structures that limit the movement of people across invented lines in space. *Dune* fundamentally opposes this view through the creation of a wall that never existed before, the opposite of a wall. The word 'wall' usually connotes a membrane that demarcates space, that divides one space in two. We think of a surface that prevents us from entering the next space. But dare to think big, allow the wall to straddle an entire continent, place the habitable spaces *inside* of it, and we get a stretch of architecture that could bind places, villages, people, even countries together. As with Rem Koolhaas's 1972 project *Exodus, or the Voluntary Prisoners of Architecture* (Koolhaas 1995), the built structure becomes an architectural refuge, an enclosure that keeps both environmental ferocity and human violence out while bringing people together on the inside.

## 9 Material Process, Construction, Structure

In his 1996 novel *Idoru,* science fiction writer William Gibson offered a premonition of a twenty-first century Tokyo in which nanomachines would weave buildings from near nothingness:

> ...you could see those towers growing at night. Rooms up top like a honeycomb, and walls just sealing themselves over, one after another (...) Like watching a candle melt, but in reverse. That's too scary. Does not make a sound. Machines too small to see (Gibson 1996).

Following a cataclysmic quake, walls are spun and joints sealed by devices that operate on an atomic or molecular scale. Gibson's dream of a built environment in which a combination of genetically engineered materials and nanotechnological processes opens up new possibilities for radically innovative architectures may seem embryonic and premature to some, but this unfledged sketch of a future society in which very small things accumulate to build and/or become very large things may not be too distant.

The use of (micro)organisms to improve the properties of ground materials in the subsurface and of construction materials is an emerging technology. To borrow yet another sentence from Gibson, "the future is already here, it's just unevenly distributed" (Gibson 1999, Johnston 1999). At Deltares, the Dutch institute for delta technology, a long-term research program called SmartSoils has been initiated, and literature searches reveal that various research groups at different institutions, such as the Delft University of Technology, are active within this rather new field. Remarkably few researchers, however, focus on architectural applications.

It has been speculated that the biblical demise of Sodom and Gomorrah (assuming the twin cities existed in the first place) was due to earthquakes and liquefaction, the ground effectively swallowing buildings and people. As a result of liquefaction, waterlogged sand can lose all of its strength, with disastrous results. As the grains move apart causing the friction and adhesion between them to disappear, the stability of the material is lost, which causes it to flow and compact, expelling the water. This is what happened when an earthquake devastated the region around Bhuj in northwestern India in 2001, killing around 20,000 people. Liquefaction often causes more damage than the tremors of an earthquake—this was the case in San Francisco (1906), Anchorage (1964), and Loma Prieta (1989). In order to counteract (or at least mitigate) this process, scientists have proposed novel engineering solutions. One way would be to inject chemicals such as epoxies into sandy soil in an attempt to improve the ground by binding the grains together to make them withstand liquefaction. These chemicals, however, may be toxic (Welland 2009a, b).

A more sustainable alternative was suggested by Professor Jason DeJong at the University of California at Davis in 2007: use *B. pasteurii*, a natural bacterium that lives between sand grains and in soils, and that causes calcite (the most stable polymorph of calcium carbonate, $CaCO_3$) to precipitate, which glues the grains together and turns loose sand into solid rock. "Starting from a sand pile, you turn it back into sandstone," DeJong explained, before making clear that there are no toxicity problems, that the treatment could be applied after the construction of a building, and that the structure of the soil does not change as the void spaces between the grains are filled in (The Engineer 2007). DeJong's method had one writer at Time magazine (which included DeJong's findings on its "Best Inventions of 2007" list) to exclaim: "Mix urea, soil and calcium, inject a little bit o' bug and voilà! The cementer bug feeds on urea and deposits calcite, which cements the soil together and turns shifting sand into sandstone" (Time 2007).

That's not an altogether bad explanation of the process. Inject sand with cultures of *B. pasteurii*, also known as *Sporosarcina pasteurii*, feed them well and provide them with oxygen, and they will solidify loose sand into sandstone. DeJong and his colleagues experimented with sterilised sand and bacteria, and were able to control and monitor nutrients, oxygen levels, and other variables to determine exactly how the bacteria hardened their sand specimens.

Basing their experiments on the notion that current methods to improve engineering properties of sand all have benefits and drawbacks, and that the need to explore new possibilities of soil improvement is particularly strong as suitable land for development becomes more scarce, the team studied how natural microbial biological processes can be used to engineer a cemented soil matrix within initially lose and collapsible sand. Microbially induced calcite precipitation (MICP) was tested through the introduction of *B. pasteurii* to sand specimens. The microbes were added in a liquid growth medium amended with urea and a dissolved calcium source (nutrients are required by microorganisms for cellular material—carbon and minerals—as well as for an energy source). Subsequent cementation treatments were passed through the specimen to test the increase in cementation level of the sand particle matrix. The results for the MICP-cemented specimens were compared with those of gypsum-cemented equivalents, and were assessed by measuring the shear wave velocity with bender elements.

The experiments indicated that the MICP-treated specimens exhibit a noncollapse strain softening shear behaviour, with a higher initial shear stiffness and ultimate shear capacity than untreated loose specimens—a behaviour similar to that of the gypsum-cemented specimens, which represent typical cemented sand behavior. SEM microscopy formation of a cemented sand matrix with a concentration of precipitated calcite forming bonds at particle–particle contacts, while X-ray compositional mapping confirmed that the observed cement bonds were comprised of calcite.

The reason why DeJong's team selected this particular species of bacterium to create their microbially cemented sand, out of the huge number that occur naturally in soils, is both that *B. pasteurii* is a common bacterium naturally occurring in the subsurface, and that it is a highly aerobic microorganism particularly good at making the waters in soils more alkaline. For MICP to be effective, a microbe must be selected that is capable of $CO_2$ production paralleled by a pH rise in the surrounding environment to an alkaline level that induces precipitation of calcium carbonate. This forces calcium and carbonate dissolved in the water to combine and form crystals of calcium carbonate—calcite—the same natural cement that binds together sandstone, as well as manmade concrete. Aerobic microorganisms capable of consuming urea as an energy source (such as *B. pasteurii*) are particularly good candidates because they provide two sources of $CO_2$: respiration by the cell and decomposition of urea. Furthermore, cells of *B. pasteurii* do not aggregate, which ensures a high cell-surface-to-volume ratio—an essential condition for efficient cementation initiation (DeJong et al. 2006).

This is a rapidly evolving research area. Another team, led by Leon A. van Paassen at the Delft University of Technology, concluded that a range of new ground reinforcement techniques are currently being developed that are based on MICP. Going beyond urea hydrolysis strategies (in which bridges are formed between the grains of sand following precipitation of calcium carbonate crystals in the presence of dissolved calcium; the process proposed by DeJong's team), this research seeked to evaluate the feasibility of other microbial processes that can lead to the precipitation of calcium carbonate (and thus to ground reinforcement). Studying evaluation factors were substrate solubility, $CaCO_3$ yield, reaction rate, as well as the type and amount of side-product. Microbial denitrification of calcium nitrate, using calcium salts of fatty acids as electron donor and carbon source, was singled out as the most promising alternative. This process leads to calcium carbonate precipitation, bacterial growth, production of nitrogen gas and some excess carbon dioxide—and could be another way forward for microbially constructed architectures (van Paassen et al. 2010).

That word—'architectures'—is of course the difference between this underlying research and the deliberate misuse of it that eventually became *Dune*. Once it had been established that this new way of creating sandstone could be used to create new architectures at a human scale, the next step in the development of *Dune* was to define potential strategies and methods for how to actually construct the microbial sandstone structures in situ. The aggregational qualities of sand were explored through the appropriation and redefinition of existing sand dunes. Instead of viewing the dunes as hills of sand built by aeolian processes, they were read as readymade building volumes. This conceptual shift opens for the process of densification, adhesion and petrification described above, binding the grains together into solid sandstone surfaces: a floor datum; a wall membrane; a spatial divider. Once excavated, the biocemented sandstone surfaces demarcate habitable spaces, turning the interior and exterior of the dune into a perfectly seamless, seemingly 'weightless' architecture of solidified sand caves inside of the initial volume. Over time, more and more loose sand is eroded away from the structure. New strands of the dune can then be precipitated with the microorganism, producing an infinitely variable structural system.

Two construction options were further developed: pneumatic balloon precipitation and injection pile precipitation.

The former involves creating a pneumatic, balloon-like structure, filling it with a mixture of *B. pasteurii* and the necessary nutrient medium, and then allowing a sand dune to migrate and wash over the vessel. Once the wind has sculpted the sand dune into an optimal shape for the solidification process to take place, the microbial solution is distributed through specifically designed apertures in the skin of the structure, solidifying the sand surrounding it into structural compressive surfaces. Some time after the solidification, once the interior of the resulting sandstone structure has reached optimal strength, the balloon structure is recycled for the construction of the next stretch of the scheme. This method could be viewed as similar to that of the application of spraycrete, the high-strength polymer coating designed predominately for the re-surfacing of old concrete and tarmac. Using air compressors and various levels of air pressure, the spraycrete can be sprayed straight onto the surface to be treated (Fig. 4).

The second alternative is to use injection piles. These piles—normally used for grouting processes—would be pushed through the dune, after which a first layer of bacteria would be distributed through the piles to solidify an initial surface within the dune. They are then pulled up, creating almost any conceivable (structurally sound) shape along the way, with the loose sand acting as a jig or mould before being excavated either by hand or by the wind to create our spaces. The procedure would be analogous to using an oversized 3D printer, solidifying parts of the dune as needed (Manaugh 2009) (Fig. 5).

Furthermore, this method would be in line with one particular system that has been developed specifically for the purpose of using calcium carbonate as a laboratory cementation agent: the Calcite In Situ Precipitation System, or CIPS. This involves injecting a proprietary chemical solution that causes the precipitation of calcite crystals within the pore fluid and on the surfaces of constituent sand grains (Ismail et al. 1999a). Studies using CIPS have shown that the solidification level and cementation rate can be altered by using different chemical formulations or multiple solution flushes (Ismail et al. 1999b).

The mechanical properties of the soil itself (cohesion, friction, stiffness, and permeability) are important parameters for engineering constructions and ecosystems in sedimentary environments. In order to control the injection through the piles, it is necessary to find ways of distributing bacterial activity homogenously in the sand bed in order to prevent clogging during injection, as well as to provide homogeneous reinforcement results. A recent study describes a methodology for distributing and fixing bacteria (with their enzyme activity) relatively homogenously in for instance a sand dune, before supplying cementation reagents. The method involves a multi-step injection procedure: a bacterial suspension is injected into the fine-grained sand body, immediately followed by a fixation fluid (a solution with high salt content). The bacteria are retarded by adsorption and filtration processes, and are permanently adsorbed to the sand grains when overtaken by the fixation fluid. Finally, the cementation fluid is introduced. One advantage of this method is that the composition of the bacterial suspension and fixation fluid can be varied to either stimulate adsorption and flocculation with high salinity, or to stimulate transport and remobilization of reversibly adsorbed bacterial cells to improve bacterial distribution with low salinity (Harkes et al. 2010).

Yet another construction method could involve solidifying just the surface of the dune rather than creating solidified subsurface spaces, using much simpler distribution methods that might perhaps include the use of machines similar to

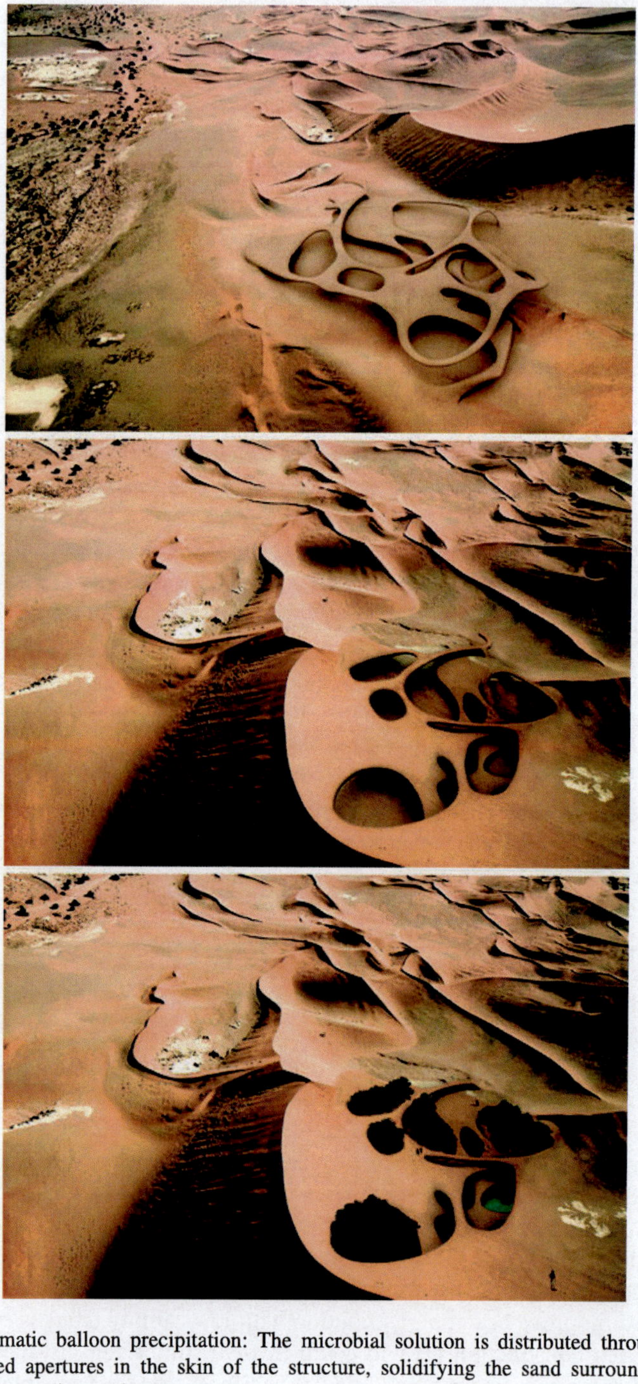

**Fig. 4** Pneumatic balloon precipitation: The microbial solution is distributed through specifically designed apertures in the skin of the structure, solidifying the sand surrounding it into structural compressive surfaces

**Fig. 5** Injection pile precipitation: This construction method would be analogous to using an oversized 3D printer, solidifying parts of the dune as needed

today's fertiliser spreaders—or even agricultural robots capable of driving themselves using GPS maps and electronic sensors—and to then connect these surfaces to the loadbearing sandstone structure. When deployed as shading devices rather than structural elements, these shell surfaces could become rather thin. Following in the footsteps of engineers such as Félix Candela, Eduardo Torroja, Pierre Lardy, Hans Hauri, Jörg Schlaich, and, in particular, Heinz Isler (Chilton 2000), these delicate shells could be designed so as to pronounce the slender character of the layers within the striated sandstone. Different methods for flushing the bacteria through the sand, as well as different sand qualities, could be investigated in order to find ways of articulating the material properties and characteristics further. The possibilities for research within this area are vast.

The five-minute theorem for masonry could be used as a basic statement to quickly assess the structural capability of such a shell in situ. This rule of thumb states that if the structure stands for five minutes once its supports have been removed, the boundaries of the masonry contain its thrust line, and will remain standing for 500 years. Admittedly, this flamboyant statement does assume static loads arising from self-weight, and dead thrust rather than resistance against live loads such as wind forces and migrating sand, while also ignoring any interaction with the foundations, but as an indicative test, it could still provide a historically proven demonstration of structural soundness (Heyman 1995).

The cementation of a previously uncemented sand dune lowers the effective void ratio of its sand mass, which in turn induces minor decreases in the porosity and increases in the density. As we solidify the dune, we also densify it. These changes occur through cementation, adding substance to the particulate mass forming around individual particles and at particle–particle contacts.

So how long would the microbial part of the construction process take? According to DeJong's study, factors critical to the success of the microbial treatment include pH, oxygen supply, metabolic status, concentrations of microbes, ionic calcium in the biological and nutrient treatment flushes, and the timed sequence of injections. While the treatment time is dependent on numerous factors such as microbial concentration, reaction kinetics, and soil characteristics, DeJong found the maximum shear wave velocity versus time value to be 1,700 min, and the relative amount of cementation resulting from each treatment to reach its peak towards the fifth or sixth injection. This would indicate that an initial sandstone surface could be created within the dune in approximately 28 h, and that the structure could reach its optimal structural strength in a single week ($1,700 \times 6 = 10,200$ min) (DeJong et al. 2006). In private conversation with the author, however, DeJong has indicated that variations are likely to occur, and that "the more patient you can be, the better" (DeJong 2008).

Soils are ecosystems, geoenvironments in which microorganisms are ubiquitous due to their great diversity and fast reproduction rate. For microbial activity to occur, a source of carbon for cell mass is required, as is a source of energy to sustain life activity, water, and a favourable environment. All of these requirements could be artificially controlled during the construction process, negating arguments about these strategies being recipes for disasters due to uncontrolled solidification.

However, many aspects still remain to be researched before a final construction method can be developed. Cost is one, though initial calculations seem to indicate that microbial construction could be cheaper than concrete construction, once the necessary machinery has been developed and accounted for. The best way of assessing the economic performance may be to meditate on the alternative: losing land to the desert, watching people die in the wake of desertification. Viewed in that light, the cost of prevention is far less than that of the cure.

Another aspect is the practical issue of supplying the microbes with water in a desert environment (though any physical intervention would clearly require water at some stage). While *B. pasteurii* is a very resistant strain that replicates easily, can live under extreme conditions, and is non-pathogenic, and while the bacteria are able to grow at high ammonium concentrations, tolerate alkaline environments, and can be grown on an industrial scale in non-sterile environments, how exactly to grow them in the desert remains an open question. Resources need to be allocated so as to provide for the further testing of soil qualities, there is a demand for the reconsidering of proper sampling and laboratory testing procedures or protocols to account for microbiological activity, and a call for rescrutinising some poorly explained laboratory results and field responses (Mitchell and Santamarina 2005). To do this properly, geotechnical engineers and 'dune architects' will need a better understanding and appreciation of both geochemistry and microbiology—as well as architecture.

But there is also a need for the architectural community to open up to new research topics, new materials, new paradigms. The promise of the use of biological treatments has already been demonstrated within a wide range of other fields. The study by DeJong et al. (2006) mentions microbes being used in environmental applications including the stabilisation of metals, the development of biological shields for zonal remediation, encapsulation of hazardous and other contaminants in natural soils, microbially enhanced oil recovery, bacteriogenic mineral plugging, and the remediation of cracks in concrete structures. What it does not mention are architectural, constructional applications, presumably since very few have been conceived of. Whereas future architects obviously need not become microbiologists, it is the author's hope that the present project stimulates interest in seeking further architectural implementations of the here described or similar biological processes. Multiple new opportunities for both soil-based and other innovative "green" construction technologies can be envisioned. Opening up a serious conversation between geotechnical engineers, microbiologists, other experts in the field, and architects can only advance the state of knowledge and practice in this area.

## 10 Formal Strategy: Tafoni

So we have a way of turning sand into sandstone in order to create spaces that people can live in, inside of the sand dunes. But how should those spaces function? How are they organised? What are the structural considerations given to configuration versus form? And what should they look like?

The architectural form of the first execution of *Dune* is inspired by tafoni, a kind of cavernous rock weathering/erosional patterns. Tafoni often include nested and cellular forms, found in granular rock such as sandstone, and hypothesised to be the results of salt weathering. It appears that this weathering might be due to water bringing dissolved minerals to the rock surface and drying, which makes the minerals form crystals that force small particles of the interior of the rock to flake off, leaving parts of the harder surface layer (had the crust not been harder, the whole rock surface would erode more or less evenly).

Tafoni, also known as honeycomb weathering patterns, are usually easily recognisable from their rounded entrances and smooth, concave walls: exquisite structures that typically develop in groups on inclined or vertical surfaces. The development and evolution of tafoni are puzzling, and continue to arouse curiosity (Hejl 2005). Since the late nineteenth century, more than 100 research articles have been published on this geomorphic topic (Boxerman 2009) (Fig. 6).

Tafoni are known to cause rapid coastal landscape retreat on geologic time scales (in desert landscapes, the retreat is slower). It has been estimated that weathering processes leading to the development of tafoni patterns cause 10% of all coastal retreat (Gill et al. 1981). On human time scales, tafoni destroy important sea walls and monuments: during a visit in 2008, this author noted tafoni patterns on the ruins of the otherwise exceptionally well-preserved Greek temple of Poseidon at Sounion (probably destroyed by the Persians in 480 BC) (Spawforth 2006).

Scaling up a tafone or a cluster of tafoni provides us with an interesting starting point for a spatial investigation: the hollow structure allows light to shine through

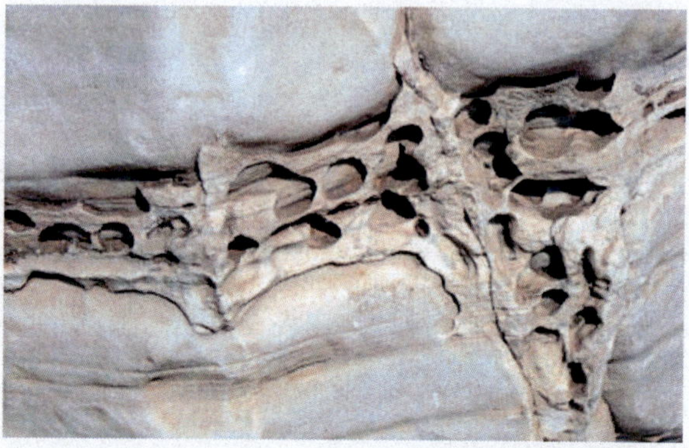

**Fig. 6** Tafoni detail: A kind of cavernous rock weathering/erosional patterns, Tafoni such as these, from the Greek temple of Poseidon at Sounion, often include nested and cellular forms, found in granular rock such as sandstone. They are hypothesised to be the results of salt weathering. (Photo: Magnus Larsson)

**Fig. 7** A certain level of formal control would be lost to nature as the bacteria solidify the sand, giving birth to a boundless beauty, the traces of *B. pasteurii* being harnessed to sculpt the desert

yet shelters from the sun, and could provide a good formal precedent for the provision of ventilation and thermal comfort. The highly differentiated and articulated section of a typical tafoni space can be viewed as a cluster of voids regulated by a highly specific yet essentially unknown growth pattern, a typological branching or stacking model that can be controlled through their structural connection points, their ratio between void and solid, their loadbearing capacity, and so on.

Basing the aesthetic, performative, and organisational strategy on tafoni gives the project a visual impact that seems surreal and continuous—the solidified dune mitigating against the migrating dune whilst offering interesting spaces; porous sandstone surfaces providing selectively shaded areas. By sculpting those surfaces according to local parameters, the wind can be controlled so that the resulting spaces shelter from sandstorms while providing ventilation and comfort. A series of pockets of sponge-like spaces could be created, voids in which to gather for a conversation, a meal, a prayer. A certain level of formal control would be lost to nature as the bacteria solidify the sand. This could give birth to a boundless beauty, the traces of *B. pasteurii* being harnessed to sculpt the desert into these habitable environments (Fig. 7).

## 11 Dune Living

One of the striking but perhaps not so obvious challenges attached to the planting of a green wall in the Sahel region is that the current level of poverty brings people to chop the trees down for firewood. This has also been a historical problem with shelterbelts, for instance in Algeria's 1970s "steampunk geoengineering project," the planting of a wall of trees up to 25 km wide and 1,200 km long that was to stretch along the length of the Sahara's edge, from the Moroccan to the Tunisian border. The planned scheme ended not only with the arrival of the pine processionary moth and the shortage of funding, but also with the local population—separated from the project's planning and planting phases—viewing the trees as a good source of building materials and firewood (Twilley 2009).

However, there is an alternative to planting trees and hope that they would not get chopped down. The habitable sandstone structure proposed in these pages essentially does three things: it adds a roughness to the surface texture of the dune in order to bind the grains to the ground and thus aid the mitigation against saltation, it provides physical support for the shelterbelt trees, and it creates habitable spaces inside of this barrier. People living inside the Green Wall could protect the trees from both humans and the forces of nature. Inside the dunes, the inhabitants have a better chance of finding shade, harvesting condensation, and beginning to green the desert from within using permacultural strategies for water harvesting and desert cultivation (Fig. 8).

Crucial to this is the temperature difference between the interior of the solidified dunes and the exterior dune surface, which makes it possible to start building an oasis-like, permacultural network, the nodal points of which could support

**Fig. 8** Performative section: The habitable sandstone structure adds a roughness to the surface texture of the dune (1), provides physical support for the shelterbelt trees (2), and creates habitable spaces inside of this barrier (3)

water harvesting and habitable thermal comfort zones. In some parts of the scheme, this would merely stabilise the shelterbelt and articulate the ground so as to provide for better local water management (adaptively storing what little rainwater there is, supplying shaded areas, creating connections to underground condensation spaces, and so on) as well as create a mitigatory shelter against wind, sand, and animals.

In nodal areas the intervention could become much more advanced, creating actual habitations or villages, with communal spaces, collaborative agriculture schemes, and large-scale water management systems; dune schools, dune mosques, dune plantations, dune wells, dune factories. The proposal is based on permacultural feedback loops. Water is a good example: from rainwater swales through to the creation of microbial sandstone tunnels down to aquifers. The latter could pick up on the history of the Garamantes tribe, a Saharan Berber-speaking people who used an incredibly elaborate network of tunnels known as *foggaras*—a subterranean water-extraction and irrigation system that allowed their part of the Sahara to bloom again—turning them into a local power between 500 BC and 500 AD (Keys 2004).

Water is obviously key. The poverty in the Sahel region means that drought can cause famine while good rains can cause drops in crop prices. *Dune* could hopefully form a stable base from which to fight back against both effects. Once the structure is in place and the permacultural network begins to support water harvesting and habitable thermal comfort zones, the economical sustainability of regions in dire need of such improvements could be increased. Social and economic conditions (poverty and food security) have a major impact on the progress and control of desertification. What we cannot do is stick our heads in the sand.

What would it be like to live in one of these sandstone structures? It would be a perfectly novel experience—or a highly familiar one. The tafoni form employed for the first iteration of the scheme could make up one part of it, but in others, *Dune* could take on traditional forms, indeed any form that could be made with more common materials, from adobe to concrete. The architecture could be orthogonal rather than sinuous, vernacular rather than contemporary, or a combination of all of them. The structure would have to support local habits and building traditions, and would need to find ways of braiding one such tradition into the next, possibly across national and religious borders. The three things the different potential adaptations would hopefully share is a connectedness within the architectural refuge, a common materiality—a seamless plasticity—as the microbes close some of the gaps in between its grains and turns it into sandstone, and a shared opportunity to use the structure to improve the local economy.

In economic terms, one aspect we must not ignore is solar power. Journalist and author George Monbiot has pointed out that we are currently experiencing an inverse relationship between human habitation and the availability of solar power, which is most concentrated and most reliable in deserts. Monbiot notes that solar electricity "generated in Sahara could supply all of Europe, the Gobi could power China, and the Chihuahuan, Sonoran, Atacama and Great Victoria Deserts could electrify their entire continents." He further mentions a calculation done by the

International Energy Agency, which shows that if solar photovoltaic panels were used to cover 50% of the land surface of the Earth's major deserts, they would produce 18 times as much energy (or 216 times as much electricity—16,742 TWh out of a suggested total of 200,000 TWh for the entire planet) as the world uses today. If people used electricity only when the sun was shining, this would mean that we only need to cover 0.23% of the land to meet demand, and "When the sun in the eastern Sahara is going down, the sun in the western Sahara is at full strength" (Monbiot 2007). The bacterially solidified sand structure proposed here could provide some of the surface area and foundation structure for such a solar panel scheme, as well as house long-distance cables to aid in the intermittency of this ambient power.

## 12 Towards Tattoine: City of Quartz

Whereas the first iteration of *Dune* was based on a system of tafoni-like habitable spaces, and future iterations may focus on the possibility of creating a novel kind of solar power facility, the very next step in the process is to bring the scheme onto the railroading scene.

On 5 March 2010, fashion designer Ozwald Boateng announced his plans to initiate a massive development plan that involves a $200 billion bond and a high-speed rail network connecting countries in Western Africa.

The scheme is the latest initiative by Boateng's Made In Africa organisation, aimed at introducing innovative ideas and new capital to help Africa go from being a developing continent towards achieving an emerging status. To date, the organisation's projects have included a mobile banking scheme in east Africa, plans for a low-carbon city in the centre of Kampala, and the planting of highly effective bio-energy trees on a 20,000-acre site (Reuters 2010).

The new pan-African railway is intended to function as a catalyst for economic and social growth. Envisioned as a high-speed connection, the railway is currently suggested to link the port of Tripoli on Libya's northern coast to the port of Takoradi in Ghana. Land would be cultivated within a strip on either side of the rail track, and state-of-the-art sustainable cities would be built at the stops.

Whereas it is still too early to go into any further detail about the scheme, the next version of *Dune* is currently being planned as part of this ambitious undertaking. At the point where the line of the cross-Saharan shelterbelt crosses the line of the new railway, a sandstone city could be built that would both function as an urban centre in its own right, and open up for tourism to this new, sustainable, infrastructural, and cultural desert node. Like Tattoine, the fictional desert planet and setting for many scenes in the *Star Wars* saga (Lucas 1977), this new city could allow moisture farmers to live in subterranean dwellings and possibly even grow crops in subterranean, hydroponic labs—though it would also aim for a more contemporary, Masdar-like quality. An environmentally ambitious, pedestrian-friendly, self-sustaining development encouraging urban growth in what is today a landscape in which it is near-impossible to feed or house human populations. An ecological desert city made from microbial sand, future home of the Sahelian Dream, the true City of Quartz.

## 13 Closing Remarks: Everything and Nothing

Estimates of recent losses of productive land suggest that the world had lost close to one-third of its arable land in the 40 years leading up to 1995 (Pimentel et al. 1995). Most specialists seem to agree that we will continue along this trajectory into the future.

As always with complex situations, action against desert-related challenges must not await complete knowledge of the circumstances. Immediate efforts need to be made using our existing understanding, not only to stop the physical processes of desert encroachment, but also to create opportunities for the people of the Sahel to get educated in how to minimise the harm done to fragile ecosystems—and to provide them with better standards of living. I believe that *Dune* could be part of such a programme. And I think the first stretches of the scheme could be created following a rather brief period of research.

In the longer run, extensive research is needed to delineate the full impact of the conceptual outline sketched above. While *Dune* underlines the need for urgent short-term relief measures, it can also be viewed as a long-term programme to prevent further desertification. This should not be delayed either: again, the cost of prevention is bound to be far less than that of the cure.

Architects create spaces that accommodate human activity. As opposed to many of its contemporary counterparts, *Dune* is not so much focused on the *styling* of that activity, as on the *supporting* of it. While designed to visually seduce, *Dune* is not primarily a formal exercise, but a social, ecological, cultural one. How are we to live with the desert, in the desert, within the desert? As architects, we have the mechanisms to understand and respond to such questions, the ability to allow them to adjust the various forms and accommodations of functionalities within the systems we design. Hopefully such adjustments turn *Dune* into a conceptually striated system, built up in layers like a sedimentary stone, possible to read on several different levels.

Let's have a closer look at one of those layers by returning one final time to the notion of *aggrerosion.* The self-organising, gregarious grains of sand almost seem to enjoy taking on different shapes as time passes, patterning the ground into three-dimensional traces of the paths they have been propelled along across the desert floor in the short and shallow trajectories known as saltation: rocky surfaces, flat sand sheets, vast dune fields. Though stratigraphically positioned within a much longer geologic timescale, sand also offers us the prospect of a rapid construction (and almost instant destruction) cycle carrying inherent possibilities of constant formal and physical invention and reinvention. In between solid and liquid, sand is an endlessly fascinating material. Its defining feature is that it aggregates—indeed, this is the mysterious heap-of-sand quality of the Sorites Paradox (Fisher 2000)—and as we have seen, all design is fundamentally about aggregation and erosion: we add and we take away.

As with other natural systems, aggregations of granular materials are self-stabilising optimisation machines. Changes in the internal or external environment have direct consequences on their emergent organisations and forms, many of which are breathtakingly beautiful: solemnly sweeping ridges of sand dunes in the desert, honeycomb rock weatherings created through salt crystallisation: the intricate geometries and endless variety of the natural phenomena at the heart of *Dune's* formal precedents.

If the morphologies of granular media are inspiring, some of their properties and potentials when interacting with biological agents such as microorganisms seem to be astonishingly underdeveloped in architectural terms—and even, to an extent, in the world of material sciences. The biocementation of sand is an example of such a granular material process. In the rare instances when it has been suggested that microorganisms be employed as a construction element in our built environment, the aim has usually not been architectural—exploring formal, performative, algorithmical, material, and similar properties—but rather related to large-scale, mitigatory engineering projects.

The novel process of engineered *architectural lithification,* creating from a pile of loose sand a solid sedimentary rock structure, a sandstone *building,* effectively involves gluing one grain of sand to the next on a microscopic level, thereby having the material shift from sand to stone. The resulting architecture becomes an infinitude of binary bits, a system of either/or relationships: solidified mass or excavated void, sand that's been precipitated or left untouched, that is grounded or fluid, sand or sandstone, very light or very heavy. The building speaks of the chronology of the sand, the vast rhythms of geological history, the evolution of villages and cities built on sand, from sand, covered in shimmering grains, forgotten in a sea of sand.

Such are the radical tectonics of similarly-sized building blocks, some of which are readily available to us today, but which constitute a conceptual virgin territory for which there is still very few maps. A cartography that *Dune* seeks to expand.

A single grain of sand is almost nothing, yet at the same time almost everything. At the end of his short story *A Passion in the Desert,* Honoré de Balzac concluded: "In the desert, you see, there is everything and nothing … It is God without mankind" (Balzac 1830). He was almost right. As we have seen, despite the Biblical admonition against building a house on sand (Matthew 7:26), close to a billion people live in arid or semiarid environments, often in substandard dwellings, but in close proximity to massive volumes of sand that we now know how to turn into pioneering buildings. Mankind has already moved into the desert. To extend the Biblical analogy, the question now is how to bring God closer to them in a landscape in which sandstorms are far more common than brainstorms.

The answer, my friend, is blowing in the wind.

**Acknowledgments** The author wishes to thank Alex Kaiser and Fredrik Nordbeck for their invaluable assistance in producing the images accompanying this text. Thanks to Steve Hardy and Jonas Lundberg, unit masters of Dip 16, as well as technical tutors John Noel, Wolf Mangelsdorf, and Michael Weinstock, all at the Architectural Association in London. At University College London: Adrian Jones, Claire Cousins, and particularly John Ward. Stefano Ciurli at the University of Bologna. Jason DeJong at University of California at Davis. Jonathan Boxerman for enthusiastic tafoni guidance. Tosin Oshinowo and Papa Omotayo in Nigeria. Geoff Manaugh at Bldgblog. Chris Anderson and Bruno Giussani at TED. Steve Hardy and Stefano Ciurli, again, for generously agreeing to referee this chapter. And finally Madelaine Levy, the most vigilant and occasionally acerbic editor on the planet, for tirelessly supporting the seemingly impossible. Any remaining errors and omissions are, of course, my own.

## References

Adeel Z, Bogardi J, Braeuel C, Chasek P, Niamir-Fuller M, Gabriels D, King C, Knabe F, Kowsar A, Salem B, Schaaf T, Shepherd G, Thomas R (2007) Overcoming one of the greatest environmental challenges of our times: re-thinking policies to cope with desertification. In: A policy brief based on the joint international conference: desertification and the international policy imperative, Algiers, Algeria, 17–19 December 2006

African Union (2006) Launching of the Green Wall Sahara Programme. Division of communication and information, press release 6. http://www.africa-union.org/root/ua/Actualites/2006/decembre/green%20wall%20launch.doc. Viewed 8 April 2010

AU/CEN-SAD (2009) Plan of action for the implementation of the Great Green Wall of the Sahara and Sahel Initiative. Draft for submission to the AU Executive Council. Addis Ababa, Ethiopia, 1–3 February 2009

Bagnold RA (1941) The physics of blown sand and desert dunes. Methuen, London

Bak P (1996) How nature works: the science of self-organised criticality. Copernicus Press, New York

Balmond C (2002) Informal. Prestel, London, p 119

Balzac H (1830) A passion in the desert. English translation appeared in The Strand Magazine, February, 1891, 1:2. http://gaslightmtroyal.ca/passion.htm. Viewed 8 April, 2010

Borges JL (1974) Fragments from an "Apocryphal Gospel", from the collection "In Praise of Darkness", Dutton, New York. Original quote: "Nada se construye sobre piedra; todo se construye sobre arena, pero debemos construir como si la arena fuese piedra."

Boxerman JZ (2009) Tafoni.com. http://www.tafoni.com. Viewed 8 April 2010

Burke M, Miguel E, Satyanath S, Dykema J, Lobell D (2009) Warming increases risk of civil war in Africa. PNAS 106:20670–20674

Burns WC (1995) The international convention to combat desertification: drawing a line in the sand? Mich J Int Law Mich 16:831

Campbell-Purdie W (1967) Woman against the desert. Gollancz, London

Chilton J (2000) Heinz Isler. The Engineer's Contribution to Contemporary Architecture series: RIBA Publications/Thomas Telford, London, pp 20–29

DeJong JT (2008), private communication, 12 March 2008

DeJong JT, Fritzges MB, Nüsslein K (2006) Microbially induced cementation to control sand response to undrained shear. J Geotech Geoenviron Eng 132:1381–1392

Dell'Amore C (2009) Africa-wide "Great Green Wall" to Halt Sahara's Spread? National Geographic News, 28 December 2009. http://news.nationalgeographic.com/news/2009/12/091228-great-green-wall-trees-senegal-sahara-desert.html. Viewed 8 April 2010

Desanker PV (2002) The impact of climate change of life in Africa: climate change and vulnerability in Africa. WorldWide Fund for Nature, Washington, DC

Desanker PV, Magadza C (2001) Africa. In: Adaptation Vulnerability, McCarthy JJ et al (eds) Climate change 2001: impacts. Cambridge University Press, Cambridge, pp 487–531

Dollo M, Sen P (2007) Afforestation: an option for combating desertification. Arunachal Times 19(12):2

Ezigbo O (2009) Nigeria: Desertification—country loses 600 m of land mass yearly. http://allafrica.com/stories/200901260201.html. Viewed 8 April 2010

Fisher P (2000) Fuzzy Logic. In: Openshaw S, Abrahart R (eds) Geocomputation. Taylor & Francis, London, pp 162–163

Fisman R, Miguel E (2008) Economic gangsters–corruption, violence and the poverty of nations. Princeton University Press, Princeton

Gibson W (1996) Idoru. Viking Press, London

Gibson W (1999) The Science in science fiction. Radio interview with William Gibson and Anne Simon on NPR. 30 November 1999, timecode 11:55. http://www.npr.org/templates/story/story.php?storyId=1067220. Viewed 8 April 2010

Gill ED, Segnit ER, McNeill NH (1981) Rate of Formation of Honeycomb Weathering Features (Small Scale Tafoni) on the Otway Coast, S.E. Australia. Proc Roy Soc Vic 92:149–154

Grainger A (1990) The threatening desert–controlling desertification. Earthscan, London

Grove AT (1977) Desertification. Prog Phys Geogr 1:296–310

Hall P, Bierut M, Kalman T, Andersen K, Heller S, Poynor R (1998) Tibor Kalman: perverse optimist. Princeton Architectural Press, New York

Hare FK (1983) Climate and desertification–a revised analysis. World climate applications programme report no. 44. World Meteorological Organization/UNEP, Nairobi

Hare FK (1984) Recent climatic experiences in the arid and semi-arid lands. Desertif Control Bull 10:15–22

Harkes MP, van Paassen LA, Booster JL, Whiffin VS, van Loosdrecht CM (2010) Fixation and distribution of bacterial activity in sand to induce carbonate precipitation for ground reinforcement. Ecol Eng 36(2):112–117

Hejl E (2005) A pictorial study of Tafoni development from the 2nd Millennium BC. Geomorphology 64:87–95

Heyman J (1995) The Stone Skeleton–Structural Engineering of Masonry Architecture. Cambridge University Press, Cambridge, p 24

Hulme M (2001) Climatic perspectives on Sahelian desiccation: 1973–1998. Glob Environ Chang 11(1):19–29

Hurt RD (1995) Forestry on the Great Plains, 1902–1942. Lecture presented in 1995 at Kansas State University; http://www-personal.ksu.edu/~jsherow/hurt2.htm. Viewed 8 April 2010

IPCC (2007a) Climate change 2007: Impacts, Adaptation and Vulnerability. Contribution of working group II to the fourth assessment report of the IPCC. http://www.ipcc.ch/ipccreports/ar4-wg2.htm. Viewed 8 April 2010

IPCC (2007b) Climate change 2007: mitigation of climate change. Contribution of working group III to the fourth assessment report of the IPCC. http://www.ipcc.ch/ipccreports/ar4-wg3.htm. Viewed 8 April 2010

Ismail MA, Joer HA, Randolph MF, Kucharski E (1999a) Cementation of porous materials using calcite precipitation. University of Western Australia, Geomechanics Group, Geotech. Rep. G1422

Ismail MA, Joer HA, Randolph MF, Kucharski E (1999b) CIPS, a novel cementing technique for soils. University of Western Australia, Geomechanics Group, Geotech. Rep. G1406

Jauffret S, Woodfine A (2009) Scope and pre-feasibility study on the Great Green Wall for the Saharan and Sahel Initiative (GGWSSI). Hernel Hempstead, Hertfordshire

Johnston A (1999) William Gibson: all tomorrow's parties: waiting for the man. Spike magazine, August 1999. http://wwwSpikemagazine.com/0899williamgibson.php. Viewed 8 April 2010

Kelly K (2007) Everything that doesn't work yet. http://www.kk.org/thetechnium/archives/2007/02/everything_that.php. Viewed 8 April 2010

Keys D (2004) Kingdom of the sands. Archeology 57(2):24–29

King FH (1911) Farmers of forty centuries: or, permanent agriculture in China, Korea and Japan. Madison, Wisconsin. http://www.gutenberg.org/etext/5350

Koechlin J (1997) Ecological conditions and degradation factors in the Sahel. In: Raynaut C, Grégoire E (eds) Societies and nature in the Sahel. Routledge, London, p 12

Koolhaas R (1995) S, M, L, XL (2nd edition 1998). The Monacelli Press, New York, pp 2–21

Krech S, McNeill JR, Merchant C (eds)(2004) Russia and the Soviet Union. Encyclopedia of world environmental history. Routledge, New York

Le Metayer-Levrel G, Castanier S, Orial G, Loubiere JF, Perthuisot JP (1999) Applications of bacterial carbonatogenesis to the protection and regeneration of limestones in buildings and historic patrimony. Sediment Geol 126:25–34

Lucas G (1977) Star Wars (later retitled Star Wars Episode IV: A New Hope): Lucasfilm/20th Century Fox, 25 May 1977

Manaugh G (2009) Sand/Stone. Blog entry, Bldgblog, April 2009. http://bldgblog.blogspot.com/2009/04/sandstone.html. Viewed 8 April 2010

Martin M (2004) Deserts of the Earth. Thames & Hudson, London

Middleton N, Thomas D (eds) (1997) World atlas of desertification. Arnold, Hodder Headline plc, London

Mitchell JK, Santamarina JC (2005) Biological considerations in geotechnical engineering. J Geotech Geoenviron Eng 131(10):1222–1233

Monbiot G (2007) Heat–how we can stop the planet burning. Penguin, London, pp 105–107

Nemati M, Voordouw G (2003) Modification of porous media permeability, using calcium carbonate produced enzymatically in situ. Enzyme Microbiol Technol 33:635

Pettijohn FJ, Potter PE, Siever R (1987) Sand and Sandstone, 2nd edn. Springer, New York

Pimentel D, Harvey C, Resosudarmo P, Sinclair K, Kurz D, McNair M, Crist S, Shpritz L, Fitton L, Saffouri R, Blair R (1995) Environmental and economic costs of soil erosion and conservation benefits. Science 267:1117–1123

Reuters (2010) Visionary African creative: Ozwald Boateng Announces Cross Sahara Railway. Video clip, Reuters, 5 March 2010. http://www.reuters.com/news/video?videoId=52377278. Viewed 8 April 2010

Spawforth T (2006) The complete greek temples. Thames & Hudson, London, pp 145–146

St. Barbe Baker R (1944) I planted trees. Lutterworth Press, London, Redhill

Stern N (2007) The economics of climate change. The stern review. Cambridge University Press, Cambridge

Takahashi G (2006) Aggregates 03. In: Hensel M, Menges A (eds) Morpho-ecologies. Architectural Association, London, pp 286–295

The Engineer (2007) Editorial, bacteria help protect from quakes. 23 February 2007. http://www.theengineer.co.uk/news/bacteria-help-protect-from-quakes/298382.article. Viewed 8 April 2010

Time (2007) Mighty microbe. The best inventions of 2007. November 2007. http://www.time.com/time/specials/2007/article/0,28804,1677329_1678027_1677996,00.html. Viewed 8 April 2010

Twilley N (2009) The Great Green Saharan Wall Redux. Edible geography. Blog entry. 31 December, 2009. http://www.ediblegeography.com/the-great-green-saharan-wall-redux/. Viewed 8 April 2010

United Nations (1977) Draft Plan of action to combat desertification. UN conference on desertification, Nairobi, 29 August–9 September 1977, Document A/CONF.74/L36, UNEP, Nairobi

United Nations Conference on Desertification (UNCOD) (1978) Round-up, plan of action and resolutions. United Nations, New York

United Nations, Department of Economic and Social Affairs, Population Division (2007) World population prospects: the 2006 revision, highlights, working paper no. ESA/P/WP.202

van Paassen LA, Daza CM, Staal M, Sorokin DY, van der Zon W, van Loosdrecht MCM (2010) Potential soil reinforcement by biological denitrification. Ecol Eng 36:168–175

Weinstock M (2010) The architecture of emergence—the evolution of form in nature and civilisation. Wiley, Chichester, pp 81–87

Welland M (2009a) Sand–the never-ending story. University of California Press, Berkeley

Welland M (2009b) Sandstone-making microbes, Tafoni: and an extraordinary design idea. Blog entry, Through the Sandglass, 29 April 2009. http://throughthesandglass.typepad.com/through_the_sandglass/2009/04/index.html. Viewed 8 April 2010

Whiffin VS, van Paassen LA, Harkes MP (2007) Microbial carbonate precipitation as a soil improvement technique. Geomicrobiol J 24:417–423

# Dune Sand Fixation: Mauritania Seawater Pipeline Macroproject

**Viorel Badescu and Richard B. Cathcart**

## 1 Introduction

Wide-spreading actively migratory sand dune fields are mainly found in the Earth's climatically designated desert regions—"hot deserts" cover ~14.2% of Earth's land (Peel et al. 2007; Parsons and Abrahams 2009). Some eremologists suspect that "global desertification", a persistent decline of ecosystems' benefits for humans—loss of utility or potential utility of land—in already dry regions, is occurring and will increase as the twenty-first century unfolds (Yizhaq et al. 2007). "Drylands cover about 41% of Earth's land surface and are home to more than 38% of the total global population of 6.5 billion" (Reynolds et al. 2007). Here, however, we focus only on certain active sand dune fields located in the northern Africa coastal nation of Mauritania where few people live and work today (Badescu and Cathcart 2008).

Mauritania's sand dune fields are well developed depositional landforms of increasingly better known accumulation history (Lancaster et al. 2002; Martinez and Psuty 2008). The geographically extensive and massive Mauritanian sand dunes contain a large volume of aeolian-moved sediment and are situated at three major depositional physiographic sites: (1) Akchar; (2) Aouker and at (3) Majabat al Koubra, the farthest inland (Fig. 1). The active sand dune fields of Mauritania exist in a region that nowadays receives <150 mm of precipitation per year. According to psammology, dune sand is an inert "soil" without any positive characteristics for flora due to its coarse particles and big pore spaces that do not retain water for plant growth, the high permeability and leaching that removes

V. Badescu (✉)
Polytechnic University of Bucharest, Bucharest, Romania
e-mail: badescu@theta.termo.pub.ro

R. B. Cathcart
Geographos, Burbank, CA, USA

V. Badescu and R. B. Cathcart (eds.), *Macro-engineering Seawater in Unique Environments*, Environmental Science and Engineering,
DOI: 10.1007/978-3-642-14779-1_21, © Springer-Verlag Berlin Heidelberg 2011

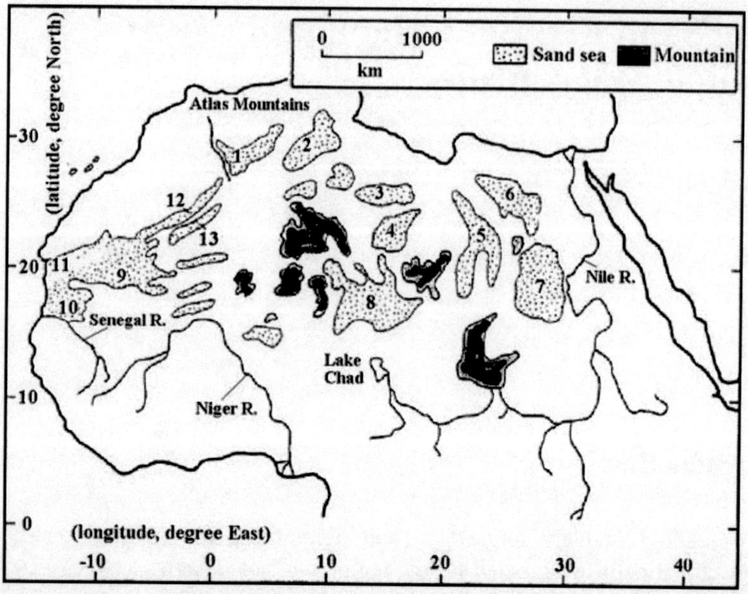


**Fig. 1** Major ergs of the mapped Sahara. *1* Grand Erg Occidental, *2* Grand Erg Oriental, *3* Ubari, *4* Murzuk, *5* Calanscio, *6* Great Sand Sea, *7* Selima, *8* Fachi-Bilma and Ténéré, *9* Majabat al Koubra, *10* Aouker, *11* Akchar, *12* Iguidi, *13* Chech. Adapted from Fig. 4. of Ba et al. (2001)

plant nutrient elements, and the wind erodibility of sand. We estimate that there are, approximately, $1.3 \times 10^9$ grains of 1 mm-diameter sand per cubic meter of any voluminous sandy dune in Mauritania. Water evaporation from immobilized wetted sand, compared to a still pool of standing freshwater, is ~12% less under identical climatic circumstances (Pavia 2008; Barclay 2009).

Based on the 1958 fieldwork experiments of O. Wipplinger in southern Africa's desertic terrain, Helga Besler asserted the permeability of 0.3 mm-diameter sand grains allows freshwater to infiltrate ~20 m/day and, in the case of larger sand particles (such as 1 mm-diameter grains) as much as ~60 m/day. Freshwater evaporation becomes ineffective ~1 m below the dune sand surface (Besler 2008). Wind erosion is the main land degradation process since plant nutrient particles are removed by saltation (Goudie 2008); eroded mineral dust components increase sand dune instability (Yair and Verrecchia 2002). However, it is worth noting that whilst Mauritania is mostly affected by disasters such as drought, epidemics, floods, insect infestations and mineral dust storms, there is some slight chance of future major earthquake activity. An earthquake of Modified Mercalli Intensity VII (estimated equivalent to Richter "local magnitude" scale 5.5–6.1) causes sand-banks to collapse gravitationally. After a lot of seawater has been added to the seawater-irrigated dune sand, perhaps saturating pockets of sand in isolated locales, then under ground shaking conditions imposed by strong temblors some liquefaction process-events may occur which could bog-down temporarily any heavy mobile machines working in the ergs of Mauritania.

Aeolian movement by saltation of surface sand grains and other dry granular media can be reduced or stopped by application of chemical stabilisers such as “Nano Clay” (Desert Control 2007) or “Biochar” (Lehmann 2007), installation of fences to trap windblown sand grains, and planted vegetation to prevent sand grain deflation. It is known that natural and artificial biological soil crusts do enhance surface stabilization (Wang et al. 2007). The present-day and future movement of sand dunes, ground surface erosion of drought-reduced lakes (White and Mattingly 2006), and accumulation of mineral dust clouds, including halite and other salts, challenge macro-engineering practice as well as seriously affect the lives and life-styles of people downwind from major active and inactive dune fields (Thomas et al. 2005). So far, macro-engineering has only a few field-tested techniques to geographically fix contemporaneous migratory sand dunes: (1) remove the dune sand mechanically; (2) disperse the dune sand by mechanical reshaping and (3) immobilize the dune sand with planted vegetation (Duran and Herrmann 2006), fences, trenches, additives and other means. Such soil fixation macroprojects, of course, involve Nature’s further hands-on management by ambitious, sometimes desperate, humans (Kareiva et al. 2007). Landscape rehabilitation is based on ecological planning and design that comprehensively examines degradational event-processes caused by global and regional climate alterations as well as global and regional biodiversity changes; every extrapolated local landscape rehabilitation macro-engineering undertaking must be judged in the encompassing context of its unique geographical, economic, temporal, cultural and directly relevant domestic environmental context. And, societal adaptation to environment-changing macroprojects are always limited, in this particular case, by Mauritanian existing societal values and ethics, geoscientific knowledge and wise respect for a panoply of geohazard risks, as well as various mutable societal cultural contructs (Adger et al. 2009).

Bodele depression (Engelstaedter and Washington 2007) in Chad is Earth’s largest single source of atmospheric mineral dust, about 40% of which is deposited ultimately in the Basin of the Amazon River (Koren et al. 2006); Mauritania’s eastern drylands and desert is Africa’s second largest source of mineral dust. The mineral dust clouds emanating from northern Africa, triggered when near-surface boundary layer wind speeds are >10 m/s, quickly change the Earth’s albedo and can strongly suppress eastern North Atlantic Ocean hurricane formation (Evan et al. 2006; Schwanghart and Schutt 2008). In other words, the Sahara’s uncontrolled or mitigated mineral dust storms do have major consequences for humans (Goudie and Middleton 2001). If mineral dust clouds were eliminated technologically, then northern Africa might become a well-vegetated region. Vegetated regions have greater dust deposition as a consequence of the filtering effect of the aerial plant parts (reduction of drag forces on sand particles). Deposition of particulate minerals by any natural or unnatural process also contributes to stabilization of sand dunes. Summarizing, we propose to foster plant growth—both wild and cultivated species—on the sandy dunal surfaces of Mauritania, and eventually Chad and Libya, by the land’s intensive irrigation with seawater pumped inland. The seawater withdrawal from the world’s ocean would also serve to induce a (measurable but modest) lowering of global sea level to prevent a worldwide "rising sea level crisis" as suggested by James E. Hansen and many other experts. The seawater intake for the Mauritanian macroproject pipeline we propose here will not even approach the macro-engineering difficulties involved with laying and operating the world's deepest (914 m below sea-level) large-diameter (1,397 mm) high-density polyethylene seawater intake pipeline at Keahole Point, Hawaii, which conveys 102,190 L/min (Shuster 2003).

## 2 The Active Dune Sand Macro-Problem

Under natural circumstances, a tenfold reduction in aeolian sand migration can be induced by a mere 3% moisture increase and the deposition of sea-spray halite particles on subaerial seashore dune sand increases the angle of repose of coastal sand dunes (Jackson and Nordstrom 1998). Shallow ground mass sand grain slides, in generally saturated sandy soil at ground surface level, made unstable by water seepage, depend on the watery fluid's gravitational flow direction and the hydraulic gradient (Ghiassian and Ghareh 2008). Injection water seepage effects on the lee-side slope of sand dunes generally acts to reduce the angle of repose while water suction acts generally to increase the angle of repose (Lu and Chiew 2007). Nowadays, it is well-known that sand saturated with hypersaline solutions does retain more moisture than sand saturated with moderately and slightly hypersaline solutions (Germann and Hagrey 2008).

Sand moves by creep, saltation and aerial suspension. Seawater regularly sprinkled onto the surface of mobile sand dunes would deposit minerals in the space between granular materials, especially after the freshwater is evaporated during the daytime (Table 1) (UN Industrial Development Organization 2005). Ground penetrating radar data could be used to assess the sand's hydraulic properties during seawater infiltration (Lambot et al. 2009). In the future, one could imagine that this surface-deposited particulate material may be harvested from the topmost crust of artificially stabilized sand dunes by specialized commercial-scale mineral harvesting machines (modified for desert use and equipped with wider tyres, Caterpillar 657E Wheel Tractor Scrapers might suffice). After

**Table 1** Chemical substances deposited from seawater onto sand dunes

| Element | Fraction of deposited mass |
|---|---|
| Chlorine | 0.0194 |
| Sodium | 0.0108 |
| Magnesium | 0.001292 |
| Sulfur | 0.00091 |
| Calcium | 0.0004 |
| Bromine | 0.000067 |
| Carbon | 0.000028 |

evaporation, the mineral-rich layer and/or encrustation will contain many useful materials and its mining could obviate some current "polluting" mine excavation and processing operations. Re-irrigation with imported seawater would simply restock the mining region and continue the effort towards sand dune field immobilisation. The irrigation potential for Mauritania using naturally renewable freshwater resources is considered quite negligible (FAO 1997); sand dune seawater sprays, imitating nature's strand seawater spray, are a new form of irrigation but for the single purpose of mining seawater elements (Walker et al. 2006). In a sense, one could mine an artificial "ore" emplaced during a very short period of time. Instead of industrialised "solar ponds", people in Mauritania would be harvesting valuable minerals from "solar dunes".

## 3 Stabilisation of Sand Dune Fields in Mauritania

Mauritania (Fig. 2) is mostly desert—it is constantly hot, dry and dusty and mostly barren, with flat western Sahara plains [elevation extremes: lowest place is Sebkhet Te-n-Dghamcha (−5 m), highest place is Kediet Ijill (915 m)] and a vast encroaching sand dune field that now threatens to inundate the nation's post-1960 capital of Nouakchott (18°07′N latitude by 15° 05′W longitude). The ~10,000 ha urban region, with a population of >700,000, is situated upon a reddish sand dune-field (Chenal and Kaufmann 2008). The yearly average solar insolation level for Nouakchott is approximately 6.55 kWh/(m$^2$ day).

Mauritania's 2009 human population of >3.1 millions is distributed very discontiguously over an area of 1.03 million km$^2$; most persons are concentrated in

**Fig. 2** Mauritania

the near-sea level capital, a seaport since 1987, in the seaport of Nouadhdibou and along the Senegal River in the southern part of Mauritania. Nouakchott may, in the near-term future, have some of its below sea level strand region submerged by adjacent ocean water influx owing to significant post-construction erosion south of the new deepwater harbour jetty (Elmoustapha et al. 2007). Mauritania's territory includes approximately 25% of the Senegal River basin ($\sim$75,500 km$^2$). Mauritania and Senegal are the only two countries in northern Africa with all their agricultural production located within drylands. Freshwater is so valuable a vital commodity to those nations sharing the Senegal River's surface runoff that only $\sim$3% of the river's outflow ever reaches the North Atlantic Ocean. Constructed in the Senegal River delta by 1986, the Diama Dam, originally designed solely as a salt-wedge barrier nowadays has an additional main function as a reservoir dam that blocks tidal or oceanic storm surge seawater intrusions upriver; the Senegal River's outlet to the northern Atlantic Ocean was also changed, in 2003, by the excavation of an artificial navigation channel through Langue de Barbarie (Isupova and Mikhailov 2008; Wu 2007). Less than 50,000 ha of Mauritania's land territory is irrigated with freshwater. Of a total annual production of 191 million kWh, only 14% of Mauritania's electricity production derives from hydropower while 86% is manufactured from fossil fuel combustion.

Mauritania is seriously affected by sand dune field migration. For instance, desert sands invaded the ancient city of Chinguetti and homes on the edge of that settlement were abandoned by its impoverished and distraught residents (see Fig. 2 of Berger (2006)). Unfortunately, this is an all-too-common present-day situation in Mauritania (Fig. 3).

## 4 The Solar-Powered Sahara Seawater Pipeline (SSSP)

Flexible solar power assemblies include a flexible photovoltaic device attached to a flexible thermal solar collector (Jeon et al. 2007; Hamakawa 2005; Kurokawa 2007). Our macroproject invokes a seawater pipeline-integrated photovoltaic flexible solar power module or membrane that will not ever—or, at least, for a period of several decades—debond under very harsh desert climate conditions of intense Saharan sunshine and strongly windblown particle abrasion. The pliable photovoltaic coating will need to cover part of the upper hemisphere of the (steel, plastic, concrete or other material) pipe, which is the basic support structure, a true *infrastructure*, for the proposed electricity-generation system. Henceforth, we will abbreviate the title of this Mauritania-sited macroproject as the SSSP.

We noticed that the map in Fig. 1 of Sakurai and Sakurai (1992) devoted to irrigation of arid zones really seemed, fortuitously, to resemble northern Africa and, in particular, Mauritania's unique geography. The identity of shape can be compared with Schluter (2006). The Sakurai's placement of the inland artificial lake seems even to represent Mauritania's world-famous 38 km-wide "Richat Structure" (21°04′N latitude by 11°22′W longitude), with a central depth elevation

**Fig. 3** Satellite image of Nouakchott (Mauritania's primary city), showing erg dunal materials, arriving from the northeast, threatening engulfment of the city with sand. Date: 9 January 2001. Source: http://earthobservatory.nasa.gov/Newsroom/NewImages/images.php3?img_id=17168 (Author: Laura Rocchio, Landsat Project Science Office, using data obtained from the University of Maryland's Global Land Cover Facility. Permission: US government, public domain)

of 400 m and surrounding walls nearly 100 m higher. The Richat Structure exposes a vast flat-lying limestone in the Maur Adrar Desert (Matton et al. 2005). The Richat Structure sits at an elevation of nearly 600 m, yet the center part of the crater-shaped geomorphologic feature is ~400 m above present-day sea level. For maximum use as a seawater pool, it may be necessary to build a small dam on the Richat Structure's southwestern edge.

Roger Henri Charlier, during 1991, suggested a route for a freshwater-carrying steel or concrete pipeline starting from the seaport of Nouakchott, which currently depends entirely on a single inadequate groundwater resource, and heading eastward into the remote interior of the Sahara, finally changing direction to meet Libya (Charlier 1991) (Fig. 4) [Charlier's pipeline would, of course, supplement depleting fresh groundwater supplies obtained by Libya from its Great Man-Made River macroproject (Elhassadi 2008). If completed, Charlier's freshwater pipeline would have inaugurated a unique supply-driven approach for freshwater management in the Sahara.].

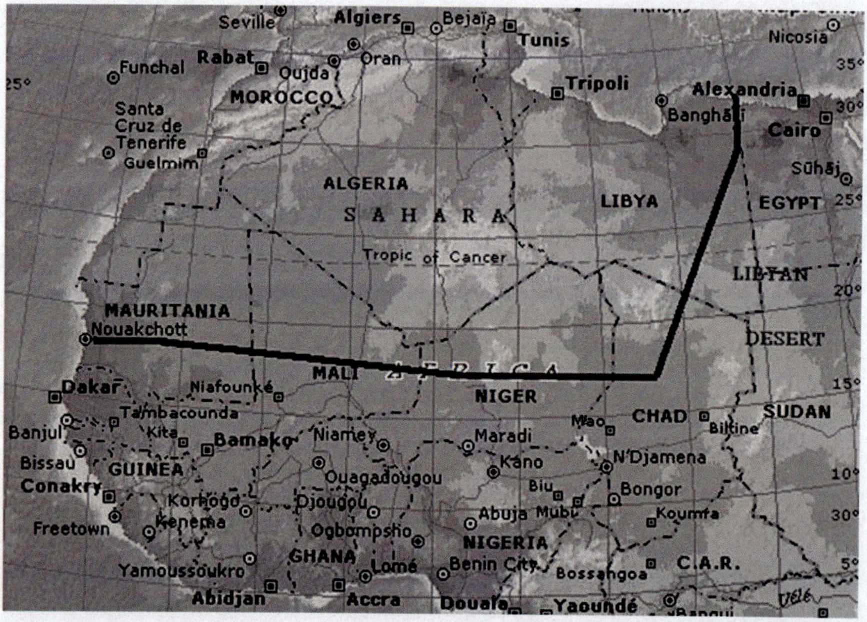


**Fig. 4** Proposed overland route of the west-to-east freshwater flow trans-Sahara pipeline (*thick black line*). Adapted from Fig. 1 of Charlier (1991)

We propose to utilize (in part) nearly the same route for the solar-powered Sahara seawater pipeline—that is, from the Capital Nouakchott to Tidjikdya (18°27′N latitude by 11°27′W longitude) the distance is ~486 km whilst we would build a seawater conveying pipeline section leaving Tidjikdya and ending near Ouadane (20°51′N latitude by 11°37′W longitude), a distance of ~265 km. Such a routing would allow use of the Richat Structure as a pooled seawater resource base from which other agricultural and industrial activities may be carried out in accord with the Sakurai's patented suggestions. Whether the Sakurai's knew it or not, this facility will not contaminate the fresh groundwater held in the "Continental Terminal" formation first discovered in 1931 (Kogbe and Dubois 1980). Therefore, construction and operation of the SSSP will not require the authoritative allocation of values in global society with respect to either freshwater or seawater (international hydropolitics).

Deserts seem likely to become prominent landscape features in our world in the near-term future years of the twenty-first century. Of course, if there is a revolutionary technical improvement in solar photovoltaic cell technology, then less land would have to be dedicated to electricity generation (Schaller et al. 2006; Mahowald 2007).

In the proposed solar-powered Sahara seawater pipe case, we assume the use of reinforced concrete or steel pipes/tubes. The presumed big hydraulic pipeline (with a large tube diameter) is expensive but such a tube has the advantage of decreasing greatly the pressure loss and increases significantly the efficiency. The SSSP pipes/

tubes may have high interior pressure so they must be composed of steel or something even stronger—perhaps from some twenty-first century-invented composite fiber material. The absence of metals would, of course, mean the absence of rusting and, perhaps, reduced interior fouling. Such composed material has higher maximum stress (up to 600 kg/mm$^2$, steel has only ~120 kg/mm$^2$) and low specific-density (~1,800 kg/m$^3$, steel is ~7,900 kg/m$^3$). Due to expected manufacturing efficiencies, it may be, or become, cheaper quite soon during the twenty-first century. In other words, no particular materials or techniques will need to be specially developed for the SSSP because we simply orchestrate the extant panoply of macro-engineering's techniques, weaving them around particular geographical and social conditions, to create a twin-goal macroproject which, in its numerous ramifications, raises a somewhat grand vision. As a critical issue for design we would expect to be longitudinal buckling of the pipeline due to thermal changes and gradients along the section due to daily and seasonal temperature variations as well as due to fluid inside the pipe having a different temperature than the surrounding environment.

# 5 Case Study

## 5.1 Dune Field Stabilization Macroproject

As an example, here we quantify the stabilization of the dunes on erg Akchar, which is one of the three major ergs of Mauritania (see Fig. 1) (Ba et al. 2001). The erg surface is about $S_{erg} = 300 \times 300 = 9 \times 10^4$ km$^2$ $= 9 \times 10^6$ ha. The distance from the ocean to the middle of erg Akchar is approximated to $L_{duct} = 150$ km, which is also the design value of the duct length accepted here. To estimate the mean altitude of the erg's dunes we used topographic data for the geographic zone 18–22°N latitude and 15–10°W longitude. We used digitized data downloaded from the Global Topography project (Smith and Sandwell 1997) (Fig. 5). The mean altitude of the erg we obtained by numerically processing the topography data is $H_{erg} = 279$ m.

The total volume of seawater $Q_{sw}$ necessary to stabilize the erg's sand dunes is given by

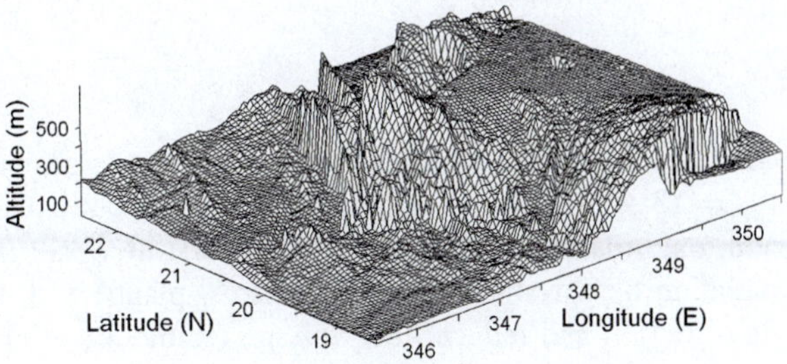


**Fig. 5** Geographical location and topography of erg Akchar

$$Q_{sw} = S_{erg} q_1 \quad (1)$$

where $q_1$ is the amount of seawater necessary to stabilize a unit erg surface. Here we accept $q_1 \approx 0.5\ \mathrm{m}^3$ water per square meter of dune surface. Now we assume the dune stabilization macroproject will be completed in $N$ years. An uniform (in time) seawater flow rate $q_{sw}$ is assumed. Then, in the volume flow rate in cubic metre per second is

$$q_{sw} = \frac{Q_{sw}}{365 \times N \times 24 \times 3,600} \quad (2)$$

The speed $w_{sw}$ of seawater in the duct of diameter $D_{duct}$ is given by:

$$w_{sw} = \frac{4 q_{sw}}{\pi D_{duct}^2} \quad (3)$$

The power $P_{pump}$ required to move the seawater in the duct is:

$$P_{pump} = g \rho_{sw} q_{sw} H / \eta_p \quad (4)$$

In Eq. 4, $g$ (9.78 m/s$^2$) is gravitational acceleration, $\rho_{sw}$ (1,030 kg/m$^3$) is the density of the seawater, $H$ is the hydraulic head and $\eta_p$ ($\approx$0.75) is the efficiency of the pump. The hydraulic head is obtained by summing the mean height of the erg with the lost pressure height $\Delta H$ due to friction:

$$H = H_{erg} + \Delta H \quad (5)$$

Only linear pressure losses are considered next and:

$$\Delta H = \lambda \frac{L_{duct} w_{sw}^2}{D_{duct}\, 2g} \quad (6)$$

where $\lambda$ is the linear pressure loss coefficient given by:

$$\lambda = \begin{cases} \dfrac{1}{\sqrt[4]{100 Re}} & \text{for} \quad Re < 10^5 \\ 0.0032 + \dfrac{0.211}{Re^{0.237}} & \text{for} \quad Re > 10^5 \end{cases} \quad (7)$$

where the Reynolds number is defined by

$$Re = \frac{w_{sw} D_{duct}}{\nu_{sw}} \quad (8)$$

where $\nu_{sw}$ is the kinematic viscosity of seawater. The constant value $\nu_{sw} = 13 \times 10^{-4} / \rho_{sw}$ is adopted in this study. We also assume, conservatively, a non-toxic marine anti-fouling agent coats the SSSP interior to prevent the accumulation of unwanted micro-organisms, plants, and animals that would increase flow friction and reduce flow volume (Efimenko et al. 2009).

The energy consumed with pumping $E_{pump,year}$ is obtained (in J/year) from:

$$E_{pump,year} = 365 \times 24 \times 3,600 \times P_{pump} \tag{9}$$

The yearly averaged daily solar global irradiation on a horizontal plane ground surface in Nema, a city in southeastern Mauritania, is $G_{day} = 24.46$ MJ/(m$^2$ day) (Badescu 2006). The yearly solar global irradiation $G_{year}$ is of course

$$G_{year} = 365 G_{day} \tag{10}$$

The energy provided per unit surface area by PV cells during a year, $E_{PV,year,1}$ is given by:

$$E_{PV,year,1} = G_{year} \eta_{PV} \tag{11}$$

where $\eta_{PV}$ is PV cell efficiency (yearly average). In computations we have used a rather high (optimized) value $\eta_{PV} = 0.15$ which applies for Nema (Badescu 2006). The energy provided by the whole PV cells surface during a year, $E_{PV,year}$ is obtained from:

$$E_{PV,year} = E_{PV,year,1} D_{duct} L_{duct} \tag{12}$$

The energy consumed per year with pumping, $E_{pump,year}$, is provided in part by the PV cells. The remaining part, $E_{classic,year}$, should be provided from classical energy sources. One has:

$$E_{classic,year} = E_{pump,year} - E_{PV,year} \tag{13}$$

To evaluate the financial magnitude of the Akchar erg dune stabilization macroproject we have to estimate the cost of the duct, of the PV cells and of pumps enabling seawater movement, respectively. The cost $c_{duct}$ of the duct is given by:

$$c_{duct} = c_{duct,1} L_{duct} \tag{14}$$

where $c_{duct,1}$ is the cost of a unit length of duct. Similarly, the cost of installing the tube, $c_{inst,duct}$ is given by:

$$c_{inst,duct} = c_{inst,duct,1} L_{duct} \tag{15}$$

where $c_{inst,duct,1}$ is the cost of installing a standard unit length of duct. The unitary costs depend of course on various factors, such as the duct diameter $D_{duct}$ as well as the material of the duct. Table 2 in Macro-Engineering Lake Eyre with Imported Seawater shows the input values used in this work.

The cost of the pumping installation, $c_{pump}$, is obtained from:

$$c_{pump} = P_{pump} c_{pump,1} \tag{16}$$

where $c_{pump,1}$ is the cost a pump of unit power. Table 3 in Macro-Engineering Lake Eyre with Imported Seawater shows statistically average values for $c_{pump,1}$, as a function of pumping device quality. The gold-plated cost in Table 3 in Macro-Engineering Lake Eyre with Imported Seawater is for a custom-built installation. The inexpensive cost is for an off-the-shelf, commercially mass-produced fluid pump. The maintenance cost per year for the pumping system, $c_{pump,maint}$ is

$$c_{pump,\mathrm{maint}} = f_{pump,\mathrm{maint}} c_{pump} \tag{(17)}$$

where $f_{pump,maint}$ is a given fraction. Here $f_{pump,maint} = 0.01$ has been adopted.

The cost $c_{PV}$ of the PV cells is obtained from the product of the cost of a unit surface PV cell, $c_{PV,1}$ and the total surface covered by installed PV cells:

$$c_{PV} = c_{PV,1} L_{duct} D_{duct} \tag{18}$$

Using the estimate for future PV Cells technology, the cost per square meter of flexible PV cells sheeting is $c_{PV,1} \approx 100$ USD/m$^2$ (Liu et al. 2007). The maintenance cost per year for PV cells, $c_{PV,maint}$ is

$$c_{PV,\mathrm{maint}} = f_{PV,\mathrm{maint}} c_{PV} \tag{19}$$

where $f_{PV,maint}$ is a given fraction. Here $f_{PV,maint} = 0.01$ has been adopted.

The cost of the classical energy consumed during a year, $c_{classic,year}$, is given by:

$$c_{classic,year} = E_{classic,year} c_{classic,year,1} \tag{20}$$

where $c_{classic,year,1}$ is the cost of a standard classical energy unit. Here electricity is considered and $c_{classic,year,1} \approx 0.1$ USD/kWh (PelletHeat 2007). The cost of classical energy during $N$ years, $c_{classic}$, is obtained from:

$$c_{classic} = N c_{classic,year} \tag{21}$$

The initial investment cost $c_{invest,stab}$ for duct, PV cells and pumping installation, is given by

$$c_{invest,stab} = c_{pump} + c_{PV} + c_{duct} + c_{inst,duct} \tag{22}$$

The maintenance cost $c_{maint,stab}$ for $N$ years of the PV cells and pumping installation is

$$c_{\mathrm{maint,stab}} = N\left(c_{pump,\mathrm{maint}} + c_{PV,\mathrm{maint}}\right) \tag{23}$$

The total costs, $c_{tot,stab}$, after $N$ years of the dune stabilization macroproject, is

$$c_{tot,stab} = c_{invest,stab} + c_{\mathrm{maint,stab}} + c_{classic} \tag{24}$$

### 5.2 *Seawater Agriculture Macroproject*

It is estimated ~10% of Earth's land is already affected by salt deposition. Of ~5,000 food and fiber crops that are cultured by farming humans, only a few can survive with water that contains >0.5% salt, and most suffer serious yield reductions at ~0.1% salt. Still, the use of saline waters and even seawater for halophytic crop cultivation is an attractive option for farmers in some dryland regions (Lieth and Mochtchenko 2004). The world's first commercial food to be grown entirely on poor soil irrigated by seawater is *Salicornia bigelovii* (Glenn et al. 1997). The increasing salinisation of inland waters (Williams 2001), and the calls for "reversing the flow of water and nutrients from the ocean to the land" (Hodges et al. 1993), combined with the prospects for progressive genomic manipulation of photosynthetic plants by "Metabolic Engineering", means that saline water may soon have a greater value to humans than it had during the twentieth century. Cropping of vast additional segments of the Sahara (Glenn et al. 1998; Rozema and Flowers 2008), or the Sahara's near-term future coverage by a Sahara Tent Greenbelt (Cathcart and Badescu 2004), might curtail or even terminate the natural suppression of North Atlantic Ocean tropical cyclones. Emplacement of the Sahara Tent Greenbelt would also effectively prevent further erosion of radiation contaminated dust, and its deposition on southern Europe, that is derived from former nuclear weapons testing regions in Algeria (Menut et al. 2009). For lack of sufficient available geoscientific data, we remain uncertain of our Sahara Tent Greenbelt's likely fertilizing-effect diminishment on the Mauritanian Upwelling Zone offshore in the Atlantic Ocean (Schlosser and Croot 2009).

In an attempt to recover part of the costs of the erg Akchar dune stabilization macroproject, an additional project has been considered. After the dune stabilization macroproject is completed, the duct and all existing infrastructure may be used for seawater agriculture. Seawater irrigation does not require special farming equipment. The large existing test farms have used either flood irrigation of large basins or moving seawater to the region in which it is used. Seawater agriculture needs approximately 35% more irrigation fluid volume when grown using seawater than conventional crops grown using freshwater. Generally, there are no insurmountable macro-engineering problems associated to commercial seawater agriculture.

Computation of the crop surface that may be irrigated per second, $S_{irrig,sec}$, is made by

$$S_{irrig,sec} = \frac{q_{sw}}{S_{irrig,1}} \tag{25}$$

where $S_{irrig,1}$ is the surface of crops that may be irrigated with one cubic meter of seawater. The cycle of the irrigation procedure is denoted $t_{irrig}$. Then, the total irrigated surface $S_{irrig}$ is given by:

$$S_{irrig} = t_{irrig} S_{irrig,sec} \tag{26}$$

Input values were suggested from the mini-project "Ras al-Zawr" where *Salicornia bigelovii*, an estuarine plant found in many parts of the world, is cultivated (SaudiAramCo 2007). There, computer-controlled pivot-irrigation arms—

one for each 50 ha circle—sprayed seawater pulled in by three diesel pumps at a rate greater than 28 $m^3$/min. It took 6.5 h for the arms to complete a single circuit. From these data, one finds that $S_{irrig,sec} = 0.763$ $m^2$ land per cubic meter of water and $t_{irrig} = 23{,}400$ s.

The cost of the irrigation installation $c_{irrig}$ is of course proportional with the irrigated surface:

$$c_{irrig} = S_{irrig} c_{irrig,1} \tag{27}$$

where $c_{irrig,1}$ is the cost of irrigating a unit surface of cultivated crops. Depending on the crops, $c_{irrig,1}$ ranges from 200 to 2,000 US \$/0.4 ha (Farm Management 2007). Here we accept 750 US \$/0.4 ha which then yields $c_{irrig,1} = 0.185$ USD/$m^2$. The maintenance cost per year for the irrigation installation, $c_{irrig,maint,1}$ is

$$c_{irrig,maint,1} = f_{irrig,maint} c_{irrig} \tag{28}$$

where $f_{irrig,maint}$ is a given fraction. Here $f_{irrig,maint} = 0.05$ has been adopted. The maintenance cost $c_{irrig,maint}$ for $N_1$ years of the irrigation installation is

$$c_{irrig,maint} = N_1 c_{irrig,maint,1} \tag{29}$$

The cost associated to the irrigation installation, $c_{tot,irrig}$, after $N_1$ years of operation is

$$c_{tot,irrig} = c_{irrig} + c_{irrig,maint} \tag{30}$$

The economic gain per year from the crops irrigated with seawater, $g_{irrig,year}$, is given by

$$g_{irrig,year} = S_{irrig} g_{irrig,1} \tag{31}$$

where $g_{irrig,1}$ is the economic gain per unit surface of cultivated crops, per year. The economic gain after $N_1$ years from the cultivated crops, $g_{irrig}$ is given by

$$g_{irrig} = N_1 g_{irrig,1} \tag{32}$$

During six years of intensive agricultural field trials in Mexico, *Salicornia bigelovii* produced an average annual crop of 1.7 kg/$m^2$ of total biomass and 0.2 kg/$m^2$ of oilseed (Imaz et al. 1998; Troyo et al. 1994). It is expected that the benefits from the cultivated, harvested and processed *Salicornia bigelovii* consists of food products like cooking oil and "sea asparagus", a gourmet food delicacy of fresh shoot tips eaten as a salad vegetable during summertime that sells in southern Europe in a niche market for US \$20 per 0.4536 kg (Global Warming 2007). In calculations we considered as possible products to sell, oil (0.75 US \$/kg) and sea asparagus (40 US \$/kg). In the two cases, $g_{irrig,1}$ is evaluated to about 0.15 and 8 US \$/$m^2$, respectively.

The economic profit $p_{irrig}$ of the irrigation system after $N_1$ years is

$$p_{irrig} = g_{irrig} - c_{tot,irrig} \tag{33}$$

The total cost, $c_{tot}$, of the erg Akchar dune stabilization macroproject and dune irrigation macroproject, after $N + N_1$ years of operation, is obtained from:

$$c_{tot} = c_{tot,stab} + c_{tot,irrig} \tag{34}$$

## 5.3 Results

Generally, the energy consumed with pumping the seawater, obtained from the Atlantic Ocean, inside the duct is decreased by increasing the number of years $N$ chosen to finish the Akchar erg stabilization macroproject. When $N$ is of the order of 10 or 20 years, the macroproject is intensively consuming energy and money. Here we show results for $N = 50$ years (Fig. 6). An inexpensive pump in Table 3 in Chap. 25 has been considered. The seawater speed in the duct $w_{sw}$ and the pumping power $P_{pump}$ become reasonably small only at duct diameters $D_{duct}$ larger than 3 m (Fig. 6a). Thus, it is obvious that at smaller duct diameters, the surface covered by PV cells is rather small and most of the energy required to shift the seawater is provided from classical energy sources (Fig. 6b). However, for

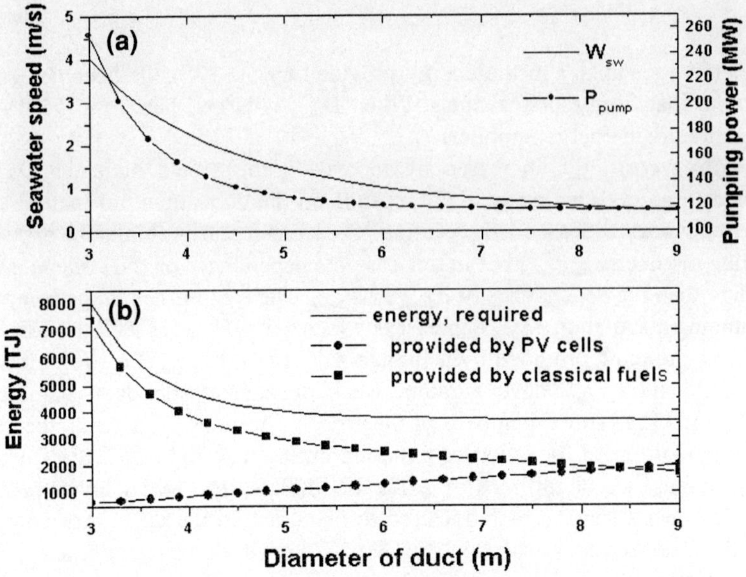


**Fig. 6** **a** Seawater speed $w_{sw}$ and pumping power $P_{pump}$ and **b** various energies (the energy consumed during a year with pumping, $E_{pump,year}$, the energy provided by PV cells and from classical fuels, $E_{PV,year}$ and $E_{classic,year}$, respectively) as a function of duct diameter $D_{duct}$. Computations performed for $N = 50$ years

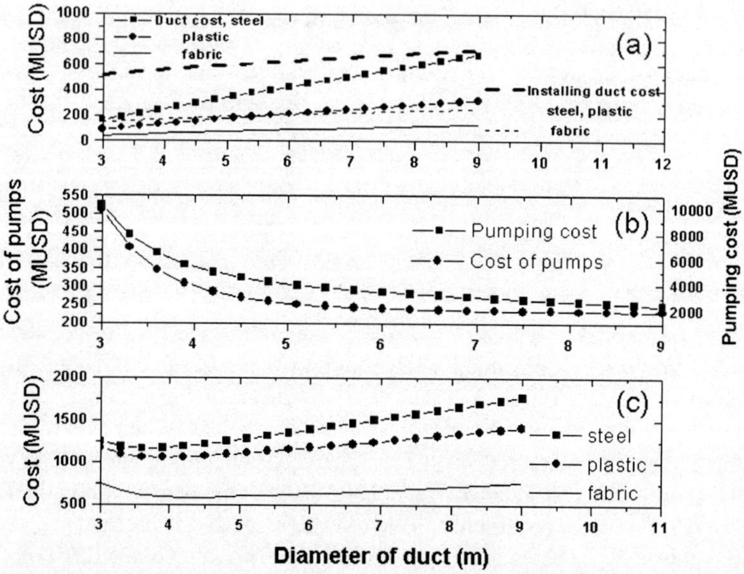


**Fig. 7** Various costs as a function of duct diameter $D_{duct}$. **a** Cost of duct $c_{duct}$ and cost of duct installing $c_{inst,duct}$ for various duct materials; **b** cost of pumping installation $c_{pump}$ and cost of pumping during $N$ years by using classical fuels, $c_{classic}$; **c** initial investment cost $c_{invest,stab}$ for various duct materials. Computations performed for $N = 50$ years

diameters larger than 7 m the energy provided by the PV cells becomes comparable with that from classical fuels. When $D_{duct}$ is larger than 9 m, the PV cells become the main energy supplier.

The duct cost $c_{duct}$ increases by increasing the pipe's diameter $D_{duct}$, as expected (Fig. 7a). Obviously, $c_{duct}$ depends on the duct structural material, with composed fabric the least costly solution. The same feature exhibits the expense of installing the duct, $c_{inst,duct}$, but in this case the dependence on duct diameter $D_{duct}$ is weaker than the dependence of $c_{duct}$ on $D_{duct}$. The cost of the pumps comprising the pumping installation $c_{pump}$ decreases by increasing $D_{duct}$, as well the cost of the energy for pumping provided by using classical fuels, $c_{classic}$ (Fig. 7b). However, both $c_{pump}$ and $c_{classic}$ have a rather weak dependence on duct diameter for $D_{duct} > 6$ m. The cost of pumps is of the order of millions US \$ while the cost of the energy consumed for seawater pumping during a period of 50 years is of the order of thousands of millions of 2,009 US dollars. The initial investment cost $c_{invest,stab}$ depends on the material used to manufacture the duct, with composed fabric the least expensive solution (Fig. 7c). Whatever the duct material, $c_{invest,stab}$ has a minimum that is about 4 m for steel and plastic ducts and about 4.5 m for composed fabric ducting. The minimum initial investment in the erg Akchar sand dune field stabilization macroproject is about 1,100 millions US \$ in the case of

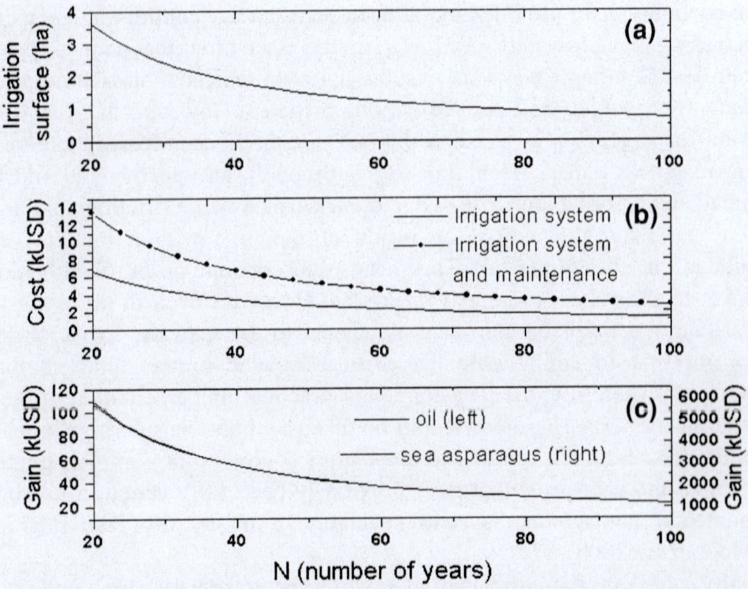


**Fig. 8** Various quantities as function of the number of years $N$. **a** Irrigated surface $S_{irrig}$; **b** cost of irrigation system $c_{irrig}$ and cost of irrigation and its maintenance for $N_1$ years, $c_{tot,irrig}$; **c** economic gain from the irrigated surface for $N_1$ years, $g_{irrig}$, for two possible products. Computations performed for $D_{duct} = 4.5$ m and $N_1 = 20$ years

pipes composed of steel and plastic and about 600 millions US $ in the case of ducts fabricated with composed fabric.

The seawater irrigation macroproject associated to the Mauritanian erg stabilization macroproject was studied next. The results are shown in Fig. 8, which refers to a duct diameter of 4.5 m, associated to minimization of the initial investments in the stabilization macroproject in case of fabric ducts (see Fig. 7c). The irrigated surface $S_{irrig}$ decreases by increasing the number of years $N$ to complete dune stabilization (Fig. 8a). This is easily understood: increasing $N$ decreases the seawater mass flow rate used by the dune field stabilization macroproject. For all computations $S_{irrig}$ ranges between 1 and 4 ha, and for $N = 50$ years (the case treated in Figs. 6, 7) the irrigated surface is about 1.5 ha. The cost of irrigation system $c_{irrig}$ and the cost of irrigation and its maintenance for $N_1 = 20$ years, $c_{tot,irrig}$, decrease by increasing the number of years $N$ (Fig. 8b). These costs are of the order of thousands of 2,009 US $, well below the investment costs associated to the dune stabilization macroprojects (see Fig. 8b). The economic gain $g_{irrig}$ from the crops irrigated with seawater is shown in Fig. 8c for two different products. This economic gain decreases by increasing $N$, as expected. In the case of cooking oil, $g_{irrig}$ is less than 0.1 million 2,009 US $ but the economic gain is of the order of a few millions US $ in case of sea asparagus. In practice $g_{irrig}$ will probably fall within the range of these extreme values.

The economic gain provided by induced seawater agriculture may be increased by increasing the seawater speed $w_{sw}$ in the duct after the dune stabilization macroproject is completed. This ensures a larger seawater mass flow rate and, consequently, a larger land surface may be irrigated. However, there is a limited provision for increasing $w_{sw}$, due to the fact that the friction losses increases with the square power of $w_{sw}$. Doubling $w_{sw}$ is probably an upper bound and this is associated to doubling the irrigated ground surface $S_{irrig}$ (friction loss may be reduced markedly by coating the inside of the pipe with a low-cost carbon nanotube fabric liner or plastic). This means the economic profit might be doubled in respect to the values presented in Fig. 8c. However, even in these most optimistic situations, the economic gain associated to the seawater irrigation macroproject goes far to compensate the costs associated to the dune stabilization macroproject. Therefore, the decision about practical implementation of the dune field stabilization macroproject should be taken without considering (immediate) economic reasons. Even so, an eco-audit must follow-up any implementation of these nested macroprojects that have previously been fully vetted in a structured environmental management system evaluation done by dry land degradation experts (Barrow 2006).

Finally, note that the computations reported here are very rough and, in practice, the costs of the macroproject may be significantly larger.

## 6 Effects on Global Sea Level Rise

*Circa* 3900 BC—about 1,100 years after global sea level stabilised following the Last Glacial Maximum—civilisation commenced with the human use of newly abundant coastal margin resources; *circa* 2300 BC, urban governments commenced construction of monumental infrastructures (Day et al. 2007). Contemporaneous destabilisation of global sea level impacts urban governments and infrastructures on present-day coastal margins (Ericson et al. 2006). For example, some major cities—places such as Lagos, Karachi, Mumbai, Kolkata, Bangkok, Jakarta, Manila and Shanghai—presently find their fresh groundwater reservoirs being contaminated by saltwater intrusions that may be further aggravated by a postulated greater, yet unproved, future global sea level rise. "Global Warming" appears to be a real existential risk, but its impact during the twenty-first century and beyond could plausibly range from nil to negligible to severe. Computer models of future atmospheric realities are still extremely simplistic as compared to the actual phenomenon of Earth's air (Tyrell et al. 2007).

Worldwide, since $\sim$10.5% of the world's population reside on land that is <100 km from the shoreline at elevations <10 m above sea level ($\sim$2.2% of all land), the international and intranational legal implications of coastal zone adjustments instigated by the alleged impending global sea level rise are profound (Caron 1990). Hydrogeology may have a central role to play in a macroproject solving any prospective future global sea level rise because that profession already has a burgeoning role in the underground sequestration (injection and storage) of aerial $CO_2$ gas in deep saline aquifers (Celia 2002). For example, during this century, BP PLC operates an isolated natural-gas processing plant in the Sahara near In Salah (27°12′N latitude by 2°28′E longitude), Algeria (Ball 2005).

The SSSP macroproject herein proposed involves a massive anthropogenic redistribution of Spaceship Earth's natural seawater cargo. Whatever the cause of a future global sea level rise, whether it is "global warming" or some other phenomenon or aggregation of phenomena, James E. Hansen defines a substantial global sea level rise to mean "a total sea level rise of at least two meters, because that would be sufficient to flood large portions of Bangladesh, the Nile Delta, Florida, and many island nations, causing forced migration of tens to hundreds of millions of people" (Hansen 2005). More than 20 years ago, Walter Stephenson Newman (1895–1978) and Rhodes Whitmore Fairbridge (1914–2006) speculated that humans could, using macro-engineering tactical technologies, manage any future global sea level rise by "...diverting sea water into continental depressions" filled to present-day global sea level (Newman and Fairbridge 1986). By our estimation, the Caspian Sea region could store ~13,000 $km^3$, the Aral Sea region ~1,000 $km^3$, the Qattara Depression ~3,200 $km^3$, the Dead Sea ~1,260 $km^3$, Lake Eyre region ~200 $km^3$ and the Salton Sea region ~400 $km^3$ of seawater. Total global land depression storage capacity: ~18,060 $km^3$ and all of the seawater is stored in the open air, subject to solar evaporation. In other words, these seawater storehouses must be constantly replenished at some undetermined financial and energy cost. The area of the Earth's ocean is $\sim 3.62 \times 10^8$ $km^2$. If the ocean rose by J.E. Hansen's 2 m, the volume of that increase would be ~725,000 $km^3$, meaning that removal of ~18,060 $km^3$ is only ~2.5% of the volume that must be shifted to the Earth's land from the ocean's basin in order just to maintain present-day global sea level. Where can the remaining 97.5% (~705,940 $km^3$) be stored, withheld from our world's ocean, for an indeterminate time period?

First, a maximum of 7,200 $km^3$ of freshwater, about 20% of the Earth's total annual river runoff is retained in artificial reservoirs created by anthropic dams (Oki and Kanae 2006). Second, the total natural unused, unsaturated volume of pore space beneath the world's land is huge—to a depth of ~2,000 m, if only unconsolidated sands, sandstones and carbonates are considered, ~25,000,000 $km^3$ exists which might be filled artificially with seawater. "Sands retain most of their original porosity down to a depth of 1 km. Porosities of approximately 48 percent at the surface show little change for the initial 100 m of burial, and then begin to decrease slightly with depth: to 45 percent at 300 m and 37 percent at 1 km" (Hay and Leslie 1990). The Kalahari sand sheet situated in southern Africa, with a 2.5 million ha area, is very probably the world's largest continuous surface of loose sand. Water of any purity can be forced to accumulate rapidly in the unsaturated vadose zone (Anderson 2007). Oilfield repressurization with deliberately injected seawater has been used for many years to halt widespread land subsidence caused by oil and natural gas mining.

J.E. Hansen's global climate warming scenario entails a global sea level rise, "...a process that will shift the interface between land and sea, resulting in the inland extension of maritime related flooding and elevated soil salinities" (Greaver and Sternberg 2007).

The SSSP macroproject installation considered in this single tactical SSSP macro-engineering assessment can pump the volume $V_{sw} = q_{sw}t$ of seawater, where $t$ is time. For the duct considered in Figs. 6 and 7, the volumetric seawater flow rate is $q_{sw} = 28.5\ \mathrm{m^3/s}$ and the amount of seawater extracted from our world's ocean and deposited on Mauritania's sand dune fields during a year is about 0.9 $\mathrm{km^3}$. This is a very small quantity if we compare with that associated to the 2 m global sea level rise expected by Hansen. However, the Intergovernmental Panel on Climate Change's latest projection for future global sea level rise during the twenty-first century is 18–59 cm. Global sea level rise was estimated at 1.6 mm/year during the twentieth century (Wöppelmann et al. 2009). Therefore, any sand dune seawater saturation effort need only be adequate to compensate for each year's potential global sea level rise. Stated differently, each year the single SSSP sited in Mauritania could transport inland 0.166% of the yearly 541.5 $\mathrm{km^3}$ increase in the world-ocean's volume. So, we conclude that the northern Africa Mauritania SSSP macroproject may contribute to, but it is not a solution for, the expected future rise of global sea level.

## 7 Conclusion

We have proposed that an SSSP macroproject sited in Mauritania can be both economic and developmental, with extensive future applicability worldwide. The SSSP's macro-engineering concept of extraction and long-term storage of excess seawater in city-threatening active coastal sand dune fields via simple and effective tactical technologies (artificially duplicating, in part, the Earth's natural Hydrologic Cycle) is revolutionary.

The Sahara includes large expanses of sand dunes called ergs. These dunes are formed and constantly reshaped by the prevailing winds. Our SSSP macroproject is intended to foster development of region-wide soil moisture heterogeneity in Mauritania's inland ergs for the purpose of stabilization by irrigating the dunes with unadulterated seawater pumped from the nearby North Atlantic Ocean. Future irrigation seawater may be infused with slightly cleansed urban sewage, which could fertilize other irrigated vegetable and fruit field crops (Heinonen-Tanski 2007; Bouvy et al. 2008).

**Acknowledgments** The authors thank Dr. Stephen Salter (University of Edinburgh) for his truly inspirational idea for the sand dune field sequestration of unwanted excess seawater and Dr. J. Marvin Herndon (Transdyne Corporation, San Diego, California) for his chemistry advice.

## References

Adger WN, Adger WN, Dessai S, Goulden M, Hulme M, Lorenzoni I, Nelson DR, Naess LO, Wolf J, Wreford A (2009) Are there social limits to adaptation to climate change? Clim Change 93:335–354

Anderson MP (2007) Introducing groundwater physics. Physics Today 60:42–47

Ba MB, Nicholson SE, Robert Frouin R (2001) Satellite-derived surface radiation budget over the African continent. Part II: Climatologies of the various components. J Clim 14:60–76

Badescu V (2006) Simple optimization procedure for silicon-based solar cell interconnection in a series-parallel PV module. Energy Convers Manage 47:1146–1158

Badescu V, Cathcart RB (2008) Sand dune fixation: a solar-powered Sahara seawater pipeline macroproject. Land Degrad Develop 19:676–691

Ball J (2005) Deep in the Sahara, BP tries to put dent in global warming. Wall St J CCXLV:1–2

Barclay DR (2009) On the shapes of natural sand grains. J Geophys Res 114:B02209

Barrow CJ (2006) Environmental management for sustainable development. Routledge, London, p 198

Berger AR (2006) Abrupt geological changes: causes, effects, and public issues. Quat Int 151:3–9

Besler H (2008) The Great Sand Sea in Egypt. Elsevier, London, p 130

Bouvy M, Briand E, Boup MM, Got P, Leboulanger C, Bettarel Y, Arfi R (2008) Effects of sewage discharges on microbial components in tropical coastal waters (Senegal, West Africa). Mar Freshw Res 59:614–626

Caron DD (1990) When law makes climate change worse: rethinking the law of baselines in light of a rising sea level. Ecol Law Q 17:621–653

Cathcart RB, Badescu V (2004) Architectural ecology: a tentative Sahara restoration. Int J Environ Stud 61:145–160

Celia MA (2002) How hydrogeology can save the world. Ground Water 40:113

Charlier RH (1991) Water for the desert—a viewpoint. Int J Environ Stud 39:11–35

Chenal J, Kaufmann V (2008) Nouakchott. Cities 25:163–175

Day JW, Gunn JD, Folan WJ, Yáñez-Arancibia A, Horton BP (2007) Emergence of complex societies after sea level stabilized. EOS, Transactions, American Geophysical Union, vol 88, pp 169–170

Desert Control (2007) http://www.desertcontrol.com. Accessed 10 October 2007

Duran O, Herrmann HJ (2006) Vegetation against dune mobility. Phys Rev Lett 97:188001

Efimenko K, Finlay J, Callow ME, Callow JA, Genzer J (2009) Development and testing of hierarchically wrinkled coatings for marine antifouling. ACS Appl Mater Interfaces 1:1031–1040

Elhassadi A (2008) Views on Libyan national plan (LNP) to resolve water shortage problem (WSP). Part Ib: Great Man-Made River (GMMR) project—capital costs with interest rates. Desalination 220:184–188

Elmoustapha AO, Levoy F, Monfort O, Koutitonsky VG (2007) A numerical forecast of shoreline evolution after harbour construction in Nouakchott, Mauritania. J Coastal Res 23:1409–1417

Engelstaedter S, Washington R (2007) Atmospheric controls on the annual cycle of North African dust. J Geophys Res 112:JD007195

Ericson JP et al (2006) Effective sea-level rise and deltas: causes of change and human dimension implications. Global Planet Change 50:63–82

Evan AT et al (2006) New evidence for a relationship between Atlantic tropical cyclone activity and African dust outbreaks. Geophys Res Lett 33:GL026408

FAO (1997) FAO Land and Water Development Division, Irrigation potential in Africa: a basin approach. FAO Land and Water Bulletin 4, United Nations Organization, New York

Farm Management (2007) http://www.cook.rutgers.edu/~farmmgmt/ne-budgets/methodology.html. Accessed 10 October 2007

Germann PF, Hagrey SA (2008) Gravity-driven and viscosity-dominated infiltration into a full-scale sand model. Vadose Zone J 7:1160–1169

Ghiassian H, Ghareh S (2008) Stability of sand slopes under seepage conditions. Landslides 5:397–406
Glenn E, Miyamoto S, Moore D, Brown JJ, Thompson TL, Brown P (1997) Water requirements for cultivating *Salicornia bigelovii* Torr with seawater on sand in a coastal desert. J Arid Environ 36:711–730
Glenn EP, Brown JJ, O'Leary JW (1998) Irrigating crops with seawater. Sci Am 279:76–81
Global Warming (2007) http://yosemite.epa.gov/oar/globalwarming.nsf/Unique.KeyLookup/SHSU5C3HJU/$File/aij.mexico.pdf. Accessed 10 October 2007
Goudie AS (2008) The history and nature of wind erosion in deserts. Annu Rev Earth Planet Sci 36:97–119
Goudie AS, Middleton NJ (2001) Saharan dust storms: nature and consequences. Earth Sci Rev 56:179–204
Greaver TL, Sternberg LSL (2007) Fluctuating deposition of ocean water drives plant function on coastal sand dunes. Global Change Biol 13:222
Hamakawa Y (ed) (2005) Thin-film solar cells: next generation photovoltaics and its applications. Springer, New York
Hansen JE (2005) A slippery slope: how much global warming constitutes 'dangerous anthropogenic interference'? Climatic Change 68:274
Hay WW, Leslie MA (1990) Could possible changes in global groundwater reservoir cause eustatic sea-level fluctuations? In: Geophysics study, chap 9. Committee, National Research Council, Sea-Level Change, National Academy of Sciences, Washington, DC, p 165
Heinonen-Tanski H (2007) Pure human urine is a good fertilizer for cucumbers. Bioresour Technol 98:214–217
Imaz M, Gay C, Friedmann R, Goldberg B (1998) Mexico joins the venture: joint implementation and greenhouse gas emissions reduction. Lawrence Berkeley National Laboratory, Environmental Energy Technologies Division, November 1998, Work supported by the U.S. Environmental Protection Agency through the U.S. Department of Energy under Contract No. DE-AC03-76SF00098
Hodges CN et al (1993) Reversing the flow: water and nutrients from the sea to the land. Ambio 22:483–490
Isupova MW, Mikhailov VN (2008) Hydrological and morphological processes in Senegal River mouth area. Water Resour 35:30–42
Jackson NL, Nordstrom KF (1998) Aeolian transport of sediment on a beach during and after rainfall. Geomorphology 22:151–157
Jeon J-H, Keum J-H, Song JH, Kim JS (2007) The fabrication and characterization of organic photovoltaic microcell into the polymer membrane using template synthesis of conductive polymer. Sol Energy Mater Sol Cells 91:645–651
Kareiva P, Watts S, McDonald R, Boucher T (2007) Domesticated nature: shaping landscapes and ecosystems for human welfare. Science 316:1866–1869
Kogbe CA, Dubois D (1980) Economic significance of the 'Continental Terminal'. Int J Earth Sci 69:429–436
Koren I, Kaufman YJ, Washington R, Todd MC, Rudich Y, Martins JV, Rosenfield D (2006) The Bodele depression: a single spot in the Sahara that provides most of the mineral dust to the Amazon forest. Environ Res Lett 1:1–7
Kurokawa K (2007) Energy from the desert. Earthscan, London
Lambot S, Slob E, Rhebergen J, Lopera O, Jadoon KZ, Vereecken H (2009) Remote estimation of the hydraulic properties of a sand using full-waveform integrated hydrogeophysical inversion of time-lapse, off-ground GPR data. Vadose Zone J 8:743–754
Lancaster N, Kocurek G, Singhvi A, Pandey V, Deynoux M, Ghienne J-F, Lo K (2002) Late Pleistocene and Holocene dune activity and wind regimes in the Western Sahara Desert of Mauritania. Geology 30:991–994
Lehmann J (2007) A handful of carbon. Nature 447:143–144
Lieth H, Mochtchenko M (eds) (2004) Cash crop halophytes—recent studies: ten years after Al Ain meeting (tasks for vegetation science). Springer, Dordrecht

Liu J, Namboothiry MAG, Carroll DL (2007) Fiber-based architectures for organic photovoltaics. Appl Phys Lett 90:063501

Lu Y, Chiew Y-M (2007) Seepage effects on dune dimensions. ASCE J Hydraulic Eng 133: 560–563

Mahowald NM (2007) Anthropocene changes in desert area: sensitivity to climate model predictions. Geophys Res Lett 34:L18817

Martinez ML, Psuty NP (2008) Coastal dunes: ecology and conservation. Springer, London

Matton G, Jebrak M, Lee JKW (2005) Resolving the Richat enigma: doming and hydrothermal karstification above alkaline complex. Geology 33:665–668

Menut L, Masson O, Bessagnet B (2009) Contribution of Sahara dust on radionuclide aerosol activity levels in Europe? The 21–22 February 2004 case study. J Geophys Res Atmos 114:D16202

Newman WS, Fairbridge RW (1986) The management of sea-level rise. Nature 320:319–321

Oki T, Kanae S (2006) Global hydrological cycles and world water resources. Science 313: 1068–1072

Parsons AJ, Abrahams AD (2009) Geomorphology of desert environments, 2nd edn. Springer, London

Pavia EG (2008) Evaporation from a thin layer of wet sand. Geophys Res Lett 35:GL033465

Peel MC, Finlayson BL, McMahon TA (2007) Updated world map of the Koppen-Geiger climate classification. Hydrol Earth Syst Sci Discuss 4:439–473

PelletHeat (2007) http://www.pelletheat.org/3/residential/compareFuel.cfm. Accessed 10 October 2007

Reynolds JF, Stafford Smith DM, Lambin EF, Turner BL II, Mortimore M, Batterbury SPJ, Downing TE, Dowlatabadi H, Fernández RJ, Jeffrey E, Herrick JE, Huber-Sannwald E, Jiang H, Leemans R, Lynam T, Maestre FT, Ayarza M, Walker B (2007) Global desertification: building a science for dryland development. Science 316:847–851

Rozema J, Flowers T (2008) Crops for a salinized world. Science 322:1478–1480

Sakurai M, Sakurai C (1992) Irrigation system and irrigation method. United States Patent 5,160,214, issued on 3 November 1992

SaudiAramCo (2007) http://www.saudiaramcoworld.com/issue/199406/samphire-from.sea.to.shining.seed.htm. Accessed 10 October 2007

Schaller RD, Sykora M, Pietryga JM, Klimov VI (2006) Seven excitons at a cost of one: redefining the limits for conversion efficiency of photons into charge carriers. Nano Lett 6:424–429

Schlosser C, Croot PI (2009) Controls on seawater Fe(III) solubility in the Mauritanian upwelling zone. Geophys Res Lett 36:L18606

Schluter T (2006) Geological atlas of Africa with notes on stratigraphy, tectonics, economic geology, geohazards and geosites of each country. Springer, Heidelberg

Schwanghart W, Schutt B (2008) Meteorological causes of Harmattan dust in West Africa. Geomorphology 95:412–428

Shuster LA (2003) Thinking deep. Civil Eng 73:46–53

Smith WHF, Sandwell DT (1997) Global sea floor topography from satellite altimetry and ship depth soundings. Science 277:1956–1962

Thomas DSG, Knight M, Wiggs GFS (2005) Remobilization of southern African dune systems by twenty-first century global warming. Nature 435:1218–1221

Troyo DE, Ortega RA, Maya Y, Leon JL (1994) The effect of environmental conditions on the growth and development of the oilseed halophyte *Salicornia bigelovii* Torr. in arid Baja California Sur, Mexico. J Arid Environ 28:207–213

Tyrell T, Shephard JG, Castle S (2007) The long-term legacy of fossil fuels. Tellus B 59:664–672

UN Industrial Development Organization (2005) Extraction of chemicals from seawater, inland brines and rock salt deposits. University Press of the Pacific, San Diego, 88 pp

Walker S, Thompson R, Ellis R (2006). Mauritania. Min J Special Publication (London), 1–12

Wang X-P, Young MH, Yu Z, Li X-R, Zhang Z-S (2007) Long-term effects of restoration on soil hydraulic properties in revegetation-stabilized desert ecosystems. Geophys Res Lett 34:GL031725

White K, Mattingly DJ (2006) Ancient lakes of the Sahara. Am Sci 94:58–63

Williams WD (2001) Anthropogenic salinization of inland waters. Hydrobiologia 466:329–337

Wöppelmann G, Letetrel C, Santamaria A, Bouin M-N, Collilieux X, Altamimi Z, Williams SDP, Miguez BM (2009) Rates of sea-level change over the past century in a geocentric reference frame. Geophys Res Lett 36:L12607

Wu W (2007) Coastline evolution monitoring and estimation—a case study in the region of Nouakchott, Mauritania. Int J Remote Sens 28:5461–5484

Yair A, Verrecchia E (2002) The role of the mineral component in surface stabilization processes of a disturbed desert sandy surface. Land Degrad Develop 13:295–306

Yizhaq H, Ashkenazy Y, Tsoar H (2007) Why do active and stabilized dunes coexist under the same climatic circumstances? Phys Rev Lett 98:188001

# Treeing the CATS: Artificial Gulf Formation by the Chotts Algeria–Tunisia Scheme

**Nicola M. Pugno, Richard B. Cathcart and Joseph J. Friedlander**

## 1 Introduction

The Chotts Algeria–Tunisia Scheme (CATS) is a grand design to form a man-excavated new Mediterranean Sea gulf to be shared by Tunisia and Algeria. It is well known that closed and nearly closed seawater volumes fluctuate in response to changes in local aerial regional evaporation and precipitation rates. The Zone of Chotts, bordering the Sahara at 32–35 North Latitude by 5–10 East Longitude, shared by Algeria and Tunisia, is an aggregation of playas that currently exist under a high-evaporation/low-rainfall regime (often >20:1); nowadays these particular playas are subjected to sporadic ephemeral floods. The mean annual rainfall is ~90–120 mm in southern Tunisia and Algeria, with high inter-annual variability. The rainy season extends from November to February, with summertime winds emanating from the northwest, and from the southwest during wintertime and Spring—that is, blowing from the aridified, desolate and mostly uninhabited Sahara landscape. The mean regional evaporation is ~2,500–3,100 mm/year in both southern Tunisia and Algeria (Jean-Pierre et al. 2010). The terrain of the Chott Gharsa and Chott el Jerid is undergoing some ground subsidence (~0.001–0.027 mm/year) caused by local tectonic movements. (The word "terrain" is cognate with "terrace", a relationship of some vital importance in our chapter's exposition.) We intend the CATS to be a landscape recovery macro-project, a grand

N. M. Pugno (✉)
Politecnico Di Torino, Turin, Italy
e-mail: nicola.pugno@polito.it

R. B. Cathcart
Geographos, Burbank CA, USA

J. J. Friedlander
Shave Shomron, Israel

V. Badescu and R. B. Cathcart (eds.), *Macro-engineering Seawater in Unique Environments*, Environmental Science and Engineering,
DOI: 10.1007/978-3-642-14779-1_22, © Springer-Verlag Berlin Heidelberg 2011

design of possibly future international significance in the Basin of the Mediterranean Sea. The CATS has been under public consideration for more than a century—the novelist Jules Verne first mentioned a fictional scientific version of it in his published works in 1877s *Hector Servadac*. Tunisia, in area, is one of the smallest ecosystem-nation on the African continent. Yet, much of its national terrain is salt-encrusted and salt-saturated arid soil wasteland that is unfit for human settlement at this time. If we do connect the Zone of Chotts with the Mediterranean Sea, then Tunisia and Algeria will gain a new seacoast caused by the anthropogenic creation of a new gulf. Northern Africa had previously experienced a similar transformational geomorphic event: the 14 March 1869 commencement of the now ~200 km$^2$ Great Bitter Lake in Egypt (32.39°E Long. by 30.37°N Lat.), when the Mediterranean Sea's water first began to flow into the salty dry basin during the construction of the Suez Canal. The first, basic iteration of the Suez Canal landscape shaping effort required the removal of ~74,000,000 m$^3$; if dug, the CATS would require, minimally, the rapid and largely automated industrial excavation of ~85,500,000 m$^3$. Earth-moving machinery is larger and more efficient today than equipment used a 100 years ago and we expect the CATS macro-project to be physically doable at reasonable cost (Haycraft 2000). Ultimately, it is possible that our artificial gulf will become a major twenty-first century northern African industrialization seaport, Port Tritonis, potentially almost equivalent to the Mainport Rotterdam (51°55′51″N Lat. by 4°28′45″E Long.) logistics node in The Netherlands, that is close to existing and developed future markets. In Europe, Mainport Rotterdam alone, as the largest seaport and major industrial complex there, accounts for >100 million tonnes of containerized and liquid cargo yearly. However, whereas Mainport Rotterdam is served by canal and river barges, highway carrying motor vehicles, airports and multiple railroad networks throughout Europe, the prospective Port Tritonis could be served only by automobiles, trucks, trains and aircraft.

## 2 The Big Picture

By ~2100, our world's ocean could elevate by as much as 1 m relative to its present-day level thereby directly affecting the world's populated coastline (Harff et al. 2007); ~81,000 years ago the Mediterranean Sea's level was ~1 m above its current level (Dorale et al. 2010). Sea level is a major factor influencing the role of sedimentary processes of the Mediterranean Sea Basin (Fischer and Garrison 2009). Still, to mid-2010 AD, there is little unambiguous oceanographic evidence that a truly catastrophic global sea level rise is in the offing (Leuliette and Miller 2009). Minimally, the Mediterranean Sea's ~13,000 km-long Basin strand could be markedly affected by global sea level rise; maximally, the Basin's strand could be as much as 46,000 km (Lejeusne et al. 2010). The crudest monetary estimate of the minimal total cost in 2010 value USA dollars [USD] for coastal protection against a 1 m rise in sea level (2010 USD costing one million/lineal kilometer) tells us that plan's cost–benefit ratio is excellent because if could off-set 13 trillion dollars of possible future coastal damage (Anthoff et al. 2010; Kuleli 2010). Is it really, then, so difficult for extra-basin human populations to fathom that geographically important region's *weltanschauung* of *weltschmerz*? (We decline to adopt *weltanschauung* with the connotation of "visionary"; instead, we adopt the more prosaic "concept of the world" conveyed by the term which still is not precisely encompassed by any English-language word.) We agree with John Brinckerhoff Jackson (1909–1996) that a "landscape" is "...a composition of man-made or man-modified spaces to serve as infrastructure or background for our collective existence" (Jackson 1984, p. 8).

Submarine archaeological sites, such as inundated harbors, ancient fishponds and crumbled buildings, indicate that the Mediterranean Sea's level during Roman times was ~1 m or more below the region's present-day sea-level (Lambeck et al. 2004). To accommodate an expected future 1 m ocean rise during the twenty-first century caused by climatic forcing, tectonic activity, anthropogenic effects and glacio-isostatic adjustment (Fig. 1), Greek experts have boldly proposed a 4.5 km-long solid causeway-barrier be emplaced to isolate the inner and outer Thessaloniki Bays at enormous, yet uncalculated, monetary expenditure and environmental cost (Perissoratis and Georgas 1994; Poulos et al. 2009). Such ugly localized

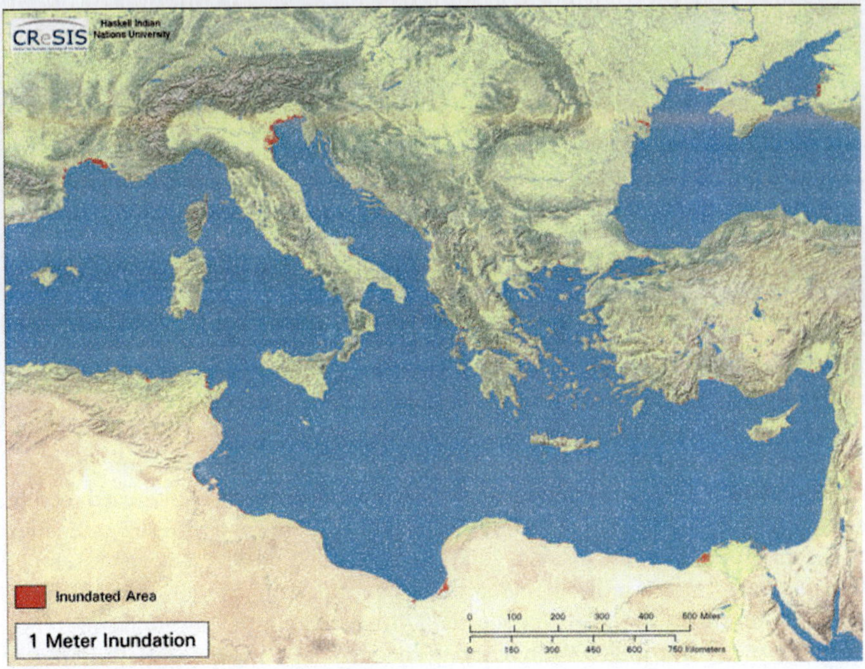


**Fig. 1** Hypothetical 1 m ocean rise in the Mediterranean area. Credit: Center for Remote Sensing of Ice Sheets. Haskell Indian Nations University; http://www.cresis.ku.edu/sites/default/files/sea-level-rise/images/mediterranean/mediterranean_1m.jpg; https://www.cresis.ku.edu/data/sea-level-rise-maps ©Copyright 2010 CReSIS

blocking macro-project constructions—planned emergency "techno-fixes"—may not be undertaken if the future earth–ocean sea-level rise is physically excluded from affecting all Mediterranean Sea Basin nations (at modest regional or spread global cost of less than 2010 USD 10 billion).

According to Bohannon (2010), a 1-m future rise of the Mediterranean Sea's level will submerge ~30% of the Nile River's delta. The land adjacent to Venice, Italy, is even more obviously affected, as is the famous vacationer's strand zone of southern France's popular beach communities.

The ultimate form of "landscape architecture", a term coined *circa* 1858, is Macro-engineering; as overtly implied during 1984 by J.B. Jackson, "landscape" combines civil engineering and landscape architecture to honor the first century of the profession's formal existence, the American Society of Landscape Architects (organized 1899) designated the year 1964–1965, beginning June 28, the Centennial Year of Landscape Architecture. However, it was not until *circa* 1964 that "Macro-engineering" was neologized and found immediate professional acceptance worldwide amongst civil and military engineers. Herein, we apply macro-engineering to remaking the seacoast and inland landscapes of Tunisia and Algeria, at least the alluvial plains containing desert saline playas which border the southern foothills of the Atlas Mountains and the northwestern margin of the Grand Erg Oriental. Our proposed macro-project, of course, has a firm linkage to the ongoing and announced macro-projects discussed thoroughly at the "Impacts of Human Activities on Dynamic Coastal Processes, Tokyo, Japan, 7–11 September 2009" ("Coastal Dynamics 2009" is the sixth international specialty conference on this subject). Tunisia, ~0.1% of the earth's land area (163,610 $km^2$), was one of the first North African ecosystem-states to actively resist the encroachment of the Sahara, desertification and salinization (Borowiec 2003; Goudie 2003).

A comprehensive technical definition of any macro-project should be judiciously phrased to highlight the concept that it is of unusually large scale, both geographically and economically, has an inherently unique character, provides a variety of useful localized earth–biosphere effects, and changes fundamentally the conditions for a society–ecosystem's growth and development (in human terms). Because it may have some of the local aura of "magic" by making something or some activity possible that was impossible before its execution, the particular macro-project can also be defined as "super important", "super costly", yet inexpensive at any price in terms of its overall effects, affective for an entire ecosystem-nation's human population, and possibly never really tried before anywhere on Earth during the twenty-first century. There are useful sub-categories of macro-projects—some that are good, beneficial and positive or even essential; there are other macro-projects that are bad, harmful, damaging or undertaken for purposes enhancing individual or societal prestige amongst all aware peoples. For example, in ancient times, "…modifying the signs of nature, especially through the construction of canals, was an arrogant act which later classical authors usually associated with tyrants" (Dora and della 2007).

Francois Charles Marie Fourier (1772–1837), in 1808, cogently remarked that human armies of industrious tool-armed persons would someday boldly subjugate the Sahara: "They will execute works the mere thought of which would freeze our mercenary souls with horror. For instance, the combined order will undertake the conquest of the great desert of Sahara; they will attack it at various points by ten and twenty million hands if necessary; and by dint of transporting earth, cultivating the soil and planting trees everywhere, they will succeed in rendering the land moist, the sand firm... They will construct canals navigable by vessels, where we cannot even make ditches for irrigation, and great ships will sail [on them]..." (Gide 1970). (After its dedication, and when it first opened to international oceanic shipping on 17 November 1869, the Suez Canal was immediately found to be too narrow and shallow to efficiently accommodate the marine traffic then desirous of passage. Obviously, this was a Suez Canal macro-project planning shortfall.)

By 1877, Donald Mackenzie published *The Flooding of the Sahara*, a geographical tome proposing an excavation from northern Africa's North Atlantic Ocean strand to the "...below sea-level central Sahara", thus permitting an [impossible] submersion with canal-imported seawater of a large region of that infamously hostile hot desert (Koger 1999). (To inspect Mackenzie's rare tome describing this potential macro-engineering fiasco, a scanned copy of his complete book can be conveniently accessed by the worldwide public: http://www.ia310843.us.archive.org/0/items/floodingsaharaa01mackgoog/floodingsaharaa01mackgoog.pdf. Therein, the most important particulars of the Sahara flooding scheme, as Mackenzie assessed them, can be perused at pages 255–263; 279–287 and 304–308 of this "portable document format", invented in 1993, and since 1 July 2008, an open standard for personal computers.) Before Mackenzie, however, heroic French macro-engineers had contemplated a similar transformative project macro-plan, but for another northern Africa locale; their speculated artificial "la mer interieure" was to be situated entirely within modern-day Algeria and Tunisia (Dora and della 2007) and was promoted by the adventurous macro-engineer Francois-Elie Roudaire (1836–1885).

In 1950, the population of the Mediterranean Sea Basin was ~170 million Europeans (73%) and 63 million North Africans (27%); by 2025, due to a profound demographic shift, there could be 305 million Europeans (44%) and 381 million North Africans (56%). world war II causalities in Europe, which caused a labor shortage needed to service Europe's obsolete industries, and post-world war II Muslim immigrants to France from Algeria as well as a prospering Germany's 1950s Muslim guest-worker national labor policy, together with high birthrates amongst European Muslims and low birthrates amongst traditional Europeans are the chief causes of the remarkable demographic shift (Caldwell 2009). A rising European standard of living, although not a precise measurement of how people live, was a social concept that first became widely used *circa* 1902, and certainly affected everyone in postwar Europe. The African Union was proclaimed on 1 March 2001 and on 29 October 2004 Europe's leaders signed a European Union Constitution. Currently, a demographic shift in Europe seems to presage an epoch—occurring, perhaps, sometime *circa* 2010 to 2050—that will alter Europe's still distinctive culture: post-world war II Europe has been colonized by Muslims mainly from North Africa. By 2010–2050, Muslims in southern Europe

(Spain, France and Italy) may form ~25% of the total population and working Muslim adults may total ~40% of the available labor force. This means, apparently, that the Gibraltar Strait Textile Barrier and its associated infrastructure developments discussed below in Sect. 3 may find decisive future acceptance amongst voting citizens of southern Europe and northern Africa since by 2050 North Africa's population should exceed southern Europe's by close to 100 million persons.

Signatories to the Barcelona Convention, an international agreement to protect the Mediterranean Sea, including Tunisia and Algeria as parties in good standing to the 1976 convention, declared in their January 2008 "Almeria Declaration" to ban any further construction on the 100 m of land nearest the present-day Mediterranean Sea's saltwater. (Marine power, particularly wave and tidal energy developments would not need to be transmitted great distances (with losses) if the human population lives with <100 km of the ocean, thus conforming to the mantra of efficiency, since efficiency is associated with increased profits or productivity, individual discipline and superior macro-management.) This declaration is a strong social response to the repulsive fact that ~40% of the Basin's strand is already disfigured by highways, railways and buildings. And, there are significant seawater and seabed pollution macro-problems lurking below the Mediterranean Sea's surface (Dachs and Mejanelle 2010; Mukerjee 2010). On 13–14 July 2008, the President of France inaugurated a new political/economic initiative for uniting the peoples and economies of the Mediterranean Sea Basin with practical macro-projects via a "Barcelona Process: Union for the Mediterranean" (Zewail 2008), with headquarters located in Barcelona, Spain. The Mediterranean Solar Plan, for example, is a planned macro-project encouraged by the Mediterranean Union to install a vast array of concentrating solar power facilities in the Sahara (Pearce 2009); some of the long-distance power transmission facilities (cables) will pass through Tunisia, across the Mediterranean Sea, then meeting the power grid of Sicily. We do not anticipate that CATS will interfere with the planning of this power-line system, due to be completed by 2050.

## 3 Mediterranean Sea Terracing

About 4,300 years ago, urban governments commenced construction of monumental edifices and massive infrastructures following the post-Ice Age natural stabilization of our world's ocean (Day et al. 2007). About 2.2% of this planet's land, that is 10 m or less in elevation above the world-ocean's current level, probably supports 10% of all humans and about 13% of all human population designated as "urban" (McGranahan et al. 2006). Humanity's activities (to make and earn a living) will "globalize" the Mediterranean Sea—its seawater, organic and inorganic contents, and periphery (Blondel 2006; Dora and della 2007).

Twenty-first century Mediterranean Sea Basin state-ecosystems still have several expensive twentieth century-conceived ameliorative macro-project options available—possibly, someday, including even Atlantropa's institutional realization as basically contrived by Herman Sorgel (1885–1952) after world war II (Vleuten and van der Kaijser 2006). However, recent material properties R&D, and newer industrial products stemming directly from advanced material technologies—particularly, technical textiles and flexible impermeable films exhibiting high-performance, purely functional, and precisely woven or non-woven fabrics—offers teamed artists/macro-engineers the prospect of a cheap Mediterranean Sea Basin anti-level rise sea barrier macro-project hung underwater by a very strong super-rope (Cullen 2005). A fabric artwork and barrier, the Gibraltar Strait Textile Barrage (GSTB), could supplement or replace eventually the MOSE (Modulo Sperimentale Elettromeccanico) macro-project to protect Italy's historic Venice from ocean inundation with an installed facility of costly, operationally complex-to-manufacture-and-maintain storm movable surge gates currently scheduled for completion *circa* 2014.

The Strait of Gibraltar (Fig. 2) connects the North Atlantic Ocean and the Mediterranean Sea, making it inevitable that the unobstructed Mediterranean Sea will rise as our world's ocean elevates caused by climate regime disruptions (Beddoe et al. 2009). The GSTB will likely be dependently draped on a 20 km-long horizon-to-horizon alignment between Tarifa in Spain and Ksar e' Sghir in Morocco, creating an aerial and submarine fabric artwork somewhat imitative of Christo's "Valley Curtain, Rifle, Colorado, 1970–1972", which was designed structurally by Earnest C. Harris (1915–1998) (Vaizey 1990). If and when constructed, its sole purpose will be to ensure the maintenance, for a long period of historical time, the existing

**Fig. 2** Gibraltar Strait, appearing as if gazed at by a person flying in an aircraft looking eastward from above the North Atlantic Ocean. The strait is about 13 km wide and is the only natural gap in the topographic barriers that separates the Mediterranean Sea from the earth's world-ocean. Source: http://www.earthobservatory.nasa.gov/IOTD/view.php?id=3926

Mediterranean Sea Basin's common sea-level; in other words, this artwork would preclude any future erosion of valued land-based artworks encompassed by the urban fabrics of North Africa and southern Europe. The global reinsurance industry, almost 75% of which is "…underwritten by companies in Germany, Switzerland and the United States [of America]" (Sturm and Oh 2010), will likely restructure their Mediterranean Sea Basin risk exposure after the Gibraltar Strait Textile Barrier macro-project we propose is constructed. The GSTBs structural, mechanical and hydro-dynamical physics was first preliminarily demonstrated in 2007 (Cathcart and Bolonkin 2007). Basically, the seawater-impervious Gibraltar Strait Textile Barrage replicates, in a crucial Mediterranean Sea Basin setting during a critical climatic change event-process, Christo's temporary suspended fabric curtain in Colorado. [Here, it is interesting to note, that Christo intends to complete a horizontally laid 9.5 km-long fabric covering of the State of Colorado's Arkansas River valley by 2012 (Christo 2008). Such a covering might be assumed to have some moisture-retention capabilities—that is, containment of evaporated freshwater, rather like the underlying impermeable lining of major freshwater irrigation canals.] Emplacement of the GSTB will not, in any significant way whatsoever, hinder the future tunnel boring machine [TBM] excavation of a Gibraltar Strait Tunnel should that become internationally desirable and economically affordable within the continuously evolving geopolitical context of the Mediterranean Union (Pliego 2005; Silva et al. 2006).

## 4 Gibraltar Strait Textile Barrage, the World's First Sea Art

"Valley Curtain" was punctured regularly at numerous places to prevent its being torn asunder by strong up-and-down valley winds. [Christo's famous "Running Fence" of 1976 in Marin County, California, had commenced and terminated in seawater (O'Doherty 2010).] Coincidentally, a watery version of "Valley Curtain", concocted by the UK expert Andrew Noel Schofield, was offered as a Thames River Storm Surge Barrier during 1971–1972. The GSTB will be impervious to seawater, safely sealed to the rocky sidewalls and sedimentary seafloor of Gibraltar Strait. Consequently, the GSTB will bow or "billow" like a ship's sail eastwards from the selected construction, emplacement, installation site because of marine (differences in sea elevations on a two-sided, bottom-anchored and virtually vertical suspended membrane and natural currents such as North Atlantic Ocean tidal solitons) and aerial (seasonal winds) pressures acting directly on the air-exposed GSTB segment. Indeed, prevailing seasonal winds flowing generally west-to-east along the fetch of the Gibraltar Strait will pile approximately 5–6 mm of seawater on the GSTBs west face. To cope with these natural environmental forces, macro-engineer planners must draw on the installation experience with heavy wire nets, floatation systems and their seabed moorings derived from world war II anti-submarine net installation in strategic harbors and that documented previous experience offered by the ~100 km-long World War I anti-submarine Otranto Strait Barrage (1915–1919).

From its western approaches, the GSTB will have the characteristic of an architectural deception resembling an English Garden or zoo landscape architect's geo-textile "ha-ha" (also known as a "sunken fence") in that—absent warning light-buoys and effective radar reflectors—ship navigators will visually misapprehend the true nature of the plotted sea-route ahead. Those mariners, such as small-scale, private-sector fishermen and yachtsmen, piloting their boats without benefit of up-to-date navigational sea-charts that indicate clearly the GSTBs presence, will have no inkling *via* normal optical clues that a 1 m drop in sea-level occurs at the Gibraltar Strait. Mariners without radar readouts using the eastern approaches will visually spy a 1 m-high tensioned fabric wall spilling some seawater caused by wave over-wash, which if made of clear or aquamarine-colored barrier material might be almost invisible until closely sighted.

The total area of the vertically draped, pendulous, GSTB is about 200 km$^2$ but only approximately 20,000 m$^2$ will be constantly exposed to the air-corrupting and material-degrading UV-strong sunshine on its eastern face while the GSTBs submerged western face will be required to continually resist a 1 m chemically complex hydraulic head of seawater. The world-ocean's seawater acidification persistent, if slow, trend must be monitored carefully. The GSTB artwork—if built, the world's first true commonly visible Sea Art—will be mechanically lifted by shore-based winches gradually only as much as the North Atlantic Ocean actually rises—it will, therefore, act as an active, scalable compensation mechanism to accommodate the real-time twenty-first century ocean changes in measured volume and surface area.

## 5 GSTB Structural and Material Design

The GSTB textile barrage structure and material have to be strong enough in order to support the nearly 1 m water hydraulic pressure difference, which remains constant along the height *h* (Fig. 3). The brittle fracture of the membrane would

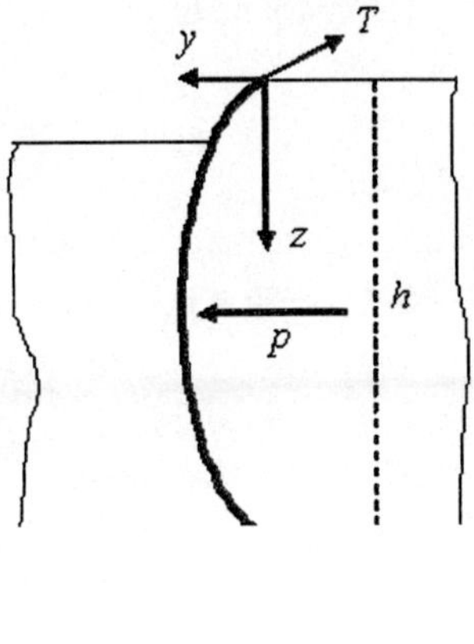


**Fig. 3** Vertical sectional scheme of the tensioned GSTB (side view)

cause the propagation of a tsunami wave and has, thus, to be avoided by a proper barrage structural and material design.

Considering the macro-engineering scheme to be installed at the geographical place imaged in Fig. 2 and reported in Fig. 3, and indicating with $y(z)$ the barrage deflection, from classical structural mechanics we have:

$$\frac{d^2y(z)}{dz^2} = -\frac{pL}{T_H} \tag{1}$$

where $p$ is the pressure difference, $L$ is the barrage length (along $x$) and $T_H$ is the horizontal component of the tension $T$. Equation (1) can be easily integrated with the boundary conditions $y(z=0) = y(z=h) = 0$. The real height $H$ of the barrage can be calculated as:

$$H = \int_0^h \sqrt{1 + \left(\frac{dy}{dz}\right)^2} dz \approx \int_0^h \left\{1 + \frac{1}{2}\left(\frac{dy}{dz}\right)^2\right\} dz = h + \frac{p^2L^2h^3}{6T_H^2} \tag{2}$$

Accordingly we find:

$$T_H = \frac{pLh}{\sqrt{6\varepsilon}}, \quad \varepsilon = \frac{H-h}{h} \tag{3}$$

The mean tension in the GSTB will, thus, be

$$T = \frac{1}{h}\int_0^h \frac{T_H}{\cos\left(\frac{dy(z)}{dz}\right)} dz \approx \frac{T_H}{h}\int_0^h \left\{1 + \frac{1}{2}\left(\frac{dy}{dz}\right)^2\right\} dz = T_H\frac{H}{h} \tag{4}$$

Consequently, noting that the ultimate tension per unit length $T_u/L = \sigma_u t$ is the product of the material strength $\sigma_u$ and GST Barrage thickness $t$, we expect a minimum thickness:

$$t = \frac{pH}{\sigma_u\sqrt{6\varepsilon}} \tag{5}$$

The mass $M$ of the GST Barrage can thus be calculated as:

$$M = \rho LHt = \frac{\rho p H^2 L}{\sigma_u\sqrt{6\varepsilon}} \tag{6}$$

where $\rho$ is the material density. The optimal design can be deduced to minimize its weight, according to:

$$\frac{\partial M}{\partial H} = 0 \quad \Rightarrow \quad \varepsilon = \varepsilon^* = \frac{1}{3} \tag{7}$$

At the optimal shape $\varepsilon^*$ corresponds to a thickness:

$$t^* = \frac{pH}{\sqrt{2}\sigma_u} \tag{8}$$

Equation (8) shows that to further reduce the weight of the barrage, and thus its thickness, we can reduce its size $H$ and/or increase the material strength $\sigma_u$.

The material strength $\sigma_u$ for a barrage having micro-structural (e.g., texture) size $q$ (the "fracture quantum") and containing an elliptical hole of half-axes $a$, perpendicular to the applied tension, and $b$ can be determined according to Quantized Fracture Mechanics (Pugno 2006) as:

$$\frac{\sigma_u}{\sigma_{ideal}} = \sqrt{\frac{1 + 2a/q(1 + 2a/b)^{-2}}{1 + 2a/q}} \tag{9}$$

where $\sigma_{\mathrm{ideal}}$ is the ideal (defect-free, theoretical) strength ($\sim$100 GPa for industrially-produced carbon nanotubes). Equations (9) and (8) show that plausible defects can reduce the strength and increase the thickness by 1–3 orders of magnitude.

Due to the huge physical size $H$, which must even accommodate the Earth's natural curvature, the GSTB composing material has to be sufficiently strong, e.g. steel, Kevlar or carbon nanotubes, or alternatively the structure itself has to be sufficiently thick throughout it full extension. Graphene sheets are ideal candidates in this geographical and structural context, thanks to their great mechanical strength, impermeability and natural two-dimensionality.

Considering $H \approx 1\,\mathrm{km}$, $\varepsilon \approx \sigma_u/E \approx 0.1 - 0.01$ ($E$ is the material Young's modulus) and $p \approx 10\,\mathrm{KPa}$, we deduce for realistic macroscopic, thus inevitably defective, graphene sheets ($\sigma_u \approx 10\mathrm{GPa}$, see Pugno 2006) $t \approx 1 - 4\,\mathrm{mm}$ (for defect-free graphene sheets, $\sigma_u \approx 100\,\mathrm{GPa}$, one would expect $t \approx 100 - 400\mu\mathrm{m}$); for comparison, for steel membranes ($\sigma_u \approx 1\,\mathrm{GPa}$) $t \approx 1 - 4\,\mathrm{cm}$.

We conclude that Kevlar and even graphene sheets are ideal as tensioned textile barrage material; their astounding super-strength, and consequently reduced thickness, consequently minimizes material consumption and maximizes flexibility, e.g. allowing rolling (expected to be useful during transportation to the work-site).

## 6 Artwork (Southern Europe and North Africa) Preservation Paramount

The proposed Gibraltar Strait Textile Barrage is a deformable physical construction intended to preserve extant as well as near-term future social artworks within the Basin of the Mediterranean Sea that are near seawater. It also is a means to improve humanity's ability to apply macro-engineering principles which skirt or correct a possible near-term future global oceanographic macro-problem impairing the future economic usefulness of southern Europe and northern Africa's low-elevation coastal lands. It is a practical and low-cost example of Sea Art for the twenty-first century that preserves humanity's ancient and modern cultural heritage situated in the multi-cultural Mediterranean Sea Basin, which is subject to earthquakes, tsunamis and even hurricanes (Bilham 2009; Weisse and von Storch 2009).

The two most essential reasons why the GSTB ought to be built are: (1) all European Union and Mediterranean Union members now hope to link Europe's and North Africa's super-smart 2050 electric power grids (Strahan 2009) and (2) construction of solar power concentration electricity generation facilities in North Africa will quickly modernize ordinary life-styles as well as rapidly change traditional national geopolitical outlooks of the peoples living in that generally arid region (Trieb and Muller-Steinhagen 2008). A third necessarily vital reason (3) to build the GSTB is to steady the micro-tidal Basin's sea-level in order to permit reliable macro-engineering assumptions about the CATS macro-project elucidated in this chapter.

## 7 The Mediterranean Sea Basin CATS Macro-Project

Disregarding the ultimate cause of future effective sea-level rise, a practical subject of great geoscientifical controversy, one prospect looms menacingly obvious: marine inundation of the Mediterranean Sea Basin's present-day strand poses a major threat to the long-term welfare of its permanently resident human populace. Besides freshwater reservoir storage, it is practicable to channel large quantities of seawater into some of our world's great interior drainage basins that lie below present-day sea-level in order to dynamically control our world-ocean's volume (Newman and Fairbridge 1986); for example, a naturally capacious potential sink adjacent to the Mediterranean Sea is Egypt's Qattara Depression. An artificial potential sink is envisioned at another low-elevation region, the most eastern part of the Zone of Chotts, which extends westward from Tunisia into Eastern Algeria and consists of Chott Gharsa (Fig. 4, 34.08°N Lat. by 10.05°E Long.; 620 km$^2$ with a maximum depth below present-day world sea-level of $\sim$10–25 m), Chott el Jerid (34.00°N Lat. by 10.00°E Long.; 5,400 km$^2$ with a maximum positive elevation of $\sim$10–25 m ASL) and Chott Melrhir (31.00°N Lat. by 7.00°E Long.; 1,100 km$^2$ surface with a maximum negative sea-level elevation of $\sim$40 m below global sea-level (BSL).

Chott el Gharsa would overflow the watershed, flowing into Chott Melrhir at a level of minus 5 m. Water levels >25 m would result in linkage with the Chott el Jerid (Fig. 5) a halite saltpan bordered by a remarkably concentric region of gypsum flats encircled by a sandy sabkha, and the Chott el Fejej, and a water level exceeding 37 m above extant world sea-level would result in the inter-connection of the entire seawater-filled Zone of Chotts (Figs. 5, 6) with the nearby Mediterranean Sea. A "Chott" is a salt flat within a hydrological basin separated from the Mediterranean Sea. The mean annual temperature for the Zone of Chotts is 20.9°C.

**Fig. 4** Chott el Gharsa. Source: http://www.upload.wikimedia.org/wikipedia/commons/c/c5/Chottelgharsa1.jpg; Wikipedia Commons by Jaume Ollé. Creative Commons Attribution 3.0 Unported License

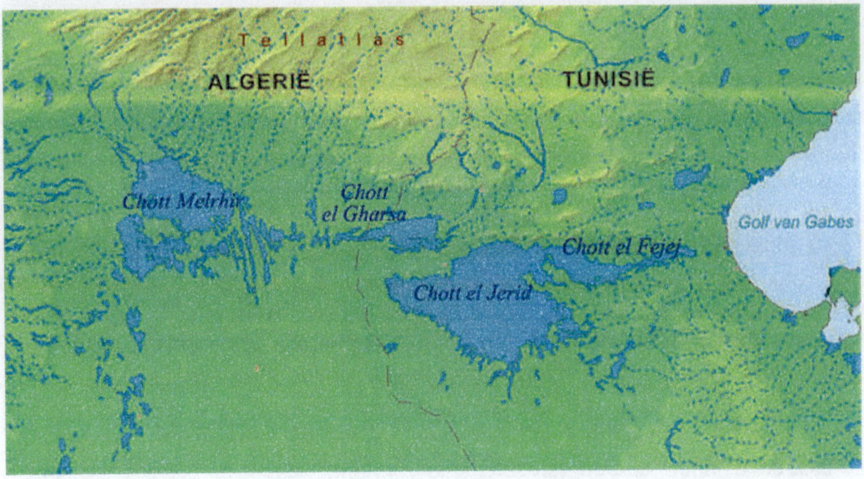


**Fig. 5** Map of northern Africa's famous Zone of Chotts. Source: http://www.upload.wikimedia.org/wikipedia/commons/7/72/Chott_el_Jerid.jpg Released under the GNU Free Documentation License

The mean yearly rainfall ranges between 89 and 150 mm. Evaporation in the Zone of Chotts is at a maximum during the dry season (between April and November), reaching between 2,500 and 3,100 mm/year. It is a closed topographic region distinguished by high evaporation and low rainfall regimes. Under some UNO Climate Change models projected changes in yearly temperature move

**Fig. 6** Chott el-Jerid with naturally channeled brine accumulation. http://www.commons.wikimedia.org/wiki/File:Chott_el-Jerid_1117.jpg Source: By Gorik Francois. Licensed under Creative Commons Attribution ShareAlike 2.0 License

upward (by +3 to +4°C) and annual rainfall is reduced (by 10–20%). Should those event-processes ever happen, it would greatly benefit our offered CATS macro-project.

With a population of >10 million humans, sovereign Tunisia, independent since 1956–1957, is bordered by the Mediterranean Sea (1,148 km-long coastline), Libya (459 km) and Algeria (965 km).

Filling Chott Melrhir (Fig. 7, −40 m BSL) in Algeria with seawater by harnessing, sequentially, the major Chotts existing in Tunisia is the goal of the "Chotts Algeria–Tunisia Scheme" (CATS) macro-engineering effort. Treeing the CATS means, simply, that we recognize this particular macro-project grand-scale design (Kaya 1991) for artificially enclosed seas to be really difficult of accomplishment, rather like getting all herded contentious wild cats into one peaceful, or at least controllable and containable, group holding pen.

During the historic period from 1957 until about 1988, nuclear energy researchers in the USA and the former USSR considered the mundane purposes achievable using mundane tools—peaceful nuclear explosives (PNEs) (Hess 1962). Super-computerized simulations of crater dimensions resulting from 200, 400 and 800 kiloton yield nuclear explosions have been done to study the feasibility of connecting the Mediterranean Sea with the Qattara Depression by a PNE-excavated channel (Toman 1980). Earthwork comprises two operations: (1) the cutting down of the elevations projecting above the level of the proposed surface and (2) the filling up of the hollows lying below the proposed surface. CATS will involve, mostly, the

**Fig. 7** Chott Melrhir imaged by earth-orbiting satellite. Source: http://www.upload.wikimedia.org/wikipedia/commons/0/0b/Chott_Melrhir_SPOT_1195.jpg

former activity. Tunisia is one of the smallest in area of all independent states in Africa. Nevertheless, sometime before 1962, Tunisian scientists comprehensively proposed a "Chotts Depression Scheme" to serially blast huge craters in the Chott el Fejej and Chott el Jerid, both wastelands which are above present-day world sea-level, and subsequently inundate the resulting depression with 27.5 to 38.5% salinity seawater (Feistel 2008) channeled from the close-by Mediterranean Sea (Fig. 8). Minimally, ~5,400 km$^2$ plus the 620 km$^2$ Chott Gharsa west of the small-scale seaport of Gabes was planned for future unnatural ocean water inundation. It was also then foreseen and contemplated that Algeria's below world-ocean sea-level Chott Melrhir (1,100 km$^2$) could, eventually, be connected to form a linked chain of synthetic mini-seas covering 7,120 km$^2$ of previously air-exposed Zone of Chotts landscape. (That would be ~35.5 times bigger in area than Egypt's Great Bitter Lake.) Planted along the 34° North parallel of latitude, the colossal channel-depression was to have been formed inexpensively *via* multiple PNEs and the Chotts Depression Scheme, which we have renamed CATS, was to result, finally, in a real-world recreated Lake Tritonis (Fig. 9) of ancient Mediterranean Sea Basin mythology. While the transforming Chotts are being filled by inflowing seawater

**Fig. 8** One possible configuration of a Zone of Chotts seawater submergence macro-project—perhaps the foundational first iteration of a future Port Tritonis? Original artwork by Joseph J. Friedlander. Base image from Orbiter

**Fig. 9** Artist-author Joseph J. Friedlander's space image-like concept of ancient mythology's supposed Lake Tritonis. Port Tritonis realized and materialized?

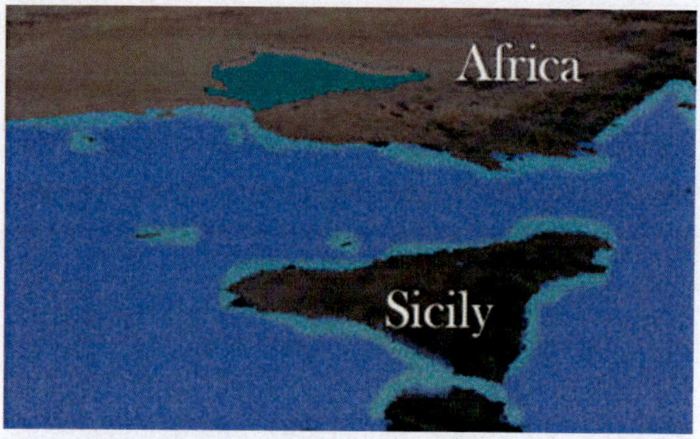


from the micro-tidal Gulf of Gabes, where Spring tides have amplitude of $\sim 1$ m, a large amount of hydropower, which entails a true cascade of seawater spinning turbines, can be generated. However, that will only be a temporary phenomenon.

Subsequently, because of the dry inland climate, these shallow bodies of introduced seawater, brought to Mediterranean Sea level, would endure rapid aerial evaporation, increasing the salinity to +200% salts, thus producing a continuous flow of seawater punctuated by tidal seawater movements through the excavated connecting channel; Tunisian macro-engineers naturally visualize that cheap and reliable hydrokinetic electricity would thereby be produced as a direct result of this constant current, a steady inflow much like a river's flow (Kahn 2009). Of course, the continuously incoming seawater's flow would not come close to the current speed within the Strait of Messina, the gap in southern Europe's land between Calabria on mainland Italy and Sicily. If the natural Tunisian terrain has a water flow resistance of over 0.6 kg/m$^2$ for an unlined canal with a 20 cm/km slope—that is, a gradient of 2/10,000—unlined bank erosion will be unlikely to occur (Litrico and Fromion 2009). Since the 3.2 Richter-scale Earthquake in southern Tunisia, with an epicenter in the mining basin near Gafsa (Karaloui et al. 2009) on 28 December 2009, it is obvious that unlined canals would endure longer without maintenance than more complicated structures; nevertheless, concrete-lined channels are best from the perspective of restriction of geographical effects. Today, hydropower-generated electricity accounts for barely 1% of overall energy production in 2010 Tunisia. The wind energy resource present in the Gulf of Tunis is promising for future development and some dredged material might be shipped northwards to be used to form artificial islands bearing masts equipped with wind turbines. Such a facility could efficaciously complement the anticipated solar power concentrators in the desert to the south of Tunis (Orstein et al. 2009).

How best to produce hydrokinetic electricity in the existing wastelands of Algeria and Tunisia? First, a man-made channel (100 m vertical bank to vertical bank width by ~175,000 m-long with a constant depth of 5 m) must be excavated by (1) yet-to-be-built floating nuclear-powered dredgers (Cathcart 2008) or automated Bucketwheel excavators similar to that once used to partly dig the Sudan's incomplete Jonglei Canal. The now idle and damaged Sudan Bucketwheel dug at a maximum consistent rate ~3,000 m$^3$/h (Collins 1990). Self-propelled canal-construction machinery is custom-built to specific job requirements and can be designed to perform in a number of applications. Canal builder's use of continuous concrete slip-form pavers can be labor and cost-saving and prevents most infiltration of canal-carried fluids into the soil and groundwater underneath the canal. Very roughly, approximately 85,500,000–100,000,000 m$^3$ of rock fragments and loose quaternary sediments (aeolian, lacustrine and sabkha deposits) must be removed to join Gabes (human population ~250,000) with Algeria's Chott Melrhir. Supposing the need to redistribute 85,500,000 m$^3$ of material using a highly automated Bucketwheel-type mechanical self-moving digger that can shift ~280,000 m$^3$/week, then excavation for the CATS might require a constructive time period of between 5 and 6 years to complete. (A floating nuclear-powered dredger could, probably, remove 350,000 m$^3$/week and compete the initial phase in 0.5 and 1.5 years.) We think it is reasonable to suppose that CATS' macro-engineers will assume that the elongated arm of the Chott el Jerid, the Chott Fejej that extends eastward towards the coastal city of Gabes, should be utilized as a direct route for the concrete-lined seawater channel. The cost of the CATS' initial digging phase ought not to exceed 2010 USD 1 billion. This estimated cost might be reduced somewhat if beneficial uses can be found for dredged material (US Army Corps of Engineers 2005) in, say, nearby salt-marshes (Allen and Pye 2009).

During the middle construction phase, the spoils from the channel mining, heaped in usefully sited mounds (artificial earth sculpture, landscape as artwork) by design, form gigantic geometrical berm-bordered ponds wherein seawater might be deposited temporarily to create permanent pumped storage electrical power plants (Hiratsuka 1993), perhaps employing a hydraulic turbine deriving energy from an inlet within the reservoir (Kouris Paul 2000), after the reduction of the Chott el Jerid ground surface elevation above sea-level by 20 m above sea-level (ASL), making it $\sim$3 m below the present-day Mediterranean Sea level ultimately. Basically, that part of the gulf-creation macro-project necessitates removal of $1.072 \times 10^{11}$ m$^3$ of material. Full excavation of the Chott el Jerid might be profitably postponed until after the channel reaches the first topographical place below the present-day sea-level Chott, namely Chott Gharsa. The GSTB in Sect. 5 will, thus, maintain a stable macro-engineering planning context for the channel and the trained seawater inlet facing the micro-tidal Gulf of Gabes since Tunisia's strand will be mostly unaffected by the generally anticipated, but nevertheless, uncertain future global sea-level rise macro-problem. The two trickiest dredger working-life moments will happen when the automated nuclear-powered dredger meets the two places where the dug channel will eventually enter the below sea-level Chotts since a gushing seawater flow could either wedge the dredger in the gap between the elevation-reduced Chotts—that is, Chott el Fejej and Chott el Jerid—and the below sea-level Chotts—Chott Melrhir and Chott Gharsa—or the unmanned dredger could be wrecked or severely damaged by simply falling over unnatural waterfalls. Not to be ignored is the likelihood of seasonal gusting winds affecting the handling of the unanchored floating dredgers.

## 8 Hydrokinetic Electricity Production, Navigation and Pollution

Several prefabricated floating railroad bridges with pre-installed tidal stream energy machines ought to be towed into their ideal final installation sites, and thereafter sunk to form a ship-passable pierced land transportation causeway. Rotation of the energy-generation mechanisms within this permeable barrier will thereby produce an interminable hydrokinetic electricity supply that the poorest people living in western region of central and southern Tunisia may wish to sell out-of-country, use themselves and/or share with neighboring folks situated in southern Algeria.

The introduction of a new source of income will certainly help improve the nutrition of the people through diet diversification. On 22 April 2009, Tunisia ratified the "UN Convention on the Non-navigational Uses of International Watercourses", which is an international legal framework for more specific bilateral and regional treaties relating to the use, management and preservation of trans-boundary water resources such as the Complex Terminal Aquifer (Edmunds 2003; Kamel 2008). CATS would seem to qualify for future discussion and inclusion. Improved harmonization between Tunisia and Algeria's anti-Sahara migration and desertification control strategies and the legitimate socio-economic development of local rural communities living and working (pastoral and mining) in the Sahara to the south will be a major challenge for CATS macro-engineers because the CATS could instigate a land-rush episode. It is easy for us to imagine a new public relations program that touts CATS as a "resurrection of the granary of Rome" (Davis 2007) combined with a concerted public-government organization effort toward regional industrialization. Such reinvigoration of a region—in fact, the macro-engineering-induced conversion of J.B. Jackson's "anti-landscape", which he defined in 1984 as a man-modified space that once served as infrastructure for collective existence but that has, over recorded history, ceased to do so—making the CATS a truly resurrected northern African "landscape". Port Tritonis, somewhat resembling mainport Rotterdam in layout, is the ultimate possible expression of such a macro-management effort.

There is a single Mediterranean Sea Basin precedent—although not fully comparable—at Italy's Ischia Island. The unique man-made Harbour of Ischia's (Fig. 10) historically fascinating macro-engineering story is fully recounted elsewhere in this book (see Chap. 10.1007/978-3-642-14779-1_1). A small boat and yacht harbor, Port d'Ischia, was dug during the 1850s by macro-engineers who flooded a land-locked volcanic crater that could be connected to the Tyrrhenian Sea economically. Harbor formation workers needed about 2 years for digging the very short channel with hand tools, pony carts and wheelbarrows. The Romans had an expression: "Multorum minibus magnum levatur onus" ["by the hands of many a great load is lightened"]. Luckily, the construction site work leaders and laborers followed the surface trace of a <10,000 year-old volcanic eruptive fracture that made their excavation effort simple and physically less difficult.

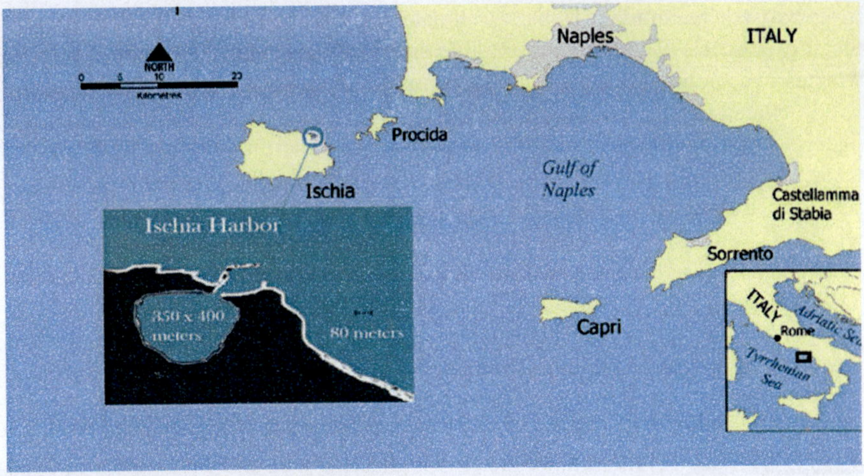


**Fig. 10** Ischia Harbor. Source: *Inset* art by Joseph Friedlander; background art by Norman Einstein, modification under the GNU free documentation license. Courtesy Wikipedia Commons at http://www.upload.wikimedia.org/wikipedia/commons/7/73/Capri_and_Ischia_map.png

Excavation and submersion by imported seawater of Chott el Fejej-Chott el Jerid and Chott Melrhir facilitates profitable commercial coastal and high-seas shippers, encouraging them to serve new seaports built along the new far-inland strand, perhaps mineral and agricultural exportation would flourish. It will be incumbent on Tunisian authorities to become vigorously involved in the promotion, design, construction, procurement and operation of these new port facilities at all strata, with particular attention to the most recent developments in applicable design and building methods and how these must be economically integrated with operational requirements and, of course, environmental considerations. Channel walls and embankments will become places of new riparian vegetation and ship-generated wake washes will cause sediment and seawater movements unrelated to natural hydrodynamic processes, causing anthropogenic shoreline erosion, for instance (Garel et al. 2008). For example, fast ferries—some are easily capable of speeds of ~40 knots in water depths <10 m—produce very-long-period low-amplitude waves which are very energetic and can build in height rapidly close to the shore and then break suddenly; this nature of confined wash waves generated by fast ferries is such that their presence would endanger beach-goers and, probably, should not be allowed to ply the artificial bay produced by CATS. The effect of living channel vegetation will be evident on the seawater's velocity, turbulence and longitudinal mixing because emergent vegetation reduces the magnitude of longitudinal shear dispersion while submerged vegetation has a wake zone. And, barge-mounted drilling oilrigs could float unobstructed from one exploration work site to another drill site rather easily.

Essentially, what will be created by Tunisian macro-engineers and others is a semi-enclosed marine system bounded by land with only a regulated inlet (Tagliapietra et al. 2009); the inland coastal ocean-connected CATS marine system will be incomparable as not even the 75 km-long Arabian Canal now under construction in Dubai will be similar. Tunisia's climate regimes will change, perhaps unpredictably (Enger 1991). Too, it is possible there may be some macro-project planner worries over potential future hydro-seismology owing to seawater loading of the wind-deflated depression's earth-crust surface since seawater is less dense than the materials that have been forcefully removed by human in the Chott el Fejej and Chott el Jerid and redistributed. Seawater circulation, the long-term pattern of seawater motion remaining after the irregular seawater movements involved in wind drift, seiches, and other short-term phenomena are averaged, in the CATS will likely be counter-clockwise just like Egypt's Great Bitter Lake (30°20′N Lat. by 32°23′E Long., Fig. 11) (Touliabah and Taylor 2004; El-Bassat 2008).

One unique import, fertile foreign silt, would be obtained and carried ~2,700 km by slurry pipeline to the vicinity of Gabes, Tunisia, from the new freshwater reservoir submarine delta created by the Aswan High Dam at the border of Egypt and Sudan or, perhaps, from some closer place (Abulnaga and El-Sammany 2004); this silt importation macro-engineering idea was first bruited in 2004 (Cathcart and Badescu 2004). Large-scale reservoir desiltation serves two interests because (1) such lake sediment mining effort prolongs the operational period of the Aswan High Dam Reservoir and (2) the mineral-rich silt extracted, especially if

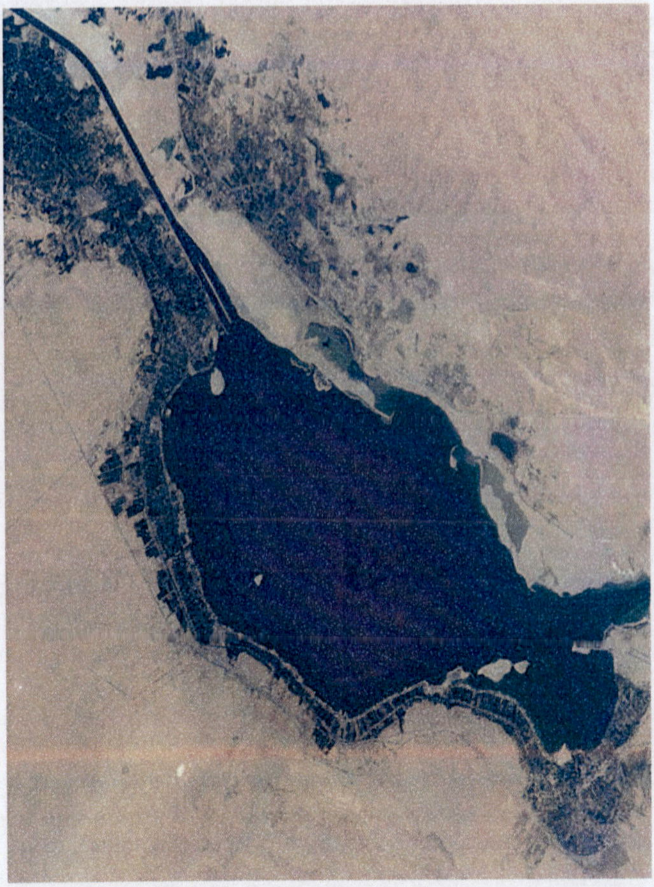

**Fig. 11** Great Bitter Lake, Egypt. Source: NASA earth observatory. This saline lake is a possible physical analogue for CATS. The Great Bitter Lake accommodates north–south/south–north international shipping using the Suez Canal

widely spread on arid land to appropriate fresh-soil profiles, could provide suitable soil base material for covering extant barren salt-flats in Tunisia and Algeria (Brandt 2000). Macro-engineers have the useful example of hundreds of lakes, including the Veeranam in Cuddalore, India, that are planned to be de-silted during the early-twenty-first century just to increase their holding capacities for vitally-needed collected normal rainwater runoff.

Eventually, the budding academic and industrial technopole of Gabes, Tunisia, could become the home of an all-digital electronic Second Library of Alexander to serve well the employed and settled citizenry of a newborn major industrial complex. Ships that use ballast water, and once used stones, can be adapted to use appropriately formulated thick mud slurry consisting of Nile River freshwater and Aswan High Dam Reservoir sediment that is stabilized for transport in slightly modified ship holds. In other words, there are at least two means to transfer material from Egypt to Tunisia and Algeria. FCM Fourier's "hands" can be supplemented with robust solar-powered robotic outdoor earth-moving equipment, off-spring of those semi-automated machines NASA R&D has devised and roughly constructed for the exploration of the far-distant planet of Mars (Badescu 2009).

Realistically, the materialized CATS inland oceanographic creation would not be easily predictable in its hydraulic behavior. CATS will certainly have many of the distinguishing characteristics of the USA's Great Salt Lake in Utah and the Salton Sea in California (O'Connell 2001) as well as Urmia Lake in Iran. Tunisia's Gulf of Gabes is the major marine region of energy dissipation for present-day—that is to say, known and experienced—Mediterranean Sea tides (Tsimplis 1995). Local tides in today's Mediterranean Sea generally have a small range of $\sim$1 m. Strand conditions have changed greatly from those of ancient times, as close examination by geological expert Roland Paskoff (1933–2005) has indisputably revealed (Paskoff 1991). If saltwater aquaculture were to be developed in the submerged Chotts, then the clean-up of incoming seawater is, seemingly, a mandatory subsequent industrial activity (see http://www.ars.usda.gov). The presence of CATS may cause further tourist development on the 514 km$^2$ Djerba Island—the most famous vacation attraction situated in southern Tunisia—as an international yacht racing locale (Anane et al. 2008), especially if Port Tritonis becomes actualized during this century, to facilitate the industrialization of North Africa and the northern Sahara.

Today's Mediterranean Sea level, 1 m higher than in olden Roman times, masks a rubbish-strewn (underwater) seascape and a badly contaminated volume of seawater (Guillaumont 1995) laden with floating plastic wastes. It is a saddening geographical fact that to our species' almost everlasting shame, vast regions of the Mediterranean Sea Basin's submarine continental shelf is burdened with rotting man-made marine debris. A strong ocean seawater current moving finally towards Chott Melrhir through the long artificial channel and other capacious seawater lagoons that were once unpopulated Chotts will, of course, redistribute this junk, garbage and other unidentifiable stuff unless remedial measures are simultaneously undertaken to prevent any pollution of Chott Melrhir. The way in which these various materials may be brought to Chott Melrhir is by the primary wave system built up in the form of a pressure maximum at the bow and the stern of a moving watercraft and a pressure minimum develops along the hull of the vessel. This distribution of pressure will cause a seawater level elevation at the bow and a drop amidships. As a consequence of the pressure distribution of the primary wave system, a secondary wave system builds up with shorter wave periods compared with the long wave periods of the primary system. The whole process results from the complex interaction of both wave systems. Every vessel, whether streamlined or not, passing through the Gabes-Chott Melrhir Channel of the CATS will generate ship waves and return currents that hit the vertical banks of the channel, causing slosh and propelling suspended and floating debris forward towards the watercraft's destination. As a result, some bank damage doubtlessly will occur. However, since the vertical banks of the channel are concreted there

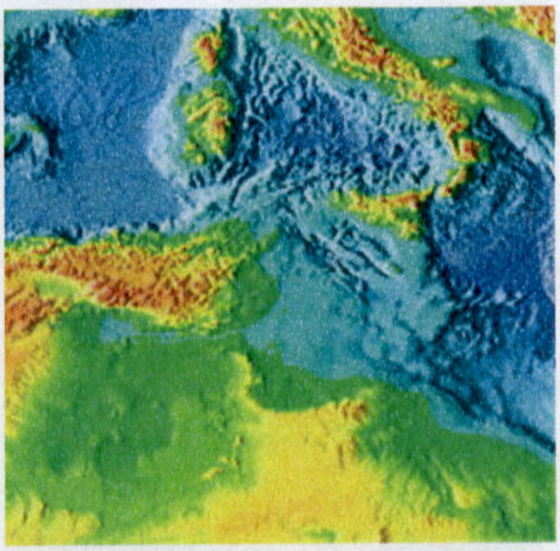

**Fig. 12** Two views of a canal-connected flooded Zone of Chotts. Above, an expansive view of a near-term future possibility, base image courtesy NOAA. Available at http://www.ngdc.noaa.gov/mgg/image/2minsurface/1350/ 45N000E. jpg. below, Fig. 13, conceivable areas of rain enhancement after Enger and Tjernström (see text)

**Fig. 13** Two views of a canal-connected flooded Zone of Chotts. Above, an expansive view of a near-term future possibility, Base Image courtesy NOAA. The *green-color* edging to the completed CATS indicates where increased rainfall may be expected to fall with some regularity. Notice that southern Algeria is benefited too by the increased rainfall effect

will be no bank slumping caused by soil instabilities, nor any need for costly time-tabled maintenance dredging. Like the Suez Canal, scheduled ship and barge convoys will traverse the CATS channel one-way, with the Chott Melrhir serving as a safe large vessel turning basin (Figs. 12, 13).

Ships, probably balanced using foreign ballast water, and entering the completed CATS, can be expected to transfer and deposit plants and marine animals from their places of origination (http://www.invasions.si.edu/ballast.htm; http://www.invasivespecies.gov and http://www.globallast.imo.org). Consequently, there are likely to be algal blooms and exotic mineral interactions producing, in effect, a horizontal bubbly lamp a la the still popular "Lava Lite®" effect; this unique graduated coloration effect ought to be quite detectable in earth-orbiting satellite images of North Africa. Marine mucilage and microbial pathogens may become established in the CATS (Danovaro et al. 2009) and the invasive sea-grasses and algae of the *Caulerpa* family could strongly affect the hydrodynamics and sedimentation rate in CATS (Hendriks et al. 2010).

A familiar pong at the beaches of CATS will be caused by odiferous DMS (dimethylsulfide) produced by microbes. Alternatively, all CATS seawater inputs could be filtered so as to resemble the world's largest seawater pool, the USD 3.5 million hotel swimming pool built at San Alfonso de Mar resort in Chile, and completed in 2007 that is 8.9°C warmer than the Pacific Ocean (Anon. 2010).

## 9 Political Aspects

Disputes of a geopolitical nature are bound to arise with the inundation by seawater of this large low-lying arid region. Since CATS will be an artificial, virtually enclosed new marine system, lengthening Tunisia's present-day 1,148 km-long shoreline, and increasing its offshore 8,250 $km^2$ continental shelf national territory, how ought it to be apprehended and categorized by the experts of extant international law? And, Algeria's new inland seacoast bordering the former Chott Melrhir must be contiguous with Tunisia's, subject to foreign control of trade just like Africa's disadvantaged, landlocked states (Faye 2004). [The place furthest from the ocean in Africa is located at 26.17°E Long. by 5.65°N Lat. (Garcia-Castellanos and Lombardo 2007).] Is it possible that the national economy of Chad, once the site of a Holocene Lake Mega-Chad (Bouchette et al. 2010) could become positively affected, induced by a new source of income caused by CATS-fostered ecotourism? Development of Port Tritonis will entail the same kind of considerations as that being made for Mainport Rotterdam—that is, environmental impacts caused by routine shipping operations which often pose threats to natural habitats and economic losses to adjacent tourist regions. Port management at Port Tritonis will, of course, necessitate aspects of sustainable shipping and efficient seaport management in the form of pollution controls (van Gils and Klijn 2007). Like Mainport Rotterdam, Port Tritonis will require some land reclamation from the ocean (Hommes et al. 2009). Figure 14 is a speculative rendering of one landfill pattern becoming the literal foundation for massive future harbor and industrial development.

The Caspian Sea is the "...largest totally enclosed water body on earth..." (Ibrayev et al. 2010); nowadays, there are strong arguments over the Caspian Sea's

**Fig. 14** A speculative vision: Tunisia's Chotts as the landfill enhanced "Mainport Rotterdam" of a future industrially developed North Africa. Note the strategically convenient natural geography of Djerba Island

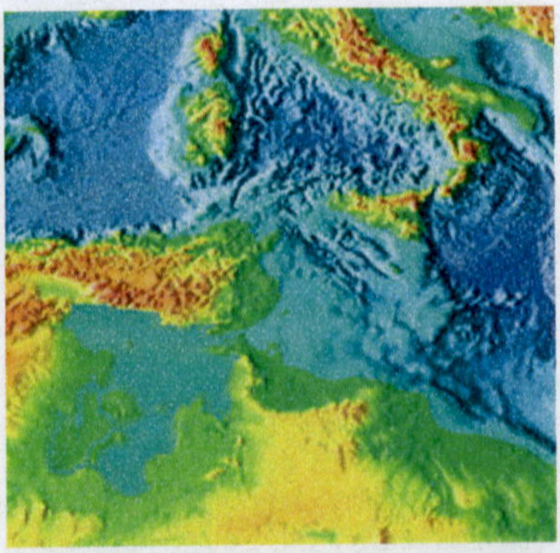

**Fig. 15** A vertical view of a possible future island-spotted saline Lake Tritonis created to supplant our Zone of Chotts CATS macro-project by magnificent spatial enlargement. It might be created by a coastal dam/seawater pumping facility. Artwork by author-artist Joseph J. Friedlander from a base image courtesy NOAA. Source: http://www.ngdc.noaa.gov/mgg/image/2minsurface/1350/45N000E.jpg

divided geopolitical status. Likewise, Tunisia shares a 965 km-long co-terminus border with Algeria. Logically, there must be a strong bi-lateral treaty and binding UNO-brokered international legal accords before a grossly revamped Chotts Algeria–Tunisia Scheme (CATS) ever becomes a fabulous Lake Tritonis-like facility publicized by the worldwide leisure industry (Fig. 15).

## 10 Conclusions

We outline the macro-engineering properties of our proposed CATS, which are intended to transform favorably the traditional livelihoods of persons chiefly living and working in western Tunisia and southern Algeria. If the imported seawater can be processed for cleanliness, then seafood, derived from local aquaculture may be a sustainable industry (Corbin 2007). Processing of seawater is, after all, considerably easier than the creation of seawater from a batch of chemicals poured from laboratory glass-jars, as the chemist Thomas Alvin Boyd (1888–1989) elucidated with his charming popular scientific essay published more than 80 years ago (Boyd 1929). Basically, with the physical realization of CATS, we create new seaports and tourist beaches that might resemble, but will be far more voluminous and extensive, than those existing at Algarroba, Chile, at the San Alfonso del Mar coastal hotel complex (Anon. 2007; Valdemoro and Jimenez 2006). Various floating structures, including oil drilling rigs may intrude on the CATS region to explore for new national wealth that could benefit markedly the citizens of two North Africa nations. Clearly, as macro-engineers, we do not apprehend the CATS as a vain national fantasy quest for valuable natural resources (Allan 1983). Instead, we think and believe CATS is a doable dream macro-project with positive economic consequences for Tunisia and Algeria. And, furthermore, that Port Tritonis is a worthy final act of stimulated macro-engineering imagination.

## References


Abulnaga BE, El-Sammany MS (2004) De-Silting Lake Nasser with slurry pipelines. Critical transitions in water and environmental resource management proceedings of world water and environmental resources congress 2004, p 284

Allan JA (1983) Natural resources as national fantasies. Geoforum 14:243–247

Allen JRL, Pye K (eds) (2009) Saltmarshes: morphodynamics, conservation and engineering. Cambridge University Press, Cambridge, 196 pp

Anane M, Kallali H, Jellali S (2008) Ranking suitable sites for soil aquifer treatment in Jerba Island (Tunisia) using remote sensing, GIS and AHP-multicriteria decision analysis. Int J Water 4:121–135

Anon. (2007) The Inland Sea. Wired. p. 124

Anon. (2010) Big dipper. The National Geographic 217, np

Anthoff D, Nicholls RJ, Tol RSJ (2010) The economic impact of substantial sea-level rise. Mitig Adapt Strategies Global Change 15:321–335

Badescu V (ed) (2009) Mars: prospective energy and material resources. Springer, Heidelberg, 695 pp

Beddoe R, Costanza R, Farley J, Garza E, Kent J, Kubiszewski I, Martinez L, McCowen T, Murphy K, Myers N, Ogden Z, Stapleton K, Woodward J (2009) Overcoming systemic roadblocks to sustainability: the evolutionary redesign of worldviews, institutions, and technologies. Proc Natl Acad Sci 106:2483–2489

Bilham R (2009) The seismic future of cities. Bull Earthq Eng 7:839–887

Blondel J (2006) The 'design' of Mediterranean landscapes: a millennial story of humans and ecological systems during the historic period. Human Ecol 34:713–729

Bohannon J (2010) The Nile Delta's sinking future. Science 327:1444–1447

Borowiec A (2003) Taming the Sahara: Tunisia shows a way while others falter. Praeger, NY, pp 168

Bouchette F, Schuster M, Ghienne JF, Denamiel C, Roquin C, Moussa A, Marsaleix P, Duringer P (2010) Hydrodynamics of Holocene Lake Mega-Chad. Quat Res 73:226–236

Boyd TA (1929) Synthetic seas. Sci Mon 29:61–65

Brandt DA (2000) A review of reservoir desiltation. Int J Sediment Res 15:321–342

Caldwell C (2009) Reflections on the revolution in Europe: immigration, Islam, and the West. Doubleday, NY, 432 pp

Cathcart RB (2008) Kra Canal (Thailand) excavation by nuclear-powered dredges. Int J Global Environ Issues 8:248–255

Cathcart RB, Badescu V (2004) Architectural ecology: a tentative Sahara restoration. Int J Environ Stud 61:145–160

Cathcart RB, Bolonkin AA (2007) Ocean terracing. arXiv.org > physics > physics/0770110

Christo Jeanne-Claude (2008) Over the River. Taschen, Hong Kong

Collins RO (1990) The waters of the Nile. Markus Wiener Publishers, New York, p 348

Corbin JS (2007) Marine aquaculture: today's necessity for tomorrow's seafood. Mar Technol Soc J 41:16–23

Cullen MN (2005) Tension membrane water retaining structures. Trans Build Environ 79:427–436

Dachs J, Mejanelle L (2010) Organic pollutants in coastal waters, sediments, and biota: a relevant driver for ecosystems during the anthropocene? Estuaries Coasts 33:1–14

Danovaro R, Umani SF, Pusceddu A (2009) Climate change and the potential spreading of marine mucilage and microbial pathogens in the Mediterranean Sea. PLos One 4:1–8
Davis DK (2007) Resurrecting the granary of Rome. Ohio University Press, Athens, 296 pp
Day JW, Gunn JD, Folan WJ (2007) Emergence of complex societies after sea level stabilized. EOS Trans Am Geophys Soc 88:169–170
della Dora V (2007) Geo-strategy and the persistence of antiquity: surveying mythical hydrographies in the eastern Mediterranean, 1784–1869. J Hist Geogr 33:514–541
Dorale JA, Onac BP, Fornos J J, Gines J, Gines A, Tuccimei P, Peate DW (2010) Sea-level highstand 81,000 years ago in Mallorca. Science 327:860–863
Edmunds WM (2003) Groundwater evolution in the continental intercalaire aquifer of southern Algeria and Tunisia: trace elements and isotopic indicators. Appl Geochem 18:805–822
El-Bassat RA (2008) Composition and abundance of the zooplankton community in the Bitter Lakes, Egypt, in relation to environmental factors. Afr J Aquat Sci 33:233–240
Enger L (1991) Estimating the effects on regional precipitation climate in a semiarid region caused by an artificial lake using a mesoscale model. J Appl Meteorol 30:227–249
Faye M (2004) The challenges facing landlocked developing countries. J Human Dev 5:31–68
Feistel R (2008) Mutually consistent thermodynamic potentials for fluid water, ice and seawater: a new standard for oceanography. Ocean Sci 4:275–291
Fischer AG, Garrison RF (2009) The role of the Mediterranean region in the development of sedimentary geology: a historical review. Sedimentology 56:3–41
Garcia-Castellanos D, Lombardo U (2007) Poles of inaccessibility: a calculation algorithm for the remotest places on earth. Scott Geogr J 123:227–233
Garel E, Fernandez LL, Collins M (2008) Sediment resuspension events induced by the wake wash of deep-draft vessels. Geo-Mar Lett 28:205–211
Gide C (1970) Design for Utopia: selected writings of Charles fourier. Shocken Books, New York, p 180
Goudie AS (2003) Enhanced salinization. Dev Water Sci 50:287–293
Guillaumont B (1995) Pollution impact study in Gabes Gulf (Tunisia) using remote sensing data. Mar Technol Soc J 29:46–58
Harff J, Hay WW, Tetzlaff DM (2007) Coastline changes: interaction of climate and geological processes. The Geological Society of America, Washington
Haycraft WR (2000) Yellow Steel: The Story of the earthmoving equipment industry. University of Illinois Press, Urbana, 463 pp
Hendriks IE, Bouma TJ, Morris EP, Duarte CM (2010) Effects of seagrasses and algae of the *Caulerpa* family on hydrodynamics and particle-trapping rates. Mar Biol 157:473–481
Hess WN (1962) New horizons in resource development: the role of nuclear explosives. Geogr Rev 52:1–24
Hiratsuka A (1993) Seawater pumped-storage power plant in Okinawa Island, Japan. Engineering 35:237–246
Hommes S, Hulscher SJMH, Mulder JPM, Otter HS, Bressers HTA (2009) Role of perceptions and knowledge in the impact assessment for the extension of mainport Rotterdam. Mar Policy 33:146–155
Ibrayev RA, Ozsoy C, Sur HI (2010) Seasonal variability of the Caspian Sea three-dimensional circulation, sea level and air–sea interaction. Ocean Sci 6:311–329
Jackson JB (1984) Discovering the vernacular landscape. Yale University Press, p 8
Jean-Pierre L, Philippe G, Mohamed A, Abdelmatif D, Larbi B (2010) Climate evolution and possible effects on surface water resources of North Algeria. Curr Sci 98:1056–1062
Kahn MJ (2009) Hydrokinetic energy conversion systems and assessment of horizontal and vertical axis turbines for river and tidal applications: a technology status review. Appl Energy 86:1823–1835
Kamel S (2008) The hydro geochemical characterization of ground waters in Tunisian Chott's region. Environ Geol 54:843–854
Karaloui F, Touzi S, Tarhouni J, Bousselmi L (2009) Improvement potential of the integrated water resources management of the mining basin of Gafsa. Desalination 248:157–163

Kaya H (1991) Ideal grand design for enclosed coastal seas. Mar Pollut Bull 23:463–467
Koger G (1999) The Great Sahara Sea: an idea whose time has come? Mercator's World 4:18–23
Kouris Paul S (2000) USA patent 6,114,773. Hydraulic Turbine Assembly
Kuleli T (2010) City-based risk assessment of sea level rise using topographic and census data for the Turkish coastal zone. Estuaries Coasts 33:640–651
Lambeck K, Anzidei M, Antonioli F, Benini A, Esposito A (2004) Sea level in Roman time in the Central Mediterranean and implications for recent change. Earth Planet Sci Lett 224:563–575
Lejeusne C, Chevaldonne P, Pergent-Martini C, Boudouresque CF, Perez T (2010) Climate change effects on a miniature ocean: the highly diverse, highly impacted Mediterranean Sea. Trends Ecol Evol 25:251
Leuliette EW, Miller L (2009) Closing the sea level rise budget with altimetry, Argo, and GRACE. Geophys Res Lett 36:Gl036010
Litrico X, Fromion V (2009) Modeling and control of hydrosystems. Springer, NL, p 409
Mukerjee M (2010) Poisoned shipments. Sci Am 302:14–15
Newman WS, Fairbridge RW (1986) The management of sea-level rise. Nature 320:319–321
O'Connell KA (2001) The forgotten sea. Landsc Architect (February), pp 50–54
O'Doherty B (2010) Christo and Jeanne-Claude: remembering the running fence. University of California Press, Berkeley, p 178
Orstein L, Aleinov I, Rind D (2009) Irrigated afforestation of the Sahara and Australian Outback to end global warming. Clim Change 97:409–437
Paskoff RP (1991) Modifications of coastal conditions in the Gulf of Gabes (Southern Tunisia) since classical antiquity. Zeitschrift fur Geomorphologie SB 81:149–163
Pearce F (2009) Sunshine superpower. New Sci 204:38–41
Perissoratis C, Georgas D (1994) The role of the earth scientist in assessing the impacts of climatic changes due to the greenhouse effect: two case studies of 'prognostic geology'. Terra Nova 6:306–312
Pliego JM (2005) Open session—the Gibraltar strait tunnel an overview of the study process. Tunn Undergr Space Technol 20:558–569
Poulos SE, Ghionis G, Maroukian H (2009) The consequences of a future eustatic sea-level rise on the deltaic coasts of Inner Thermaikos Gulf (Aegean Sea) and Kyparissiakos Gulf (Ionian Sea), Greece. Geomorphology 107:18–24
Pugno N (2006) Dynamic quantized fracture mechanics. Int J Fract 140:159–168
Silva PG et al (2006) Neotectonic fault mapping at the Gibraltar Strait Tunnel area, Bolonkia Bay (South Spain). Eng Geol 84:31–47
Strahan D (2009) Green grid. New Sci 201:42–45
Sturm T, Oh E (2010) Natural disasters as the end of the insurance industry? Scalar competitive strategies, alternative risk transfers, and the economic crisis. Geoforum 41:154–163
Tagliapietra D et al (2009) A review of terms and definitions to categorize estuaries, lagoons and associated environments. Mar Freshw Res 60:497–509
Toman J (1980) Technical report UCID-18531, 78 pp
Touliabah HE, Taylor WD (2004) The phytoplankton of Great Bitter Lake, Egypt, including the impacts of nutrient-laden and heated effluents. Afr J Aquat Sci 29:259–264
Trieb F, Muller-Steinhagen H (2008) Concentrating solar power for seawater desalination in the Middle East and North Africa. Desalination 220:165–183
Tsimplis MN (1995) A two-dimensional tidal model for the Mediterranean Sea. J Geophys Res 100:16223–16239
US Army Corps of Engineers (2005) Beneficial uses of dredged material. University of the Pacific, San Diego, 200 pp
Vaizey M (1990) CHRISTO. Rizzoli, New York
Valdemoro H, Jimenez J (2006) The influence of shoreline dynamics on the use and exploitation of Mediterranean tourist beaches. Coast Manage 34:405–423
Van Gils M, Klijn E-H (2007) Complexity in decision making: the case of the Rotterdam harbour expansion. Connecting decision, arenas and actors in spatial decision making. Plan Theory Pract 8:139–159

Vleuten E, van der Kaijser A (2006) Networking Europe: transnational infrastructures and the shaping of Europe, 1850–2000. Science History Publications, Sagamore Beach
Weisse R, von Storch H (2009) Marine climate and climate change: storms, wind waves and storm surges. Springer-Praxis Books, NY, p 200
Zewail A (2008) Mediterranean scientopolitics. Science 321:1417

# Mapping of the Qattara Depression, Egypt, using SRTM Elevation Data for Possible Hydropower and Climate Change Macro-Projects

**Ragab A. Hafiez**

## 1 Introduction

The Qattara Depression, which has a nearly triangular shape with a vertex at about 67 km distance from the Mediterranean Sea, is the largest natural closed land depression (19,605 km$^2$) of the Eastern Sahara. It forms the most significant geomorphologic feature in the northern part of Egypt's Western Desert. The deserted periphery of the depression is taken at the present sea level contour, while the lowest point in the depression is 134 m below mean sea level (b.m.s.l.). The large area of the depression, and the fact that it falls to a depth of 134 m b.m.s.l., has led to several proposals of major hydropower projects, to generate a huge hydroelectric power by conveying seawater from the Mediterranean Sea in an open channel or tunnel (Ball 1933). Recently, there is a serious concern to use the Qattara Depression as a basin to discharge the extra ocean water possibly resulting from Earth's climate change. The transformation of Qattara Depression into isolated anthropogenic inland sea could provide some ocean level adjustment, as well as generate energy, induce rainfall over some of the adjacent desert, reduce hottest desert daytime and nighttime air temperatures, and permit new local-use fisheries (aquaculture) as well as international tourism resorts. Persons visiting the Aswan High Dam will surely also be drawn to view the proposed seawater canal and the enormous anthropogenic desert seawater lake located so close to Alexandria's great Library.

R. A. Hafiez (✉)
DASCO, Cairo, Egypt
e-mail: ragabr@gmail.com

V. Badescu and R. B. Cathcart (eds.), *Macro-engineering Seawater in Unique Environments*, Environmental Science and Engineering,
DOI: 10.1007/978-3-642-14779-1_23, 

## 2 Geomorphology

The Qattara Depression forms one of the most significant features of the northern part of the Western Desert of Egypt (Libyan Desert in older literatures). The depression is a closed inland basin (Fig. 1), bounded by steep escarpments along its northern and western sides, with an average escarpment elevation of 200 m above mean sea level (a.m.s.l.). To the south and east, the arid floor of the depression rises gradually to the general desert level at 200 m a.m.s.l. (Figs. 2, 3). The periphery of the depression is usually taken at the present zero Mediterranean Sea level contour line, and, by definition, the area of the region of the depression is estimated to be 19,605 km$^2$ .The lowest negative elevation datum point in the depression is 134 m b.m.s.l. at the western end. Its maximum length extends 300 km in an east-west direction, and a maximum geographical width of 145 km north–south direction. The closest point to the Mediterranean Sea located at 55 km south of El Alamein.

The floor of the Qattara Depression below the contour line of 60 m b.m.s.l. is mainly covered by sabkha and evaporite materials resulting from the evaporation of seeped saline groundwater, while salt-beds cover the western end of the lowest portion of the depression. Within the depression, cones, towers, mushrooms and plateau-like hill landforms, ranging in height from 5 to 30 m, are common, especially, near the western scarp of the depression. Sinkholes and caves are

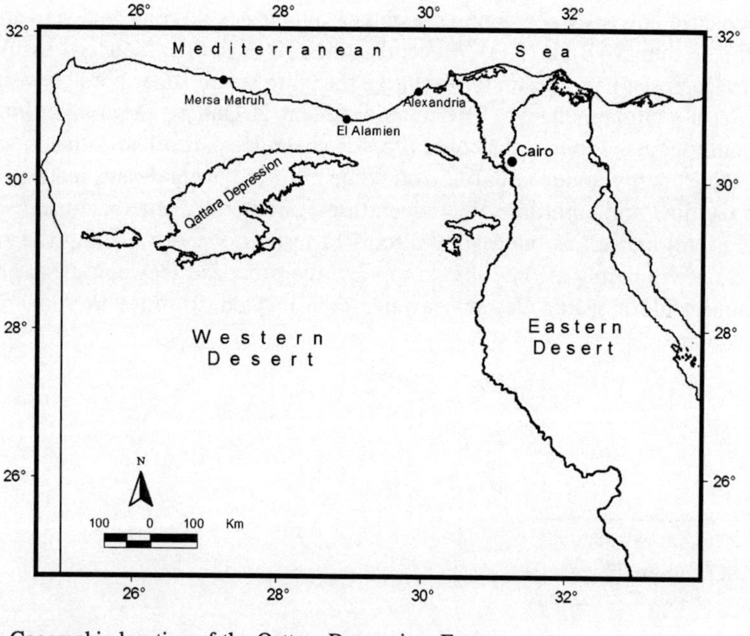


**Fig. 1** Geographic location of the Qattara Depression, Egypt

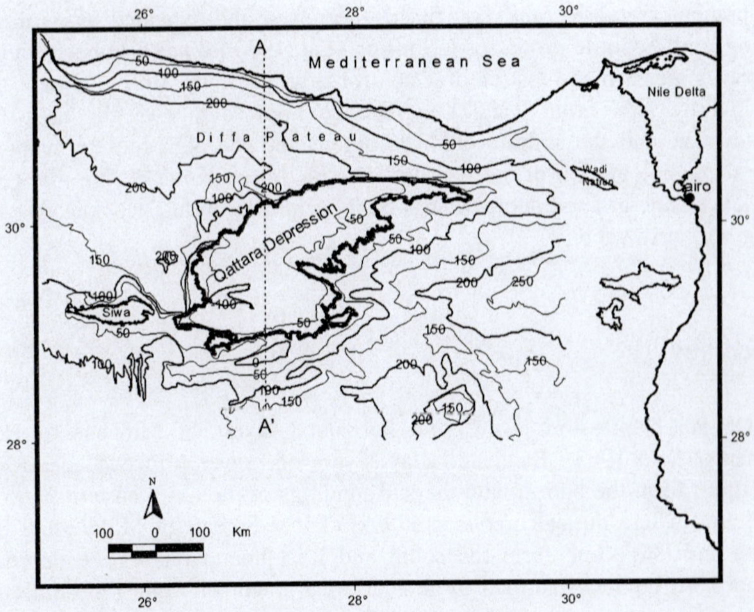


Fig. 2 Topography of lands surrounding the Qattara Depression

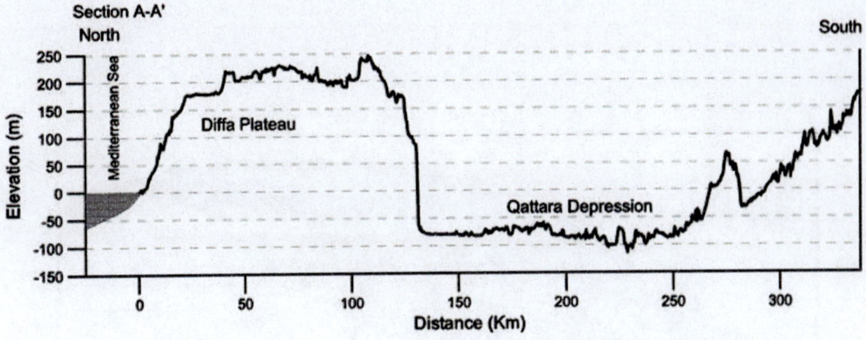


**Fig. 3** Cross section: A–A′ from north to south crossing the Mediterranean Sea shore, Diffa plateau, Qattara Depression and the southern highlands

common in the northern Diffa Plateau. The northern and western escarpments are dominated by large mass-wasted blocks (Aref et al. 2002).

The geological origin of the Qattara Depression remains an unresolved Earth Science puzzle. The first hypothesis was proposed by Ball (1933), he suggested that the Qattara Depression originated as the result of wind deflation to base-level controlled by the groundwater table. Other explanations include solution, mass-wasting followed by wind deflation (Said 1960), or the depression was originally excavated as a stream valley that was subsequently dismembered by karsts development processes, and was further deepened and extended by earth mass-wasting, and fluviotile processes (Albritton et al. 1990). It has also been suggested that the depression is of structural control origin (Gindy 1991). A new theory proposed by Aref et al. (2002) is that the high salinity of the near-surface groundwater, and the sodium chloride nature and the high rate of evaporation cause the disintegration of the bedrock. Dissolution of halite during the wet episode leaves fine-grained debris that is easily removed during dry episodes by the strong northerly winds.

# 3 Geology

The Qattara Depression is cut into horizontal layers of Miocene to Eocene sediments (Said 1962). Sand and clay layers of Lower Miocene age (Moghra Formation) form the bottom and the surroundings of the north-eastern part of the depression, where the calcareous sand and clay sediments of Middle and Upper Eocene and Oligocene form the south and west boundaries of the depression. The northern border is formed by a steep escarpment of white limestone of the Middle Miocene Marmarica Formation (Fig. 4).

Over large areas of the floor, the bedrock is covered with younger deposits including sand dunes, sabkha and Quaternary evaporite sediments.

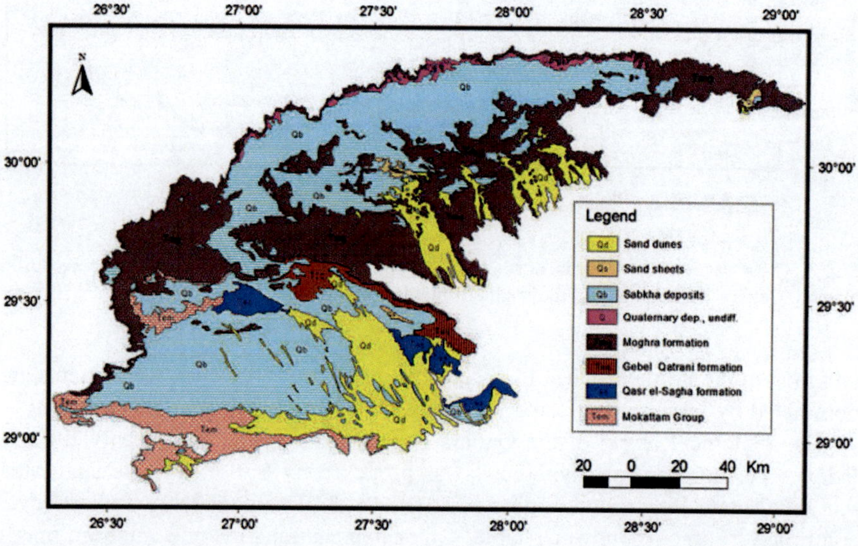


**Fig. 4** Geologic Map of the floor of the Qattara Depression (adapted from EGPC/CONOCO-Coral (1987); Geological map of Egypt)

## 4 Climate

The Qattara Depression is a part of the Great Sahara and located within that hyper-arid northern Africa region. The average temperature ranges from 36.2°C during the Northern Hemisphere summer months of June, July and August, to 6.2°C during the coldest winter month of January. The annual precipitation ranges from 25 to 50 mm on the northern rim to less than 25 mm on the southern part of the depression. The humidity value ranges from 71% during December and January, to 17% during April, May and June. The average annual evaporation rate is 1,700 mm. The prevailing wind is generally northerly but varies at times from northeasterly to northwesterly. The wind speed reaches a recorded maximum of 11.5 m/s in March and a minimum of 3.2 m/s in December (Aref et al. 2002).

## 5 Qattara Depression Macro-Project History

The utilization of the Qattara Depression for hydro-power generation was suggested for the first time by the Dutch geographer professor Penck in 1912, and later by Ball (1927) (De Martino 1973). Ball studied in particular the possibility of utilizing the Qattara Depression for hydro-power purposes by the formation of three lakes at final levels of 50 m, 60 m, and 70 m below sea level. After examining the effect of climatic changes, evaporation, seepage, minor seawater conveyance losses, geological and topographical considerations Ball, in 1933, finally recommended that the artificial lake of 50 m below sea level with a supply system along line route D is most convenient (Fig. 5). He calculated the surface area of the lake at that level would be 13,500 km$^2$, and the required quantity of seawater to be channeled into the lake is 656 m$^3$/s. The suggested route is 72 km length, an open channel to be constructed for the first 20 km, and that three circular tunnels with a diameter of 11 m should be built to the prospective lake's site at a level of −50 m.

Ball concluded that the resulting firm base-load power potential would be 175 MW, and he anticipated the possibility of using a power surplus during periods of low consumer demand to pump some parts of the inflowing seawater into a high-level reservoir atop the fringing escarpment of the depression, and using the 200 m head created to generate hydropower to meet peak-load requirement.

In 1964, a commission of experts took a favorable point of view of the possibility of a power plant to produce peak load energy, and F. Bassler was appointed to study possible development schemes in detail (De Martino 1973). Bassler concluded that the most convenient solution is the formation of an anthropogenic lake at a level of −60 m, with a corresponding seawater area of 12,000 km$^2$. In this case, the volume of water that would disappear annually due to the uncontrolled solar evaporation was estimated at 19,000 million cubic meters. It was

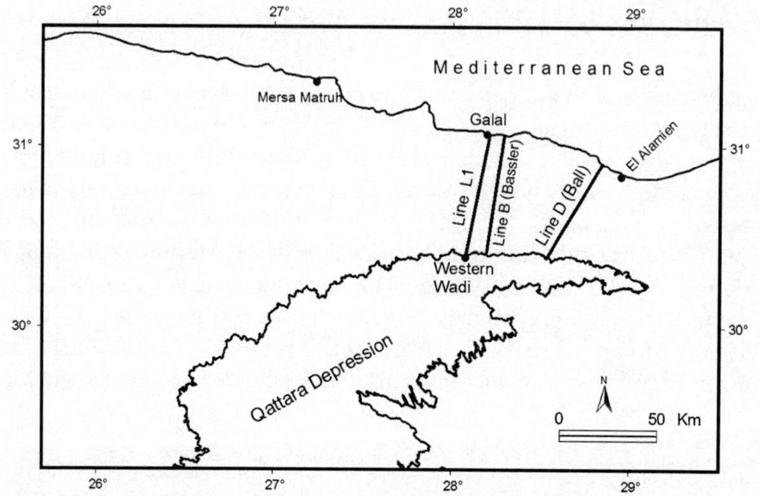


**Fig. 5** The different routes chosen over the years by various investigators to conduct seawater to the Qattara Depression

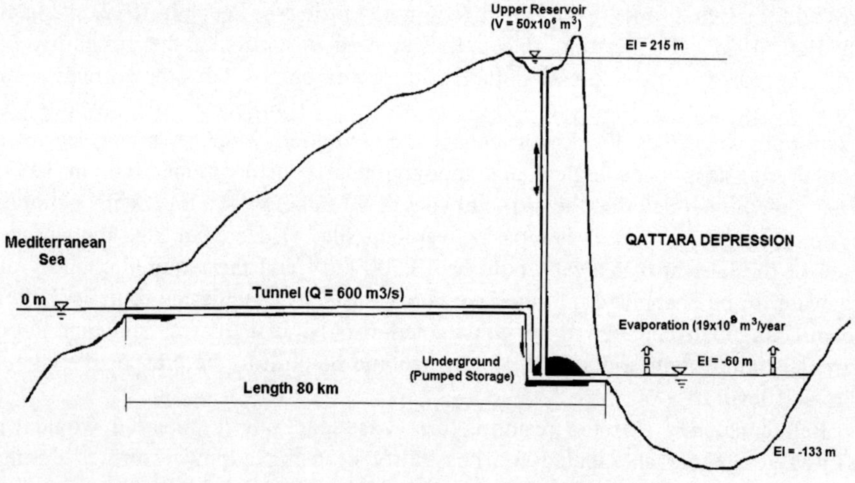


**Fig. 6** Schematic profile of the Mediterranean Sea-Qattara Depression hydro-power scheme with pumped storage proposed by F. Bassler

anticipated that such a lake would have to be fed by flow of ~600 $m^3$/s from the Mediterranean Sea to fully compensate for the evaporated water. The feeding seawater flow would be carried through tunnels having a length of about 80 km to an underground power plant at level −50 m (Fig. 6). During the hours of low load demand, the energy could be used to pump the seawater into a high-level natural reservoir having a capacity of about 50 million cubic meters. Therefore, during peak load hours, a valuable head of water would be available, and in addition, by not allowing the pumps to work during these periods, the low-head system would be able to generate much useful power. To sum up, this scheme is a combined high- and low-head electricity generating system, capable of producing peak power up to 4,000 MW and adaptable to a varying consumer demand pattern.

In 1981 The Ministry of Electricity and Energy, Arab Republic of Egypt published a comprehensive study for the macro-engineering geological investigations of the Qattara Depression Project prepared by Joint-Venture Qattara. This comprehensive study examines the geological conditions of the land region between the Mediterranean Sea and Qattara Depression with respect to a tunnel or peaceful nuclear explosion excavated canal, the determination of the soil and rock mechanical characteristics, the assessment of the foundation conditions of the proposed civil infrastructures and the slope stability conditions at the northern depression rim.

Three main alternatives had been proposed for the outlined route (Joint Venture Qattara 1981) as follows:

1. Nuclear Headrace Solution: Nuclear Headrace Canal between the Mediterranean Sea at Galal (west of El Alamein) and the Qattara Depression, terminating at the Western Wadi area (Fig. 5), with maximum capacity of 1,200 $m^3$/s. At the depression rim, a dam is envisaged, a run-off power house (600 MW) and a pumped storage power hose (4,800 MW) to be constructed, a tailrace canal will serve as lower pumpage reservoir, the upper reservoir to be located northwest of an elevation of 208 m.
2. Headrace Tunnel Solution: Two tunnels with diameter of 15.3 m on shore between Galal and the Western Wadi, the capacity of the two tunnels will be 800 $m^3$/s. In the Western Wadi the tunnels end in a forebay which will also serve as a surge pond, a run-off powerhouse (320 MW) and a pumped storage power hose (4,800 MW) as well as the tailrace canal and the upper reservoirs are similar to those of alternative 1.
3. One Way Pumped Storage Solution: At the Mediterranean Sea west of zawyet Harun a conventional intake canal led to a pump station which pumps seawater through a penstock into a high headrace canal running to the Western Wadi area and being terminated by a reservoir, by an intake, penstocks and peak power station the water will run through the tailrace into the depression.

## 6 Materials and Methods

The Shuttle Radar Topographic Mission elevation data product (SRTM) is one of the most valuable global resources of topographic data to-date (Rabus et al. 2003). The SRTM data product was developed for about 80% of the Earth's surface. The processed SRTM 90 m digital elevation model (DEM) for the entire planet was compiled by Consultative Group for International Agricultural Research Consortium for Spatial Information (CGIAR-CSI), and made available to the public via Internet mapping format (Gorokhovich and Voustianiouk 2006). The absolute geo-location accuracy for the continent of Africa is 11.9 m, whereas the vertical (absolute) accuracy is quoted at 5.6 m (Rodriguez et al. 2005)

This chapter is based on SRTM data acquired by the National Aeronautics and Space Administration (NASA) and National Imagery and Mapping Agency (NIMA) in February 2000 aboard of the shuttle spacecraft Endeavour. The data used herein is version 3.0 SRTM data, which resulted from substantial post-processing of the original release SRTM 90-m resolution (3 arc seconds) data and are provided by the Consortium for Spatial Information (CSI) of the Consultative Group for International Agricultural Research (CGIAR). The data are distributed in a geographic (Latitude/Longitude) projection, with the WGS84 reference ellipsoid and datum, and are currently available from the CGIAR-CSI SRTM database: http://srtm.csi.cgiar.org; where the following procedure has been followed:

1. The CSI-SRTM DEM 3-arc-second (90 m) tiles srtm_42_06 and srtm_42_07 were downloaded from CGIAR site: (http://srtm.csi.cgiar.org).
2. Mosaicking the tiles by joining the geo-referenced images together to form a larger image.
3. Generate a zero level base contour line of the entire Qattara Depression.
4. Subset the DEM file to the zero contour line.
5. DEM was re-projected to UTM zone 35 with both reference ellipsoid and datum to WGS84, resampled by nearest neighbor and a DEM with Spacing 92 m was derived.
6. The re-projected DEM was imported into Golden Software Surfer for further calculations (Golden Software 1993–2002).
7. The volume and surface area of the Qattara Depression at different altitude values starting from 0 to 120 m b.m.s.l. has been calculated in Surfer.

## 7 Results

The desert periphery of the Qattara Depression has been delineated at the present 2010 sea level (zero contour), with a surface area of 19,605 $km^2$ and a total volume of 1,213 $km^3$. The deepest point in the depression reaches 134 m b.m.s.l. and is located at the depression's western corner (29°23′33″N. latitude by 26°43′57″E. longitude).

The maximum depression length from east to west is 300 km at approximately latitude 29°45′N, where the maximum width from north to south is 145 km at longitude 27°30′E. The shortest distance to the Mediterranean Sea shore is 55 km at (30°20′45″N by 28°52′46″E). A new model for both the spatial volume and imagined surface area of the Qattara Depression were calculated below certain altitudes at intervals of 10 m (Table 1).

**Table. 1** The volume ($km^3$) and surface area ($km^2$) of the Qattara Depression below certain altitudes

| Altitude (m) | Volume ($km^3$) | Percent of volume (%) | Surface area ($km^2$) |
|---|---|---|---|
| 0 | 1,213 | 100 | 19,605 |
| −10 | 1,021 | 84 | 18,692 |
| −20 | 839 | 69 | 17,646 |
| −30 | 668 | 55 | 16,573 |
| −40 | 508 | 42 | 15,405 |
| −50 | 360 | 30 | 14,065 |
| −60 | 227 | 19 | 12,510 |
| −70 | 113 | 9.3 | 9,740 |
| −80 | 39 | 3.2 | 4,652 |
| −90 | 15 | 1.3 | 1,314 |
| −100 | 6.3 | 0.5 | 526 |
| −110 | 2.8 | 0.2 | 236 |
| −120 | 0.8 | 0.07 | 153 |

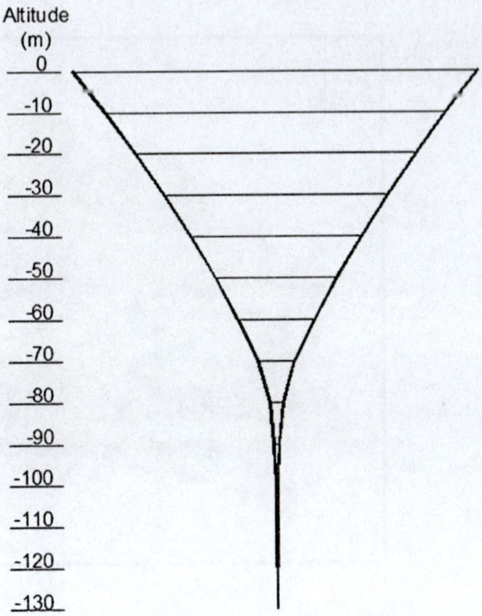


**Fig. 7** Graphical relationship between the contour levels and the capacity of the Qattara Depression

The graphical representation of the depression capacity with the change of contour level is presented in Fig. 7.

The calculated volume and surface area of the Qattara Depression were plotted against the contour levels starting from 0 to −130 m (Fig. 8). It is apparent from the model that 80% of the depression volume (986 $km^3$) is contained between contours 0 and −60 m, where the rest of the depression volume (287 $km^3$) exists below contour −60 m; the volume of the depression becomes dramatically negligible below contour line −70 m.

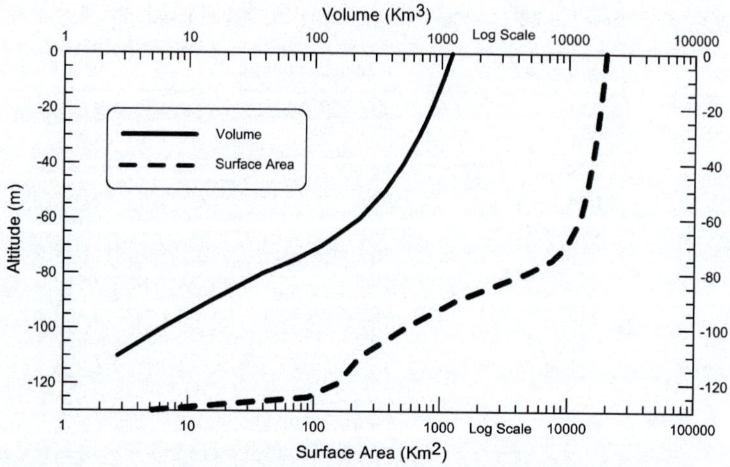


**Fig. 8** Volume, surface area—altitude model of the Qattara Depression, Egypt

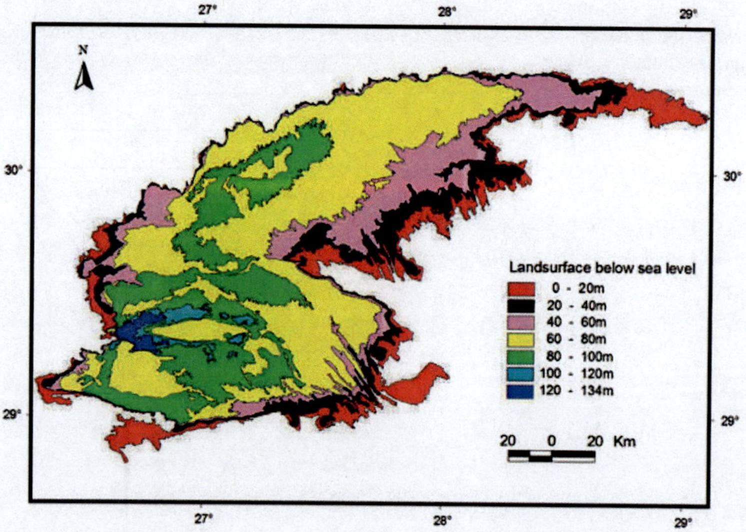


**Fig. 9** Topography of the Qattara Depression generated from SRTM elevation data

A set of contour lines for the un-flooded Qattara Depression has been generated from SRTM elevation data (Fig. 9); which highlights the surface variations of the depression floor as if the same as simulated stages of filling the depression with the imported seawater.

Three topographic profiles crossing the Qattara Depression has been generated from the SRTM data to demonstrate the topographical variations of the depression. The topography of the eastern and southern parts of the depression is relatively flat, while that of the western side is rough and irregular (Fig. 10).

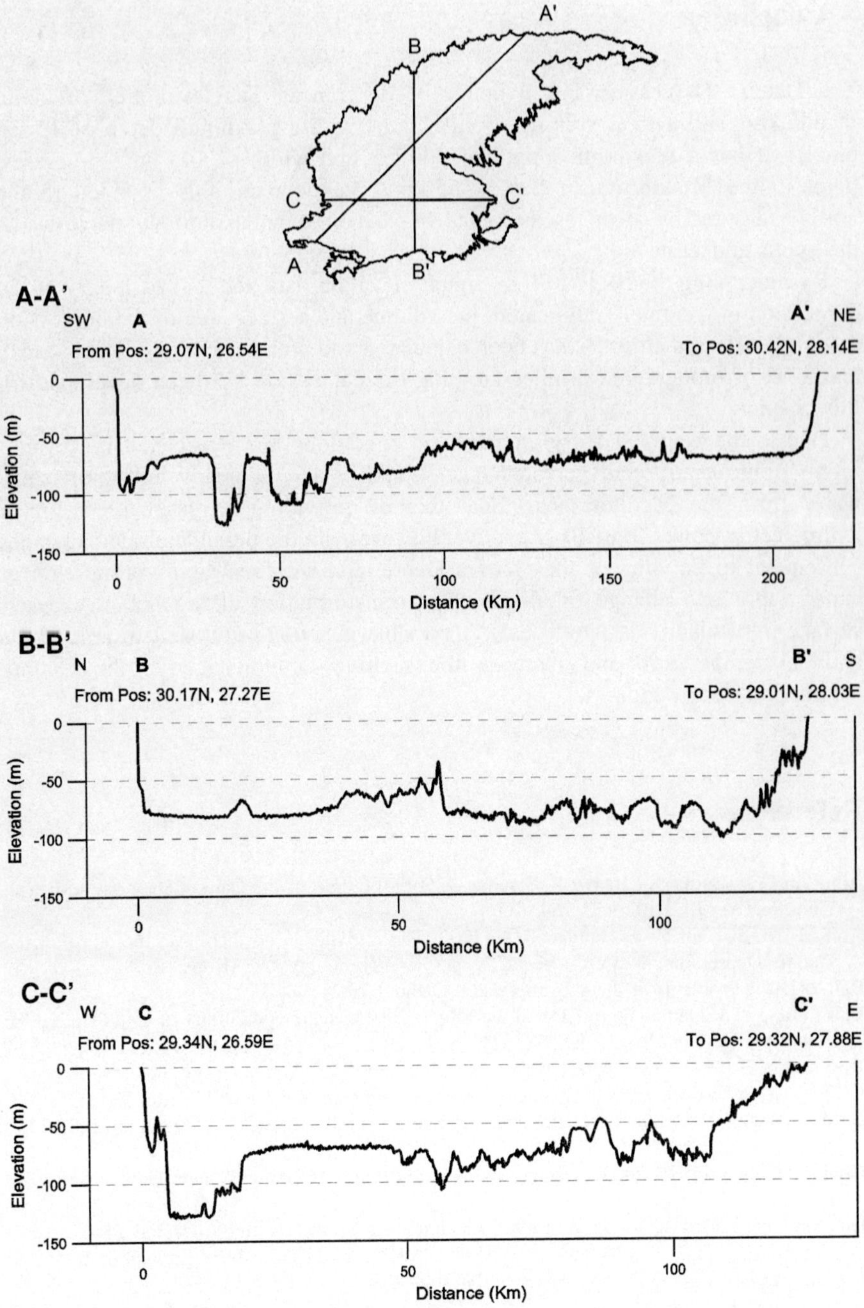


**Fig. 10** Topographic profiles generated from SRTM data across the Qattara Depression where: A–A′ cross section in the long axis of the depression extend from south west to north east, B–B′ show the topographic gradient existing between the northern and southern side of the depression, and C–C′ is the cross-section from west to east traversing the lowest point in the Qattara Depression

## 8 Conclusion

The Qattara Depression has a below Mediterranean Sea level area of about 19,605 $km^2$ and a total volume of 1,213 $km^3$ with a maximum depth of 134 m b.m.s.l. It has a maximum length of 300 km and width of 145 km, the closest point to the Mediterranean Sea at 55 km. The depression is bounded to the north and west by steep escarpments but becomes comparatively flat towards the south and east.

By processing the SRTM-90 m digital elevation data, the real periphery of the depression is accurately delineated, the volume and surface area of the depression at certain selected altitudes has been calculated and presented, a new contour map and a set of topographic profiles crossing the depression has been developed for this chapter.

Due to the warm and dry atmospheric conditions, the expected large evaporation from an artificially created lake should balance a steady inflow of ocean water from the Mediterranean Sea, thereby providing a reasonable gain of hydroelectric power. Equally attractive, however, are the possible climate changes subsequent to flooding of the Qattara Depression with seawater—production of water vapor and change of albedo. The transformation of the hot, dry desert surface into inland sea would have a possible evaporation impact to reduce the ambient temperature and increase the relative humidity above the Qattara Depression's surroundings.

## References

Albritton CC, Brooks JE, Issawi B, Swedan A (1990) Origin of the Qattara Depression, Egypt. Geol Soc Am Bull 102:952–960

Aref MAM, El-Khoriby E, Hamdan MA (2002) The role of salt weathering in the origin of the Qattara Depression, Western Desert, Egypt. Geomorphology 45:181–195

Ball J (1927) Problems of the Libyan Desert. Geogr J 70:21–38

Ball J (1933) The Qattara depression of the Libyan Desert and the possibility of its utilization for power-production. Geogr J LXXXII:289–314

De Martino G (1973) The Qattara depression. Reprinted from Water Power, January 1973

EGPC/CONOCO-Coral (1987) Geological map of Egypt, scale 1:500000, Cairo, Egypt

Gindy AR (1991) Origin of Qattara depression: discussion. Geol Soc Am Bull 103:1374–1375

Golden Software (1993–2002) Surfer Version 8.00-Surface Mapping System. Golden Software, Inc.

Gorokhovich Y, Voustianiouk A (2006) Accuracy assessment of the processed SRTM-based elevation data by CGGIR using field data from USA and Thailand and its relation to terrain characteristics. Rem Sensing Environ 104:409–415

Joint Venture Qattara (1981) Study Qattara Depression, Feasibility Report, Volume III part 2. General Egyptian Electricity Corporation, Arab Republic of Egypt

Rabus B, Eineder M, Roth A, Bamler R (2003) The shuttle radar topography mission—a new class of digital elevation models acquired by spaceborne radar. ISPRS J Photogramm Rem Sensing 57(4):241–262

Rodriguez E, Morris CS, Belz JE, Chapin EC, Martin JM, Daffer W, Hensley S (2005) An assessment of the SRTM topographic products. Technical Report JPL D-31639. Jet Propulsion Laboratory, Pasadena, California, 143 pp

Said R (1960) New light on the origin of the Qattara Depression. Societe de Geographie d'Egypte Bulletin 33:37–44

Said R (1962) The geology of Egypt. Elsevier, Amesterdam, 377 pp

# Assessing the Climate Response to Major Surface Inundation: Lake Eyre, Australia

**Pandora Hope, Andrew B. Watkins and Robert L. Backway**

## 1 Introduction

A World Meteorological Organization assessment of freshwater resources noted that approximately 1.7 billion people, or one-third of the world's population, live in countries that are water-stressed (Stockholm Environment Institute 1997) (defined as using more than 20% of their renewable water supply, a commonly used indicator of water stress), and that this number is projected to increase to around 5 billion by 2025. As a result, projects to 'drought-proof' regions vulnerable to water stress are clearly of interest to many people and governments.

While weather modification, primarily through cloud seeding, has been seen as one possible solution (Committee on the Status of and Future Directions in US Weather Modification Research and Operations and Board on Atmospheric Sciences and Climate Division on Earth and Life Studies National Research Council of the National Academies, 2007; and references therein), the flooding of large regions of land with a view to encouraging rainfall 'down-stream' has also been a popular, though less well explored, theme. Any such major project will have significant and unavoidable costs: financial, environmental, ecological and social. Determining the efficacy of any proposed scheme to produce useable rainfall in a downstream target region is imperative.

P. Hope (✉)
Centre for Australian Weather and Climate Research, Australian Bureau of Meteorology, Melbourne, Victoria, Australia
e-mail: p.hope@bom.gov.au

A. B. Watkins
National Climate Centre, Australian Bureau of Meteorology, Melbourne, Victoria, Australia

R. L. Backway
Lake Eyre Yacht Club, Belgrave Heights, Victoria, Australia

V. Badescu and R. B. Cathcart (eds.), *Macro-engineering Seawater in Unique Environments*, Environmental Science and Engineering,
DOI: 10.1007/978-3-642-14779-1_24, 

This chapter aims to describe methods of assessing the climate response—in particular rainfall enhancement—to the macro-engineering of wide-spread inundation. Lake Eyre, a large ephemeral playa in Australia, will be used as an example, following on from the work of Hope et al. (2004). This chapter will examine the potential impact that flooding Lake Eyre would have on the rainfall in the agricultural regions to its east.

Cloud-seeding is a well established science and can provide guidance as to how such an a *priori analysis* could be achieved. It is common practice to analyse the background weather conditions in the proposed region to determine if they are conducive to a positive response to cloud-seeding (e.g. Kuleshov et al. 2008). The response can also be modelled using fine-scale models of the weather over the region (e.g. Ćurić et al. 2007).

Studies of the climate response to man-made extensive flooding that is already in place will also guide an analysis of future projects by highlighting the expected responses. Around the globe, regions of inundation have been achieved via extensive irrigation. India, Israel, China and the USA all have wide-spread irrigation on a scale that would be likely to modify at least the local climate, and possibly the climate of regions remote from the source (e.g. Simmonds et al. 1999).

The USA has seen an extensive increase in irrigation over the last century. A statistical study of the conditions in the central USA by Barnston and Schickedanz (1984), extending an earlier study by Schickedanz (1976), suggested that irrigation enhanced rainfall from naturally occurring systems. Thus stationary fronts, that entrain surface moisture as they persist over the region, resulted in increased rainfall due to the irrigated surfaces. The overall climatic response to irrigation was decreased temperature, particularly during the day, and increased water vapour content. These changes were evident at stations up to 160 km from the main irrigated regions. Moore and Rojstaczer (2001) examining the same region, found that for the period 1950–1997 there is "at best, slight evidence that irrigation induces rainfall" and that "If irrigation-induced rainfall exists, its impact is only minor relative to the natural determining factors of plains climate". Moore and Rojstaczer (2001) also suggested that the apparent irrigation effects noted by Barnston and Schickedanz (1984) may simply have been due to a large-scale increase in precipitation rather than an irrigation effect.

The USA rainfall response to irrigation was further considered from both a conceptual viewpoint and using a fine-scale, weather resolving (mesoscale) model by Segal et al. (2001). In their conceptual model they suggested that moisture evaporated from irrigated areas in dry regions may act as a source for rainfall downwind. With a dynamical mesoscale model they assessed normal, drought and flood conditions with irrigated and non-irrigated surfaces. They found that irrigation did indeed alter existing rainfall regions but concluded that "the impacts of irrigation on the resolvable rainfall systems were generally so weak that hardly any new rainfall areas were generated; rather, irrigation only altered existing rainfall fields".

In some contrast, increased irrigation and a doubling of the area of rice paddies in the south of West Bengal, India, since the 1970s has been accompanied by a decrease in local rainfall. Lohar and Pal (1995) on the basis of a mesoscale model study, suggested that the decrease in temperature that accompanied the irrigation hampered the local sea breeze, and enhanced the extent of the land breeze. Although there is more actual evaporation from the surface, the amount of moisture brought by the sea breeze far outweighs that from the surface, so the enhancement of the land breeze leads to an overall decrease in moisture in the air. The decreased temperature and moisture limits the energy available to convective thunderstorms decreasing the potential rainfall from those systems also. Thus, given the mesoscale climate in this region, atmospheric modifications caused by the irrigation may have actually decreased local rainfall.

Inland from the coastal region of West Bengal, a westward shift in rainfall patterns and storm tracks has been noted by Singh and Sontakke (2002) who suggested this shift was probably due to large-scale circulation changes, rather than associated with changes in irrigation. They did, however, suggest that "A combined effect of regional (expansion and intensification of agricultural activities and spreading irrigation network) and global environmental changes… appears to be the main cause of cooling" in the region from 1958.

Increases in irrigation in the south of Israel since the mid-1960s have coincided with increased annual rainfall, discernible decreases in the diurnal temperature range and an increase in early wet season convective rainfall. Such changes were reproduced in a simple model experiment of de Ridder and Gallee (1998) leading to the conclusion that irrigation could have lead to enhanced moist convection. Ben-Gai et al. (1994) agree that the increased irrigation may have contributed to the observed increase in annual rainfall totals for 1961–1990, though they also suggested that the early 1960s was a time when the global sea surface temperatures were changing, affecting large-scale systems. Steinberger and Gazit-Yaari (1996) go further and state that the change in Israeli rainfall was related to large-scale circulation changes, and discount possible influences from increased irrigation.

Finally, in Australia's far north-west Kimberly region, there lies Lake Argyle, an artificial lake of some 700 $km^2$ (i.e. <10% of Lake Eyre when Lake Eyre is full) that was built in 1971, as part of the large Ord River irrigation scheme. Over the latter half of the twentieth century rainfall has increased over the east Kimberly (Rotstayn et al. 2007), leading some to suggest that the lake and irrigation scheme might have influenced rainfall in the region. However the timing of the rainfall increase pre-dates the lake being built and a number of studies have proposed other possible reasons for this broadscale rainfall increase (Wardle and Smith 2004; Rotstayn et al. 2007).

In most of the cases mentioned above, increased irrigation leads to a local decrease in temperature and sometimes changes in precipitation (although the evidence regarding the precipitation effect is less clear). Any increases in rainfall tend to occur in regions where there was rainfall prior to the irrigation. Responses distant to the moisture source are not as clear as those in the local region. Thus the evidence from international studies suggests that maintaining a constantly full Lake Eyre might reduce temperatures locally, and perhaps modify convective storms in the Lake Eyre region. However, on balance, these international studies do not suggest that a constantly full Lake Eyre would result in major, widespread increases in precipitation distant from the water surfaces.

In Australia, there have been a number of schemes proposed to 'drought-proof' the key agricultural regions. One recurring suggestion to achieve this aim is to 'water the inland' by forming a large water expanse in central Australia, with the assumption that moisture evaporating from such an expanse would later precipitate over agricultural regions in eastern Australia. Proposed methods included widespread irrigation, damming of rivers or flooding a low-lying region or pre-existing playa, in particular, Lake Eyre. Many of the early schemes are documented in Towner (1955). Most famous was the Bradfield Scheme, described in Warren (1945). The hydrological estimates behind this scheme have since been found to be flawed (Kotwicki 1986). As consideration for the hydrological and engineering limitations develop, new schemes have been proposed, particularly connecting Lake Eyre to the sea (Osborne and Dunn 2004 and Macro-Engineering Lake Eyre with Imported Seawater of this book). It must be noted that an enhanced rainfall response over agricultural regions is not always the desired outcome of these schemes, as described in Macro-Engineering Lake Eyre with Imported Seawater.

A Committee of meteorologists from the University of Melbourne and the then Commonwealth Meteorological Bureau (the predecessor of today's Australian Bureau of Meteorology) examined the Bradfield proposal and their conclusions are reported in Warren (1945): "The best that could be hoped for would be a slight amelioration of the climate, more particularly of temperature in the immediate neighbourhood of the storage area". This conclusion was supported by Hope et al. (2004), who explored the question of whether permanently flooding Lake Eyre would enhance rainfall using climate model experiments and found a localized cooling, but no consistent rainfall response.

From the studies described above and the guidance from the tests on the potential for success from cloud-seeding projects, two methods stand out as the dominant ways to assess the usefulness of a particular proposed scheme prior to its initiation. One is to impose an artificial region of extensive flooding in a climate model and assess whether the resultant climate is different to that when the flooding is not present (e.g. the studies of Segal et al. 2001; Hope et al. 2004). The other method examines the climate data in the region to be flooded and the 'target' region to determine, given a strong conceptual understanding of the regional weather, whether a large water expanse would firstly provide additional moisture, and secondly, if that moisture would have an impact on rainfall totals in the 'target' region. These two approaches will be explored for Lake Eyre, drawing on previously published works.

Section 2 provides a brief overview of the geography and climate of Lake Eyre and its catchment, while Sect. 3 describes the agricultural importance and climate variability of the 'target' region. Results from an analysis of the relationship between observed variability of the volume of Lake Eyre and rainfall in the target region are described in Sect. 4 while the results from the climate modelling studies of Hope et al. (2004) and further proposed model experiments are detailed in Sect. 5. A discussion of our findings and the methods used follows in Sect. 6.

## 2 Climate and Description of Lake Eyre

Lake Eyre is a large, shallow, and at times dry, salt lake sitting at Australia's lowest point, approximately 15 m below sea level. It drains a catchment of 450,000 km$^2$, as shown in Fig. 1, an area that is approximately one sixth of the contiguous Australian land mass, and is the fourth largest internal basin in the world. Lake Eyre consists of two lakes—Lake Eyre North and Lake Eyre South, linked by the Goyder Channel. The total lake covers 9,500 km$^2$, and when full its deepest point has a depth of only 6 m (Bye et al. 1978). The bathymetry changes during and after a filling due to the solution and re-deposition of the surface salts. Figure 2 shows the catchment, Lake Eyre North and South (shaded dark), and some of the surrounding rivers and towns.

All rivers that flow into Lake Eyre are ephemeral. The majority of inflow to Lake Eyre is from the rivers to the north, with an annual inflow of approximately 2.4 km$^3$. Due to the gentle slope along much of the river courses, there are many tributaries and channels, at times making the river courses unclear. The Georgina, Diamantina and Warburton make up one system, with a catchment of 365,000 km$^2$, while Cooper Creek has a catchment of 306,000 km$^2$ (Kotwicki 1986; McMahon et al. 2005). Water from the Diamantina rivers flows into Lake Eyre approximately every second year, while flow from other sources is more sporadic. West and south, the rivers include the Macumba, Neales, Frome,

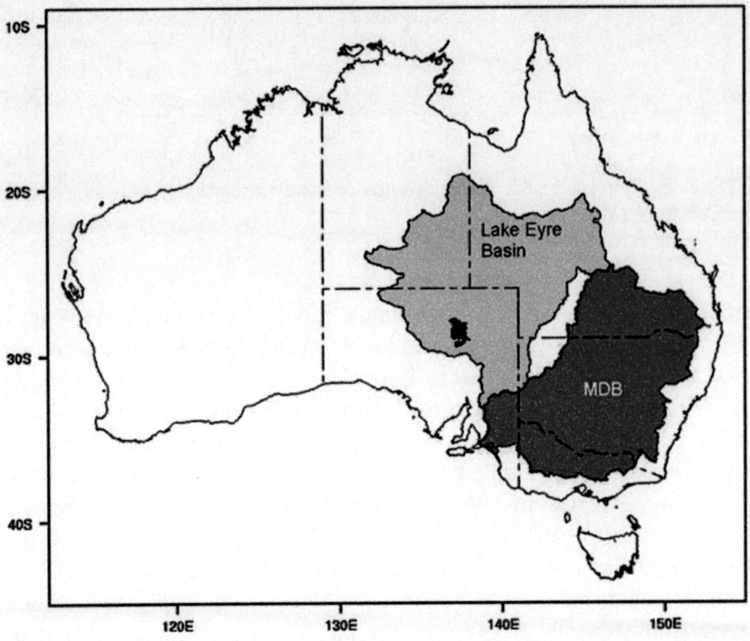


**Fig. 1** Map of Australia showing the location of Lake Eyre (*shown black*) and Lake Eyre and Murray Darling (MDB) basins

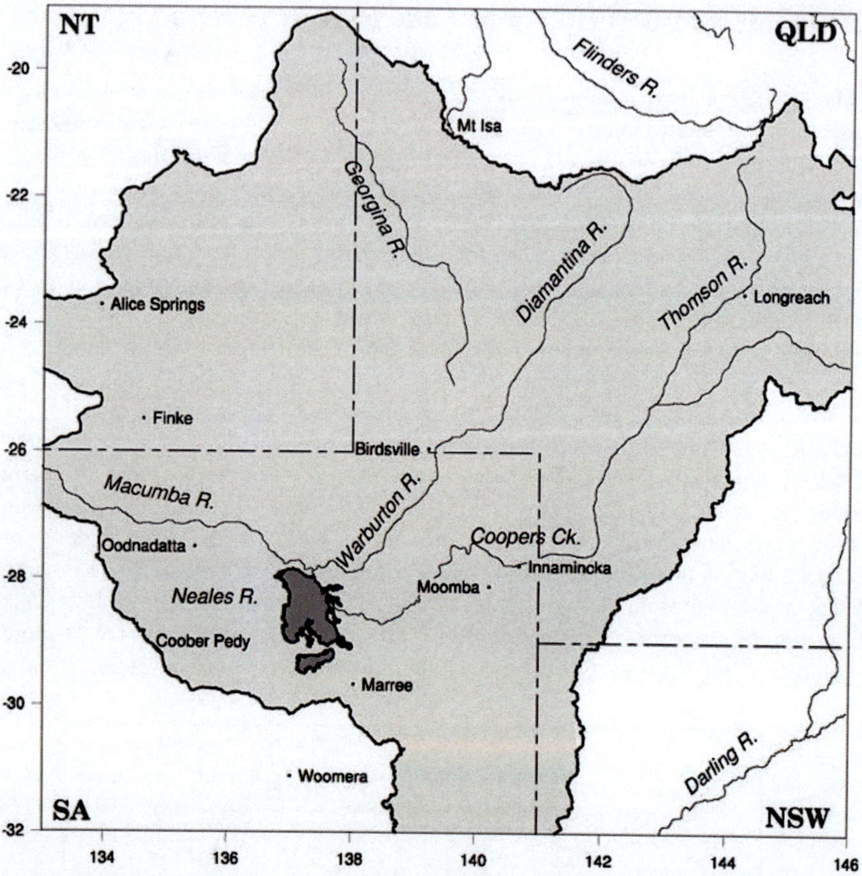


**Fig. 2** The Lake Eyre catchment, showing some of the major rivers and towns, the states (*NT* Northern Territory, *QLD* Queensland, *NSW* New South Wales, *VIC* Victoria) and state boundaries (from Hope et al. 2004)

Margaret and Warriner, with a mean annual input of 0.74 km$^3$. The water in the rivers that feed into Lake Eyre is often lost to evaporation and infiltration into the flood-plains and dunes well before it reaches the lake (Knighton and Nanson 1994). For instance, the rivers to the northwest, including the Hay, Plenty and Finke, traverse and usually end in the Simpson Desert (Kotwicki 1986).

Early recordings of the fillings of Lake Eyre date from Edward John Eyre's expeditions in the region in 1840, but have been sporadic since then due to the limited population in the vicinity of the lake. The first verified extensive flooding of Lake Eyre was in 1949 (Bonython and Mason 1953; Peake-Jones 1955; Allan 1985). Unfortunately, there can be problems with the quality of the evidence of Lake Eyre filling events. Because of the shallowness of the lake, if it is only partially full (as in 2000) no water may be visible from the edge if the wind is blowing away from the shore. An increase in the number of aircraft in the region have increased the number of sightings of water in the lake in the second half of the twentieth century, although there are limitations with estimates of the extent and depth of the filling from the air, since extremely shallow water can give the appearance of a great deal of water (http://www.lakeeyreyc.com). Satellites have provided a basis for quantitative values of the extent of fillings used in this study, as described in detail in Sect. 4.

Evidence over the last 50 years indicates that portions of the Lake partially fill approximately every 3 years, that the lake floor is completely covered about every 8 years, with major floods about every 20 years. The largest known filling since European settlement occurred from 1974 to 1976, during the big wet of the 1973–1976 La Niña event (Dulhunty 1989). Contrary to popular perception, Lake Eyre is rarely completely dry. It also supports extensive wetlands and ecosystems (Roshier et al. 2001). The fact that Lake Eyre often has some water in it means that the current local climate is already somewhat tuned to the presence of water in Lake Eyre.

Lake Eyre and much of its catchment is located in the driest part of Australia (Fig. 3), with the average rainfall over the catchment of about 250 mm/year. Most of the catchment is also very hot during the day (Fig. 4). Gentilli (1971) describes the air mass over much of the catchment as being all of continental origin during both summer and winter. The northeast of the catchment receives the largest rainfall totals, mostly associated with the summer monsoon (Gentilli 1971; Roshier et al. 2001; Peel et al. 2007). The rest of the catchment is at the geographical

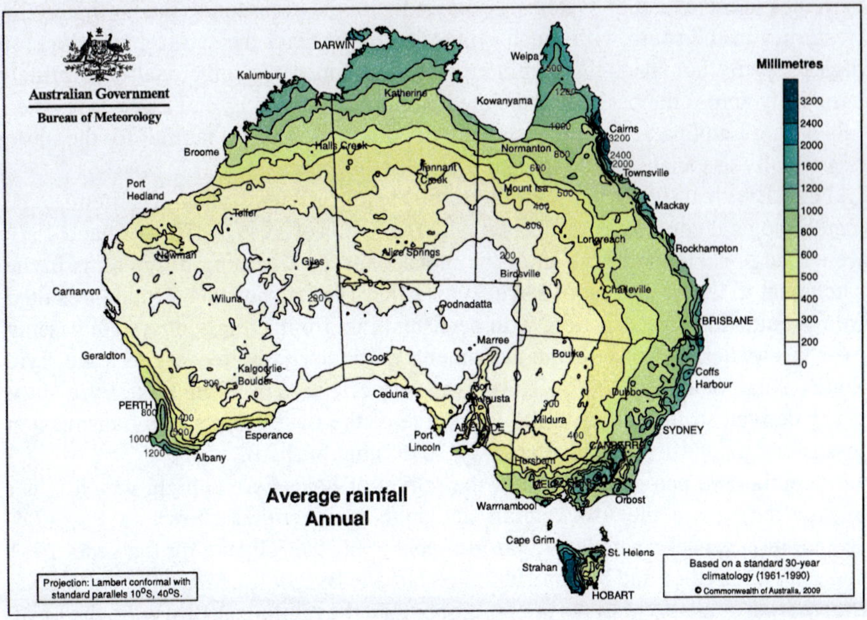


**Fig. 3** Australia's mean annual total rainfall. From Australian Bureau of Meteorology, http://www.bom.gov.au

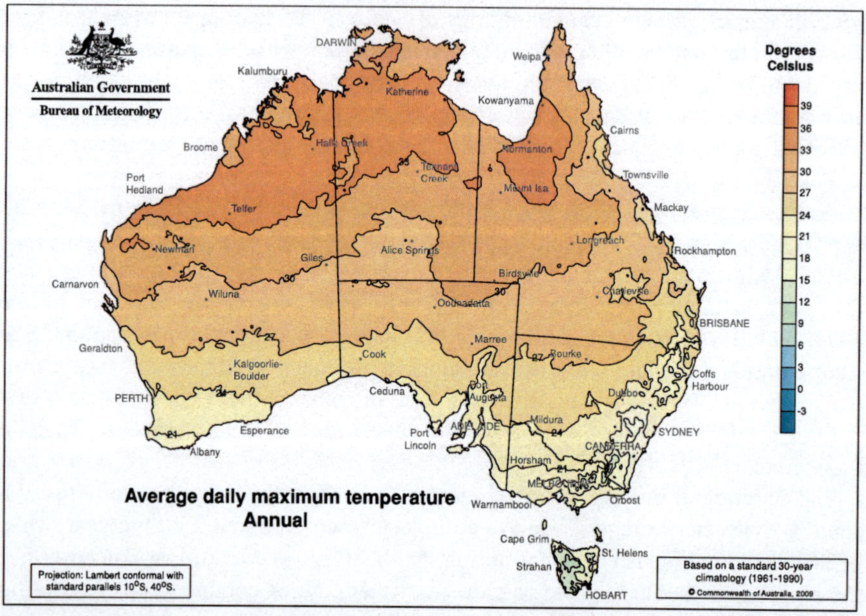


**Fig. 4** Australia's annual mean daily maximum temperature. From Australian Bureau of Meteorology, http://www.bom.gov.au

extremes of the extent of systems associated with the monsoon to the north and the westerlies to the south with much lying directly beneath the sub-tropical ridge of high pressure that spans the continent, and thus suppresses precipitation. Rainfall variability across the basin is associated with La Niña events, and Lake Eyre often fills during La Niña years (McMahon et al. 2008). As a result, rainfall for the basin is generally associated with sporadic large-scale events.

Lake Eyre can flood via rainfall from several different types of systems. The meteorological conditions associated with the floods of 1949, 1955, 1974 and 1976 were a large-scale modification of the monsoonal flow, causing heavy rains in the catchment to the north-east of Lake Eyre (Allan 1985). The 1984 flooding resulted from a quite different situation, with persistent rain from a transient system passing over the western portion of the catchment, particularly the feeders to Lake Eyre South (Allan et al. 1986). This storm was very effective in filling Lake Eyre since water derived from rainfall in the local area of the Lake suffers less transmission loss (Tetzlaff and Bye 1978; Backway 2009, unpublished).

Potential and pan evaporation in the region of Lake Eyre is high, which is not unexpected, given the low rainfall and high temperatures shown in Figs. 3,4. Evaporation was estimated at 2,000 mm year$^{-1}$ off the full lake for the years 1951 and 1975 (Bonython and Mason 1953; Tetzlaff and Bye 1978). Measurements of pan evaporation at nearby stations are far higher than the rainfall: at Birdsville the mean annual pan evaporation is 3,358 mm, whereas the rainfall is 169 mm. A reduction in the potential and pan evaporation has been noted when Lake Eyre is full.

When Lake Eyre was full in 1974 the pan evaporation at nearby stations reduced to 1,800 mm year$^{-1}$ (Tetzlaff and Bye 1978). With this potential for high levels of evaporation, to maintain a full lake would require an enormous increase in both the volume and consistency of the inflow. Kotwicki (1986) estimated that to off-set the evaporation alone, 18.9 km$^3$ year$^{-1}$ inflow would be required, about six times the current average annual inflow, and equivalent to the annual discharge of the Murray River (http://www.k26.com/eyre), or 3 days of Amazon discharge.

The information gained by examining the local climate and geography of the region to be inundated provides a good background for an heuristic estimate of what the likely outcomes of flooding might be. The fact that climate over Lake Eyre is dominated by the sub-tropical ridge suggests that, even if a filled lake provided more moisture to the atmosphere, there might be little mechanism for uplift to create rainfall.

## 3 Target Region—The Murray Darling Basin

The Murray Darling Basin (MDB—see Fig. 1) not only lies 'downstream' of Lake Eyre (its mid-point is about 1,200 km east of the lake), but is also Australia's chief agricultural region, often described as Australia's "food basket". Despite being only 14% of Australia's land area, it produces 80% of the countries citrus, grape and stone fruit, over 50% of its grain production, and 28% of its vegetables. The basin houses 32% of Australia's dairy cattle, 45% of its sheep herd and 62% of its pigs. The MDB also produces 90% of the countries cotton. In total, the relatively small MDB generates over 40% of Australia's gross value of agricultural production [http://www.mdba.gov.au/encyclopaedia/agriculture].

Agriculture in the MDB relies upon both irrigated and dry land farming, with the main crop and pasture growing season in the south being from April to October, while in northern areas of the basin a summer crop is possible. Much of the MDB, and in particular the wheat-sheep belt, receives far less than 600 mm/year, resulting in a basin-wide 1961–1990 average annual rainfall of only 489 mm. Large parts of these relatively low rainfall areas are also of moderate to high variability [http://www.bom.gov.au/jsp/ncc/climate_averages/rainfall-variability/index.jsp], making crop yields, overall, somewhat unreliable from one year to the next.

For every full year between 2001 and the time of writing (2009), below average rainfall fell in the MDB, with little to no relief from the particularly severe drought years of 2002 and 2006. Furthermore, since the mid-twentieth century, each drought has been warmer than the last, adding to the water stresses experienced during dry years (Nicholls 2004).

Climate change projections suggest winter and spring (i.e. southern cropping season) rainfall will be between 2.5 and 7.5% below the 1961–1990 average by 2030, while in the northern summer cropping regions, rainfall may increase by up to 2.5% (Bureau of Meteorology and CSIRO 2007). Eastern Australia is also projected to experience up to 20% more drought months per year by 2030 (Mpelasoka et al. 2007).

It is clear that rain which does fall upon the basin is of extremely high value both for Australian agriculture as well as for MDB farmers, including those farming in irrigated areas.

## 4 Analysis of Observed Data

The dry land farming in the MDB during the austral winter season (April–October) is the most sensitive activity in the MDB to drought. Hence it is this season that is of most interest if the aim is to reduce the severity of future droughts through such schemes as flooding Lake Eyre. The discussion of the climate and geography of the target region (MDB) and Lake Eyre and its catchment provide insight into whether filling Lake Eyre might have a positive outcome for winter agriculture across MDB.

For this study, rainfall averaged over the MDB was calculated from the Australian Water Availability Project (AWAP) dataset; the official Australian climate dataset of the Australian Bureau of Meteorology (http://www.bom.gov.au/jsp/awap/). AWAP grids are produced by interpolating station rainfall percentage of normal data onto a 5 km grid, then adding in a high resolution climatology. This effectively allows for topographical effects. Although this method includes a varying number of stations through time, the method of interpolation means that, even as stations move in and out of the record, the values at any one point should be a good estimate of the actual climate variability through time.

The example of Australia's Lake Eyre as a potential site for inundation provides the, perhaps unusual, possibility of examining periods when the Lake is flooded naturally as a proxy for artificially flooding the lake. As described in Sect. 2, Lake Eyre fills every 8 years or so. In this section we aim to explore the associations between the dry and wet periods of the lake bed and rainfall across the MDB.

The monthly volume data for Lake Eyre South and a number of basins across the south of Lake Eyre North (Madigan Gulf, Belt Bay and JackBoot Bay) were estimated from sporadic depth measurements compiled by Robert L Backway, Commodore of the Lake Eyre Yacht Club. These depths were estimated by careful analysis of Landsat satellite information (Images available from the Auslig website http://www.auslig.gov.au). Images were first "calibrated" using published depth diagrams checked by manual depth measurements. Experience has shown depths can be estimated within 100 mm in the navigation critical 1–3 m depth range. Distinct colouration provides accurate readings: when the Lake is either white or very light blue it means it is dry, and when it is very dark blue it has a depth of greater than 50 mm. The proximity of water to certain features in the Lake was used to guide the final depth result. There were still some concerns using this method, but allowances were made for these issues by careful consideration of the timing of inflow and weather events. For instance: recent local rain can make the whole Lake appear dark and give a misleading indication of depth. Measurements of lake depth were compiled from satellite images from September 1979. Landsat images were available from NASA since 1972, but due to technical problems and satellite failures they did not become reliable until 1979. The record was extended back to 1973 using data from the Australian Geographic Society booklet on the 1973–1978 flood (Bonython and Fraser 1989) and images from the book 'The spell of Lake Eyre' (Dulhunty 1975). Records of the source of the inflow to Lake Eyre suggest that local rains contribute to Lake Eyre depth more than may have been previously thought.

Depth measurements were multiplied by the area of each bay to create an approximate volume. Volumes were then summed across the bays. A monthly time-series was created by interpolating between the dates of each depth reading and averaging those when more than one reading was taken during a month (Fig. 5).

The work of Hope et al. (2004) demonstrated that Lake Eyre's extensive catchment has a high level of moisture recycling—i.e. when the surface is wet, subsequent rainfall falling over the catchment has been enhanced by that surface water. This recycling of moisture can cause rainfall anomalies to persist for some months (Simmonds and Hope 1997, 1998). However, Hope et al. (2004) found that the lake itself has little impact on the level of rainfall enhancement across the catchment through providing a surface moisture source. Hope et al. (2004) did not, however, assess the association with the MDB to Lake Eyre's east.

Relating the month to month variation in the volume of Lake Eyre with the MDB rainfall from 1974 to 2009 indicates a small, but positive and significant (not allowing for autocorrelation), relationship between the two events. Correlations between these variables were $r = 0.10$. Thus, if water in Lake Eyre were the only factor influencing MDB rainfall, it could be said that there was an association between these two entities, explaining approximately 10% of the variance. Lagging

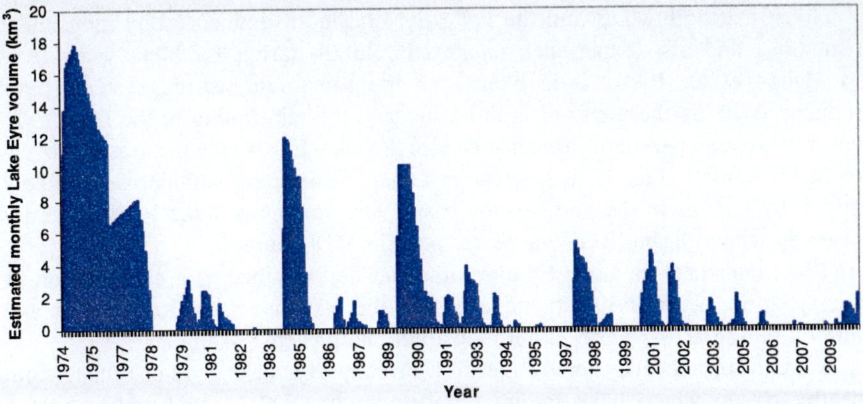


**Fig. 5** Estimated monthly volume of Lake Eyre South and the southern basins of Lake Eyre North ($km^3$)

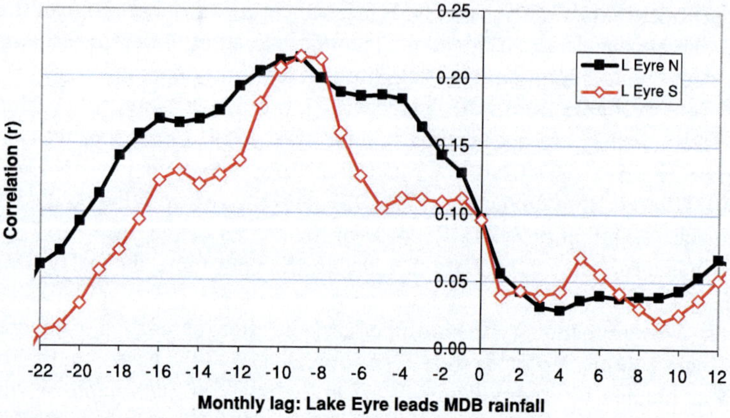


**Fig. 6** Lagged correlations between monthly MDB rainfall and Lake Eyre volume. Positive lags mean that Lake Eyre volumes lead MDB rainfall while negative mean that MDB rainfall lead Lake Eyre volumes

the correlations (i.e. looking at the delayed relationships between the water volume in the lake and rainfall over the MDB) revealed that, as the Lake Eyre fillings led the MDB rainfall, month by month, there were no further significant correlations (see the values for positive lags in Fig. 6). Thus, at first glance, there appears to only be a small, simultaneous association between water in Lake Eyre and rainfall across the MDB, and no link when the volume of Lake Eyre leads MDB rainfall.

When the situation was reversed, and MDB rainfall led the fillings of Lake Eyre, each correlation was significant (not accounting for autocorrelation) and positive, out to a lead of 16 months for Lake Eyre south and 19 months for Lake Eyre north, with a peak of $r = 0.21$ at 9 months lead (Fig. 6). This finding suggests some sort of physical relationship between rain falling in MDB, and subsequent fillings of Lake Eyre.

Given that both MDB and the Lake Eyre basin are influenced by monsoonal incursions and also experience enhanced rainfall during La Niña years (e.g. McMahon et al. 2008), it is likely that the same rain-bearing systems have influenced the northern end of both basins, and the rain falling in the Lake Eyre basin has, over a period of time, flowed into the lake. Examining the seasonal cycle of MDB rainfall (Fig. 7), it is clear that there is a strong summer peak, likely driven by regions in the north of the basin. The volume of Lake Eyre generally peaks in winter. There is also a winter peak in MDB rainfall.

The winter peak in Lake Eyre volume can be explained as a combination of winter storms from the south raining over the lake itself and local rivers and rainfall falling across north eastern Australia the preceding summer.

Rainfall from local systems either falls directly over Lake Eyre or runs off into nearby rivers, which have minimal transition time before contributing to the volume of the lake. In contrast, streamflow resulting from rain falling during the preceding summer wet season across the north of the basin has great transmission

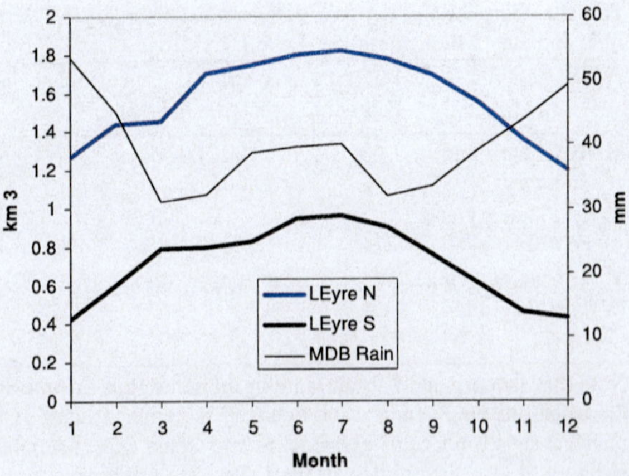


**Fig. 7** Mean monthly values of Lake Eyre volume in the lower regions of Lake Eyre North (*thick grey line*), Lake Eyre South (*thick black line*) and the mean monthly rainfall across the Murray Darling Basin (*thin black line* mm on right axis)

**Fig. 8** Simultaneous correlation between monthly Lake Eyre volume and gridpoint rainfall across Australia for the years 1974–2009. Correlations of magnitude greater than 0.10 are statistically significant

times, and thus contributes to Lake volumes many weeks to months after the rain fell (Knighton and Nanson 1994).

The maps of correlations between Lake Eyre filling and monthly rainfall across Australia (Fig. 8) reveal that, at zero lag, the greater, positive, correlations are with regions that are influenced by winter storms as evident in Köppen climate zones (Peel et al. 2007). It is particularly this correlation that would be of most interest to the proponents of a scheme to permanently flood Lake Eyre in order to enhance rainfall across the MDB.

**Table 1** Correlations between January and February MDB rainfall or Lake Eyre fillings and June and July MDB rainfall or Lake Eyre fillings.

| | MDB January and February | L. Eyre January and February | MDB June and July | L. Eyre June and July |
|---|---|---|---|---|
| MDB January and February | 1.00 | *0.56* | −0.07 | 0.51 |
| L. Eyre January and February | | 1.00 | −0.10 | 0.77 |
| MDB June and July | | | 1.00 | *−0.14* |
| L. Eyre June and July | | | | 1.00 |

Note that January and February are in the same year as June and July, thus correlations between these seasons have January and February (summer) leading June and July (winter), which makes physical sense under this example, as one would expect summer rainfall across the MDB (which is likely to also have fallen in the Lake Eyre catchment) to contribute to resultant winter-time Lake levels. Highest correlation is between Lake Eyre volume in summer and the following winter ($r = 0.77$), suggesting a high degree of autocorrelation within the time-series
Correlations within the same season are highlighted in italics

Dividing the monthly time-series into series for summer (January and February) and winter (June and July) allows focus to be placed on the winter correlations (see Table 1). Unexpectedly, Lake Eyre volumes and MDB rainfall in June and July are actually negatively correlated ($r = -0.14$). A far larger, positive, correlation is found in summer ($r = 0.56$). This summer relationship may be due to both Lake Eyre volumes and MDB rainfall responding to La Niña (Puckridge et al. 2000; Power et al. 2006; McMahon et al. 2008).

Seasonally, an El Niño or La Niña typically develops through the austral autumn to become a clear signal, typically by the early winter. Once established, these events clearly influence northern and eastern Australian rainfall for the winter and spring. Into summer, the relationships tend to break down somewhat, although a La Niña signal may be apparent through into the new year (Puckridge et al. 2000; Power et al. 2006).

Lagging the relationships, so that summer (January and February) leads winter, it can be seen that there is also minimal association between Lake Eyre volumes in summer and MDB rainfall the following winter ($r = -0.10$).

The seasonal correlations also reveal that there appears to be a high degree of autocorrelation and persistence in the Lake Eyre volumes ($r = 0.77$ between summer Lake Eyre volumes and volumes the following winter). There is no such relationship for MDB rainfall.

It must be noted though, that the volumes of Lake Eyre are not a smoothly varying measure, and closer inspection on a filling-by-filling basis might be warranted before a full understanding of these correlations could be gleaned.

This simple correlation analysis reveals that there is little association between Lake Eyre being full in either summer or winter and winter (June and July) rainfall in MDB. These results do not suggest a strong association between Lake Eyre fillings and MDB winter rainfall in the observed record, which significantly weakens the likelihood that a permanent filling of the Lake would have any significant impact upon MDB rainfall.

## 5 Modelling Experiments

Once a clear understanding of the drivers of climate variability and rainfall type in both the flooded region and the 'target' region have been identified, a suite of experiments can be performed using an appropriate climate model. There are a range of climate models to use, and the nature of the problem, in particular, the scale of the regions considered, will drive the decision on which is appropriate. Climate model simulations are almost always constrained by the available computing power and availability of super-computing time, thus, in general, the smaller the region considered, the greater the detail that can be simulated.

In some instances a very fine-scale, mesoscale model is appropriate if the region is small and the expected response is a modification of the convection (e.g. Lohar and Pal 1995; Segal et al. 2001). These types of models can provide a great deal of information about the local responses (e.g. Yan and Anthes 1988; Physick and Tapper 1990). Given the large-scale of the regions under consideration in the Australian example used in this chapter, a larger-scale model is appropriate. A global climate model could be used, which does not focus specifically on the region, but can capture the physics of atmospheric (and sometimes oceanic) processes around the globe. Hope et al. (2004) used such a model (MUGCM, Noone and Simmonds 2002) at two different resolutions (approximately 300 by 200 km or 500 by 300 km at Australian latitudes). Another approach is to focus the model on the local region with a regional model. A stretched grid regional model was used by Hope et al. (2004) (C-CAM, McGregor and Dix 2001) resulting in a fine mesh over Australia (65 km spacing) and very coarse resolution on the other side of the globe (800 km spacing) (see Fig. 9). Another type of regional model simulates the climate over the region of interest and receives global information at the edges of the region from either data (reanalyses) or a global climate model. Very fine-scale areas can be embedded within fine-scale regions, as is the case for a recently proposed study of the climate response to a flooded Lake Eyre where a grid with a mesh of 10 km spacing will be placed over south-east Australia, and a mesh of 40 km spacing over the whole continent (see Fig. 10 D. Abbs, CSIRO, pers. comm.).

Experimental design is also important, as the time length of simulations can also be limited due to computing constraints. The experiments in Hope et al. (2004) included an artificially permanently flooded lake in the location of Lake Eyre, and each simulation was forced by ocean temperatures for the years 1979–1989. These results were compared with simulations where Lake Eyre was left dry. This period includes many dry years, so it is an appropriate period to choose to simulate as the factors that drive both rainfall in the MDB and the filling of Lake Eyre would not be present, thus any signal of enhanced rainfall associated with the artificially full lake might be clearer in the otherwise dry conditions. The new experiments that are

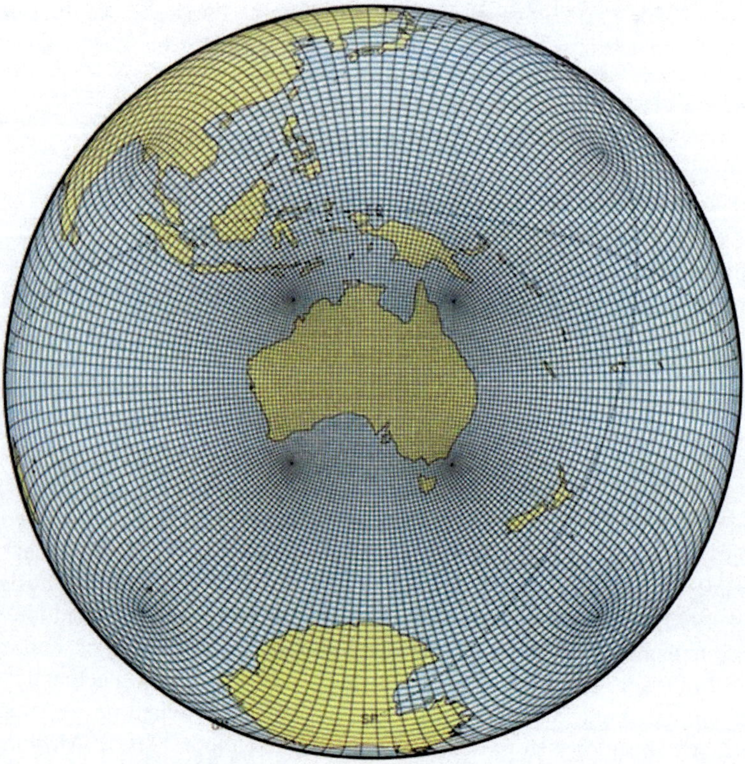

**Fig. 9** The grid used by the CCAM model. The grid is fine over Australia, extending to broad spacing on the other side of the globe (IOCI, Indian Ocean Climate Initiative Stage 2: Unabridged Reports of Phase 1 Theme 2 Activity, July 2003–December 2004, available from http://www.ioci.org.au)

currently underway are far more computationally expensive, and thus only a limited number of months can be simulated. The agriculturally important winter months for 5 years when Lake Eyre was particularly dry were chosen: 1982, 1995, 1999, 2006, and 2008. Simulations where a permanently inundated Lake Eyre was imposed and simulations with observed conditions will then be compared to assess what impact the flooding has on the wider climate.

The physics and parameterisations within the model mean that defining the flooded region requires some knowledge of the way the surface is modelled. Hope et al. (2004) discovered that the roughness length of the imposed water surface has a strong impact on the rate of evaporation. When the surface roughness length was left as that over land (e.g., 29.63 cm in the MUGCM, Simmonds 1985) but the surface was saturated, the resultant evaporation was 2.6 times greater than if the standard roughness length over water had been used. These results suggest that the evaporation from an irrigated region with crops (which will increase the roughness length) will be greater than that from an unvegetated water expanse.

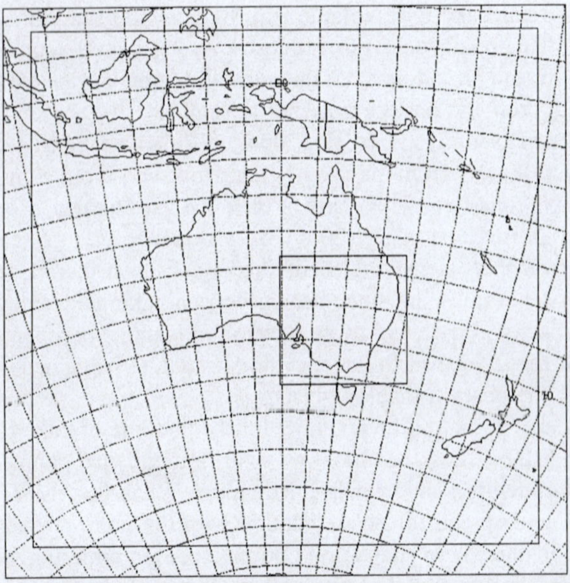

**Fig. 10** The embedded region over south-east Australia over which a very fine-scale (10 km) regional model (RAMS) will simulate the climate response to filling Lake Eyre. The grid over all of Australia is the region where a climate model at 40 km resolution will by used. The lines of latitude and longitude are placed at the spacing of the global data used to force the regional model (approximately 250 km resolution) (*Source:* D. Abbs, CSIRO, pers. comm.)

The modelling experiments completed by Hope et al. (2004) all suggest that there was no consistent rainfall response in the MDB to a water expanse imposed in the Lake Eyre region, even when that region was reasonably large. However, there was a local response—in particular a moistening of the air and a cooling and drying. These results align with the findings of de Ridder and Gallee (1998).

## 6 Discussion and Conclusions

Two methods of determining the potential of enhancing rainfall in the MDB were detailed and briefly explored. One described the underlying rainfall variability and drivers in both the region of Lake Eyre and in the target region. The other aimed to decide the appropriate modelling tool to use and devise appropriate experiments that would minimise the computing resource requirements while still providing a robust answer to the question.

Both the Lake Eyre basin and the MDB have highly variable climate, dominated by the ENSO cycle (e.g. Puckridge et al. 2000; Power et al. 2006). In the north of the basins, both are influenced by summer rainfall, while in the south, winter storms dominate. Lake Eyre often has water in at least part of its area, and is sometimes dry. Extensive wet conditions provide a proxy for artificial fillings. The correlation between Lake Eyre volume and MDB winter rainfall is weak—in fact it is negative. This result does not suggest a strong causal relationship between filling Lake Eyre and providing a reliable rainfall regime for the MDB, particularly during its main crop and pasture growing season.

Modelling results of Hope et al. (2004) revealed no consistent, widespread increase in rainfall away from the lake region. However, it did simulate a localised cooling, which is consistent with many of the cited international modelling studies. The models used in Hope et al. (2004) were reasonably large-scale, which was appropriate given the extensive size of the regions considered, though, a more focussed, finer scale study has since been proposed, taking advantage of increased computing capacity. However, even now, intelligent experimental design will allow the most robust answer to our specific question to be gained—building upon the findings of the observational analysis.

All these results point to little potential for the flooding of Lake Eyre to enhance rainfall in the MDB. These results align with the guidance gained from international studies of regions of extensive irrigation. What may be a more sensible approach than widespread inland inundation with the hope of enhancing 'downwind' rainfall, is one where adaptation options, focused on living with natural and imposed climate, are explored in the regions that currently, or are likely in the future to, suffer drought, high rainfall variability and growing demand for water (Prabhakar and Shaw 2008).

**Acknowledgments** The authors would like to thank Julie Arblaster and Debbie Abbs for their advice on experimental design.

## References

Allan RJ (1985) The Australasian summer monsoon, teleconnections, and flooding in the Lake Eyre basin. S Aust Geogr Pap 2:47

Allan RJ, Bye JAT, Hutton P (1986) The 1984 filling of Lake Eyre South. Trans R Soc S Aust 110:81–87

Barnston AG, Schickedanz PT (1984) The effect of irrigation on warm season precipitation in the southern Great Plains. J Clim Appl Meteorol 23:865–888

Ben-Gai T, Bitan A, Manes A, Alpert P (1994) Long-term changes in annual rainfall patterns in Southern Israel. Theor Appl Climatol 49:59–67

Bonython CW, Fraser AS (1989) The great filling of Lake Eyre in 1974. Royal Geographical Society of Australasia (South Australia Branch) Inc. pp 119

Bonython CW, Mason B (1953) The filling and drying of Lake Eyre. Geogr J 119:321–330

Bureau of Meteorology and CSIRO (2007) Climate change in Australia. Technical report. Commonwealth Scientific and Industrial Research Organisation, Canberra, pp 148

Bye JAT, Dillon PJ, Vandenberg CJ, Will GD (1978) Bathymetry of Lake Eyre. Trans R Soc S Aust 102:85–89

Committee on the Status of Future Directions in US Weather Modification Research, Operations, Board on Atmospheric Sciences, Climate Division on Earth, Life Studies National Research Council of the National Academies (2003) Critical issues in weather modification research. The National Academies Press, Washington, DC, p 123

Ćurić M, Janc D, Vučković V (2007) Cloud seeding impact on precipitation as revealed by cloud-resolving mesoscale model. Meteorol Atmos Phys 97:179–193
de Ridder K, Gallee H (1998) Land surface-induced regional climate change in Southern Israel. J Appl Meteorol 37:1470–1485
Dulhunty R (1975) The spell of Lake Eyre. Lowden Publishing
Dulhunty JA (1989) Levels, salt crusts and lake bed instability. In: Bonython CW, Fraser AS (eds) The great filling of Lake Eyre in 1974. Royal Geographical Society of Australasia, (South Australia Branch) Inc, pp 56–59
Gentilli J (1971) Climates of Australia and New Zealand: World Survey of Climatology, vol 13, pp 405
Hope PK, Nicholls N, McGregor JL (2004) Rainfall response to permanent inland water in Australia. Aust Meteorol Mag 53:251–262
Knighton AD, Nanson GC (1994) Flow transmission along an arid zone anastomosing river, Cooper Creek, Australia. Hydrol Process 8:137–154
Kotwicki V (1986) Floods of Lake Eyre. Government Printer, South Australia
Kuleshov Y, Biadeglgne B, Chambers L, Cook G, Coughlan M, de Hoedt G, Fawcett R, Hollis L, Hope P, Jakob D, Seed A (2008) Western Australia cloud seeding study report: scientific reviews and analysis. Bureau of Meteorology, Australia, p 127
Lohar D, Pal B (1995) The effect of irrigation on premonsoon season precipitation over South West Bengal, India. J Clim 8:2567–2570
McGregor JL, Dix MR (2001) The CSIRO conformal-cubic atmospheric GCM. In: Hodnett PF (ed) IUTAM symposium on advances in mathematical modelling of atmosphere and ocean dynamics. Kluwer, Dordrecht, pp 197–202
McMahon TA, Murphy RE, Little P, Costelloe JF, Peel MC, Chiew FHS, Hayes S, Nathan R, Kandel DD (2005) Hydrology of Lake Eyre Basin. SKM. http://www.lebmf.gov.au, pp 131
McMahon TA, Murphy RE, Peel MC, Costelloe JF, Chiew FHS (2008) Understanding the surface hydrology of the Lake Eyre Basin: Part 1—Rainfall. J Arid Environ 72:1853–1868
Moore N, Rojstaczer S (2001) Irrigation-induced rainfall and the Great Plains. J Appl Meteorol 40:1297–1309
Mpelasoka F, Hennessy K, Jones R, Bates B (2007) Comparison of suitable drought indices for climate change impacts assessment over Australia towards resource management. Int J Climatol 27:1673–1690
Nicholls N (2004) The changing nature of Australian droughts. Clim Change 63:323–336
Noone D, Simmonds I (2002) Associations between $\delta^{18}$O of water and climate parameters in a simulation of atmospheric circulation for 1979–1995. J Clim 15:3150–3169
Osborne M, Dunn C (2004) Talking water: An Australian guidebook for the 21st Century. Farmhand Foundation. http://www.farmhand.org.au, pp 113–115
Peake-Jones K (1955) Lake Eyre, South Australia—the great flooding of 1949–50. Royal Geographical Society of Australasia, (South Australia Branch) Inc
Peel MC, Finlayson BL, McMahon TA (2007) Updated world map of the Köppen–Geiger climate classification. Hydrol Earth Syst Sci 11:1633–1644
Physick WL, Tapper NJ (1990) A numerical study of circulations induced by a dry salt lake. Mon Weather Rev 118:1029–1042
Power S, Haylock M, Colman R, Wang X (2006) The predictability of interdecadal changes in ENSO activity and ENSO teleconnections. J Clim 19:4755–4771
Prabhakar SVRK, Shaw R (2008) Climate change adaptation implications for drought risk mitigation: a perspective for India. Clim Change 88:113–130
Puckridge JT, Walker KF, Costelloe JF (2000) Hydrological persistence and the ecology of Dryland Rivers. Regul Rivers Res Manage 16:385–402
Roshier DA, Whetton PH, Allan RJ, Robertson AI (2001) Distribution and persistence of temporary wetland habitats in arid Australia in relation to climate. Aust Ecol 26:371–384
Rotstayn LD, Cai W, Dix MR, Farquhar GD, Feng Y, Ginoux P, Herzog M, Ito A, Penner JE, Roderick ML, Wang M (2007) Have Australian rainfall and cloudiness increased due to the remote effects of Asian anthropogenic aerosols? J Geophys Res D Atmos 112

Schickedanz PT (1976) The effect of irrigation on precipitation in the Great Plains. Illinois State Water Survey, Urbana
Segal M, Pan Z, Turner RW, Takle ES (2001) On the potential impact of irrigated areas in North America on summer rainfall caused by large-scale systems. J Appl Meteorol 37:325–331
Simmonds I (1985) Analysis of the 'spinup' of a general circulation model. J Geophys Res 90:5637–5660
Simmonds I, Hope P (1997) Persistence characteristics of Australian rainfall anomalies. Int J Climatol 17:597–613
Simmonds I, Hope P (1998) Seasonal and regional responses to changes in Australian soil moisture conditions. Int J Climatol 18:1105–1139
Simmonds I, Bi DH, Hope P (1999) Atmospheric water vapor flux and its association with rainfall over China in summer. J Clim 12:1353–1367
Singh N, Sontakke NA (2002) On climatic fluctuations and environmental changes of the Indo-Gangetic plains, India. Clim Change 52:287–313
Steinberger EH, Gazit-Yaari N (1996) Recent changes in the spatial distribution of annual precipitation in Israel. J Clim 9:3328–3336
Stockholm Environment Institute (1997) Comprehensive assessment of the freshwater resources of the world. UN/UNDP/UNEP/FAO/UNESCO/WMO/WB/ WHO/UNIDO
Tetzlaff G, Bye JAT (1978) Water balance of Lake Eyre for the flooded period January 1974–June 1976. Trans R Soc Aust 102:91–96
Towner ET (1955) Lake Eyre and its tributaries. Qld Geogr J
Wardle R, Smith I (2004) Modeled response of the Australian monsoon to changes in land surface temperatures. Geophys Res Lett 31:16205
Warren HN (1945) Bradfield scheme for 'watering the inland'. Meteorological aspects: a) Possibilities of climatic amelioration; and b) Rainfall characteristics of river basins proposed to be harnessed. Commonwealth Meteorological Bureau
Yan H, Anthes RA (1988) The effect of variations in surface moisture on mesoscale circulations. Mon Weather Rev 116:192–208

# Macro-Engineering Lake Eyre with Imported Seawater

**Viorel Badescu, Richard B. Cathcart, Marius Paulescu, Paul Gravila and Alexander A. Bolonkin**

## 1 Introduction

By conduction and thermal radiation (~11.7%) and rising air (~7%), approximately 18.7% of the Earth's energy budget is used for direct heating of the air. Whatever the causes—perhaps, in part, an expansion of the width of the Earth's Southern Hemisphere atmospheric Hadley cell (Johanson and Fu 2009; Liu et al. 2007a, b)—a widening of the Tropical Zone air circulation and a southward shift of the Southern Hemisphere's tropospheric jet stream and Australia's associated subtropical arid lands appears to be taking place (Fu et al. 2006; Evans et al. 2009). It is worth noting, too, that since the "Geocentric Datum of Australia 1994" became Australia's newest country-wide geographical coordinate system, it has been discovered and documented subsequently that Australia, an isolated landmass, is moving to the Northeast at ~3 cm/year. The geographical shift of the continent's supporting tectonic plate bodes a future change in Australia's vital hydrologic cycle. Australia is the world's driest permanently inhabited continent and it has the most variable precipitation, with periods of widespread drought (Sohn 2007).

Subject to a testing climate dominated by the subtropical high-pressure belt which encloses high-pressure air systems that move from the West Coast to the East Coast above the ocean-isolated terrestrial ecosystem, Australia is a vast

V. Badescu (✉)
Polytechnic University of Bucharest, Bucharest, Romania
e-mail: badescu@theta.termo.pub.ro

R. B. Cathcart
Geographos, Burbank, CA, USA

M. Paulescu and P. Gravila
West University of Timisoara, Timisoara, Romania

A. A. Bolonkin
C&R, New York, USA

V. Badescu and R. B. Cathcart (eds.), *Macro-engineering Seawater in Unique Environments*, Environmental Science and Engineering,
DOI: 10.1007/978-3-642-14779-1_25, 

dry-land region; ~75% of the continent—approximately ~3,030,000 km$^2$—is classed as either arid or semi-arid. Naturally, the boundaries of these continental climate regions are neither geographically static nor obviously abrupt as physically represented by the continental landscapes beneath the superincumbent air. During the coolest period of the Southern Hemisphere year (Winter: May to October), the high-pressure systems pass slowly over central Australia's land in a belt of air extending from 29° to 32°S latitude and often remain stationary for several consecutive days, causing orographic precipitation on the mountains near the East Coast. Studies have shown that precipitation for the nation varies from Summer-dominant rainfall in the north to Winter-dominant rainfall in the south, making the resulting available flowing freshwater resource exceedingly problematic (Fleming 1995). During the warmest period of the year (Summer: November to April), the subtropical high-pressure systems belt normally shifts offshore, flowing over the Southern Ocean—at 37° to 38°S latitude. Pigram (2006) offers a reliable comprehensive weather and climate assessment.

Currently, the ephemeral Lake Eyre is Australia's largest saline playa and it is also that island-continent's lowest land elevation. The Lake Eyre Basin is a vast landscape, endorheic, a topographically depressed region, spanning the Tropic and the Temperate Zones, with extremely variable drainage system types. Its arid topographical relief is extremely low (about 8,500 km$^2$ of the Basin is below present-day global sea level); almost throughout the entire Basin and the rivers are characterized by ephemeral, anastomozing river channels and large-scale floodplains (Webb et al. 2009). The inland Lake Eyre Basin receives high-intensity monsoonal rains from northern Australia that, rather suddenly by flash flood overland flows, can connect the rare waterholes and lakes remnant from previous pulsative inundations caused by similar intense precipitation. These torrential rainfalls are variable (in magnitude and duration) and, thus, the waterhole and lake connections formed are not persistent long-term (Box et al. 2008). The annual rainfall on Lake Eyre amounts to ~125 mm with an average annual pan-evaporation of ~3,800 mm (Prata 1990); northeast of Lake Eyre the rainfall averages ~500 mm and pan-evaporation is ~2,400 mm. The Lake Eyre Basin, ~1,140,000 km$^2$, or ~15% of Australian land territory, exhibits ephemeral stream channels, aeolian dunes, gibber plains and bare bedrock. Early-twenty-first Century human residents of the Lake Eyre Basin number fewer than 60,000 persons. Lake Eyre is comprised of two depositional basins, the North Basin, which is the deepest, has an estimated plane surface area of ~8,400 km$^2$ and the South Basin is ~1,200 km$^2$ (Fig. 1).

The brackish water level in Lake Eyre is unregulated and, owing to rare and variable strength rainstorms, the depressed land region's water level can rise rapidly in a matter of months with flows of 5,000 to 10,000 m$^3$/s; during 1974 water flowed from the −15 m-deep North Basin into the South Basin but during 1984 the South Basin overflowed into the North Basin. Present-day Lake Eyre (28.3°S Lat. by 137.2°E Long.) is situated in the "hot dry summer, cold winter" Australian Climatic Zone—meaning the average January maximum air temperature is >30 °C.

The three o'clock afternoon water vapor pressure at Lake Eyre is <2.1 kPa and the average July mean air temperature exceeds 14 °C. Kingsford (2006) offered a

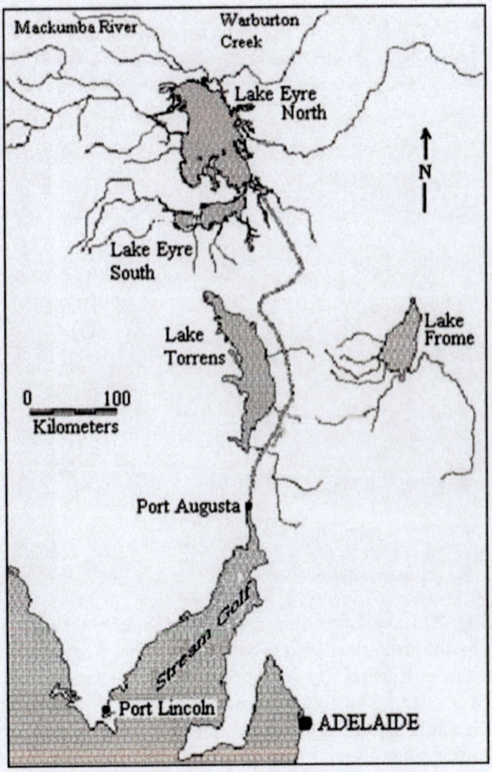


**Fig. 1** Lake Eyre and Port Augusta to Lake Eyre Seawater Pipeline Corridor. A slightly different course might prove to be more efficacious if only the South Basin is to be filled at the Lake Eyre terminal south of the proposed Goyder Channel Tensioned Textile Dam. [Map courtesy of Fereidoun Ghassemi and Ian White (Ghassemi and White 2006, Fig. 7.5 at p. 146)]

comprehensive overview of watered arid lands worldwide, including the Lake Eyre Basin. Figure 2 shows the evolution of Lake Eyre during most of the year 2009.

In this chapter-length planning document, we have undertaken merely a preliminary assessment of a Macro-engineering project—i.e., seawater flooding of Lake Eyre—that could instigate a remarkable increase in South Australia's land area that can be profitably grazed by currently domesticated ranch animals. Bringing seawater to Lake Eyre has been discussed publicly since at least 1883–1902; the idea was again broached publicly by a civil engineer in Hungary, Joseph Holibal, in 1933. But here, for the first time ever, we offer useful technical and economical macro-engineering details. The macro-project will utilize ~0.11% of Australia's territory that is below present-day global sea level, most of which is entirely unsettled, with small annual GDP economic impact, even though it is also often described as picturesque (Gibbs 2006).

Anthropogenic flooding of Lake Eyre with a large volume of seawater could increase to ~50% (from its current 45%) the volume of Earth's inland waters that are saline (Hahn 2006). This study does not give final macro-project solutions; there are additional macro-problems that must be investigated before any political and financial decision-making can be undertaken to bring about a successful completed macro-project. Some of these macro-problems are enumerated in Sect. 5.

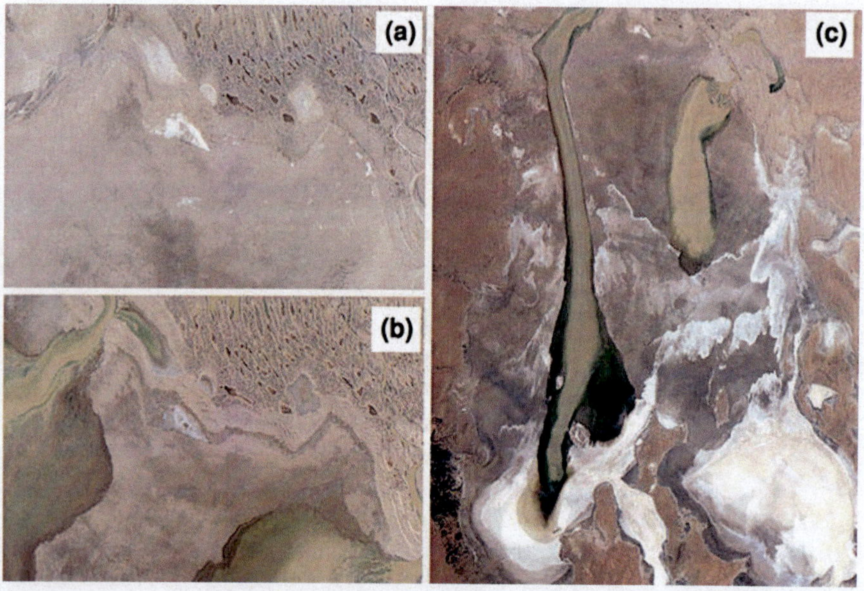


**Fig. 2** Lake Eyre during 2009; **a** 18 February 2009. The image provides a view of normal conditions in the desert. The landscape has been carved by the occasional flow of water across the surface. Even in dry conditions, the river channel and alluvial fan are imprinted on the land; **b** 9 May 2009. The image provides a view of water pouring into the lake through one of many channels that drain the desert during the rainy season. The muddy water disperses into the lake in a triangular alluvial fan**; c** 10 June 2009. Lake Eyre at its fullest in 2009. Shallow and episodic water covers most of the impermanent lake's bed. The lake's southwestern lobe is darkest in color. Source http://earthobservatory.nasa.gov/IOTD/view.php?id=38994&src=iotdrss; Author: Jesse Allen; Images created using Landsat data provided by the United States Geological Survey

## 2 Previous Lake Eyre-Filling Macro-Projects

Twentieth century macro-engineers have put forward many schemes for augmenting desert rainfall by flooding Africa's desert topographical depressions (Schwarz 1923; Letolle and Bendjoudi 1997). In twenty-first century Australia, Macro-engineering proposals for the flooding of Lake Eyre with salt water from the Southern Ocean (possibly via Spencer Gulf's Port Augusta) continue to be a staple of conversation in ordinary and national political circles (Boia 2005).

Filled to present-day ocean level, the Eyre Seawater Reservoir (ESR)—North and South basins combined—would hold >200 km$^3$ of imported seawater "By contrast, the deepest historical filling held 30 km$^3$" (DeVogel et al. 2004). Ghassemi and White (2006) offer the best historical account of some major macro-engineering schemes for the creation of an artificial inland sea in arid central Australia. Excavation of a canal from the general vicinity of Port Augusta (32°S Lat. by 137°E Long.) at the head of Spencer Gulf to Lake Eyre, a distance of ~320 km with a maximum above global sea-level intervening ridge of land elevation of <60 m will exceed 2009 USD 50 billions. Port Augusta is a seaport which faces Spencer Gulf and has daily tides that are relatively moderate in size, with amplitudes of the four major tidal components ($M_2$, $S_2$, $F_1$ and $O_1$) of 0.29 to 0.74 m. It is currently the proposed site of the largest solar-powered desalination plant in the Southern Hemisphere. The ecological effects—especially brine discharge—of the potential desalination plant on the upper Spencer Gulf's seawater is being examined (Dupavillon and Gillanders 2009). By 2013, South Australia may have eight completed major urban desalination plants, with two plants each producing more than 100 ML/day at Whyalla (33°02′S Lat. by 137°34E Long.) and Adelaide (34°55′S Lat. by 138°36′E Long.) (Hoang et al. 2009).

## 3 Interlinking the ESR Macro-Project Characteristics

Long-term flooding of Lake Eyre may have important consequences for the sparse nearby human population. In terms of aims, this macro-project proposes the ESR as a center of a region of biosaline agriculture. It is this subject which will be treated in more detail here. Other possible benefits are briefly enumerated in Sect. 5.

In terms of tools, the macro-project is based on three interlinked macro-engineering ideas. First, seawater is transported by flexible pipes rather than by a single excavated canal. Second, use of PV cells is envisaged to ensure the electricity necessary to pump the seawater inland. Third, artificial covering of the created permanent Lake Eyre will greatly diminish evaporation. All of these tools were previously proposed and studied by various groups of researchers. Some details and critical particular notes follow on this "tame problem" that can be handled through normal technologies (Lach et al. 2005).

### *3.1 Seawater-Transporting Pipeline*

There is an Australian national precedent macro-engineering plan for a seawater pipeline (Allan et al. 2001). These authors mention that the Water Corporation of Western Australia proposed a 348 km-long Esperance to Kalgoorlie seawater pipeline as a major water supply option for Western Australia. More than a century ago, wood-fired stills were employed by industrial firms to desalt water in that region's famous goldfields.

Our proposal is to use low-pressure and cheap pipes—we characterize them as "hoses"—by having multiple seawater pumping solar power stations along the ~320 km-long route. If a conveying pipe should ever happen to rupture, whether by accident or by human political protest or outright daffy terrorism, the piped seawater delivery disruption and the landscape damage should be easily repaired, probably short-term, and without creating any major or truly significant long-lasting post-break environmental macro-problems. We suggest the pipe walls could be composed mainly of inexpensive waterproof tensioned textile material. Any tensioned textile pipeline will, in fact, have many of the characteristics of a very large diameter garden hose or standard-size Australian city fire-fighting hose. Therefore, it is possible to imagine multiple tubes instead of merely one, two or more "pipelines" transporting seawater to Lake Eyre in a special zoned land-use exclusion corridor, the ESR Pipeline Corridor. Such hoses would be aggregated, but not bundled. A wide-diameter hydraulic tube is more expensive but such a pipe, hose or tube has the advantage of decreasing greatly the pressure loss and increases significantly the efficiency and decreases the wall interior friction loss as well as the needed pump power. Tubes near the fixed pumping stations have high pressure (up 30 atmospheres). They must be made from something even stronger—perhaps from some composite fiber material. Such composed material has higher maximum stress (up to 600 kg/mm$^2$) and low specific-density (1,800 kg/m$^3$). Such materials may be, or become, cheaper quite soon. Below, is the computation equation for tube-wall thickness:

$$\delta = \frac{pD}{2\sigma} \tag{1}$$

where $\delta$ is tube-wall thickness; $p$ is seawater pressure, $D$ is tube's diameter and $\sigma$ is safety tensile stress. The computation's result is presented in Fig. 3.

The ESR Pipeline Corridor would pass east of Lake Torrens (31°02′S Lat. by 137°51′E Long.) (see Fig. 1), just skirting an episodic ~6,000 km$^2$ saline playa lake first filled in recorded history, entirely by natural rainfall, during March 1989.

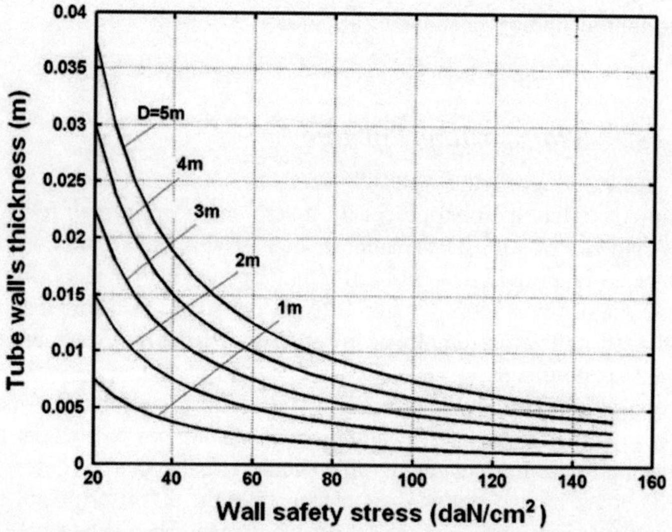


**Fig. 3** Tube-wall thickness via wall safety tensile stress for various tube diameters and a water pressure of 30 atmospheres

Subsequently, it dried up during 1990. It was "discovered" in 1839, amidst a then occurring severe meteorological drought, by Edward John Eyre (1815–1901).

The tension textile tube must be supported along its entire length by a special chute, a cradle-like trough or discrete air-inflated pillows. The overall seismicity of the Earth-crust foundation of South Australia is low, with most low Richter Scale-intensity earthquakes concentrated near the Flinders Ranges and on the salt-lake studded Eyre Peninsula (Timms 2009); we assume that no great threat is posed to the ESR by temblors and that structural designers will optimize the ESR's seismic resilience (for a given construction cost) because all Macro-engineering projects must have careful, proactive macro-project failure analyses.

## 3.2 Photovoltaic Cells

Kurokawa (2006) touts the benefits of carpeting desert regions with photovoltaic (PV) cells. Essentially, his macro-engineering proposals differ little from those made by TREC Australia (http://www.trec.net.au). Both TREC Australia and Kurokawa recognize that Australia's deserts are perfect places for such vast and fixed power-generating solar installations, just like the proposed Northern Hemisphere Sahara (North Africa) and Ordos City (Inner Mongolia) solar-power installations (AP 2009). However, Lake Eyre's Basin is vegetated, making it unsuitable, or at least less suitable, for fulfillment of Kurokawa's plans or TREC Australia's plans. Also, while progress is reportedly being achieved in the industrial development of solar PV cells, vast permanent solar energy collecting surfaces inevitably will result in some heating with atmospheric effects, require considerable maintenance effort to maintain in good working condition, coat large swathes of sometimes picturesque desert landscape, affect natural vegetation and wild animals living beneath, photovoltaic cell manufacture frequently results in toxic chemical releases to Earth's air and standing waters, needs electricity storage facilities and entails costly human labor for construction and maintenance in non-urban remote locales Here we shall use PV cells, factory installed and placed only on the uppermost surface of the seawater-conveying tensioned textile tubes. We admit that the seawater-conveying pipeline, whilst shading the ground beneath it, will have the slight net effect of directly heating the air immediately above it. However, we do not expect to affect local weather trends in significant way, and certainly not as effectively as 26,000 ha of ground-cooling glassy reflective greenhouses measurably have in southeastern Spain (Campra et al. 2008).

## 3.3 Lake Eyre's Future Man-made Lid

Alvia Gaskill and Charles E. Reese, working at Environmental Reference Materials, Inc., during 2003, proposed "Global Warming Mitigation by

Reduction of Outgoing Longwave Radiation Through Large-Scale Surface Albedo Enhancement of Deserts Using White Plastic Polyethylene Film—the Global Albedo Enhancement Project (GAEP)". They desired to cover allegedly useless desert with a white-colored, sunlight-reflective polyethylene–aluminum sheet intended by them to increase the mean albedo from 0.36 to 0.8, thus causing a $-2.75\ W/m^2$ reduction in radiative forcing. Instead, we wish to lid a large, saline water body, in order to decrease evaporation over the region east and northeast of the Eyre Seawater Reservoir by cooling the boundary layer. It is noteworthy that saline lakes have a tendency to emit carbon dioxide gas to the Earth–atmosphere (Duarte et al. 2008) and the world's 85,000 to 104,000 $km^3$ of saline lake water will surely respond, like the world's 125,000 $km^3$ of lake freshwater, to ongoing and future global climate change (Lerman 2009). A possible macro-problem with Lake Eyre's mat cover is that a plastically flexible floating-lid there could allow rainwater to accumulate atop the roof-like mat; a reasonable solution for this incidental problem would be to install flexible drainpipes that funnel rainwater to the middle of the ESR, carrying it through the saltwater "tank" below for deposition on land. Since freshwater is less dense than seawater, such wheel-spoke-like drainage funnels will add to the mat's basic inherent buoyancy.

A floating lid made from thin polyethylene sheeting—rather like a heavy "bubble-wrap" used by commercial shipping industry product packagers—with 100% water imperviousness would cost 10 $USD/m^2$ (in 2009). But, there are other means to cover the body of artificially emplaced liquid that are just about as useful, namely, floating white-shaded formed plastic hollow or solid balls. The material costs will be about the same as that of extensive "bubble-wrap" sheeting but deployment will be much easier, quicker to accomplish. Hollow or solid manufactured floating balls will be more affected by wind than rim-anchored flat plastic sheeting—they can be piled on the shoreline by normal deflationary event-processes such as gusty winds. However, they can be cheaply returned to the lake's imported seawater surface by diligent macro-project caretakers employing rideable mobile machines or light-weight man-portable backpack suction devices because the plastic balls will have very low mass. Caretakers may dwell in camps with low physical impact on the arid surrounding landscape (Halley 2009). Since we will be using a lot of plastic, the bulk manufacturing costs should be low since bidding will offer the prospect of worldwide suppliers; balls seem especially easy to outsource and can be tested by sampling at the separated places of manufacture or at the worksite's point of delivery. Whilst the sheeting is impermeable and causes 100% seawater retention (zero evaporation) the floating balls, which can spin unfettered and constantly remain wetted, will consequently cause only ~70% seawater retention. Added moisture above the covered Lake Eyre will be entrained by the plume of rising heated air. Floating plastic balls might be more ecologically, economically and aesthetically acceptable than a gigantic plastic sheet. Of course, a strong vertical motion of the air plume caused by the ESR could be more strongly induced if the plastic balls were of an aesthetically pleasant and scientifically suitable shade of blue coloration or merely flat black (Black and Tarmy 1963). Viewed by passengers aboard the Earth-orbiting International Space Station or by planet-bound Earthlings sent the imagery end-products of unmanned remote sensing Earth satellite databases, the ball covered ESR will generally have the appearance of a darkly-shaded "lake"; in fact, the ESR will be a spatially gigantic *trompe-l'oeil*, perhaps this feature will become the world's largest example of completed Installation Art. Such "Aesthetic landscapes" are nowadays deemed necessary for the sustainability of modern-day societies (Barrett et al. 2009).

Man-made thermal releases now commonly reach several thousand megawatts in the case of electric power plant cooling towers and cities generate heat island effects; the Port Kembla (34°25′S Lat. by 150°54′E Long.) integrated steelworks, an economical combination of industrial complex and associated seaport situated on the New South Wales coast has measurably had a "… greater than 5% on total rainfall…" increase effect (Ogden 1969) since it first commenced full-scale industrial operation in 1931. The ESR might provide moisture via Anthropic Clouds created above and downwind of the ESR moving naturally eastward to Australia's most mountainous landscape to eventually supply the city reservoirs of eastern Australia's most populous major coastal farms and ranches as well as cities that are nowadays suffering economically crippling drought-related hardships. The Murray-Darling River Basin—Australia's agricultural heartland, producing ~40% of the country's agricultural products—is still under severe drought stress. Extending across southeastern Australia from Brisbane to Adelaide, the Murray-Darling River Basin is ~14% of the nation's area but easily accounts (during non-drought periods) for ~70% of its freshwater use for crop irrigation (Draper and Mills 2008; Cai and Cowan 2008). The floating ESR balls, whatever their coloration, will cost about USD 5/m$^2$, maybe markedly less if global competitive plastics factory production is massively revved-up by the prospect and existence of the ESR macro-project. Some ball manufacturers will be retained on long-term fixed-price contract to provide replacements for lost balls blown off the ESR by the wind and a tasked conscientious clean-up crew will constantly be working mostly downwind of the filled Eyre Seawater Reservoir to remove all plastic litter that it is ever feasible to gather from the countryside that is obviously caused by the installed ESR macro-project's existence. Also, since the balls can easily, quickly, and cheaply be collected and removed from the ESR by human fiat, this sublime fact means that reversibility is an attractive key structural feature of the recommended ESR macro-project.

## 4 ESR Macro-Project Case Study

Our study involves some macro-engineering and economic calculations. However, these evaluations are very rough, not taking into account the many unknown and missing parameters involved for a unique macro-project. This is always the case with site-specific macro-projects.

## *4.1 Flooding Lake Eyre Macro-Project*

### 4.1.1 Technical Aspects

Lake Eyre stood at +10 m above sea level (asl) and covered $\sim$35,000 km$^2$, more than three times the current saline playa area, at its peak high-stand $\sim$125,000 years ago during the late Pleistocene. (Lake Eyre's theorized mega-lake incarnation is named by geologists "Lake Dieri".) The ESR macro-project assumes Lake Eyre becomes permanently filled with seawater. When Lake Eyre is full, and its asl is +5 m, short of its "Lake Dieri" presence, then Lake Eyre could cover an area of $\sim$19,600 km$^2$ and contain $\sim$215 km$^3$ of saline liquid. After some consideration, we suggest that it might be best to envision Lake Eyre with an artificial level of −3.5 m asl, an area of 9,920 km$^2$ and a seawater volume of 75 km$^3$. Our vision offers a future man-created Lake Eyre where both the North and South basins are filled constantly, instead of intermittently as under the influence of only Nature. The seawater volume and the surface area of a full Lake Eyre is denoted $V_{\text{lake}}$ and $S_{\text{lake}}$, respectively. Here we accept $V_{\text{lake}} = 75$ km$^3$ and $S_{\text{lake}} = 9{,}920$ km$^2$.

The source of incoming freshwater into the future artificial lake may be non-continuous (massif rains, occurring at a random time interval $t$). Lake Eyre South is known to have filled in 1938, 1955, 1963, 1968, 1973, 1974, 1975, 1976, 1984 and even filled partially by mid-2009 as illustrated by reference to Fig. 2, an image obtained from an Earth-orbiting satellite's imagery database. In 1984 Lake Eyre South overflowed to Lake Eyre North. In 1974 water flowed from Lake Eyre North to Lake Eyre South between March and October when an equilibrium level was obtained. Groyder Channel is a 15 km-long topographic depression that links Lake Eyre North and South (Fig. 3). The width as well as lowest elevation of the Goyder Channel—named to honor George Woodroffe Goyder (1826–1898)—change with each significant natural freshwater flooding event. In this study we accept $t = 8$ years as an average value of the observed data (McMahon et al. 2005).

The source of seawater for Lake Eyre may be a continuous or non-continuous macro-engineered addition from the Southern Ocean. In this chapter we assume a single duct connecting Lake Eyre's southernmost natural basin with the ocean situated south of Australia. Lake Eyre's South Basin lies 356 km from Port Augusta. Cost-free seawater can be extracted from Spencer Gulf in the general vicinity of Port Augusta. The duct length and diameter is denoted $L_{\text{duct}}$ and $D_{\text{duct}}$, respectively, while its maximum elevation above sea level is $H_{\text{duct}}$.

The seawater volume flow rate through the duct is $Q_{\text{sw}}$. The duct includes some pumps. Part of the necessary power is provided by PV cells.

The seawater-filled Lake Eyre surface is assumed to be partially covered, diminishing natural evaporation.

A land region of surface $S_{\text{irrig}}$ adjacent to the subject proposed artificial lake is prepared to allow biosaline agriculture. The necessary volume flow rate of water for irrigation is $Q_{irrig}$. One denotes by $q_{irrig,1}$ the necessary volume flow rate of seawater to irrigate one square meter [in $m^3/(m^2\ s)$]. Then, one has:

$$S_{irrig} = \frac{Q_{irrig}}{q_{irrig,1}} \tag{2}$$

The volume balance for the seawater in the trial ESR during the period $t$ is:

$$yV_{lake} + Q_{sw}t_{sw} = Q_{irrig}t + Q_{evap}t \tag{3}$$

Here $0 \leq y \leq 1$ is a parameter showing how much of the water in the lake will be used during the time $t$. $y = 1$ means that all the seawater is used (i.e. after the time $t$ the lakebed is empty). Also, $t_{sw}$ denotes the time period with seawater pumped from the ocean. Finally, $Q_{evap}$ is the volume rate of evaporated freshwater.

We denote by $q_{evap,1}$ the rate of evaporated freshwater [units: $m^3/(m^2\ s)$]. Then:

$$Q_{evap} = q_{evap,1}(1 - x)S_{lake} \tag{4}$$

Here $x$ is the fraction of lake surface covered ($x = 1$ means the whole surface is covered and consequently, no evaporation occurs).

Earlier estimates show that the annual rainfall on Lake Eyre amounts to ~125 mm with an average annual pan-evaporation of ~3,800 mm. The difference between these two quantities gives the net specific evaporation rate. The measure of evaporation used in a recent report (McMahon et al. 2005) is the mean annual area potential evapotranspiration (APET). APET values vary from ~1,000 mm/year immediately south of the Lake Eyre Basin to >1,500 mm/year directly north of the whole Lake Eyre Basin. For an average value of 1,250 mm/year, we found $q_{evap,1} = 0.396 \times 10^{-7}\ m^3/(m^2\ s)$.

We assume that:

$$Q_{sw} = zQ_{evap} \tag{5}$$

where $z$ is an under-unitary fraction ($z = 1$ means that the water flow rate coming from the ocean equals the evaporated water flow rate). The speed $w_{sw}$ of seawater in the duct of diameter $D_{duct}$ is given by:

$$w_{sw} = \frac{4Q_{sw}}{\pi D_{duct}^2} \tag{6}$$

The applied impelling power $P_{pump}$ required to move the seawater within the special duct is obtained from:

$$P_{pump} = g\rho_{sw}q_{sw}H/\eta_{pump} \tag{7}$$

In Eq. (7), $g$ (=9.78 m/s$^2$) is gravitational acceleration, $\rho_{sw}$ (=1,030 kg/m$^3$) is the density of the seawater, $H$ is the hydraulic head and $\eta_{pump}$ ($\approx$0.75) is the efficiency of the pump. The hydraulic head is obtained by summing the highest elevation of the duct $H_{duct}$ ($\approx$60 m) with the lost pressure height $\Delta H$ due to friction:

$$H = H_{duct} + \Delta H \tag{8}$$

Only linear pressure losses are considered next and:

$$\Delta H = \lambda \frac{L_{duct}}{D_{duct}} \frac{w_{sw}^2}{2g} \tag{9}$$

where $\lambda$ is the linear pressure loss coefficient given by:

$$\lambda = \begin{cases} \dfrac{1}{\sqrt{4100RE}} & \text{for} \quad Re < 10^5 \\ 0.0032 + \dfrac{0.211}{Re^{0.237}} & \text{for} \quad Re > 10^5 \end{cases} \tag{10}$$

where the Reynolds number is defined by

$$Re = \frac{w_{sw} D_{duct}}{\nu_{sw}} \tag{11}$$

where $\nu_{sw}$ is the kinematic viscosity of seawater. The constant value $\nu_{sw} = 13 \times 10^{-4}/\rho_{sw}$ is adopted in this study.

The energy consumed per year with pumping $E_{pump,year}$ is obtained (in J/year) from:

$$E_{pump,year} = 365 \times 24 \times 3,600 \times P_{pump} \tag{12}$$

The average annual temperature varies from 21 °C south of the Lake Eyre Basin to 24 °C north of it, and the average maximum temperatures are 18 °C and 24 °C respectively in July, and 36 °C and 39 °C in January. The annual hours of sunshine vary from 3,250 to >3,500. Surprisingly, the literature is poor on detailed solar radiation data in the vicinity of the extant, humanly unaltered, Lake Eyre. In Table 1 there are some data that could be a starting point for enlightening future computations (Solar Irradiation Database 2007). Here the following yearly averaged daily solar global irradiation on a horizontal plane around Lake Eyre is adopted $G_{day} = 21.6$ MJ/(m$^2$day), which is close to the value estimated in Table 1 for Moomba, Australia. The yearly solar global irradiation $G_{year}$ is of course

$$G_{year} = 365 G_{day} \tag{13}$$

The energy provided per unit surface by PV cells during a year, $E_{PV,year,1}$ is given by:

$$E_{PV,year,1} = G_{year} \eta_{PV} \tag{14}$$

**Table 1** Monthly mean of daily global solar irradiation [kWh/(m$^2$day)] (P-measured by pyranometer, E-estimated by model)

| Location | Dry-Creek | Moomba | Woomera | Gladstone |
|---|---|---|---|---|
| Lat. and Long. | 34.83°S, 138.58°E | 28.10°S, 140.20°E | 31.52°S, 137.17°E | 33.28°S, 138.37°E |
| Type | P | E | E | E |
| Jan | 8.03 | 8.25 | 7.42 | 6.86 |
| Feb | 6.98 | 7.90 | 6.97 | 6.53 |
| Mar | 5.66 | 6.97 | 6.00 | 5.81 |
| Apr | 4.95 | 5.00 | 4.81 | 5.33 |
| May | 2.71 | 4.18 | 3.53 | 4.39 |
| Jun | 2.29 | 3.78 | 3.08 | 4.00 |
| July | 2.43 | 3.75 | 3.36 | 4.28 |
| Aug | 3.32 | 4.18 | 4.14 | 4.86 |
| Sep | 4.61 | 5.52 | 5.28 | 6.06 |
| Oct | 5.57 | 7.10 | 6.36 | 6.50 |
| Nov | 6.97 | 7.85 | 7.08 | 7.03 |
| Dec | 7.69 | 8.14 | 7.50 | 6.81 |
| Yearly mean | 5.10 | 6.06 | 5.47 | 5.69 |

where $\eta_{PV}$ is PV cell efficiency (yearly average). In computations we have used a rather high (optimized) value $\eta_{PV} = 0.15$ (Badescu 2006). The energy provided by the whole PV cells surface during a year, $E_{PV,year}$, is obtained from:

$$E_{PV,year} = E_{PV,year,1} D_{duct} L_{duct} \tag{15}$$

This takes into account that the PV cell area, being mounted on the duct, is equal to the duct's projected area. The energy consumed per year with pumping, $E_{pump,year}$, is provided in part by the PV cells. The remaining part, $E_{classic,year}$, should be provided from classical energy sources. One has:

$$E_{classic,year} = E_{pump,year} - E_{PV,year} \tag{16}$$

### 4.1.2 Economical Aspects of the ESR Macro-Project

For each of these macro-engineered components (i.e., covering the lake, ducts and pumping and biosaline agriculture) there are two obvious associated costs:

Costs of construction (investments). Costs are proportional with the main extensive quantity of each component. These costs refer to covering the surface $xS_{lake}$ of the lake, building the conveying duct to the Spencer Gulf-Lake Eyre Basin, including the seawater pumps and the PV cells (this cost is a function of $L_{duct}$, $H_{duct}$, $D_{duct}$, $Q_{sw}$ and the material of the duct) and preparing the irrigation land surface;

Costs of completed macro-project's continuous operation and general facility upkeep. They are proportional with the main extensive quantity of each component. They are related to seawater pumping, maintenance of the floating covering system and operation of irrigation surface.

Also, for the irrigated land region, one may have an economic gain from farming and ranching, from selling the harvested agricultural as well as the rounded up and corralled living animal grazing end-products. This economic gain is proportional to $S_{irrig}$. The enterprise's profit, which is the difference between economic gain and costs, is one of the economic performance indicators. One objective function may be system profit, after $t$ years of operation. This may be seen as a temporally cyclic process, because $t$ can be taken as a period. That would be a measurable economic gain. Other, immeasurable economic gains are listed in Sect. 5.

To evaluate the financial magnitude of the macro-project, it is a sheer necessity to estimate the cost $c_{cover}$ of partially covering Lake Eyre:

$$c_{cover} = xS_{lake}c_{cover,1} \tag{17}$$

where $c_{cover,1}$ is the cost of covering the unit surface area. Here we adopt $c_{cover,1} = 5\ \mathrm{USD/m^2}$. The maintenance cost per year, for the covering system, $c_{cover,maint,year}$ is

$$c_{cover,maint,year} = f_{cover,maint}c_{cover} \tag{18}$$

where $f_{cover,maint}$ is a given fraction. Here $f_{cover,maint} = 0.01$ has been adopted.

The cost of the duct, the PV cells and of the electric pumps enabling steady seawater pumping, are estimated. The cost $c_{duct}$ of the conducting tube is given by:

$$c_{duct} = c_{duct,1}L_{duct} \tag{19}$$

where $c_{duct,1}$ is the cost of a unit length of duct. Similarly, the cost of installing the tube, $c_{inst,duct}$ is given by:

$$c_{inst,duct} = c_{inst,duct,1}L_{duct} \tag{20}$$

where $c_{inst,duct,1}$ is the cost of installing a unit length of duct. The unitary costs depend, of course, on various factors, such as the desired duct diameter $D_{duct}$ as

**Table 2** Approximate 2009 costing of the duct and its installation on ground

| Diameter of duct, $D_{duct}$ (m) | Cost of steel duct (USD/m) | Cost of plastic duct (USD/m) | Cost of fabric duct (USD/m) | Cost of installation (steel, plastic) (USD/m) | Cost of installation (fabric) (USD/m) |
|---|---|---|---|---|---|
| 0.1 | 19 | 9 | 4 | 900 | 100 |
| 0.5 | 45 | 21 | 10 | 1,500 | 400 |
| 1 | 380 | 200 | 90 | 2,000 | 600 |
| 2 | 600 | 300 | 140 | 2,700 | 800 |
| 3 | 1,100 | 600 | 290 | 3,500 | 1,000 |
| 5 | 2,300 | 1,200 | 510 | 4,000 | 1,300 |
| 10 | 5,100 | 2,400 | 990 | 5,000 | 1,800 |

well as the optimum material composing the duct. Table 2 shows the input values used in this chapter.

Thickness of steel duct wall is about 1% of the diameter, thickness of plastic duct wall is about 10% of the duct's overall diameter. Fabric ducts are inflatable and flexible, comprised of exceptionally strong artificial fibers.

The cost of the pumping installation, $c_{pump}$, is obtained from:

$$c_{pump} = P_{pump} c_{pump,1} \tag{21}$$

where $c_{pump,1}$ is the cost a pump of unit power. Table 3 shows statistically average values for $c_{pump,1}$ as a function of pumping device quality. The gold-plated cost in Table 3 is for a custom built pump. The inexpensive cost is for an off-the-shelf, mass-produced pump. The maintenance cost per year for the pumping system, $c_{pump,maint,year}$ is

$$c_{pump,maint,year} = f_{pump,maint} c_{pump} \tag{22}$$

where $f_{pump,maint}$ is a given fraction. Here $f_{pump,maint} = 0.01$ has been adopted.

The cost $c_{PV}$ of the PV cells is obtained from the product of the cost of a unit surface PV cell, $c_{PV,1}$ and the total surface covered by installed PV cells:

$$c_{PV} = c_{PV,1} L_{duct} D_{duct} \tag{23}$$

Using the estimate for future PV Cells technology, the cost per square meter of flexible PV cells sheeting is $c_{PV,1} \approx 100\ \text{USD/m}^2$ (Liu et al. 2007a, b). The maintenance cost per year for PV cells, $c_{PV,maint,year}$ is

$$c_{PV,maint,year} = f_{PV,maint} c_{PV} \tag{24}$$

where $f_{PV,maint}$ is a given fraction. Here $f_{PV,maint} = 0.01$ has been adopted.

The cost of the classical energy consumed during a year, $c_{classic,year}$, is given by:

$$c_{classic,year} = E_{classic,year} c_{classic,year,1} \tag{25}$$

where $c_{classic,year,1}$ is the cost of a standard classical energy unit. Here electricity is considered and $c_{classic,year,1} \approx 0.1$ USD/kWh (PelletHeat 2007). The cost of classical energy during the period $t$ (years), $c_{classic}$, is obtained from:

$$c_{classic} = c_{classic,year} t \tag{26}$$

**Table 3** Cost of a pump of unit power $c_{pump,1}$

| Pump quality | Cost 2009 (USD/W) |
|---|---|
| Cutting corners | 2 |
| Inexpensive | 4 |
| Average | 8 |
| Expensive | 16 |
| Gold plated | 32 |

Adapted from Axion (2007)

if $t$ is expressed in seconds. The initial investment cost $c_{\text{invest,Eyre}}$ for the covering system, the duct, PV cells and pumping installation, is given by

$$c_{\text{invest,Eyre}} = c_{\text{cover}} + c_{\text{pump}} + c_{\text{PV}} + c_{\text{duct}} + c_{\text{inst,duct}} \tag{27}$$

The maintenance cost $c_{\text{maint,Eyre}}$ during the time period $t$ (years) of the covering system, PV cells and pumping installation is

$$c_{\text{maint,Eyre}} = (c_{\text{cover,maint,year}} + c_{\text{pump,maint,year}} + c_{\text{PV,maint,year}})t \tag{28}$$

The total costs, $c_{\text{tot,Eyre}}$, after the time period $t$, is

$$c_{\text{tot,Eyre}} = c_{\text{invest,Eyre}} + c_{\text{maint,Eyre}} + c_{\text{classic}} \tag{29}$$

## 4.2 Seawater Irrigation Macro-Project

Biosaline irrigation requires no special equipment. Existing test farms have tried either normal dug ditch flood irrigation of large basins or aerially broadcast seawater (Glenn et al. 1998). Generally speaking, seawater agriculture needs ~35% more irrigation fluid when grown using seawater than conventional farm crops grown using only freshwater. The main macro-problem is that the land fosters evaporation of freshwater—it then becomes escaping water vapor, a part of the moving air overhead—and the land's soil salinity will increase year by year (Beresford 2004). However, generally, there are no insurmountable macro-engineering problems associated with biosaline agriculture (Ozturk et al. 2006).

Input values were suggested from the Project Ras al-Zawr where *Salicornia bigelovii* is cultivated (SaudiAramCo 2007). There, computer-controlled pivot-irrigation arms sprayed seawater sucked in by three diesel pumps at a rate exceeding 28 $m^3$/min, each unit watering a 50 ha circle, requiring 6.5 h for the revolving arms to complete one full circuit. From these data we found that $q_{\text{irrig},1} = 0.0093 \times 10^{-4}$ $m^3/(m^2$ s).

The irrigation installation cost $c_{\text{irrig}}$ is, of course, proportional with the irrigated land surface:

$$c_{\text{irrig}} = S_{\text{irrig}} c_{\text{irrig},1} \tag{30}$$

where $c_{\text{irrig},1}$ is the cost of irrigating a unit surface of cultivated crops. Depending on the crops, $c_{\text{irrig},1}$ ranges from 200 to 2,000 USD 2007/acre (Farm Management 2007) (one acre equals about 4,048 $m^2$). Here we accept 750 USD/acre which then yields $c_{\text{irrig},1} = 0.185$ USD/$m^2$. The annual maintenance cost for the irrigation installation, $c_{\text{irrig,maint,year}}$ is

$$c_{\text{irrig,mai,year}} = f_{\text{irrig,maint}} c_{\text{irrig}} \tag{31}$$

where $f_{\text{irrig,maint}}$ is a given fraction. Here $f_{\text{irrig,maint}} = 0.05$ has been adopted. The maintenance cost $c_{\text{irrig,maint}}$ for the time period $t$ (years) of the irrigation installation is:

$$c_{\text{irrig,maint}} = c_{\text{irrig,maint,year}} t \tag{32}$$

The cost associated to the irrigation installation, $c_{\text{tot,irrig}}$, after the time period $t$ of operation is

$$c_{\text{tot,irrig}} = c_{\text{irrig}} + c_{\text{irrig,maint}} \tag{33}$$

The economic gain per year from the crops irrigated with seawater, $g_{\text{irrig,year}}$, is given by

$$g_{\text{irrig,year}} = S_{\text{irrig}} g_{\text{irrig,1}} \tag{34}$$

where $g_{\text{irrig,1}}$ is the economic gain per unit surface of cultivated crops, per year. The economic gain after the time period $t$ (years) from the cultivated crops, $g_{\text{irrig}}$, is given by

$$g_{\text{irrig}} = g_{\text{irrig,year}} t \tag{35}$$

During 6 years of rigorous field trials in Mexico, *S. bigelovii* produced an average annual crop of 1.7 kg/m$^2$ of total biomass and 0.2 kg/m$^2$ of oilseed (Imaz et al. 1998). It is expected that the economic benefit consists of food products such as cooking oil and "sea asparagus", a vegetable delicacy that sells in some parts of Europe for USD 40/kg. In calculations, we considered as possible enterprise sale products: oil (0.75 USD 2009/kg) and sea asparagus (40 USD 2009/kg) but only results for this last (more advantageous economically) product are given herein. In the two cases examined, $g_{\text{irrig,1}}$ are evaluated to about 0.15 USD 2009/m$^2$ and 8 USD 2009/m$^2$, respectively.

The economic profit $p_{\text{irrig}}$ of the land seawater irrigation system after $t$ years is given by

$$p_{\text{irrig}} = g_{\text{irrig}} - c_{\text{tot,irrig}} \tag{36}$$

The total cost, $c_{\text{tot}}$, of the Lake Eyre flooding and adjacent irrigation macro-project, the ESR, after the time period $t$, is obtained from:

$$c_{\text{tot}} = c_{\text{tot,Eyre}} + c_{\text{tot,irrig}} \tag{37}$$

### 4.3 Results

Covering half of the artificially seawater-filled Lake Eyre ($x = 0.5$) to diminish evaporation is the first choice. One assumes that during the time period $t$ the volume of liquid in the lake is halved ($y = 0.5$). Also, the seawater income from the Spencer Gulf is half of the evaporated water ($z = 0.5$). This method of choosing the values of the parameters *x*, *y* and *z* in the middle of their range of variation may be a good evaluative strategy when virtually nothing is known with certitude yet about the actual dependence of various quantities on these parameters. For these input data, the seawater evaporation flow rate is $Q_{evap} = 196.4$ m$^3$/s and the Southern Ocean seawater flow rate income is $Q_{sw} = 98.2$ m$^3$/s. The seawater flow rate $Q_{irrig} = 50.43$ m$^3$/s will irrigate a biosaline agricultural area $S_{irrig} = 54.23$ km$^2$.

An inexpensive pump in Table 3 has been considered. The seawater speed in the duct $w_{sw}$ and the pumping power $P_{pump}$ become reasonably small only at tube diameters $D_{duct}$ larger than 3 m (Fig. 4a). It is obvious that at smaller duct diameters, the surface covered by PV cells is rather small and most of the energy required to shift the seawater is provided from standard classical energy sources (Fig. 4b). However, for diameters larger than 7 m the energy provided by the PV cells becomes comparable with that from classical sources. When $D_{duct}$ is larger than 8 m, the PV cells become the main energy supplier.

The duct cost $c_{duct}$ increases by increasing the pipe's diameter $D_{duct}$, as expected (Fig. 5a). Obviously, $c_{duct}$ depends on the duct material, with composed fabric the least expensive solution. The same feature exhibits the cost of installing the duct, $c_{inst,duct}$, but in this case the dependence on duct diameter $D_{duct}$ is weaker than the dependence of $c_{duct}$ on $D_{duct}$. The cost of the pumps comprising the pumping installation $c_{pump}$ decreases by enlarging $D_{duct}$, as well the cost of the energy for pumping provided by using classical fuels, $c_{classic}$ (Fig. 5b). However,

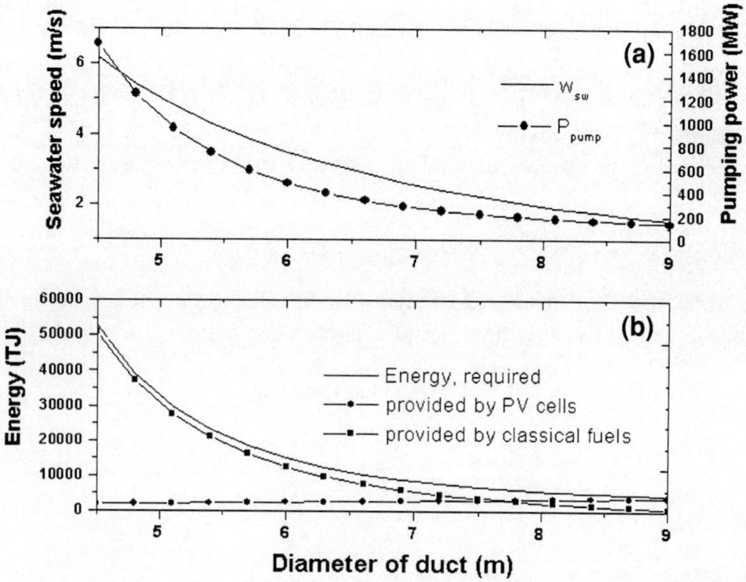


**Fig. 4** **a** Seawater speed $w_{sw}$ and pumping power $P_{pump}$ and (**b**) various energies consumed and generated during one year (the energy consumed with pumping, $E_{pump,year}$, the energy provided by PV cells and from classical fuels, $E_{PV,year}$ and $E_{classic,year}$, respectively) as a function of duct diameter $D_{duct}$

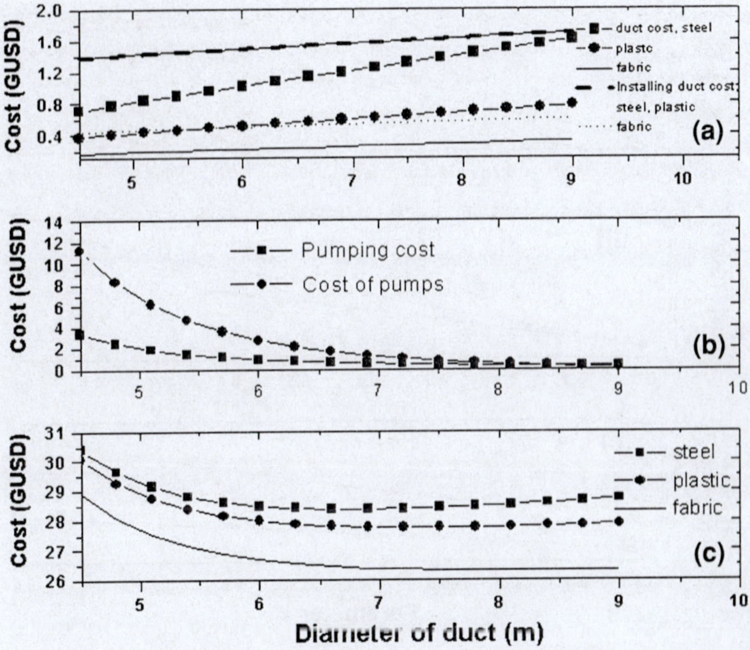


**Fig. 5** Various estimated 2009 USD costs as a function of duct diameter $D_{\mathrm{duct}}$. **a** Cost of duct $c_{\mathrm{duct}}$ and cost of duct installation $c_{\mathrm{inst,duct}}$ for various associated duct materials; **b** cost of pumping installation $c_{\mathrm{pump}}$ and cost of pumping during $t$ years by using classical fuels, $c_{\mathrm{classic}}$; **c** initial investment cost $c_{\mathrm{invest,stab}}$ for various duct materials. Computations performed for $t = 8$ years

both $c_{\mathrm{pump}}$ and $c_{\mathrm{classic}}$ have a rather weak dependence on the duct's diameter for $D_{\mathrm{duct}} > 7$ m. The cost of electric seawater pumps is of the order of a few 2009 GUSD while the cost of the energy consumed for seawater pumping during time $t$ (years) is of the order of one GUSD (i.e. 2009 USD).

The initial investment cost $c_{\mathrm{invest,Eyre}}$ depends on the material used to manufacture the duct, with composed fabric the least expensive solution (Fig. 5c). However, it is obvious that the larger contribution to the macro-project cost is provided by covering Lake Eyre. Indeed, compare the total costs in Fig. 5c with the cost of the duct in Fig. 5a and costs of the pumps in Fig. 5b. The difference between the costs in Fig. 5c, on one hand, and the sum of the cost in Fig. 5a and Fig. 5b, on the other hand, is the cost associated to the covering phase of the proposed ESR macro-project. Whatever the duct material, $c_{\mathrm{invest,Eyre}}$ has a minimum that is about 6.5 m for steel and plastic ducts and about 7.5 m for tubing comprised of fabric. The minimum initial investment is ~28 GUSD in the case of pipes composed of steel and plastic and about 26.5 GUSD in the case of ducts fabricated with composed fabric.

The seawater irrigation component associated to the ESR macro-project was studied next. The results of that study are shown in Fig. 6, which refers to a pipe diameter of 6 m, which is rather near the optimum values minimizing the initial

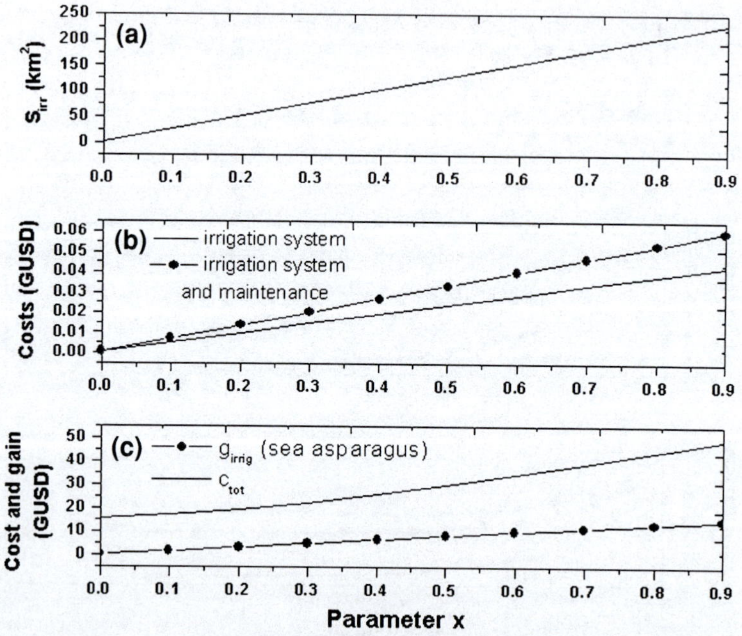


**Fig. 6** Various quantities as function of the cover parameter *x*. **a** Irrigated surface $S_{irrig}$; **b** cost of irrigation system $c_{irrig}$ and cost of irrigation and its maintenance for *t* years, $c_{tot,irrig}$; **c** economic gain from the irrigated surface for *t* years, $g_{irrig}$, for one possible product and the total cost, $c_{tot}$, of the Lake Eyre seawater flooding project and the biosaline irrigation project, after the time period *t*. Computations performed for $D_{duct} = 6$ m and $t = 8$ years. Also, $y = 0.8$ and $z = 0.4$

investments in the seawater flooding component in the case of a fabric duct use (see Fig. 5c). The covering parameter *x* is a variable while $y = 0.8$ and $z = 0.4$.

The irrigated surface $S_{irrig}$ increases by increasing the invoked parameter *x*, as expected (Fig. 6a). Understandably, increasing *x* decreases the evaporated freshwater rate and more seawater may be directed toward the irrigated land surface. For all computations, $S_{irrig}$ ranges between 0 and 250 km$^2$. The cost of the irrigation system $c_{irrig}$ and the cost of irrigation and its continuous maintenance for $t = 8$ years, $c_{tot,irrig}$, increases by increasing the parameter *x* (Fig. 6b). These money costs are of the order of tens of millions of 2009 USD, well below the investment costs associated strictly with the covering of Lake Eyre (see Fig. 6c). The economic gain $g_{irrig}$ from the crops irrigated with seawater is shown in Fig. 6c for one product (sea asparagus). This economic gain increases by increasing *x*, as expected. The economic gain is up to 10 GUSD. The total cost, $c_{tot}$, of the Lake Eyre seawater flooding macro-project and irrigation macro-project, after the time period *t* is shown in Fig. 6c. It increases by increasing the parameter *x*. A short comparison between $g_{irrig}$ and $c_{tot}$ may be useful. At large values of the parameter *x*, the total cost $c_{tot}$ is about five times the economic gain $g_{irrig}$. Thinking roughly, this means that in about $5t = 40$ years the ESR macro-project will become

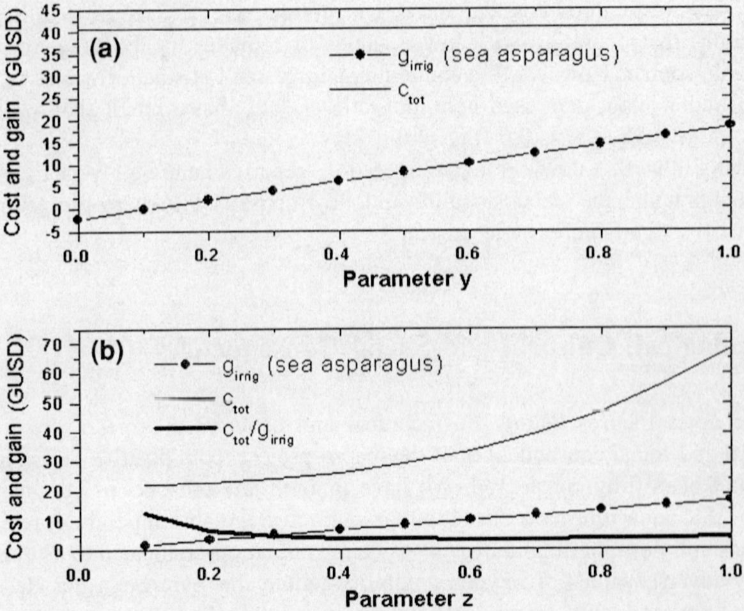


**Fig. 7** The economic gain from the irrigated surface for $t$ years, $g_{irrig}$, for one possible product and the total cost, $c_{tot}$, of the Lake Eyre seawater flooding project and irrigation macro-project, after the time period $t$. **a** Dependence on the established parameter $y$, for $x = 0.6$ and $y = 0.8$; **b** dependence on the parameter $z$, for $x = 0.4$ and $y = 0.8$. Here the ratio $c_{tot}/g_{irrig}$ is also shown. Computations are performed for $D_{duct} = 6$ m and $t = 8$ years

economically profitable. At smaller values of $x$, the time interval spent until the macro-project becomes profitable increases.

The total cost, $c_{tot}$, of the Lake Eyre seawater flooding and adjoining region biosaline irrigation macro-project is very slightly increased when the parameter $y$ increases from zero (Lake Eyre is kept at a constant level during the time period $t$) to unity (Lake Eyre is completely emptied after the time period $t$) (Fig. 7a). The economic gain $g_{irrig}$ from the biosaline crops increases significantly by increasing parameter $y$. This has important consequences on the strategy of operation. For instance, in case $y = 0.3$, then $c_{tot} \approx 50$ GUSD and $g_{irrig} \approx 5$ GUSD. The ratio between the two quantities is eight and the macro-project will become economically profitable in about $8t = 64$ years. But when $y = 1$, then $c_{tot} \approx 40$ GUSD and $g_{irrig} \approx 15$ GUSD which reduces the time interval until the macro-project operates profitably to about 40 $t$/15 $\approx$ 21.3 years. To conclude, operation with higher values of $y$ induces the ESR macro-project profitability period to arrive sooner.

The total cost, $c_{tot}$, of the Lake Eyre seawater flooding macro-project and an adjacent biosaline irrigation macro-project increases significantly by increasing the parameter $z$ related to the amount of seawater incoming from the Southern Ocean (Fig. 7b). The increasing rate is important and non-linear at higher values of $z$. The comparison of the total cost $c_{tot}$ and the economic gain $g_{irrig}$ seems to indicate (at first sight) that operation with lower values of $z$ makes the ESR macro-project profitable sooner. However, the economic gain is small at smaller values of $z$. For the particular input data used here, the ratio $c_{tot}/g_{irrig}$ has a much shallower minimum value 3.36 at $z \approx 0.6$ (Fig. 7b).

Finally, note that the working computations reported herein are very rough and, in actual practice, the real costs of this unique proposed ESR macro-project may be higher by a significantly large amount.

# 5 Ecological, Cultural and Social Consequences

This chapter focuses mainly on technical and economical aspects. Ecological, cultural and social consequences of the macro-project (both positive and negative) have not been fully considered. We have ignored any prospect of commercially viable ESR aquaculture because the maximum average air temperature, related to the constantly evaporating imported seawater (salination), indicated to us that truly fishery-adverse values of oxygen solubility within the Eyre Seawater Reservoir would unquestionably occur (Debelius et al. 2009). To give an over-arching perspective and closure, a few additional important ESR macro-project considerations are presented next.

An important macro-problem is that, with continuous imports of seawater from the Southern Ocean, the salinity of Eyre Lake will increase relentlessly year by year (except when freshwater floods occasionally occur). A rather similar problem occurs in instances of closed seas, such as the lake-like Caspian Sea (Millero et al. 2008). But that large body of water in Central Asia has a rather constant salinity because it fills by markedly seasonal freshwater river inflows. The difference is that, in the case of Australia's Lake Eyre converted to the ESR, one has an anthropogenic salinization of an episodic inland body of brackish water (Williams 2001).

The accumulation of sea-salt and its effects on Lake Eyre have to be examined, and solutions to diminish the consequences of sea-salt accumulation need to be proposed. For example, an artificial evaporative southern gulf of the ESR might be created, playing the same hydrological role that Kara-Bogaz Gol does for the Caspian Sea (Kosarev et al. 2009). In this way, the ESR's salinity might be kept within desirable limits. Some sea-salt deposited in the southern gulf of the ESR could be removed afterwards by profitable commercial mining, used then for anti-global warming purposes. Electrolysis of mined ESR fresh sea-salt, making strong bases (sodium hydroxide) and strong acids (hydrochloric acid), may become commercially saleable chemical end-products because when strong bases are dissolved in oceanic seawater, with a normal "basic" pH of 8.2, it causes aerial $CO_2$ to be stored in such treated oceanic seawater as $HCO_3$, making the industrially doctored impure seawater (at its multiple places of point discharge and an unknown distance beyond) more alkaline and, thus, helping in a small, yet obviously significant way, to combat the adjacent greater world-ocean's ongoing acidification thought nowadays to be almost exclusively caused by the anthropogenic $CO_2$ gas buildup in the world's atmosphere (House et al. 2007). Specifically, at the Great Barrier Reef, it has been shown that the coral to macro-algal phase shift induced by global atmospheric warming and local seawater $CO_2$ enrichment occurring between 1996 and 2006 "...increased slightly.... (Bruno et al. 2009). Barring anthropogenic intervention, by 2100 it is predicted that the Earth's ocean surface seawater will have a pH of 7.8. Ocean acidification, locally expressed, is bound to adversely affect Australia's world-famed Great Barrier Reef (extending from 8° to 27°S Lat. in the Coral Sea); sodium hydroxide manufactured at the ESR, piped to Port Augusta and loaded aboard distributor barges which then discharged their cargoes directly into the natural seawater between the Great Barrier Reef and Australia's mainland—i.e., inside Australia's territorial waters—could beneficially increase the alkalinity of the seawater inside the reef and elsewhere nearby, thus bettering the oceanographic conditions of growth for the reef-forming biota (coral polyps). So, we speculate, a sodium hydroxide ESR to Port Augusta pipeline/Port Augusta barge conveyance system to Queensland's not-far-distant coast may be both possible, and necessary, someday to preserve the celebrated 1981 UNESCO World Heritage Site, which is Queensland's official state icon, as well as a 344,400 $km^2$ national Great Barrier Reef Marine Park since 1975 (Hutchings et al. 2008). Excavation and processing of fresh sea-salt at the ESR will prolong the operational design life of the ESR macro-project by an as yet uncalculated period of time. The ESR-based electrolyzers will be solar-powered—the ESR situated solar energy facility will be "stranded", unconnected by power-lines to any distant urban market—and will of necessity require a local market within the enormous Lake Eyre Basin. The effectiveness risks of the sea-salt processing operation and exportation system need not be detailed herein because the ordinary post-2009 world economic reality, as well as the common twenty-first Century Australian legal certification process, will work eventually to sort out all such effectiveness risks.

Also, we think it is very worthwhile to mention herein, an interesting new mechanical solution does exist to extract freshwater from the Earth–atmosphere (Bolonkin 2007). This locally-derived aerial freshwater might be used to decrease the ESR's seawater importation need by some unknown quantity. But the implementation of these (and other) macro-engineering solutions should, in future, be studied in much more detail.

The present ESR macro-project should be underpinned by an inter-basin freshwater transfer Macro-engineering plan renewal—namely the diversion of Cooper Creek (Walker et al. 1997; Kingsford et al. 1998). Cooper Creek is one of the last 'wild' (i.e. unregulated) river systems in Australia and is protected by the Lake Eyre Basin Agreement. But the valuable and infrequent flash-flood river runoff may be used to dilute from time to time the Eyre Seawater Reservoir, replacing for short time intervals the industrialized importation of seawater withdrawn from Spencer Gulf because of ESR operational demands. The potential unrestrained pulsative freshwater inflow to Lake Eyre contributed by the poorly-monitored Diamantina River must also be dealt with somehow. Given that these, and the other rivers that flow into Lake Eyre, have the potential to fill the lake in times of flood, the diversion of these additional rivers would constitute a macro-project in itself. More simply stated, upstream reservoirs should be established. In addition to the large and irregular flood events that fill the present-day lake on occasion, brackish water also covers about half of the lake every three years and more than half the lake approximately every decade (Kingsford and Porter 1993). All these contributory important aspects should be studied in detail and solutions to keep the salinity of the ESR at a useful level should be proposed.

Lake Eyre is not a very brackish lake and the land surrounding the present-day terminal lake can hardly be rated as an unforgiving barren desert. There are abundant known and documented plant and animal species established in this remote region. Indeed, Lake Eyre is one of Australia's largest ephemeral wetlands and a major breeding ground for a variety of desirable water-birds. When Lake Eyre floods it supports great numbers of birds, fish and invertebrates. Kingsford and Porter (1993) estimate more than 100,000 water-birds use the lake while Roshier et al. (2001) allege the Lake Eyre Basin has the highest habitat availability for migratory water-birds in Australia, with interconnected wetlands providing broad pathways to the wetter regions of currently drought-stricken southeastern Australia. Also, unpredictable rainfalls produce regions that support a high diversity and abundance of wildlife (Stafford Smith and Morton 1993). The present macro-project could, no doubt, disturb the extant Lake Eyre Basin eco-system. Lake Eyre acts as an ephemeral breeding ground for numerous bird species; covering any part of it with floating man-made material could upset this biological activity. Floating plastic balls covering the lake could be manufactured too large for all visiting bird species to ingest by swallowing. Varying Lake Eyre's salinity could have adverse effects too.

Australia is a highly urbanized country and >50% of Australians live within 10 km of the world-ocean (Chen and McAneney 2006). In southern Australia, South Australia is classified as mostly "remote", "other regional-rural" and "small regional" in terms of human settlement characteristics. Only Adelaide is a "large city" (Rofe and Oakley 2006). The desert surrounding Lake Eyre is a region of very low human population density and with a low future population growth demographic projection (Taylor 2003); human demographic projections for the region in question are unlikely to change radically during the early twenty-first century, even when Australia's national population reaches ~35 million persons by circa 2049. It is preferable that economic development occur only after the Eyre Seawater Reservoir is emplaced and indisputably proven to be a benefic optimum solution by several supercomputer modeling outcomes (Dean et al. 2007). Provision for insignificant cultural heritage damage to take place ought to be assured. Also, the capping and increased salinization of Lake Eyre would have consequences on the aesthetics of the landscape, which supports some tourism. The lake is protected as a National Park.

How groundwater flow beneath Lake Eyre might be changed by the ESR's presence should also be thoroughly studied, even if no important or surprising negative effects are really expected by confident, but definitely not over-confident, macro-engineers (Holzbecher 2005). There is a real possibility the sub-aerial topography and existing soils will endure some landscape contamination during future ESR operational emergencies. Design solutions should be given for the visual impact to be tolerable as the hoses/pipelines/tubes can be camouflaged, made generally inconspicuous almost everywhere, that the region's groundwater and surface waters would not be polluted, that air quality may improve (less mass of blowing dust), that noise from enclosed electric pumps would not provide a nuisance to anyone with normal hearing abilities, that construction wastes will be removed and/or discretely entombed, perhaps in some entertaining form of twenty-first Century Land Art (Hogan 2008).

Briefly, further research is necessary on the ESR macro-project's possible effects on flora and fauna, groundwater, surface water and soil salinity and certain aesthetic and cultural values. A useful initial reference along this line is Williams (2001).

There are many benefits of the ESR macro-project such as a beneficial rising air plume, possible climate modification in the region and farther downwind by reduction of aerosols, provision of photovoltaic energy, potential nearby human settlement in the region and increased areas for lightly-grazing livestock. Also, there would be immeasurable economic gains stemming from the removal of dry-land as a ground source of dust storm aerosols (Shao et al. 2007)—it would be immersed by imported seawater—and locally enhanced rainfall will help curtail/diminish such economically damaging weather events in the downwind region. Reddish mineral aerosols, dust eroded from the arid landscape of Central Australia, and blown southeast towards the coastal city of Sydney (33°51′S Lat. by 151°12′E Long.) caused a city-enveloping daytime "vermillion haze" sky there during September 2009. (Previously, intense storms have caused identical shroud-like aerial urban hazes in Sydney during January 1942, December 1957, September 1968, April 1994 and October 2002.) All of these bruited ESR benefits ought to be included in any assessment of the probable value of the ESR, but studied in much more detail by qualified Australian experts.

A full-of-seawater Lake Eyre may change the circumferential weather and climate regime, making the region more favorable for additional and larger human settlements. Such a postulated change has been studied, with rather disappointing preliminary conclusions so far, and is still under controversial public scrutiny and public debate (Hope et al. 2004). The mechanisms by which this macro-engineering ESR macro-project may induce rainfall inland of the country's eastern seaboard or alter the climate regime above and surrounding Lake Eyre, as well as the supposed adjacent benefic results, should also be studied. For example, hygroscopic sodium chloride (50 nm particles sea-salt particles that activate at 0.33% super-saturation) is an excellent cloud condensation nucleus (Feather 1971).

We have proved that biosaline agriculture in the ESR macro-project region might be profitable. However, a much smaller geographical scale experiment undertaken closer to the coast (even at the physical scale of a single Australian citizen landholder) might be appropriate before undertaking an ESR macro-project. Indeed, this would eliminate the costs of both ducted pipelines to the lakebed and costs of transporting the produce from such a remote region that is not served by any built dedicated railroad or spur line. Also, it is obvious that sea asparagus gains its high market price per kilogram from the fact that it is somewhat of a "novelty' food product, and the mass production of sea asparagus could lower the local market prices to unsustainable levels whilst an export market is definitely unwarranted at this time. These pertinent aspects of the ESR macro-project should also be intensely studied.

## 6 Conclusive ESR Macro-Project Summary

The present work proposes a macro-engineering project, aimed at stimulating economic activity and human settlement in the Lake Eyre Basin. This territory, belonging to South Australia, often described as picturesque, is semiarid and practically unsettled.

Historically, human settlement was dependent on desert waterholes and traditional agriculture as a basis, eventually followed by urbanization and industrial activities. In the past 150 years, a number of proposals were made, but none convinced or demonstrated the possibility of improving regional arid conditions of inland central Australia. It may be that chronic lack of freshwater and salt-ridden soils in the Lake Eyre Basin do not allow for the traditional agriculture approach.

But, during the early twenty-first century, it may be opportune to change this previous tame Science paradigm during a time of record-breaking meteorological drought. The all-important quantity in this new Macro-engineering thinking is energy, instead of rich soils and precipitation. The Lake Eyre Basin is suitable for clean, large-scale photovoltaic energy production. Large quantities of seawater can be transported without losses on the ground, using big diameter, cheap tensioned textile tubes. Saltwater from the Southern Ocean can be pumped for input into Lake Eyre, which can be used as seawater reservoir for biosaline agriculture on its encircling shore. Increasing and stabilizing the level of the lake is of crucial importance in the macro-project. For this, apart from the relatively modest saltwater additional input, evaporation from the lake has to be reduced; the only means to do this is by lidding the lake with a floating, impervious, plastic mat or with buoyant white hollow plastic balls. The exact composition and desired optical properties (reflectance) of this cap remains to be seen. Suppressing evaporation of the freshwater/brackish water is an important factor designed to raise the level of the lake and has the beneficial consequence of preventing salt concentration. In a first transformational stage, during the initial period of artificially raising the level of the ESR, the salinity is expected even to decrease.

A crude economic cost estimate is done in this chapter. We do believe that all the methods and elements presented in this expose will work and using them together may even have an effectual synergistic result. However, further research is necessary to provide a broader perspective on both the benefits and the negative consequences of this macro-project. The future mainlines of macro-engineering R&D, as we currently think are probable, are briefly described in Sect. 5 of this chapter.

## References

Allan GL, Banens B, Fielder S (2001) Developing commercial inland saline aquaculture in Australia: part 2. Resource inventory and assessment. FRDC Project No. 98/335; NSW Fisheries Final Report Series, No. 31, pp 62–63

AP (Associated Press) (2009) First solar wins approval for massive China field. The Wall Street Journal CCLIV: B7 (news report dated September 9, 2009)

Axion (2007) http://members.axion.net/~enrique/tenmegawattprice.html. Accessed 10 Oct 2007

Badescu V (2006) Simple optimization procedure for silicon-based solar cell interconnection in a series-parallel PV module. Energy Convers Manage 47:1146–1158

Barrett TL, Farina A, Barrett GW (2009) Aesthetic landscapes: an emergent component in sustaining societies. Landscape Ecol 24:1029–1035

Beresford Q (2004) Salinity crisis: landscapes, communities and politics. University of Western Australia Press, Perth

Black JF, Tarmy BL (1963) The use of asphalt coatings to increase rainfall. J Appl Meteorol 2:557–564

Boia L (2005) The weather in the imagination. Reaktion Books, Chicago

Bolonkin A (2007) Extraction of freshwater and energy from atmosphere. http://arxiv.org/ftp/arxiv/papers/0704/0704.2571.pdf

Box JB, Duguid A, Read RE, Kimber RG, Knapton A, Davis J, Bowland AE (2008) Central Australian waterbodies: the importance of permanence in a desert landscape. J Arid Environ 72:1395–1413

Bruno JF, Sweatman H, Precht WF, Selig ER, Schutte VGW (2009) Assessing evidence of phase shifts from coral to macroalgal dominance on coral reefs. Ecology 90:1478–1484

Cai W, Cowan T (2008) Evidence of impacts from rising temperature on inflows to the Murray-Darling Basin. Geophys Res Lett 35:L07701

Campra P, Garcia M, Canton Y, Palacios-Orueta A (2008) Surface temperature cooling trends and negative radiative forcing due to land use change toward greenhouse farming in southeaster Spain. J Geophys Res 113:D18109

Chen K, McAneney J (2006) High-resolution estimates of Australia's coastal population. Geophys Res Lett 33:L16601

Dean SM, Flowerdew J, Lawrence B, Eckermann S (2007) Parameterization of orographic cloud dynamics in a GCM. Clim Dyn 28:581–597

Debelius B, Gomez-Parra A, Forja JM (2009) Oxygen solubility in evaporated seawater as a function of temperature and salinity. Hydrobiologia 632:157–165

DeVogel SB, Mageeb JW, Manleya WF, Millerc GH (2004) A GIS-based reconstruction of late Quaternary paleohydrology: Lake Eyre, air central Australia. Palaeogeogr Palaeoclimatol Paleoecol 204:1–13

Draper C, Mills G (2008) The atmospheric water balance over the semiarid Murray-Darling River Basin. J Hydrometeorol 9:521–534

Duarte CM, Prairie YT, Montes C, Cole JJ, Striegl R, Melack J, Downing JA (2008) $CO_2$ emissions from saline lakes: a global estimate of a surprisingly large flux. J Geophys Res 113:JG000637

Dupavillon JL, Gillanders BM (2009) Impacts of seawater desalination on the giant Australian cuttlefish *Sepia apama* in the upper Spencer Gulf, South Australia. Mar Environ Res 67:207–218

Evans AD, Bennett JM, Ewenz CM (2009) South Australian rainfall variability and climate extremes. Clim Dyn 33:477–493

Farm management (2007) http://www.cook.rutgers.edu/~farmmgmt/ne-budgets/methodology.html. Accessed 10 Oct 2007

Feather OR (1971) Method of increasing the likelihood of precipitation by the artificial introduction of sea water vapor into the atmosphere windward of an air lift region. USA Patent 3,601,312 awarded 24 August 1971

Fleming M (1995) Australian water resources are different. Austral Sci 16:8–10

Fu Q, Johanson CM, Wallace MJ, Reichler T (2006) Enhanced mid-latitude tropospheric warming in satellite measurements. Science 312:1179

Ghassemi F, White I (2006) Inter-basin water transfer: case studies from Australia, United States, Canada, China and India. Cambridge University Press, Cambridge

Gibbs LM (2006) Valuing water: variability and the Lake Eyre Basin, central Australia. Aust Geogr 37:73–85

Glenn EP, Brown JJ, O'Leary JW (1998) Irrigating crops with seawater. Sci Am 279:76–81

Hahn MW (2006) The microbial diversity of inland waters. Curr Opin Biotechnol 176:256–261

Halley C (2009) CAMPS: a guide to 21st century space. MIT Press, Cambridge

Hoang M, Bolto B, Haskard C, Barron O, Gray S, Leslie G (2009) Desalination in Australia. CSIRO: National Research Flagships, Water for a Healthy Country. 26 pp

Hogan E (2008) Spiral Jetta: a road trip through the land art of the American West. University of Chicago Press, Chicago

Holzbecher E (2005) Groundwater flow pattern in the vicinity of a salt lake. Hydrobiologia 552:233

Hope PK, Nicholls N, McGregor JL (2004) The rainfall response to permanent inland water in Australia. Austral Meteorol Mag 53:251–262

House KZ, House CH, Schrag DP, Aziz MJ (2007) Electrochemical acceleration of chemical weathering as an energetically feasible approach to mitigating anthropogenic climate change. Environ Sci Technol 41:8464–8470

Hutchings P, Kingford M, Hoegh-Guldberg O (eds) (2008) The Great Barrier Reef: biology, environment and management. CSIRO Publishing, Melbourne

Imaz M, Gay C, Friedmann R, Goldberg B (1998) Mexico joins the venture: joint implementation and greenhouse gas emissions reduction, Lawrence Berkeley National Laboratory, Environmental Energy Technologies Division. Work supported by the U.S. Environmental Protection Agency through the U.S. Department of Energy under Contract No. DE-AC03-76SF00098

Johanson CM, Fu Q (2009) Hadley cell widening: model simulations versus observations. J Clim 22:2713–2725

Kingsford R (2006) Ecology of desert rivers. Cambridge University Press, Cambridge

Kingsford RT, Porter JL (1993) Waterbirds of Lake Eyre, Australia. Biol Conserv 65:141–151

Kingsford R, Boulton AJ, Puckridge JM (1998) Challenges in managing dry land rivers crossing political boundaries: lessons from Cooper Creek and the Paroo River, Central Australia. Aquat Conserv Mar Freshw Syst 8:361

Kosarev AN, Kostianoy AG, Zonn IS (2009) Kara-Bogaz-Gol Bay: physical and chemical evolution. Aquat Geochem 15:223–236

Kurokawa K (2006) Energy from the desert: practical proposals for very large scale photovoltaic systems. Earthscan, London

Lach D, Rayner S, Ingram H (2005) Taming the waters: strategies to domesticate the wicked problems of water resource management. Int J Water 3:1–17

Lerman A (2009) Saline Lakes' response to global change. Aquat Geochem 15:1–5

Letolle R, Bendjoudi H (1997) Histoires D'une Mer au Sahara. L'Harmattan, Paris

Liu J, Namboothiry MAG, Carroll DL (2007a) Fiber-based architectures for organic photovoltaics. Appl Phys Lett 90:063501 (Also, derived from a broad survey of commercial solar power sites on the World Wide Webb during the period August–September 2007)
Liu J, Vecchi GA, Reichler T (2007b) Expansion of the Hadley cell under global warming. Geophys Res Lett 34:L06805
McMahon TA, Murphy R, Little P, Costelloe JF, Peel MC, Chiew FSH, Hayes S, Nathan R, Kandel DD (2005) Hydrology of Lake Eyre Basin. Department of Civil and Environmental Engineering, University of Melbourne and Sinclair Knight Merz, Malvern VIC, Australia
Millero F, Mirzaliyev A, Safarov J, Huang F, Chanson M, Shahverdiyev A, Hassel E (2008) The equation of state for Caspian Sea Waters. Aquat Geochem 14:289–299
Ogden TL (1969) The effect of rainfall on a large steelworks. J Appl Meteorol 8:585–591
Ozturk M, Waisel Y, Khan MA, Gork G (2006) Biosaline agriculture and salinity tolerance in plants, Birkhauser, Basel (Switzerland)
PelletHeat (2007) http://www.pelletheat.org/3/residential/compareFuel.cfm. Accessed 10 Oct 2007
Pigram J (2006) Australia's water resources: from use to management. CSIRO Publishing, Collingwood
Prata AJ (1990) Satellite-derived evaporation from Lake Eyre, South Australia. Int J Remote Sens 11:2051–2068
Rofe MW, Oakley S (2006) Constructing the port: external perceptions and interventions in the making of place in Port Adelaide, South Australia. Geogr Res 44:272–284
Roshier DA, Robertson AI, Kingsford RT, Green DG (2001) Continental-scale interactions with temporary resources may explain the paradox of large populations of desert waterbirds in Australia. Landsc Ecol 16:547–556
SaudiAramCo (2007) http://www.saudiaramcoworld.com/issue/199406/samphire-from.sea.to.shining.seed.htm. Accessed 10 October 2007
Schaller RD, Sykora M, Pietryga JM, Klimov VI (2006) Seven excitons at a cost of one: redefining the limits of conversion efficiency of photons into charge carriers. Nano Lett 6:424
Schwarz EHL (1923) The Kalahari or Thirstland Redemption. Oxford University Press, Oxford
Shao Y, Leys JF, McTainsh GH, Tews K (2007) Numerical simulation of the October 2002 dust event in Australia. J Geophys Res 112:D08207
Sohn E (2007) The big dry. Science News 172:266
Solar Irradiation Database (2007) http://energy.caeds.eng.uml.edu/fpdb/Irrdata.asp. Accessed 10 Oct 2007
Stafford Smith DM, Morton SR (1990) A framework for the ecology of arid Australia. J Arid Environ 18:255–278
Taylor J (2003) Population futures in the Australian desert. Aust Geogr 34:355–370
Timms BV (2009) A study of the salt lakes and salt springs of Eyre Peninsula, South Australia. Hydrobiologia 626:41–51
Walker KF, Puckridge JT, Blanch SJ (1997) Irrigation development on Cooper Creek, Central Australia: prospects for a regulated economy in a boom-and-bust ecology. Aquat Conserv Mar Freshw Syst 7:63
Webb NP, McGowan HA, Phinn SR, McTainsh GH, Leys JF (2009) Simulation of the spatiotemporal aspects of land erodability in the northeast Lake Eyre Basin, Australia, 1980–2006. J Geophys Res 114:F01013
Williams WD (2001) Anthropogenic salinization of inland waters. Hydrobiologia 466:329

# Can Geoengineering Sustain Critical Ocean Currents?

**Peter C. Flynn and Jason Songjian Zhou**

## 1 Introduction

The oceans play a major role in climate through many mechanisms. Two critical roles are the storage of carbon, with transfer of carbon dioxide from atmosphere to ocean, and the transport of heat through ocean currents. Any consideration of a response to climate change through global warming from the buildup of greenhouse gases (GHGs) must examine both of these roles. The structure of the ocean and the mechanisms of key ocean currents are critical to both processes.

## 2 The Structure of the Ocean

The ocean can be thought of as two independent components, the shallow and deep ocean, that have very limited circulation between them. The barrier to mixing between the two is the pycnocline, a sharp change in density associated with a change in temperature: the deep ocean is colder and denser that most shallow ocean water. Figure 1 shows a typical profile for both temperature and density.

The sharp increase in density means that transport of a bulk amount of water vertically between the shallow and the deep ocean requires a high degree of energy; in practice, such currents are limited. In two places on earth surface water is sufficiently dense to flow into the deep ocean. The first is the Weddell Sea off the coast of Antarctica, and the second is the Greenland Iceland Norway (GIN) area of the North Atlantic Ocean. In the Weddell Sea, the driving force is low temperature:

P. C. Flynn (✉)
University of Alberta, Edmonton, Canada
e-mail: peter.flynn@ualberta.ca

J. S. Zhou
Weatherford, Edmonton, Canada

V. Badescu and R. B. Cathcart (eds.), *Macro-engineering Seawater in Unique Environments*, Environmental Science and Engineering,
DOI: 10.1007/978-3-642-14779-1_26, © Springer-Verlag Berlin Heidelberg 2011

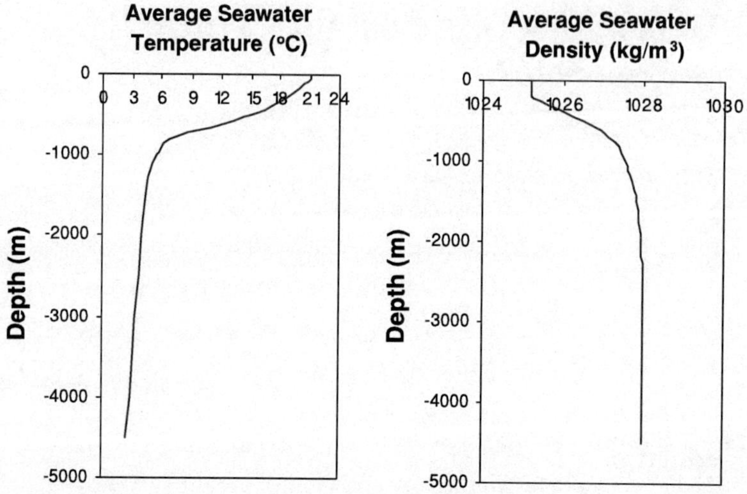


**Fig. 1** Typical low latitude seawater temperature and density profiles (Source: Ingmanson and Wallace 1979)

surface water gets cooled to a point where its density causes it to sink in large volumes. This downwelling current, called the Antarctic Deep Water (AADW) averages about 50 Sverdrup (a Sverdrup is $10^6$ cubic meters per second). In the GIN area, the driving force is a combination of higher salinity and lower temperature: a current coming from the Mediterranean Sea to the western Atlantic Ocean is high in salinity due to the high net evaporation rate in the Mediterranean. This water flows north as the Gulf Stream, cooling as it goes north, until an area of high salinity cold water reaches a density greater than the deep ocean. Flow from this downwelling current, called the North Atlantic Deep Water (NADW) averages about 15 Sverdrup. These two currents are the primary circulation mechanism from the shallow to the deep ocean. The North Pacific does not have a downwelling current corresponding to NADW because relative to the North Atlantic it is low in salinity.

The two downwelling currents flow along the bottom of the ocean, and are thought to reach the Indian and Pacific Oceans. Flow into this area is offset by rising warm water that flows along the surface oceans back into the regions at which the downwelling currents originate (Broecker 1991). The combined flows are known as thermohaline circulation, since both heat and salinity are redistributed by the paired currents. The Gulf Stream, a warm current flowing up the east coast of North America before crossing the Atlantic towards Northern Europe, is a surface current that offsets the NADW. The name "ocean conveyor belt" is given in general to the circulating pairs of currents, and often specifically to the coupling of NADW and the Gulf Stream.

Given the volume of the deep ocean, about $1.2 \times 10^{18}$ m$^3$, the flow into the ocean would have an average residence time of about 600 years. In practice, some areas of the ocean may be isolated from the circulation pattern, somewhat reducing the residence time for downwelling water.

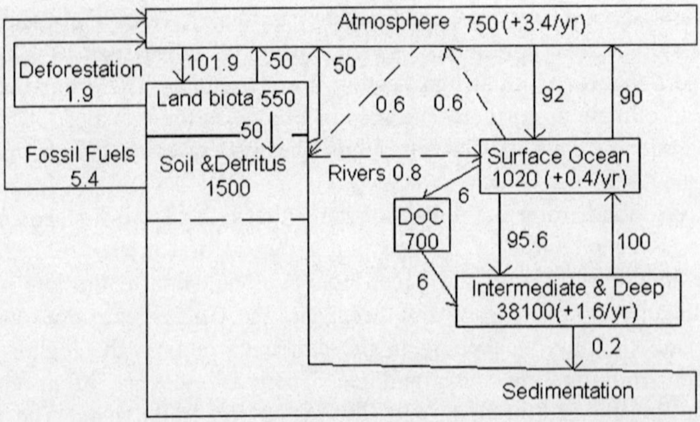


**Fig. 2** Estimated carbon amounts (GtC) and fluxes (GtC/year) in the 1980s. DOC is dissolved organic carbon, including detritus. Data from Siegenthaler and Sarmiento (1993)

## 3 The Ocean as a Storehouse of Carbon

The low temperature and large volume of the deep ocean mean that it can hold a great deal of carbon, in the form of dissolved $CO_2$, bicarbonate, carbonate, and dissolved organic carbon (DOC). Carbon flows into the deep ocean from two key mechanisms, the flow of biological material, typically dead plants and animals, and the carbon carried in downwelling currents. The former is called the biological pump, the latter the solubility pump. Figure 2 shows an estimated amount and flux of carbon for the earth in the 1980s.

Two observations can be made:

- Of the carbon at the surface of the earth, the amount held in the deep ocean is more than an order of magnitude larger than all other reservoirs combined.
- The annual increase in carbon in the atmosphere, about 3 GtC, if transferred to the ocean would result in an increase of less than 0.1% of the concentration.

The deep ocean has a significantly higher concentration of total carbon than shallow oceans, approximately 2,200–2,400 μmol/kg at depth versus an average of 1,900–2,100 μmol/kg at the surface (Takahashi 1981), depending on surface temperature and seawater chemistry. Despite this, the deep ocean is not saturated in carbon. Given the long residence time of the deep ocean, it is a suitable candidate for the storage of anthropogenic carbon.

## 4 The Role of the Ocean Conveyor Belt in European Climate

Europe maintains large centers of population at higher latitudes than any other continent. The city of Edmonton in Canada is the northernmost urban center in North America with a metropolitan population of one million, and lies at 54° north latitude. Belfast, Glasgow, the Scandinavian capitols, St. Petersburg, and Moscow are all located further north. East Asian "northern" cities such as Vladivostok, Beijing and Sapporo are all at significantly lower latitudes. The prime reason cited for the mild climate at northern latitudes in Europe is the transport of heat to the North Atlantic by the Gulf Stream: air is warmed by contact with this surface current.

It is a profound irony of global warming that one of the first potential consequences is the plunging of Europe into a period of extreme cold. The cause would be an interruption of the ocean conveyor belt due to the interruption of downwelling NADW; if the NADW weakens, the Gulf Stream does as well. In turn, the cause of this weakening in the paired currents is the higher influx of fresh water from melting snow and ice. There is evidence of a catastrophic cooling in Europe 12,800 years ago that lasted for 1,300 years. The period is known as the Younger Dryas (Segar 1997); the name arises from an alpine/tundra flower that reestablished in much of northern Europe. The most accepted cause of the rapid cooling is an interruption in thermohaline circulation in the North Atlantic due to a cessation of NADW. The proposed cause was the flow into the North Atlantic of Lake Agassiz, a large body of fresh water in north central North America that formed from melting glaciers, presumably when an ice dam released. Flow was large enough to cause a noticeable increase in the level of the ocean, and this event may be the source of flood myths in many cultures. The flow of a large volume of fresh water into the North Atlantic interfered with downwelling currents because a surface layer of low density fresh water formed, preventing the formation of high salinity cold water that was sufficiently dense to penetrate the pycnocline. The resumption of NADW required the dissipation of the layer of fresh water.

There are concerns today about the weakening of NADW due to global warming (Segar 1997). The ultimate cause is the same as occurred at the onset of the Younger Dryas: increased fresh water influx due to melting ice, although the rate is more gradual today. Continued reduction of the Arctic ice cap and melting of Greenland glaciers are seen as threats to NADW, and hence to the Gulf Stream, with a corresponding threat of a period of rapid cooling for Northern Europe.

## 5 Geoengineering and Ocean Currents

From the above, two opportunities arise for studying geoengineering related to climate change. The first is whether human intervention can add additional carbon dioxide to downwelling currents, in essence enhancing the existing solubility pump that carries carbon from the atmosphere into surface waters and then into the deep ocean. The second is whether human intervention could enhance downwelling currents in the North Atlantic in the event that a weakening in NADW is observed.

**Table 1** Saturation of high latitude seawater

| Location | $pCO_2$ | Sat. TC (μmol/kg)[a] | % Saturation (Carbon) | $\frac{\partial TCO_2}{\partial T}$ (μmol/kg°C) |
|---|---|---|---|---|
| 61°N, 27°W–64°N, 36°W | 337 | 2,124 | 99.48 | −8.2 |
| 61°N, 19°W–65°N, 34°W | 294 | 2,102 | 98.81 | −8.6 |
| 62°N, 18°E–78°N, 11°W | 333 | 2,162 | 99.90 | −8.0 |
| 70°S, 17°W–72°S, 20°W | 353 | 2,156 | 99.95 | −7.6 |

[a] Saturation TC was calculated using Lewis' program with Roy's constants and total pH values (Lewis and Wallace 1998)

Adding additional carbon dioxide to downwelling water requires that it not be saturated at the point at which it sinks. Two theoretical analyses suggested that as the Gulf Stream moves north and cools in exchange with the atmosphere it would not have a chance to reach equilibrium saturation of carbon, due to the slow rate of absorption from atmosphere to surface water. Broecker and Peng (1982) estimated that it would take on the order of 1 year for ocean water to equilibrate with the atmosphere, and Volk and Hoffert (1985) did an elegant analysis of the efficiency of the solubility pump and predicted that uptake of carbon was well below equilibrium.

As elegant as this theoretical analysis is, it is not supported by actual measurements of surface sea water in the North Atlantic. Zhou and Flynn (2005) analyzed data from four series of samples of sea water at northern latitudes (Talley et al. 1997; Brewer and Takahashi 1981; Clarke et al. 1984; Heywood and King 1995). The results are shown in Table 1. No samples of NADW have been analyzed for carbon content, but based on adjacent surface water the potential to add incremental carbon to NADW by enhancing the exchange of the atmosphere to the sea water is negligibly small.

The second geoengineering opportunity is technically feasible: since thermohaline circulation is created by the downwelling of cold salty water, it is technically possible to increase downwelling currents by transferring heat from surface waters to the atmosphere. A number of industrial methods are in use to transfer heat, with or without the transfer of mass, from fluids to the atmosphere: condensing of steam in power plants and cooling of liquid streams during petroleum refining are examples. Using a basis of cooling one Sverdrup of sea water from 6 to 0°C during winter months, seven methods of cooling sea water were compared. The assumption of one Sverdrup is arbitrary, since the extent of weakening of downwelling NADW currents, if any, is unknown. The basis represents a large scale intervention achieving full economy of scale that could be replicated if the creation of more downwelling current was desired.

The average ambient temperature of air over the northern ocean in winter was assumed to be −10°C (Jones and Otaegui 1997; Martin and Munoz 1997), and −20°C over land (NCDC 2002), with an average wind speed of 10 m/s (MOST 2002). Land based power requirements are met by a dedicated nuclear power plant [installed cost $1,500/kW, average operating cost $0.015/kWh (EIA 2002)]. Barge based power requirements are met by wind generators [installed cost $1,000/kW (Cadogan et al. 1996) and an operating cost of $0.01/kWh (DWIA 2002; NREL 1996)]; the installed cost for wind includes 4 h of battery backup storage. Equipment would be shut off in the event of longer periods of no or low wind. A 30-year life was assumed for all equipment, and capital cost was annualized by setting the pre-tax return on total investment at 10%.

## 6 Seven Methods of Cooling Sea Water

Seven schemes for cooling large volumes of sea water were identified and subjected to preliminary screening (Zhou and Flynn 2005). The methods are:

1. Forced draft heat exchangers. Units identical to those in use in power plants and refineries would be located onshore at Greenland. The total power requirement is 250 GW, based on an elevation gain of 8 m to allow the water to flow down the cooling tower and the requirement to operate air fans. $10^5$ cooling tower cells would be required, at a cost of $5.2 M per cell (all costs are in 2005 US$).
2. Cooling ponds. A process comparable to that in use in many power plants was envisioned, a large area with passive heat transfer from water to air, but that would be filled and drained daily. An area of 1–3.6 × $10^5$ $km^2$ was calculated to be required to provide sufficient cooling. This alternative was dropped from further evaluation based on the massive land requirement relative to available low lying areas in either Greenland or Iceland.
3. Injection of cold air. Air would be slightly compressed by a blower to 120 kPa, cooled by a finned exchanger to remove the heat of compression, then injected into the ocean at a depth of 1.5 m below the surface. Given the low density of air relative to water, a massive amount of air compression is required, with a total power requirement of 21 TW.
4. Circulation of warmer layers of the ocean to the surface. The existence of a warmer layer of water below a surface layer of cooler water is sometimes observed in the ocean, and heat transfer could be enhanced by a very low head ship mounted pump that moved warmer layers to the surface. This alternative was dropped from further evaluation because of the limited extent of such inversions containing warm layers of water below colder surface waters.
5. A barge based finned heat exchanger. Sea water would be pumped up through a finned heat exchanger mounted on a barge, then returned to the ocean. Because there is no net change in head, the requirement for pumping power is low; the total power required is 15 GW. 4 × $10^4$ barges would be required at a cost of $3 M per unit.
6. Barge mounted heat pumps that use a circulating coolant to transfer heat from the ocean to the atmosphere. Coolant could circulate by gravity (condensation of evaporated coolant at an elevated heat exchanger and return to a coil

located under the barge). $4 \times 10^4$ barges would be required at a cost of \$3 M per unit.

7. Formation of thicker sea ice. Barges with low lift screw pumps would be locked in areas of annual sea ice. Initially low volume high lift pumps would spray sea water in the air prior to the formation of sea ice, to advance its formation, after which low lift pumps would put ocean water onto the surface of sea ice. In winter cold ambient temperatures would thicken the ice; in spring and early summer the ice would be melted by contact with sea water, cooling 12 times the mass of water by 6°C per unit mass of melted ice because of the high latent heat of crystallization of ice. $8.1 \times 10^3$ barges would be required at a cost of \$3 M each.

Capital and operating costs were developed for each of the five alternatives that remained after the preliminary feasibility screening (for details, see Zhou and Flynn 2005). Air injection was by far the most costly alternative, with a calculated cost of \$4 $\times 10^{-1}/m^3$ of incremental current; the high cost reflects the high energy required for compression of air. Incremental current from forced draft heat exchange cost \$8 $\times 10^{-3}/m^3$, due in part to high power requirements to lift the sea water to the top of the cooling tower and move air through the cells. Incremental current from barges using finned heat exchange or a refrigeration cycle had a cost of \$1 $\times 10^{-3}/m^3$, with the lower cost reflecting in part that there is no net head gain or loss for the seawater in either case. The lowest cost case was the formation of thicker sea ice, with a cost of \$4 $\times 10^{-4}/m^3$ of incremental current. The low cost reflects the fact that only 8% of the total volume of cooled seawater is pumped, and the pumping is low head, about 1 m, enough to allow it to flow on the surface of the ice.

Based on this analysis a more detailed scope was developed for thickening of sea ice. Equipment, operation and maintenance cost estimates were based on a remote Arctic location. Barges were assumed to be unmanned but remotely monitored. Necessary maintenance would be by helicopter, with barges equipped with emergency accommodations in the event of the rapid onset of severe weather.

Relative to finned heat exchange or refrigeration barges, one other benefit, although small, is that water melting sea ice in the spring and summer would have the ability to absorb some carbon from the atmosphere. As noted above, the effect is small, but in the case of the finned heat exchange or refrigeration barges there is no contact between sea water and the atmosphere during cooling.

## 7 Formation of Thicker Sea Ice as a Means of Creating Additional Downwelling Current

Naturally forming ice over a body of water forms at the bottom of the ice sheet. There are two consequences of this: over time the thickening ice becomes its own insulation, slowing the rate of formation of ice. Also, for sea water the forming ice rejects the brine component as water molecules are preferentially formed into ice along the bottom of the ice sheet. As a result, naturally formed sea ice is low in salt.

A longstanding practice in northern environments is the thickening of ice by pumping water onto the surface of existing ice: ice is quickly formed since the insulation of the ice sheet itself is no longer a barrier to heat transfer. Applications have included the formation of roads for winter access in northern communities (including access to the City of Leningrad during the long siege in World War II), and the formation of ice islands to support drilling in the shallow Beaufort Sea. This same concept can be used to enhance the thickness of areas of existing annual sea ice. Barge spacing would be optimized by comparing the availability of low power low lift pumps with the rate of heat transfer on the surface of the ice.

A second key issue for future resolution is whether the salt in ice formed on the surface of a sea ice sheet remains in the ice. If concentrated brine finds its way to the ocean via micro-channels in the ice, then what would remain is a low salinity ice block, which when melting would not create a body of cooled high salinity water that could enhance downwelling current.

Barges for the formation of sea ice would have to be designed to withstand the ice pressure of being locked in annual sea ice. They would be unmanned and serviced as required during the winter season by helicopter, from bases in Greenland and Iceland. Power would come from a barge mounted wind turbine, with battery backup. Design would have to include an emergency power supply for heat tracing and instrumentation support in the event of long periods of low or no wind, although wind over the ocean is more reliable than over land. Barges would be towed into ports in Greenland or Iceland in mid-summer for maintenance and towed back out in early fall.

## 8 Is Sea Ice Formation Economic? Should it Proceed?

Creating one Sverdrup of incremental NADP, about 7% of the existing flow, would have a price tag of \$45 G and an annual operating cost of \$1.3 G. Tables 2 and 3 summarize capital and operating costs (Zhou and Flynn 2005).

These numbers are daunting on first review, but become more tolerable when the population affected by a cooling of Europe is considered. The extent of impact of a mini ice age in Europe is unknown, but conservatively, 100 million lives would become untenable in a Europe plunged into a period of extreme cooling, and the number may be as high as 500 million. The capital cost would thus range from \$100 to \$500 per affected person; given the GDP of Europe, the cost is certainly within the range of practical expenditures.

Any preliminary conceptual study has a high degree of uncertainty, and for this kind of study a range of at least ±50% would typically be expected. However,

**Table 2** Capital cost of sea ice formation

| Category | Element | No. of units | Unit cost w/o learning factor ($000) | Learning factor | Cost ($M) |
|---|---|---|---|---|---|
| Direct costs | | | | | |
| Barge | Material | 8,100 | $1,900 | 0.8 | $28,000 |
| | Labor | 8,100 | $1,100 | | |
| | Overhead and profit | 8,100 | $1,300 | | |
| Wind power | Installed wind power system | 8,100 | $450 | 0.8 | $3,500 |
| | Battery bank | 8,100 | $88 | | |
| Pumping | Low volume pump | 8,100 | $30 | 0.8 | $3,500 |
| | High volume pump | 16,200 | $200 | | |
| | Heat tracing and insulation | 8,100 | $120 | | |
| Transportation | Helicopter | 32 | $1000 | 1 | $32 |
| Land based construction | Control center and instrumentation | 1 | $70,000 | 1 | $990 |
| | Harbor and harbor machinery | 4 | $100,000 | | |
| | Air base | 4 | $130,000 | | |
| Indirect costs | | | | | |
| Engineering and supervision | 5% of direct costs | 1 | $1,700 | 1 | $1,700 |
| Contingency | 20% of direct costs | 1 | $6,900 | 1 | $6,900 |
| *Total* | | | | | $45,000 |

none of the components of sea ice thickening represent new technology: artificial ice formation in marine environments has been practiced in the Beaufort Sea to support exploratory drilling for oil and gas. Hence the risk of a huge error in estimate arising from an unknown technology is relatively low.

With any geoengineering scheme there is a risk of unknown impacts on the complex interactions of the earth's ecosystem. Relative to other geoengineering systems, the risk for thickening sea ice is relatively low, since ice is being thickened in areas that it already forms on an annual basis. The impact of incremental ice on both Northern European and world climate could be monitored after the start of ice formation. Because the barges would be dispatched on an annual basis, the project could be cancelled at any time if unexpected adverse consequences were observed.

Sea ice formation remains a possible response to the weakening of the NADW. Given the magnitude of the cost, it should not be implemented unless more convincing evidence of the weakening of the current is observed. The critical piece of additional research to confirm the validity of the proposal is determining whether salt remains in sea ice that forms from sea water pumped onto the surface of an ice sheet.

**Table 3** Operating and maintenance cost

| Category | Element | No. of units | Estimate basis | Unit cost ($k/year) | Total cost ($M/year) |
|---|---|---|---|---|---|
| Barge | Tug rent | 8,100 | $300/h; Deployment and retrieval: 1,440 h 1 tugboat per 20 barges | $22 | $790 |
| | Maintenance | 8,100 | 1% of direct costs | $34 | |
| | Insurance | 8,100 | 1% of total costs (barge, wind power and pumping system) | $42 | |
| Wind power | Wind turbine and battery O&M | 8,100 | $0.015/kWh | $15 | $190 |
| | Battery replacement | 8,100 | 10 years lifespan | $9 | |
| Pumping | Pump O&M | 8,100 | 3% of capital costs | $13 | $300 |
| | Heat tracing O&M | 8,100 | 20% of capital costs | $24 | |
| Transportation | Helicopter maintenance | 32 | 15% of capital costs | $150 | $26 |
| | Fuel consumption | 32 | $0.2/l 230 l/flight hour | $70 | |
| | Labor and overhead* | 32 | | $580 | |
| Land based construct-ion | Control center | 1 | Building and equipment maintenance: 5% of capital costs Man-hours*: 5,400/year | $3,000 | $27 |
| | Harbor | 4 | Docks and machinery maintenance: 5% of capital costs Man-hours*: 151,200/(year·harbor) | $3,300 | |
| | Air base | 4 | Runway, control tower, and machinery maintenance: 5% of capital costs Man-hours*: 3,240/(year·base) | $2,700 | |
| *Total* | | | | | $1,300 |

## References


Brewer PG, Takahashi T (1981) Carbon-related and hydrographic data from transient tracers in the ocean (TTO) Leg 5 (WOCE NDP004/R1, 1981). Carbon Dioxide Information Analysis Center (CDIAC), Oak Ridge National Laboratory, US Department of Energy, Oak Ridge, Tennessee

Broecker WS (1991) The great ocean conveyor. Oceanography 4(2):79–89

Broecker WS, Peng T-H (1982) Tracers in the sea. Lamont-Doherty Geological Observatory, Columbia University, Palisades, pp 149–158

Cadogan J, Parsons B et al (1996) Characterization of wind technology progress. Annual Conference and Exhibition of the American Wind Energy Association, Denver, Colorado (June 23–27, 1996). Power 96, pp 163–172

Clarke RA, Reid JL, Swift JH (1984) CSS Hudson Cruise 82-001 data report, a joint venture of Bedford Institute of Oceanography and Scripps Institution of Oceanography. vol 1 (Physical and Chemical Data)

DWIA (Danish Wind Industry Association) (2002) Guided tour on wind energy. http://www.windpower.org/tour/index.htm

EIA (Energy Information Administration) (2002) International energy outlook 2002. US Department of Energy. http://www.eia.doe.gov/oiaf/ieo/tbl_18.html

Heywood KJ, King BA (1995) Carbon-related and hydrographic data from the WOCE hydrographic program cruises section A23 (March 20–May 6, 1995). Carbon Dioxide Information Analysis Center (CDIAC), Oak Ridge National Laboratory, US Department of Energy, Oak Ridge, Tennessee

Ingmanson DE, Wallace WJ (1979) Oceanography—an introduction. Wadsworth Publishing Company Inc., Belmont, pp 79–119

Jones ISF, Otaegui D (1997) Photosynthetic greenhouse gas mitigation by ocean nourishment. Energy conversion and management, 38 supplement:s367–s372

Lewis E, Wallace D (1998) Program developed for $CO_2$ system calculations, ORNL/CDIAC-105, Carbon Dioxide Information Analysis Center (CDIAC), Oak Ridge National Laboratory, US Department of Energy, Oak Ridge, Tennessee

Martin S, Munoz EA (1997) Properties of the Arctic 2-meter air temperature field for 1979 to the present derived from a new gridded dataset. J Clim 10:1428–1440

NCDC (National Climatic Data Center) (2002) Online surface data. US National Oceanic and Atmospheric Administration. http://lwf.ncdc.noaa.gov/oa/climate/climatedata.html

NREL (National Renewable Energy Laboratory) (1996) Wind energy information. US Department of Energy. Report 1996. No. DOE-GO-10095-83, pp 11–15

Segar DA (1997) Introduction to ocean science. Wadsworth Publishing Company, Belmont, p 235

Siegenthaler U, Sarmiento JL (1993) Atmospheric carbon dioxide and the ocean. Nature 365:119–125

Takahashi T (1981) Supplement to the alkalinity and total carbon dioxide concentration in the world oceans. In: Bolin B (ed) Carbon cycle modelling, Wiley, pp 159–199

Talley L, Wallace D, Millero F, Takahashi T (1997) Carbon-related and hydrographic data from the WOCE hydrographic program cruises section A24 (May 30–July 5). Carbon Dioxide Information Analysis Center (CDIAC), Oak Ridge National Laboratory, US Department of Energy, Oak Ridge, Tennessee

MOST (Marine Observing System Team) (2002) Ocean surface winds. NOAA. http://orbit212.wwb.noaa.gov/doc/oceanwinds1.html

Volk T, Hoffert MI (1985) Ocean carbon pumps: analysis of relative strengths and efficiencies in ocean-driven atmospheric $CO_2$ changes. Geophys Monogr 32:99–110

Zhou S, Flynn PC (2005) Geoengineering downwelling ocean currents: a cost assessment. Clim Chang 71:203–220

# Calling upon Neptune: Ocean Energies as "Renewables"

**Roger H. Charlier**

*Progress is nothing but the accomplishment of utopias*
Oscar Wilde (1854-1900)

## 1 Introduction

In the broad area of ocean engineering, there is one domain wherein technology poses, in at least some of its fields, little or no problem and, even is mostly environment-friendly. And yet, notwithstanding proven technology, implementation is hesitant. The field is harnessing the huge resources of ocean energies. Several studies have been made, and some new ones are in process, but except for one instance, in Korea, no large project is actually in gestation or at an advanced stage; there is, on the other hand, no shortage, of test runs or pilot trials. Yet, no doubt Neptune, or his eponym Poseidon, is ready to make his contribution to efforts to counter climate change and yet provide the energy that is needed.

Ocean sources of energy can and should be put to work; they are non-polluting, and minimally environment impacting. Their extraction is impeded less by technology than by the onerous capital intensiveness (Hilsop 1992; Kristoferson and Bokalders 1991). Some have been tapped, with unequal success though, such as tides, waves, marine winds, others remain currently more engineers' dreams like marine currents, salinity differential, conversion of biological products, marine geothermal energy use and some others. OTEC, ocean thermal energy conversion, is being used in some modest way and in very specific sites. Though proven technically possible to be put to work, economically it remains unattractive.

According to the IEA-OES, the International Energy Agency-Implementing Agreement on Ocean Energy Systems, by the end of 2008, more than 25 countries were involved in ocean renewable energy technology development activities. These include the deployment of multi-unit wave technology in Portugal, and the start of construction of the 260 MW tidal power plant in the Republic of (South)

---

R. H. Charlier (✉)
Vrije Universiteit Brussel, Brussel Belgium
e-mail: rocharli@vub.ac.be

V. Badescu and R. B. Cathcart (eds.), *Macro-engineering Seawater in Unique Environments*, Environmental Science and Engineering,
DOI: 10.1007/978-3-642-14779-1_27, © Springer-Verlag Berlin Heidelberg 2011

Korea. But, lack of targeted national priorities and policies remains a major barrier. Yet, as examples of the stakes at play the accessible resource in United Kingdom waters could be as much as 700 TWh/year and the North American marine energy resource could supply some 10% of US electricity demand.

## 2 Marine Winds

Offshore marine wind turbines (OWEC, ocean wind energy conversion) dot today many landscapes, from the Utgrunden (Scandinavia) marine wind farm to the very modest Zeebrugge (Belgium) structure, and more developments are planned. Of all the ocean energies, marine winds have known the most important development during the last decades. They are a "renewable" which was easy to harness and which required only relatively modest capital investments. Sites are numerous, and a judicious choice permits to dampen the objections voiced because of the noise they cause. Marine wind "farms" have been implanted in numerous locations particularly in Northern and Western Europe. However, environment-related objections to existing and planned farms are being raised, spurring engineers to devise new approaches. A claim is made that by 2013 the largest offshore wind farm in the world will have been established off the Belgian shore (De Doncker and Loos 2009).

There is a trend towards taller towers and larger turbine capacity. Over the past 20 years, the wind energy turbine tower has grown from 40 to 90 m with turbine sizes reaching 2.5 MW. The 4.0 and 8.0 MW turbine is on the horizon along with towers exceeding 100 m in height. It is cheaper to construct fewer wind towers with larger capacity because this reduces total infrastructure cost. Towers are today tubular steel structures.

Offshore, where the highest efficiency curves are obtained, the minimum turbine sizes reach 5.0 MW and increasing towards 7.0 MW. In its web site, the American Wind Energy Association (AWEA) equates the 2009-added 8,500 MW of wind power generation to 8.5 GW or about nine medium-sized nuclear power plants.

Wind tapping for electricity generation is still a "young" industry and surrounded by unresolved problems; twelve issues that confront currently structural engineering in tower design are turbine over-speed, E-stop, frequency-rotational stiffness, fatigue failure of welds, blade failure, turbine eccentricity loads, rotor imbalance, residual stresses induced by internal welds, foundation cracking from fatigue, foundation softening due to poor drainage, corrosion of foundation bolts from aggressive soils, seismic and/or earthquake loads; some of these concern only on-shore structures. On the other hand for offshore projects, issues include additional technical complexities such as hydrodynamic wave loading on the tower foundation, frequency response of tower-foundation with complexity of the wind and water wave action on the structure, fatigue and corrosion from the water-wave action, submarine access to the foundation for monitoring and maintenance, geotechnical performance of the sub-sea floor foundation behaviour (Khatri 2010).

## 2.1 Offshore Winds Use on US East Coast and Great Lakes

Not with standing forceful opposition from coastal dwellers, such as on Cape Cod in Massachusetts, but equally loud support in Rhode Island and other parts of Massachusetts, offshore winds tapping seems to be on its way. The marine winds development is also hampered by capital intensive costs. Why is offshore wind so much more expensive? And, more importantly, will it stay high?

Offshore facilities require more complex design to bear up under storms, waves and tides. Turbines represent the greatest expense for land-based wind but foundations, towers, transmission and installation tend to account for offshore project's larger costs various studies show. Among these, foundations for offshore facilities require more steel and concrete. Additionally water depth is a large cost factor with each meter of tower height representing €1200 (about US-$2000).

The marine winds panorama (2010) on the US East coasts can be summarized as follows:

- Maine targets to develop 300 MW of offshore wind by 2020.
- Massachusetts targets to develop 2,000 MW, most of it offshore, by 2020.
- Rhode Island has formed a partnership to assist in developing 28.8 MW offshore.
- In New York, a collaboration of industry and government agencies seeks a project developer to build up to 700 MW of wind power off the coast of Long Island. The New York Power Authority issued a solicitation for projects totaling 120–500 MW to be built in the Great Lakes. Across the border Ontario has set up a distinct set of rules for offshore projects and offers them priority connection rights to the transmission grid.
- New Jersey struck a deal to build a 350 MW offshore wind farm. The goal is to attain production of 1,000 MW by 2012 and 3,000 MW by 2020.
- In Delaware, Bluewater Wind won the right to build an offshore wind farm. A 450 MW project is planned. The state joined Maryland and Virginia to develop offshore wind power.
- Duke Energy and the University of North Carolina plan testing of demonstrator turbines at 11–16 km from the Outer Banks.
- According to the "Trillium Report", Great Lakes wind farms would be an economic bonanza for Canada: 20,790 MW of offshore wind projects costing Can$83.2 billion in capital investment would generate C$253.5 billion in gross economic activity, and create 66,362 jobs.

## 2.2 Wave Energy Conversion

J. B. Multon and H. benAhmed of the Brittany SATIE-CNRS ENS cachan-site proposed to quantify wave energy according to the sea-state (Table 1); with height

**Table 1** Energy from waves depending on sea-state

| Sea-state | $H$ (m) | $T$ (s) | $P$ (kW m$^{-1}$) |
|---|---|---|---|
| Agitated | 0.6 | 5 | 0.9 |
| Choppy | 9 | 11 | 437 |
| Heavy | 18 | 15 | 2380 |

$H$ in m, $T$ period in seconds and $P_w$ in watts, they proposed to express it mathematically by Eq. 1:

$$P_w = \frac{1}{8}\rho g H^2 \cdot \frac{gT}{4\pi} \cdot \frac{1}{2} \tag{1}$$

Thus, the incoming (incident) power equals the surface energy density times the propagation velocity and an adaptation coefficient.

Waves are a concentrated form of wind energy, even of solar energy. Their very nature requires a large number of small devices for its energy extraction. Waves make more energy available as energy is extracted, due to the inefficiency at which energy is transferred from the wind to the sea at highly developed sea states. The potential along the US West Coast could match all renewable energy systems presently deployed. The Mavericks Surf Contest at Half Moon Bay, CA, in February 2010 illustrated the sea's power when waves breaking on shore swept dozens of spectators from a man-made jetty. Years ago a 25 ton block of concrete was brought inland several kilometres from the southeast English coast by waves from a particularly heavy storm. The Electric Power Research Institute (EPRI) suggests that generation of electricity from wave energy may be economically feasible in the near future (*TreeHugger*; Feb. 7, 2007. A BBC program reported an Australian scheme in regular production and a Norwegian one functioned for 6 years).

Energy companies and coastal cities like New York and San Francisco hope to tap ocean waves, and tides as well. Ocean waves would likely minimally impact the environment. Wave action has been said to possibly power as much as 10% of the world's electrical demand. Pelamis$^{TM}$, a wave energy converter was tried out off the Portuguese mainland coast. (The device has received so much notoriety from the installation offshore Portugal, that the name of project Pelamis has been for some time commonly used to designate the converter itself and has become a trade mark of the manufacturer). Portuguese have been trying to establish a wave conversion scheme for many years in the Azores (Cf. Proceedings of the European wave and tidal power conferences: yearly 2004 through 2009. Indexes available on Google).

This device, anchored down on one end, floats along the surface of the water. The rolling motion of the waves is resisted by hydraulics which pump oil through a hydraulic motor that converts the wave motion to electricity by powering electrical generators. The 750 kW, "snake-like" Pelamis, tested at the European Marine Energy Centre (EMEC) in Orkney—which is to get 'biggest' wave farm—makes up in Portuguese waters, near Póvoa do Varzim, the Aguçadoura wave farm *Agucadoura Wave Park*. The world's first commercial wave farm, of three Pelamis

P-750 machines generating 2.25 MW ran, however, into technical difficulties and plans to bring installed capacity to 21 MW with a further 25 Pelamis machines seem stalled. The wave farm in Scotland, the largest in the world (capacity of 3 MW) will use four Pelamis devices (Figs. 1, 2).

Another system, OPT, (Ocean Power Technologies) has buoys anchored to the seabed and floating beneath the surface; they capture and convert wave energy into a controlled mechanical force that drives a generator, linked by an undersea cable to the shore. A smart sensor optimizes power in differing wave conditions, switching the generator off when wave activity is too strong, to avoid damaging the equipment. Severe storms and flat calm weather therefore mean downtime for the buoys, though the buoys still offer between 80 and 90% availability, comparable with conventional fossil fuel generators, and enjoy a key advantage over

**Fig. 1** Pelamis deployed in a choppy sea

**Fig. 2** Pelamis device deployed. (company photographs) ©

wind (30–45%) and solar (20–30%) power generation. Like Pelamis™, OPT machines are designed to be highly resistant to the forces of the open ocean.

A near shore floating device facing incoming waves, similar to the oscillating wave column without a turbine, was patented by Indian researchers in 2007. The rise and fall of the waves is converted to mechanical motion by a heavy buoyant piston driving an overhead crankshaft that in turn drives a gearbox and alternator.

Numerous buoys designs have been developed. A far from complete list includes projects of the universities of Oregon, Uppsala, Lancaster, the Trident, OPT, Aqua Buoy, Seavolt, Manchester Bobber, Wave Reaper, Wave Hub, Seadog, etc. Wave energy converters developed can be grosso modo placed into three groups, viz. pneumatic/or oscillating water column devices (OWC), mechanical or float devices and overtopping-tapered devices (Falcao AF de 2010).

Another wave energy tapping scheme functioned for some time in Toftestallen, Norway (Charlier 2010). Unfortunately it was severely damaged during a particularly heavy storm and has not been restored (Fig. 3). The Musée Océanographique de Monaco used wave (*énergie houlomotrice*) power and so did a private party—among the first—near Royan, France [Area where a new island has appeared in this century (reported on French Television)].

A pier in California was lit by wave energy early in the 20th century. A wave energy tapping project is the topic of a doctoral thesis at the University of Ghent (Gand, Gent). The challenges slowing down development of implementation of wave energy extraction are akin to those hampering other ocean energies: funding, global opportunities, grid connection, environmental permits, maintenance strategies, public interest beyond mere curiosity, site selection. In the matter of sites, a recent study examined criteria for North Sea locations (Beels et al. 2008).

**Fig. 3** The Toftestallen, Norway, wave energy capturing scheme (Photograph Kvarner, Inc.)

Oscillating water column (OWC) technology has been explored by several companies. Orecon's wave-powered buoy is based on a multi-resonant chamber (MRC) oscillating water column and joined forces with HydroAir developer of a bi-directional air impulse turbine and with Portuguese developer Eneólica. The Joint Venture company is to build and deploy Orecon's first full-scale 1.5 MW MRC buoy in a grid-connected installation. Additional units will increase output to 4.5 MW making it the world's largest operating wave farm.

Another OWC design prototype, the Danish Wave Dragon, was installed in Nissum Bredning in 2003, a 7 MW Wave Dragon is to be deployed off the coast of Milford Haven (Wales) in 2010, ten devices in Portugal between 2011 and 2012 and an additional ten dragons in an array off Wales in 2013.

New Jersey's, Ocean Power Technologies (OPT) developed the PowerBuoy, a device that uses waves to move the buoy up and down and converts the resultant mechanical stroking via a power take-off to drive an electrical generator. PowerBuoy has projects in Reedsport (Oregon), Victoria, Australia and the UK. The design is similar in concept to that of Wavebob Ltd of Ireland. Trident Energy's device also uses the linear motion of waves to generate energy. It is solidly anchored, does not use inertial forces like OPT and Wavebob, and floats are used to drive linear generators. A test rig will generate 20 kW from eight full scale linear generators.

Then there are the Archimedes Wave Swing of Ocean Energy, Voith Siemens Hydro Power Generation's WaveGen the CETO device that uses a submerged piston to deliver high pressure water to shore where itch is in conventional hydro technology.

Aquamarine Power's Oyster (Fig. 4) consists of an oscillating flap, which pumps high pressure water through an on-shore turbine to generate electricity. Green Ocean Energy Ltd Wave Treader machine, attaches to offshore wind farm

**Fig. 4** Oyster wave machine (Aquamarine Power company photograph)

monopiles and shares infrastructure, Seaenergy Ltd's Checkmate reminds of a rubbery submarine-like tube that is the Anaconda.

## 3 Ocean Termal Energy Conversion (OTEC)

Sometimes referred to as *thalassothermal* energy, more commonly called OTEC, uses the difference of temperature prevailing between different ocean waters layers to produce electrical power. Statisticians eager to impress the amount of energy available from the ocean, stress that in the waters between the tropics the quantity of heat stored daily by the surface water layers in 1 $km^2$ is equal to burning 2,700 barrels of oil.

The proposal and pilot projects of Arsène d'Arsonval (1851–1940) and Georges Claude (1871–1961) have been abundantly and repeatedly described; they date back to the first half of the last century [French television program of February 22 (Channel A2, 20h00)]. Claude sank his entire personal fortune in his efforts to implement OTEC technology; his best known attempt was offshore Abidjan (Republic of Ivory Coast) (1930). Following the oil crisis of 1973, there was a new flurry of interest for OTEC and "Mini-OTEC" and "OTEC-1" were launched, respectively, in 1979 and 1980. In 1981, Japanese researchers built a close-circuit central on Nauru that delivered 31.5 kW/h; they had placed the cold-water conduits on the ocean floor at a depth of 580 m. It was a result that went way beyond the most optimistic expectations.

The sun's energy heats the surface water of the ocean. In tropical regions, the surface water can be much warmer than the deep water and this temperature differential can be harnessed to produce electrical power. An OTEC system requires a temperature difference of at least 20–25°C (77°F) to operate, limiting, in principle, its use to tropical regions.

Hawaii has, since the 1970s, and still is experimenting with OTEC. Among the major engineering challenges are water adduction (pumping) and economic transmission to land of electricity generated by the system. It will probably be 10–20 years before the technology is available to produce and transmit electricity from OTEC systems. The US Department of Energy's Office of Energy Efficiency and Renewable Energy is currently funding research and development on OTEC cold water pipe manufacturing techniques to help create a more cost-effective OTEC system.

Several technical improvements have been introduced into the plans of proposed schemes. Energy conversion reaches an efficiency of 97%, water exchanges are no longer made of titanium, but of the far less expensive aluminium, corrosion and bio-fouling have been considerably reduced, and the closed circuit system is far more ecologically benign than the open circuit one. The 1993 ducts closed-circuit prototype set up at Keahole Point (Hawaii) delivered 50 kWh net. Turbine improvements are under scrutiny. Closer to the present, a Pacific Ocean luxury resort hotel installed an OTEC system that reportedly provides the electricity needed for lighting, cooking, heating, "spa" facilities, etc.

## 4 Ocean Currents Energy

Often discussed, ocean currents hold immense potential, however, they have come repeatedly under severe criticism on the part of ecologists and environmentalists. Harnessing the Gulf Stream or placing a barrage in the English Channel are viewed as unconceivable projects. Merely 1/1,000th of the available energy from the Gulf Stream would supply Florida with 35% of its electrical needs (Finkl 1980). Ocean currents flow in complex patterns affected by the wind, water salinity and temperature, topography of the ocean floor, and the earth's rotation. Surface ocean currents are generally wind driven and deep ocean currents are driven by density and temperature gradients. The ocean currents are driven by wind and solar heating of the waters near the equator, though some ocean currents result from density and salinity variations of water. These currents are relatively constant and flow in one direction only, in contrast to the tidal currents closer to shore. The Coriolis Project received some publicity already more than a quarter of a century ago. A scheme is reported to have functioned even before that in Northern Iceland [Personal communication at the Second Symposium, on Coasts (Høfn, Iceland, 2005)].

Some examples of ocean currents are the Gulf Stream, Florida Straits Current, and California Current. Far fewer objections are raised against tidal stream schemes. The matter has been discussed again recently by Finkl and Charlier in regards the coast of southern Florida (Finkl and Charlier 2009). No commercial grid-connected turbines are currently in operation though a few prototypes and demonstration units have been tested (Myers and Bahaj 2005; Smith 2005). The US Minerals Management Service has shown interest in an ocean current energy harnessing systems (Fig. 5).

Harnessing marine currents' energy could be achieved by using submerged water turbines similar to wind turbines that would have rotor blades, a generator for converting the rotational energy into electricity, and a means of transporting

**Fig. 5** Ocean Current Project (Source: Minerals Management Service)

the electrical current to shore to connect with the electrical grid. Posts or cables anchored to the sea floor would keep the turbines stationary, and concentrators put around the blades could increase the flow and power output from the turbine.

A marine current "farm", with a predicted density of up to 37 turbines per $km^2$ is conceivable in open spaces provided space is foreseen between the water turbines to eliminate wake-interaction effects and to allow access by maintenance vessels.

Other approaches involve a barge moored in the current stream with a large cable loop to which "parachutes" are fastened. The parachutes would be pushed by the current, then closed on their way back, forming a loop similar to a large horizontal water wheel. It reminds of the bygone days of floating tide mills on the lower Danube River and of those in use during a siege of Rome.

Technical challenges include avoidance of bubble formation (cavitations), prevention of marine growth build-up, reliability, corrosion resistance, bio-species protection from turning turbine blades. Consideration of shipping, fishing, recreational uses, current flow slow-down, temperature and salinity modifications as they might impact species in estuaries.

## 5 Other Sources

Some ocean energies need technological refinement and drastic capital costs reductions; it does not mean that they must be written off, rather that their use is further away than wind, wave, OTEC or tide. In that group fall salinity differential, marine bio-energy, submarine geothermal energy and some others. But perhaps a successful future is in store for tidal power; it has been on and off the list of renewable ocean energy tapping schemes for decades.

## 6 Tidal Energy

Tidal energy has been used for centuries with both tidal current and rise and fall of tides put to work. They provided power for flour mills, saw mills, breweries, etc. Tide mills that dotted several geographic areas regions of Europe from The Netherlands to Spain and from Wales to England, but also coastal strips of the United States and Canada. They may well be considered the forerunners of the 20th century power-generating tidal power stations (Charlier et al. 2003). The *centrales marémotrices*, tidal power stations, *Gezeitenkraftwerke*, are not numerous—except for the mini plants in China—nor are any but one (Rance River) large electricity producers. Besides the Rance River plant in Brittany (France), facilities were built in Russia and Canada. [TPPs that obviously had a short life span, functioned in Boston (Massachusetts), Bristol (UK), Busum (Germany), Lake Van Bemmelen (Suriname), (Frau 1993)].

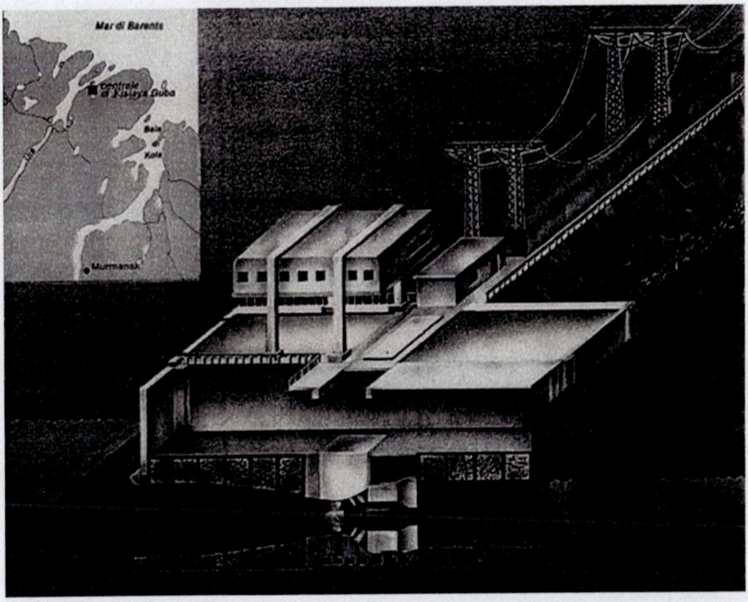


**Fig. 6** Stylized view of Kislaya, Russia plant with location map

Tides provided the dismal Russian North with the electricity needed to develop a rather desolate region (Fig. 6) (Charlier 1982, 2010; Chaineux and Charlier 2008).

The Canadian plant, in Nova Scotia (Fig. 7), is more a trial run than a badly needed plant (De Lory 1987). Nonetheless, it was a convincing trial run because the provincial governments involved have decided to install an additional, bigger, facility.

Barrage systems use the potential energy in the difference in height (or *head*) between high and low tides. Barrages are dams closing off entirely a tidal estuary, and suffer from civil infrastructure costs, and debatable environmental fall-out (Fay1982). Tidal barrage systems provide electricity at ebb tide, or flood tide, or both.

Modern advances in turbine technology may eventually see large amounts of power generated from the ocean, especially tidal currents using the tidal stream designs but also from the major thermal current systems such as the Gulf Stream, which is covered by the more general term marine current power. Tidal stream turbines may be arrayed in high-velocity areas where natural tidal current flows are concentrated such as the west and east coasts of Canada, the Strait of Gibraltar, the Bosporus, and numerous sites in Southeast Asia and Australia. Such flows occur almost anywhere where there are entrances to bays and rivers, or between land masses where water currents are concentrated.

Hammerfest, in Norway, the most northerly city (Fig. 8), has been one of the leading sites on the foreground of tidal power utilization experiments in the 21st century. Indeed, while the British Marine Current Turbines company has

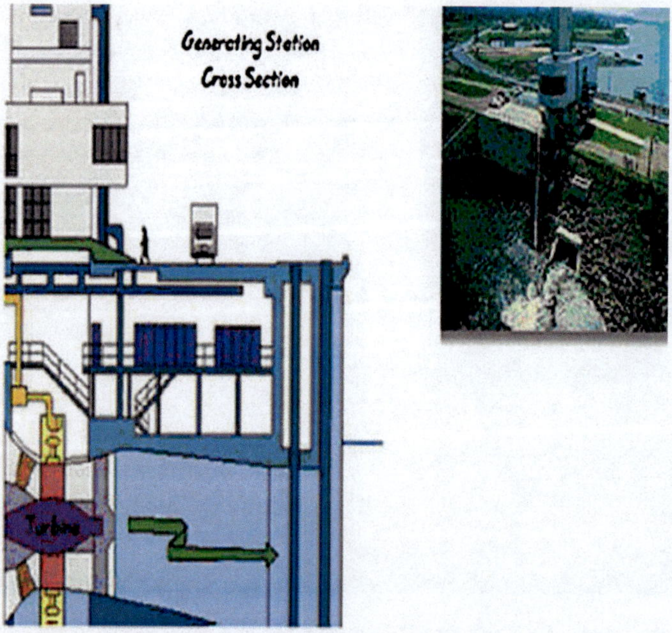


**Fig. 7** The Annapolis Royal TPP. This plant is equipped with a Straflo turbine

**Fig. 8** View of Hammerfest Harbour

conducted several tests in Great Britain, which hold great promise, the Norwegian company, Hammerfest Strøm connected its 300 kV tidal generator machine to the town of Hammerfest's grid thus becoming one of the first grid connected tidal turbine schemes in the world. The company believed that it would have its first tidal farm of over 20 s generation devices operational before the end of 2008. It would have been the 3rd phase of the "Blue Concept" project and would result in a tidal farm that would produce 10 MW of renewable electricity. Hammerfest Strøm AS developed the Lànstrøm tidal turbine, which is due for tests in Scottish waters.

Originally geographically limited to coasts with large tidal ranges, the development of very-small-head turbines permits the implantation of tidal power plants in many more locations (Baker 1997; Charlier 2001).

The development of the tidal power plant went hand-in-hand with, or at least was boosted by, that of the bulb turbine (France, Russia) (Charlier 1982; Gibrat 1976; Gibrat and Auroy 1956) and later of the Straflo turbine (Canada) (Charlier and Justus 1993; De Lory 1987). Tidal power schemes have been considered around the world (Charlier 1982; Charlier and Justus 1993; Charlier and Finkl 2009). See also the planisphere shown in Fig. 9 and Table 2 (compiled from the New World Encyclopedia). In Table 2 "–" means lack of, and "?" indicates information not determined. The huge Korean Lake Shiwa project, under construction since 2008, the Wales (Severn River) TPP plans, some pilot projects in Scotland, and the Hammerfest, Norway (functioning) plants are not included in Table 2, nor are plants, at work, of France, Canada, Russia and China. For France

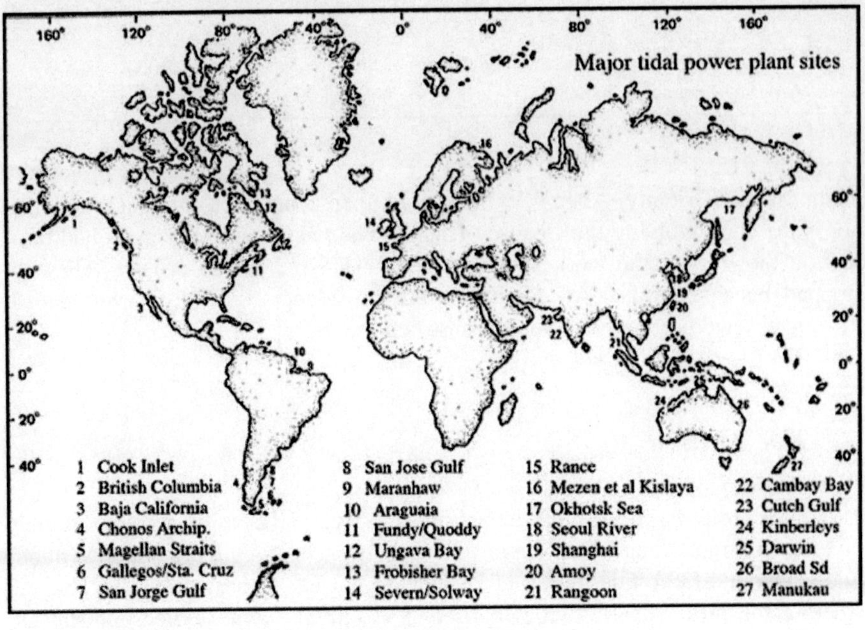


**Fig. 9** Planisphere showing suitable sites for large TPPs

**Table 2** Survey of sites once considered for the implant of a TPP (see Fig. 9)

| Country | Place | Mean tidal range (m) | Area of basin ($km^2$) | Maximum capacity (MW) |
|---|---|---|---|---|
| Argentina | San Jose | 5.9 | – | 6,800 |
| Australia | Secure Bay | 10.9 | – | ? |
| Canada | Cobequid | 12.4 | 240 | 5,338 |
| | Cumberland | 10.9 | 90 | 1,400 |
| | Shepodya project | 10.0 | 115 | 1,800 |
| | Passamaquoddy | 5.5 | – | ? |
| India | Kutch | 5.3 | 170 | 900 |
| | Cambay | 6.8 | 1,970 | 7,000 |
| Korea | Garolim | 4.7 | 100 | 480 |
| | Cheonsu | 4.5 | – | – |
| Mexico | Rio Colorado | 6–7 | – | ? |
| | Tiburon | – | – | ? |
| United Kingdom | Severn | 7.8 | 450 | 8,640 |
| | Mersey | 6.5 | 61 | 700 |
| | Strangford Lough | – | – | – |
| | Conwy | 5.2 | 5.5 | 33 |
| United States | Passamaquoddy Bay, Maine | 5.5 | – | ? |
| | Knik Arm, Alaska | 7.5 | – | 2,900 |
| | Turnagain Arm, Alaska | 7.5 | – | 6,501 |
| | Golden Gate, California | ? | – | ? |
| Russia | Mezen | 9.1 | 2,300 | 19,200 |
| | Tugur | – | – | 8,000 |
| | Penzhinskaya Bay | 6.0 | – | 87,000 |
| South Africa | Mozambic Channel | | 68,000 | |

neither is the (Brittany) Aber W'rach (started, then abandoned) project mentioned nor is the one involving the Chausey Islands. The Passamaquoddy plant had been seen at the start of the work as a bi-national—USA/Canada—project. All work stopped because of lack of funding. Both presidents F. D. Roosevelt and J. F. Kennedy were proponents of its construction.

## 6.1 Tide Mills

Tide mills are of course not different from run-of-the-river mills, except that they include an impounding basin where the water brought in by the incoming (flood) tide is stored; at ebb tide the water is released but has to pass through a channel wherein the mill wheel is set (Charlier et al. 2003). Some more sophisticated mills even captured power from both ebb and flood tides. And still others captured the energy of the horizontal movement from tide (tide current). A variant relies on air contained in a metal or concrete conduit that is compressed by an incoming tide. The air, so compressed, becomes air power. An artificial basin, made by damming part of the sea, fills at incoming tide and is then released at low tide through turbines back to the sea, like at the Rance River Plant today, or to another basin. The tide mills' demise in man's industrial arsenal was slow but their numbers declined rapidly and abruptly, as newer technology unfolded (Wailes 1941).

Though power stations have thus been constructed during the second half of the 20th century in France (Rance River estuary) (Fig. 10), Canada (Bay of Fundy, Hog Island), on Kislaya Guba, and "mini" stations in China, interest in building more plants waned, as fossil fuels remained abundant and their prices fell (Bernshtein and Usachev 1957).

**Fig. 10** *Upper* aerial view of Rance River Power Plant (Brittany), *Lower left* plant under construction "in the dry", cofferdams, *Lower right* bulb turbine

### 6.2 Harnessing Tidal Energy

Some views on the future of ocean energies use and particularly of the future of tidal energies published by Pike Research in the 2nd Quarter of 2009 might be appropriately recalled here. "The high capital costs associated with renewable energy technologies is largely avoided with marine kinetics. The capital costs of marine renewable energy systems will be 50–100 times smaller than investments required to create the same amount of electricity from either wind or solar." Gulf Stream Turbines LLC in Fort Lauderdale has patented a submersible turbine similar to a wind turbine in price but 4–10 times as effective as wind or solar in its capacity factor and needs no site preparation, is invisible, causes no $CO_2$ emissions, nor harms marine life, operates at 85% capacity factors equal to fossil fuel plants but carries no fuel costs and each turbine has a carbon offset of 10,000 tons of $CO_2$ each year. If the FPL Group realizes that 10,000 turbines can supply the state with 8.94 GW of electricity and at the Euro Carbon Offset futures price of January 2011 is US $30/ton, that is US $3 billion in $CO_2$ offsets alone, not counting the revenues of electricity or hydrogen that can be realized by 2012–2013. Rowen Negrin, President of Gulf Stream Turbines LLC would like to begin a dialogue with FPL Group to explore bringing this technology to market" (see http://www.gulfstreamturbines.com).

If the largest plant ever built thus far, and that operated successfully for 40 years, is in France on the Rance River, plans have been afoot in Britain for decades. (In Great Britain tidal power stations could usefully be built, in several locations including the Severn, Dee, Solway and Humber estuaries.)

There have been plans for a "Severn Barrage", a facility that could provide as much power as 3–12 nuclear centrals, depending on site and project. Such a plant would in addition to furnishing power protect a large stretch of coastline against high storm tides, and depending on site provide a road bridge. On the other side of the coin drastic changes to the currents in the estuary could strongly impact the ecosystem, and cut off estuarine feeding grounds (mud flats) of birds. The Rance and most other projects are all on-shore power stations. Their construction is very costly and some experts back therefore the idea of off-shore centrals.

A plant with offshore turbines would be much cheaper to build, and would avoid environmental problems of a barrage. The idea has been taken up by Marine Current Turbines (Fig. 11).

The University of Wales Swansea's "Swanturbines" differs from several other devices significant in that it is direct drive, where the blades are connected directly to the electrical generator without a gearbox between, uses a "gravity base", a large concrete block to hold it to the seabed, (no drilling into the seabed is needed), and its blades are fixed pitch, rather than actively controlled (http://www.swanturbines) (Fig. 12). Severn is merely one project site and Swanturbines only one energy extracting system. There are many others.

Tidal energy has an efficiency of 80% in converting the potential energy of the water into electricity. It is thus very efficient, compared to other energy resources

**Fig. 11** The offshore scheme developed by Marine Currents Turbines Ltd. Artist's view

**Fig. 12** Swanturbines-artist's view

such as solar power or fossil fuel power plants. It is also more regular than wind power.

Capturing and putting to work—to generate electricity—the energy of the tides is achieved by damming a stream when it enters an estuary. A water storage pond can be created by enclosing an area in a tidal creek. Water flowing from one level to a lower one produces "work": the force is gravity, the work is potential energy. Figure 9 shows the geographical distribution of suitable sites where a tidal power plant could be implanted. Today, far more sites are suitable because of the low head turbines that have been developed. The numerous Chinese mini-plants are not shown on this map, some have been abandoned due to silting, and, information on location being scant or unavailable (Charlier 2008).

A plant in China's Zhejiang Province supposedly predates the French Rance River plant; it is claimed to have been built in 1959. No details concerning the type of turbine(s) used in the facility are available to this author. By 1969, plants were at work in the provinces of Zhejiang, Jiangsu and Shandong, but according to a Chinese author some stations were not producing power, some were even abandoned. More plants were constructed in 1980 (Jian Xia). Only the plants at Jian Xia and Liu He Kuo are double-effect centrals.

## *6.3 The TPP*

The "modern" tidal power plant consists of a barrage, in which turbines are installed, a retaining basin and sluices. A tidal current facility has no barrage, and is consequently less onerous to build and has still lesser environmental impact. Tides re-timing has thus been a concern. Compressed air schemes have been proposed; another approach is the two-basin schemes wherein one is filled at high tide and the other is emptied at low tide. In such schemes turbines are placed between the basins. The system allows thus generation time to be adjusted with high flexibility and to generate almost continuously. However, two-basin schemes involve high capital expenditures to construct as they require extra length of barrage, except in some rare sites; that approach had been proposed for the Passamaquoddy plant.

Tide mills did not need a barrage but had a water wheel. Tidal power stations, such as the Rance, can provide valuable co-generation but the other existing stations are all very modest electricity producers (Fig. 13). The bulb turbine, once hailed as revolutionary, has not been maintained as a "must" in contemporary TPPs. However, it permitted a variety of methods of producing electricity. The Rance River plant can produce power with or without pumping and can generate at both the incoming and outgoing tide: this feature, often called dual power, was installed in some tide mills (Table 3). It is used in at least two Chinese TP plants but has, however, not been included in the Canadian Bay of Fundy pilot plant where Straflo turbines were installed instead; indeed some doubt whether the extra cost it carries is compensated by resulting production.

However, utilizing the tidal processes has received considerable attention during the last few years as oil reserves, and especially prices, and climate changes cause increasing concern. China is considering building a plant and has received feasibility study funds from the World Bank and, on Lake Shiwa, the Republic of Korea has undertaken the construction of the largest ever tidal power station (260 MW) (Fig. 14). Tidal lagoons, can play the role of retaining basin with "natural" barrages, and entrain lower cost and have less impact as self-contained structures do not fully cut across an estuary. Under some circumstances they can be configured to generate continuously. Barrage schemes need multiple basins or additional technical arrangements to adjust to solar cycle.

The Shiwa "lake" communicates with the sea and will act as the retaining basin and only ebb-tide generation is foreseen. Korea has kept a lively interest in barrage tidal power for decades, as it possesses several attractive sites ranging from

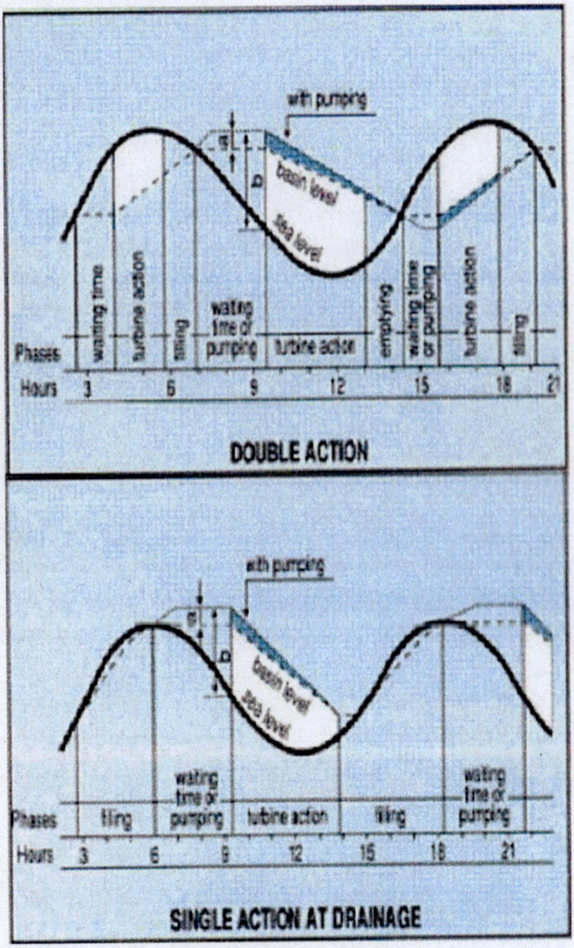


**Fig. 13** Tidal power plant, modes of operation

**Table 3** Some characteristics of operating TPPs

| TPP | Generating capacity (MW) | Barrage length (m) | Impounded estuary area ($km^2$) | Number of turbines | Net power output (gW/y) |
|---|---|---|---|---|---|
| Rance | 240 | 720 | 22 | 24 | 480 |
| Annapolis | 20 | | | 1 | 30 |
| Kislaya | 400 | | | 1 | |
| Jangzhu (East China Sea) | 500 | | | 1 | |

Yanhkakto in North Korea to the Seoul River and Jinsen in South Korea, on the west coast. On the east coast, excellent locations exist in Chonsu and Inchon bays. At Inchon 1,300,000 kW could be generated.

**Fig. 14** View of Lake Shiwa tidal power plant (Korea) that was projected to be completed by the end of 2009, but still is being worked on according to information available in early 2010

## *6.4 Mathematical Modeling of Tidal Schemes*

In mathematical modeling of a tidal power scheme design, the basin is broken into large segments (1-D), squares (2-D), or cubes (3-D). The complexity and accuracy of the model increases with dimension segments, each maintaining its own set of variables. Time is advanced in steps; every step, neighboring segments influence each other and variables are updated.

The simplest type of model is the flat estuary model, in which the whole basin is represented by one segment. The surface of the basin is assumed to be flat. This model gives rough results and is used to compare many designs at the start of the design process. Thus, quantitative information is gathered for a range of parameters, including water levels (during operation, construction, under extreme conditions), currents, waves, power output, turbidity, salinity and sediment movements. Modeling has been called upon for some tidal stream projects.

## *6.5 Tidal Stream or Tide Current*

Tidal stream systems use the kinetic energy of moving water to power turbines. The method has received increased attention lately because of the lower cost in civil engineering and lesser ecological impact. The benefits of tidal flow, exploited by some tide mills of bygone times, have been largely ignored. Indeed, compared to other sources of renewable energy, tidal flow is reliable, predictable, of known gravity and sustainable. Furthermore, acceptable from an aesthetic viewpoint, being out of sight, contrary to wind power, such schemes create no noise. In San Francisco, a sub-sea tidal power station has been under consideration for some time; no moving parts are involved, except on land, advantage would be taken of a venturi, and conventional air-driven turbines could be installed. If maintenance is a flaw in the system, there are probably ways to solve this, and it does not justify the low profile it has been given. Cathodic corrosion protection, another "warning" often uttered, has been improved, witness a study at the Huaneng Dandong Power Plant in the PRC, (China). This "huge source of energy should be elevated through research to a much higher profile". Financial requirements remain an impediment but bureaucratic barriers are to be overcome as well.

Location is critical for a tidal stream power generator and tidal stream tapping requires sites with fast currents where natural flows are concentrated between obstructions, such as entrances to bays and rivers, around rocky points, headlands, or between islands or other land masses. Such locations exist at the Pentland Firth (Scotland), the Channel Islands near the French coast, Cook Strait (New Zealand), the Bosporus, the Bass and Torres straits (Australia), etc.

Tidal current power, used for hundreds of years by rather unsophisticated yet clever technology (tide mills, a.k.a sea mills), received limited attention until relatively recently. In 1980, a small research program was launched with the purpose of evaluating the potential of tidal and marine currents. Countries involved were the United Kingdom, the United States and Japan. Government funding, particularly in Europe, is currently more common as several new approaches as getting refined.

The first utilization of tidal power in the United Kingdom dates back to a Bristol 1931 scheme, on the heels of the Aber W'rach aborted attempt (Brittany) and just before the United States' Passamaquoddy project got the go-ahead signal. Work came rapidly to a halt at "Quoddy" under pressure from traditional electrical power providers. In fact it took another half century before, as the anecdote goes, French President Charles de Gaulle, irritated by the dilly-dallying, ordered construction of the Rance River TPP started. As the French dug and built, the British looking at the Severn Estuary kept talking, as they still do. (The British Government and the regional government of Scotland have in recent years shown keener interest in tidal power and several projects, tests and the like have been carried out.)

Russians in Kislaya Bay and the Canadians in Fundy Bay, decided to fall in step with the French and constructed pilot plants (Chaineux and Charlier 2008; De Lory 1987). The expected resurgence of interest in TPPs (Wilson 1977) after completion of the Annapolis-Royal (Nova Scotia) plant, did not materialize, yet discussions of such mega-schemes as the Chausey Islands Plant (France) found their way in print (Banal 1997a, b; Bonnefille 1976).

Various try-outs, studies, plans, projects are on-going; in China the technology of the LonWorks Fieldbus digital communication network has been applied in the 70 KW tidal current electricity generating system.

In Singapore, four sites—Lighthouse, Pasir Panjang, Sembawang, and Tanjong Fagar—have been considered for the implant of a tidal power plant, locations that are rather remote; thus these facilities would provide small quantities of power, ranging from 1 to 10 MW. The limited information available about Chinese TPPs hampers the view of China; the Kianghsia facility, however, was briefly described in the (French) *Revue de l'Énergie,* and the recent call for specialists for a pre-feasibility study for a TPP in Zhijiang Province bears witness of continued interest (Anonymous 1981, 1998; Zhenzia et al. 1989). The project is buttressed by funding from the World Bank (Ch'iu Hou-Ts'ung 1958). A tidal stream plant is being built in China on the Yalu estuary.

Even though 2006 dated information pertaining to tidal energy use in Russia indicates a current disinterest, Usachev et al. (2004) had published a review of its large-scale use and the observations made over 35 years of operation of the Kislaya Bay TPP. The TPP has operated successfully, with moving parts underwater, under extreme climatic conditions. The plant has been used as a biological test site in a nearly closed-off basin, including fisheries and mariculture observations. A variety of units were tested among which sorption and electrolytic, and for cathodic behaviour. The reporters confirmed that a TPP can be integrated into a power system, in basic and peak periods of the load diagram, allowing for the specific conditions of tidal energy generation.

There is general agreement that one of the problems of commercialization of tidal current schemes is to solve involve installation and subsequently maintenance of the device, but so are loading conditions and power transmission. Though one may consider that there is a lesser environmental impact than with a barrage plant the systems are not free of environmental matters to be adjusted.

The World Energy Council sees considerable opportunities between the present and the end of the next decade for renewable energy sources. A rapid turnover is excluded because of investments in traditional facilities. The choices that are made today, however, will determine whether the cortege of ill consequences of climate warm-up, will be blocked by the development and implementation of better technologies, or will proceed with its scientific uncertainties, an Absalom sword.

## 6.6 Environmental Impact

In brief, environmental impact of barrage plants include fish populations, other marine life, water quality and in some instances they impede maritime traffic. The Rance River TPP's construction phase proved particularly environment damaging, yet a different and diversified flora and flora took over once the plant began operating. Migratory organisms are able to pass through sluices and turbines, and even a female seal passed through; the animal was removed repeatedly from the retaining basin and released to the open sea, but she stubbornly returns, well decided to establish its habitat in the basin (Fig. 15).

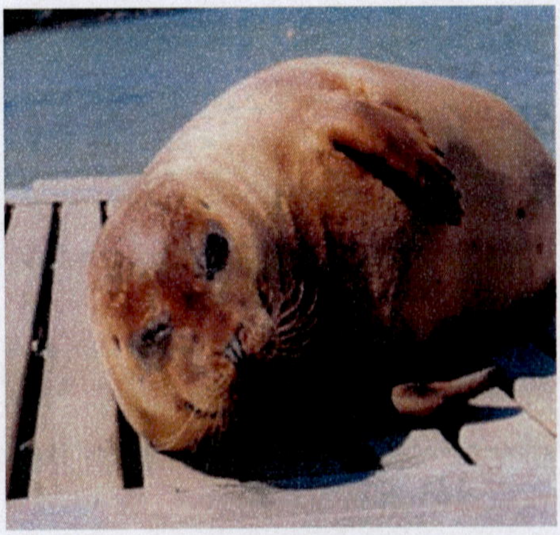

**Fig. 15** EDF photograph of the seal that elected domicile in the TPP Rance River retaining basin

The present author does not share the viewpoint that environmental concerns are the reason for the small number of TPPs—rather the situation should be ascribed predominantly to economic motives—but shares the view that EIAs (environmental impact assessment) are useful tools to identify impacts, and that applications of modern appropriate technologies might help abate the objectionable effects of a tidal power plant. Already in 1993, additional research had been suggested pertaining to the interface of the TPP output with national grids, that a sound assessment of such plants, their economic interest, design, and implementation according to the site, and environmental effects was both desirable and urgent (Frau).

Environmental effects have been again re-examined by Van Walsum in 2003. Interestingly, that study addressed two-basin schemes, an approach rarely still considered today, though such scheme had been repeatedly proposed for the Bay of Fundy and nearby sites such as Quoddy.

An unusual reaction was triggered by the release of the UN Environmental Panel on Climate Change report that held—for the second time—that human activity strongly impacts the climate, with the combustion of fossil fuels playing a major part (2000). The International Energy Agency predicts, if no measures are taken to reduce emissions from transportation and power sectors, a still increasing release into the atmosphere of greenhouse gases. It holds that population growth and GNP are proportional to such increase. At the 2002 meeting of the IEEE (Institute of electrical engineers) Power Engineering Society, a diametrically opposed view was presented by a group of scientists who, through the paper authored by Meisen, dispute these findings.

Company long-range planners, suggest the use of renewable energies. The price of these, however, must be made competitive with that of conventional fuels. Be it mentioned here that each time the cost of oil rises, automatically the price gap narrows. Another condition is that means be found to secure systems' reliability, as intermittent resources, such as tidal power among others, expand into the daily load mix. And there is a psychological factor as well, linked to the final recognition by governments—particularly the US—to wit acceptance by the public at large and the governments' willingness to address the problem of climate change.

While Gibrat proposed a formula to rate the suitability of sites for location of a TPP, Sergey (2003) established a semi-quantitative rating of environmental disturbance brought about by power generating schemes. The thermal power industry is, according to that scale, the most damaging (coefficient 74) and the wave energy capturing industry the most benign (coefficient 31). Second to best is tidal power with a score of 42, just one point "better" than that of solar power (43).

Similarly a formula has been calculated to assess the energy "available" as different turbine designs have varying efficiencies, hence varying power output. When the turbine efficiency ($\eta$) is known, the power available from a kinetic system will be:

$$P = \frac{\eta \rho A V^3}{2} \tag{2}$$

where $P$ = power generated (in Watt), $\rho$ = water density (seawater is 1,025 kg/m$^3$), $A$ = turbine sweep area (in m$^2$), $V$ = flow velocity.

Relative to an open turbine in free stream, depending on the geometry of the shroud, shrouded turbines are capable of as much as 3–4 times the power of the same turbine rotor in open flow.

### 6.6.1 Tidal fences

In order to circumvent, even to remove, environmental objections, alternatives have been developed to the "straight" tidal barrage approach. They include tidal fences and tidal reefs.

A tidal fence has vertical axis turbines mounted in a fence. All the water that passes is forced through the turbines. Tidal fences can be used in areas such as channels between two landmasses, which is in fact not different from tidal plant barrages. The fences are cheaper to install than tidal barrages—a cost considerably reduced since the use of cofferdams can be avoided. An additional advantage is that fences have less impact on the environment than tidal barrages, although they can disrupt the movement of large marine animals.

Tidal fences (Fig. 16) are effectively barrages which may but do not necessarily completely block a channel. They may thus not be an impediment to navigation nor to passages of animals. If established across the mouth of an estuary they can be very environmentally destructive. However, in the 1990s their deployment in channels between small islands or in straights between mainland and island was frequently considered as a viable option for generation of large amounts of electricity.

The advantage of a tidal fence is that all the electrical equipment (generators and transformers) can be kept high above the water. Furthermore, by decreasing the cross-section of the channel, current velocity through the turbines is significantly increased.

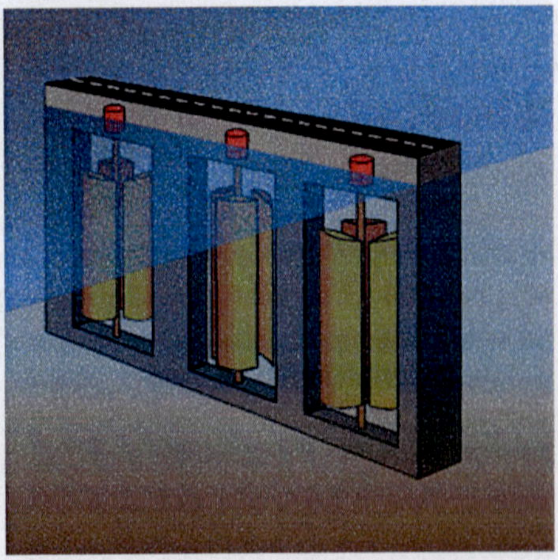

**Fig. 16** Tidal fence

### 6.6.2 Tidal Reefs

Again in connection with plans revolving around the Severn River, still another idea has surfaced: the tidal reefs. Reefs are equipped with reversible turbines attached to the estuary floor and theoretically can, like at the Rance and Kislaya installed bulb turbines, generate power both with ebb and flood tides. Thus far no prototype exists and hence no test has ever been made of this technology. It could have lower environmental impact but would extract less energy than a barrage. The British government recently allocated £500,000 (€550,000) to accelerate research.

Some proponents claim that not only would a reef be cheaper to build, and be less environmentally damaging, it would even generate more power. The reef, in their opinion, would hold back about 2 m of water and not alter tidal patterns as much as a barrage would.

Supporters of a proposal for a 16 km barrage between Cardiff and Weston-super-Mare argue it could generate some 17,000 GW hours (GWh) of electricity each year, the equivalent of almost 5% of the UK's needs. The reef alternative, it is claimed by its proponents, involving a 19 km reef across the Severn between Aberthaw (Glamorgan) and Minehead (Somerset) would cost about £2bn (€2.2) less and produce 20,000 GWh.

The reef would keep intact most of the estuary's bird feeding ground and its slower-moving turbines would reduce danger to migrating salmon and eels, could generate electricity for longer periods than the barrage, and therefore be more able to meet power needs at peak demand times.

## *6.7 Modus Operandi Modifications at the Rance*

Rance River TPP operators are assisted since the 1990s by a sophisticated computer enabling to carry out simultaneously automatic control and human shift reduction. Automatization insures higher reliability and a more sensitive response to external conditions. Alternators were rebuilt in 1997, the bulb turbines renovated and underwent design changes.

The AGRA program was put in use at the Rance TPP in 1970. The special features of the AGRA new operational model for the Rance, introduced in 1982, encompass close optimization by dynamic programming, modular program structure, and developed graphical outputs (Merlin et al. 1982; Sandrin 1980). As for other French projects, the Rance Plant has followed the trend for automation (Ferdinand).

## 6.8 New Turbines

In operating tidal power plants bulb (Rance, Kislaya) or straflo (Hog Island) turbines have been installed. Low heads being a limiting factor for tidal power plants, Darrieus turbines might offer a solution. Shiono et al. (2000) developed a Darrieus turbine specifically for use in tidal current energy harnessing. Darrieus turbines, originally designed for windmills and adapted for use in water, can always revolve in a same direction without being influenced by tidal flow direction. Difficult to start, these turbines, now equipped with spiral blades, have overcome that problem. Efficiency of tidal current power generation is influenced by the characteristics of the turbine, and Darrieus turbines are affected by the solidity and the number of blades. The experimental turbine was tested in ad hoc channels to ascertain the best values for rotor solidity and the number of blades.

Tide variations in the Amazon River basin provided an opportunity to study turbine rotor-blades with simple curvature for use in small hydro-power plants with maximal generation of 10 kW. Application in tidal power plants seemed promising and although rotor yield proved to be low, decentralized power generation would be viable (Mauad and Mariotoni 2000).

A shroud or duct housing a turbine can produce up to four times the energy output of the same turbine for little extra project cost. Shrouded turbines can produce the same energy from a much smaller turbine while larger turbines if housed in a shroud have significant attraction to commerce where "Rate Of Return" on investment may be up to 15 years on some sites where open turbines are used.

Australian Tidal Energy Pty Ltd undertook successful commercial trials of shrouded tidal turbines on Queenland's Gold Coast in 2002. Commercial trials undertaken on such turbines were tested in a remote Australian community in northern Australia with some of the fastest flows ever recorded (11 m/s); two small turbines provide 3.5 MW. A larger 5 m diameter turbine, capable of 800 kW in 4 m/s of flow, was planned for deployment as a tidal powered desalination showcase near Brisbane in 2008.

Still another device, the Hydro Venturi, was to be tested in San Francisco Bay and along the Canadian west coast. Some flows reach 8.5 m/s; a 1 $m^2$ turbine in such speed flow could produce around 0.5 megawatt (MW) and a shrouded turbine would provide nearly 2 MW; and average flow is about 3–5 m/s. These shrouds can be mounted horizontally or vertically.

### 6.8.1 Axial Turbines

The number of turbines developed to harness tidal stream's energy is surprisingly high, even though all fit into one of two types: those with a horizontal axis—parallel to the direction of water flow—and those with a vertical axis—perpendicular to the direction of water flow. Gorlov's and Tidal Fence Davis Hydro are the only leading systems involving vertical axis turbines.

Leading schemes with horizontal axes are Evopod (Fig. 17), Delta Stream, Lunar Energy Tidal, Aquamarine's Neptune Tidal Stream, Verdant's Free Fow, Atlantis' Nereus and Solon Turbine, Tidal Stream Turbine. But there are others such as TidEl Stream Generator, the all but abandoned Stingray, SeaGen, Open Hydro's Open Center that was taken for tests to France, and Pulse Turbine Hydrofoil. Evopod, close in concept to traditional windmills "operating under the sea", is a semi-submerged floating platform approach tested in Strangford Lough, the same area where SeaGen completed in 2010 successful tests. A prototype is undergoing testing at 1/10th scale. The advanced hull form maintains optimum heading into the tidal stream and is designed to operate in the peak flow of the water column.

SeaGen, made up of two axial flow rotors, each of which drives a generator, is currently the only commercial scale device installed anywhere in the world. The turbines are capable of generating electricity on both the ebb and flood tides because the rotor blades can pitch through 180°.

Prototypes currently operating include the one connected to the grid in 2003 at Kvalsund, near Hammerfest in Northern Norway. The turbine has a reported capacity of 300 kW.

**Fig. 17** Evopod, the scheme constructed at Hammerfest

Seaflow, an 11 m diameter turbine generator fitted to a steel pile driven into the seabed is a 300 kW Periodflow marine current propeller type turbine manufactured by Marine Current Turbines, the makers of SeaGen. It was installed off the coast of Lynmouth, Devon, England (in 2003).

Since April 2007 Verdant Power has been running the first major tidal-power project in the United States. It is a prototype project in the East River between Queens and Roosevelt Island in New York City. Reinforced turbines were installed in 2008.

Following the Seaflow trial, a full-size prototype, called OpenHydro is a prototype tested at the European Marine Energy Centre (EMEC), in Orkney, Scotland. Nova Scotia Power has selected OpenHydro's turbine for a tidal energy demonstration project in the Bay of Fundy, Nova Scotia.

### 6.8.2 Cross-flow Turbines

Cross-flow turbines, invented by Darrieus (1923), can be deployed either vertically or horizontally. Neptune Renewable Energy has developed Proteus which uses a barrage of vertical axis crossflow turbines for use mainly in estuaries. A crossflow variant was developed by Gorlov; it features a helical design being commercially tried out in S. Korea, with a 1 MW plant (2009) scheduled to be expanded to 90 MW by 2013 (Gorlov 2003).

Ocean Renewable Power Company, LLC (ORPC) successfully completed 3 years ago the testing of its turbine-generator unit (TGU) prototype at ORPC's Cobscook Bay and Western Passage tidal sites near Eastport, Maine. The TGU is the core of the OCGen technology and utilizes advanced design cross-flow (ADCF) turbines to drive a permanent magnet generator located between the turbines and mounted on the same shaft. ORPC has developed TGU designs that can be used for generating power from river, tidal and deep water ocean currents.

Trials of the Kobold concept in the Strait of Messina, Italy, started 10 years ago. A Kobold turbine is able to convert kinetic energy from marine currents into mechanical; the turbine is a rotor mounted on a vertical shaft which produces mechanical energy. It rotates independently of the direction of the current; its high torques permit spontaneous starting even under intense conditions without the need of an ignition device. It has been patented and is developed by Ponte di Archimede Inc. (Erikson et al. 2008).

### 6.8.3 Oscillating Devices

Oscillating devices use aerofoil sections pushed sideways by the flow. More commonly considered among wave harnessing systems, such oscillating stream power extraction was proven with the omni- or bi-directional Wing'd Pump windmill. During 2003, a 150 kW oscillating hydroplane device, the Stingray, was tested off the Scottish coast. The Stingray uses hydrofoils to create oscillation, allowing it to create hydraulic power, in turn used to power a hydraulic motor, that then turns a generator.

Pulse Tidal operates an oscillating hydrofoil device in the Humber estuary. Backed by European Union funding, Pulse is developing a commercial scale device to be launched in 2012. This uses a shroud to increase the flow rate through the turbine. These can be mounted horizontally or vertically. Pulse Tidal are designing a commercial device with seven other companies The consortium was awarded an €8M EU grant to develop the first device, which will be deployed in 2012 and generate enough power for 1,000 homes.

Modern advances in turbine technology may eventually see large amounts of power generated from the ocean, especially tidal currents using the tidal stream designs but also from the major thermal current systems such as the Gulf Stream, which is covered by the more general term marine current power. Tidal stream turbines may be arrayed in high-velocity areas where natural tidal current flows are concentrated such as the west and east coasts of Canada, the Strait of Gibraltar, the Bosporus, and numerous sites in Southeast Asia and Australia. Such flows occur almost anywhere where there are entrances to bays and rivers, or between land masses where water currents are concentrated.

### *6.9 Tide Current (Tidal Stream)*

Over 20,000 GWh per year could be generated on the West Coast of Canada by harnessing tidal current energy. Tidal stream (current) has also been considered in Ireland. Tidal power has been investigated at Bull's Mouth: Achill, the largest Irish island, could be electrically self-sufficient using winds and tides as energy sources, provided the network be upgraded (Ottewell 2003).

Besides assessing the resource, estimated to have a mean output of 130 MW in several sites, Bryans' studies (2004, 2005) looked at methods of deployment and control inclusive of down rating of the generator in regards turbine size and operational output reduction to reduce capital costs, capacity factor increase and reduction of impact on the grid (Bryden et al. 2004; Charlier 2004). Two tidal energy devices involved horizontal-axis turbines and one had an undulating wing design.

Relatively little research has been conducted to determine the characteristics of turbines running in water to convert kinetic energy. The fundamental issues likely to play important roles in MCT (Marine Current Turbines™) systems implementation include particularly harsh marine environments—though already examined in the case of Kislaya Bay TPP—cavitations phenomena, and high stresses (Bahaj and Myers 2003). Myers and Bajah submitted the design of a horizontal axis turbine; they used tidal data from Alderney Race to run simulations to show a potential annual energy output of 1,340 Gwh. Such "races" exist also near Scotland's west coast. Land mass constrictions in these geographical locations are at the origin of high current velocities at depths suitable for the placement of turbines. In a preceding paper (2004), the same authors had emphasized that the power density—for marine currents—for a horizontal axis turbine is similar, in form, to that of a wind turbine and depends on the cube of the speed and the fluid density (water ds = 1,000 × air ds). A tidal current turbine can thus be smaller than the wind turbine. Energy yields exceeding 7.4 TWh could—theoretically—be realized, which would satisfy 2% of the demand of the United Kingdom (2000).

Marine Current Ltd used a single turbine variant of one originally developed by IT Power. Two axial flow rotors with diameters 17–23 m approximately in diameter drive a generator via a gearbox, mounted on a tubular steel monopile about 2.7–3.3 m in diameter, set into a hole drilled in the seabed (Sanford 2003).

SeaGen, the prototype tidal energy turbine of Marine Current Turbines Ltd (MCT) has, in early 2010, exceeded 1,000 h of operation in Northern Ireland's Strangford Lough, the first tidal current or wave energy system that achieved such performance (Figs. 18, 19).

**Fig. 18** SeaGen device (Stanford Lough)

**Fig. 19** SeaGen under construction

The 1.2 MW tidal current turbine, achieved a capacity factor of 66% and delivered 800 MWh into the National Grid, thus a totally predictable output delivered on average at the same rate as a wind turbine of approximately twice the rated power.

Marine Current Turbines installed SeaGen, the world's largest ever tidal current device (1.2 MW). The turbine is a prototype to be replicated on a large scale over the next few years. Rotors turn slowly about 10–20 rpm or ten times slower than a ship's propellers, making risks of impact small. Prospects for tidal current commercial tidal farms, of up to 10 MW in UK waters are thus realistic.

Gorlov of Boston's Northeastern University has been active for decades in developing systems for capturing tidal energy (Gorlov 1982, 2000). A few decades ago he proposed a "removable barrage" scheme and energy storage with pneumatic chambers. He recently (2003) introduced a helical turbine that could be used in harnessing tidal as well as wave energy. The scheme dispenses with dams and can operate in free or ultra-low head water currents. The cross-flow unidirectional rotation machine is usable in reversible tidal streams and estuaries, for instance.

In line with the philosophy that even small quantities of power are worth harnessing, Bryden et al. (2004) proposed to link tidal current turbines with energy storage mechanisms even if small. They furthermore (Bryden and Melville 2004; Bryden et al. (2004)) developed a simple model of tidal current energy extraction and showed that flow in a simple channel is altered. Ten per cent of the flux ought to be the limit to assess the extractable amount. Admittedly the limitation may be less stringent where sea lochs are concerned.

Using a two-dimensional tide driven hydrodynamic model, Blunden and Bahaj (2006) assessed the Portland Bill (Dorset, GB) proposed site for a tidal current (stream) TPP. Their results could eventually be used if a TPP was indeed implanted there, a site close to population centers and thus economically attractive.

The capability of tidal currents to provide firm power was defended recently, stressing concomitantly the finite nature of wind power expansion (Clarke et al. 2006).

Tidal current tapping must take into consideration that increasing the number of turbines will not only have an unfavorable environmental effect but will also block the flow and thereby reduce power generated (Charlier 2003; Garrett and Cummins 2004). The latter is substantially less than the average kinetic energy flux observed. According to Garrett and Cummins (2004, 2005) the maximum average power lies between 20 and 24% of the peak tidal pressure head (from one end of the channel to the other) in the undisturbed channel. Garrett and Cummins also showed that an array of turbines placed at the entrance of a bay, gulf, inlet, is quite effective if uniformly distributed across the entrance. (The maximum power available from a quadratic drag law occurs when the tidal range inside the bay is reduced to 0.74 of the original amplitude.)

The continuous operation in a bay's entrance of current turbines is not less productive than accumulating the high tide waters in a retaining basin for release at outgoing tide, and power generation is compatible with flushing and purposes other than power generation.

### 6.9.1 Lagoons

Lagoons have been proposed as sites for tidal power plants. They are, it is claimed, less environmental damaging. One may consider the Shiwha Lake plant, now under construction in the Republic of Korea as a lagoon site. In the Republic of (South) Korea a lagoon type tidal plant has been functioning for some time (Jindo Uldolmok Tidal Power Plant).

Offshore Lagoons

The "perennially examined" proposal for a tidal power plant involving the Severn River in Wales, has been the subject of a Tidal Electric Company proposal that rated "Friends of the Earth" endorsement; the project would generate electricity by using thirteen estuary tidal lagoons built up on the estuary floor from rock. They would trap water at high tide and release it later through electricity-generating turbines.

Coastal Lagoons

The lagoons proposed in Wales at Fleming and Bridgewater Bay, similar to the offshore lagoon but sited with one side on land pose more risk to the birds on adjacent mudflats. Tide-generated coastal currents are part of the nearshore ocean resources that governments are eyeing to provide relief from global warming (Jones and Westwood 2005). Studies of tidal current harnessing were launched as collaborative public/private project early in 2005. The EPRI (Electric Power Research Institute) plays a leading role in the project (Siddiqui and Bedard 2005). The promised technical–economic results for tidal current power plants and several sites have not yet been disclosed. A keen interest has also been manifested by San Francisco for a hydro-venturi approach which contrary to the Rance River one, would involve no moving parts placed underwater; the value of Blue Energy or Verdant Power vertical or horizontal axis-type propellers might be determinant in any implementation here. Conventional air driven "turbines are to be placed on land and generate the electricity from the venturi suction. [In a Venturi tube the velocity is increased and the pressure lowered for a fluid that is conveyed through it, by a constricted passage. Named after the Italian G. B. Venturi (1746–1822).]

Indeed, this system causes the least disturbance for the fauna of the Bay area (Hammons 2005; O'Donnell 2005). Arlington (Virginia) based Verdant Power (Free Flow) is a major cog in the project to place tidal current turbines in New York City's East River (Grad 2004).

San Francisco city officials passed a resolution committing the city to the development of a 1 MW TPP during the 2003–2006 time span. But 2006 is long gone, and no further news concerning this proposal has been gleaned.

Conversion of marine currents' kinetic energy, different from tidal currents', has been proposed for many decades, witness the pharaonic Coriolis Project. The technology is similar to that for wind turbines but we are still quite far from a cost-effective large-scale system. Marine Currents Turbine Ltd (London, UK) has been involved in the development of such system for some time (Fraenkel 2002). Marine Current Turbines Ltd (MCT) (Basingstoke, UK) participated in the Sea-flow Project whose aim was the development of a tidal power plant based on the windmill concept. Such a scheme was established off the Devon Coast in the English Channel (Lang 2003). A 130 Mt unit was cemented to the sea-floor at about 1.1 km from the shore, protruding a few meters above sea-level. "Tidal currents turn the eleven-meter-long rotor, but as they reverse direction, the rotor's blades can be pitched to accept flow from the opposite direction. The rotor turns slowly in water, (but) at 17 rpm the speed is sufficient, with appropriate gearing, to harness the tide's energy and drive a turbine". Rotor speed varying, a power-conditioning system (with Ac–Dc–Ac conversion) allows to get a current output of 50 Hz (grid frequency). Mean power reaches 100 kW with a peak of 300 kW, fulfilling the needs of about 500 homes.

Japan, actively searches ways to extract energy from the ocean: researchers there examine the possibility of adapting reciprocating flow turbines developed for tapping wave energy to tidal energy harnessing (Takenouchi et al. 2006). This would evidently widen the field of tidal power use with extra-low head and time-varying energy characteristics. They concluded the output of the plant depends (evidently) on the tidal (range) and a pond inundation area. Others developed a new type of two-way diffuser suitable for a fluid flow energy conversion. Power from such fluid flow is proportional to the cube of the flow's velocity, thence by increasing velocity results in more power. Based upon experience with wind turbines, a system was developed for tidal streams (Kaneko 2004).

## 6.10 New Approaches, New Technologies

A new approach has been proposed by Venturigen in its tidal power generating concept (Venturigen 2006). Tidal power installations have been traditionally conceived as coastal project. Offshore tidal power has now been suggested by Ullman (2002). The installation would involve a self-contained impoundment structure. Likely sites include southwest Alaska, Puerto Mont in Chile and Swansea Bay in the UK.

The specific materials requirements for a tidal power plant have been discussed by Gooch (2000) with a view on capital costs reduction, harsh environment, longevity prolongation, generation efficiency increase.

In its effort to remain at the forefront of wave and tidal power development, the United Kingdom decided to support facilities at Blyth and in Scotland's Western Isles that aim to switch from research to implementation. The Severn River is probably the tidal river that has been most thoroughly studied, and the most often as well, with an eye on building a TPP. It has also been shown that had a plant been built after WW II, it would have quite long ago paid for itself. Because of environmental effects of a barrage, and no doubt the economic aspects of any project, no TPP has ever been even started. Fry (2005) has described alternatives to a TPP-with-barrage. Among these tidal lagoons schemes have been proposed and so has the modest utilization of the tidal stream (tidal current).

The same concern about environmental effects of a barrage led to the Hastings (2005) proposal. The time difference of tides between the two sides of the Cotentin Peninsula (France) has retained the attention of geographers and engineers alike. The tidal phase difference, or tidal delay, could be put to use, and the landform constitutes a natural "barrage". The Australian company Woodshed Technologies proposes to utilize the phenomenon to capture tidal energy. Still another scheme to dispense with the TPP barrage has been put forth by Appleyard (2005). The hydro-electricity generator of Hi-Spec Research (Fowey, UK) is an off-shore system that has a series of four channels created from two outer walls and three rows of chambers housing turbines.

A new perspective is perhaps added to utilization of tidal stream (current) by Mueller and Baker (2005) in their analysis of conversion to electricity of energy produced by marine energy converters, using direct drive electrical power take-off, without use of complex pneumatic, hydraulic or mechanical linkages. To this end, the linear vernier hybrid permanent magnet machine and the air-cored tubular permanent machine were examined. Taking into account the electrical properties of the topologies of these machines, problems of sealing, lubrication and corrosion were studied.

An attempt at defining the viability of tidal energy generated power was made in Wood (2005), however, it failed to take fully into account the rise in oil prices so that a revised schedule is timely. Tidal current tapping remains thus far economically not interesting. This opinion expressed in prior writing (Charlier 2003; Wood 2005) is reconfirmed by O'Rourke et al (2010): "Kinetic energy can also be harnessed from tidal currents to generate electricity and involves the use of a tidal current turbine. This is the more desired method of capturing the energy in the tides. However, tidal current turbine technology is currently not economically viable on a large scale, as it is still in an early stage of development."

The Delta Stream Turbine, Lunar Energy Tidal Turbine, Neptune Tidal Stream Device, Pulse Tidal, Hydrofoil and Tidal Fence Davis Hydro Turbine are all at the design stage, and to-date (2010) nothing has been built. Several scale-models have been built and tested including the Nereus and Solon Tidal Turbines, Evopod Tidal Turbine (Fig. 17). Gorlov Helical Turbine, TidEl Stream Generator and Stingray Tidal Energy Converter. The Stingray Tidal Energy Converter was installed and removed and is no longer under development. The SeaGen (Fig. 18) and Seaflow, Open Centre Turbine, Tidal Stream Turbine and Free Flow Turbines are the only full-scale operational tidal current turbines, which are generating electricity. All of these demonstration devices operate with a horizontal axis of rotation, which suggests that this may be the optimum configuration for tidal current turbines. At the rate at which tidal current turbine technology is developing, it is expected that other high potential tidal current sites will become available which were previously uneconomical for energy extraction (O'Rourke et al. 2010).

The European Commission funded CAOC (Co-ordinated Action on Ocean Energy), and put in place a platform enabling devices developers, wave and tidal energy researchers, and standard agencies to share information and knowledge in order to facilitate transition of wave and tidal energy from an energy research technology to one approaching commercial competitiveness with a medium-term time frame (Johnstone et al. 2006).

## 7 Quo vadis, Neptune?

This chapter has provided a condensed survey of the ocean energies which could be provided by Neptune, however, only a very limited amount can, considering current technology and present capital demands, be transformed into electrical power. Nevertheless, the contribution is not negligible and should be called upon. Of the several sources only a few have been or could be tapped. Marine wind parks have been multiplied during the last decade and are functioning satisfactorily. The noise they make and the un-sightliness have not been too disturbing and the feared for birds' hecatomb has not materialized. Bird casualties will be further reduced when slower rotating turbines will be generalized. Commercially viable and price competitive, WECS, however, are less reliable in delivering power than for instance TPPs.

Offshore development of winds in Europe, particularly in Germany and Britain, grew on a large scale after 2007. US market growth is strongly influenced by federal tax incentives to promote wind power. Europe continues to be the driving force in the wind power industry and accounted already for 66.5% of global megawatts installed in 2003 (Fig. 20). The five biggest wind turbine makers are Vestas and NEG Micon (both in Denmark), GE Wind (US), Enercon (Germany), and Gamesa (Spain) (Fig. 20).

The British government pledged to reach the targets of the 1997 Kyoto Agreement, (reduction of fossil energy use) can meet it by use of renewable energy. Some 82 wind farms were operating in the UK in 2005 already. There has been a lot of public support for renewable energy on environmental grounds but further expansion of wind use might lead to mass protest from people living on the coast as the future lies most probably in large wind energy farms out to sea with fears that wind farms of hundred of turbines could disturb birds habitats and is demanding an urgent study about the impact of wind farms on bird populations [Saturday Times (London) 13 February and 4 March 2004]. Yet, the same sensitivity does not prevail in Germany because at the same time that the British commissioned an ad hoc study the Germans granted permission for the offshore-windpark project "Borkum Riffgrund" in Germany's Exclusive Economic Zone in the North Sea (Plambeck Neue Energien AG). It entails the establishment of 77 wind mills with a capacity of 277 MW. Ecological aspects were taken into consideration when granting the permission, specifically targeting birds (Reichenbach 2003). Wind waves have been tapped for over a 100 years for power supply of independent buoys, beacons, scientific devices, in mariculture.

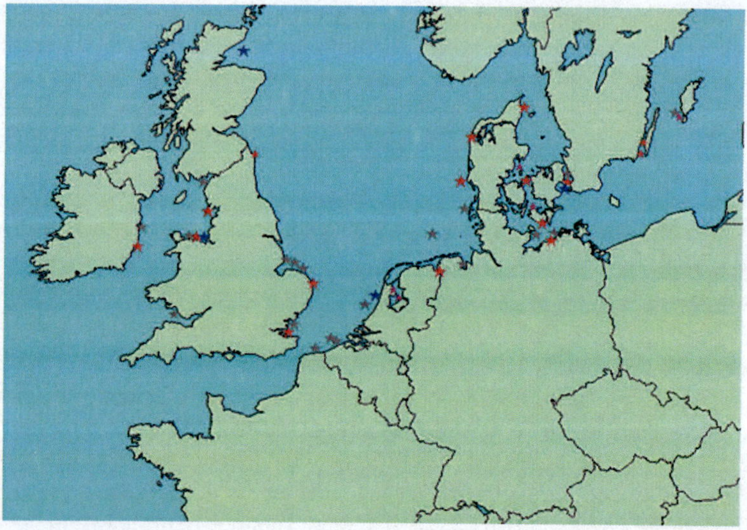

**Fig. 20** Existing (vivid*) and planned (dull*) locations of "farms" to tap marine winds (Thys 2006, Oceans Conference, Tokyo-Marinouchi)

At Orkney, Scotland, EMEC, the European Marine Energy Centre, is readying for the arrival of more devices capable of generating electricity from waves or tidal currents. It operates open-sea, grid-connected test facilities for prototype wave and tidal energy technologies.

Atlantis Resources Corporation is testing off Orkney Islands the world's biggest tidal turbine. Irish OpenHydro is testing its tidal turbines and generated electricity to the grid. It has successfully deployed a commercial turbine at the Bay of Fundy (Nova Scotia). A second test site for tidal devices off the island of Eday is functioning.

Scottish Aquamarine Power, active in both the waves and tides fields, is testing its Oyster wave energy converter (Fig. 4). Pelamis Wave Power generated electricity to the National Grid from its deep water floating device. Ocean Power Technologies that began construction off Oregon's coast of a commercial US wave energy farm, planned to install a wave project at the test site.

If ocean energy trial projects are successful in the next few years, the industry could represent a large source of renewable electricity generation capacity by 2025 [Pike Research Report, 2010, "Hydrokinetic and Ocean Energy"]. The Pike study assessed market opportunity for four marine energy technologies: ocean wave, tidal stream, ocean current, and ocean thermal. Regretfully tidal barrage and similar (fences, reefs) was left aside.

Building of wave power stations is determined by an optimum choice of water area of ocean with a stable reserve of wave energy, an effective construction of station with devices of smoothing of a non-uniform regime of agitation built in. From a commercial viewpoint, wave stations can work at use of power about 80 kw/m but the electric power delivered is from doubly to triply more expensive than from a "traditional" source.

Some plants have worked in Madras (India),Vinga (Gotheborg, Sweden) and in Toftestallen (Norway). Ocean Power Technologies Inc. (OPT), USA, and Iberdrola (Spain) embarked several years ago on a pilot project involving ten power generating buoys that were to generate 1.25 MW. This was to be a prelude to a 100 MW wave farm to be in place by 2006. These buoys take up less space per megawatt than either wind farms or conventional shore-based generators. Refinement in the area of turbines is still needed (Brooke 2003; Cruz 2008). Falcao and collaborators published about a decade ago a thorough survey of wave energy utilization in the Sustainable and Renewable Energy Review and Falcao updated the survey of the ad hoc technologies in 2010 (Falcao AF de 2010).

Thermal differences of water layers can provide power but as long as construction and operation may lead to use of simpler and less expensive turbines. The worst impact on life in the estuary of the Rance was observed during the first 3 years and construction; it took 10 years to reach a biological balance. Environmental objections may become fewer, considering that even at the so vocally criticized Rance River plant only one animal species disappeared (the sea-snail) and a female seal, brought back repeatedly to open sea, unharmed, stubbornly returns to the basin where she elected domicile each time passing through the system (Fig. 15).

Setting aside tidal power stations and WECS, already at work, ocean energy conversion may boost of some schemes having reached an advanced development stage. With a large and avid world market, and as proven in Kislaya suitable to modular construction, these renewable energies ought to be supported in their research and development as they can significantly contribute to quench the demand for electricity.

## References

Anonymous (1981) China looks at tidal power. Electr Rev 3: 208

Anonymous (1998) Experience in the topographical survey for a tidal power plant: survey for a tidal power plant. Design Institute of the Yao-Ning Province Administration of Water Management and Power Construction, Beijing (a Russian translation of the Chinese version)

Appleyard O (2005) High tide (tidal power). Int Dam and Power Const 57(4):36–37

Bahaj AS, Myers L (2003) Fundamentals applicable to the utilization of marine current turbines for energy production. Renew Energy 28(14):2205–2211

Baker AC (1997) Tidal power. Peregrinns, London

Banal M (1997a) History of tidal power in France La Houille Blanchde. J Int de l'Eau 52(3):14–15

Banal M (1997b) The technical origins of the tidal power station. La Houille Bl. J Int de l'Eau 52(3):16–17

Beels A, De Rouck J, et al. 2008, The impact of several criteria on site selection for wave energy conversion in the North Sea: Scientific Commons, University of Ghent

Bernshtein LB, Usachev IN (1957) Utilization of tidal power in Russia in overcoming the global and ecological crisis. La Houille Blanche-Rev Int de l'Eau 52(3):96–102

Blunden BS, Bahaj AS (2006) Initial evaluating of tidal power resource in Portland Bill, UK. Renew Energy 31(2):121–132

Bonnefille R (1976) Les réalisations de l'Electricité de France concernant l'énergie marémotrice. La Houille Blanche 31(2):87–149

Brooke J (2003) Wave energy conversion. vol 6. Elsevier, Amsterdam

Bryden IG, Melville GT (2004) Choosing and evaluating sites for tidal power development. Proc Inst Mech Eng 219(A-3):235–247

Bryden IG, Grinsted T, Melville GT (2004) Assessing the potential of a simple tidal channel to deliver useful energy. Appl Ocean Res 26(5):198–204

Ch'iu Hou-Ts'ung (1958) The building of the Shamen TPP. Tien Chi-Ju Tung-Hsin 9:52–56

Chaineux M-CP, Charlier RH (2008) Women's tidal power plant: Kislaya Guba's 40 candles. Renew Sustain Energy Rev 12(8):2508–2517

Charlier RH (1982) Tidal energy. Reinhold-Van Nostrand, New York

Charlier RH (2001) The view from China—small is beautiful. Renew Sustain Energy Rev 5(3):403–409

Charlier RH (2003) A "sleeper" awakes: tidal current power. Renew Sustain Energy Rev 7:515–529

Charlier RH (2004) A sleeper awakes. Power from tidal currents. Ren Sust Energy Rev 8(6):34–46

Charlier RH (2008) Nothing new under the sun: tidal power status and perspectives. J Mar Des Oper N°B-13:15–24

Charlier RH (2010) Power from Arctic waters. (Book of Extended Abstr. Int. Polar Year, in press)

Charlier RH, Finkl CW (2009) Ocean energy: tides and tidal power. Springer, Heidelberg

Charlier RH, Justus JR (1993) Ocean Energies-Environmental, and Economic and Technical Aspects of Alternative Power Sources. Elsevier, Amsterdam

Charlier RH, Menanteau L, Chaineux M-CP (2003) The rise and fall of the tide mill. In: Proc of the 6th Intern Conf on Hist Oceanography, Qingdao, PR China, UNESCO-Paris and 1st Inst Oceanogr Qingdao, PR China, Part III, chapter 39, pp 315–340

Clarke JA et al (2006) Regulating the output characteristics of tidal power current stations to facilitate better base load matching over the lunar cycle. Renew Energy 31(2):173–180

Cruz J (ed.) (2008) Ocean wave conversion. Springer, Heidelberg

De Doncker A, Loos J (2009) Kust-en zeegids. Willy Ibens, Mechelen (Mechlin, Malines)

De Lory RP (1987) Prototype power plant achieves 99% availability. Sulzer Techn Rev 1:3–8

Erikson H, Fiorentino A, Moroso A (2008) The vertical axis turbine Kobold in the Strait of Messina—a case study of a full scale marine current energy prototype. In: Abstr./Proc. World Maritime Technology Conference, London, March 2008, no pp. nbrs

Falcao AF de O (2010) Wave energy utilization: a review of the technologies. Renew Sustain Energy Rev 14:899–918

Fay JA (1982) Design principles of horizontal-axis tidal current turbines. In: Proc Intern Conf New Approaches to Tidal Power, Bedford Institute of Oceanography, Dartmouth, Canada

Finkl CW (1980) Le strada del mare. Jacques Cousteau Planeta Mare Enciclopedia di Scienza e di Avventura, Gruppo Editoriale Fabbri, Milano, 1(4):81–96

Finkl CW, Charlier RH (2009) Some environmental considerations of electrical power generation from ocean currents in the Straits of Florida. Renew Sustain Energy Rev 13:2597–2604

Frau JP (1993) Tidal energy—promising projects. La Rance, a successful industrial-scale experiment. In: IEEE Trans on Energy Conversion, pp 552–558

Fraenkel PL (2002) Power from marine currents. Proc Inst Mech Eng J Power & Energy 216(A-1):1–14

Fry C (2005) All at sea (tidal power). Power Eng 19(5):24–27

Garrett C, Cummins F (2004) Generating power from tidal currents. Proc Coast Ocean Eng 130(3):114–118

Garrett C, Cummings F (2005) The power potential of tidal currents in channels. Proc Roy Soc London A 461(2060):2560–2572

Gibrat R (1976) The current revival of tidal power. Proc of Association pour l'avancement des sciences et technologies de la documentation (ASTED), Montreal

Gibrat R, Auroy F (1956) Problèmes posés par l'utilisation de l'énergie des marées. World Power Conf, Vienna: 12, 111, H/22, pp 4299–4328

Gooch DJ (2000) Materials issues in renewable power generation. Int Materials Rev 45(I):1–14

Gorlov AM (1982) Hydropneumatic approach to harnessing tidal power: New Approaches Tidal Power. In: Proc Conf Bedford Inst Oceanog, Bedford, Nova Scotia, Canada, pp 8–16

Gorlov AM (2000) Tidal power. Encyclopedia of Ocean Science, p 2955

Gorlov AM (2003) The helical turbine and its application for tidal and wave power. Abstr Oceans 4:1996–1997

Grad P (2004) Changing tide of power generation. Eng Australia 76(11):51–52

Hammons TJ (2005) Energy potential of the oceans: tidal, wave, currents and OTEC. In: Proc 40th Int Univ Power Conf, Orlando, Florida, 1, 2, pp 1047–1057

Hastings S (2005) Working with Nature: tidal power. Int Power Gener 28(3):27–28

Hilsop D (ed) (1992) Energy Options: an introduction to small-scale energy technologies. Intermediate Technology Publications, Rugby

Johnstone CM, Nielsen K, Lewis K, Sarmento A, Lemonis G (2006) European Commission coordination action on ocean energy: a European platform for sharing technology information and research outcomes in wave and tidal energy systems. Ren Energy 31(2):191–196

Jones AT, Westwood A (2005) Recent progress in offshore renewal energy technology development. Proc IEEE Gen Meet Power Engng Soc 2:2017–2022

Khatri D (2010) Economics of Taller Wind Towers. Renew Energy World North Am Mag 2(1):3–10

Kristoferson LA, Bokalders V (1991) Renewable energy technologies: their applications in developing countries. Intermediate Technology Publications, Rugby

Mauad FF, Mariotoni CA (2000) Reaction to simple curvature blades, an option for the generation of energy in small energy plants. Informacion Tecnologica 11(5):11-16 [in Spanish]

Merlin A, Sandrin P, Goes JM, Hillariet M (1982) Agra, the new operational model for the La Rance tidal power plant. IEEE Trans Power Apparatus & Syst 101(2):290–294

Mueller MA, Baker NJ (2005) Direct Drive Wave Energy Converters. I Mech E Journal of Power and Energy 219(A3):223–234

Myers L, Bahaj AS (2005) Simulated electrical power potential harnessed by marine current turbine arrays in the Alderney Race. Renew Energy 30(11):1713–1731

O'Donnell P (2005) Update '05: ocean wave and tidal power generation in San Francisco. In: Proc Gen Meet IEEE Power Engng Soc, Orlando, Florida, N°2, pp 1990–2003

O'Rourke F, Boyle F, Reynolds A (2010) Tidal energy update 2009. Appl Energy 87:398–409

Ottewell S (2003) Ireland's renewable island. Power Eng 17(3):10–11

Reichenbach M (2003) Auswirkungen von Windenergieanlagen auf Vögel—Ausmass und planerische Bewältigung. N° 123 in series Landschaftsentwicklung und Umweltforschung Technische Universität, Abt Publikationen, Berlin

Sandrin P (1980) Agra: s new operational management system for the La Rance tidal power plant. Bull Dir Et & Rech Sr B 3:29-40

Sanford L (2003) Winning the tidal race. Modern Power Syst 23(7):11–12

Sergey F (2003) Winning the tidal race. Mod Pow Syst 3(7):11–12

Shiono M, Suzuki K, Kiho S (2000) An experimental study of the characteristics of a Darrieus turbine for tidal power generation. Electr Eng Jpn 1, 32, 3, 38–47

Siddiqui O, Bedard R (2005) Feasibility assessment of offshore wave and tidal power production: a collaborative public/private partnership. Proc Gen Meet IEEE Power Engng Soc 2:2004–2010

Smith D (2005) Why wave tide and ocean current promise more. Peregrinns, London

Takenouchi K et al (2006) On applicability of reciprocating flow machines developed for wave power to tidal power conversion. Renew Energy 31(2):209–223

Ullman PW (2002) Offshore tidal power—beyond the barrage. Modern Power Syst 22(6):38–39 (See also Int Water Power and Dam Constr 54(9):24–27)
Usachev IN et al (2004) Performance control of a marine power plant in the Russian Arctic coast and prospects for the wide-scale use of tidal energy. Power Technol Engng 37(4):201–206
Van Walsum E (2003) Barriers against tidal power. Int Water Power Dam Constr 55(9):34–41
Venturigen (2006) Tidal flow energy system. Lewis Eng Technol 5(10):19–20
Wailes R (1941) Tide mills in England and Wales. Jun Inst Eng J Record Trans 51:91–114
Wilson EM (1977) Tidal energy and system planning. Cons Eng London 41(4):25
Wood J (2005) Marine renewables face paperwork barrier. UK Inst Electr Eng Soc 51(4):26–27
Zhenzia L, Donking X, Mingzuo F (1989) Evaluation on sited of two tidal power stations. In: Coastal Zone '89, Charleston, North Carolina, pp 2203–2209

# Coastal Ecosystem Management Using a Wave Energy-driven Seawater Pump: Some Mega-engineering and Environmental Aspects

**S. P. R. Czitrom, I. Penié and G. de la Lanza**

## 1 Introduction

Mexico has an extensive coastline and a great number of coastal water bodies where intense economic activities take place. Some of the most important and famous touristic sites are located along the coast, and most commercial products are moved through ports. In addition, an increasingly greater proportion of marine protein is produced by aquaculture in coastal water bodies.

More than 130 coastal water bodies are found in Mexico, of which the total area corresponds to a 1.25 km wide strip spanning the whole coast of the country (11,000 km). A significant part of this ecosystem is threatened by the intense economic activity it supports, that in turn generates important resources for the country. Ironically, the buffering capacity and low dynamics of this ecosystem, characteristics that favor many port and biological activities, also render it vulnerable to the accumulation of sediments and exogenous substances. The inputs to the seawater and sediment, both systematic and accidental, constitute the main factors that affect the environment (Xue et al. 2004; Guerra-García and García-Gómez 2005; Borja and Elliott 2007; Cornelissen et al. 2008). Port pollutants pose negative externalities to natural habitats and are generated by routine shipping operations in Mexico's ports.

A considerable part of the Mexican coastal ecosystem is threatened by a deficient planning of anthropogenic activities, and studies on which to judge the feasibility of coastal management and development plans currently are insufficient. This has

S. P. R. Czitrom (✉) and I. Penié
Instituto de Ciencias del Mar y Limnología, Universidad Nacional Autónoma de México, Circuito Exterior S/N, Ciudad Universitaria, 04510, Mexico D.F., Mexico
e-mail: stevenczitrom@gmail.com

G. de la Lanza
Instituto de Biología, Universidad Nacional Autónoma de México, Circuito Exterior S/N, Ciudad Universitaria, 04510, Mexico D.F., Mexico

V. Badescu and R. B. Cathcart (eds.), *Macro-engineering Seawater in Unique Environments*, Environmental Science and Engineering,
DOI: 10.1007/978-3-642-14779-1_28, 

resulted in an inadequately controlled growth of economic activities as well as environmental damage that becomes evident in the medium term, by which time the damage can be so economically costly as to be practically irreversible (De la Lanza-Espino et al. 1998; Espejel et al. 2004). Additionally, local and global climate variations favor sedimentation that promotes geomorphologic changes and impacts on coastal ecosystems (Ortiz and de la Lanza 2006).

The rehabilitation of these coastal areas has usually included the treatment of effluents and periodic sediment dredging. However, these procedures are costly, often result in a slow rehabilitation, or lose effectiveness on the long term (Dauvin 2008; Shin et al. 2008). An adequate environmental rehabilitation of these seawater bodies should include decreasing the input of pollutants, favoring an increase in biodiversity, and increasing the hydrological dynamics.

Examples of coastal ecosystems that have been affected may be found along the Mexican Pacific, the Gulf of California, the Gulf of Mexico and the Caribbean Sea where agro-industrial, touristic, oil production and port activities have caused serious pollution problems. Enormous oil spills in the Gulf of Mexico (Ixtoc I, gushing uncontrollably from 3 June 1979 until 23 March 1980 and BP/Deepwater Horizon on 20 April 2010) presage future coastal pollution caused by offshore economic development. In some areas, the domestic and industrial residues that have been poured into coastal seawater bodies over long periods of time have caused them to remain permanently in a polluted state despite treatment now being given to some land-based point discharges. Considering that it is practically impossible to eradicate or limit the uses to which coastal ecosystems are subjected, apart from decreasing the inflow of pollutants, it is necessary to implement coastal management measures that favor a renewal of the seawater and contribute positively to the rehabilitation of the coastal ecosystem.

A Wave Energy-Driven Seawater Pump (SIBEO) has been developed at the National University of Mexico (UNAM). It may be used to clean polluted coastal seawater bodies by injecting "clean" and oxygen-rich seawater from the surf zone into stagnated areas. Since it has no moving parts, it allows the passage of marine organisms and so may also be used in the management of coastal lagoon fisheries. The SIBEO uses the renewable and freely available energy of waves, leading to remarkably low economic and energetic operational costs.

In an initial proposal for a national coastal management plan, we propose here to install SIBEOs to inject seawater into three coastal ecosystems representative of the country: Bojórquez Lagoon (BL) in the State of Quintana Roo, Lagartero Lagoon (LL) in Oaxaca, and the Port of Ensenada (PE) in Baja California (Fig. 1).

In the case of BL located by the resort of Cancún, one of the coastal ecosystems most affected by tourism development, the SIBEO will be combined with measures to rehabilitate the mangroves. The SIBEO will enhance the dynamics and oxygenation of the water in the lagoon thus decreasing the trend to eutrophication and promoting the decomposition of organic matter both in the water and sediment. These effects will be enhanced by the mangrove forest rehabilitation, and various degrees of ecosystem health improvement may be expected in the short, medium and long terms.

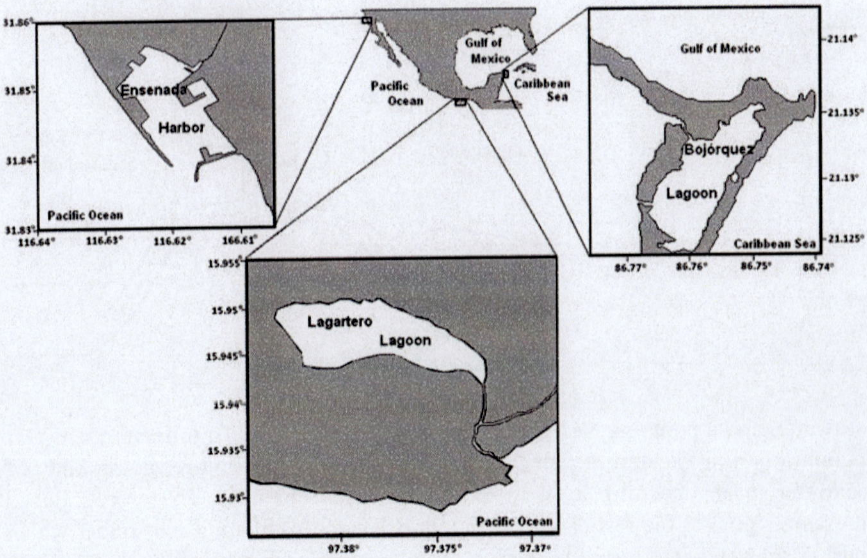


**Fig. 1** Initial sites proposed for an initial coastal management plan with the SIBEO

LL is typical of the coastal lagoons of the Mexico's Pacific Ocean coast. Its inlet closes by sediment accumulation and the SIBEO, apart from stabilizing its hydrological regime, will favor the development of an extensive shrimp fishery. In the case of the PE, the SIBEO will be used to dilute and oxygenate the water and sediments that have received and harbor domestic and seaport activity-derived pollutants over many decades.

The installation of the SIBEOs for the rehabilitation of the proposed ecosystems will be the first permanent application on a real scale of this technology invented and developed at UNAM. The experience acquired by the installation of the SIBEO in the selected ecosystems will serve as a research and development base for its use in other areas of the national and international coastal ecosystem of the Earth's ocean.

## 2 The SIBEO

SIBEO (Fig. 2) is the Spanish acronym of "Sistema de Bombeo por Energía de Oleaje" (Wave energy-driven seawater pump). It was developed and patented by the Instituto de Ciencias del Mar y Limnología of UNAM, to reduce the residence time in coastal ecosystems (Czitrom 2002; Czitrom et al. 2003). The SIBEO acts by transporting "clean" and oxygen rich seawater from the adjacent ocean into the affected coastal water body, to promote its ventilation and cleaning. Since it has no moving parts and uses wave energy only, the SIBEO is a sustainable coastal management tool with low operational and maintenance costs.

The system is composed of a resonant duct, which is exposed to the action of the ocean's waves, an exhaust duct, which discharges into the receiving body of

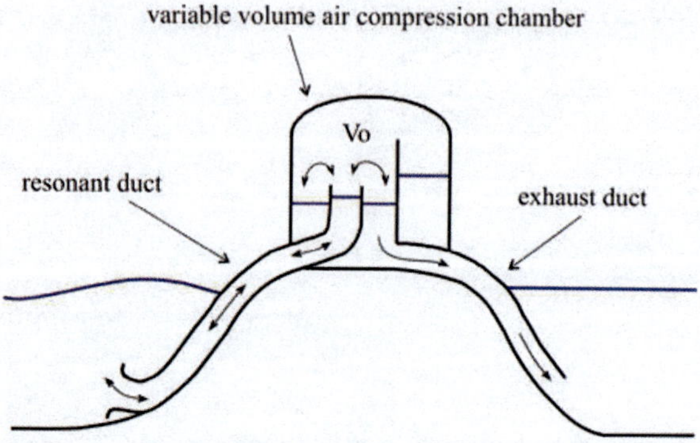


**Fig. 2** Schematic diagram of the SIBEO

seawater, and a pumping element connecting the two ducts. To prime the seawater pump, a partial vacuum elevates seawater from the ocean and receiving body of seawater to an operating level in the compression chamber. The wave-induced pressure signal at the resonant duct mouth drives an oscillating movement which pushes seawater into the compression chamber with each passing wave. The spilt seawater cannot return into the resonant duct since it protrudes above the seawater level in the compression chamber, and it accumulates therein, driving a gravity flow to the receiving body of water through the pouring discharge duct.

The SIBEO is basically a system composed of two seawater masses (in the resonant and exhaust ducts), coupled by an air spring (the gas in the compression chamber). Any system with springs and masses has an intrinsic natural frequency of oscillation that depends on the size of the masses (in this case the amount of seawater in the ducts) and the hardness of the spring (for the SIBEO, this depends on the volume of air). When the natural frequency of oscillation of the SIBEO coincides with the frequency of the driving ocean waves, the system becomes resonant, the seawater oscillations increase in size, and seawater pumping is maximized. This is similar to a child's suspended swing in which the oscillations increase when it is amplified at the appropriate time during the swing's pendulum-like movement.

The development of the SIBEO has gone through several stages, from the conceptual design to the deduction of the differential equations that describe it and tests with a scale model in a wave channel (Czitrom et al. 2000a). A 1:4 scale prototype of the SIBEO was installed on the coast of Oaxaca, near LL to test it under marine conditions, to demonstrate the feasibility of installing it in a high wave-energy area, even by hand, and to show that marine organisms such as shrimp post-larvae could be pumped without suffering damage for extensive aquaculture. The prototype was successfully installed from land with the help of fishermen from a cooperative (Czitrom et al. 2000b) and proved its potential as a seacoast management tool. The prototype was dismantled after the tests.

Outstanding among these tasks is the performance of the numerical model that integrates the differential equations of the SIBEO. Figure 3 shows the flow measured with a 1:25 scale model ($Q_{\mathrm{Exp}}$) in wave channel experiments, and the

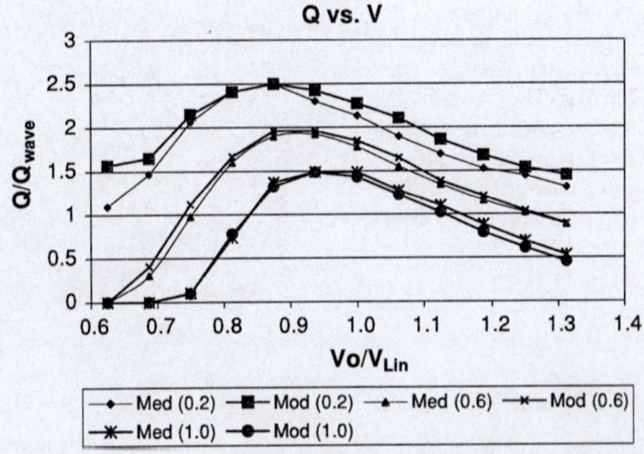


**Fig. 3** Flow of the scale model ($Q_{Exp}$) and the numerical model ($Q_{Mod}$) of the SIBEO, for several tuning volumes and various collector heights (Czitrom et al. 2000b)

flow predicted by the numerical model ($Q_{Mod}$), for a variety of collectors (the outstanding height of the resonant duct above the seawater level in the compression chamber) and volumes of air in the compression chamber. It is clear that the numerical model successfully reproduces the experimental data, showing it is a trustworthy tool for fluid flow design and, significantly foretelling in the case of real-world applications.

In the ocean, wave frequencies normally change over time, so that the SIBEO will not always be exactly at resonance. If the range of wave frequencies is small, the SIBEO can be designed to resonate with an intermediate frequency and its performance will decrease only marginally when it is slightly out of resonance.

For cases with a wide range of ocean wave frequencies, a system was invented, and patented, to tune the natural frequency of oscillation of the system so that it coincides with that of the most energetic waves at all times (Czitrom 2002). Tuning is attained by changing the hardness of the air spring in the SIBEO's compression chamber, by modifying its volume of air through the interchange of seawater and air with an adjacent container (this container is not shown in the schematic diagram above). Tuning is controlled by a programmed electronic circuit that analyses the wave signal to locate the frequency with the greatest energy. This same circuit controls an air compressor and a seawater pump to regulate the volume of air in the compression chamber to maintain a state of resonance with the most energetic waves.

Figure 3 shows how tuning can increase SIBEO efficiency, particularly when the ocean wave frequency is very different from the SIBEO's natural frequency of oscillation. This is to say, tuning makes it possible to move along the curves shown in Fig. 3 to maximize the pumped seawater flow.

## 2.1 Design of the SIBEO

From the above basic description of the pump operation, it is clear that the SIBEO performance hinges importantly on the efficiency with which the ocean wave energy drives the seawater oscillation in the resonant duct. The oscillations

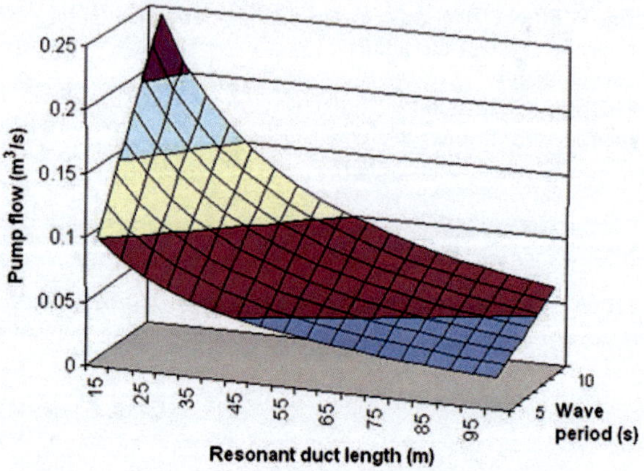


**Fig. 4** Pumping flow for various resonant duct lengths and wave periods for Cancún, Mexico. The plotted values were obtained using the solution to the linearized equations of the SIBEO (Czitrom and Prado 1999)

produced by the wave energy, and therefore the pumping, will be greater when the resonant duct is shorter and there is a smaller seawater mass to be moved.

Figure 4 shows the maximum flow that can be obtained with the SIBEO at resonance, as a function of the driving wave period and the resonant duct length, for a particular case in the Mexican Caribbean. This plot was obtained using the solutions to the linearized equations of the SIBEO (Czitrom and Prado 1999). In agreement with the previous paragraph, it can be seen that the shorter the resonant duct (tens of meters), the greater the pumping, up to 10 or more times greater than when the resonant duct is long (hundreds of meters). It is clear that one of the main criteria when designing the SIBEO for a particular site is to strive for as short a resonant duct as possible, to maximize the flow.

Another design criterion is that, in order to diminish the silting problems, the intake of the SIBEO should be placed seaward of the surf area to avoid the region with most wave-suspended sediment. The ideal case is when the waves break very close to the seacoast (e.g., against a port breakwater), which makes it possible to place the pumping element on land and the intake mouth behind the breaking waves, even with a short resonant duct (as in Fig. 2). As will be seen further on, this is the situation in the PE where it will be possible to place the pumping element on the breakwater that protects the port, and the mouth of the resonant duct will lie behind the breaking waves. In this case, the short resonant duct will guarantee maximum pumping, and the pumping element on land will facilitate the control and management of the SIBEO and its tuning system.

This, however, is not the case in BL or LL where the waves break at least a hundred meters offshore. In these cases, a maximum pumping efficiency (short resonant duct) and a seawater intake behind the breakwater are only possible by submerging the compression chamber of the pumping element and anchoring it to the seafloor (see Fig. 5).

The operational principle of the SIBEO is the same whether the pumping element is placed on land or submerged in the ocean close to shore. The difference between these two cases is that, when on land, a partial vacuum elevates seawater to the

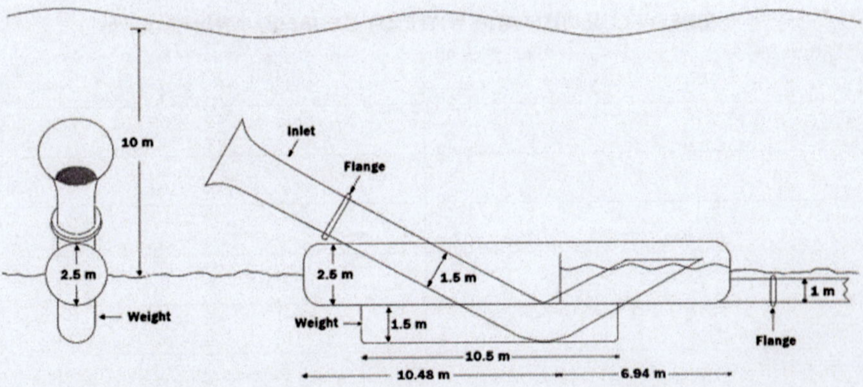


**Fig. 5** Pumping element of a submerged SIBEO designed to resonate with 7 s period waves

operating level in the compression chamber, whereas air must be injected to prime the pump when it is submerged. The resonant duct reaches above the seawater level in the compression chamber in both systems so that, once seawater has cascaded, it cannot return to the resonant duct, thus establishing the pumping mechanism. In both systems, the resonant duct is communicated with the sea-level while the exhaust duct is communicated with the lagoon and, in agreement with the principle of communicating vessels; the difference in the levels between the ocean and lagoon is reproduced in the levels in the compression chamber. At present, wave-channel tests are being carried out with a submerged scale model, which confirm this description.

Placement of the pumping element underwater to increase luquid flow, however, results in serious difficulties when implementing the tuning system. The installation of electro-mechanic valves, compressors and seawater pumps underwater, hundreds of meters offshore, complicates the design, operation and maintenance of the SIBEO and increases both the cost and the probabilities of system-wide failure. In these cases, it may be simpler to abstain from implementing the tuning system and to design the SIBEO to resonate at a fixed intermediate frequency. The decrease in flow resulting from not implementing the tuning system must be evaluated and compared to that obtained when the pumping element is placed on land with a long resonant duct to avoid silting.

An analysis for Cancún, Mexico, is presented as a representative case of the situations in which the submerged SIBEO will be used. For this analysis, the flow that can be expected with the SIBEO illustrated in Fig. 5 (without the tuning system) was compared with the expected flow using the same design but with the dynamic tuning system implemented. Both alternatives were computed using the numerical model of the SIBEO driven by the ocean waves and tides off Cancún (CFE 2004).

These ocean wave conditions indicate that the period varies mainly between 4 and 11 s, with a maximum incidence of 27% between 7 and 8 s. A period of 7 s was chosen for the design of the submerged pumping element shown in Fig. 5.

Figure 6 presents the annual average SIBEO flow driven by ocean waves with significant periods and heights observed off Cancún, plotted against the tidal level

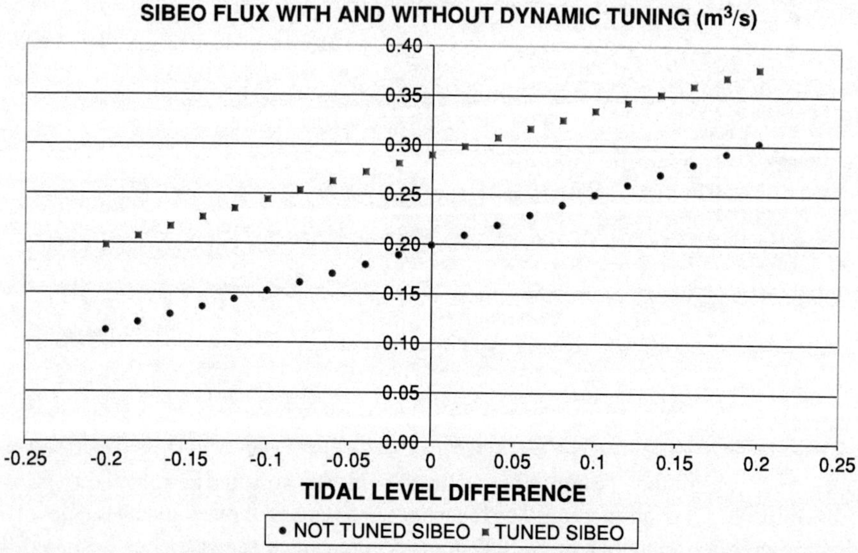


**Fig. 6** Average annual SIBEO flow, both with and without dynamic tuning, plotted against the tidal level difference between the sea and BL. The flows calculated with the numerical model for each wave period (*Tp*), significant height (*Hs*) and tidal level difference were averaged for each tidal level difference, weighing with the percentages of time for each *Tp* and *Hs* present (CFE, Comisión Federal de Electricidad 2004)

difference, with and without dynamic tuning. The line of higher values in Fig. 6 corresponds to the SIBEO flow with dynamic tuning. The flow varies with the tide from 0.2 to 0.38 $m^3/s$ with an annual average of 0.29 $m^3/s$, whereas the SIBEO tuned at a fixed period of 7 s pumps an annual average of 0.2 $m^3/s$. That is, the SIBEO numerical model predicts an average 30% drop in flow when the dynamic tuning system is not implemented.

The effect of the ocean tide on the SIBEO flow is linear, increasing when the ocean level is higher than the lagoon and vice versa. Since the mean level in the lagoon and the ocean are practically the same, no significant net effect of the tide on the SIBEO flow throughout the year is to be expected.

The SIBEO shown in Fig. 5 was designed to pump optimally when the most energetic waves off Cancún have a period of 7 s. In order to analyze the performance when other periods predominate, the flow of the SIBEO when excited by 1 m high waves and various wave periods is presented in Fig. 7. The flow increases rapidly with wave periods from 3 to 7 s up to a maximum of 0.275 $m^3/s$, at resonance. For periods above resonance, the flow decreases, albeit more slowly.

The same Fig. 7 shows the fraction of time each wave period is present in the deep waters off Cancún. From this it can be seen that the efficiency of the SIBEO will be greater than 85% during 70% of the time (20 + 27 + 23), whereas it will be around 98 during 47% of the time (20 + 27).

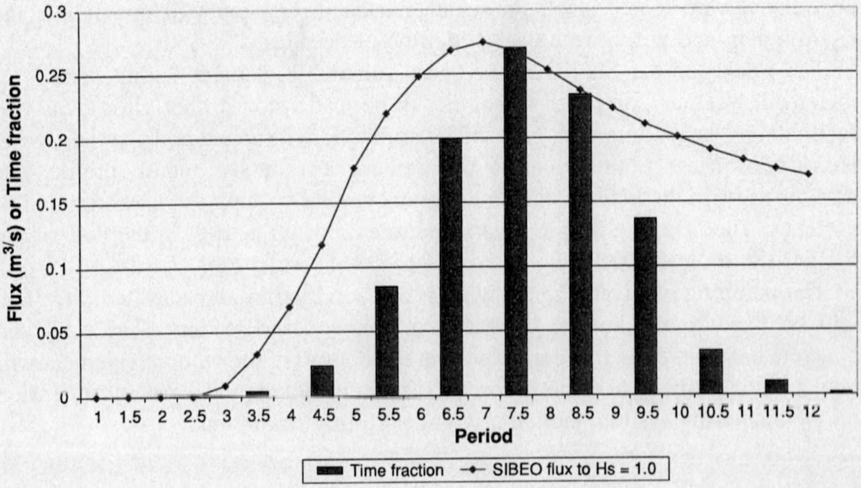


**Fig. 7** The fluid flow pumped by the submerged SIBEO, when driven by a 1 m high wave and various periods. The vertical bars correspond to the fraction of time during which waves of the various periods are present (deep-water data in the ocean offshore of Cancún, taken from the United Kingdom Meteorological Office webpage)

It is clear that the efficiency of the SIBEO approaches a maximum when the most common ocean waves are present and, although the efficiency is lower when less common periods occur, the SIBEO pumps seawater with a very reasonable efficiency for the whole range of frequencies present. This explains why, in spite of not implementing the dynamic tuning system, the net annual flow of the SIBEO decreases only 30% with respect to the tuned SIBEO.

In the case of Cancún, the increase of some 10 times the flow when placing the pumping element underwater, together with the simplification in design and reduction in costs resulting from not implementing the dynamic tuning system, more than compensates for the 30% loss of flow when using the fixed tuning system.

We have not included in this present analysis, however, the additional cost of installing and mooring the pumping element to the ocean floor and the additional vulnerability to extreme conditions when submerged, vis a vis the pumping element installed on land. Nevertheless, the ten-fold increase in efficiency still leaves much room to maneuver before a cost-benefit analysis determines the SIBEO should be placed on land.

## *2.2 Other Considerations with Respect to the Design of the SIBEO*

The pumping element of the SIBEOs will be manufactured in carbon steel with cathodic protection, and the exhaust duct will be made from high density polyethylene or coated steel, for which reason the system will not pollute the environment and will have a useful lifetime of decades.

The intake of the SIBEO is the most vulnerable element during periods of inclement weather, for which reason it will be made of cast fiber-glass. This will facilitate its substitution in case of destruction by, for example, a hurricane. Hurricanes affect both seacoasts of Mexico. The intake mouth profile was developed from theoretical and experimental studies to suppress the formation of vortexes. The resulting profile favors an increase of up to 20% in the flow of the SIBEO (Stern et al. 1999).

The pumping element (Fig. 5) will be partly buried in an excavated ditch and will be fixed to the bottom by stakes driven into the seafloor. The steel and concrete ballast weight that can be seen at the bottom of the pumping element will help maintain it fixed to the seafloor, decreasing its flotation and easing the tension on the anchoring system, particularly under stormy conditions.

## 3 Management Areas

Three seawater bodies representative of the Mexico's coastal ecosystems (Fig. 1) were selected as the starting point for a coastal management plan based on the use of the renewable energy of ocean waves. In two of them, BL and PE, the SIBEO will be used to rehabilitate the ecosystem by increasing the circulation and ventilating the water. In the third, LL, it will be used to enhance the shrimp fishery.

Swell waves arriving at the coast along the Mexico's Pacific Ocean shore have periods typically in the 10 to 20 s range and an energy density of about 25 kW/m (Cruz 2008). By contrast, along the Mexico's Caribbean Sea shoreline swell waves have shorter periods and an energy density of 5 kW/m. Despite the difference, SIBEO is designed to operate on both seacoasts can pump similar flows (typically 0.2 $m^3/s$ for a 1.5 m diameter resonant duct). This is due to the fact that shorter ocean wave periods require shorter resonant ducts, so that an SIBEO designed for the Caribbean Sea, for example, will use the available wave energy more efficiently and can thus pump similar flows to one designed for the Pacific Ocean.

### *3.1 Environmental Rehabilitation: Bojórquez Lagoon, Cancún*

BL (Fig. 8) in the state of Quintana Roo, a seawater body that forms part of the Nichupté Lagoon System, is an example of a location where the SIBEO can be used for environmental rehabilitation. The loading capacity of this lagoon is reduced as a result of its limited seawater exchange with the rest of the lagoon system and the adjacent ocean. Merino et al. (1992) estimated a hydraulic residence time of between two and four years, testifying to the reduced loading capacity of BL.

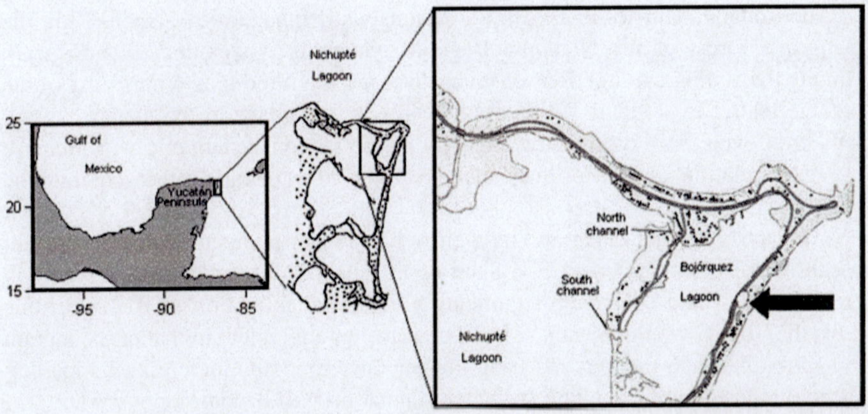


**Fig. 8** Bojórquez Lagoon in the Nichupté Lagoon System, Quintana Roo, Mexico. The arrow indicates the most convenient point for the installation of the SIBEO. It has the shortest distance to the ocean, is located mid-way along the island barrier of the lagoon, and does not affect the highway infrastructure

**Fig. 9** Floating conglomerates of organic matter (*left*) and mangrove trees affected by Hurricane "Wilma" in BL

Carbajal-Pérez (2009) and Herrera-Silveira (2006) recorded abnormally high nutrient contents of nitrates (18 μM) and ammonium (10 μM) in BL at various times during the year. They associated these values with inputs of both underground freshwater and the residual freshwater discharged by the hotels (Fig. 9).

The resort tourism development of Cancún, starting in the 1970s, is the main cause of its environmental decay. Treated and untreated residual freshwaters have been discharged into the lagoon for decades so that the lagoon now presents a significant decrease in biodiversity, an accumulation of organic matter in the seawater and sediment, and the production of foul-smelling air even in the light hours of the day (González-López 1989; Reyes and Merino 1992; Herrera-Silveira 2006). These symptoms seriously affect the fundamentally touristic objective to which the BL ecosystem has been dedicated.

Additionally, Hurricane "Wilma" seriously affected more than 90% of the mangrove forest of the Nichupté Lagoon System in November 2005 (Fig. 9). Among the ecological functions of mangroves are the filtering and purifying of the organic matter in wetlands (Páez et al. 2007). Any damage to the mangrove will thus limit even more the loading capacity of the lagoon system and will increase the accumulation of the residues that are even now being discharged into the system.

A 0.2 $m^3/s$ flow of clean and oxygen rich seawater pumped from the adjacent ocean at the point shown in Fig. 8, in conjunction with a mangrove restoration programme, would contribute significantly to the rehabilitation of BL. With this flow, the SIBEO would inject a volume of seawater equivalent to that of the lagoon in approximately 6 months, reducing its residence time and increasing its loading capacity considerably. Lagoon seawater stagnation will become impossible.

Circulation in the Nichupté Lagoon System is mainly wind-driven (Carbajal-Pérez 2009). Even moderate winds generate seawater speeds of at least 5 cm/s, reaching up to 10 cm/s in certain areas of the lagoon. A seawater speed of 5 cm/s over a period of 12 h corresponds to a linear displacement of 2.16 km, which is approximately the largest dimension of BL. From this fact of physics it can be inferred that wind generates a significant exchange of water throughout the lagoon on a time scale of days. Thus, the seawater pumped by the SIBEO can be re-distributed to most of the lagoon on a short time scale.

The pumped seawater would dilute the lagoon, decrease its tendency to eutrophication, and promote the decomposition of organic matter in the seawater and seabed sediment. As the seawater quality improves, the lagoon is expected to recover the natural composition and structure of its flora and fauna that existed before the tourists became a major local environmental impact factor. In the short term (1 month), the influence of the SIBEO will be detected near the discharge area, while reaching the rest of the lagoon in the medium and long term (6 months to 1 year), aided by the ocean's tidal currents and the ambient wind.

It is quite likely that a healthy ecosystem in BL will encourage the return of fisheries species which might have been exploited commercially in the past. Nevertheless, in view of the brisk activities related to tourism, it seems likely that only sport fishing might develop significantly once the lagoon is rehabilitated.

## *3.2 Environmental Rehabilitation: The Port of Ensenada*

PE (Fig. 10), one of the most important in Mexico, has received considerable anthropogenic impact over several decades. As a result of the limited loading capacity of the port, pollutants have accumulated in the sediment, from where they can again become available to the water column. Despite the treatment that effluents receive in the port nowadays, the water circulation pattern within the port, characterized by closed gyres in the northern and southeastern areas, favors the accumulation of pollutants and prevents the water from being renovated from

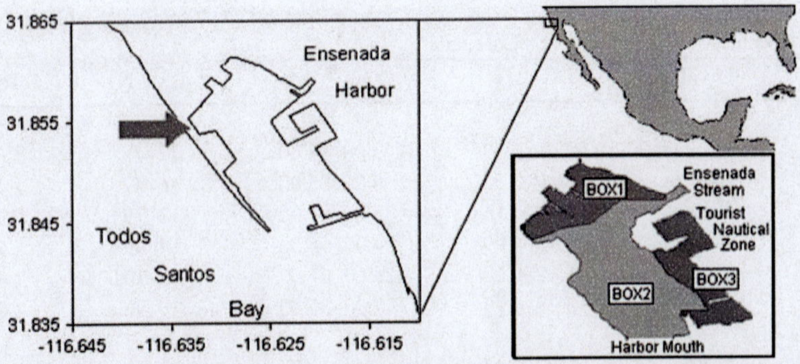


**Fig. 10** The Port of Ensenada, Baja California, Mexico. The *arrow* indicates the place where the SIBEO can be installed. The subdivision into *boxes* responds to the geomorphologic characteristics and hydrologic dynamics of the seaport (De la Lanza-Espino et al. 2010)

the adjacent sea (Czitrom et al. 2003). On 2 October 1858 the only known hurricane to make landfall on Mexico's west coast so far north of the usual tropical cyclone track adversely affected the still undeveloped Port of Ensenada (Chenworth and Landsea 2004). Such a future destructive ocean storm event is, thus, to be anticipated and accommodated by a prudently-designed SIBEO installation thereabouts.

In general, PE presents an oxygen content gradient that spans from the port entrance, under the influence of oxygen rich water (5–6 mL/L) from Bahía de Todos Santos and the California Current, to the northern region of the port. The activities in this area have a strong impact as discharged organic matter consumes oxygen and results in hypoxic (<3 mL/L) or even anoxic (0.8 mL/L) conditions. This is particularly the case along the sea-bottom, although oxygen concentrations there also depend on the season of the year, increasing when strong wind mixing is present (González-de Zayas 2004).

Usually, nutrient concentrations in PE are higher in the northern section of the port (14.5 μM ammonium and 5 μM nitrate), as a result of the closed circulation patterns there (Czitrom et al. 2003). In the southern section, close to the harbor entrance, the influence of the outside seawater maintains lower values.

In 2005 (Table 1), however, the Chemical Oxygen Demand (COD) and total Nitrogen (TN) increased dramatically throughout the port (maximum 440 mg$O_2$/L and 843 μM respectively), coinciding with an extraordinary toxic red tide (Almazán-Becerril et al. 2007).

The TN recorded in 2004 presented, on average, high values of 11–75 μM. Even more so, the values recorded in 2005 (Table 1) increased approximately 10-fold up to an average of 321 μM throughout the port, confirming the presence of an extraordinary event, a red tide, together with an increase in port activities and organic discharges.

The total phosphorus content was generally high in 2004. Its origin may have been organic considering the port activities, the dredging and resulting movement of this nutrient from the sediment to the seawater, and the peripheral settlements and

**Table 1** Dissolved Oxygen DO and Chemical Oxygen Demand COD (mg/L), total N and total P (μM) concentrations in the Port of Ensenada

| Box | Level | DO | COD | Total N | Total P |
|---|---|---|---|---|---|
| October 2004 | | | | | |
| 1 | Surface | 4.50–6.53 | 29.00–97.00 | 17–75 | 1.6–9.7 |
| | Bottom | 3.89–6.32 | 35.00–85.00 | 6–47 | 1.1–16.3 |
| 2 | Surface | 4.50–6.16 | 5.00–56.00 | 13–79 | 0.2–17.7 |
| | Bottom | 3.73–4.99 | 24.00–56.00 | 4–49 | 1.1–17.7 |
| 3 | Surface | 5.19–6.20 | 26.00–61.00 | 18–101 | 1.4–8.5 |
| | Bottom | 3.28–4.90 | 7.00–55.00 | 10–96 | 1.8–17.7 |
| October 2005 | | | | | |
| 1 | Surface | 0.99–4.47 | 240–390 | 93–529 | 0.5–1.1 |
| | Bottom | 1.57–4.47 | 240–330 | 79–843 | 0.5–2.3 |
| 2 | Surface | 3.27–5.58 | 220–330 | 79–664 | 0.4–4.1 |
| | Bottom | 2.07–5.54 | 280–370 | 100–757 | 0.5–3.1 |
| 3 | Surface | 3.80–4.09 | 230–440 | 100–543 | 0.5–2.8 |
| | Bottom | 3.13–4.77 | 180–330 | 107–536 | 0.6–1.9 |

clandestine discharges. A slight decrease was recorded in 2005 (Table 1), although the values were still high in comparison with coastal seawater bodies that receive agricultural and domestic residual freshwaters (De la Lanza-Espino et al. 2010).

Two events clearly established the differences between 2004 and 2005. Ensenada received the most important rains of the previous 50 years in October 2004 (SMN 2004) with a maximum contribution of surface runoff to the port, together with clay sediments from the surrounding area. In October 2005, the effects of the red tide which occurred mainly between April and August, including Bahía de Todos Santos (Almazán-Becerril et al. 2007), were still present in PE. A few days before the 2005 sampling, there was a massive die-off of anchovies (approximately 10 metric tons in a three day period) inside the Touristic Nautical Zone, that reduced DO levels to zero. It is probable that the enrichment of organic matter measured as COD and TN was due to the secreted products of the species involved in the phytoplanktonic bloom produced by the red tide and by lixiviation from dead fish. The high contents of COD and TN indicated that the environment was being affected significantly by an allochthonous source.

In the 2004–2005 study (De la Lanza-Espino et al. 2010), the sediment in the Port of Ensenada presented large percentages of organic matter, especially in the southern area (15–22%). TN presented maximum concentrations in the sediment in the southern (1596 μg/g) and northern (924 μg/g) areas of the port. TP presented its greatest values in the northern area of the port (4443 μg/g).

Czitrom et al. (2003) used a two-dimensional hydrodynamic model of the circulation driven by the $M_2$ tide to demonstrate that a SIBEO flow of 1 $m^3/s$ into the northern area of the port would be adequate to alter the gyres formed there by the ocean tides and to promote a greater seawater exchange between the different sections of the port.

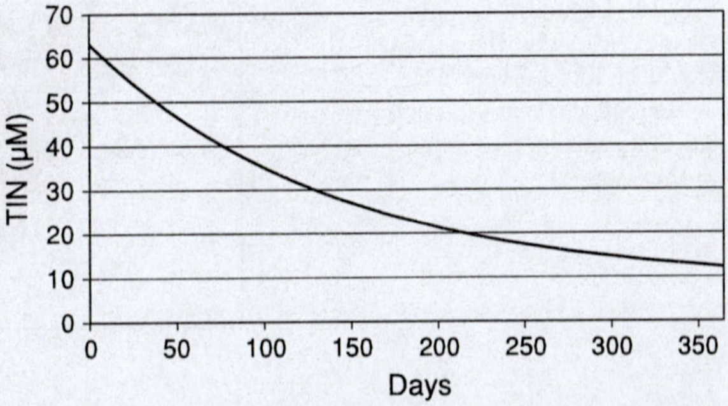


**Fig. 11** Effect of conservatively diluting the TIN using the SIBEO in the port of Ensenada (taken from De la Lanza-Espino et al. (2010))

In the following rough calculation, it was assumed that the SIBEO water is instantaneously mixed throughout the port, diluting the nutrient concentration. Also it was assumed that there is no nutrient exchange between the water column and the sediment. By mass conservation, an equivalent of the seawater being pumped by the SIBEO exits the port with the diluted concentration. It is estimated that, for a concentration of 8.24 μM of Total Inorganic Nitrogen (TIN) in the adjacent ocean (as that recorded in October 2005) and a SIBEO flow of 1 $m^3/s$, the maximum average concentration of the nutrient in the Port of Ensenada (63.16 μM) would be diluted by 80% after the first year of seawater pumping (Fig. 11), De la Lanza-Espino et al. (2010).

## 3.3 *Aquacultural Management: Lagartero Lagoon, Oaxaca*

A considerable number of lagoons along Mexico's Pacific Ocean coastline, including LL (Fig. 12), are isolated from the ocean by sandbars that form due to the accumulation of sediment transported by the littoral currents. The bars open occasionally during the rainy season when the lagoon level increases until the sandbar is overtopped. The escaping lagoon seawater scours a channel through the sandbar which remains open for as long as the freshwater discharge is sufficient to avoid the accumulation of sediment at the inlet. As a result of these episodes, the productivity of the lagoons increases substantially, following the immigration of larvae of diverse species that support important local fisheries. These episodes, however, are sporadic, which makes the planning of efficient fisheries activities difficult, added to which they are less and less frequent as a result of a decrease in river water caused by deforestation and the construction of reservoir-creating dams for agricultural purposes inland.

Channel dredging the sandbars to restore coastal lagoons as nursery grounds is a common practice in Mexico. This technique has had good results at some places, causing an increase in fishery productivity (de la Lanza-Espino and García-Calderón 2000). In other cases, however, the dredging of channels, which has a high economic cost, has resulted in negative environmental impacts such as the destruction of productive mangrove ecosystems (de la Lanza-Espino et al. 1998).

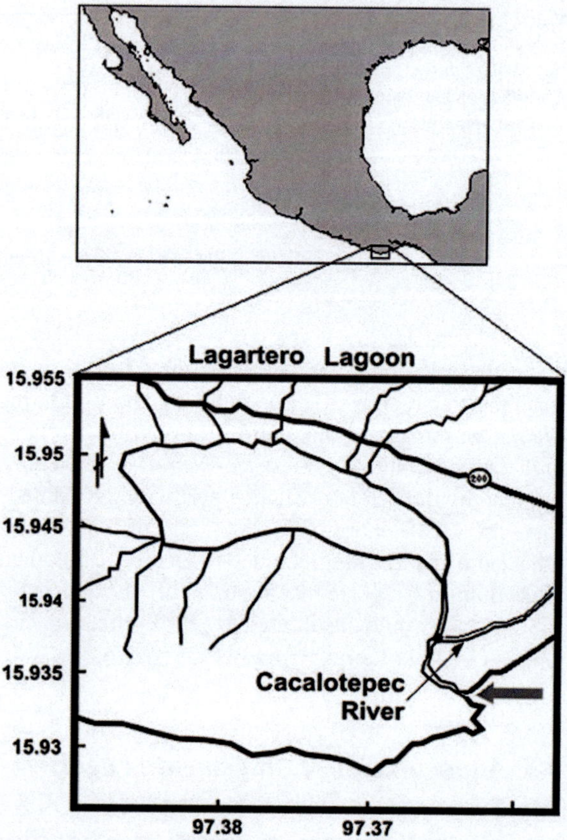


**Fig. 12** Lagartero Lagoon, Oaxaca, Mexico. The SIBEO would be located very near the lagoon inlet on the sandbar that remains closed most of the year

LL behaves essentially like an evaporation basin, with a close inverse relation between its volume and salinity during the periods when the sandbar is closed (Czitrom et al. in press). At those times, both variables are controlled by the freshwater inputs and outputs resulting from the over-lagoon precipitation/evaporation balance and land runoff deposition. This hydrologic dynamic allowed the annual volume and salinity cycles to be modeled with precision. One of the main conclusions of this chapter is that the phase of the tide at which the sandbar closes at the end of the rainy season is determinant to the amount of water and salt trapped in the lagoon, thus defining the initial conditions for the subsequent yearly cycle. A greater water volume trapped in the lagoon increases the likelihood that the sandbar will open during the following rainy season and the time lapse during which it will remain open.

TN and TP concentrations in LL exhibit annual variability from normal levels (18.32 and 8.24 μM, respectively), to levels considered to represent a significant state of eutrophication (171.4 and 35.16 μM respectively), as can be seen in Fig. 13 for the September 2002–2003 cycle. The observed variability throughout the year is the result of external contributions from sources such as the discharge

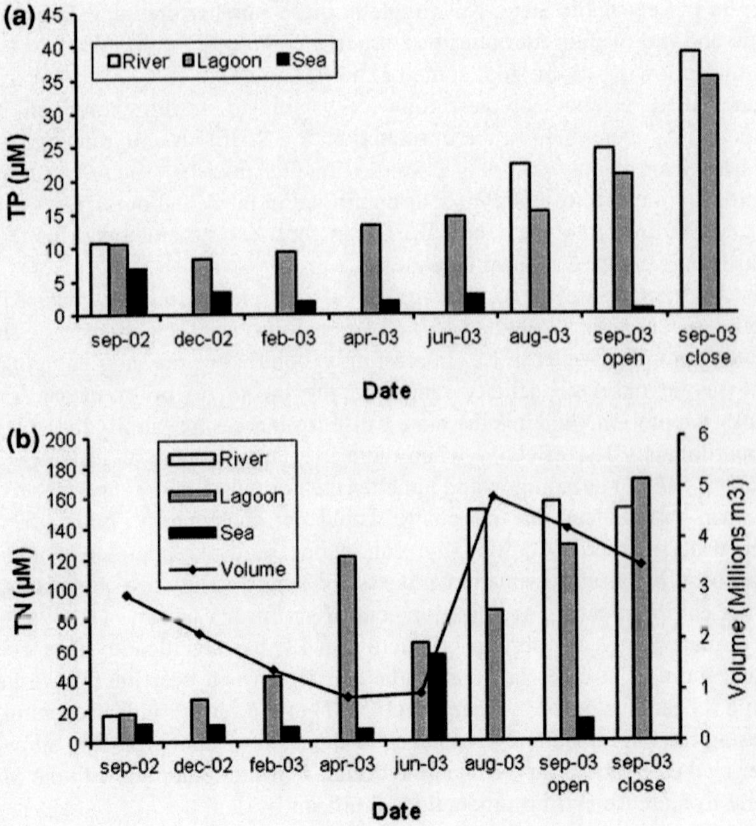


**Fig. 13** Total phosphorus (**a**) and total nitrogen (**b**) concentrations during an annual cycle in Lagartero Lagoon

from the Cacalotepec River and land seepage, as well as biophysical processes internal to the system.

Variations also occur between yearly cycles. In 1998, Penié (2005) reported TN = 6.08 μM; TP = 2.56 μM, average values throughout the year. By comparison with Fig. 13, it is clear that during the 2002–2003 cycle concentrations up to 10 times greater were registered. Although there are no data for the intervening years, or since the last sampling, it is quite likely that this observed increase is evidence of the eutrophication process which affects most coastal ecosystems today.

The eutrophication process endangers the trophic food chain, with the appearance of undesired toxic species, and negatively impacts fisheries (especially shrimp). This process has been inadvertently mitigated on occasion by the local fishermen when they dredged the sandbar, while attempting to force ocean-dwelling shrimp post-larvae into the lagoon. The interchange of lagoon and relatively low-nutrient-concentration oceanic seawater probably contributes to moderate the eutrophic state. Nevertheless, these sandbar dredgings are not systematic and the overall eutrophication process continues.

In the following discussion, some of the likely effects that seawater pumping with the SIBEO can have on the lagoon ecosystem and on shrimp productivity are considered. As mentioned in the introduction, the SIBEO has no moving parts so that, aside from simply pumping seawater for flushing, the system can transport larvae of marine species unharmed, as confirmed in the actual ocean tests with the 1:4 scale SIBEO prototype. The SIBEO can thus also be employed for the biological management of coastal ecosystems.

Another conclusion of the Czitrom et al. (in press) study is that seawater pumping increases the likelihood of sandbar openings and their persistence since it raises the volume of water in LL. Interestingly, continuous seawater pumping does not cause the observed salinity ranges in the lagoon to be exceeded. This is because evaporation, which is the main cause for increasing salinity in the lagoon, is proportionately less effective when there is a large volume of seawater. Thus, continuous SIBEO pumping would not alter the normal salinity ranges observed in the lagoon and, at least on this count, would not constitute a limiting factor to shrimp fisheries there. Additionally, continuous seawater pumping would contribute to curb the eutrophication process by simple dilution, contributing to a healthier environment for the development of shrimp.

In the last few years, shrimp production in LL has significantly decreased. A similar condition at other lagoons in the Mexico-owned coastline of the Pacific Ocean has been attributed by Páez et al. (2007) to the fewer sandbar openings. By increasing the likelihood and persistence of the sandbar openings, thus allowing a greater intake of shrimp larvae from the ocean, seawater pumping with the SIBEO is likely to increase shrimp production significantly.

Another way in which shrimp production can increase is that larvae would enter the lagoon throughout the year with the seawater being pumped by the SIBEO. Considering a concentration of 2 org/m$^3$, an average for the periods of greater concentration during April–May and November–December (Cabrera-Jiménez 1997), and an average SIBEO flow of 0.1 m$^3$/s, then potentially more than 2 million organisms would be pumped. Taking a 90% mortality rate (Edwards 1977; Gracia 1989) and a 25 g weight for pre-adult organisms (de la Lanza-Espino and García-Calderón 2000), this "sowing of larvae" would lead to annual local shrimp fisheries in excess of 5,000 kg. This is more than double the annual historic catch at LL.

This estimate, however, does not consider the intake of larvae when the sandbar opens and is therefore likely an underestimation of the potential shrimp production. At this time, however, we cannot estimate the amount of larvae that enter the lagoon when the sandbar opens. Also, other biotic and abiotic factors must be considered when estimating the production of shrimp, including growth rhythms, abundance of depredating species and the effect of all this on the trophic chain (Minello and Zimmerman 1991). Nevertheless, since LL has an abundant cover of halophyte vegetation and a high productivity, we can safely conclude that extensive shrimp fishery would be significantly enhanced by pumping seawater with the SIBEO.

## 4 Summary and Conclusions

It has been shown that the SIBEO is a viable seacoast management tool. It can be used to increase the loading capacity of enclosed coastal seawater bodies by favoring the hydrologic dynamics and interchange with the immediately adjacent ocean. In Bojorquez Lagoon, seawater ventilation and a reduction by 75 % in the residence time will diminish eutrophication and promote the rapid rehabilitation of the ecosystem. In the Port of Ensenada, as in other ports which are susceptible to toxic red tides, high nutrient concentrations in the stagnant sections can be diluted significantly by the SIBEO.

The SIBEO can also be used to manage the fisheries production of some coastal seawater bodies. In Lagartero Lagoon, pumped seawater will increase the frequency and persistence of the sandbar openings which in turn will allow shrimp post-larvae to stock the lagoon fisheries, while not affecting negatively the lagoon ecosystem. Additionally, since the SIBEO system has no moving parts, enough shrimp post-larvae can be pumped to at least duplicate present lagoon production.

**Acknowledgements** We wish to express our thanks to the following persons and institutions whose support contributed to the work presented in this chapter. Virginia Osuna Salazar, Noel Carbajal Pérez Salvador Hernández Pulido, Xavier Flores Vidal, César Coronado Méndez, Ma Antonieta Ozuna Salazar and Esteban Prado Bravo. The Cacalotepec community, the Consejo Nacional de Ciencia y Tecnología (CONACYT) (Project: Flushing of the Port of Ensenada using a SIBEO wave driven seawater pump. No G334644-T), the Fondo Sectorial de Investigación Ambiental SEMARNAT-CONACYT (Project: Ventilación del Puerto de Ensenada Mediante un Sistema de Bombeo por Energía de Oleaje (SIBEO). No. C016-2002), the Instituto de Ciencias del Mar y Limnología, UNAM (Proyect: Energía de Oleaje No. 139), the Posgrado en Ciencias del Mar y Limnología, UNAM, through an M.Sc. grant for I. Penié and the Instituto de Ingeniería, UNAM, through an Academic Commission for S. Czitrom. The John D. and Catherine T. MacArthur Foundation, through a leadership grant for S. Czitrom, the North American Fund for Environmental Cooperation (Project: 'Implementación de un Sistema de Bombeo por la Energía del Oleaje para el Manejo de Pesquerías', 1998-1999). The CICESE, the SEMAR and API-Ensenada for logistic support.

## References

Almazán-Becerril A, García-Mendoza E, Peña-Manjarrez JL (2007) Segundo Taller de Florecimientos Algales Nocivos. Resultados, análisis y conclusiones de las ponencias y mesas de trabajo. Technical Report No. 40. Departamento de Oceanografía Biológica. CICESE, novenber 21–23, 2007, Ensenada, Baja California, 16 pp

Borja A, Elliott M (2007) What does "good ecological potential" mean, within the European Water Framework Directive? Mar Pollut Bull 54:1559–1564

Cabrera-Jiménez JA (1997) Los períodos de presencia y ausencia de las postlarvas del camarón Penaeus (Litopenaeus) vannamei BOONE (crustácea, decápoda, penaeidae) en una boca litoral tropical. Revista de Investigaciones Marinas 18(3):260–267

Carbajal-Pérez N (2009) Hidrodinámica y transporte de contaminantes y sedimentos en el Sistema Lagunar de Nichupté-Bojórquez, Quintana Roo. Instituto Potosino de Investigación Científica y Tecnológica, AC. SNIB-CONABIO proyecto No. CQ063. México D. F

CFE, Comisión Federal de Electricidad 2004. Informe del oleaje medido durante 10 meses en aguas de 20 metros frente a Punta Cancún (unpublished)

Chenworth M, Landsea C (2004) The San Diego Hurricane of 2 October 1858. Bull Am Meteorol Soc 85:1689–1697

Cornelissen G, Pettersen A, Nesse E, Helland EA, Breedveld GD (2008) The contribution of urban runoff to organic contaminant levels in harbour sediments near two Norwegian cities. Mar Pollut Bull 56:565–573

Cruz J (Ed.) (2008) Ocean wave energy, current status and future perspectives. Springer, Berlin, 423 pp

Czitrom SPR (2002) Sintonizador para sistemas de energía de oleaje que operan por resonancia. Titular Instituto de Ciencias del Mar y Limnología, UNAM. Patente Mexicana No. 210329

Czitrom SPR, Prado E (1999) Design optimisation of a wave driven seawater pump. In: Proceedings of the OMAE99 18th international conference on offshore mechanics and arctic engineering. St Johns, Newfoundland, Canada

Czitrom SPR, Godoy R, Prado E, Pérez P, Peralta-Fabi R (2000a) Hydrodynamics of an oscillating water column seawater pump. Part I. Theoretical aspects. Ocean Eng 27(11):1181–1198

Czitrom SPR, Godoy R, Prado E, Olvera A, Stern C (2000b) Hydrodynamics of an oscillating water column seawater pump. Part II. Tuning to monochromatic waves. Ocean Eng 27(11):1199–1219

Czitrom SPR, Núñez I, Ramírez I (2003) Innovative uses of Wave power: environmental management of the Port of Ensenada, Mexico. J Mar Technol Soc 36:74–84

Czitrom SPR, Penié I, de la Lanza G, Osuna V (in press) Management proposal for an intermittently closed coastal lagoon using a wave-driven seawater pump and a salinity-volume box model. J Coast Res

Dauvin JC (2008) The main characteristics, problems, and prospects for Western European coastal seas. Mar Pollut Bull 57:22–40

De la Lanza-Espino G, García-Calderón JL (2000) Ambiente pesquerías y camaronicultura en una laguna costera al noroeste de México. MÉXICOA 1(2):129–132

De la Lanza-Espino G, Flores-Verdugo FJ, Wulff F (1998) Teacapan-Agua Brava- Marismas Nacionales, Sinaloa and Nayarit: N/P Budget. http://data.ecology.su.se/MNODE/mexicanlagoons/tabbud.htm

De la Lanza-Espino G, Penié-Rodríguez I, Czitrom-Baus SPR (2010) Water quality of a port in NW Mexico and its rehabilitation with swell energy. Mar Pollut Bull 60:123–130

Edwards RRC (1977) Field experiments in growth and the mortality of Penaeus vanammei in a Mexican Coastal Lagoon complex. Est Coast Mar Sci 5(1):107–121

Espejel I, León C. Fermán JL, Bocco G, Rosete F, Graizbord B, Castellanos A, Arizpe O, Rodríguez G (2004) Planeación del uso del suelo en la región costera del Golfo de California y Pacifico Norte de México. En: Arriaga Rivera et al (eds.) El Manejo Costero en México. PEOMEX, Semarnat, CETyS, Univ. de Quintana Roo, pp 321–340

González-de Zayas R (2004) Caracterización físico-química del Puerto de Ensenada, Baja California, con vistas a la instalación de un Sistema de Bombeo por Energía del Oleaje (SIBEO). Tesis de Maestría, Instituto de Biología, UNAM, 92 pp

González-López A (1989) Hidrología y nutrientes en la Laguna Bojórquez, Cancún, Q. Roo. Tesis de Maestría. Instituto de Ciencias del Mar y Limnología, UNAM, 110 pp

Gracia A (1989) Impacto de la explotación de postlarvas sobre la pesquería del camarón blanco penaeus setiferus (linnaeus, 1767). Anales del Instituto de Ciencias del Mar y Limnología 10(2):255–262

Guerra-García JM, García-Gómez JC (2005) Assessing pollution levels in sediments of a harbour with two opposing entrances. Environmental implications. J Environ Manage 77:1–11

Herrera-Silveira JA (2006) Lagunas costeras de Yucatán (E, México): investigación, diagnóstico y manejo. Ecotrópicos 19(2):94–108

Merino M, González A, Reyes E, Gallegos A, Czitrom SPR (1992) Eutrophication in the Lagoons of Cancún, Mexico. Science of the Total Environment (Supplement). Elsevier, Amsterdam, pp 861–869

Minello TJ, Zimmerman RJ (1991) The role of estuarine habitats in regulating growth and survival of juvenile penaeid shrimp. In: Deloach PF, Dougherty WJ, Davidson MA (eds) Frontiers of shrimp research. Elsevier, Amsterdam, pp 1–16

Ortiz MA, de la Lanza G (2006) Diferenciación del espacio costero en México: Un inventario regional. Geografía para el siglo XXI, Serie Textos Universitarios. No. 3, 138 pp

Páez F, Ramírez G, Ruiz AC, Soto M (2007) La contaminación por nitrógeno y fósforo en Sinaloa: Flujos, fuentes, efectos y opciones de manejo. Serie las lagunas costeras de Sinaloa. UNAM/Instituto de Ciencias del Mar y Limnología/El Colegio de Sinaloa/SEMARNAT/CONACYT, 304 pp

Reyes E, Merino M (1992) Diel dissolved oxygen dynamics and eutrophication in a shallow, well-mixed tropical lagoon (Cancún, Mexico). Estuaries 14(4):372–381

Shin PKS, Lam NWY, Wu RSS, Qian PY, Cheung SG (2008) Spatio-temporal changes of marine macrobenthic community in sub-tropical waters upon recovery from eutrophication. I. Sediment quality and community structure. Mar Pollut Bull 56:282–296

SMN (2004) Servicio Meteorológico Nacional. http://www.smn.cna.gob.mx/; accessed December, 2004

Stern C, Czitrom SPR, Godoy R (1999) Oscillating flow through a funnel. Special Section: "Gallery of Fluid Motion". Phys Fluids 11(9):S3

Xue X, Hong H, Charles AT (2004) Cumulative environmental impacts and integrated coastal management: the case of Xiamen, China. J Environ Manage 71:271–283

# Wave Powered Desalination

Stephen H. Salter, Joao M. B. P. Cruz, Jorge A. A. Lucas and Remy C. R. Pascal

## 1 Introduction

One of the wave energy devices studied in the 1970s and 80s was the Edinburgh duck. Figure 1, taken by Jamie Taylor in 1976, shows a duck under test in a narrow tank.

This photograph is a 1-s exposure of a 1/100 scale model on a fixed mounting. The wires are coming from of an electromagnetic dynamometer, which is absorbing power. Waves are approaching from the right. Drops of a neutrally buoyant tracer-fluid consisting of a mixture of carbon tetrachloride and xylene with titanium oxide pigment have been injected to show the decaying orbits of wave motion.

The amplitude of the incoming waves can be measured from the thickness of the bright band. Nodes and anti-nodes due to the small amount of reflection are evident. However the thickness of the bright band to the left of the model is largely due to the meniscus, as is confirmed by the very small orbits of tracer fluid in this region. As the energy in a wave is proportional to the square of wave amplitude we can use the photograph to do energy accounting. If reflected and transmitted waves are one-fifth of the amplitude of the input they would have one twenty-fifth, or 4%, of its energy. This means that 96% has gone into the movement of the test model. The dynamometer showed that just over 90% of the power in the full width of the tank had been absorbed by the power take-off, leaving 6% loss through viscous skin friction.

A later model-mounting with controlled stiffness and damping in heave and surge allowed us to replicate the behaviour of models on a long, compliant, yielding spine and to make the circular stern of the model behave like the highly

S. H. Salter (✉), J. M. B. P. Cruz, JorgeA. A. Lucas and R. C. R. Pascal
University of Edinburgh, Edinburgh Scotland, UK
e-mail: S.Salter@ed.ac.uk

V. Badescu and R. B. Cathcart (eds.), *Macro-engineering Seawater in Unique Environments*, Environmental Science and Engineering,
DOI: 10.1007/978-3-642-14779-1_29, 

**Fig. 1** A time exposure of a small-scale duck model with tracing fluid

efficient Evans cylinder, Evans et al. (1979). The movement generated waves that cancelled the longer ones that were escaping below the model and so widened the bandwidth of efficient operation. For very large scale electrical generation we proposed mounting many ducks—at the very least ten—on a long spine which could flex to provide the correct mounting compliance and yield to avoid extreme bending moments. While this would have given very large output from a limited sea front, the need for a minimum installation of about 60 MW made it unattractive for initial investments before confidence was established. It was therefore interesting to study a solo duck.

It was shown (Budal and Falnes 1975) and others that in a wide tank solo devices could extract *more* energy from the sea than was contained in their own geometrical width and so appear to have an efficiency of more than 100%, just as the power from a radio antenna is not set by the diameter of the wire. This led to the new terms of 'capture-width' and 'capture width ratio'.

An asymmetric device able to move in the right combination of vertical and horizontal directions should be able to get energy from a width of wave of $3/\pi$ times the wavelength. Unfortunately very wide capture widths from small devices would require very large excursions, which soon make things non-linear. Nevertheless Skyner (1987) tested a tank model on a mounting which could simulate intelligent fore and aft mooring legs and was able to achieve capture width ratios of 1.8 over nearly all the useful Atlantic spectrum.

Sadly, we found that Ducks on tension leg moorings suffer from severe snatch loads if the moorings go slack and then retighten. This would occur quite often in small but steep waves. A possible solution to the mooring problem will be described later.

## 2 Desalination Basics

Of the several techniques for desalination, reverse osmosis is the present well known favourite. A high and preferably fairly steady pressure (3–6.5 MPa) is used to pass some fraction of seawater through a semi-permeable membrane while retaining most of the salt. The pressure gradient is from the source to the product and so there is a possible route by which the product can be polluted. Desalination membranes for general use are designed to meet, but only just meet, the World Health Organization guideline for total dissolved solids below 1000 parts per million (Bartram and Callan 2006). Reverse osmosis needs careful filtration and chemical pre-treatment to give the membranes an economical life.

There is an alternative but less used technique known as vapour compression (Darwish and El Dessouky 1996). Figure 2 shows a block diagram. Seawater at

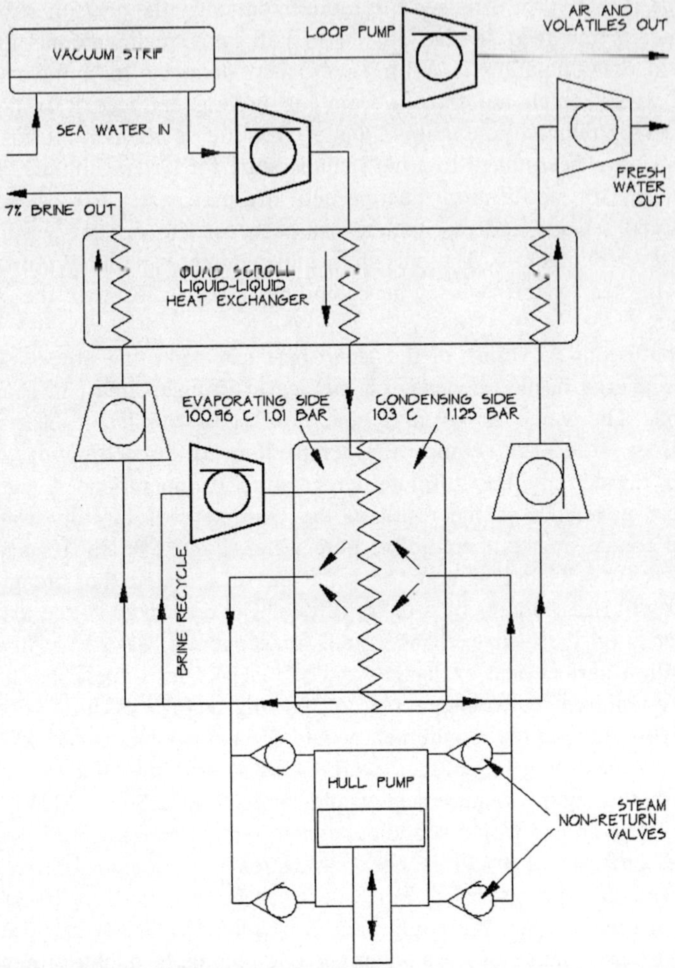


**Fig. 2** The block diagram of the vapour compression process. Most of the work done by the mechanical pump goes into raising the temperature difference across the steam heat exchanger. Latent heat is recycled

a temperature close to boiling point is trickled down one side of a hot heat-transfer surface. Pressure is reduced by a pump so that some water is converted to vapour, drawing the necessary latent heat from the heat transfer surface. The pump increases the pressure of the vapour by about 20 kPa, thereby increasing its temperature by a few degrees, and delivers it to the other side of the heat-transfer surface, which will have just been cooled by the loss of latent heat. The pure vapour condenses and releases its latent heat, which is used again and again for evaporation of more sea water. The vapour compression process can be done with pumping energy taken directly from sea waves (Salter 1985, 2005).

While the volumes of vapour are very large, the pressure drop is much smaller than for reverse osmosis. But more importantly the pressure drop is *from* the product *to* the feed so that any leaks will lose small amounts of product rather than contaminating it. With an efficient demister to remove the small drops of water which are thrown from the boiling surfaces it is possible to produce an extremely pure output, better than 0.5 parts per million. As the output will have been taken to a high temperature for several minutes there is a further means for sterilisation. Pharmaceutically pure water has been produced from hospital sewage and the method has been used to dry the mash residues from whisky manufacture. The only intractable source of polluted output will be volatiles in sea water, which have boiling points close to the operating temperature.

The continuous recycling of the latent heat can make the process very efficient. The energy needed depends on the heat exchanger area and heat-transfer coefficients. The value is given by the product of the latent heat of steam (2.256 MJ/kg at 100 °C) times the temperature difference across the heat exchanger divided by the absolute temperature of operation. A part of the temperature difference is the result of the boiling point of salty water being above the condensing temperature of pure water. For 3.5% NaCl sea water the elevation is 0.46 °C but as soon as some has evaporated the strength of the remainder will rise, perhaps by a factor of two. The remainder of the temperature drop depends on the heat exchanger and, for economic reasons, is likely to be larger. With a perfect heat exchanger and 3.5% NaCl feed operating at 373.3 K the energy requirement would be only 2.78 kJ/kg or 0.77 kWh/m$^3$. For a more realistic brine strength the requirement would perhaps double to 1.55 kWh/m$^3$ for a very large heat transfer surface working at a very low rate. For practical throughputs the energy requirement would range from 2.5 to 10 kWh or more per cubic metre and so would straddle the energy requirement of reverse osmosis plant. The hot fresh water and the hot double-strength brine contain heat which is far too precious to be discarded. Much of it can be recovered by transfer to the incoming sea water flow. We might have two volumes of 3.5% salt water being converted to one volume of fresh water and one volume of double strength brine. The energy loss will depend on the temperature difference between incoming and outgoing streams.

## 3 Fouling of Heat Transfer Surfaces

All desalination methods are bedevilled by the problem that some of the many materials dissolved in seawater are close to their limit of solubility. An increase in solution strength or a rise in temperature can make them come out of solution and form a hard scale on any suitable substrate. This will grow in thickness and will rapidly reduce the heat transfer coefficients, halving them in a few days of operation. This problem has been tackled by Maxwell Davidson who has, over the years, produced a series of heat transfer materials with steadily improving transfer coefficients resistance to fouling and very long operating life even in contact with ammonium sulphate and flue gases. The present choice is a metal mesh which has been sprayed with a continuous layer of polyvinylidene fluoride, usually known as PVDF or Kynar, filled with flakes of carbon to improve heat transfer. The PVDF layer has excellent high temperature properties and resists fouling in the same way as a non-stick Teflon-coated frying pan. It is also used for the fibres of some reverse osmosis modules.

## 4 Pumping with Waves

In conventional vapour compression systems, energy for driving the very large volumes of vapour is provided through turbo-compressors which have to rotate very fast. But it would also be possible to use a slow, positive-displacement pump formed by the body of a Duck, half-filled with water. It would have a central partition dividing a pair of steam compartments as shown in Fig. 3. While the Duck undergoes its wave-driven alternating rotations, the 'water pendulum' tends to move in the opposite direction and so the system forms an enormous, positive-displacement pump with no sliding seals or accurate machined parts and very low internal losses.

The torque opposing the wave motion of the Duck comes from the pressure-difference across the partition. This will be set by the temperature-difference across the heat-exchanger, which will be determined by throughput and so proportional to the angular-velocity of the Duck body. With the direct relationship between the pressure across the partition and the angular velocity of the Duck body we are close to the ideal damping of a wave energy device with none of the misery that this normally involves (Salter et al. 2003).

There is also a need to pump all the working fluids round the Duck and to pump the product ashore. It would be nice to avoid the use of any electrical items in the hot, wet conditions. This can be done with a mechanism known as a loop pump. We all know that a column of liquid in a vertical tube induces a pressure at the bottom with a value of fluid-density times gravitational-acceleration times column-height. We are used to situations where gravity is constant. But exactly the same thing happens in a coiled pipe which is subjected to

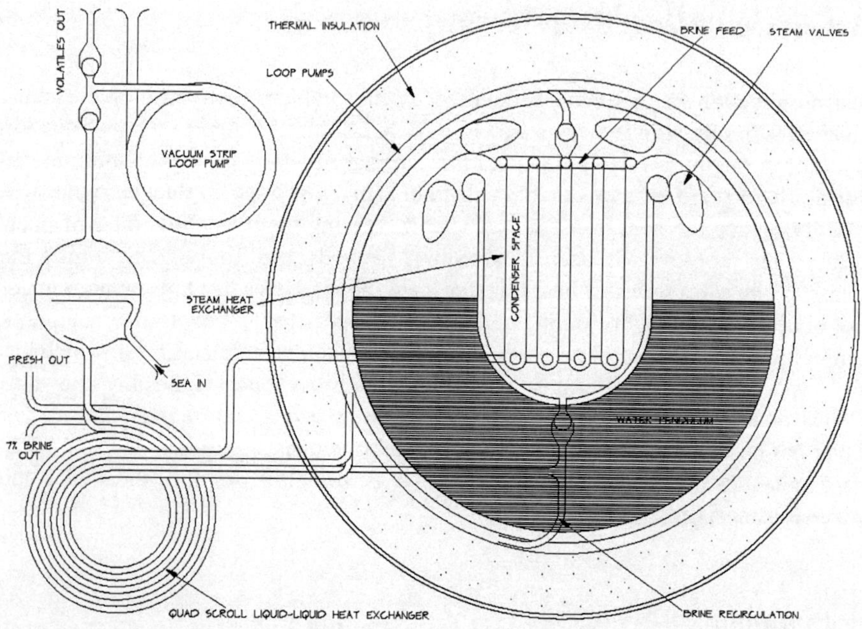


**Fig. 3** A simplified layout of the hull pump

alternating *angular* accelerations. Rectifying valves can prevent return flow. There will be room for many loops to be wrapped round the pumping chamber. Each 360° loop can easily produce a pressure of 30 kPa. With several loops in series we can pull a vacuum to remove dissolved gases, lift water from bottom to top of the steam heat exchanger, pump fresh water ashore, and even up to a respectable height of reservoir.

An updated layout of the system is shown in Fig. 4. Here the previous front shape of the duck has been replaced by offsetting the rotation axis of a circular cylinder. Numerical modelling (Cruz and Salter 2006) gives hydrodynamic coefficients and capture widths.

## 5 Heat Exchanger Design

The Maxwell Davidson heat exchanger (Davidson 1974) consists of sheets of the heat transfer material into which have been pressed corrugations that lie at 45° to the sheet edge as in Fig. 4. The press tool is used twice to make two rectangles of corrugations separated by an area of plain sheet. Other holes can be punched out for pipe work and clamping rods. When a sheet is folded into a U shape the corrugations will contact one another and make the form of an X. A large number of U-folds are clamped together with short spacing tubes which can form feed

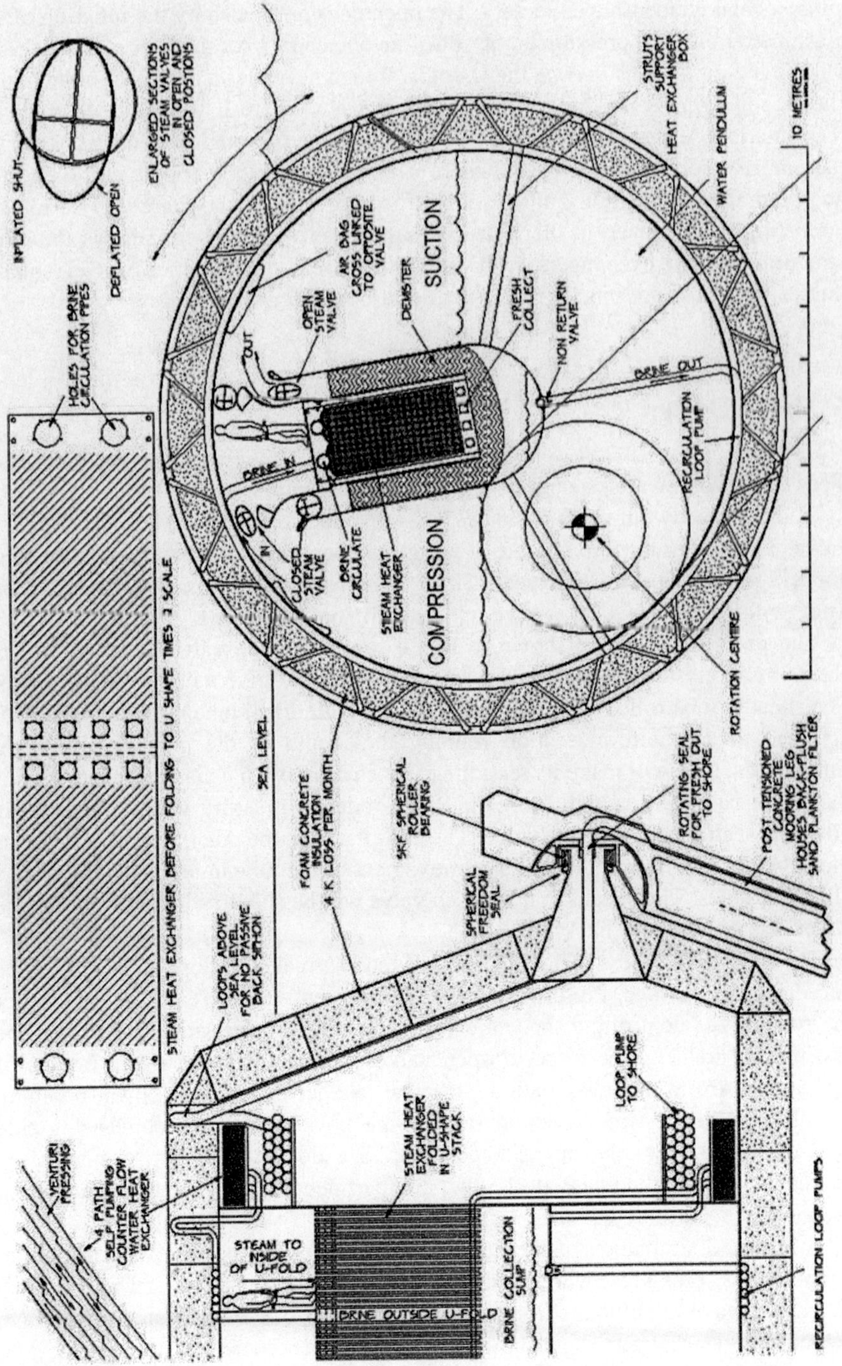


**Fig. 4** Sections through the desalinating offset cylinder. People would not be present during operation

sprayers and condensate collectors. The open ends are sealed by the inflation of an elastomeric bag to a pressure higher than the one chosen for the process. Sea water will be fed to the gaps *outside* the U-folds. It then moves as a falling film down the sheets and drips from the bottom of the U from where it will be collected for recirculation. Vapour will be drawn off and pumped through a demister stack to the *inner* surfaces of the U-folds where it will condense and collect at the bottom and be drained. The falling film is ideal for heat transfer and there will be lots of demisting taking place in the heat-exchanger itself. This design allows the construction of heat exchangers with very large areas, many thousands of square metres, at a cost not much greater than that of the raw material.

## 6 Valve Design

Very large volumes of steam have to be pumped and we need valve areas with hydrodynamically fair passage shapes so as to reduce the pressure drop through the valve. The valves have to operate at a high temperature for every pumping cycle, about 5 million operations a year. There will sometimes be impacts from water. They will be expensive to service and so must not fatigue.

The practical layout is shown in Fig. 4 top right. The valves are made from sheets of an elastomer reinforced with a wire mesh to give a controlled stiffness. The sheet will be rolled into the form of a long cylindrical tube with its long edges clamped to a cruciform section running the length of the pumping chamber. Inflating the tube will make its section nearly circular, with a diameter that just fills the valve passage and will close it. But a reduced pressure will force it to an elliptical section and open the valve. The end-plates on the cruciform section inside the valve tube will limit the range of movement of the tube in both its circular and elliptical states. This means that a closed valve will be able to withstand quite large water impacts without buckling.

The internal stiffness of the textile will tend to make it adopt a circular shape and close the passage. Preforming the reinforcement can increase the effect. For the two valves controlling the outlet from the evaporating side, closure can be assisted by gravity. The force to open valves can be increased if the insides of the valves from the evaporation side are connected to their own pumping chamber while the two valves controlling the flow back to the condensing side are cross-coupled to the opposite chamber. We do not want any water in the valves and so air-filled bags in the two compartments will generate the actuation pressures.

A Mathcad simulation of the desalination process has been written to predict temperatures, pressures, weights and productivity as a function of Duck dimensions and wave conditions. Graphs of results for 0.7 m root mean square wave height at 7 s energy period are shown in Fig. 5 and the theoretical productivity as a function of wave amplitude and period is shown in Fig. 6. Optimisation of the offset cylinder version is still in progress.

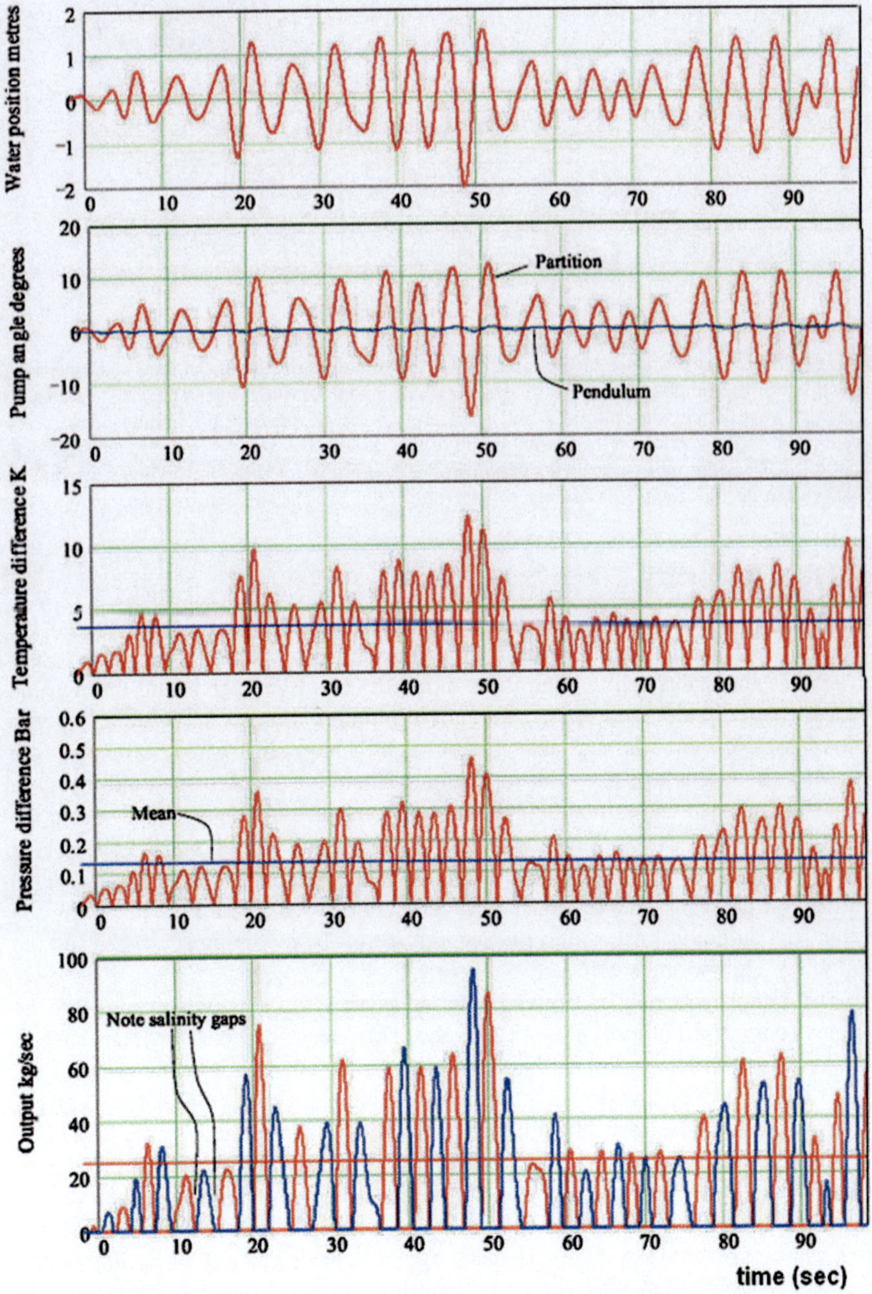


**Fig. 5** Mathcad output of water position, angles of the hub pump, temperature and pressure differences and the instantaneous output of fresh water for a 0.7 m root mean square wave height and a Pierson-Moskowitz spectrum with an energy period of 7 s. This would produce about 2000 $m^3$ of fresh water a day from a 12 m diameter duck. Note that significant wave height is now, by definition, four times the root mean square height but that other, less satisfactory, definitions exist

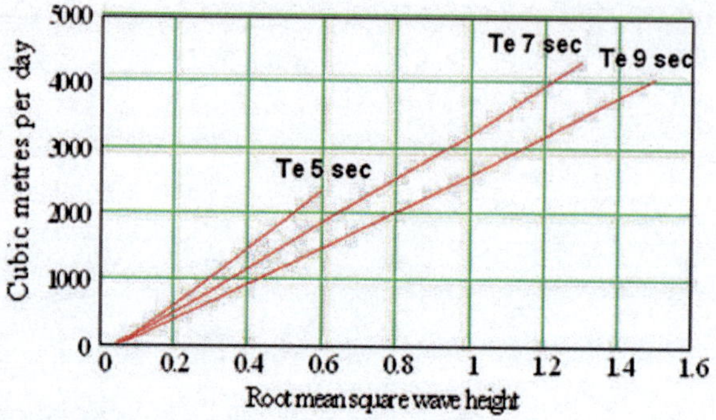


**Fig. 6** Theoretical estimates of productivity for the reference design of the duck displacer as a function of wave amplitude and energy periods of 5, 7 and 9 s. Note the nearly linear relationship at larger wave amplitudes and the small offset due to the elevation of the boiling point of sea water. This is also evident in the bottom graph of Fig. 2 for waves around 10 s into the sequence. Careful tank testing will be needed to confirm this prediction for the offset cylinder version so that we can decide if any loss of performance outweighs the cheaper construction

## 7 Heat Recovery

Other energies calculated by the Mathcad simulation software are the pressure drops of vapour through the demister, round the pipes and bends and through the valves. Next it calculates energy for moving liquids through the system, the viscous losses of the pump and heat lost through the container walls. All except the last of these losses will rise with throughput. There is a need for a liquid-to-liquid heat exchanger to warm incoming sea water from the outgoing product. Again the angular motion of the duck can be turned to advantage. We can make a cheap heat exchanger by wrapping four sheets of PVDF-clad mesh in a four-start spiral. This forms four passages. The sequence of layers will be used for half the incoming sea water, the outgoing fresh water, the second half of the incoming sea water and the outgoing, high-strength brine.

The four sheets can be pressed into an asymmetric sawtooth form with the angles equivalent to those of a Venturi tube. This means that the flow resistance in one direction will be less than that in the other. The alternating movements of the duck will induce a flow with a small direct component superimposed on a larger alternating one. This distributed pumping avoids the large pressure that would be needed to force water through long passages by a pump at one end. It is difficult to think of a more satisfactory heat transfer mechanism. Warmth and a regular supply of nutrients will encourage rich biological growth in the inlet passages. However this can be prevented by the daily discharge of a small quantity of boiling water.

Heat will be lost through the casing and through the difference in temperature between the in-going stream of seawater and the out-going streams of fresh water and extra strength brine. Heat will also be lost through the difference in chemical energies of fresh and salt water. Heat will be generated by all the internal flow losses and the work done in compressing steam. It is absolutely necessary to understand each of these mechanisms if we are to regulate the temperature swings to the chosen narrow range. Overheating and the consequent overpressure are a real possibility.

The likely dimensions of the duck mean that heat loss through the casing is surprisingly small. If we provide a metre thickness of foam concrete with conductivity about 0.1 W/mK round a 10 m diameter 20 m long pumping compartment we will lose just under 5 kW through a temperature difference of 75 K. The large water mass inside the duck means that the temperature would fall by less than 4 K a month if all flows were stopped. It would be quite possible for ducks to be filled with hot water at the factory, towed long distances to installation sites and still be at working temperature on arrival. At most places with water problems, solar inputs would be enough to maintain temperature in prolonged calms.

There are two ways to regulate temperature. The dominant heat losses are in the output streams. We can provide slightly more area than is necessary to maintain temperature but split this between two heat exchangers. By cutting the flow through one for a variable fraction of the time we can adjust their effectiveness. The second method is to gain heat by increasing the number of times that water is dropped through the steam heat exchanger before being pumped ashore.

## 8 Bearings

The bearing requirements for many renewable energy applications are often beyond those of off-the-shelf components. At the start of this project it was expected that desalinating ducks would need special hydrostatic bearings which would use filtered sea water pressurised by wave forces. It was a surprise to find that, for the tropical wave climates where desalinating ducks are like to be used, we can use a bearing from the standard SKF large-bearing product range (Anon 2006). Bearing ratings are essentially three numbers moderated by a series of application factors. There is a load that will do damage with a single application. There is another load that will start damage 10% of a large population of bearings following the application of a million rotations. An equation allows the calculation of the fraction of a population that would be damaged for other loads and other numbers of rotations. The third number is the load that is low enough never to cause failure no matter how many rotations occur.

If we have a scatter diagram for a chosen wave climate and know how to use the Rayleigh distribution to get the total number of waves in each small range of amplitudes, we can combine loading data with the second of the above bearing ratings to get the fraction of bearing life that each wave will use up. It turns out that for a moderate wave climate such as Lanzarote in the Canaries, a pair of SKF spherical roller bearings which can take the single highest wave in ten years will also have 90% survival probability over that period. Most of the life will be used up by the very large number of medium sized waves. SKF advise that far more bearings fail because of seal problems than from loading.

## 9 Moorings

A crucial problem for all marine energy sources is developing a reaction against which the device can do work. Wave powered desalination raises the following requirements.

- For a solo duck in the North Atlantic 100-year wave the forces will be about 40 MN.
- The mooring may have to rotate 360° about a vertical axis and follow movements of waves and tides in the elevation axis.
- We may have to make an attachment to a sea bed of soft sand.
- We may need to remove and replace equipment especially in the early days of the project.
- We may not be allowed to leave permanent hardware in the sea bed.
- We should try to avoid the need for heavy weights and large anchor-handling vessels.

Because of the snatching problem of ropes the system will use post-tensioned concrete tubes as rigid struts, with bearings to give spherical freedom at both ends. Fatigue happens as a result of tensile stresses and so will not occur in the compressed concrete. The wires which provide the compression will be under constant tension and so will not fatigue either. The values of mooring loads are now being measured in a test tank. Until full results are available we are using a specification appropriate to Atlantic conditions, much higher than in likely drought-stricken countries. The 40 MN force will be applied at an angle of about 45°. When this reaches the universal joint it will be split into vertical and horizontal components of about 28 MN.

The sea-bed attachment will be split into two parts with a quick disconnect between them. The permanent part shown, in Fig. 7, will be in the form of a tripod with legs made from post-tensioned concrete. The wall thickness will be about 0.9 of the diameter - thin enough for the tubes to be buoyant for easy towing when full of air but able to sink if filled with water. The top of the tripod will be a hollow

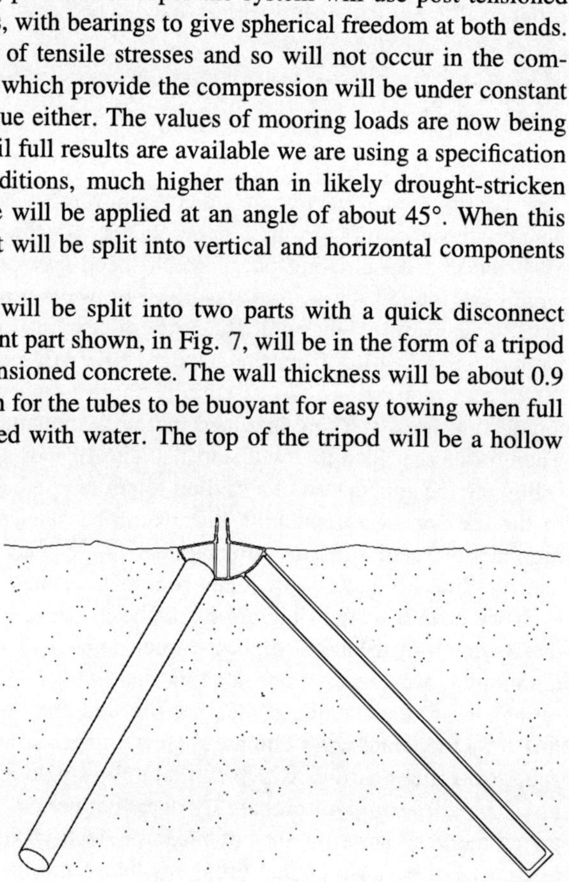

**Fig. 7** Part of the mooring is a tripod anchor made from three, slip-formed post-tensioned concrete legs. It is sunk in the sand by controlled water-jetting and retained by suction

cast-steel shell large enough for people to work inside applying tension to the post-tension strands.

The tripod will be towed to site and lowered to the sea bed. Water jets at the bottom and underside of each leg will eject sand and allow the tripod to sink in the sand. By adjustment of flow rates between legs the correct attitude can be maintained. When the tripod has reached the chosen depth, the direction of water flow will be reversed and water will be removed from the body of sand round the leg. As we remove water from around the legs the weight of the surrounding sand, increased by the pressure of the head of sea water above it, will produce a large radial force on the outer surface of each leg. The friction caused by this force will prevent the leg from sliding out of the sand. A friction coefficient of 0.3, a water depth of 50 m, a dry sand density of 1600 kg/m$^3$ and a buried length of 30 m will give a friction force of 45 MN so a single leg could resist the 100-year Atlantic wave load.

The flow rate needed to remove water in contact with the leg will depend on the grain size of the sand. With coarse sand this may initially be quite high but as pumping proceeds the passages between sand grains will be come progressively blocked by smaller grains and the flow rate will fall. With a non-return valve at the top of each leg, the inner volume can serve as a vacuum store to maintain suction. Initial pumping can be provided from a surface vessel but, later, this function can be performed by loop pumps in the desalination system. These are required anyway to strip dissolved air from the water feed.

If ever it is necessary to remove the tripod, the vacuum can be released and air pumped into the inner leg volume to give positive buoyancy. Water jets on the outside of each leg can then fluidise the sand above the tripod, which will rise through the sand and then to the surface. This design means that all installation and removal can be done with very light vessels fitted with water, air and vacuum pumping systems.

The universal joint at the sea bed is shown in Fig. 8. While the large angular movements of the displacer body are suitable for rolling element bearings they are less attractive for cases where there are many movements through angles much smaller than the gaps between rollers. The SKF catalogue shows that the GEC 1250 PSA plain bearing has a static load rating of 64 MN and a dynamic rating of 36.5 MN. This can take the horizontal loads and can provide the freedom to yaw about the vertical axis so that the system can align itself to waves from any direction. However these values apply at zero velocity and allowable loads fall at very low velocities, even a few millimetres a second. Ten millimetres per second is equivalent to a rotation period of 7 min—so the bearing can rotate fast enough about the vertical rotation axis to follow a change of weather patterns.

The velocity of the SKF GEP 480 P4S bearings for the elevation movement in a 1.5 m amplitude 8-s wave in 50 m water depth will be 7 mm/s and so the velocity restriction will often be more serious. We propose that the velocity limit can be increased by fitting one (or two) hydraulic rams at the lower end of the mooring struts. These will give oil pressures proportional to the applied load. This oil can

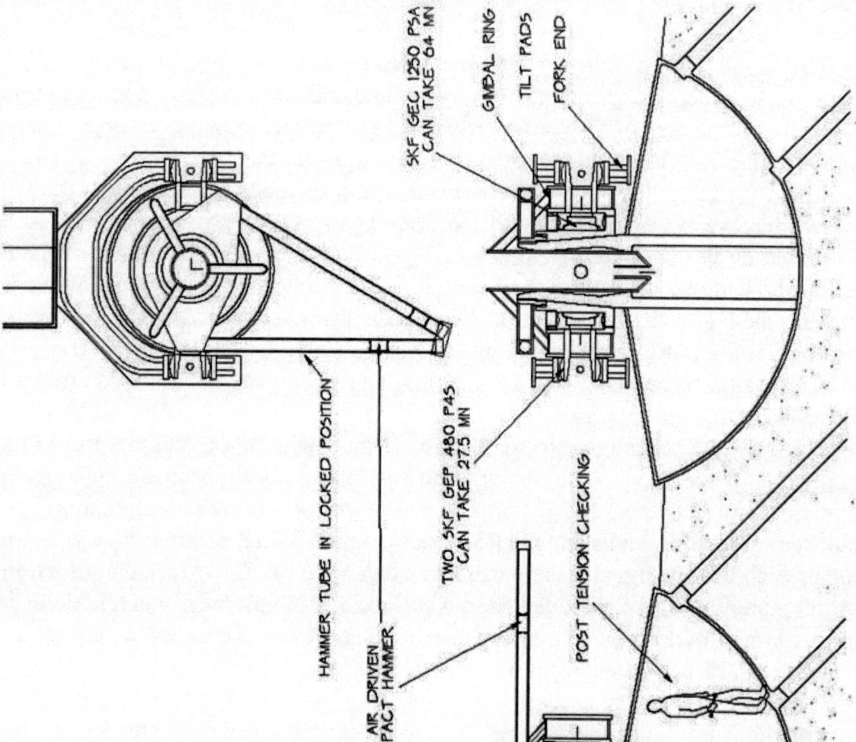


**Fig. 8** The universal joint at the bottom of the mooring legs can be disconnected from the tripod head by a hammer-driven breech screw

be fed to the loaded side of each bearing to offset a large part of the applied load as in a hydrostatic bearing, but with a pressure ratio not quite sufficient to provide complete separation. The vertical component of the strut load will be taken by pads of flat disks made from Glycodur F, mounted on spherical seatings which can tilt to give evenly-spread loading against a thrust face.

## 10 Installation

The first experiences of installing pioneer wave energy devices have not been happy, with long delays and expensive day-rates for installation vessels. We must also expect that early devices will have to be installed and removed quite frequently. The challenges are that good wave sites will have few calm days and that items weighing thousands of tonnes will often be moved by waves randomly through distances of metres but must be positioned to millimetre accuracy.

The design approach is to use modern electronic navigation methods to control the *forces* exerted by a number of very light, agile vessels in a tightly coordinated way as opposed to the traditional method of using a slow, heavy vessel with *position* control. The vessels would have a strong central box surrounded by very large air-filled tubes, much larger than those of conventional inflatables. Some of the control can be done with direct pushing. But in addition, strips of magnetic beads consisting of neodymium tiles energising steel pole pieces lying parallel to the air tubes can be attached to any ferrous plating and provide both pulling and shear forces quite large enough for moving wave devices and even large ships.

They can be quickly released by the inflation of a pair of flattened fire hoses. The vessels can also carry low inertia, *torque-controlled* winches to apply force with a vertical component up to their safe change of freeboard. Each vessel would be driven by a pair of Voith-Schneider propellers driven by compact and lightweight ring-cam Diesel engines with a power rating of 600 kW. The Voith-Schneider propeller uses vertical-axis blades with cyclical pitch-variation to give rapid changes in the magnitude of thrust and direction. Together these can produce a bollard pull of the order of 1 MN from each inflatable.

These vessels would be aligned by a mobile structure in the form of a Vee-block with four water-filled bags on its sloped faces as shown in Fig. 9. The Vee-block can be towed on the surface. Water would be pumped into it to give slightly negative buoyancy so that it can be lowered to the sea bed and then sunk at the correct distance from the tripod. Wave forces and buoyancy would be used in an intelligent way to avoid all heavy lifts. At a water depth of 50 m the amplitude of wave motions is much lower than at the surface. For an 8-s wave period, the horizontal movement at the sea bed is only 8.5% of the surface one and only 11% at the top of a Duck lying on the bed. By choosing the amount of negative buoyancy and applying judicious pulls by the installation vessels, we can make fine adjustments, say 300 mm, to the position and angle of the Vee-block relative to the tripod. More water can then be added to hold the Vee-block in place.

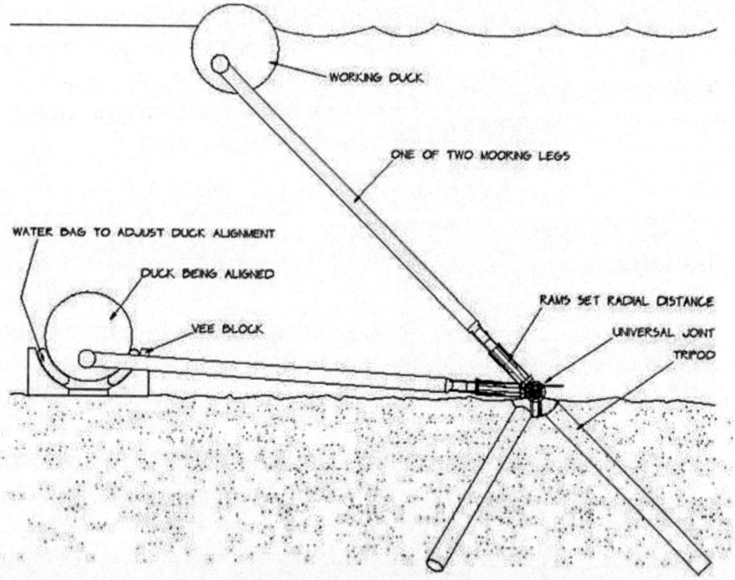


**Fig. 9** A composite view showing the fully flooded duck in the alignment position on its Vee-block and water bags and raised to the working position with half the water removed

The Duck will be filled with fresh water so that the top of its body is nearly awash but the mooring legs will be filled with air to give attachments points above the surface. It will be moved to position over the Vee-block and lowered to make contact with the water-filled bags on the slope surfaces. The bags will be flexible but have an inelastic textile reinforcement so that when the weight of the cylinder is applied they will provide an accurate and rigid support. The bag pressure will follow the pattern of wave loadings. A shoreward acceleration of a wave will increase the shoreward bag pressure and reduce that in the seaward one. By opening and shutting a valve between the bags at times chosen to suit the phase of the wave loading, we can make micro adjustments to the bag volumes to as to move the cylinder to the correct angular direction relative to the tripod. The mooring legs can be lowered with acoustic distance measurements to a datum marker on the tripod. The cylinder at the end of the mooring legs can be used to make radial adjustments. The final contact will be between a female cone on the universal joint and a male cone at the top of the tripod.

Locking of the universal joint to the tripod will be done with a mechanism known as a breech screw. These were originally used for opening and closing the chambers of large breech-loading guns. The normal threads of screw and nut are partially cut away in, say, six grooves, leaving just less than half the thread on each. The cut-aways allow the screw to be fully inserted into the nut with an axial movement. All the remaining threads can then mated with a partial turn of the screw.

A breech-screw big enough to retain a universal joint loaded with 40 MN can be turned by an inertial impact hammer. This will consist of two tubes 4 m long with a 170 mm bore. They will contain free-sliding 'pistons' of solid hardened steel running on Teflon seals. A piston will be moved to the outer end of the cylinder by air pressure. This will be abruptly reversed with vacuum on the inboard side and 10 bar air pressure on the outboard side. By the time the piston reaches the hardened plug at the far end of the cylinder, its velocity will be over 65 m/s and its momentum over 2600 kg m/s. The magnitude of the resulting force will depend on the time over which momentum is converted. With a short stiff connection to the screw threads this is likely to be fast but if necessary it can be extended by force transfer through a stack of Bellville washers. If it extended to 1 ms, the impact force would be over 2.6 MN.

Water will then be removed from the cylinder hull so that it rises to its working position at the surface. The final locking of the breech screw can be done when wave forces are pushing the system downwards and forwards. The Vee-blocks can be removed to install other systems.

## 11 Conclusions

Wave energy can be used directly for the production of very high purity, sterile fresh water even from polluted sources, provided that non-scaling heat transfer materials such as PVDF are available.

The energy requirement is set by the area of heat transfer surface. Several thousand square metres will be required for each unit but the Maxwell Davidson design gives heat exchanger costs which are little more than that of the sheet material. Energy requirement can be below 2 kWh/m$^3$ at low power levels but rises at higher ones to straddle the requirement of reverse osmosis.

Output volume is almost linear with wave amplitude, except for a small offset due to the elevation of the boiling point of salt water, and is higher at shorter wave periods. For 12-m diameter by 24-m wide Ducks, output will be of the order of one or two thousand cubic metres a day in a tropical wave climate exposed to a large ocean.

The low thermal conductivity of foam concrete means that heat losses for objects of the likely dimensions are not a problem. Heat gain is quite possible and needs careful temperature regulation.

It may be possible to build fluidised bed technology followed by suction to install the heavily loaded moorings in sandy sea beds without heavy lifting vessels.

Attachment and removal can be performed by means of a set of several light, agile, inflatable vessels with torque-controlled winches, magnet contact strips and tightly coordinated vector thrust propulsion.

**Acknowledgements** This research project has been carried out within the WAVETRAIN Marie Curie Research Training Network of the European Community's Sixth Framework contract MRTN-CT-2003-7505166.

## References


Anon. (2006) SKF bearing catalogue. http://www.skf.com/portal/skf/home/ products?maincatalogue = 1&newlink = 1&lang = en

Bartram J, Callan P (2006) Guidlines for drinking water quality, 3rd edn. WHO Geneva; http://www.who.int/water_sanitation_health/dwq/gdwq0506.pdf

Budal K, Falnes J (1975) A resonant point absorber of ocean-wave power. Nature 256:478–479

Cruz JMBP, Salter SH (2006) Numerical and experimental modelling of a modified version of the Edinburgh Duck wave energy device. I Mech E J Power Energy 220:129–147

Darwish MA, El Dessouky H (1996) The heat recovery thermal vapour compression desalting system: a comparison with other thermal desalination processes. Appl Thermal Eng 16:523–537

Davidson MW (1974) Multistage Evaporator, vol 921, p 2042. US Patent 3808104

Evans DV et al (1979) Submerged cylinder wave energy device: theory and experiment. Appl Ocean Res 1:3–12

Salter SH (1985) Wave-powered desalination. In: Twidell J (ed) Proceedings of conference on energy for rural and island communities, inverness, September 1985. Pergamon, Oxford, pp 235–241

Salter SH (2005) High purity desalination using wave-driven vapour compression. World Renewable Energy Conference, Aberdeen

Salter SH, Taylor JRM, Caldwell N (2003) Power conversion mechanisms for wave energy. Proc Inst Mech Eng J Eng Maritime Environ 216:1–27

Skyner D (1987) Solo Duck linear Analysis. Report to UK Wave Energy Steering Committee

# A Novel Macro-Engineering Approach to Seawater Desalination

**Alexander A. Bolonkin, Shmuel Neumann and Joseph J. Friedlander**

## 1 Introduction

No life resource is needed in such enormous volumes as freshwater—most living biomass in the Earth's biosphere consists of 60–95% water, and the absence of freshwater limits individual human survival to just a few miserable days, as those who have ever found themselves cast mercilessly upon the ocean's vast undulating surface, or lost in the wilderness on land, can testify—if, of course, they have lived to relate to others their life-threatening experience. New natural resources of potable water are being depleted worldwide (Rizzuti et al. 2007).

Desalination refers to any of several processes that remove salt and other minerals from water (ocean water, ground water). Water is compatible with more substances found in the Earth-biosphere than any other known solvent; all water, even just fallen rain water, contains dissolved salts. The minerals known to exist in seawater are fully dissolved, which means these minerals exist in free-flowing, non-bound, ionic form; they are not microscopic metallic suspended particles in seawater. About 84 basic chemical elements have been identified in the ocean's water as either macro or trace mineral ions. Sometimes, seawater has been defined as a weak complex solution of almost everything, which includes mineral salts and decayed biologic matter. A significant oceanographic fact regarding seawater's elemental content, first discovered *circa* 1865–1884 by William Dittmar

A. A. Bolonkin (✉)
C&R, NJIT, 1310 Avenue R, suite 6-F, Brooklyn, NY, 11229, USA
e-mail: abolonkin@juno.com, abolonkin@gmail.com

S. Neumann
Michlala Jerusalem College, Jerusalem, Israel

J. J. Friedlander
Shave Shomron, Israel

V. Badescu and R. B. Cathcart (eds.), *Macro-engineering Seawater in Unique Environments*, Environmental Science and Engineering,
DOI: 10.1007/978-3-642-14779-1_30, © Springer-Verlag Berlin Heidelberg 2011

(1833–1892) and J.G. Forchhammer (1794–1865), is that while the total concentration of dissolved salts varies from place to place in the Earth's ocean, the ratios of the more abundant components actually stay almost constant. Even so, the ocean's variance of seawater content is caused by river inflow, melting of Polar Zone glacial ice, evaporation over $\sim$70% of Earth's watery surface, rainfall and snowfall, wind, ocean waves and seawater currents. If all the salt in the ocean's seawater could be removed—certainly not by any possible human desalination macro-projects—and spread evenly atop the Earth's land ($\sim$30% of the planet's area), then it would form a homogeneous solid material stratum $\sim$152 m thick. Annually, about 10,000,000,000 metric tons of sea-salt is wafted upwards from our world's ocean into the Earth's encompassing atmosphere (Thornton et al. 2010).

The Earth's ocean undoubtedly is the most voluminous "storehouse" of minerals available to industrial exploitation by humans and it probably contains every known natural element and those that have been manufactured artificially. The recovery of certain valuable minerals from virtually inexhaustible seawater concentrates has attracted the attention of industrial innovators. First, mineral resources on land are becoming more difficult to find due to the political and economic constraints on ore exploration. Second, the remarkable ongoing shortages of freshwater of a globally increasing human population and food animals fosters the macro-engineering concept that this shortage of a vital liquid (freshwater) can be solved only by large-scale desalination of seawater. Such desalination plants will, of course, produce (as a by-product) enormous quantities of brines having mineral concentrations $\sim$3 times that of seawater. Obviously, these brines can be assessed as large, reliable and concentrated resources for minerals. Third, some eco-regions of our planet's biosphere are endowed sufficiently with the best climatic regime to produce great quantities of common salt by solar evaporation of seawater in confined ponds (Nunez 2010). The separation of salt from water has an extensive history, dating from ancient times when salt, not freshwater, was the most precious commodity to be derived from seawater. Seawater desalination was first industrialized on passenger, cargo and warships that left ports for long periods of time, at isolated military bases and on freshwater-short islands populated with few humans.

The focus of this chapter is on seawater desalination (Escobar and Schafer 2009), with a particular focus on the coupling of renewable energies with desalination processes (Cipollina 2009). Salty water is desalinated in order to be converted to freshwater suitable for animal consumption, irrigation and if almost all of the salt is removed, for drinking by humans along with other consumptive practical uses. Sometimes, desalination produces tasty and healthy table-salt as a commercial by-product. Most modern interest in seawater or brackish water desalination is focused on developing cost-effective industrialized method of providing freshwater for human use in eco-regions where the availability of freshwater is naturally limited, such as on various types of ships and aboard long-submergence voyages by military nuclear submarines. In instances in which vast land eco-regions are separated from a freshwater source, sometimes large-scale desalination is required for human development of the eco-region. Typically, large-scale desalination requires

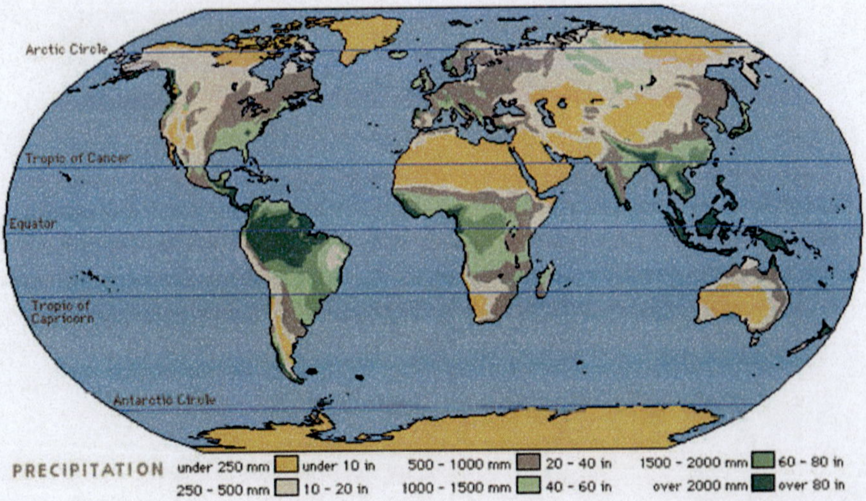


**Fig. 1** The world's vast drylands illustrated here by mapped average precipitation (rain and snow) on all land

substantial amounts of energy as well as specialized, expensive infrastructure, making the factory process costly when juxtaposed to the cost of freshwater taken directly from rivers or groundwater (Semiat 2008). Nonetheless, Middle Eastern countries' large energy reserves of oil and natural gas, coupled with their profound freshwater scarcity have led to an extensive construction of desalination plants there (Fig. 1). Saudi Arabia's industrial-scale desalination plants, for instance, account for ~24% of the world's total 2010 freshwater-making capacity. For example, five major urban regions (the seaport of Yanbu, Medina, Jeddah, Makkah/Taif and Assir) account for most of the desalination capacity facing the Red Sea. (Hoepner and Lattermann 2002). The growing phenomena of desalination is substantiated by a vividly worded 2008 press-release from the International Desalination Association: "World-wide, 13,080 desalination plants produce more than 45,000,000 $m^3$ of freshwater daily".

## 2 Methods of Seawater Desalination, Associated Macro-Problems

As of 2010, globally, the two leading desalination methods are reverse osmosis (~47% of installed capacity) and multi-stage flash (~36%). The usual industrial process used in these particular desalination operations is vacuum distillation, which entails boiling the containerized seawater at reduced atmospheric pressure and at a much lower temperature than is normal at sealevel. Energy is saved because (1) liquid boils when the vapor pressure equals the ambient pressure, and (2) vapor

**Fig. 2** Ashkelon seawater reverse osmosis desalination plant, Israel. As of circa 2006 it was manufacturing 50,000,000 $m^3$ yearly. Ultimately, it will manufacture 100,000,000 $m^3$ annually

pressure increases with temperature. Since the start of the 21st Century, most new large-scale urban and industrial desalination facilities use reverse osmosis technology (Fig. 2) and membrane processes which utilize semi-permeable plastic membranes as well as gas pressure to separate sea-salt from seawater and brackish water (Einav and Lokiec 2003).

These processes typically use less energy than thermal distillation commonly used on ships, which has led to a reduction in overall desalination costs over the past decade. (Shipboard desalination plants usually replicate, in miniature, the planetary hydrological cycle.) Desalination remains energy intensive and future costs will continue to depend on the price of both available energy and "perfected"—no technology is ever truly perfected, merely improved as human experience increases—desalination technology, examples of which include: capacity and facility type, absolute and relative geographical location, utilized feed-water, human labor costs (i.e., salaries, wages, insurance), employed energy, financing and its terms (amortization), as well as post-production disposal of wasted concentrate materials.

## 3 Desalination's Demerits

Some critics of large-scale seawater desalination assert: (1) the high monetary cost of currently available desalination technologies is especially onerous for impoverished developing states; (2) desalination leads to a by-product of brine, which some Green environmentalists claim is a major cause of marine pollution when it is dumped into the oceans at high temperatures by the desalination plant; (3) transporting or piping of massive volumes of freshwater (made from desalinated seawater) in the interiors of spacious dryland countries is impracticable and cost intensive.

In response to the first criticism, desalination is certainly more suitable for affluent, coastal regions than poor regions situated in the interior of an arid continent, or at high topographic elevation where costly pipeline pumping would be required. The latter region requires transport of freshwater from another region, as opposed to desalination. Indeed, one needs to lift the useful freshwater by 2,000 m, or transport it over more than 1,600 km for transport costs to equal desalination costs (Dore 2005). In regard to the second criticism, the reverse osmosis technology that is generally used to desalinate seawater typically does not produce heated brine as a by-product. Additionally, depending on the prevailing currents of the receiving ocean, the seawater concentrate by-product (brine) can be diluted and dispersed to match background levels within relatively short distances of the desalination plant's ocean outfall (Einav et al. 2003). In light of the third criticism, geography plays a strong role in desalination costs. In places far from the ocean, such as New Delhi (India) or in mountainous places, like Mexico City (Mexico), high transport costs add to the high desalination costs. Desalinated water is also expensive in places that are both far from the ocean and high in land elevation, such as Riyadh (Saudi Arabia) and Harare (Ethiopia). In places closest to the ocean, the dominant economic cost is actual desalination, not transport cost; therefore, desalination would be less expensive in places like Beijing (China), Bangkok (Thailand) and, obviously, major coastal cities situated near sealevel.

Desalinations' cost-effectiveness compared with other freshwater consumer supply options, which include mandatory installation of rainwater tank reservoirs or storm water harvesting infrastructure, have led many major coastal cities in industrialized nations to consider seriously the feasibility of large-scale seawater desalination. Two such countries are Israel, which now desalinizes water at a cost of 2010 USD 0.53/$m^3$ and Singapore, which is currently desalinizing seawater at USD 0.49/$m^3$. Israel's Ashkelon plant is slated, ultimately, to produce $\sim$320,000 $m^3$/day (Fig. 2). A prime example of a country that has tapped into the technological resource of industrial desalination is Australia. Engineering studies have shown that desalination is among the most cost-effective options for boosting the freshwater supply in major Australian states. (Australia is currently enduring a dreadful drought condition almost nationwide.) The city of Perth (Western Australia) has been successfully operating a reverse osmosis seawater desalination plant since circa 2006, and the national Australian Government has announced that more desalination plants are soon to be built in Australia's largest city, Sydney, in the near future.

Australia's decentralized desalination plants are the paradigm of environmental-friendliness; the Perth desalination plant, for example, is powered partially by renewable electric energy from the nearby Emu Downs Wind Farm (Garcia-Rodriguez et al. 2001).

In addition, the Sydney desalination plant will be powered entirely from renewable energy sources, thereby eliminating harmful greenhouse gas emissions to the atmosphere, a common argument used against seawater desalination due to the great energy requirements of the transformational technology.

Renewable resources, although ecologically desirable, add to the base financial operating costs of seawater desalination plants. However, recent experience in Western Australia (notably, the City of Perth) indicates that the additional cost is acceptable to local participating human communities because it allows the communities to augment their freshwater supply without doing marked environmental harm to the air by way of smoke and gas emissions from fossil-fueled powerplants (Fig. 3). As a result, The Gold Coast Desalination Plant, a water factory soon to be built in northeastern coastal Australia, will be powered entirely by fossil fuel burning. At a rate of over 4 $kWh/m^3$ to produce, this plant will be the most expensive domestic source of potable freshwater in all of Australia. The Gold Coast is Australia's most significant tourist destination because of its sunny climate, clean beaches; it attracts both adjacent residents and even foreign tourists mainly because of its agreeable weather and scenic landscape, as well as the Great Barrier Reef. One way to fund the high cost of desalination is by raising freshwater retail sale prices; for instance, the seawater desalination plant near the City of Adelaide, Australia, located at Port Stanvac will be funded entirely by increasing freshwater rates to local consumers in order to achieve full-cost recovery early-on.

Summarizing, as illustrated by the examples of the rising number of desalination plants in the Middle East, the USA and Australia, countries are very interested in public- and private-sector monetary investments in desalination plants. In order to better understand the weighty costs of the desalination plants, it is necessary to discuss the process by which seawater is typically desalinized,

**Fig. 3** The Tampa Bay, Florida (USA) desalination plant. By circa 2007 it was producing 34 million cubic meters of freshwater per year

reverse osmosis. Reverse osmosis is a separation process that uses pressure to force a solution through a semi-permeable membrane; the solute is retained on one side where-as the pure solvent is allowed passage to the other side. In reverse osmosis, a solvent moves from a region of high solute concentration through a membrane to a region of low solute concentration. As the name implies, reverse osmosis is the reversal of the normal osmosis process, the natural movement of solvent from an area of low solute concentration through a membrane, to an area of high solute concentration when no external pressure is applied.

The membranes that are used for reverse osmosis have a dense barrier layer in the polymer matrix where most separation occurs. In most cases the membrane is designed to only allow water to pass through a dense layer while preventing the passage of solutes (such as salt ions). This process requires that a high pressure be exerted on the high concentration side of the membrane, usually 2–17 bar for fresh and brackish water and 40–70 bar for seawater, which has around 24 bar natural osmotic pressure that must be overcome.

The desalinated water that is produced is very corrosive and thus is "stabilized" to protect downstream pipelines and storages, usually by adding lime and carbon dioxide to prevent the corrosion of concrete or cement lined surfaces. Moreover, liming material is used in order to adjust the pH from 6.8 to 8.1 to meet the potable water specifications, primarily for effective disinfection and for corrosion control. Additionally, although desalination processes are very effective barriers to pathogenic organisms, disinfection by the addition of chlorine is often used to ensure a "safe" water supply to the consuming public.

Reverse osmosis units sold for residential purposes offer water filtration at the cost of large quantities of waste water. For every 5 gallons of output, a typical residential reverse osmosis filter will send around 10–20 gallons of water down the drain although it may be captured and used for watering plants and lawns. The substantial waste, however, can be minimized greatly through using our efficient method of desalinization that is outlined below.

## 4 Creative Macro-Engineering: The Solar-Powered Desalination System

Our proposed solar seawater desalination system has two layers of film tubes divided by partitions into separated sections connected by back valves for air (vapor) and water (Fig. 4). The top layer of the tubes contains sea water and air; the lower layer of the tubes contains freshwater and air (Fig. 5). Solar radiation (70–80°C) heats the seawater in the top tube layer which evaporates the fresh water, creating a vapor. The moist air (vapor) is then forced into a lower tube layer, which is cooled by seawater (up to 25°C), and then condenses into freshwater which is pumped to the seacost (Neumann et al. 2010).

The tubes are separated into sections (Fig. 5) connected by unidirectional back valves on top (14) and on bottom (15). In the top tube layer, the top valves push the

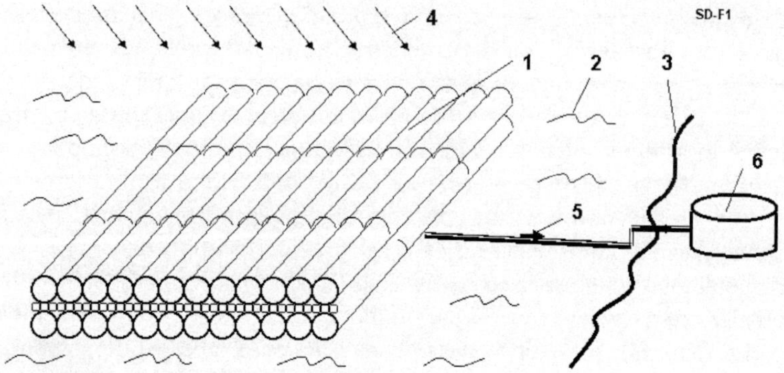


**Fig. 4** Solar seawater distiller. Notation *1* film tubes; *2* sea; *3* coast; *4* solar radiation; *5* fresh water tube and pump; *6* fresh water storage

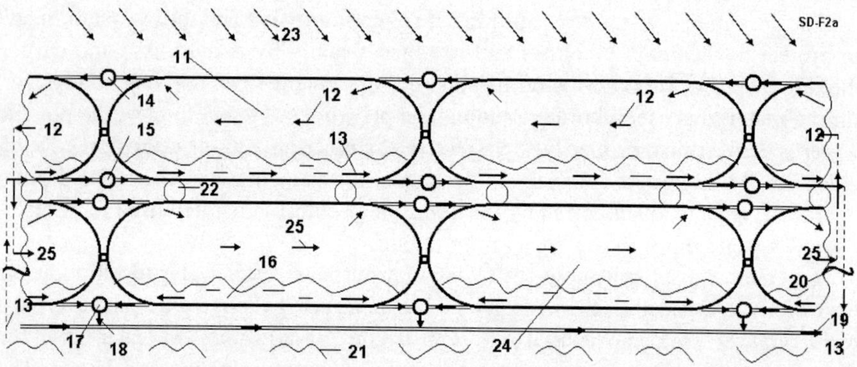


**Fig. 5** Longitudinal cross-section of solar seawater Distiller. *Notations 11* section of flexible tube; *12* air wet (vapor) flow; *13* seawater flow; *14, 15, 17* unidirectional fluid valves (other names: one direct (direction) valves, back valves, inverted valves, return valves, holding valves); *16* fresh water; *17* back valve; *18, 19* freshwater flow; *20* sea surface; *21* ocean; *22* hear protection; *23* solar radiation; *24* lower tube; *25* air flow in lower tube layer

wet air vapor (12) in one direction. The lower valves pass the sea water (13) in the opposite direction. The lower valves (17) pass the condensate (fresh water) into under water tubes and to ground or fresh water storage (16). The sea waves then compress and decompress the inflatable tube sections and pump the wet air, sea and fresh water. In addition, the wet air (vapor) and water may use small electric pumps (ventilators) for circulation.

Rain water (37) then flows into tubes (38) to the fresh water storage (40) (Fig. 6). The float installation has special protection on the top tube layer against the seawater (Fig. 9). In order to increase the useful absorption of the incidental solar radiation, the lower part 34 of the top tubes (Fig. 6) are colored black. The material of the top tube passes the solar radiation but reduces heat losses upwards. In addition, the top part of 33 is transparent. Below the top tube the heat reflector

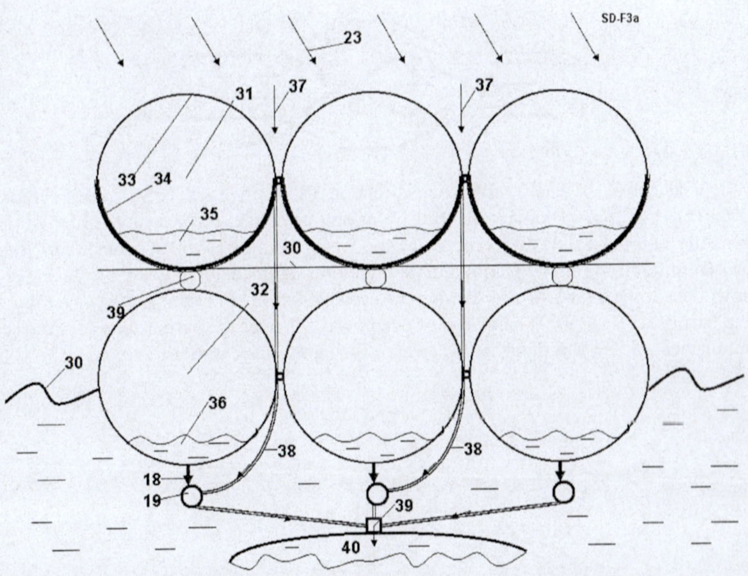


**Fig. 6** Diametrical cross-section of macro-engineered solar seawater distiller. *Notations 30* heat reflector; *31* top film tubes; *32* lower film tubes; *33* transparent top part of top tube; *34* black lower part of top tube; *35* seawater; *36* freshwater in lower tubes; *37* rain; *38* rain fresh water tube; *39* back valve; *40* freshwater storage; *30* sea surface; *23* solar radiation; *18, 19* tubes for freshwater and rainwater

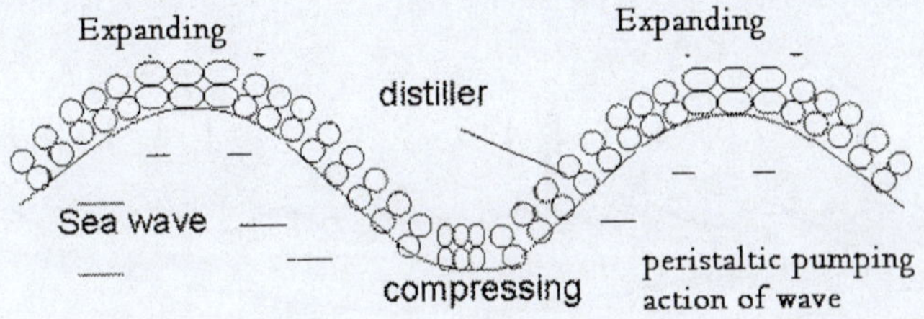


**Fig. 7** Explanation for the pumping of air and seawater through tubes of distiller by sea waves. Ocean waves alternately compress and expand the separate sections of the flexible inflatable tubes. The back valves allow the air and seawater into tubes to flow only in desired direction

30 is located (Fig. 6). A reflector blocks the lower tubes from the incident heat radiation from the upper heat tubes.

The ocean's waves periodically expand and compress the inflatable sections of tubes and the moving air (vapor) and flowing sea water (Fig. 7). Back valves (Fig. 8) allow the flow to move only in a selected, given direction. Air and seawater, for example, flows into the top layer in opposing directions. The air (vapor) flows from the top layer and moves into the lower layer where it is cooled by the ocean through thin-film lower tubes, where it then condenses, and freshwater flows into the freshwater storage (40) (Fig. 6).

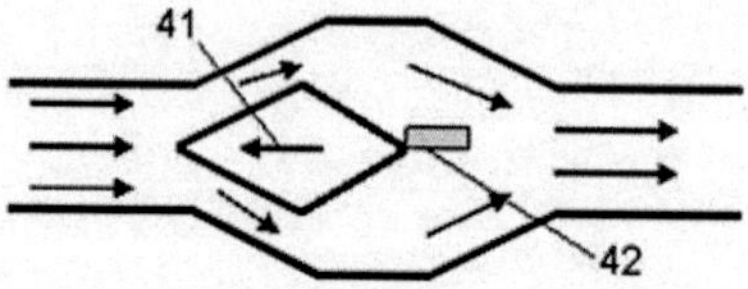


**Fig. 8** Typical action of back valve (one direction valve, inverted valve, (non) return valve, holding valve). *Notation 41* valve, *42* string. Working principle: the pressure is high on left side that opens the internal valve *41* (compresses the weak spring *42*) and substance (air, seawater) overflows from left to right. When pressure is higher on right side, it closes the hole and stops the back flow. The top surface of the distiller is protected from the strong sea waves by special inflatable barrier *43* (Fig. 9). When a top of ocean wave is over the barrier tube (*43*) the seawater flows into space *45* and through the back valve (*44*) goes back into the ocean

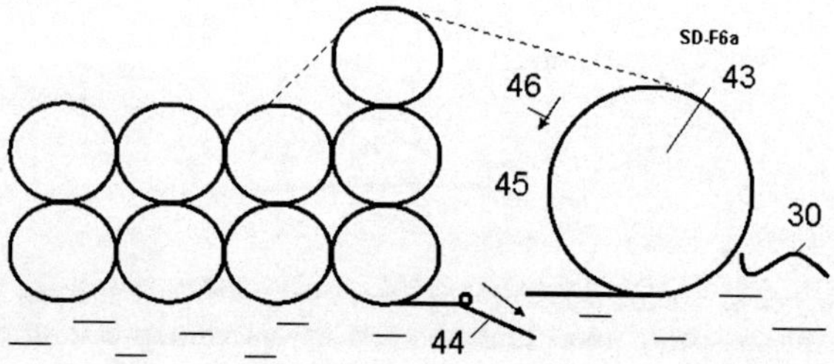


**Fig. 9** Barrier against surge of seawater over surface of distiller. *Notations 43* inflatable barrier tube; *44* back valve; *45* internal space between the barrier tube and distiller; *46* water of a sea wave. *Work:* when a top of sea wave is over the barrier tube *43*, the sea water flows into space *45* and through the back valve *44* goes back to the sea

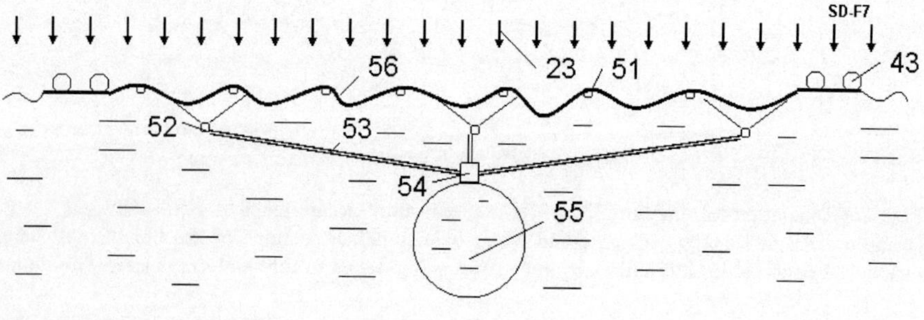


**Fig. 10** Simplest collector of rain freshwater on ocean and dew on land. *Notations 51* support float (optional); *52* small plumb; *53* rain water tube; *54* back valve, *55* inflatable storage of rain and dew freshwater; *56* thin film; *43* barrier against sea waves; *23* rain. *Work:* the rain and dew water flows to the freshwater storage and pumps to customer

The solar desalinization system can also work on land. In this instance, the lower tube layer is cooled by air. However the productivity significantly decreases. An inexpensive portable version of the offered system may be included in soldier or travelers accoutrements for production of fresh water in the desert so they would able to distill fresh water from bad water or urine.

The simplest (rain and dew) version of our system is shown in Fig. 10. Thin film is located over the ocean's surface and provides a barrier against big waves. It can also collect rain and (early morning and night) dew freshwater. The rain and dew water flows to the freshwater storage (55) from small pump (52) through rain tube (53) and back valve (54). If wanted or needed, the film can have a high reflectivity surface; if this technical improvement covers a large area, the seawater will have significantly lower temperature (solar radiation does not heat the seawater). Thus, the quantity of the night-deposited dew freshwater is increased.

## 5 Summation of Project's Innovations and Advantages

- Located on the unpopulated ocean, the system does not require expensive coastal land surface;
- The devise is made from inexpensive thin film, making the system more affordable than the conventional reverse osmotic or multi-flash plants;
- Installation uses solar energy which produces freshwater at a cost close to zero;
- Sea waves are used for all needed water and damp air circulation;
- The system may be delivered anyplace on Earth by watercraft and easily transported so that it can be an at-sea source of freshwater to ships along busy shipping lanes;
- Preassembled or disassembled, it is very compact and easily stored in conventional warehouses;
- The installation may be also used for collection of the local rainwater and dew;
- The Solar Sea Distiller is environmentally friendly as it does not dump brine in limited geographical places which can be hazardous to cherished and valuable marine life forms.

## 6 Computation and Estimates

An informed reader can derive the equations below from well-known physical laws. Therefore, no detailed explanations or reassurances of these laws are furnished herein.

### *6.1 Estimation of a Maximal Theoretical Productivity of a Solar Distiller*

The energy flux of a solar radiation out of atmosphere is 1366.1 W/m$^2$. After atmospheric losses, the average solar radiation in a clear day at the middle latitude $0 \div$ (30°N–40°N) is about $1{,}000 \div 700$ W/m$^2$. Let us take as average value

$q = 700$ W/m$^2$, 8 h of insolation per day and 300 unclouded days every year. Then a square meter of the Earth's surface will receive solar energy in one day of

$$E = 700 \times 8 \times 60 \times 60 \approx 20 \times 10^6\,\mathrm{J/\left(m^2 day\right)}\ \text{or}\ E_y \approx 6 \times 10^9\,\mathrm{J/\left(m^2 year\right)} \tag{1}$$

For heating 1 kg water from 20 to 100°C and its evaporation it is necessary the amount of thermal energy

$$Q = c_p \Delta T + r = 4.18 \times 80 + 2,260 = 2,595\,\mathrm{kJ/kg} \tag{2}$$

where $c_p = 4.18$ kJ/(kgK) is thermal capacity of water; $\Delta T$ is change of temperature, K; $r = 2{,}260$ kJ/kg is vapor capacity of water.

Theoretically, one square meter of Earth surface water can evaporate during a solar day about:

$$m = \frac{E}{Q} = \frac{2 \times 10^7}{2.595 \times 10^6} = 7.7\ \mathrm{kg/(m^2 day)} \quad \text{or} \quad m_y = 2.3\ \mathrm{ton/(m^2 year)} \tag{3}$$

That means:

- The area 100 × 100 m$^2$ will produce 77 m$^3$/day or 23,100 m$^3$/year;
- The area 1 × 1 km$^2$ will produce 7,700 m$^3$/day or 2.3 × 10$^6$ m$^3$/year and
- The area 10 × 10 km$^2$ will produce 770,000 m$^3$/day or 230 × 10$^6$ m$^3$/year

(That is the productivity of a big desalination plant). If one person, on the average, requires 10 L of water per day, that is enough for 640,000 people, or for the irrigation of 23,000 ha ground (1 ha = 100 × 100 m$^2$ area).

Here we must add the rain freshwater which also is accumulated by the operating seacoast-sited desalination plant. The average annual world-wide rainfall is ~1 m$^3$/(m$^2$year). That means from 0 to 43% additional freshwater will be produced by a near-ocean installation.

## *6.2 Amount of Water in the Earth-Atmosphere*

The amount of water in Earth's air depends entirely upon temperature and humidity. For relative humidity 100%, the maximum partial pressure of freshwater vapor is shown in Table 1.

The amount of freshwater in 1 m$^3$ of air may be computed by equation:

$$m_W = 0.00625\ [p(t_2)h - p(t_1)] \tag{4}$$

where $m_W$ is the mass of water, kg in 1 m$^3$ of air; $p(t)$ is vapor (steam) pressure from Table 1, $h = 0 \div 1$ is relative humidity. Results of computation using Eq. (4) are presented in Fig. 11. Typical relative humidity of atmospheric air is 0.5–1.

**Table 1** Maximum partial pressure of water vapor in atmosphere for given air temperature

| $T$ (°C) | −10 | 0 | 10 | 20 | 30 | 40 | 50 | 60 | 70 | 80 | 90 | 100 |
|---|---|---|---|---|---|---|---|---|---|---|---|---|
| $P$ (kPa) | 0.287 | 0.611 | 1.22 | 2.33 | 4.27 | 7.33 | 12.3 | 19.9 | 30.9 | 49.7 | 70.1 | 101 |

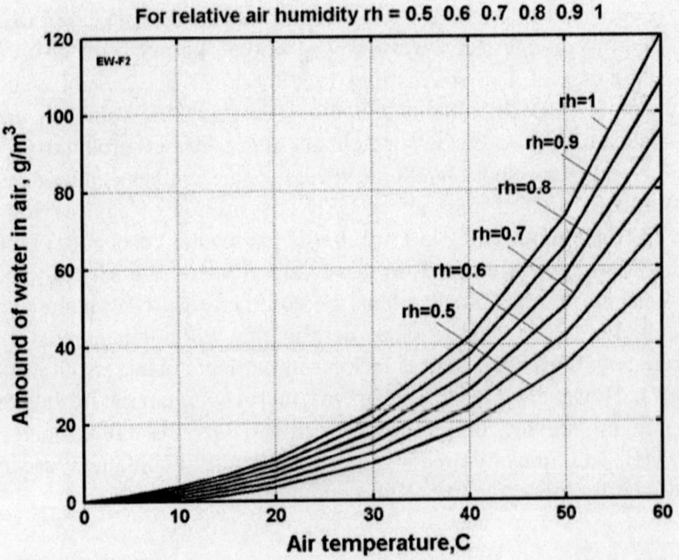


**Fig. 11** Amount of freshwater in 1 m$^3$ of air versus air temperature and relative humidity (rh). $t_1 = 0^\circ C$

## 6.3 Heat Transfer from Damp (Humid) Air to Seawater

Let us assume that the temperature in upper solar tubes is 75°C, and in the seawater it is 25°C. The average change of temperature is $\Delta T = 0.5(75 - 25) = 25°C$.

The average transfer of heat flow from air to water is (tubes made of thin film):

$$q \approx \lambda \Delta T = 100 \times 25 = 2.5 \text{ kW/m}^2. \tag{5}$$

Here $\lambda = 100$ W/(m$^2$K) is coefficient of heat transfer from air to water. The value 2.5 kW/m$^2$ is larger than the solar flux energy density of 1 kW/m$^2$. That means the water will cool down the wet air into lower tubes.

In our construction the air is moved by the wavy ocean surface movements. If the air is moved by a fan, the computation show the requested energy is very small indeed.

## 6.4 Cost Estimation

### 6.4.1 Cost of Installation

Assume the film used for tubes has thickness $\delta = 0.07$ mm. Then the volume of two tubes film with surface area $S = 1$ m$^2$ is $V = 2\pi\delta S = 2 \times 3.14 \times$

$7 \times 10^{-5} = 4.4 \times 10^{-4}$ m$^3$. If the specific mass density of the film is $\gamma = 1{,}500$ kg/m$^3$, the mass of 1 m$^2$ film is $m = 0.66$ kg/m$^2$. If the cost 1 kg film is USD 3/kg, the cost of 1 m$^2$ of distiller is USD 2/m$^2$. The cost of a surface area $100 \times 100$ m$^2$ is USD 20 K, the cost of $1 \times 1$ km$^2$ is USD 2 million, and the cost of a seawater still $10 \times 10$ km$^2$ which has a freshwater productivity of about 270 million m$^3$ a year (about five large reverse osmosis plants equivalent) is about USD 200 million of film, plus about USD 250 million of engineering and construction, the total about USD 450 million. If freshwater costs $\sim$0.5 USD per m$^3$, then payback impressively occurs in less than 3.5 years. We suggest that, at start-up, the freshwater output of our plants be confined to use by humans and stock animals only. Pure freshwater made by desalination will not meet most agricultural needs (crop irrigation) because it is lacking in nutrients plants require (Yermiyahu et al. 2007). However, traceable elements in such unnatural freshwater can be useful in tracking the flow of man-made freshwater in soils, streams and rivers and ground water and, thus, indicating leaks in poorly-maintained urban potable freshwater supply infrastructure (Kloppmann et al. 2008).

## 7 Conclusion

The solar desalinization system is the most advanced and efficient seawater desalination system invented to-date. The solar desalinization system is (1) inexpensive, (2) does not require covering picturesque and expensive real-estate at the humanly attractive coastline since its location can be on the offshore ocean surface, even providing a new local tourist attraction, (3) does not require energy except for pumping the freshwater to purchasing customers on land and (4) needs minimal amount of service or maintenance worker crews on site. Moreover, the minimal production costs qualify our efficiently macro-engineered system as by far the least expensive freshwater production method. In addition, the pre-assembled construction kits may be conveniently stored in conventional storage facilities, and may be easily moved (transported, delivered) to anyplace in the world. Related works can be found in Bolonkin (2006a, b, 2007), Krinker (2009).

## References


Bolonkin AA (2006a) Extraction of freshwater and energy from atmosphere. http://arxiv.org/ftp/arxiv/papers/0704/0704.2571.pdf

Bolonkin AA (2006b) AB method of irrigation without water (closed-loop water cycle). http://arxiv.org/ftp/arxiv/papers/0712/0712.3935.pdf

Bolonkin AA (2007) Cheap artificial AB-mountains, extraction of water and energy from atmosphere and change of regional climate. http://arxiv.org/ftp/arxiv/papers/0801/0801.4820.pdf

Cipollina A (2009) Seawater desalination: conventional and renewable energy processes. Springer, Heidelberg, p 306

Dore MHI (2005) Forecasting the economic costs of desalination technology. Desalination 172:207–214

Einav R, Lokiec F (2003) Environmental aspects of a desalination plant in Ashkelon. Desalination 156:79–85

Einav R, Harussi K, Perry D (2003) The footprint of the desalination processes on the environment. Desalination 152:141–154

Escobar I, Schafer A (eds) (2009) Sustainable water for the future, vol 2. Elsevier, Dordrecht, 444 pp

Garcia-Rodriquez L, Romero-Ternero V, Gómez-Camacho C (2001) Economic analysis of wind-powered desalination. Desalination 137:259–265

Hoepner T, Lattermann S (2002) Chemical impacts from seawater desalination plants—a case study of the northern Red Sea. Desalination 152:133–140

Kloppmann W, Vengosh A, Guerrot C, Millot R, Pankratov I (2008) Isotope and ion selectivity in reverse osmosis desalination: geochemical tracers for man-made freshwater. Environ Sci Technol 42:4723–4731

Krinker M (2009) Review of space towers. http://www.scribd.Com/doc/26270139, http://arxiv.org/ftp/arxiv/papers/1002/1002.2405

Neumann S, Bolonkin AA, Friedlander J (2010) Sea solar distiller method and installation. USPTO Patent 12856665, 15 August 2010

Nunez M (2010) Assessing potential desertification environmental impact in life cycle assessment. Int J Life Cycle Assess 15:67–78

Rizzuti L, Ettouney HM, Cipollina A (eds) (2007) Solar desalination for the 21st century. Springer, NL, 380 pp

Semiat R (2008) Energy issues in desalination processes. Environ Sci Techno 42:8193–8201

Thornton JA, Kercher JP, Riedel TP, Wagner NL, Cozic J, Holloway JS, Dubé WP, Wolfe GM, Quinn PK, Middlebrook AM, Alexander B, Brown SS (2010) A large atomic chlorine source inferred from mid-continental reactive nitrogen chemistry. Nature 464:271

Yermiyahu U, Tal A, Ben-Gal A, Bar-Tal A, Tarchitzky J, Lahav O (2007) Rethinking desalinated water quality and agriculture. Science 318:920–921

# The Ike Dike: A Coastal Barrier Protecting the Houston/Galveston Region from Hurricane Storm Surge

**William J. Merrell, Lyssa Graham Reynolds, Andres Cardenas, Joshua R. Gunn and Amie J. Hufton**

## 1 Introduction

On the morning of 13 September 2008, Hurricane Ike made landfall near the east end of Galveston, Texas. Ike's strong Category 2 winds and near Category 5 equivalent storm surge, created sufficient devastation to be ranked as the third costliest hurricane to make landfall in the United States. One factor that somewhat mitigated the damage and loss of life in Galveston and adjacent areas was the presence of an existing Seawall constructed after the 1900 Galveston storm. In this chapter we provide a rationale and conceptual framework for a next-generation coastal protection system necessary to reduce the threats posed by future hurricane surges in the Houston–Galveston urban-industrial region.

Despite the initial lack of public and governmental attention, Hurricane Ike may well come to be viewed as an epic storm. It has already changed how the National Oceanic and Atmospheric Association (NOAA) will classify hurricanes by giving more credence to surge potential. Moreover, as we will discuss in this chapter, the devastation caused by Ike clearly pointed out the vulnerability of the Houston/Galveston area to hurricane surge and triggered ideas on regional approaches to suppressing surge for this urbanized region.

Because of the significant threat of major hurricanes on the Texas Coast, there is a need for strategic approaches to reduce the region's vulnerability. There are two quite different strategies that significantly reduce a region's vulnerability to storm surge:

W. J. Merrell (✉), A. Cardenas, J. R. Gunn and A. J. Hufton
Texas A&M University at Galveston, Galveston, TX, USA
e-mail: merrellw@tamug.edu

L. G. Reynolds
Green Light Multimedia, Galveston, TX, USA

V. Badescu and R. B. Cathcart (eds.), *Macro-engineering Seawater in Unique Environments*, Environmental Science and Engineering,
DOI: 10.1007/978-3-642-14779-1_31, © Springer-Verlag Berlin Heidelberg 2011

- Limit the presence of human infrastructure in the first place or reduce it where it exists, or
- Build and maintain comprehensive coastal defenses to protect regions where we have chosen to have large human settlements.

Both approaches work. The question is when and where should they be applied?

At first glance, the Texas coast does not seem like a likely place for arguing the need for comprehensive coastal defenses. Unlike most United States coastal states, which have populations concentrated near the coast, only 25% of the Texas population lives within its 18 coastal counties. There are large areas of the Texas coast that are rural or even pristine—places where it might be wise to apply a policy of discouraging or limiting development in order to reduce vulnerability to surge.

On the other hand, 75% of the Texas coastal population lives in the counties that surround Galveston Bay, the nucleus of a concentrated human settlement on the Texas coast, which includes not only residential areas but also a very robust industrial base. Moreover, population projections call for the Houston/Galveston region to steadily increase over the next decades (http://www.hgac.com/community/socioeconomic/forecasts/archive/documents/2035_regional_growth_forecast.pdf). Given these factors, we believe a regional structural approach to surge protection makes sense for this region.

This chapter will focus on the need for and the possible design of a coastal barrier—the Ike Dike—that would protect the Houston/Galveston area from hurricane-induced storm surge. In the rest of this chapter, we will briefly discuss Hurricane Ike and its impacts, which were the catalyst for action; examine what we want to protect in the Houston/Galveston region; look at the nature of the threat; explore aspects of existing coastal barriers and finally build a concept for and examine the Ike Dike.

## 2 Hurricane Ike

Hurricane Ike was a devastating blow to the Upper Texas Coast and will likely be recorded as the most expensive and destructive storm to hit Texas (Hurricane Ike Impact Report 2008). The magnitude of Ike's impact, predominantly caused by storm surge, was unexpected and has become the catalyst of an increase in political will to rebuild the area to a less vulnerable state. Ike originated off the west coast of Africa on 28 August 2008. On 10 September, the storm entered the Gulf of Mexico and loitered in the Gulf for 3 days before making landfall on the Texas coast. Ike made landfall on Galveston Island on 13 September as a category two hurricane with a maximum recorded wind speed of 175 km/h (108 mph). The recorded maximum wave height off the Galveston Coast was 6 m (19.6 ft). Ike's slow movement across the Gulf of Mexico caused an excessive amount of water to be pushed onto the coast. The storm surge had the greatest impact on Bolivar Peninsula, Galveston Island and the counties surrounding Galveston Bay.

Because of the intensity of the storm surge many gauges were destroyed, but water marks indicate that water levels on Bolivar Peninsula reached nearly 6 m (20 ft) and as high as four and a half meters (14.9 ft) on Galveston Island. Storm surge along the upper regions of Galveston Bay was between 2 and 5 m (7 and 16 ft). The inundation from Hurricane Ike's storm surge is shown in Fig. 1. The 2008 hurricane season was the most expensive in Texas history, and Hurricane Ike the third most expensive storm in US history (DeBlasio 2008), causing nearly $30 billion in damages. Twelve fatalities directly related to Hurricane Ike were reported in Galveston and Chambers counties and the storm caused another 67 indirect deaths throughout Texas (Berg 2009). As of February 2010, 23 people from Galveston and Bolivar Peninsula were still listed as missing, of these 20 are unaccounted for but not the subjects of active searches and three people are still the subject of searches (The Laura Recovery Center 2010).

The region's population, utilities and major employment sectors were also severely affected. By 1 February 2009, over 730,000 disaster assistance requests, indicative of the number of homes that were damaged or destroyed, had been made to the Federal Emergency Management Agency (FEMA), (Hurricane Ike Devastation Report 2009). For the first time in Texas history, a disaster declaration was issued in every coastal county (Governor's Commission on Disaster Recovery and Renewal).

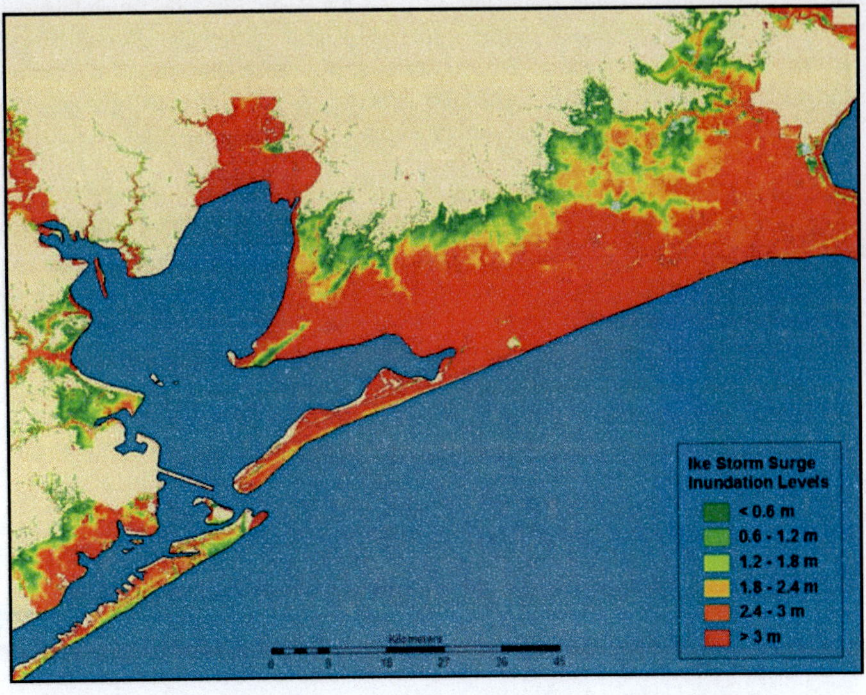


**Fig. 1** Map of the Houston/Galveston area showing inundation levels during the storm surge from Hurricane Ike. *Reproduced by permission of Kyle Johnson*

# 3 The Houston/Galveston Region

## *3.1 What We Wish to Protect*

The Galveston Bay area has a large human population, a thriving economy that is vital to Texas and the US and contains the most biologically productive ecosystem in the state. The Bay itself is approximately 1,554 square km (600 square miles) in area; 48 km (30 miles) at its largest north-south extent and 27 km (17 miles) at its largest east-west extent, and has an average depth of a little over 2 m (7 ft). Almost 64,750 square km (25,000 square miles) of land drain into the Bay, along with the Trinity and San Jacinto Rivers. The average Bay salinity is 11 ppt.

Protecting the Bay from the Gulf of Mexico is Galveston Island, a barrier island approximately 44 km long (27 miles) and up to 5 km (3 miles) at its widest, and the Bolivar Peninsula which extends from the eastern mainland. Bolivar Roads, the major opening between Galveston Bay and the Gulf, separates Galveston Island from the Bolivar Peninsula. Just above Bolivar Roads, the Houston Ship Channel and the Gulf Intracoastal Waterway cross, both of which support considerable water transportation.

The Houston–Galveston metropolitan area had an estimated population of over 5,728,143 people as of 2009 (U.S. Census Bureau 2010) making Galveston Bay the most heavily populated watershed in the Gulf Region and the seventh most populated in the U.S. The Houston ship channel runs through Galveston Bay, serving the ports of Galveston, Texas City and, of course, Houston. The Port of Houston is the busiest port in the nation generating $3.7 billion of revenue and $117 billion of total economic impact annually (Port of Houston Authority 2010). The economic status and national strategic importance of Galveston Bay's industrial complex, especially its refining capacity, is readily illustrated by ship call statistics. In 2008, ports on Galveston Bay had 7981 ship calls or 13% of all ship traffic into the U.S. Of that, 5157 calls were by tankers, almost 25% of all tanker calls in the U.S. If we add Freeport and the Sabine ports to Galveston Bay ports, the greater region saw almost 32% of U.S. tanker traffic. Loss of the upper Texas coast refining capacity or access to it would be devastating nationally.

The Gulf Intracoastal Waterway (GIWW) also runs through Galveston Bay. Fifty-eight percent of GIWW traffic is within the state of Texas. In 2006, over 74 million short tons of cargo, valued at over $25 billion, moved through this portion of the GIWW. Eighty-seven percent of the goods moved were petroleum or chemical products (Texas Department of Transportation 2009).

A report by the Perryman Group for the Independent Insurance Agents of Texas found that a Katrina-like storm that hit the Port of Houston area would result in the losses of $73 billion of gross product, over 800,000 permanent jobs and about $2.5 billion in state revenue (Perryman Report 2006, p. 3).

Galveston Bay is also home to the largest and most biologically productive estuary in Texas (http://www.gbep.state.tx.us/about-galveston-bay/geography.asp). The waters, flora and fauna of the Bay drive commercial and recreational activity and provide valuable ecosystem services, including commercial and recreational fisheries, ecotourism, research, waterfowl hunting and boating. The Galveston Bay Estuary Program estimates that over 55,847 ha (138,000 acres) of coastal wetlands exist around the Bay. These coastal wetlands serve as habitat for the many harvestable species found in the Bay. Shrimp accounts for about half of the value of seafood harvested in the area and about one-third of the Texas blue crab catch is harvested in Galveston Bay (http://www.gbep.state.tx.us/about-galveston-bay/economics.asp).

### 3.2 *Hurricane Threats and Evacuation Issues*

The Houston–Galveston region experiences a major (category three or higher) hurricane about every 15 years. Fortunately no hurricane has actually hit the coast as a category five although some have achieved that status over open waters. The potential worst case hurricane used when considering the Ike Dike is Hurricane Carly, a category four hurricane moving at eight knots and hitting west of Galveston Bay such that its maximum winds cause a massive surge up the Houston Ship Channel. To date, a hurricane with Carly's strength and path has not occurred in Galveston Bay but it is certainly a possibility because similar hurricanes have made landfall on the upper Texas coast. Should it hit, a hurricane with Carly's characteristics would cause a storm surge of 7.6 m (25 ft) in the upper reaches of the channel with a maximum surge of about 4.7 m (15.5 ft) at the coast. This storm surge is truly a worst-case scenario and represents the surge from a storm that would occur in the Houston/Galveston area only about every 10,000 years.

As the region's population grows, the Galveston Bay area becomes increasingly difficult to evacuate in a timely manner. In 2005, the National Hurricane Center designated Galveston as one of the five most vulnerable places in the United States because of its low elevation and limited evacuation routes (Galveston Long Term Recovery Plan 2009). These difficulties present another reason to suppress storm surge with a regional structural approach. The evacuation of heavily populated areas strains modes of transportation and can create massive traffic jams. In addition, heavily populated areas often have higher numbers of elderly and poor who have a more difficult time evacuating.

Evacuating from Ike was difficult because of the unusually large changes in its forecast path. Figure 2 shows official path forecasts by the National Weather Service. The Tuesday (9 September 2008) and Wednesday (10 September 2008) forecasts showed Ike making landfall in south Texas, far from Galveston. By Thursday (11 September 2008) it was forecast to hit on the Texas mid-coast, still far from Galveston/Houston. But by early Friday (12 September 2008), Ike was forecast to hit just west of Galveston Bay close to its actual landfall on Galveston Island very early on Saturday (13 September 2008).

The infamous 2005 Hurricane Rita evacuation of 2.5 million people in the Houston area discouraged many coastal inhabitants from participating in the

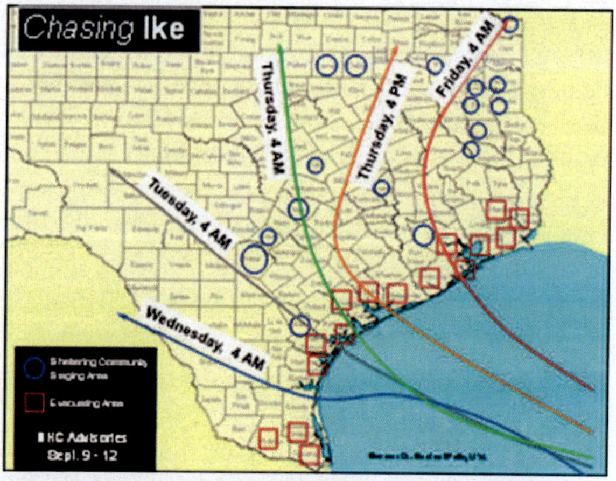


**Fig. 2** Ike's official forecast paths 9–12 September 2008. Modified from Gordon Wells Presentation "Lessons Learned from the State Response to Hurricane Ike". *Reproduced by permission of Gordon Wells, Center for Space Research, The University of Texas at Austin*

**Fig. 3** The Rita evacuation: even with contraflow lanes open, the highway system cannot accommodate evacuating traffic through and out of Houston

mandatory evacuation as Hurricane Ike approached. Figure 3 powerfully illustrates the transportation problems experienced by many coastal residents during the Rita evacuation. Of the 111 deaths associated with Hurricane Rita, 90 were related directly to the evacuation process (Zachria and Patel 2006). Evacuation of the large populations surrounding Galveston Bay can be extremely costly, stressful and dangerous.

Evacuation problems are also exacerbated by the fact that the upper Texas Coast is especially vulnerable to quick-forming and rapidly intensifying storms.

Between 1851 and 2006, 66% of the storms hitting the upper Texas coast formed in the Gulf of Mexico. These storms are particularly threatening because of their rapid intensification and fast forward speeds. An example is the 1932 Gulf hurricane which hit near Galveston after forming off Yucatan about two and one-half days prior. On August 13 around midnight local time, the 1932 Storm reached the status of a category one hurricane about 240 miles away from the coast. The hurricane rapidly gained intensity, becoming a category two a scant 6 h later and a major hurricane (category three) 6 h after that when it was about 100 miles off the coast. The 1932 Storm made landfall as a category four hurricane the evening of August 14, killing about 40 people and causing extensive damage to Freeport, Angleton and Galveston.

Upper Texas Coast storms can also make landfall within a few days of forming in the Gulf of Mexico. Hurricane Humberto, a minimal storm initially, intensified faster than any other tropical cyclone on record and is an excellent example of a hurricane that hit soon after its formation and intensified as it hit. Developing on 12 September 2007, off the Texas coast the cyclone rapidly strengthened and struck High Island with winds of about 150 km/h (90 mph) early on 13 September. This type of storm creates a major problem for public officials and citizens as the decision to evacuate must be made and communicated quickly. The City of Galveston attempts to call mandatory evacuations 72 h before landfall although this guideline was not followed during Hurricane Ike in 2008 because of the storm's erratic path (Rice 2008).

In addition to the potential for rapid intensification, the number of Gulf-formed storms hitting the Texas Coast peaks in June, while the number of Atlantic storms hitting the Texas Coast peaks in August and September (Islam et al. 2009). Hurricanes that hit early in the season have the potential to surprise individuals who are not fully prepared. Hurricane Audrey, a very fast-moving Gulf hurricane, made landfall near the Texas-Louisiana border in June of 1957. Audrey was category one or two as it initially moved through the Gulf and then suddenly gained intensity to a category four hurricane about 201 km (125 miles) off the coast. Six hours later Audrey made landfall as a category four hurricane causing catastrophic damages in eastern Texas and western Louisiana. An estimated death toll of between 416 and 550 people makes Audrey the sixth deadliest hurricane to hit the United States.

### *3.3 Social Costs Vulnerabilities*

In addition to the monetary value of the physical damage that a major hurricane or flood causes, there are also hidden costs that are difficult or impossible to quantify. Although there is little data yet available concerning the social costs of Hurricane Ike, it is apparent that vulnerable populations (the young, elderly, non-English speaking and minorities) within the region were negatively impacted and will have a more difficult path to recovery. These vulnerable populations have less access to resources and can become permanently displaced or disadvantaged after a severe disaster.

Soon after Ike, in October 2008, FEMA reported that 716,000 households had registered for individual assistance and 228,000 of those applicants had verified housing losses. Healthcare facilities, including hospitals and nursing homes, were damaged. Chambers County lost 45.5% of its certified nursing home beds; Galveston County was forced to close indefinitely two facilities that previously provided 314 beds (Hurricane Ike Impact Report 2008). Many nursing homes and assisted living facilities in Orange County, Jefferson County, Brazoria County and Harris County sustained enough damage to force temporary or permanent closure. The closure of the University of Texas Medical Branch, a major regional facility providing care for the poor and indigent, affected public health in the entire region.

Many schools and childcare facilities closed temporarily and some permanently after Ike. The Galveston Independent School District (GISD) enrollment figures provide solid evidence of the displacement of schoolchildren. In December 2008, GISD reported a 33% decrease in enrollment in elementary schools from September 2008, with middle school enrollment down 27% and high school enrollment down 15%.

## 4 Design Premises

The human settlement near Galveston Bay has reached such a size and importance and with hurricane surge such a significant threat that a comprehensive regional structural approach to surge suppression makes sense. The structural approach that we propose, the Ike Dike, is a coastal spine which would keep Gulf surge out of Galveston Bay. The physical geography of the Bay lends itself to the approach. Because the Bay is protected from the Gulf of Mexico by a barrier island and a peninsula, land barriers can be used and gated sea-barriers are only needed in the passes on each end of Galveston Island.

## 5 Coastal Barriers

The Ike Dike concept uses existing, proven technology to demonstrate the feasibility of a Galveston Bay coastal spine. Dutch coastal barrier technologies were chosen as examples because of a history of proven effectiveness as well as innovative design. Other designs obviously exist and should be considered but the work in the Netherlands is particularly relevant because of the navigation and environmental considerations in Galveston Bay. The present work in New Orleans is also particularly relevant in that it represents a large-scale attempt to suppress surge near a heavily populated and socially significant U.S. city as well as an expedited process that has enabled major works to be build very quickly.

## 5.1 Barriers in the Netherlands

The Dutch have a long and ultimately successful history of using structures to deal with flooding and surge suppression. As early as the thirteenth century Netherlands inhabitants were using dikes to protect the region from flooding. A ruling by Count Floris V in 1280 determined that, "the monastery, the knight, the priest, the common man, everybody alike," was responsible for funding dike maintenance (Bijker 2007).

In response to frequent river flooding, the Netherlands established the Rijkswaterstaat, a national water management agency, in 1798 (Toussaint 2009). Before this time, water management had been executed by isolated regional water boards without any sort of collaborative approach. A changing political mood and the frequency of flooding damage drove the initiative to centralize. Now located at the Ministry of Transport, Public Works and Water Management, the three main tasks of the Rijkswaterstaat are river and coastal defense and "interior" water management (Toussaint 2009).

In 1937, a Dutch public works report stated that phenomena that cause high sea levels could endanger the Dutch people living within close proximity to open waters. Initial protective measures began in 1950 but a major flood in 1953 caused a temporary stall in the construction. However, the nearly 2,000 lives lost and 150,000 ha (370,000 acres) of flooded land in 1953 confirmed the belief that the project needed to continue. The following years saw the creation of the largest network of water-control structures in the world. In addition to protecting the population, the Deltawerken (Deltaworks) system provides safe transportation, preservation of key ecosystem elements and water for human uses (http://www.deltawerken.com). Through the years, advances in technology and a better understanding of the natural environment have allowed the network to evolve, improve services and better protect human and natural habitats.

A lesson learned first-hand from the Dutch experience is that universal design rules are unfeasible. The hydraulic, ecological and geographical conditions of each project are unique and all must be considered together in order to attain success. The Deltaplan included closing the tidal outlets of the Maas and Rijn rivers but, in other areas, gates were used (Bijker 2007) to allow for water circulation and ecological function.

This use of circulation and navigation gates has transformed thinking about approaches to surge protection because coastal barriers can now be constructed to not only effectively suppress surge when the gates are closed but also to allow the passage of large ocean-going vessels and appropriate estuarine circulation when the gates are open. The Deltaworks contain ample examples of both navigation and circulation gates. Figure 4 shows a map of major features of the Deltaworks with inserts to illustrate the Maeslantkering, a navigation gate near Rotterdam, and the Oosterscheidekering, a series of gates which allow for water communication between the North Sea and the Eastern Scheldt preserving the ecosystem function of that estuary. Also shown on the map are the system's earlier-constructed solid

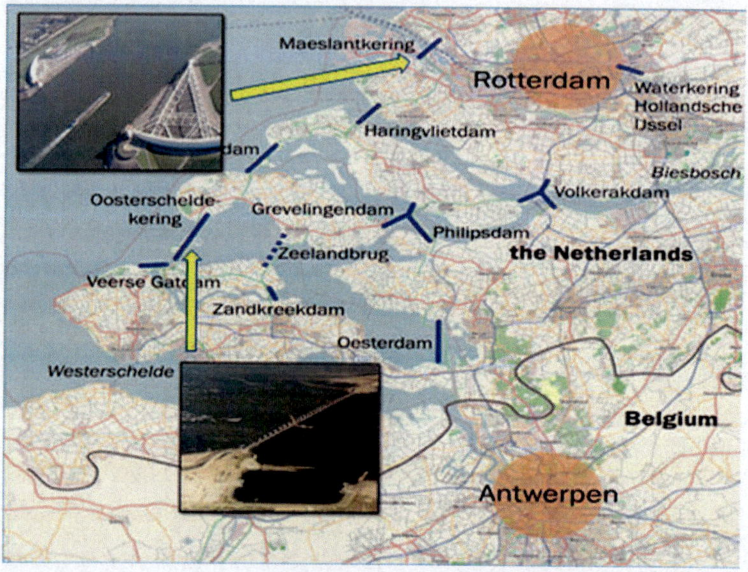


**Fig. 4** Dutch Delta Works Project, a comprehensive flood control system featuring a coastal spine with conventional dikes as well as flood gates

coastal barriers, the Brouwersdam and the Haringvlietdam. These solid barriers are effective against coastal surge but have transformed formerly salt water estuaries into fresh water lakes. Although these dams were constructed early in the Delta-works program, the Dutch made their natural appearance a design priority. Figure 5 shows the attention to a natural appearance as one looks back toward the fresh water lake standing on top of the coastal barrier.

Because we will use Dutch surge gates as existing, proven examples of components needed to successfully build the Ike Dike concept, we'll describe three of them in more detail. Near Rotterdam there are two types of gates designed for use in navigable channels, the Maeslantkering or Maeslant barrier and the Hartelkering or Hartel barrier. Farther south at the mouth of the Oosterschelde (Eastern Scheldt estuary) is the Oosterscheldekering, a surge barrier that is designed to allow tidal flows and water communication between the open ocean and the estuary.

The Maeslantkering (see Fig. 6) is located in the port of Rotterdam. The surge barrier is a huge double swinging gate that allows even the largest ocean going vessels easy transit. One of the largest moveable structures in the world, each gate is 22 m (72 ft) high and 210 m (689 ft) long with a swing arm of 237 m (718 ft). Each arm is as long as the Eiffel Tower is tall with each weighing twice as much as the Tower, yet the design allows the gates to swing shut in a little more than an hour to provide surge protection for a channel 360 m (1,181 ft) wide and 22 m (72 ft) deep. An important design feature is that the barrier imposes no restrictions on vessel height.

**Fig. 5** View from the dam portion of Dutch coastal barrier looking inland. *Reproduced by permission of William Seitz*

**Fig. 6** The Maeslantkering gates swing shut to provide surge protection for a channel 360 m (1,181 ft) wide and 22 m (72 ft) deep and imposes no restrictions on vessel height when open

Also located in the port of Rotterdam complex, the Hartelkering (see Fig. 7) is composed of two large lift gates with horizontal dimensions of 98 m (322 ft) and 49.3 m (161 ft) respectively. When open for ship traffic, the barrier has 14 m (46 ft) of vertical clearance vessel height, allowing the passage of small ocean-going ships and barges. When closed, the Hartelkering can handle a three meter storm surge. In the Ike Dike concept, a smaller version of this gate could be used at San Luis Pass, the relatively narrow, shallow opening at the far west end of

**Fig. 7** The Hartelkering is a large lift gate that allows passages 98 m (322 ft) and 49.3 m (161 ft) wide with 14 m (46 ft) clearance

Galveston Island, or on the Intracoastal Waterway if gates are needed there. Lift gates could also be incorporated into the surge barrier at Bolivar Roads.

Farther south on the mouth of the Eastern Scheldt estuary sits the Oosterscheldekering, a Dutch example of storm surge gates specifically designed to allow for water communication between the open ocean and an estuary. Deemed one of the Seven Wonders of the World by the American Society of Civil Engineers, the storm gates extend, in three sections, a total of 4 km (2.5 miles). The barrier is composed of 62 individual steel doors each 42 m (138 ft) wide set between concrete pillars each weighing 18,000 tons (19,842 short tons). Figure 8 shows the two northern sections of the Oosterscheldekering while Fig. 9 shows the southern section with the floodgates down during a surge event.

## *5.2 Surge Barriers in the United States*

Natural disasters, particularly hurricanes, storm surges and floods, have played a key role in the practices of American coastal engineering (Bijker 2007). Because of the adverse effects on lives and property, natural disasters typically result in elevated public awareness of regional vulnerability (Wiegel and Saville 1996). Historically, the damage caused by specific hurricanes has motivated several

**Fig. 8** Two of the three sections of the Oosterscheldekering, a surge barrier which allows tidal flows and water exchange between the Eastern Scheldt estuary and the North Sea. The total length of floodgates is 4 km (2.5 miles) composed of 62 individual steel doors each 42 m (138 ft) wide

**Fig. 9** The Oosterschelde-kering with flood gates down during a surge event

concrete coastal barrier projects: for example, the Galveston Seawall (1904) and the New England Hurricane and Coastal Storm Barriers (1960s).

In response to the damage done by Katrina, over $14 billion has been dedicated to a network of structures in the New Orleans, LA area designed to protect the city. The Inner Harbor Navigation Canal (IHNC) project is an example of major project being completed much earlier than if traditional methods had been used in its design, permitting and construction. The IHNC is designed to reduce storm surge in parts of New Orleans to the 100-year flood plain level and includes a floodwall and gates with a gated barrier on the Gulf Intracoastal Waterway. Katrina hit in August of 2005 and the U.S. Army Corps of Engineers issued a Request for Proposals 2 years later in August of 2007. Possible designs were submitted at the five percent level in Jan 2008 and a contract awarded in April of 2008. Construction began in December of 2008 and the estimated completion date of the project is June 2011 less than 6 years after Katrina hit.

This accelerated approach will assure the citizens of New Orleans hurricane surge protection much earlier than conventional methods and should be used in the Houston/Galveston region to provide timely protection from surge.

## *5.3 Structures in the Houston/Galveston Region*

The Netherlands, New Orleans and the Houston/Galveston region in Texas share a need to protect human and natural habitats from recurring natural threats So it is not surprising that the Galveston Bay region, in particular the City of Galveston and Texas City, has a long history of using structures to provide protection from hurricane surge.

### 5.3.1 Galveston Seawall

In the 1900 hurricane, nearly 8,000 people perished from the storm surge that covered the island. After that devastating storm, Galveston residents decided to construct a Seawall to protect the island from future storms. The original 3 miles of the remaining Seawall from 6th Street west to 39th Street was completed in 1904. Five additions to the Seawall were constructed between 1904 and 1963. The original Seawall was a concrete gravity wall built on treated timber piles, a reinforced concrete sheet pile cutoff wall and riprap toe protection backed by a landward sand-fill embankment. The modern Seawall now has paved surfaces in most locations with a broad sidewalk and major 4-lane thoroughfare, Seawall Boulevard, on its crest.

Another major hurricane hit the Galveston area in 1915 but the Seawall prevented a recurrence of the damage that was seen in the 1900 storm. Since then, the Galveston Seawall has been tested many times, notably by major storms in 1932, 1942, 1945, 1961, 1983 and 2008 (Fig. 10).

### 5.3.2 Texas City Circling Flood Barrier

The Texas City Flood Barrier, completed in 1987, provides hurricane flood protection to Texas City and La Marque, Texas, located on the southwest shore of Galveston Bay, about 14.5 km (9 miles) northwest of Galveston, Texas. The barrier consists of about 2 km (1.3 miles) of concrete walls and 25.3 km

**Fig. 10** City of Galveston Aerial showing the modern Galveston Seawall. *Reproduced by permission of Robert Mihovil, copyright 2010*

(15.7 miles) of levees, a tide control and navigation structure in Moses Lake and two pumping stations. This system encircles approximately 50,000 residents as well as refineries and chemical plants (Fig. 11).

### 5.3.3 Freeport Hurricane Flood Protection

Freeport is located in the southern part of Brazoria County about 6.3 km (4 miles) from the Gulf of Mexico and about 69 km (43 miles) southwest of Galveston, Texas. Its levee system extends for 85.3 km (53 miles) and encircles portions of the cities of Freeport, Lake Jackson, and Oyster Creek. It protects 60,000 residents and nine industrial plants.

### 5.3.4 Previous Plans for Mega-structures in the Galveston Bay Region

There have been previous plans for hurricane protection mega-structures in Houston/Galveston area. An article titled "Similar plan suggested a century before

**Fig. 11** The Moses Lake Flood Gate is part of an encircling flood barrier protecting Texas City and La Marque, Texas. *Reproduced by permission of the Galveston County Daily News*

"Ike Dike" idea" in the *Houston Chronicle* (March 18, 2010) reported that a 1902 plan was recently unearthed in the Galveston District Clerk's office showed a Seawall that would have protected the bay side of Galveston Island as well as the Gulf side. Although it is unclear why this proposed approach wasn't followed, a similar 1979 study was deemed to be too expensive. This comprehensive report by the U.S. Army Corps of Engineers, entitled "Feasibility Study—Texas Coast Hurricane Study," contains plans for a series of levees protecting the entire region from catastrophic overflows that look similar to the Ike Dike concept, although without the floodgates and adding a circular levee to protect downtown Galveston.

## 6 The Proposed Ike Dike Concept

The Ike Dike structure would include three components: the existing Galveston Seawall, land extensions of the protection afforded by the Seawall and sea barriers in the passes connecting Galveston Bay with the Gulf of Mexico.

### *6.1 Ike Dike Component 1: The Galveston Seawall*

The first component of the Ike Dike is Galveston's present Seawall. Standing 5.18 m (17 ft) tall and 16 km (9.9 miles) long, the Seawall is located on the east side portion of the island. The Seawall was constructed after the Great Storm of 1900 to protect the city from hurricane storm surges. Galveston's Seawall has stopped devastating storm surges coming from the Gulf-side; however, it offers no protection on the Bay side. Moreover, because the Island was raised to Seawall levels at the wall and gradually tapered down a few feet to the bay side to assure drainage, much of the Island including the historic business district is vulnerable to bay side flooding.

Indeed, as reported by the local news after Ike, "The Seawall did its job, but water simply went around its edges" (Barajas 2010). Because the Ike Dike when completed will keep water out of Galveston Bay, no barriers will be needed on the Bay side. And, because the existing Seawall is effective against surges from the Gulf, it is a cost-effective and proven component of a coastal spine. Figure 12 shows the Seawall and the groin field designed to keep sand covering the Seawall's base.

**Fig. 12** The Galveston Seawall stretches along the east end of Galveston Island, with intermittent groins that jut out into the Gulf of Mexico. *Reproduced with permission of the Galveston County Daily News*

## 6.2 Ike Dike Component 2: Land Barriers

The second component consists of land-based extensions of the protection afforded by the current Seawall. The Ike Dike proposal envisions construction of a 5.18 m (17 ft) wall extending from the existing Galveston Seawall out to San Luis Pass.

This new extension would cover an area of 28.9 km (18 miles) and would be constructed close to the beach. Alternatively, raising the existing highway (FM 3005) by 3.6 m (12 ft) would also create a sea barrier but move its location back from the beach. On the Bolivar Peninsula, construction of a 5.18 m (17 ft) tall coastal barrier would be required or Highway 87 could be raised 3.6 m (12 ft). The completed Bolivar coastal barrier would cover a total area of 56.3 km (35 miles) going from Bolivar Roads to High Island.

The extension of the Seawall protection could be a quite different design than the current Seawall. An example is the revetment design shown in Fig. 13. (Edge and Garrett, personal communication) The revetment design includes the placement of sand and vegetation on both sides of the wall to resemble a sand dune. The design also includes the installation of walkovers to facilitate beach access and to prevent the disturbance of the sand dunes. The proposed revetment would give a more natural look to the beach area than the current seawall.

If the coastal barrier's land extensions do not continue along the coast past the edges of Galveston Bay, inland extensions or "wrap-arounds" would be needed at both ends of the dike to complete the construction. One wrap-around would be located near San Luis Pass and the other one near High Island. These wrap-arounds

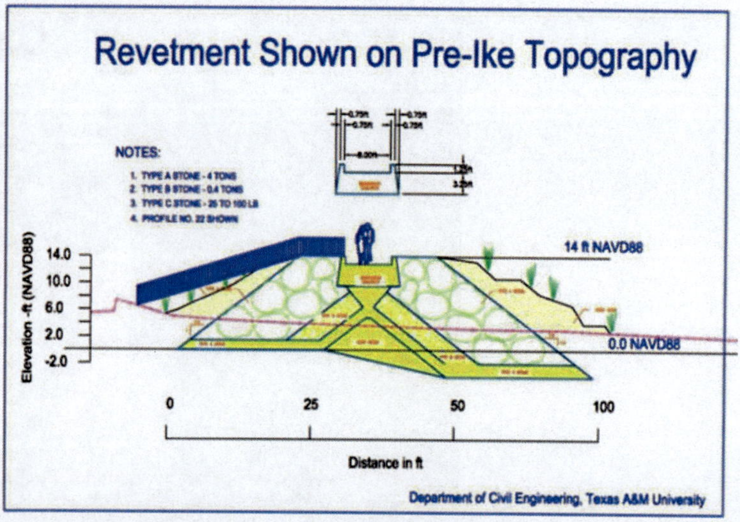


**Fig. 13** Possible revetment design for land barriers on west Galveston Island and the Bolivar Peninsula. *Reproduced by permission of Billy L. Edge and Marie Garrett*

will prevent surge from entering Galveston Bay by flowing around the ends of the barrier.

## 6.3 Ike Dike Component 3: Sea Barriers

The third component of the Ike Dike is a pair of sea barriers, a large one located at Bolivar Roads and a smaller one on San Luis Pass. The proposal calls for two sets of gated barriers which would only be closed in the case of a severe storm event. The costliest component of the Ike Dike would be the barrier located on Bolivar Roads. The proposed Bolivar sea barrier must allow for both navigation and water circulation as well as create a 5.18 m (17 ft) barrier above sea level when closed.

Appropriate Dutch examples of barriers allowing navigation and barriers allowing circulation may need to be combined here. Such a combination has already been proposed for a barrier in the Verrazano Narrows near New York City (Jansen, P.L.M. and Dircke, P.T.M.) Figure 14 shows this design in place. It combines Maeslantkering type swinging gates as well as lift gates similar to those in the Hartelkering handle ship and boat traffic. Tidal circulation is assured by the addition of a series of sluice gates similar to those in the Oosterscheldekering. A similar design for Bolivar Roads would work. These gates would also create a major tourism venue.

The proposed gates in San Luis Pass would much smaller because the boat traffic in that area is primarily smaller, recreational craft and the pass is much

**Fig. 14** Verrazano Narrows storm surge barrier design has gate combinations to assure both navigation and water circulation. *Reproduced by permission of P.L.M. Jansen and P.T.M. Dircke, ARCADIS*

shallower. The San Luis Pass gates should also be designed to allow for appropriate water circulation in the Bay.

It is possible that surge gates would need to be constructed on the Intracoastal Waterway but their need or even the requirements if they are needed are not clear without the results of surge models determining the height and timing of surges on the Intracoastal Waterway with coastal spines present.

## 7 Other Design Considerations

The Ike Dike will suppress the storm surge from entering the Bay, preventing disastrous storm surges and reducing the need for individual property hardening.

Unlike examples in the Netherlands or New Orleans, parts of which are below sea level, the Texas coast can tolerate some "leakiness" along the coastal spine, meaning that some water could be allowed through the barrier without causing severe damage. The Ike Dike can allow some of the storm surge to leak through the gates without major consequences to either the Bay ecosystems or man-made structures. With a fast moving hurricane, storm surge only needs to be held for a few hours. Water will flow out naturally when the gates are opened after the storm moves through.

## 8 Preliminary Surge Modeling

Scientists at The Institute for Computational Engineering and Sciences at The University of Texas at Austin designed storm surge models for the Houston/Galveston area to assess the problems associated with storm surge suppression. The purpose of the models was to compare the effects of the Hurricane Ike surge with and without the proposed Ike Dike. The intensity of the storm surge used in the models was the same as the one created by Hurricane Ike. The results of the models show that the Ike Dike would suppress the storm surge and prevent inundation in areas adjacent to the bay.

The model without the Ike Dike, Fig. 15, shows 4.57 m (15 ft) of water going across Bolivar Peninsula and the west end of Galveston Island. The water level is so high that Bolivar Peninsula and a portion of Galveston are not visible. The only part of Galveston Island not completely inundated is near the Seawall on the east end of the island. The current Gulf-facing Seawall runs from 6th Street to 95th Street. The entire Bolivar Peninsula was overrun by the storm surge. During Ike, the water level in the Bay would rise as much as 5 m (16 ft), according to the UT Austin model.

In the model runs with an Ike Dike, Fig. 16, the storm surge is effectively suppressed. Water levels in the Bay are lower and Bolivar Peninsula and Galveston Island are not inundated by surge. The model results show a significant

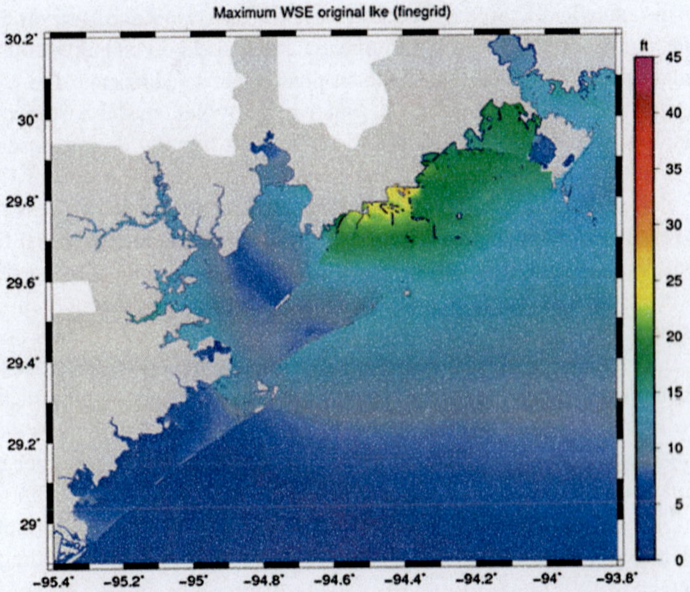


**Fig. 15** Storm surge model results without the Ike Dike. *Reproduced by permission of Gordon Wells, Jennifer Proft and Clint Dawson; The University of Texas at Austin*

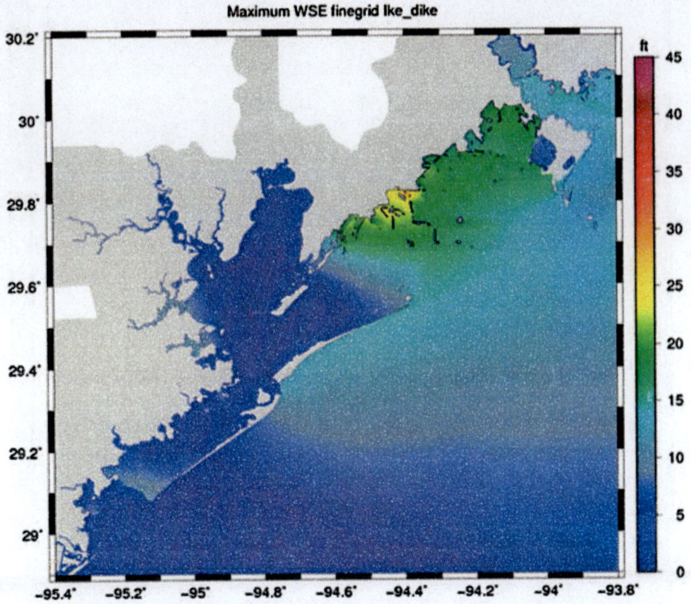


**Fig. 16** Storm surge model results with Ike Dike in place. *Reproduced by permission of Gordon Wells, Jennifer Proft and Clint Dawson; The University of Texas at Austin*

decrease in water levels around the Bay. Galveston Island and Bolivar Peninsula are visible, suggesting that Bolivar Peninsula and Galveston Island would have had little damage if the Ike Dike had been in place prior to Hurricane Ike.

According to the model results, the water level by the coastal barrier in Bolivar Peninsula, could reach up to 4.57 m (15 ft) without topping the proposed barrier, keeping storm surge from covering the peninsula. On the west end of Galveston water levels reach 3.05 m (10 ft) and the barrier is again able to repress the storm surge. Overall water levels in the Bay rise 1.52 m (5 ft)–3.05 m (−10 ft) difference from levels reached in the previous model runs without the Ike Dike. The water level in the Bay could have been reduced further—by about 1 m (3 ft)—if the gates were closed at low tide. Further surge reduction would have occurred if the west end gates were opened as the storm passed. Opening of the west end gates would have allowed for water to be blown out of the Bay, reducing the water level.

Figure 17 shows the difference between the two models. On the east side of the Ike Dike—the yellow zone—the difference in the water level was higher by 0.3 m (1 ft). This increase could be attributed to the coastal barrier. The suppression of the storm surge caused water to accumulate, increasing the water level outside the barrier. The water level increment was not high enough to overcome the Seawall. The water level difference in the bay ranges from −1.52 m (−5 ft) to −1.83 m (−6 ft) which would reduce flooding drastically. Water level on the west end of

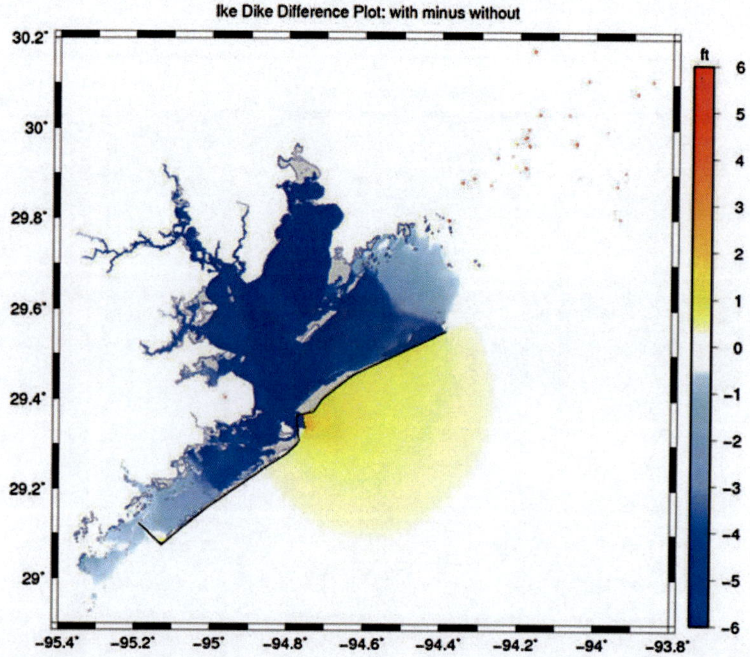


**Fig. 17** Model showing difference of storm surge heights with and without the Ike Dike. *Reproduced by permission of Gordon Wells, Jennifer Proft and Clint Dawson; The University of Texas at Austin*

Galveston island is from −0.91 m (−3 ft) to −1.22 m (−4 ft), also a significant reduction in the water level.

If the Ike Dike modeling had included active flood gates on Bolivar Roads—gates that would have been closed at low tide when water level is at its lowest in the bay—the water level rise would have been about 1 m (3 ft) lower. The gate design could allow for the dike to leak some water into the Bay without causing an inundation. Using the proposed dike plans, gates at San Luis pass would have been opened to allow the outflow water circulation from the Bay after the storm passed over Galveston Island.

## *8.1 Benefits of the Ike Dike*

The Ike Dike concept would work to significantly reduce hurricane surge in Galveston Bay, providing comprehensive protection from storm surge for people, properties and an industrial base with nationally strategic importance. It also prevents surge damage to the Bay's natural resources. Because vulnerability would be significantly reduced, its presence will encourage investment in and commitment to the region.

The construction of a coastal spine costs much less than a single hurricane recovery and may well cost less than individually armoring the entire Bay complex. Additionally, the Ike Dike may also prove to be a more environmentally sound option than armoring the Bay.

Structural surge reduction is the best and perhaps only way to protect the regions less resilient citizens, especially the elderly and the poor who do not have the means to recover. It would help protect the lives of those who cannot or do not evacuate as well as during difficult evacuations from hurricanes that quickly change path or intensity.

# 9 Status of the Ike Dike Concept

The Ike Dike concept was introduced to the public on 11 November 2008 in a Guidry News article by one of the authors of this chapter (Merrell) and regional decision-makers began to consider the idea as a possible solution to storm surge in the Gulf Coast region. Media sources such as the Guidry News, Houston Chronicle, the Galveston Daily News, and the Wall Street Journal latched onto the Ike Dike, using the catchy name freely in articles as locals and experts debated the issue.

Galveston Mayor Lyda Ann Thomas took the idea before the US Congress in February 2009. The Galveston delegation to Congress reported that several Hurricane-Ike related concerns were met with a receptive hearing from Congressmen, including regional shoreline protection such as the "Ike Dike" concept (Thomas 2009).

In response to Hurricane Ike, Texas Governor Rick Perry issued Executive Order RP 69 creating the Commission on Disaster Renewal and Recovery (Governor's Commission Preliminary Report 26 March 2009). The 23-member commission, comprised of private and public sector experts, was charged with the objective of creating a "plan to assist Texas communities with recovery efforts after a natural disaster (http://governor.state.tx.us/news/appointment/11723/)". The three specific tasks of the Governor's commission were to help communities recover from the 2008 storms, reduce the impact of future storms and renew coastal communities (Governor's Commission Preliminary Report 26 March 2009). After holding four public hearings, the commission made several recommendations to the 81st Texas Legislature, with a focus on improving coastal communities' ability to restore housing, infrastructure and enterprise. The commission noted that the areas in Galveston County that were not protected by a Seawall experienced 15–19 foot storm surge and consequently suffered catastrophic damage.

On 26 January 2010, the Harris County Commissioners Court approved the revised Articles of Incorporation and Bylaws of the Gulf Coast Community and Recovery District, Inc. This action joined Harris County with Brazoria, Chambers, Galveston, Jefferson and Orange Counties to "aid, assist and act on behalf of" the sponsoring counties to form a six-county public corporation, Gulf Coast Community and Recovery District, Inc, for the purpose of planning storm surge, flood, wind and disaster mitigation. The nine directors on the board will consider possible solutions and make recommendations regarding their feasibility. While this newly-formed public corporation has many startup decisions to make, the collaboration of the six counties represents a significant movement towards a coordinated approach to regional storm response and surge protection.

While discussions continue over the advantages and disadvantages of the Ike Dike, both advocates and detractors agree that benefits and impacts studies must be conducted to determine the feasibility of such a mega-structure on the Texas coast. Preliminary modeling and the proven outcomes in the Netherlands indicate that the Ike Dike design would significantly reduce storm surge and allow Galveston Bay's ecosystem services to be maintained.

## 10 Other Potential Benefits

There are other benefits that a coastal spine or an "Ike Dike" could provide beyond storm surge protection. The active gates are a potential tool to manage the Bay's ecosystem. Salinity might be controlled during particularly dry or wet seasons by raising or lowering some of the gates, which in turn could benefit the growth of certain shellfish, fish or plants within the Bay. Coastal spines with gates would also serve as a precautionary measure against future sea level rise because of climate change. In particular, as base sea level increases, smaller storms, which occur more frequently, will generate surges that reach damaging levels in the Bay so the gates will need to be closed more often. Eventually, sea level rise could reach appoint where gate systems with locks would be necessary to maintain freshwater balances in Galveston Bay and allow ship traffic into bay ports.

## 11 General Applicability of the Approach

Worldwide, the number of people living within 100 km (62 miles) of a shoreline and 100 km (62 miles) of sea level is three times higher than the population found inland (Martinez et al. 2006). As of 2003, the coastal counties of the United States were home to 53% of the human population as well as ten of the 15 most populated economic centers (http://oceanservice.noaa.gov/programs/mb/supp_cstl_population.html). By 2025, it is estimated that U.S. coastal population sizes will be comparable to the rest of the world with 75% of the population living within 125 km (77 miles) of the coast (CT2020 Exec. Sum.; NOAA).

Population in coastal areas will continue to increase because of the coasts' numerous economic and recreational advantages. Much of this population increase will be concentrated in coastal cities. As human populations and associated infrastructure grow in coastal areas, the effects on the natural environment often experience a parallel growth. Hence both the human population and natural environments become more vulnerable to natural disasters. When coastal settlements are concentrated, a regional structural approach to suppressing the impacts of storm-induced surge and flooding makes sense to protect both the built and natural environments.

## References


Barajas E (2010) "Ike Dike" prevent storm surge? ABC 13 KTRK-TV. http://abclocal.go.com/ktrk/story?section=news/local&id=6913610. Accessed 5 Mar 2010

Berg R (2009) Tropical cyclone report Hurricane Ike. National Hurricane Center, 55 pp

Bijker W (2007) American and Dutch coastal engineering: differences in risk conception and differences in technological culture. Soc Stud Sci 37(1):143–151

DeBlasio SM (2008) Foreword by federal coordinating officer. Hurricane Ike Impact Report by the State of Texas, Federal Coordinating Officer, DR-1791-TX U.S. Department of Homeland Security Federal Emergency Management Agency. http://www2.fema.gov/pdf/hazard/hurricane/2008/ike/impact_report.pdf. Accessed 12 Jun 2010

Islam T, Merrell W, Seitz W, Harriss R (2009) Origin, distribution and timing of Texas hurricanes: 1851-2006. Natural Hazards Review, pp 136–144

Martinez ML, Gallego-Fernandez JB, Garcia-Franco JG, Moctezuma C, Jimenez CD (2006) Assessment of coastal dune vulnerability to natural and anthropogenic disturbances along the Gulf of Mexico. Environ Conserv 33:109–117. doi:10.1017/S0376892906002876

Port of Houston Authority (2010) General Information. http://www.portofhouston.com. Accessed 12 Mar 2010

Rice H (2008) Ike's changing course delayed Galveston's evacuation call, city says. Houston Chronicle. 11 September 2008

Texas Department of Transportation (2009) Gulf Intracoastal Waterway. Legislative Report to the 81st Legislature of Texas

The Laura Recovery Center (2010). http://www.lrcf.net/Ike/display.names.cgi. Accessed 20 Mar 2010

The Perryman Group (2006) An economy at risk: our vulnerable coast and its importance to the Texas economy. Report commissioned by the Independent Insurance Agents of Texas, December 2006

Thomas LA (2009) Written testimony presented to the U.S. House Committee on Homeland Security Subcommittee on Emergency Communications, Preparedness and Response. http://homeland.house.gov/SiteDocuments/20090303102053-20881.pdf

Toussaint B (2009) Training the rivers, exploring the coasts. Knowledge evolution in the Netherlands in two engineering fields between 1800 and 1940. Phys Chem Earth 34:132–141

U.S. Census Bureau (2010) State and County Quickfacts: Texas. http://quickfacts.census.gov. Retrieved 12 Mar 2010

Wiegel RL, Saville T (1996) History of coastal engineering in the USA. In: Kraus NC (ed) History and heritage of coastal engineering. American Society of Civil Engineers, pp 513–600

Zachria A, Patel B (2006) Deaths related to Hurricane Rita and mass evacuation. University of Texas Health Science Center-Houston, Houston, TX

# Kalpasar: Potential Coastal Impacts for India of a Mega-Engineering Project "Fulfilling All Wishes"

D. Venkata Subba Rao

## 1 Introduction

The Gulf of Khambhat, (also known as the Gulf of Cambay, Gujarat), an inlet of the Arabian Sea is located between 20° 35′–22° 20′N and 72° 05′–72° 55′E on the West Coast of India (Fig. 1). A high tidal energy environment this Gulf receives freshwater from 12 rivers including the Shetrunj, Mahi Sagar, Narmada, Tapi, Sabarmati, Kim and the Dhadhar, that flow through Gujarat State (Fig. 2). These rivers have traditionally supported thriving human settlements (Karanth et al. 2005). Known as the 'lifeline of Madhya Pradesh' Narmada has $34.537 \times 10^6$ m$^3$ of utilizable freshwater to be shared by Madhya Pradesh (65%), Gujarat (32%), Rajasthan (1.78%) and Maharashtra (0.89%).

Gujarat, essentially both an agrarian and an industrial growth state, accounts for the following levels of India's economic activities: industrial output 39%, mineral production 10%, salt production 80% (10.4 million tons per year), exports 20%, textile production 25%, pharmaceutical products 40% and petrochemical production 67%. However, Gujarat experiences recurrent droughts and spends nearly 40% of its domestic energy generation just to extract sufficient groundwater to meet its freshwater requirements. The unfettered development of industrial activities and unhindered labor exploitation in this state are attributed to the lack of bio-diversity, bio-resources and minimal social opposition (Khanna et al. 2000).

To alleviate the freshwater shortage, Dr. Anil Kane, the former Vice-chancellor of M.S. University, Baroda, envisioned a mega-engineering project as 'Kalpasar' a reference to a mythological wish fulfilling tree (Kane 1995). Kalpasar is considered a harbinger of prosperity to Gujarat state by creating a freshwater reservoir

D. V. Subba Rao (✉)
Dartmouth Nova Scotia, Canada
e-mail: seshu35@gmail.com

V. Badescu and R. B. Cathcart (eds.), *Macro-engineering Seawater in Unique Environments*, Environmental Science and Engineering,
DOI: 10.1007/978-3-642-14779-1_32, 

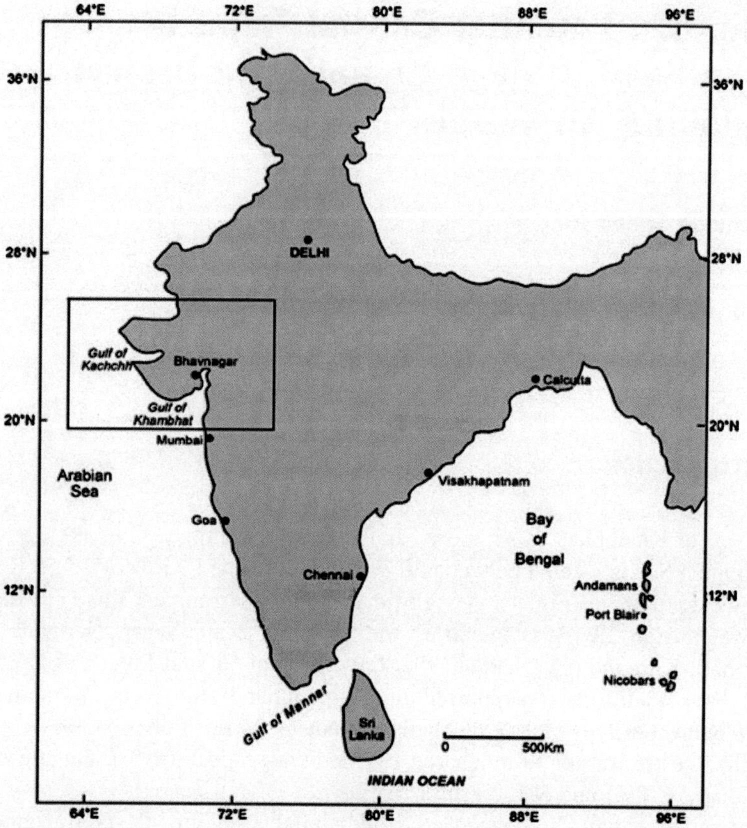


**Fig. 1** Map of India showing the state of Gujarat

with the assurances of "no human displacement" or "no involuntary human displacement".

In the year 1975 Eric Wilson identified the Gulf of Khambhat (also known as "Gulf of Cambay") as a promising site for tidal power generation. A Reconnaissance Report was prepared in 1988 by the Haskoning Foreign Consulting Firm, with experts from "Rijkwaterstaat"—the Dutch Public Work Department which recommended transformation of the tidal power mega-project to a tidal power mega-project cum freshwater supply mega-project for large multipurpose benefits (Haskoning 1998). This survey was followed by a pre-feasibility study in February 1998 that addressed six specific mega-engineering studies regarding its techno-economic feasibility and environmental aspects. Projected benefits from Kalpasar include increased freshwater supply for domestic use and industries, increase in irrigation, generation of hydropower, land reclamation, road communication, waterborne transport and fisheries development. The basic development scenario was to transform the region into a nationally and internationally recognized 21st Century export base for agro-industrial products. Objectives of this

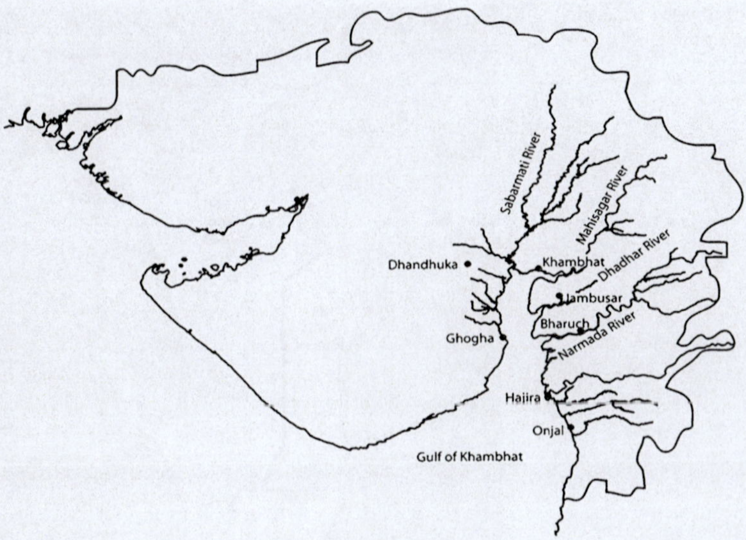


**Fig. 2** Current map of Gujarat showing Kalpasar region

multifaceted development mega-project have been expanded, now to include overall socioeconomic development to bring economic, social, and cultural changes affecting not only the local region but also India.

Consistent with this, the Government of Gujarat, Narmada Water Resources, Water Supply and Kalpasar Department has proposed creation of the 2,000 km$^2$ freshwater coastal reservoir by closing the Gulf of Khambhat across Ghoga, in the Bhavnagar district, and Hansot in Baharuch district through construction of a 64.16 km-long dam; reducing the distance between the two places by 225 km (Fig. 3) (Government of Gujarat 1999). Recently, the Gujarat Government approved two changes to the Kalpasar plan: (1) shift the closing dam 15 km northwards between Aladar in Bharuch and Kalatalav in Bhavnagar that would decrease the length from the proposed 64–29 km and also cuts the mega-project cost to USD $\sim 7.5 \times 10^9$. This design also includes a six-lane road and a railway track on the dam structure, and (2) a 30 km-long freshwater canal to divert Narmada waters into the Kalpasar dam created reservoir.

The mega-project also includes development of a 660 km inland canal system that would irrigate 1.05 million hectares of arable territory in coastal Saurashtra. The tidal fluctuations outside the Kalpasar Reservoir are to be harnessed for generating 5,880 MW of tidal power. The Government of Gujarat seeks national and international private sector investment in this mega-scale renewable energy project proposed for completion by 2020. Consistent with the Kyoto protocol mechanisms emission credits could be applied to the mutual benefit of investors.

It should be noted that Narmada runoff appears to be the main source of freshwater for Kalpasar. However, earlier the Narmada Control Authority approved plans for construction of a series of hydroelectric dams on the river

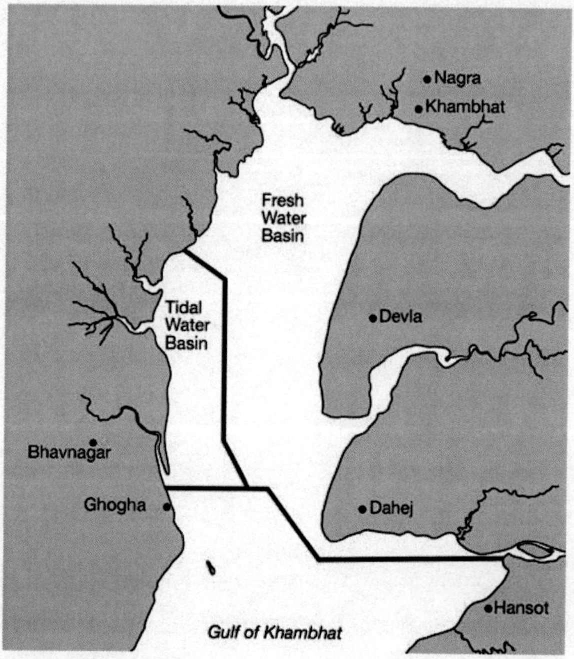


**Fig. 3** Proposed tripartite division of the Gulf of Khambhat

Narmada. Of the proposed thirty large dams on Narmada, the Sardar Sarovar Project (SSP) is the largest multipurpose project to irrigate more than 18,000 km$^2$ in the drought prone Kutch and Saurashtra areas.

The physiographical, seismological, geological, chemical, and biological features of this area including its fisheries are reviewed in this paper. Stressors on the ecology of the Gulf of Khambhat are identified with a view to examining the possible environmental impacts of the Kalpasar mega-engineering project on the overall estuarine and marine ecosystems.

## 2 Physiography

The funnel-shaped Gulf of Khambhat is 129 km long and occupies 6,346 km$^2$ (Kadam et al. 2008). It constitutes a shallow gulf of the Arabian Sea lying between Kathiawar peninsula and Gujarat, varying between 5 m at its land origin and 40 m in the mouth region facing the Arabian Sea. The Gulf of Khambhat receives directly 600–800 mm rainfall yearly, mostly during the south-west monsoon. The Gulf receives annually 38 km$^3$ of freshwater carrying $74 \times 10^9$ kg sediment; suspended particulate matter in the Gulf waters ranges between 600 and 4,000 mg/l. The Gulf environment is highly influenced by the fluxes by Narmada estuary where the spring and neap tides ranges are 8.7 and 4.6 m, respectively. Several channels are formed due to tidal bores and discharge of large volumes of river water during the monsoons. Post-monsoon, a low salinity plume exits the Gulf (Rao 2009). Of various channels, the Grant and Sutherland Channels are safe navigational routes. During flood-tide large, areas are submerged that form vast mudflats along the eastern shore. On the eastern bank the rivers Sabarmati and Mahi form wide estuaries while other rivers Dhadhar, Narmada and Tapi form wide and shallow estuaries. Several minor seasonal rivers such as Utavali Nadi, Shetrunji and Dhantarvadi are on the western bank.

Erosion and depositional processes due to strong tidal forces, construction of dams, destruction of the mangrove vegetation has caused notable geomorphologic changes in the Gulf's shoreline. These changes extend to the estuaries, mudflats, islands, mangrove swamps, cliffs, dunes, and floodplains (Shaikh et al. 1989). Additional changes include the westward shift of the channel of Sabarmati followed by its narrowing, a shift in the Mahi river southwards to carve out two islands. New shoals in Narmada and Tapti resulted in the advancement of the shoreline west of Khambhat, and sedimentation around Dholera. The erosion of the southern bank of Dhadhar and joining of a large island between Sonari creek and Kalubhar resulted in the gradual expansion of the mudflat which eventually connected the Aliya Bet Island to the mainland.

The Gulf of Khambhat has extensive areas of inter-tidal mud and sand flats, coastal salt marshes, and degraded mangrove associations, particularly in the deltas of the Mahi and Sabarmati rivers. It has a large number of estuaries, islands, and salt pans and some coral reefs around small inlets in the western part of the Gulf. This is one of the largest salt producing states in India. The salt pans have been expanded by 53.91 $km^2$ between 1995 and 2000, 22.56 $km^2$ between 2000 and 2001 and 35.73 $km^2$ in 2001–2003 which has caused a devastation of natural habitats (Charatkar et al. 2005).

Several industries contribute to pollutants. On the eastern shore the industrialized cities Surat, Bharuch and Vadodara are located as well as industrial complexes at Hazira, Ankleshwar and Dahej. Bhavnagar is the only city on the western shore and has a soda ash factory. There are cement factories at Jafarabad and Pipavav. At Alang- Sosiya Ship-Breaking Yard, one of the largest such yards in the world, is located on a Gulf beach. This particular ship breaking yard may recycle as much as 50% of the ships salvaged worldwide. Super-tankers, car ferries, container ships, since June 1983, have been torn apart on the beach, contaminating the strand and near-offshore regions with toxic substances such as asbestos. An estimated 15,000 workers are employed in this deconstructive commercial task.

## *2.1 Discussion*

Throughout the past 350 years of human history, societies have dammed rivers for irrigation, transport and other major industrial applications; only a few inland seas have been used for similar purposes. Unfortunately the consequences of such large-scale engineering systems are not fully understood nor do we have the ability to predict accurately, or fully, changes in complex ecosystems. On a global scale, large global freshwater reservoirs retained behind 45,000 dams intercept >40% of global freshwater discharge and also retain <30% global river sediment flux (Vörösmarty et al. 1997, 2003), which will impact the environment. Similarly, damming the strongly active seismic Gulf of Khambhat zone, that receives annually 38 $km^3$ freshwater carrying $74 \times 10^9$ kg sediment, and its future tripartite division into: (a) freshwater reservoir, (b) a brackish water tidal basin, and (c) the open ocean would alter the region's environment drastically.

Anthropogenic pressures, coming mostly from industrial and domestic activities, exert strongly negative environmental impacts upon the shallow euphotic seawaters of the Gulf. ICMAM (2002) observed that "the carrying capacity of the Gulf and the coastal regions seems to be exceeding the limit, causing resource conflicts". Some of the conflicts include oil and natural gas exploration, encroachment onto the Gulf by way of land reclamation, and destruction of mangrove forests (Fig. 4).

Tripartite division of the Gulf would have strong environmental impact on the fluvial ecosystem as well as on the inter-basin estuarine and marine ecosystems and may lead to overall deterioration of the environment (Fig. 4). Kalpasar mega-project would result in the development of transitional water ecosystems which are open systems, ecotones between land and the Arabian Sea, seawater and freshwater that is subjected to additional anthropogenic pressure (Orfanidis et al. 2008). Based on lessons learned from general geoscience and strictly environmental studies elsewhere, I provide here perceptions of several potential environmental impacts that could result from future realization of the Kalpasar mega-project.

## 2.2 Dams

Dams have been built for hydropower generation and barrages were built in estuaries for water storage and recently for hydrokinetic tidal power generation. The planning for these major civil engineering projects however does not include the potential impacts on the marine environment (Drinkwater and Frank 1994). In the Sea of Azov, anthropogenic interference with natural ecosystems by way of construction of large number of dams caused a diversion of freshwater flow to allow heavy industry, energy production, and irrigated agriculture. These activities contributed heavily polluted runoff which, in turn, has negatively impacted the reproduction of commercially important fish and resulted in reduced catches (Volovik et al. 1994; Volovik 2001). Elsewhere, construction of dams has resulted in the loss of ecosystems and loss of fisheries. Examples include the Aswan High Dam on Nile River that reduced the fishery by 95%, dams on Colorado River that decreased the shrimp catch in the Gulf of California, the Kotri Barrage, Pakistan decreased the fish take by a factor of three, and in the China Sea off Bohai reduction of the shrimp catch by 85% was observed following marked changes in total sediment discharge of the Yellow River.

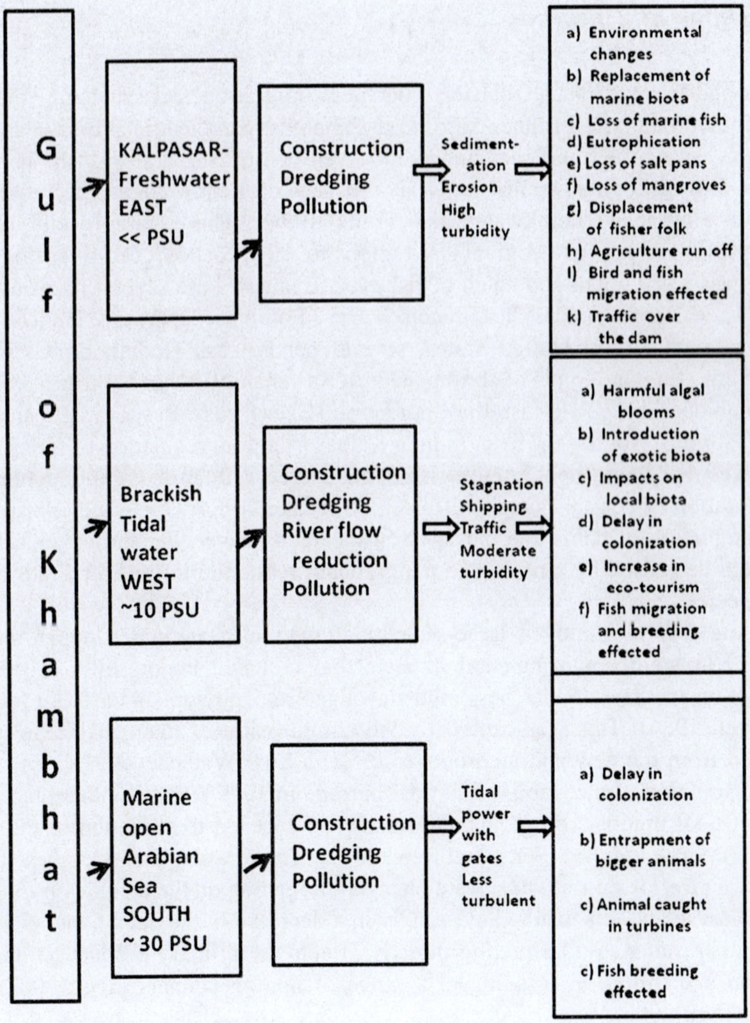


**Fig. 4** Flow diagram of the possible impacts of the mega engineering project Kalpasar

In Koljö fjord, Sweden to provide shipping with a faster and safer route to Uddavalla, a new channel was constructed in a shallow area around 1940 and the old connection was closed (Filipsson et al. 2005). This major mega-engineering work altered the fjord's bottom topography, internal surface circulation and caused substantial changes in the phytoplankton abundance and composition.

Fearside (1995) studied the impact of 79 Brazilian dams with an installed capacity of 85,900 MW as sources of "Greenhouse" gases and concluded that decomposition of organic matter such as the submerged underwater biomass, macrophyte beds would release approximately $0.26 \times 10^6$ tons of methane and $38 \times 10^6$ tons of carbon dioxide.

## 2.3 Dams and Fauna

Dams and barrages act as barriers and block migration and spawning routes of fishes. Hydroelectric turbines and the associated hydroelectric pressure and turbulence can cause significant injury and even death to migratory fish. The tidal power barrage of Annapolis River in the Bay of Fundy contributed to >46% mortality of shad (Drinkwater and Frank 1994). These authors cite several examples from various geographical regions to show that reduced river discharge influences recruitment and catch of fisheries. Examples are several fisheries from Norway, eastern Canada, San Francisco Bay, Potomac, Delaware, Hudson estuaries in northeastern United States, several pelagic fish (India), brown shrimp (Southern England), pink shrimp (Florida), shallow water shrimp (central Mozambique), prawns (Australia), blue crab (Chesapeake Bay), oysters and blue crab (Florida). Not too far from India's Kalpasar region, regulation of Indus River discharge led to nutrient impoverishment, decrease in primary production and deterioration of coastal mangrove forests (Snedaker 1984; Quraishee 1988). With the completion of dams and barrages on the Indus River, the shrimp industry in Pakistan decreased by tenfold and a significant reduction in the catch rate for all fish species.

In the Bay of Fundy a large-scale tidal mega-project in the upper reaches resulted in major environmental impacts that included major shifts in primary production in the salt marshes, mud flat algae and phytoplankton (Gordon and Dadswell 1984). This was caused by the compression of inter-tidal zone which resulted from the new and anthropogenic Mean Low Water level.

Construction of the Cardiff Bay tidal barrage in 1999 resulted in a replacement of inter-tidal mudflats by permanent freshwater that led to reduction of the shelducks *Tadorna tadorna* (Ferns and Reed 2009). The foraging habits of these birds, mainly on *Nereis diversicolor*, were changed to feeding on the benthic chironomid midge larvae and this resulted eventually in a decrease in the duck populations. In the British estuaries tidal mixing directly affects the primary production and the location and timing of foraging of a range of marine animals (Scott 2007) and birds (Clark 2006).

The Yellow Sea Large Marine Ecosystem survey showed that the construction of a tidal power plant in Gororim Bay, South Korea, could destroy tidal flats which are critical spawning grounds of many fishes.

## 2.4 Tidal Power

Around the world several tidal power plants exist which make use of the daily rise and fall of the tides. These experiences should be taken into consideration by mega-engineers planning the Kalpasar in the Gulf of Khambhat. In the Bay of Fundy, construction of a major tidal power plant in the upper reaches could elevate the tidal levels of the head pond that would have major impacts on inter-tidal ecology (Gordon 1994). A drop in the mean high seawater level by about 0.7 m, compression of inter-tidal zone, conversion of mud and sand flats to sub-tidal habitat, and major shifts in the primary production by salt marshes and in the abundance of inter-tidal organisms and fish are predicted.

The ecological impact of La Rance tidal power plant after its 20 years of operation was reevaluated (Retiere 2008). On the northern coast of Brittany on the Rance River during the construction period ecological degradation was more (Bershtejn 1993) than during post-construction. During the post-construction period the ecosystem experienced variable degrees of adjustments to the new environmental conditions. The new environmental equilibrium is linked to the degree of stability of abiotic conditions, which in turn relates to the operating conditions of the tidal plant.

## 2.5 Inter-Basin Changes

Changes in river fluxes may eliminate distinctive habitats as in the Ebro River and delta (Ibanez and Prat 2003, Ibanez et al. 1996). Changes due to the inter-basin transfer of freshwater include: (1) increase in salinity in the delta and estuary due to restriction of river flow; (2) changes in geomorphology caused by changes in the fluvial system; (3) decrease in biological productivity following decrease in nutrient input and (4) affect all species distribution, aquaculture and fisheries in the bays.

Tidal inlets respond to sediment infilling and are controlled by the impinging wave energy, local tidal range, tidal prism, sediment supply, and direction and rates of sediment delivered to the inlet. As siltation would be a major problem, a complete inventory of the existing sedimentary conditions, volumetric estimate of the sediment, grain size, sources and transport should be made and utilized in a hydrodynamic predictive model. To develop hydrodynamic predictive models of siltation and understanding of existing hydrodynamic conditions, their relation to the sedimentation character of the region, the mode and direction of transport, and the available supply of material are required (Amos 1979).

## 2.6 Algal Blooms and Other Biological Alterations

Ecosystem mega-engineering facilitates invasions by exotic biota. Disturbances to the resident communities are common and often cause partial removal of dominant species (Badano and Marquet 2008). In the Kalpasar mega-project, the brackish water transformed from the open-to-ocean marine environment to an enclosed nearly stagnant lagoon may favor development of harmful algal blooms. The anticipated increase in the commercial shipping activity within this proposed mega-engineered artificial lagoon may bring in exotic biota. Human-induced changes resulting from mega-engineering construction in Asia Pacific harbors—for example, in Hong-Kong, Manila Bay, and Jakarta Bay caused significant degradation of the coastal ecosystem, resulting in significant loss of biodiversity as well as development of harmful algal blooms (Wolanski 2006). Harmful algal blooms are known to occur elsewhere in the Arabian Sea (Thresiamma et al. 2008, Karunasagar et al. 1989) and may spread to the Kalpasar mega-project region. Development of a major seaport, one intermediate port and seven minor ports for watercraft would increase shipping traffic, both commercial and naval, and associated with it more discharges of untreated ballast water and introduction of more alien species.

The question asked by Cloern (2007), "How do the changes in eutrophication interact with other stressors and how are the multiple stressors linked?", is relevant in the Gulf of Khambhat. A highly stressed environment the Gulf of Khambhat acts as a sink to nutrients from agricultural runoff, and for industrial pollutants and its carrying capacity is in excess of its limit, causing resource conflicts (ICMAM 2002).While planning a mega engineering project of the magnitude of Kalpasar, an environmental strategy should be mandatory to respond to new developments or unforeseen events associated with the project. A comprehensive baseline survey over multi-site, multi-year will provide a better understanding of the structure, functioning and response and recovery of the ecosystem to perturbations. The existing baseline data for this Gulf lack the robust baseline data of the biology. A multi-level approach to monitor the freshwater, brackish and marine ecosystems and to evaluate the risks needs still to be carried out. Christie et al. (2009) stressed the need to slow down the dramatic decline of the ocean environment; they propose the tropical Marine ecosystem-based management (EMB). This proposal merits consideration in Kalpasar—a planned mega-project in the tropical seas. For management in these areas, EMB seeks utilization of a wide variety of investigative tools using oceanographic and ecological principles and giving equal consideration of governance and social conditions. However, biological issues of the region are the least addressed. There is only one designated consultant on fisheries.

## 2.7 Societal Impacts

Although Kalpasar mega-project is wishfully envisaged as a harbinger of future economic prosperity by providing much needed freshwater and electrical power to Gujarat, we must pay attention to its several adverse societal impacts. Because of the lack of experience and expertise locally, the Government of Gujarat with all good intentions hired international experts to provide advice. There appears to be emphasis on geotechnical matters including seabed mega-engineering, geophysical geochemical and earthquake investigations including tsunamis, tidal power generation and mega-engineering design of the dam but only slight emphasis on subsequent biological and societal changes. There are 56 national consultants in geotechnical investigations, seismic studies, irrigation studies, hydrology, dam designing, seaport development, ocean engineering, coastal engineering, and tidal power generation. This is understandable because the Gulf of Khambhat located in an intense seismically active zone is potentially vulnerable to earthquakes. Large reservoirs as they fill increase the pressure on local faults and can cause seismic events (ICE 1981) as in the case of the Yangtze River Three Gorges Dam (TGD) in China. The TGD, with its almost insurmountable challenges, is the most controversial due to its growing threats from landslides, pollution and potential for catastrophic downstream floods (Oster 2007). Although the Kalpasar mega-project in India is smaller compared to China's TGD we should be specifically concerned about the probability of an earthquake damaging the dam resulting in a great fluid outburst devastation in which event firm rescue plans should be in place. Depending on when the Kalpasar dike break occurred, water (both fresh and salty) could strike the Gulf's beaches or move as an odd anthropogenic double "tsunami" outwards into the Arabian Gulf, perhaps swamping nearby boats and small ships with freshwater and salt water waves.

The biological and societal changes that should be assessed include the possible displacement of local human population, particularly the 360,000 fisher-folk population and 123,366 active fishermen. The Gujarat government hopes to generate 55.21 million man-days of employment in fisheries activities. As Caesar and Pet-Soede (2003) observed, many coastal communities, particularly in the poor and developed countries, would be trapped in the "vicious cycle of poverty, resource depletion and further impoverishment", it would be crucial to include these in planning the mega-project's overall management equation. In the USA, the Klamath River dams were built from 1918 to 1961 in Oregon and California along an upstream stretch of the river, and are operated by PacifiCorp, which uses them to generate electricity. Restriction of river flow resulted in a decline of salmon populations. With the help of environmentalists, fishermen faced with economic hardships brought pressure on the Federal and other government officials who finally agreed to a removal of these dams (beginning in 2020) at an estimated USD 450 million cost. Poor and uneducated Gujarat fishermen must be closely consulted and their collective voices heard on this mega-engineering project.

Alternatively, construction of several small dams may relieve the freshwater shortage in the State of Gujarat. Experts not in favor of tidal power generation under the Kalpasar project favor setting up eco-friendly wind farms to generate 3,000 MW at a 50% of the total Rs 54,000 crore (India, equal to about USD $11 \times 10^9$) mega-project cost (http://www.expressindia.com/latest-news32009). Nuclear power is another alternative worthy of consideration.

## 2.8 Recommendations

- Consider construction of several small dams on the various rivers to relieve the water shortage in Gujarat.
- Augment energy supply with renewable resources such as biomass energy, solar energy, wind energy and geothermal energy or consider nuclear power as an alternative source of energy.

- Institute a long-term environmental monitoring program encompassing the entire water column and the sea floor, not just once but over all seasons as populations move and habitats shift, to obtain a solid biological baseline.
- A thorough evaluation of the impending environmental impacts, similar to those addressed by a Before/After and Control/Impact (BACI model—Underwood 1992; Hewitt et al. 2001) is recommended.
- Bring together all stakeholders, from industrialists, government agencies, environmentalists and fishermen and provide assurance to the public, particularly the 851 fishing villages, with a population of 360,000 of which 123,366 are active fishermen.
- Develop a suitable warning system, disaster preparedness and management of disasters utilizing Information Technology. Plan should include pre-earthquake phase preparedness, mitigation and prevention concepts, and post earthquake relief measures including immediate relief and rehabilitation.

# 3 Technical Considerations

## *3.1 Narmada River*

Narmada River is a 1,312 km-long river with a total catchment of 98,796 $km^2$ and it meanders amongst the Mandla Hills, the Vindhya and the Satpura mountain ranges before pursuing a direct westerly course joining the Gulf of Khambhat near Luhara. The major tributaries of Narmada are Kaveri, Orsang and Karjan. July–August is the flood season when vast areas along the banks get inundated causing extensive damage to agriculture. The Sardar Sarovar Project prevents inundation. Bharuch Channel is the principal entrance to the estuary. Narmada River receives pollutants from the domestic wastewater releases from Bharuch Town, wastewaters from a fertilizer complex near Bharuch and an industrial estate [industrial park] located at Ankleshwar.

The locally major river Narmada has only 27.22 million acre-feet mean annual flow (MAF) at 75% dependability. Gujarat under its existing dam-centric development, proposes construction of four mega-dams, Sardar Sarovar Project (SSP) that includes 30 large dams, 165 medium sized dams and 3,000 small dams. With a 136.5 m height, the SSP will irrigate 18,000 $km^2$. The State of Gujarat has constructed Sipu and Dantiwada dams across Narmada River that would divert >11 $km^3$ freshwater for irrigation. (For a vivid volumetric comparison, that is equivalent to 18 times the seawater volume of Sydney Harbour, Australia). Upadhyay (2004) points out that although 80% of the projected freshwater to Kalpasar that is slated to come from Narmada River, the downstream flow due to Sipu and Dantiwada dams would decrease the freshwater flow and would, thus, impact Kalpasar mega-project.

**Table 1** Salient features of the Kalpasar project*

| Marine ecosystem | Open to the Arabian Sea |
|---|---|
| Freshwater reservoir | 2070 $km^2$ |
| Brackish water tidal basin | 872 $km^2$ |
| Total length of dam | 64.16 km |
| Actual dam (between Gogha on the eastern bank of the Gulf and the Luhara point) | 28.9 km |
| Extension (across the Bharuch Channel towards the island of Alia Bet) | 5.1 km |
| Maximum height of dam | 53 m where the bottom is at GTS-40 m |
| Top width | 35 m |
| Top level | MSL +14.5 m |
| Maximum height above | 45 m |
| Discharge openings | 65, each 17 m wide |
| Crest level | GTS + 13 m |
| First ship-lock | 200 m length, 35 m wide with sill level at GTS-10 m, allows ships ~50,000 DWT |
| Second ship-lock | Lets ships ~20,000 DWT to enter the tidal basin |
| Freshwater basins: | Sabarmati 870, Mahi 4807, Dhadhra 512, Narmada 6016 and Surashtra 112 $Mm^3$ |
| Freshwater Reservoir | 2,070 $km^2$ |
| Storage | 16,791 $Mm^3$ |
| Average inflow of freshwater | 12,317 $Mm^3$/year |
| Usage of freshwater Domestic and industrial | 1,387 $Mm^3$/year |
| Irrigation | 6,100 $Mm^3$/year |

* Govt. Gujarat Report 1999
*GTS* geotechnical and tunnel analysis system, *MSL* mean sea level

### *3.2 The Kalpasar Dam Mega-Project*

The scope and size of the mega-project could exceed any other similar construction in the world. When completed, the dam will cause a tripartite transformation of the extant large marine ecosystem into (a) a freshwater reservoir (b) a brackish water tidal basin, and (c) the open ocean of the Arabian Sea. Salient features of the Kalpasar mega-project are provided in Table 1.

Because of the absence of deep channels in the north of the Narmada estuary, the physiographic conditions are more stable than in the southern part of the estuary.

## 4 Seismic Activity

This Gulf of Khambhat is prone to severe seismic activity. In the past 500 years several earthquakes have shaken the region, including the major +8 magnitude Richter Scale event of 26 January 2001. On 16 January 1819 an 8.3 magnitude event devastated several areas nearby. These quakes have caused much subsidence and elevation at various places.

Seismic studies in the Gulf revealed a seismically active zone with an 80 km radius (Mukhopadhyay 2002) with significantly large thermal outputs and a compressive stress in east–west to NNE-WSW direction (Mukhopadhyay and Arora 1999). Directional correlation exists between thermal reservoirs and plume activity along Cambay and Narmada collision zone (Veeraswamy and Harinarayana 2006). High heat flow (max. 468 $mW/m^2$) and geothermal gradient (234°C/km max.) were observed in the Tatapani in the Narmada Son Lineament (NSL) region. Presence of volcanic plugs with seismic activities of magnitude $\leq$4 at their contacts in the Saurashtra region is of concern. Circular gravity highs and magnetic anomalies of 40–60 mGal (10–5 $m/s^2$) and 800–1,000 nT, respectively were reported in this region; three of them are related to the Gulf of Khambhat and the co-extensive Cambay rift basin (Chandrasekhar et al. 2002; Tewari et al. 2009). Veeraswamy and Harinarayana (2006) observed significantly large thermal outputs and a correlation between thermal reservoirs and plume activity along Cambay and Narmada collision zone.

## 5 Currents and Tidal Streams

Currents in the coastal waters are bimodal exhibiting prominent north-northeasterly directions during flood-tides and south-southwesterly direction during ebb tides with very short slack periods. The maximum currents for neap and spring tides are 1.4 and 3.5 m/s, respectively, that induce a coastal circulation parallel to the shoreline. Due to pronounced tides the seawater column is well mixed without any stratification, even in the presence of discharges from Narmada River and other contributory rivers.

Circulation in the Gulf is dominated by barotropic tides (Unnikrishnan et al. 1999). In the Gulf's creeks and shallow areas of the tidal amplitude is much higher than in the more open, unconstrained areas. In the open ocean of the Gulf of Khambhat, currents were 1.4 m/s and in the Gulf about 3.2 m/s The velocity of tidal streams in the northwestern region of the Gulf is high (Sanil Kumar et al. 2006).

The tidal excursions extend 25 km upstream; for example in Tapi the salinity decreases from 32.53 PSU at the mouth to 0.1 PSU within a distance of 35–40 km (Bapardekar et al. 2004). These tidal streams act especially as dynamic barriers to long-shore sediment transportation resulting in four distinct clay-mineral assemblages (Nair et al. 1982). During the southwest monsoon, the large tidal action enhances deposition onto the mud flats (Chandramohan et al. 2001). Based on a budget of non-conservative material, that is, dissolved inorganic phosphate (DIP) and dissolved inorganic nitrogen (DIN), Bapardekar et al. (2004) showed that the inner segment of the Tapti estuary serves as a sink (DIP 13.40 $mmol/m^2/day$, DIN 108.9 $mmol/m^2/day$).

## 5.1 Sediments

Along with the river waters, the Gulf received a sediment load of $0.53 \times 10^{10}$ m$^3$ over 45 years contributing to 0.03 m/year increase in thickness (Chandramohan et al. 2001). In this tidally high-impact environment fine-grained sediments are transported into the Gulf, whilst the coarse sandy sediments are transported southwestwards out of the Gulf of Cambay (Kunte 2008) where they probably contribute to the formation of bed form structures offshore. The Olpad and Cambay shales sediments and the overlying Bhavnagar and Tarapur sediments of the Gulf of Cambay are considered to have a good to excellent potential for oil and natural gas (Berry et al. 1996). On the wave dominated western coast of the Gulf of Khambhat, net transport of sediment is from north to south in the pre-monsoon season and south to north in the post-monsoon season (Rajawat et al. 2005).There was observably significant onshore-offshore transport of sediment along the west coast.

There is some compelling evidence for the transportation of Indus River-derived sediments from the north to the outer shelf of Cambay-Goa of the western India coast by a southerly surface current (Rao and Rao 1995) and with an apparent decrease from north to south (Rao and Rao 1995). The influence of these ocean currents seems to mix with the clays transported from the hinterland to result in the montmorillonite-rich zone, relatively rich in organic carbon, off the Gulf of Khambhat.

The suspended sediments, churned up by strong tidal currents, limit the euphotic layer (1% light penetration) to very close to the water surface (Varkey and Kesavadas 1976). Mud brought by the Narmada and Tapti rivers renders the waters highly turbid particularly in the interior of the estuary and hinders plankton growth (Rao et al. 1971).

The northern area (north of 20° latitude) is characterized by uneven bottom topography, with steep gradients, sand waves and mega-ripples which are likely to pose major problems for offshore marine construction and should be avoided (Siddique et al. 1981).

## 5.2 Tidal Power

On the west coast of India off Mumbai, the continental shelf is wide and leads to a converging channel situated at the Gulf of Khambhat. Because of the funnel-shaped geometry, the semi-diurnal tides amplify about threefold from mouth to channel head, resulting in the highest tides on the entire Indian coast because of sluicing (Nayak and Shetye 2003). For tidal power, areas with high tidal amplitude are most suitable; areas with low tidal amplitudes are unsuitable for costly tidal power mega-projects (Hammons 1993). In the former USSR, Penzhinskaya Cauba, Menzen, and Tugar and at San Jose (Argentina), the Severn River (UK), Turnagain

Arm (USA), Cobequid (Canada) the amplitudes correspond to 200, 50, 27, 20, 17, 17 and 14 while for the Gulf of Khambhat (Cambay) it is 15.

In the Kalpasar mega-project, a single basin with 5,880 MW production or a double basin with 1,680 MW capacity are envisaged. Assuming that thermal combustion to generate 5,000 MW of electricity would normally emit 22,660,000 metric tons of carbon dioxide gas, the Kalpasar mega-project, while sustaining clean power to Gujarat would be consistent with the legally defunct Kyoto Protocol and its Clean Development Mechanism (CDM) that is gaining the increasing acceptance for reducing global emission of carbon fluorocarbons.

## 6 Pollution

Gujarat State has established a corporate body to develop an industrial base that includes food processing, tobacco, textiles, wood, paper and pulp, leather, rubber, chemical, glass, cement, non-ferrous metals, ferrous metals, non-electricity machinery, electrical machinery, transport machinery and other kinds of equipment and several other types of small-scale industries.

Oza (1975) cautioned that the increasing populace of Gujarat should press the government to enact proper restrictive legislation to prevent construction of a 56 km masonry conduit carrying untreated industrial effluent from Baroda to Sarod, Gulf of Khambhat. The Gulf of Cambay already receives waste generated by more than 155 large- and medium-scale industries established near Vadodara city. At the confluence of Mahisagar river, near the Gulf of Cambay effluents are continuously discharged; fish caught in these polluted waters have a higher metal content than those captured further downstream (Sharma 1995, unpublished). Once common to these waters, the flat fish *Synaptura* has been absent since about 1993. Reddy et al. (2007) reported bioaccumulation of heavy metals in some commercial fishes and crabs taken from the Gulf of Cambay.

The Tapi basin, with a population of more than 15 million humans, unserved by any sewage facilities, is utilized for 93 large and medium scale industries such as fertilizers, engineering, thermal power, food and beverages, distilleries, textiles, chemical and metallurgical industries. The domestic, industrial and agricultural wastes amount to $4{,}026 \times 10^6$ $m^3$/year (Bapardekar et al. 2004). Load from the agricultural wastes in MT per year alone amounts to 24,826 nitrogen, 6,478 phosphorus and 6,848 potassium.

Gujarat Alkali and Chemicals Limited (GACL), established at Dahej in Bharuch District, manufacture phosphoric acid and soda (Kadam et al. 2008). Other products are calcium chloride, hydrogen peroxide, poly aluminum chloride and anhydrous aluminum chloride. GACL also runs a 98 MW captive power plant. Treated effluent of GACL meeting the norms of the Gujarat Pollution Control Board (GPCB) is disposed of in the coastal waters off Luhara through a 5 km long submarine pipeline equipped with a suitably designed outfall diffuser. GACL has engaged the National Institute of Oceanography (NIO), Goa to assess status of the coastal environment off Luhara for identifying the adverse impacts due to the release of treated effluent of GACL at the designated site and assess efficiency of their effluent disposal system. Accordingly, NIO investigated the marine environment off Luhara during October 2007 (Table 2).

This Gulf has one of the world's largest ship-breaking yards at Alang-Sosiya. About 2,453 tankers, cargo carriers and bulk carriers are broken for scrap and provide over 17 million metric tons of steel to the re-rolling industries (GSEAP 2000) and the yard generates 7,500 MT of industrial solid waste yearly, of which ~33% is hazardous (GSEAP 2000). This industrial activity contributes pollutants such as dangerous heavy metals, contaminants and petroleum hydrocarbons which reportedly accumulate in some of the commercially important fish *Harpodan nehereus* and crab *Metopograpsus maculates* (Reddy et al. 2007). Reddy et al. (2005a, b) reported higher concentrations of petroleum hydrocarbons and heavy metals during winter than during the monsoon season or summer.

**Table 2** Summary of values of variables off Luhara, Gulf of Khambhat, India, during October 2007*

| Variable | Near shore | Estuary mouth | Offshore |
|---|---|---|---|
| **Water Temp (°C) | 28.7 | 28 | 26.4 |
| pH | 7.9 | 7.9 | 8 |
| Suspended solids (mg/l) | 1,083 | 1,345 | 979 |
| Salinity PSU | 21.1 | 21.8 | 21.2 |
| Dissolved oxygen (ml/l) | 4.1 | 3.8 | 3.7 |
| BOD (mg/l) | 0.7 | 1 | 0.8 |
| $PO_4$ (μmol) | 4.6 | 4.9 | 4.1 |
| $NO_3$ (μmol) | 42.9 | 35.5 | 39.6 |
| $NO_2$ (μmol) | 0.6 | 0.6 | 0.5 |
| $NH_4$ (μmol) | 0.7 | 1.1 | 0.4 |
| PHc (μg/l) | 11.1 | 9.7 | 11.2 |
| Phenols (μg/l) | 7.4 | 7.6 | 7.4 |
| **Sediment -Sand | 95.4 | 34.3 | 12.9 |
| Silt | 3.4 | 63.1 | 82.7 |
| Clay | 1.2 | 2.6 | 4.4 |
| Al (%) | 1.4–5.5 | | 1.1–3.6 |
| Cr (μg/g) | 46–152 | | 75–165 |
| Mn (μg/g) | 1,018–3,014 | | 2,372–5,615 |
| Fe (%) | 3.2–11.3 | | 6.4–13.1 |
| Co (μg/g) | 17–54 | | 28–50 |
| Ni (μg/g) | 38–96 | | 68–106 |
| Cu (μg/g) | 33–113 | | 88–166 |
| Zn (μg/g) | 70–234 | | 133–279 |
| Hg (μg/g) | 0.02–0.08 | | 0.06–0.07 |

* Based on NIO/Sp-32/2008 (SSP 1963)

** % dry weight

# 7 Biology

## 7.1 Flora

In the Gulf of Khambhat, mangroves have experienced the worst vegetation decline. Once occupying an area of 338 $km^2$ in ca. 1870, the mangrove forests were reduced to 54 $km^2$ by ca. 1998. There were extensive tracts of mangrove forest, now reduced to less than 2,000 $km^2$ of scrub-type forest. Species of Rhizophoraceae are now rare. Thus, the zonation of mangroves consists of a seaward band of *Avicennia marina* leading to a back-mangal consisting of *Salicornia brachiata, Suaeda urochondra setulosa,* which appears at the extreme eastern limit of its distribution. Land-cover of the Gulf of Cambay shows mangrove mixed swamps (67.2 $km^2$), inter-tidal mud flats (3,268 $km^2$), rocky shore 5.2 $km^2$, salt pans (132.5 $km^2$) sand dunes and sandy beach (22.8 $km^2$).

## 7.2 Fauna

Khambat Gulf's newest seaport, Bhavanagar Harbour, is one of the largest colonies of *Egretta gularis* in the world, approximately 1,000 pairs (Parasharya et al. 2004). Other common breeding species include *Nycticorax nycticorax*, *Ardeola grayii*, *Bubulcus ibis*, *Egretta garzetta*, *E. alba*, *Mycteria leucocephata*, *Threskiornis melanocephalus*, *Pseudibis papillosa*, and *Platalea leucorodia*. The heron colonies in this area are of interest as it is one of the few places where *E. gularis* and *E. garzetta* nest side by side and interbreed.

The extensive mudflats and sand flats of the Gulf of Khambhat at Ghogha support very large numbers of migratory shorebirds, gulls and terns, together with large feeding flocks of *Phoenicopterus ruber* and *Phoeniconaias minor*. The most abundant shorebirds are *Recurvirostra avosetta*, *Charadrius mangolus*, *C. leschenaulti*, and species of *Tringa* and *Calidris*, and *Limicola falcinellus*. Numerous crab plovers (*Dromas ardeola*) winter in the area, and a few Indian skimmers (*Rynchops albicollis*) have been reported. Large roosting flocks of *Grus grus* and *Anthropoides virgo* are often present. *Chelonia mydas* and *Lepidochelys olivacea*, two species of marine turtles nest in large numbers along the coast and on Piram Island.

## 7.3 Biology of the Waters

Annual studies of the pelagic, nektonic and benthic fauna and flora are almost nonexistent except by Rao et al. (1971) and Kadam et al. (2008). The Gujarat Ecology Society published a bench mark survey of the Gulf of Khambhat (1998) incorporating their findings for the 20 km belt coastline of Gujarat. Besides information on the impact of industries and other developmental activities on agriculture and allied activities, coastal geology, water quality of important stations and plankton composition was given. For the Gulf of Khambhat, utilizing remote sensing and Relational Database Management System (RDBMS) data on seawater, the sediment quality and phytoplankton data were extracted from COMAPS reports (ICMAM 2002) (Table 3).

Plankton was abundant in the less turbid waters near the mouth of the Gulf but poor in the highly turbid interior water volume; the only macro fauna was the foraminifera (Kameswara Rao 1971). The eastern bank of Khambhat Gulf with a high tidal range and high turbidity caused by high concentrations of suspended matter seems to indicates a possible mobile substrate which prevents settlement of live foraminifera (Nigam 1984). In the macro-tidal estuary *Galitellia vivans*, a foraminiferan indicative of stressed and up-welled areas is carried 50 km into the estuaries (Ghosh et al. 2009). Contributions of the inter-stitial algae diatoms and dinoflagellates remains essentially unknown.

Phytoplankton surveys during December 1998 and February 1999 (ICMAM 2002) showed a maximum of 35 diatom genera, 22 dinophyceae. Cyanophyceae were also present. Phytoplankton abundance (cell numbers/l) was 168–586 for

**Table 3** Summary of values of biological variables off Luhara, Gulf of Khambhat, India, during October 2007*

| Variable | Near shore | Estuary mouth | Offshore |
|---|---|---|---|
| Chl a (μg/l) | 0.8 | 0.6 | 0.6 |
| Phytoplankton cells ($\times 10^3$/l) | 28.6 | 11.2 | 15 |
| Total genera | 5 | 4 | 7 |
| Major species | Thalassiosira, Thalassiothrix, Peridinium | Cyclotella, Nitzschia | Thalassiosira, Thalassiothrix, Peridinium |
| Zooplankton Biomass (ml/l) | 4.9 | 2.8 | 2.4 |
| Nos ($\times 10^3$/100 m$^3$) | 43.4 | 35.5 | 22.3 |
| Total groups | 12 | 12 | 12 |
| Major genera | Copepods, chaetognaths, decapod larvae, lamellibranch larvae | Copepods, chaetognaths, decapod larvae, lamellibranch larvae | Copepods, chaetognaths, decapod larvae, lamellibranch larvae |
| Benthos biomass (g/m$^2$) | 0.06 | 0.66 | |
| Population (nos/m$^2$) | 22 | 12 | |
| Total groups | 1 | 1 | |
| Major groups | Polychaetes | Polychaetes | |

* Based on NIO/Sp-32/2008 (SSP 1963)

diatoms, 14–204 dinoflagellates and 4–8 cyanophyceae. Diatoms were represented by *Coscinodiscus*, *Biddulphia*, *Surirella*, *Navicula*, and *Nitzschia* sp.

Zooplankton abundance (numbers/m$^3$) for copepods was 500–18,7500, hydrozoans 2,500–17,500, tintinnids 2,500–30,000, pyrosomids 2,500–12,500, foraminiferans 2,500, chaetognaths 5,000, cladocerans 2,500, fish eggs 2,500–87,500, decapod larvae 2,500–50,000, echinoderm larvae 2,500–7,500, and polychaete larvae 2,500–15,000.

## 7.4 Corals and Associated Ocean Fauna

A high species diversity on the coral reefs of the Gulf exists. Raghunathan et al. (2004) surveyed the oceanic coral reefs, at depths of 2–5 m, during March 10–13, 2004 and duly reported *Gorgonium* sp, *Tubinaria crater*, *Tubastrea aurea*, *Polycyathus verrilli*, and *Porites lutea*. Polyp density was 53–500 m$^2$ and a biomass range of 195 to 1,000 g m$^2$. There were 18 diatom species, 3 dinoflagellate species, contributing to a maximum cell number of $75.8 \times 10^2\ l^{-1}$, chlorophyll 1.35 μg/l. Zooplankton volumes ranged from 5.1 to 8.8 ml/100 m$^3$ and their numbers 7,280 to 15,600 individuals/100 m$^3$. The waters also contained a rich diversity of species: 18 foraminiferans, 2 tintinnids, 2 annelids, 6 copepods and one each of chaetognath, ostracod, decapod larva, and gastropod, appendicularian, salp and fish egg.

The species diversity of macrofauna associated with the corals was rich and consisted of 27 gastropods, 7 bivalves, 6 brachyuran crabs, 2 each of polychaetes, amphipods, cirripeds, hermit crabs, shrimp, nudibranchs, and one each of polyplacophoran, sea anemone, sponge and an eel species.

## 7.5 Fisheries

There are 851 fisher-folk villages along the 1,600 km coastline, with a population of 360,000 of which 123,366 are active fishermen (Somvanshi et al. 1999). Some 740 non-mechanized and 499 mechanized fishing boats fish are active in the State of Gujarat's coastal waters. Trawl-nets, gill nets, drag nets and cast nets are commonly used. The Fishery Survey of India estimated for Gujarat State a potential marine fishery of $0.703 \times 10^6$ tonnes/year constituted of 64.7% demersal and 35.3% pelagic fish. Recent estimates gave an additional potential yield of 111,400 tonnes demersal and 51,500 tonnes of other resources.

The coastal waters of Gujarat make a significant contribution to the fisheries. Of the total $0.44 \times 10^6$ tonnes catch along Gujarat coast, 67% was in the inshore waters (<50 m depth), 26% in the 50–100 m depth range and 6.5% in the 100–200 m depth zone. Bottom trawling, purse seining, mid-water trawling, pelagic trawling and tuna long-lining are employed. In the productive grounds of Kutch, Porbundar, Cambay and Dwaraka regions, Ghol, sciaenids, perches, catfish, prawns and elasmobranches are dominant. Other important fish include Bombay duck, ribbon fish, seer fish, perches, polynemids, clupeids, sharks, yellow-fin tuna, marlins, swordfish, sailfish, lobsters, squid and cuttlefish.

Coastal waters also serve as a migratory route to Hilsa (*Tennalosa ilisha*), a clupeid that migrates to Narmada River for breeding; the juveniles and some adults return to sea. Presence of a successful population of Bombay duck (*Hapadon nehereus*) in the Gulf of Cambay suggests that it provides the key factors, that is, the ideal nursery grounds, right type of forage in abundance and optimal tidal amplitude (Fernandez and Devaraj 1996).

**Acknowledgments** I am most grateful to Dr. James E. Stewart, Emeritus Scientist, Bedford Institute of Oceanography, Dartmouth, Nova Scotia, Canada and Dr. Ravi V. Durvasula, University of New Mexico, Albuquerque, NM, USA for constructive criticism of the manuscript. I thank Professor Viorel Badescu for his infinite patience. My thanks are due to Dr. P.S. Rao, Senior Scientist, National Institute of Oceanography, Goa, India for providing several publications on the Gulf of Khambhat and Mr. Francis Kelly, Technographics Bedford Institute of Oceanography for the illustrations, and Bala T. Durvasula for formatting the text.

## References

Amos CL (1979) Sedimentation resulting from Fundy tidal power. In: Severn RT, Dineley DL, Hawker LE (eds) Tidal power and estuary management. Symposium of the Colston Research Society, University of Bristol, Scientechnica, Bristol, 1978, pp 173–179

Badano EI, Marquet PA (2008) Ecosystem engineering affects ecosystem functioning in high-Andean landscape. Oecologia 155: 821-829.

Bapardekar MV, de Sousa SN, Zingde MD (2004) Biogeochemical budgets for Tapi estuary. An assessment of nutrient, sediment and carbon fluxes to the coastal zone in south Asia and their relationship to human activities, Final report for APN Project—Ref. Nos. 2001–2020 and 2002–2005 (April 2001–February 2004) submitted to APN, pp 153–161

Berry CM, Mehrotra NC, Nautiyal DD (1996) Hydrocarbon potential in the Gulf of Cambay—a palynological perspective. Contributions to XV Indian Colloquium on Micropaleontology and Stratigraphy, pp 707–710

Bershtejn LB (1993) Around the Rance a Tugursk tidal power plants. Gidrotekhnicheskoe Stroitel'stvo 1:50–53

Caesar HB, Pet-Soede L (2003) The economics of worldwide coral reef degradation (WWF and ICRAN) Arnhem, The Netherlands

Chandramohan P, Jena BK, Sanil Kumar V (2001) Littoral drift sources and sinks along the Indian Coast. Curr Sci 81:292–297

Chandrasekhar DV, Mishra DC, Poornachnadra Rao GVS, Mallikharjuna Rao J (2002) Gravity and magnetic signatures of volcanic plugs related to Deccan volcanism in Saurashtra, India and their physical and geochemical properties. Earth Planet Sci Lett 201:277-292

Charatkar SL, Mitra D, Biradar RS, Pikle M (2005) Study of saltpan increment in Gulf of Cambay using GIS. Annual ESRI user International Conference at San Diego July 25–29, 2005

Christie P, Pollnac RB, Fkuharty DL, Hixon MA, Lowry GK, Mahon R, Pietri D, Tissot BN, White AT, Armada N, Eisma-Osorio R (2009) Tropical marine EBM feasibility: a synthesis of case studies and comparative analyses. Coast Manage 37:374–385

Clark NA (2006) Tidal barrages and birds. Ibis 148:152–157

Cloern JE (2001) Our evolving conceptual model of the coastal eutrophication problem. Mar Ecol Prog Ser 210:223–253

Cloern JE (2007) Habitat connectivity and ecosystem productivity: Implications from a simple model. Am Nat 169:E21–E33

Drinkwater KF, Frank KT (1994) Effects of river regulation and diversion on marine fish and invertebrates. Aquat Conserv Freshw Mar Ecosyst 4:135–151

Fearside PM (1995) Hydroelectric dams in the Brazilian Amazon as sources of 'Greenhouse' gases. Environ Conserv 22:7–29

Fernandez I, Devaraj M (1996) Dynamics of the Bombay duck (*Harpodon nehereus*) stock along the northwest coast of India. Indian J Fish 43:1–11

Ferns PN, Reed JP (2009) Effects of Cardiff Bay tidal barrage on the abundance, ecology and behavior of shelducks *Tadorna tadorna*. Aquat Conserv Mar Freshw Ecosyst 19:466–473

Filipsson HL, Björk G, Harlan R, McQuoid MR, Nordberg K (2005) A major change in the phytoplankton of a Swedish sill fjord—a consequence of engineering work? Estuar Coast Shelf Sci 63:551–560

GESAP (Gujarat State Environmental Action Programme) (2000) Coastal and marine Environment Phase-1. March 2000, Gujarat Ecology Commission, Phase 1 report

Ghosh A, Saha S, Saraswati PK, Benerjee S, Burley S (2009) Intertidal foraminifera in the macrotidal estuaries of the Gulf of Cambay: implications for interpreting sea-level change in paleoestuaries. Mar Pet Geol 26:1592–1599

Gordon DC Jr (1994) Intertidal ecology and potential power impacts, Bay of Fundy, Canada. Biol J Linn Soc 51:17–23

Gordon DC Jr, Dadswell MJ (1984) Update on the marine environmental consequences of tidal power development in the upper reaches of the Bay of Fundy. Can Tech Rep Fish Aquat Sci 1256(vii):686

Government of Gujarat (1999) Narmada, Water Resources and Water Supply Department KALPASAR Gulf of Khambhat Development Project, specific studies, summary and reappraisal report

Hammons TJ (1993) Tidal power. Proc IEEE 81:419–433

Haskoning (1998) Khambhat Gulf development project (Kalpasar) pre-feasibility report Executive summary, Prepared by HASKONING in association with DCL, Fugro Engineering, Rijkswaterstatt, Euroconsult, Grondmechanica Delft, TU Delft, Delory, 22 pp

Hewitt JE, Thrush SE, Cummings VJ (2001) Assessing environmental impacts: effects of spatial and temporal variability at likely impact scales. Ecol Appl 11:1502–1516

Ibanez C, Prat N (2003) The environmental; impact of the Spanish National Hydrological Plan on the Lower Ebro River and Delta. Int J Water Resour Dev 19:485–500

Ibanez C, Prat N, Canicio A (1996) Changes in the hydrology and sediment transfer produced by large dams on the lower Ebro River and its estuary. Red Riv Res Man 12:51–62

ICMAM (2002) Critical habitat information for Gulf of Khambhat, Gujarat, Government of India, Department of Ocean Development, Integrated Coastal and Marine Area Management Project Directorate, Chennai, May 2002, 34 pp

Institution of Civil Engineers (ICE) (1981) Dams and Earthquake. In: Proceedings of a Conference at the Institution of Civil Engineers, Thomas Telford Ltd. London, England

Kadam AN, Sukumaran S, Mandalia AV, Naidu VS (2008) Monitoring of Coastal Waters off Luhara (Gulf of Khambhat) for GACL Sponsored by Gujarat Alkali and Chemicals Limited, Dahej NIO/SP-32/2008 (SSP1963)

Kameswara Rao K (1971) On some foraminifera from the northeastern part of the Arabian Sea. Proc Plant Sci 73:155–178

Kane AS (1995) 'Kalpasar', the only permanent solution of the perennial water problem of Gujarat, Paper presented at the seminar on The Water Problems of Gujarat: approaches to Solution, organized by the Centre for Research and Training in Rural Development (CERITA) and Gujarat Science Academy (GSA), 7 January 1995, PRL, Ahmedabad

Karanth RV, Thakkar PS, Gadhvi MS, Sant DA, Negi JG (2005) Vedic Saraswati and the Arabian Sea (Dr. Ambedkar College, Pune 411 006). In: 7th National Conference on Marine Archaeology of Indian Ocean Countries. http://www.nio.org

Karunasagar I, Segar K, Karunasagar I (1989) Potentially toxic dinoflagellates in shellfish harvesting areas along the coast of Karnataka State (India). In: Okaichi T, Anderson DM, Nemoto T (eds) Red tide: Biology, Environmental Science, and Toxicology, Elsevier, New York, pp 65–68

Khanna K, Jauhari RK, Bhattacharya SS (2000) Risk of emergence of epidemics vis-à-vis costal development of Cambay. International Workshop on Subtle Issues in Coastal Management. Sudarshana R, Mitra D, Mishra AK, Roy PS, Rao DP (eds) Indian Institute of Remote Sensing NRSA, Dehradun, India, pp 143–147.

Kunte P (2008) Sediment concentration and bed form structure of Gulf of Cambay from remote sensing. Int J Remote Sens 29:2169–2182

Mukhopadhyay M (2002) Current seismicity in Northern Maharashtra and Southern Gujarat: Implications of plume tectonics. J Geol Soc India 60:629–637

Mukhopadhyay M, Arora SK (1999) Current seismicity and tectonics near the Gulf of Cambay: evidence for the Khambhat plume induced activity. J Geol Soc India 54:23–36

Nair RR, Hashimi NH, Purnachandra Rao V (1982) Distribution and dispersal of clay minerals on the western continental shelf of India. Marine Geology 50:M1-M9

Nayak RK, Shetye SR (2003) Tides in the Gulf of Khambhat. West coast of India. Estuar Coast Shelf Sci 57:249–254

Nigam R (1984) Living benthonic foraminifera in a tidal environment: Gulf of Khambhat (India). Mar Geol 58:415–425

Orfanidis S, Reizopoulou S, Basset A (2008) Transitional states in transitional waters. Aquat Conserv Mar Freshw Ecosyst 18:S1–S3

Oster S (2007) In China, new risks emerge at giant Three Gorges Dam. Wall St J A1

Oza GM (1979) The danger of dumping wastes into sea and river waters of India. Environ Conserv 2:59–60

Parasharya BM, Borad CK, Rank DN (2004) A checklist of the birds of Gujarat. Bird Conservation Society, Gujarat

Quraishee GS (1988) Variations in the Indus River discharges and their hazards. In: El-Sabh MI, Murty TS (eds) Natural and man-made Hazards, Reidel, Holland, pp 369–375; http://www.expressindia.com/latest-news.exper–forscarppingtidalpowergeneration/3.2009.

Raghunathan CR, Sen Gupta U, Wangikar Lakhmapurkar A (2004) A record of live corals along the Saurashtra Coast of Gujarat, Arabian Sea. Curr Sci 87:1131–1138

Rajawat AS, Gupta M, Pradhan Y, Thomaskutty AV, Nayak S (2005) Coastal processes along the Indian coast- case studies based on synergistic use of IRS-P4 OCM and IRS-1C/ D data. Indian J Mar Sci 34:459–472

Rao AD (2009) Observed low-salinity plume off Gulf of Khambhat, India, during post-monsoon period. Geophys Res Lett 36:L03605

Rao VP, Rao BR (1995) Provenance and distribution of clay minerals in the sediments of the western continental shelf of India. Cont Shelf Res 15:1757–1771

Rao TSS, Rao KK, Wagh AB, Desai BN (1971) Studies on the hydrological features of the Gulf of Cambay. Part 1. preliminary account. Mahasagar 4:111–117

Reddy MS, Basha S, Joshi HV, Ramachandraiah G (2005a) Seasonal distribution of contamination levels of total PHCs, PAHs and heavy metals in coastal waters of the Alang-Sosiya ship scrapping yard, Gulf of Cambay, India. Chemosphere 61:1587–1593

Reddy MS, Basha S, Kumar VGC, Joshi HV, Ramachandraiah G (2005b) Distribution, enrichment and accumulation of heavy metals in coastal sediments of Alang-Sosiya ship scrapping yard, India. Mar Poll Bull 48:11–12

Reddy MS, Mehta B, Dave S, Joshi M, Karthikeyan L, Sarma VKS, Basha S, Ramachandraiah G, Bhatt P (2007) Bioaccumulation of heavy metals in some commercial fishes and crabs of the Gulf of Cambay. Curr Sci 92:1489–1491

Retiere C (2008) Tidal power and the aquatic environment of La Ranee. Biol J Linn Soc 51:25–36

Sanil Kumar V, Pathak KC, Pednekar P, Raju NSN, Gowthaman R (2006) Coastal processes along the Indian Coastline. Curr Sci 91:530–536

Scott BE (2007) A renewable engineer's essential guide to marine ecology. Oceans Europe 4302218:1–3

Shaikh MG, Nayak S, Shah PN, Jambusaria BB (1989) Coastal landform mapping around the Gulf of Khambhat using Landsat TM data, Photonirvachak. J Indian Soc Remote Sens 17:41–48

Sharma K (1995) Cretaceous-tertiary tectono-magmatism in the NW Indian shield: a fragmenting continent, unpublished thesis. MS University of Baroda

Siddiquie HN, Gopala Rao D, Wagle BG, Vora KH, Gujar AR, Karisiddaiah SM (1981) The continental shelf in the southern Gulf of Khambhat—an evaluation of the sea bed for construction. INCOE 81, VII, pp 35–42

Snedaker S (1984) Mangroves: a summary knowledge with emphasis on Pakistan. In: Haq BU, Milliman JD (eds) Marine geology and oceanography of Arabian Sea and coastal Pakistan. Van Nostrand and Reinhold Co, New York, pp 225–262

Somvanshi VS, Chandrapal GD, Paleri P, Patel MI (1999) Fisheries monitoring, control and surveillance in India, with special reference to Gujarat State, GCP/INT/648/NOR—Field Report C-3 (En) 57, Report of a Regional Workshop on Fisheries Monitoring, Control and Surveillance, Muscat, Sultanate of Oman, 24–28 October 1999

Tewari HC, Surya Prakasa Rao G, Rajendra Prasad B (2009) Uplifted crust in parts of western India. J Geol Soc Ind 73:479-488

Thresiamma J, Shaiju P, Laluraj CM, Balachandran KK, Nair M, George R, Nair KKC, Sahayak S, Prabhakaran MP (2008) Nutrient environment of red tide—infested waters off south-west coast of India. Environ Monit Assess 143:355–361

Underwood AJ (1992) Beyond BACI: the detection of environmental impacts on populations in the real, but variable, world. J Exp Mar Biol Ecol 161:145–178

Unnikrishnan AS, Shetye SR, Michael GS (1999) Tidal propagation in the Gulf of Khabhat, Bombay High, and surrounding area. Proc Ind Acad Sci (Earth Planet Sci) 108:155–177

Upadhyay H (2004) Narmada project: concerns over command area environment. Econ Political wkly 39(19):1879–1882

Varkey MJ, Kesavadas V (1976) Light transmission characteristics of the northern Arabian Sea during December–May. Indian J Mar Sci 5:147–151

Veeraswamy K, Harinarayana T (2006) Electrical signatures due to thermal anomalies along mobile belts reactivated by the train and outburst of mantle plume: evidences from the Indian subcontinent. J Appl Geophys 58:313–320

Volovik SP (2001) Ecological and fishery problems of the Azov Sea basin. Fishery 4:22–23

Volovik SP, Dubinina VG, Semenov AD (1994) Hydrobiology and dynamics of fisheries in the Azov Sea. GFCM Studies and Reviews Rome. FAO 64:1–58

Vörösmarty CJ, Sharma K, Fekete B, Copeland AH, Holden J, Marble J, Lough JA (1997) The storage and aging of continental runoff in large reservoir systems of the world. Ambio 26:210–219

Vörösmarty CJ, Meybeck M, Fekete B, Sharma K, Green P, Syvitski J (2003) Anthropogenic sediment retention: Major global-scale impact from the population of registered impoundments. Glob Planet Change 39:169–190

Wolanski E (2006) Increasing trade and urbanization of the Asia Pacific Coast. In: Wolanski E (ed) The Environment in Asia Pacific Harbours, Springer, The Netherlands, pp 1–13

# The Bering Strait Seawater Deflector (BSSD): Arctic Tundra Preservation Using an Immersed, Scalable and Removable Fiberglass Curtain

**Richard B. Cathcart, Alexander A. Bolonkin and Radu D. Rugescu**

## 1 Introduction

The First International Conference on Iceberg Utilization, held at Iowa State University during 2–5 October 1977 (Husseiny 1978) awakened the world-public's interest in a mega-engineering concept first advanced circa 1949 by John Dove Isaacs (1913–1980), the maverick USA oceanographer (Behrman and Isaacs 1992). More than 60 years ago few took Isaacs' mega-project proposal seriously and, until 1977, little was done by mega-engineers to rigorously evaluate the technical and economic feasibility of towing tabular icebergs, naturally and artificially calved from the periphery of the icy continent of Antarctica, to the warm dry-land regions enduring, or predicted to endure during the twenty-first century, significant consumer demand-driven freshwater shortage. During the twenty-first century, the Earth's two coldest regions, our planet's northern and southern Polar Zones, achieved their commonly accepted cultural–geographic status as the axial and symbolic polar foci of the supposed anthropogenic "Global Climate Change" (Yusoff 2005; Cameron 2005). In the Arctic, at the southern tip of Novaya Zemlya (i.e., the island of Yuzhnyj), which lies at about the same latitude as Alaska's northernmost point (latitude 71°N), the Russians demonstrated their technology by detonating on October 30, 1961 the most powerful aerial nuclear explosion in history—the Tsar Bomb yielding 50 Mt; the Arctic is contaminated by decaying

R. B. Cathcart (✉)
Geographos, Burbank, CA, USA
e-mail: rbcathcart@gmail.com

A. A. Bolonkin
C&R, Brooklyn, NY, USA

R. D. Rugescu
Polytechnic University of Bucharest, Bucharest, Romania

V. Badescu and R. B. Cathcart (eds.), *Macro-engineering Seawater in Unique Environments*, Environmental Science and Engineering,
DOI: 10.1007/978-3-642-14779-1_33, 

radioactive materials deposited by weapons test fallout, industrial nuclear accidents on land and warship mishaps at sea, and from intentional disposal of nuclear waste by marine dumping. However, and beneficially, with the advent of nuclear-powered submarines "…the entire Arctic Ocean…ceased to be remote and is open to study on a year-round basis…" (Molloy 1962). We accept the practical mega-engineering outlook that physical things of our world are never truly static, and that utopia/dystopia is a dynamic terminology and, therefore, the properly professional approach of the mega-engineer is 'if it needs to be done, it can be done'. We only have to secure a real need.

Antarctica's permanent colonization was initiated in 1954 by Australian citizens (Collis 2007) and construction on the USA's permanent McMurdo Base commenced during 1956. Thus, Antarctica has undergone more intensive scientific examination ever since human explorers discovered, in 1840, that the enormous icy mass actually represented the coast of a continent mostly covered by Nature-frozen water. The only major "industrial" activity allowed, so far, by conventional regulations under the Antarctic Treaty, which entered force on 23 June 1961, has permitted the ongoing assembly of the voluminous Cherenkov detector to search for muons from high-energy neutrino interactions, the IceCube mega-project at the geographical South Pole. IceCube is being installed, to a depth of ~2.4 km, within the glacial ice precisely at the permanent South Pole human settlement. Construction started during 2005 and is slated to be finished by 2011, possibly at a final cost of 2010 USD 271 million. Nevertheless, important scientific data on extra-planetary high-energy neutrinos are already being collected by the still incomplete IceCube facility (Abbasi et al. 2010).

Well within the climate change era, the lowest measured air temperature at our planet's surface, −89.2°C, was recorded at Vostok Station, Antarctica, at 0245 Universal Time on 21 July 1983 by brave and hardy humans employing some automatic outdoor meteorological data recorders (Turner 2009). Several fully surveyed sites on the summit of Antarctica's ice-cap plateau are still being evaluated for future astronomical bases because the incumbent free atmosphere has low amounts of obscuring water vapor and the air column's turbulence overhead is concentrated in a thin surface layer.

Intentional, direct creative mega-engineering of general seawater movement in the Southern Ocean, which naturally circulates around the continent via the unique Antarctic Circumpolar Current, was first proposed in Japan circa 1971 by Keiji Higuchi. Higuchi proposed the possibility of changing the Southern Hemisphere seawater circulation with the emplacement of an anthropogenic deflective seawater flow barrier composed of natural tabular icebergs harvested from Antarctica into the constricted Drake Passage situated between South America and the Antarctic Peninsula (Higuchi 1971). Higuichi's intention was to sink these flat-topped icebergs by loading them with rock fragments. [A few years later, Douglas R. MacAyeal proposed a mega-project to pin the West Antarctic Ice Sheet permanently to the southernmost continent by loading the ice sheet with pumped seawater to forestall a global rise of sea-level (MacAyeal 1984).]

Drake Passage is one of the Earth's seven major natural land gaps (Drake Passage, Tasman Gateway, Indonesia, Gibraltar Strait, Denmark-Strait/Iceland-Faeroe Ridge and the Bering Strait) allowing massive movements of seawater with contrasting features into the ocean (Thompson 1995). The ocean covers ~70% of the planet's surface. Drake Passage, with a sill depth of ~400 m on a 700 km-long charted line near 60°W. longitude, connects the submarine Burdwood Bank south of the Falkland Islands, yet midway to Cape Horn (55°58′47″S. latitude), and Elephant Island (61°08′S. latitude by 55°07′W. longitude) near the Antarctic Peninsula, is the only large-scale link between the world-ocean's different major basins. With the Drake Passage blocked by Keiji Higuchi's barrier, created with artificially sunken icebergs, the termination of the effect of the Antarctic Circumpolar Current on the Southern Hemisphere climate will be marked: it might become characterized by warmer surface air temperature and sea ice-free South Polar Zone conditions and, quite possibly, almost zero North Atlantic Ocean Deep Water overturning (Sijp and England 2004). In the case of geophysical theories, it is important to establish whether they are correct or incorrect, or perhaps candidates to be true; as yet, there is no peer-reviewed publication describing the results of a twenty-first century mega-engineered total blockage of Drake Passage. Within the Northern Hemisphere Polar Zone, and second only to Gibraltar Strait in width, the *Bering Strait* gateway between the Pacific Ocean and the Arctic Ocean impacts global climate when it was closed in the past (as during the most recent Ice Age) or when it is open, as it is during the twenty-first century (Gordeev and Kravchishina 2010). Northern Hemisphere Polar Zone economic development, past, present, and future is closely related to the state (temperature and salinity) and movement of seawater as currents within the Arctic Ocean.

The possibility, potentially materialized through twenty-first century mega-engineering, of changing the climate of the Arctic Ocean Basin by the deliberate manipulation of moving seawater currents has long fascinated the most imaginative mega-engineers, especially non-oceanographic mega-engineers. Russian oceanographers have several times since the late-nineteenth century suggested the artificial closing of the ~50 m-deep and 85 km-wide Bering Strait, with or without the pumping of seawater through the imposed barrier in either direction (i.e., north or south), and they have, just as often, announced the infeasibility of such a massive and energy-intensive mega-project (Fig. 1).

Nevertheless, the ratio of cross-section areas for Drake Passage/Bering Strait is ~70:1, obviously a far less onerous semi-exclusionary seawater diversion barrier mega-project construct than that proposed by Keiji Higuchi. Higuchi really offered his proposal to encourage computer modeling of the mega-problem and, obviously, Japan would not then have benefited from a Drake Passage blockage any more notably than any other industrialized Northern Hemisphere ecosystem-nation. Like so much other scientific model thinking, Higuchi's technology got ahead of his ability to clearly use it in a responsible way.

Recorded changes of the Arctic's typifying environmental parameters (as, for example, annual sea-ice coverage and the ground depth of yearly permafrost defrosting) during the late-twentieth and early-twenty-first century has the

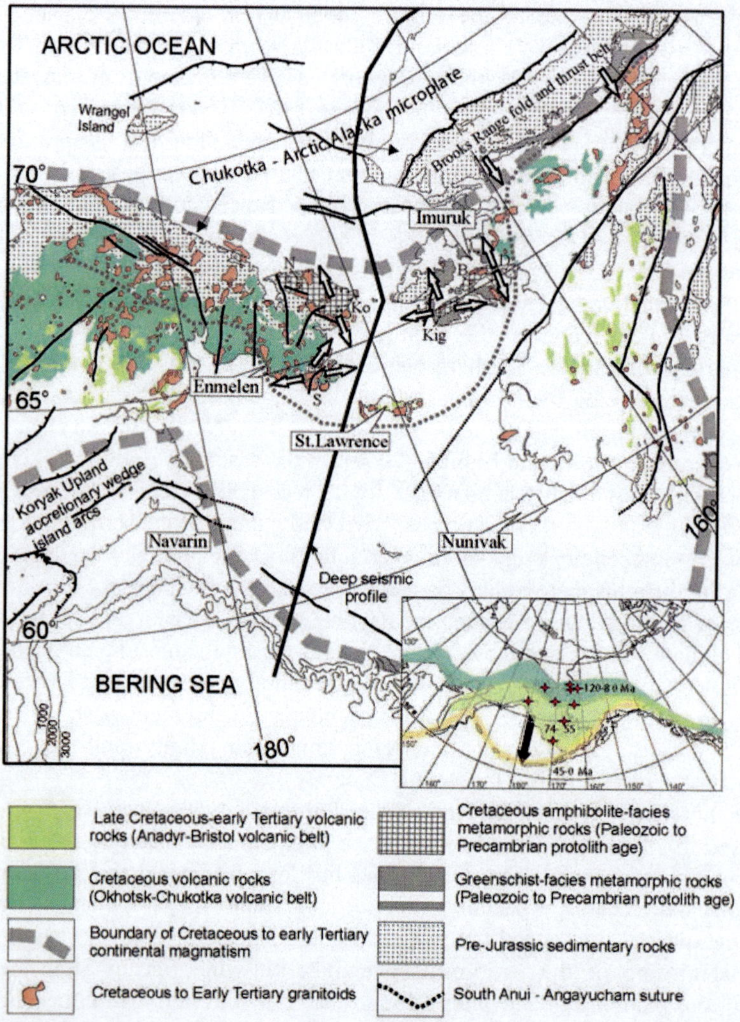


**Fig. 1** Geographical position and geology of the Bering Strait

potential to alter the Earth's climate regimes, in the Northern Hemisphere at least, because Arctic tundra landscapes and the Arctic Ocean do account for ~25% of the Earth's carbon dioxide sink (McGuire 2009; Boe et al. 2010)). The Arctic is also a source for methane ($CH_4$)—estimated at between 32 and 112 Tg $CH_4$/year—to the Earth-atmosphere primarily because of the large organic decomposition within the area of wetlands, still present in the Arctic; the East Siberian Arctic Shelf alone emits 7 Tg yearly—about equal to the amount of methane emitted from the rest of the world-ocean (Heimann 2010). Norilsk (69°20′N. latitude by 88°13′E. longitude, founded circa 1935), the Arctic's largest industrial town (Fig. 2), within the recognized permafrost zone, harbors mines and works for

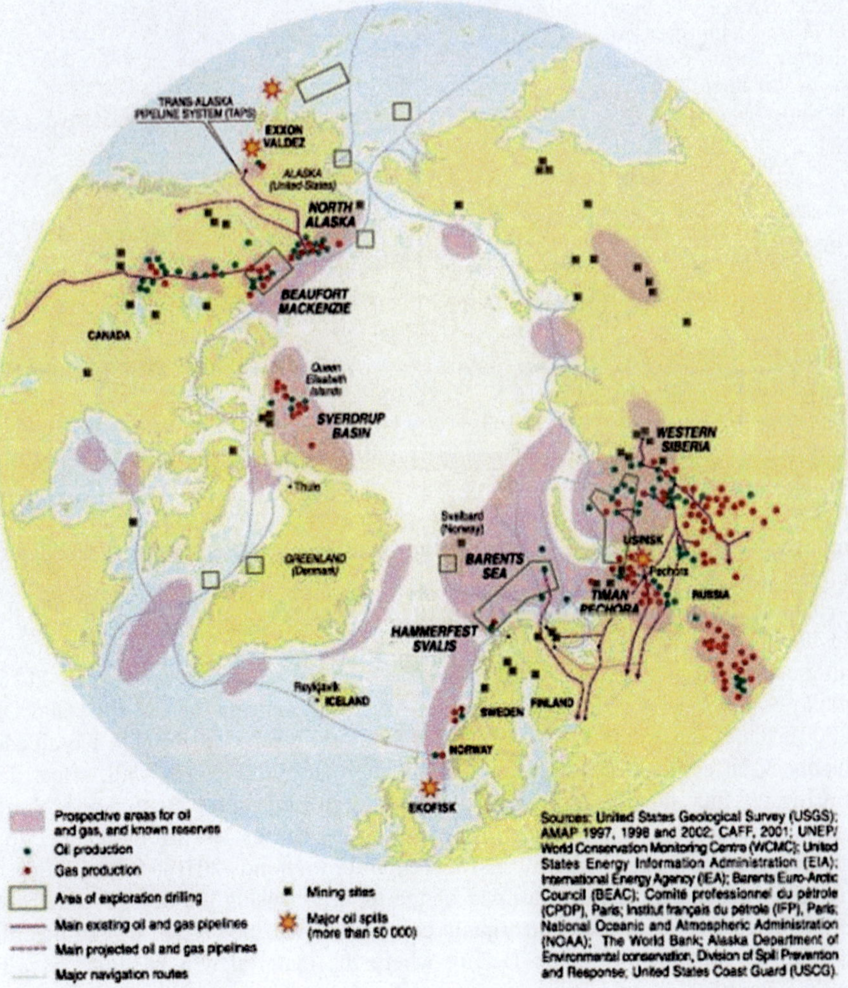


**Fig. 2** A generalized map of the Arctic's current inventory of industrial infrastructures as well as their impressive interconnections

nickel, copper, cobalt, platinum, palladium and coal, which are powered, since 1970, by a natural gas resource obtained and drawn off from beneath the ground's zero temperature annual variation permafrost line.

In addition, methane gas from decomposing organics escapes from the Arctic Ocean's Shelf seabed at its continental margin in Siberia and North America. "If this process becomes widespread along Arctic continental margins, tens of Teragrams of methane per year could be released into the ocean" (Westbrook 2009), and eventually to the Earth-atmosphere. Gaseous methane in the Earth's air is 21 times more effective as a warming agent than gaseous carbon dioxide ($CO_2$); in other words, a regional, possibly demi-global climate change could rapidly occur because of a swift upward shift in an atmospheric "greenhouse" gas concentration.

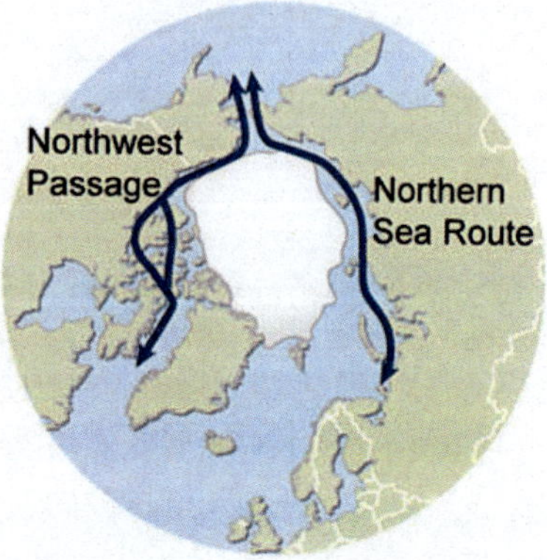


**Fig. 3** Various placements of known and mapped permafrost formations enclosing the sea basin of the Arctic Ocean

Permafrost is widespread in the Arctic (Fig. 3). There is an increasingly well-calibrated geophysical linkage between rapid sea-ice loss and tundra-permafrost instability (Lawrence 2008). Sea-ice extent is the primary parameter for characterizing the state of the Arctic sea-ice cover and Earth-orbiting satellites with microwave instruments have routinely and accurately reported on the yearly extent of sea-ice since 1979.

During summer 2009, the Arctic Ocean's sea-ice cover was >25% less than the 1979–2000 average. Future modeled projections of Arctic Ocean sea-ice cover, although varying, show with some models a seasonal ice-free conditions by 2070, while others predict virtually no change over the same time period (Boe and Qu 2009); the absence of sea-ice will foster Arctic air warming through ice-albedo feedback, affecting weather systems in the high-latitudes and, as well, affect the freshwater flux flowing out of the Arctic Ocean and, subsequently, the North Atlantic Ocean seawater circulation. Less sea-ice causes less sunlight reflection and slower tropospheric photolysis (Screen and Simmonds 2010). Furthermore, a hypothetical anthropogenic warming of the tundra, causing the tundra to defrost, might release high amounts of organic carbon that will be carried by northward-flowing rivers into the Arctic Ocean, where the material will become gaseous carbon dioxide, subsequently entering the Earth-atmosphere.

Among indigenous peoples of the Arctic approximately 50% of the metabolized calories consumed are derived from country foods (seal, walrus, whale, fish, wildfowl, caribou and moose). And, last but not least, warming of the Arctic will promote the northward advance of green vegetation that will very likely amplify that extant atmospheric temperature rise trend by the change of albedo because of the increase in shortwave absorption (Swann, et al. 2010).

Therefore, playing our proper role as mega-engineers, we sought to investigate a speculated possibility of inducing the apparent Artic warming trend to reverse toward an apparent trend of regional cooling by employing cost-effective creative mega-engineering. Specifically, we propose a beneficial mega-project blockading Bering Strait, the Bering Strait Seawater Deflector (BSSD), controllably halting most of the seawater interchange between the Arctic Ocean and the northern Pacific Ocean that occurs presently. The goal of our proposed BSSD is to help cool the Arctic in order to sufficiently maintain more of the frozen tundra existing today that would, otherwise, defrost by directly causing an influential reduction in the Arctic Ocean's advection of heat to the air. In essence, we are proposing an extension of civil engineering's commonplace ground freezing technology, used by structure constructors since circa 1883, using another means—a geographically large-scale indirect surface-level air chilling technique (Braun and Nash 1985). Artificial ground freezing, as it is employed nowadays, was originally developed by F.H. Poetsch in 1883 to control groundwater and increase soil load support capabilities (Anon. 1889); Poetsch's method involved the circulation of a refrigerated coolant through a series of sub-surface pipes to extract heat, thus converting the soil's freshwater to ice, creating a specific watertight material mass. Yet, we also plan to accommodate the global maritime shipping that may, in future, use the Arctic Ocean for safe navigation in new shipping lanes in all but perhaps the coldest months of Arctic wintertime (Howard 2009). As professional mega-engineers, it is our governing credo that the daring technologists' approach to mega-project design is more innovative and carries less risk than a mundane mega-project design approach. Obviously, no such project is without risk however, thus we deserve a short analysis to such risks in the end.

## 2 The Truly Pivotal Arctic

Large-scale climatic intervention in the Arctic Basin has, so far, occurred mostly by human inadvertence and will affect this region during the twenty-first century (Launder and Thompson 2009). Mega-engineering studies of the kind we propose to facilitate, particularly the BSSD, recognize that fixed infrastructure impacts take place in two distinctive phases:

- Phase I, time-specific and generally related to the preparation and assembly phase of the constructed mega-project;
- Phase II, open-ended, starting with the dedication of the completed mega-project for use and the coming "on stream" subsequently of all economic benefits.

Between Phase I and II an ecological phase-change takes place, as a fuzzy but barely perceivable transitional period. Marine primary productivity in the Arctic Ocean has increased measurably owing to the marked macroscopic diminishment of overall sea-ice area (Arrigo 2008; Vermeij and Roopnarine 2008); in other words, the fishery is being anthropogenically boosted positively by unnatural sea-ice reduction, and so the region is thus beginning to attract large-vessel, industrial factory fishing, commercial enterprises. No mega-engineering project is ever strictly focused on pure Science and practicable civil or military engineering—there are always additional, complicating, sometimes external issues such as aesthetics, economics, technological ethics and geopolitics.

The Bering Strait is a Northern Hemisphere regional climate keystone and whether it is open freely to Nature-driven seawater flows or not makes a profound difference in the Arctic Basin's prevailing climate regime. Unquestionably, our proposed BSSD could be an important instigative process-event factor in the near-future human civilization. Hence, we have emphasized the removable nature of the BSSD solution; the BSSD will be no more difficult to remove than a large floating harbor boom often used to suspend anti-submarine and anti-torpedo netting during periods of hostilities. However, there are other significant geophysical factors impacting the twenty-first century Arctic (Borgerson 2008).[this could happen with the BSSD as well].

Five national-ecosystems own Arctic Ocean coastline and have legal territorial claims to its seabed: the USA, Canada, Russia, Norway and Greenland. The area of the Northern Hemisphere north of the Arctic Circle—during 2010, located at 66°33′39″N. latitude, comprising ~6% of Earth's area—contains an estimated "...30% of the world's undiscovered gas and 13% of the world's undiscovered oil resources ...", mostly offshore under less than 500 m of water. Undiscovered natural gas is three times more abundant than the oil in the Arctic region and is largely concentrated in Russia" (Gautier et al. 2009). Potentially, there is greater marine access along Russia's Arctic Ocean coastline for domestic and international commercial shipping; the possibility also arises for special ship transits (North Atlantic Ocean to North Pacific Ocean and return) of hazardous wastes and other sensitive cargos because double-bottomed ships could avoid the geopolitically instable Middle East and the potentially deadly piratical attacks frequently occurring near Indonesia and Somalia (Fig. 4). Particularly, the reader should note the potentially great future importance of the Bering Strait (at 65° to 67°N. latitude and 167° to 170°W. longitude) for all significant North Pacific Ocean commercial and naval operations.

Continued human settlement of the Arctic Ocean's coast might easily promote a new form of open-ocean seawater fertilization—the removal of partially treated seaport-generated sewage from the urbanized parts of the seacoast via "solid-water disposal" (artificial bagged icebergs composed of discarded sewage, or floating "sewagebergs"). Viable Arctic Ocean shipping lanes accessed by non-nuclear powered surface vessels—that is, ships moved by fossil-fuel combustion engines—will increase the air's photo-oxidants as well as other pollutants in the Arctic (Granier 2006). Opening of the Arctic Ocean to commercial and naval traffic "...would enable a sensible cut in the transportation cost of local cargos. The Suez and the Panama canals will also lose some of their sealane chokepoint role. The discovery and exploitation of mineral resources will become much easier and possibly lead to a new pattern of raw material flows. The area suitable for forestry and some agriculture will expand northwards whereas biological marine resources face varying destinies.... Vast areas beyond territorial waters will gain economic value, challenging the relevance of international agreements" (Laulajainen 2009). Figure 5 highlights the region's infrastructure as it is publicly known. Since the Arctic mainly consists of ocean and ice and multiple ecosystem-states, it may never be possible for the politically disunited region to have a single telecommunications address, as Antarctica does [11 + 672 + Area Code + Telephone/FAX #]. Every

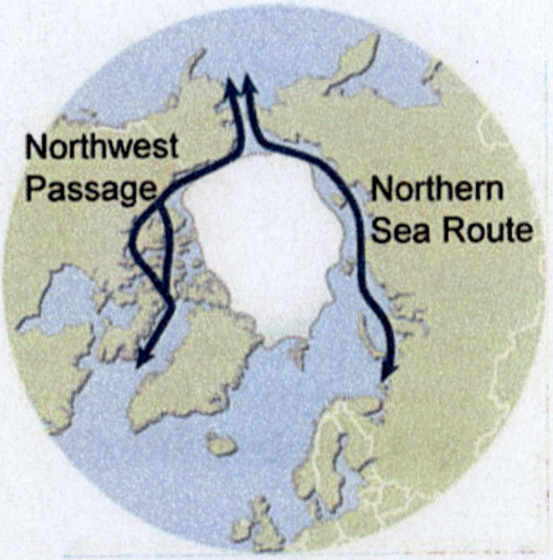


**Fig. 4** Both the northwest passage and the northern sea route have been ship-tested during the twentieth century as safe shipping lanes and powered vessel sailing conditions are likely to improve markedly during the twenty-first century. Imagine the Arctic Ocean during summer without any sea-ice. A direct shipping lane passing over the North Pole would be shorter than either of the routes mapped above. Fast container ships move at 20–25 knots, while bulk dry carriers and petroleum tankers move more slowly, at 13–17 knots on the high seas

continent is linked presently into the www by fiber-optic submarine cables except Antarctica.

Today, telecommunication networks connecting the people of all country-eco-systems consist of two basic systems: (1) wired networks on land and submarine cables and (2) wireless networks (radio stations on land and Earth-orbiting tele-communications satellites). Fiber-optic cables are continuously penetrating the market (Malecki and Wei 2009). An undersea fiber-optic cable to be installed between Tokyo (Japan) and London (UK) through the Arctic Ocean's Bering Strait would cut the transmission delay to 88 ms from its current 140 ms; such a low latency [transmission time] can be achieved via the Kodiak-Kenai Cable Co.'s 2010-proposed USD 1.2 billion "ArctiLink" mega-project. "ArctiLink" would be a twenty-first century mega-engineering project that echoes a nineteenth century mega-project then intended to complete the first global telegraphic circuit from North America, across the Bering Strait to Siberia, and thence to London (UK) surveyed during 1856–1857 by Perry McDonough Collins (1813–1900) (Richardson 1999). The first section of a submarine telegraph cable, laid on the continental shelf between Nome and St. Michael, was installed by 1900. All Alaskan submarine telegraph cables suffered chafing by ordinary sea-bottom currents as well as infre-quent turbidity currents which affected their long-term operation and period of commercial and military usefulness (Heezen and Johnson 1969). "ArcticLink", possibly completed by 2013, with a data stream capacity of 6.4 terabit per second, will have branches that extend from the main cable to service Nome, Kotzebue and Prudhoe Bay, Alaska.

The submarine BSSD would entail installation of hundreds of depending cables which would hold securely in place the impervious-to-seawater glass-fiber panel barrier which then causes a diversion of natural seawater flows into new geographical

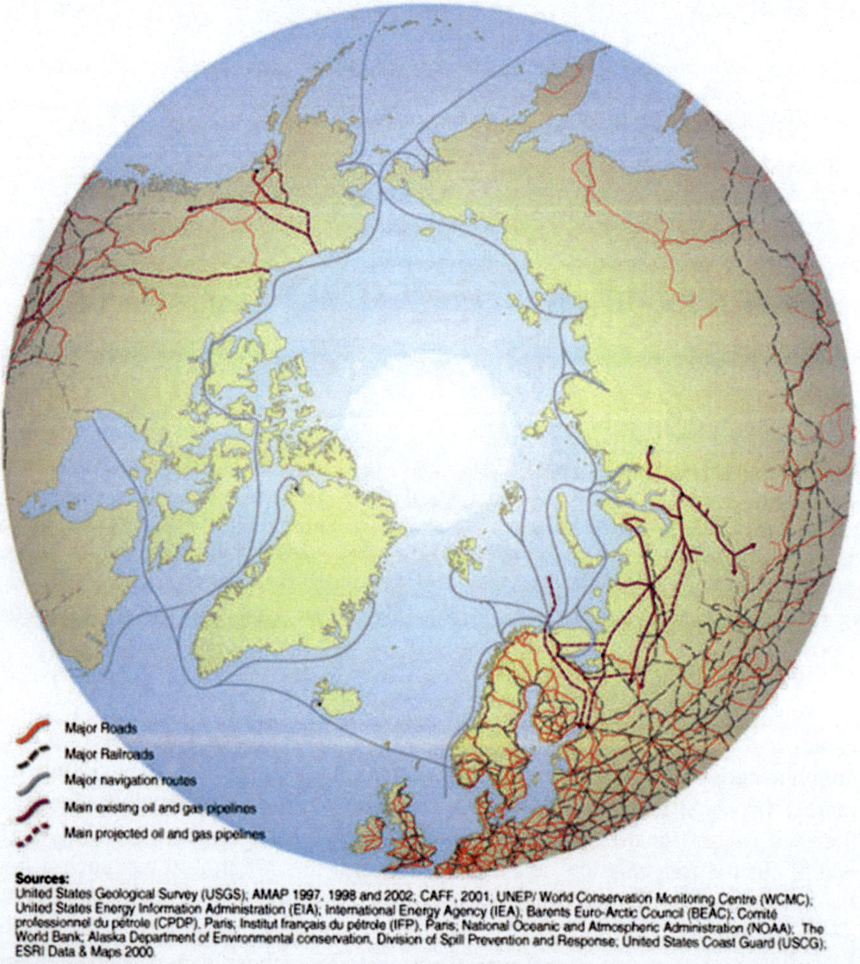


**Fig. 5** Mankind is moving northwards when geophysical conditions allow, as is clearly shown by the mapped physical connections of humanity's infrastructure in this increasingly important geopolitical world region

patterns south of the anthropogenic Bering Strait obstacle. The Deflector's hung textile skirt-like barrier ought to be installed piecemeal—that is, in convenient, readily handled, factory-made fiber-glass/cement panel sections brought to the worksite from storage yards or fabrication factories located on or very near land.

Additionally, glassy fiber-optic telecommunication cables and a flexible BSSD-suspending textile air-filled pipeline must be included in the BSSD. A light-weight stagnant air-holding pipeline at the top of the BSSD would work as a simple floatation mechanism, somewhat like a standard gas-filled floatation collar, for the sea-bottom-anchored strong glass-fiber barrier formed of many panels, the actual seawater exclusion barrier. These panels could be cheaply manufactured just as

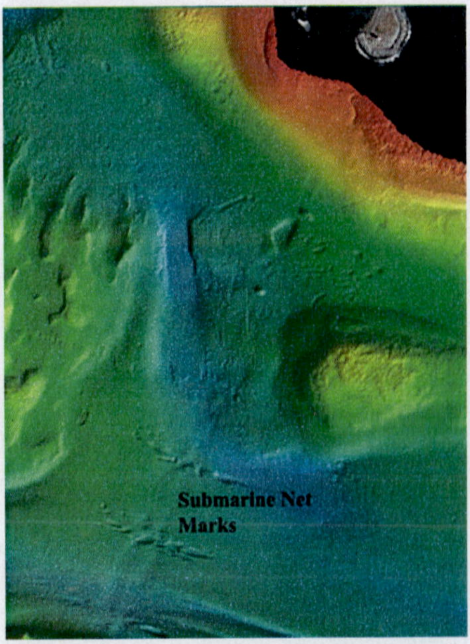


**Fig. 6** This narrow-beam sonar image clearly shows the trace remaining in the muddy/sandy seafloor since the dismantlement by Canada of the unneeded World War II Halifax Harbor-protection anti-submarine net. Knowledge of the magnitude and location of future seafloor changes is essential if we are not to put offshore infrastructures at risk or make them too costly because of over-design

concrete tilt-up warehouse walls are created virtually on-site. Too, 3-D fiber-glass textile has three directions in yarn architecture and/or textile architecture, which can be made in a one-step manufacturing process. Possibly infused with cement, such non-biodegradable panels then become cement based composite structures; the fiber-glass serving as a physical reinforcement for the cement (Roye and Gries 2007). By 2010, retailed glass-fiber cost only ~1/7 that of Kevlar; factory-direct wholesale bulk purchases would likely reduce fiber-glass cloth's cost even more. Since laying the economic groundwork for the BSSD will probably be more difficult than assembling the infrastructure itself, the necessity of including an international fiber-optic telecommunication cable, as well as the far-future potential for High Voltage Direct Current (HVDC) transfers of tidally-produced electricity across the Bering Strait (Bernshtein 1994) will increase its mega-project planner's leverage in dealing with political and economic local consumer demands. Wartime barriers, such as the lengthy anti-submarine nets stretched across Italy's Gulf of Taranto during World War I or the Gulf of Finland, or in Halifax Harbor in Canada, during World War II, indicate instructively the workload for installing our BSSD; these massive installations affected the seabed that remains visible even after the postwar complete removal of these specific defensive steel nets (Fig. 6).

## 3 Planned Bering Strait Fixed Link Crossings and the BSSD

Like a similar transition zone in the eastern North Atlantic Ocean, the Bering Sea is undergoing a northward bio-geographical shift in response to changing surface air temperatures and atmospheric forcing. Due to environmental uncertainties, all proposals for innovative fixed infrastructural links in the Bering Strait are controversial. We deal head-on with the controversial BSSD, by describing the mega-engineering options for technological, societal, and policy actions. Though on a smaller scale, that part of the Arctic Ocean immediately north of the Bering Strait is similar to the Barents Sea, which releases more energy to the air above it than any other part of the Arctic Ocean (Smedsrud et al. 2010). The warm seawater flow estimated at $\sim 2\times10^6$ m$^3$/s which passes over the Barents Sea's sill into the North Atlantic Ocean is like the Bering Strait's annual mean northward throughflow of $\sim 0.8 \times 10^6$ m$^3$/s. Oceanic heat flux, 5–6 $\times$ $10^{20}$ J/year, through the Bering Strait during 2007 was "…twice the 2001 heat flux, comparable to the annual shortwave radiative flux into the Chukchi Sea, and enough to melt 1/3rd of the 2007 seasonal Arctic sea-ice loss. It is suggested the Bering Strait inflow influences sea-ice by providing a trigger for the onset of solar-driven melt, a conduit for oceanic heat into the Arctic, and (due to long transit times) a subsurface heat source within the Arctic winter" (Woodgate et al. 2010). Perfect exclusion of the incoming Pacific Ocean heat flux should increase the chilling of the seawater resident in the Chukchi Sea; such refrigeration-effect may consequently induce a slowdown of methane release from the East Siberian Arctic Shelf. An artificially barred Bering Strait, which terminates all Pacific Ocean seawater inflow to the Arctic Ocean, should cause saltier North Atlantic Ocean surface seawater to flow increasingly into the Arctic Sea (Hu et al. 2010). Unfortunately, it is still unclear if this saltier seawater influx will significantly affect the Arctic Ocean anthropogenically, negatively or positively (Williams et al. 2010) and this item remains to be clarified by oceanographers. Certainly, the nearly 100% exclusion of the Bering Sea current normally flowing northwards will cause that deflected seawater to move laterally, with yet-to-be-mapped consequences to seawater circulation in the Bering Sea and finally the North Pacific Ocean. And, heat normally reaching the Arctic Ocean from the Bering Sea would also be retained by the Bering Sea. The possible absence of any Pacific Ocean surface seawater flow might, subsequently, induce an anthropogenically-forced stronger Atlantic Ocean overturning, thus potentially counteracting Keiji Higuchi's Drake Passage Barrier, were that ever actually emplaced.

## 4 Computations and "Blueprint" for the BSSD

A sketch of the BSSD is shown in Fig. 7. Our current deflector, 2, is factory-fabricated as thin 3-D glass-film architecture perhaps—if necessary—infused with hydrophobic cement. Fiber-glass was invented circa 1938 by Russell Games Slayter (1896–1964). Glass fiber cloth has a greater specific tensile strength than steel wire of the same diameter and a low elongation under load, generally 3% or less. Being inert, it is durable, not requiring replacement for $\sim$30 years and, unlike Kevlar, is unaffected by sunlight or bacteria. It is even tough enough to be used

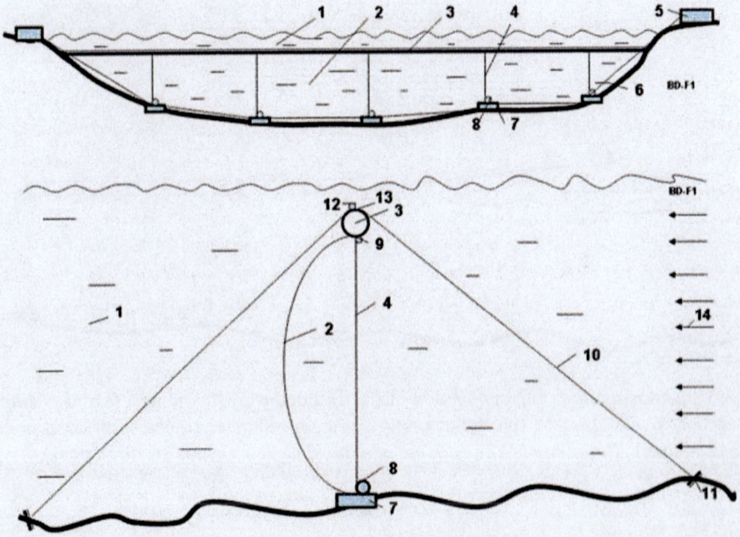


**Fig. 7** Sketch of Bering Strait Seawater Deflector (BSSD). *Top*: Long cross-section; *bottom*: radial cross-section. *Notation*: *1* Bering stream, *2* film (fabric) deflector, *3* sectional air tube supports the top part of deflector, *4* control cable, *5* control station, *6* electric power and control cables, *7* load, *8* electric engine, *10* brace, *11* anchor, *12* tube of compression air, *13* valve, *14* seawater flow

commonly in sheet piles immersed in seawater (Gellert and Turley 1999). Top part is supported by an air tube, 3, made also from the glass cloth and the control cable 4 connected to a load 7 through an electric engine, 8. The electric engine allows the BSSD's operators to change the length of cable, 4, and, thus, the height of the seawater current deflector from zero to the prevailing sea surface. Such movement is necessary to avoid the impact of floating jammed sea-ice with deep keels, deep-draft ships and to control the Arctic Basin's future climate regime.

The air tube, 3, is separated by sections and every section connected to tube, 12, of the compressed air though the valve, 13. This system allows maintenance of a certain air pressure in sections of tube, 3. This economic mega-project design allows working the system should any parts of the sections become damaged, requiring repair or replacement.

Construction of our seawater current deflector should be relatively easily and rapidly performed by skilled marine constructors familiar with the region's seasonal weather extremes (Fig. 8). The factory-fresh deflector ought to be put into carefully managed folds and then should be reeled onto a big movable spool. The spool is then shifted from the factory directly onboard the docked seaworthy installation ship. During the period of installation, the ship propelled forward at slow speed, although making steady headway, and it gradually un-reels the spooled deflector material, and the trained crew then connects to it the loads, cables and so on. This could best be done during the Arctic summertime, in our estimation.

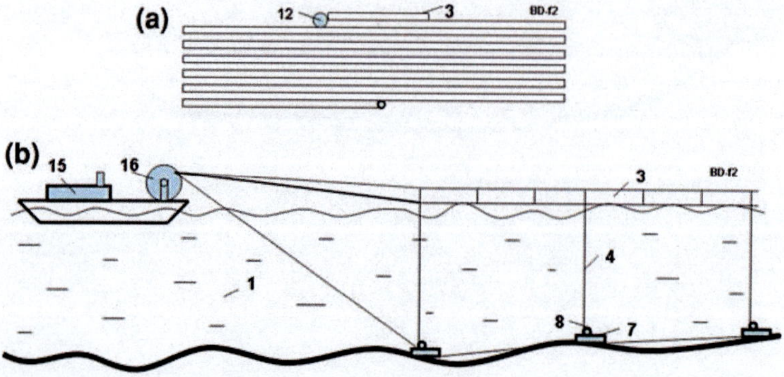


**Fig. 8** BSSD construction. **a** Cross-section of the film deflector in compact form for a shipboard storage spool, **b** Building of the deflector by sea-going ship equipped with a securely fixed rotating film-spool. *Notation*: *1–12* are same as with Fig. 7; *15* ship; *16* film spool

For repair the air tube, 3, lifts to surface. The electric engine, 8, can be also lifted to ship by cable, 4, and change the new engine and new load, 7.

The offered design and building method of BSSD have a lot of marine technical innovations. For example: the guided (controlled) height, compact form for spool, building by unreeling, etc. (Bolonkin and Cathcart 2009). That will surely decrease the monetary and material costs of the BSSD's construction by, at least, tens of times and allows, somewhat, the control of a region's weather regime.

Our computations and estimates for the BSSD follow herewith.

1. Dynamic seawater pressure on deflector is

$$p = \frac{\rho V^2}{2}, \tag{1}$$

where $p$ is the pressure, N/m$^2$; $\rho = 1{,}000$ kg/m$^3$ is the seawater density; $V$ is seawater current speed, m/s. For an average $V = 0.2$ m/s we have $p = 20$ N/m$^2$.

2. Vertical force from the dynamic pressure in 1 m of a film-like cloth BSSD having the height and distance $H = 50$ m between support cables 4 and the angle between an edge and plate of deflector $\alpha = 30^{\circ}$ degrees is

$$P = \frac{pH}{2 \sin \alpha} = 1000 \text{ N/m}, \tag{2}$$

where $P$ is the force, in N/m.

3. Need thickness $\delta$ of film-like panels of glassy fabric for the safety stress $\sigma = 10$ kg/mm$^2$ $= 10^8$ N/m$^2$

$$\delta = \frac{P}{\sigma} = 0.01 \text{ mm}. \tag{3}$$

We take $\delta = 0.05$ mm.

4. Let us take the fabric of reflector from glass-fiber having the density $\rho = 2{,}450\ \text{kg/m}^3$, maximal tensile stress of fiber-glass cloth $\sigma = 3{,}500$ MPa = $3.5 \times 10^9\ \text{N/m}^2$ and the present-day cost of a fiber $c = 0.7$ USD/kg. We take the cost of a glass-fiber fabric $c = 1$ USD/kg. Then the deflector has

$$v = \delta \times 1\ \text{m}^2 = 5 \times 10^{-5}\ \text{m}^3, \quad m = \rho v = 0.1225\ \text{kg/m}^2 \tag{4}$$

where $v$ is volume of 1 $\text{m}^2$; $m$ is mass of 1 $\text{m}^2$ of glass fabric cloth.

5. The area $S$, volume $V_d$ and mass $M_d$ of BSSD are

$$\begin{aligned} &S = HL = 50 \times 85000 = 4.25 \times 10^6\ \text{m}^2, \quad V_d = vS = 212.5\ \text{m}^3, \\ &M_d = mS - 521\ \text{tons} \end{aligned} \tag{5}$$

where $H = 50$ m is the average height of the deflector, $L = 85{,}000$ m is length of deflector.

6. The cost of our film-like fiber-glass deflector is $C = cM_d$ = USD 0.521 M. Let us take the full cost of reflector USD (20 ÷ 100) M.
7. The top support tube, upon which the cables depend, has a diameter of about 0.5 m and a thickness taken as 0.5 mm.

Our estimation of the heat flow relevant to artificial Arctic chilling follows herewith. Let us to estimate the annual heat flow which the Bering Strait stream passes from the Bering Sea to Arctic Ocean through the Chukchi Sea. The seawater speed into Bering Strait streams at $V = 0.04 \div 0.15$ m/s. For average $V = 0.1$ m/s, $S_b = 85{,}000 \times 50 = 4.25 \times 10^6\ \text{m}^2$, $\rho = 10^3\ \text{kg/m}^3$, mass $M$ of the seawater flow and heat flow $Q$ are

$$M = \rho S_b V = 4.25 \times 10^8\ \text{kg/s}, \quad Q = qM\Delta T = 1.78 \times 10^{10}\ \text{kJ/s} \tag{6}$$

where $q = 4.19$ kJ/kg · degree is seawater's heat capacity; $\Delta T \approx 10°\text{C}$ is the accepted seawater temperature. Substituting our values into (6) we receive $M = 4.25 \cdot 10^8$ kg/s, $Q = 1.78 \cdot 10^{10}$ kJ/s. This heat deposition is enough to affect a strong melting of the Arctic Ocean's sea-ice mass

$$M_i = \frac{Qt}{\lambda} = 4.6 \times 10^9\ \text{ton/day} \tag{7}$$

where $t$ is time, s (1 day $= 8.64 \cdot 10^4$ s); $\lambda = 334$ kJ/kg is the melt heat. For our data this heat is enough for melting $4.6 \times 10^9$ tons of freshwater ice in a single day. If the thickness of sea-ice is 2 m, that is area $S = 2.3 \times 10^3\ \text{km}^2$ or 48′48 km.

Let us now estimate the air temperature if the natural seawater flow is artificially discontinued at the Bering Strait (i.e., current mass not passed into the Arctic Ocean). We use the following data: time $t = 1$ day $= 8.64 \times 10^4$ s, the change of temperature of seawater (from 10°C to 5°C) and air (from 0°C to 5°C) is $\Delta T = 5°\text{C}$; air heat capacity $q_a = 1$ kJ/kg K; air density $\rho = 1.225\ \text{kg/m}^3$, height of troposphere $H_t = 4$ km; the heat flow is $Q = 1.78 \times 10^{10}$ kJ/s. The air mass, volume and area heated daily are

$$M_a = \frac{Qt}{q_a \Delta T} = 30.8 \times 10^{13} \text{ kg/day}, \quad V_a = \frac{M_a}{\rho_a} = 2.5 \times 10^5 \text{ km}^3,$$
$$S_a = \frac{V_a}{H_a} = 6.25 \times 10^4 \text{ km}^2 \tag{8}$$

This area forms a square of 250 × 250 km per day. During some months the heating may be of a geographically very large Earth area, including the southern Chukchi Sea, Alaska, and the peninsula of East Kamchatka. Our preliminary indicative calculated results follow herewith. The suggested BSSD may have the following "first observable effects" on the Arctic's climate regimes: (1) stops warming of the southern part of the Arctic Ocean closest to the Bering Strait; (2) BSSD reduces and/or preferably halts, the natural venting of subterranean methane gas from the Siberian Arctic Shelf seafloor which warms the Earth-atmosphere; (3) putatively, our BSSD improves the climate prevailing directly above the South Chukchi Sea, western part of the State of Alaska, Russia's Kamchatka and possibly even in larger areas.

## 5 BSSD Operational Aspects

Circa 1840, almost simultaneous with the determination that Antarctica is a spatially extensive and isolated continent, many naval professionals commonly inaugurated the nautical practice of adding or subtracting a day to a vessel's reckoning upon crossing the 180$^{\text{th}}$ meridian from Greenwich, UK. During 1884, however, an official International Date Line was first delimited and demarcated by international convention (chiefly because of the all-pervasive political and economic influence of the global commercial shipping industry) that divides the shallow Bering Strait and the navigationally tricky seaway between the Diomede Islands—Little Diomede (USA) and Big Diomede (Russia)—situated within the Bering Strait (Palmer 2002). In the image reproduced below, Fig. 9, Big Diomede (approximately 29 km$^2$) is the westernmost island and Little Diomede (approximately 3 km$^2$ located at 65°45′N. latitude by 168°56′W. longitude) is the easternmost island. The islands are about 4 km apart, divided by a channel rimmed with reefs and are at this time mostly uninhabited; a fluctuating population of Inupiat (community of ~150 persons) dwells on Little Diomede (USA). Little Diomede has steep cliffs that rise from the ocean to a plain at an elevation of ~350–363 m (Gualtieri and Brigham-Grette 2001). Generalizing, the islands are uplifted marine terraces composed of hard hornblende granites. The seismicity of the Bering Strait poses no obvious overwhelming threat to any small area-footprint seawater exclusion/diversionary infrastructure such as a BSSD (Mackey et al. 1997). Volcanic ash-falls from major eruptions can smoother the vegetation sporadically. The ambient air temperature in

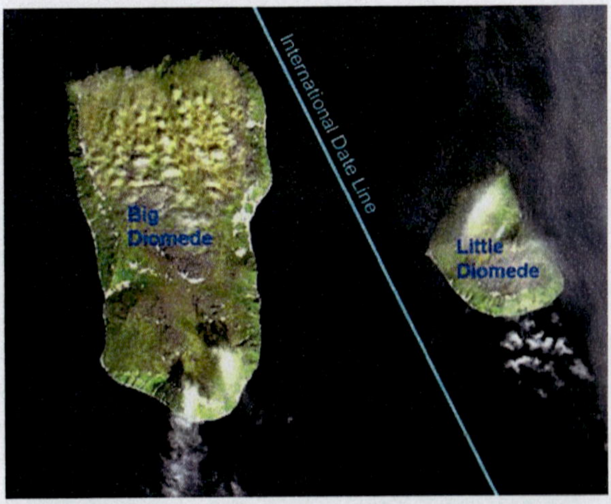


**Fig. 9** An Earth-satellite view of the Diomede Islands located in the central Bering Strait. The distance between the islands is about 4 km and the greatest depth between the islands is 45 m. Seawater current between islands flows northwards and is subject to sea-ice jamming

the Bering Strait at the Diomede Islands during summertime is ~10°C and in wintertime is about −14°C. Sea-ice cover in the Bering Strait reaches a maximum of 100% during late Winter early Spring and is lacking during July–September; the duration of sea-ice coverage in the Bering Sea ranges from a 20% coverage for 80–252 days during an unusually warm Winter to 56% coverage for 170–365 days during a severe Winter. Minimum sea-ice was recorded during September 2007 with open water north of Bering Strait persisting until December 2007. Quick emplacement of the BSSD could change these variances because its presence would keep warm surface seawater in the Bering Sea rather than letting it move northwards into the Arctic Ocean by almost halting any Bering Strait throughflow.

Sea-ice jamming in the channel between the islands often makes it possible to walk between the two separate geopolitical entities, across the International Date Line. (However, this activity will be stopped once the seawater is slightly warmed to curtail sea-ice jamming and to promote use of the Arctic's emerging sea-lanes.) This type of Arctic sea-ice jamming, but on a more massive geographical scale of course, may have happened during the most recent Ice Age to create a piled ice dam totally blocking all seawater flow within the Bering Strait (Sandal and Nof 2008). With the cessation of northward seawater flow into the southern limit of the Arctic's most dense pack ice, the fringing sea-ice would probably advance southwards by a few tens of kilometers and a termination of the Pacific Ocean inflow would likely be noticed most intensely along the northern coastline of Alaska, at least to Point Barrow (71°17′44″N. latitude by 156°45′59″W. longitude), facing the Chukchi Sea, where the summertime sea-ice limit should shift slightly southwards too.

In 1890, William Gilpin, the first territorial governor of Colorado, was the first to propose a rail tunnel under the Bering Strait, as part of a railway network on a global scale, in his book "Cosmopolitan Railway: Compacting and Fusing Together All the World's Continents", at a time when Alaska was a recently acquired territory by the USA from Russia, only since 1867. A little bit later, Joseph Baermann Strauss (1870–1938) first proposed, in his pioneering 1892 civil engineering graduation thesis at the University of Cincinnati—now lost, destroyed or misplaced—a plan to build an inter-continental railroad bridge across the Bering Strait. This was a report that evidences a definitive mega-engineering mentality that later became physical reality in the form of the Golden Gate Bridge at San Francisco, California, USA by 1937. Comparatively, the Bering Strait is about as wide as the Panama Canal is long and twice the width of the English Channel at its narrowest. Circa 1899, the USA railroad magnate Edward Henry Harriman (1848–1909), leader of the Southern Pacific Railroad Company, which had stopped the uncontrolled accidental creation of California's Salton Sea by the rampaging Colorado River, breaching of a canal diversion structure in 1905, and built the famous Lucin Cut-off Trestle across Utah's Great Salt Lake in 1904, envisioned a modified rock-fill causeway, an earthwork emplaced within the Bering Strait (Cole 1990). Subsequently, others have proposed bridges and tunnels linking two great Northern Hemisphere landmasses and fostering passenger and cargo transport with trains and other wheeled land vehicles such as cars and trucks (Kudoyarov 1993; Elliot and Howard 1994; Brezhnev et al. 2005). The tunnel would have a length of twice the Channel Tunnel and connecting approximately Ouelen (Chukchi Peninsula in Russia) and Wales (Alaska, USA). The tunnel should take between 15 and 20 years to build. According to an estimate made in the 1990s, the project would cost 2010 USD 65 billion (48 billion euros), because it would require a total of 6,000 km of tracks to be built. All of these proposed mega-projects seek to recreate a land connection, named Beringia, that naturally existed between Alaska and Siberia during the most recent Ice Age when sea-level worldwide was $\sim$120 m lower than today (Kitchen et al. 2008). Often geoscientists and anthropologists allege that Beringia's temporary existence thousands of years ago may have forced the ensuing, most recent, Ice Age and, subsequently, allowed long-distance human migration from Asia to populate North and South America.

The Bering Strait sometimes has a 3 m-thick sea-ice covering that originates from the Arctic Ocean, but never the huge icebergs like the stormy Southern Ocean surrounding Antarctica, even though seawater circulation in the Bering Strait is dominated by a northward mean flow ranging from 4 to 15 cm/s, varying 11–37% due to tidal influences (Foreman 2006; Hu and Wang 2010). Frequent seawater flow reversals are coincidental with, and contingent upon, strong seasonal weather shifts. These ocean tide reversals seem to invalidate any artificial warming of the Arctic Ocean by the Bering Strait LUNAHEAT permeable barrier of rotating tidal flaps that might be installed within the seaway proposed in 1976 by Harold Henry Anderson (1906–1994). One hundred planned LUNAHEAT tidal flaps, each with a mass of one million metric tons, were supposed to be affixed permanently into place by suspension bridge cables and seabed anchors (Anderson 1975). In other words, LUNAHEAT would form a kind of depending skirt blocking the Bering Strait. Because it would use seawater currents to generate electricity LUNAHEAT is an example of the new field of "hydrodynamics", as opposed to "hydropower". We are unsure how the seawater slowing and sifting effective of a built LUNAHEAT in the archipelago to the south might affect, either favorably or unfavorably, the daily operation of our BSSD. Lynne Cox, on 8 August 1987, became the first human known to swim from the USA's Little Diomede Island to Russia's Big Diomede Island, across a short segment of the Bering Strait (Fig. 9). The presence of sea-ice does reduce the impact of wind stress forcing (wind fetch). Nevertheless, any emplaced Bering Strait bridge piers must be capable of enduring large ice piling loads, possibly as great as 2 MPa (Hansen 2007; Timco and Weeks 2010).

In the northern Atlantic Ocean, thinner sea-ice observably present in Nares Strait no longer appears to function effectively to partly "pin" the polar ice-cap with arched sea-ice jamming (Kwok 2010); the presence, absence or unusual fragility of these obstructing sea-ice arches in straits tends to govern the volume of outflow of sea-ice leaving the Arctic's permanent ice-cap. Bridges are exposed to strong, barely forecasted, cyclical meteorological events, sea-ice pile ups on supporting piers and offering operational aerial exposure risks for trains and other vehicles (intrinsic super-structure icing) using the postulated Northern Hemisphere international transit crossing. H.H. Anderson's LUNAHEAT installation would be affected similarly. [LUNAHEAT might best be employed at the archipelago's seawater passages between the Aleutian Islands (Osafune and Yasuda 2010).] A tunnel, or tunnels, under the Bering Strait could be bored within a fairly predictable, benign granite and hard limestone submarine rock formation. The danger to people in soundly built, stable rock-lined tunnels is fire, especially in the railway rolling stock and other motorized vehicles that carry flammable fuel. However, such sound seafloor geology also bodes well for a BSSD barrier's seafloor anchoring system.

Affecting our proposed BSSD are the potential infrastructure developments that might occur during the twenty-first century in northern Canada and Greenland. When Robert Bourassa (1933–1996) revealed the first phase of the James Bay hydroelectric mega-project in the early-1970s, he called it "the Project of the Century" (Bourassa 1985; Hornig 1999). However, instead, it inspired a Green social movement call to public protest halting such development. Still, it is possible that great geographical changes, caused by both humans and Nature, will alter northern Canada. For example, it is known that, on balance, that the afforestation of northern Canada and Russia could chill the climate of the Northern Hemisphere by change of albedo and net atmospheric carbon drawdown. Too, the ice-cap of Greenland is undergoing obvious diminishment, notably glacial recession at its edge, that is sometimes beneficial, such as the uncovering of new fertile farmland used by native-born Greenlanders (Etter 2006), and sometimes non-beneficial. Such events are similar to events now well documented historically that are directly related to Norse colonization of Greenland circa 790–1070 AD (Patterson et al. 2010). Some mega-engineers have proposed harnessing Greenland's glacial melt-water to generate hydropower exportable to Iceland and Europe via undersea HVDC cables (Partl 1978a, b; Bolonkin and Cathcart 2008). Partl's collective electrical super-grid and cross-border infrastructural mega-engineering proposal becomes even more viable with the anticipated circa 2020 materialization of a 2010 USD 40 billion power grid of HVDC cables under the North Sea intended to serve nine European countries as a multinational grid designed to address the mega-problem of the fluctuating nature of Green-inspired electricity generation plants (tidal, wind, solar). Other engineers, presenting their nascent mega-engineering concept-thoughts on the World Wide Web (http://pruned.blogspot.com/2006/10/what-if-greenland-was-africas-water. html), have already suggested that as a national income-generating renewable resource, Greenland's meltwater should be exported to crowded ecosystem-nations that are willing to pay the true overseas importation costs of drinking water. Reiterated, if any of these mega-projects ever come to fruition, then our BSSD must be operationally adjusted, at least marginally.

According to a few widely published 2007 news media reports, Russia is tentatively seeking to build a 102 km-long Bering Strait Tunnel roughly estimated to cost 2010 USD 65 billion. Meantime, in Alaska during 2010, an ordinary ~804 km-long motor vehicle highway, branching from Elliott Highway near Manley Hot Springs, that would follow the Yukon River and west to Norton Sound, and eventually reaching Nome near the Bering Strait might cost 2010 USD 2.7 billion. Certainly, roads are paths of human endeavor—in fact, the infrastructure upon which almost all other infrastructure depends (Conover 2010). Simultaneously, another mega-project, the Alaska Pipeline Project, meant to transport natural gas from the North Slope to market in Alaska, through Canada and the contiguous USA, could cost as much as 2010 USD 41 billion. In other words, mega-engineered projects in the environmentally harsh Arctic are very costly in time, energy and money. A particularly good example of the mega-engineering difficulties, and the associated high economic costs, of constructing something resembling a Bering Strait dam is the proposed railroad causeway from Russia's mainland to Sakhalin Island (Zaretskii and Korchevskii 2008) where petroleum and natural gas fields are being developed. Its true financial cost is yet to be reckoned. Of course, only one mega-engineer, Piotr Mikhailovich Borisov (1901–1973) in Russia, has actually planned a materially bulky dam spanning the Bering Strait between Russia and Alaska since circa 1959 (Anon. 1961; Borisov 1973); the former USSR awarded P.M. Borisov a Patent, Number 7337, for "A Radical Climate Improvement in the Arctic and Temperate Zones of Our Earth". Our realistic contention, in this chapter, is that due to the Arctic Ocean's thermal inertia if, alone, a near 100% stoppage of warm seawater northwards through the Bering Strait was accomplished shortly, then the Arctic's current climate regime would likely not respond measurably for a decade, or several decades. Nevertheless, our proposed mega-project is designed to extend at least the period before the seasonal disappearance of Arctic sea-ice, as is now expected by experts (Wang and Overland 2009).

## 6 Bering Strait Barriers: Concrete, Rock-Fill or Textile?

P.M. Borisov's Bering Strait Dam mega-project proposal was designed to double the inflow of warm North Atlantic Ocean surface seawater by blocking the inflow of Pacific Ocean surface seawater through the Bering Strait and also by pumping

500 km$^3$/day of surface seawater from the Arctic Ocean directly into the Bering Sea which connects with the northern Pacific Ocean. This mega-engineering concept is fully described in detail by Borisov in *Can Man Change the Climate?* This 175-page book offered a rosy picture of the consequences of his mega-project. Nevertheless, P.M. Borisov's all-too-apparent ingenuity was praised by some, even though geopolitical apprehensions were aroused eventually in eco-system-nations closely neighboring the USSR. Nowadays, his idea is passé and, occasionally, he is even equated with Jules Verne (1828–1905), because the ultimate effect of his imported electricity-using Bering Strait pump-house installation might be to instigate an anthropogenic global climate catastrophe much like the climate change disaster caused fictionally by human imprudence portrayed so vividly in Verne's *San dessus dessous* (1889). It is quite fair to say that, like his predecessor Jules Verne, who was a playwright and novelist in France, the Russian mega-engineer P.M. Borisov engaged in some wishful thinking. Despite that mild condemnation, we still view Borisov as a good role model for modern mega-engineers because he successfully combined the scientist and the engineer. As the USA civil engineer Henry Petroski has cogently noted, Science does not exactly precede Engineering in the really creative process—scientists (the research in R&D) and engineers (the development in R&D) have traditionally co-operated, but when budgets in various R&D public or private-sector organizations are preponderantly front-loaded for undirected basic research, all that may be produced is new knowledge that is remarkably irrelevant and inefficacious for solving the mega-problem of, say, global and regional climate change. To prove his bold assertion, Henry Petroski cites some instructional examples of scientists finding inane mega-project "solutions" for alleged man-made Global Warming that could become a sustainability/survivability crisis for humankind (Petroski 2010). Twenty-first Century "Global Warming" alarmism—commencing circa 1988, when the World Meteorological Organization and the United Nations Environment Program established the Intergovernmental Panel on Climate Change, to the present—may be deemed a popular delusion: modern-day examples of popular alarmism are the "Y2K" scare (in 2000) and "Global Cooling" (1970–1979). It is still unrealistic computer-based models that seem to foster ever-changing "Global Warming" alarmism. A popular delusion dissipates when scientific predictions based on the theory are repeatedly and publicly contradicted by real-world event-processes (verified by repetition). It is advisable however, to minutely analyze all potential risks from the new projects prior to their establishment, as much as they concern mega-resources and could yield mega-effects.

So far, all Bering Strait dams and causeways proposed are monumental but standard barrier designs that are extremely expensive monetarily. Formed of pervious rock-fill and sunken, prefabricated impervious concrete caissons rising visibly above the Bering Strait's local sea-level, such permanent installations would be massive structures relentlessly attacked by the elements usually present in the Arctic outdoors, at least the unavoidable super-structure icing, strong winds, seawater corrosion and ocean waves. Seasonal sea-ice jamming is a particular structural integrity nuisance during times of otherwise inclement weather. No final mega-engineering design specifications for a Bering Strait Dam yet exist, so if a materially solid Bering Strait Dam of P.M. Borisov type is supposed, of 85,000 m long, 60 m high and 100 m wide, then it would total $\sim$510,000,000 $m^3$ of deliberately deposited rock-fill and/or solidified concrete bulk. And, $\sim$18% of that enormous bulk would be above local sea-level and, thus, visible to people on land and aboard nearby shipping that traverses the Arctic region, whenever clear weather conditions permit such viewing. Doubtlessly, there are profound mega-engineering aesthetic decisions involved in the planning of the architectural presentation to the world's publics of an aerial section of any Bering Strait Dam work of Art, Science and Engineering.

Our proposed removable textile glass-fiber barrier for the Alaska-Siberia region, the BSSD, would be completely underwater. Submerged it would be entirely invisible and, as well, capable of avoiding hazardous super-structure icing.It is possible, after intricate physical and computer modeling studies are done, to determine that the BSSD should extend from the sea-bottom to, approximately, 3–4 m beneath the seawater surface—that is, the BSSD would block $\sim$96% of the Bering Strait's normal northward seawater flow whereas the Bering Strait solid dams proposed are all built with the expressed and explicit intent to control 100% of that natural seawater movement. In other words, the BSSD won't affect deep-keeled natural sea-ice movement through the Bering Strait. At the 55th meeting of the International Whaling Commission in Berlin, Germany during 2003, it was first reported that, annually and worldwide, >300,000 marine mammals die unintentionally in fishing nets (gillnets, traps, weirs, long-lines and trawls). The BSSD's fiber-glass seawater-impermeable film will be "visible" to echo-locating mammals, unlike ensnaring fishnets. Bowhead whales (*Balaena mysticetus*), for example, feed in the Bering and Chukchi seas; these whales could pass above the submerged top of our BSSD without much physical impairment or collision damage to the BSSD. Toothed whales root in the seafloor sediments for food and they have, occasionally, become entangled in undersea telegraph cables because the cables were not previously detected by those large moving creatures (Heezen 1957). Branches from the "ArcticLink" or, perhaps, a different new submarine communications cable should extend East–West at the base of the BSSD for the simple reason that underwater fiber-optic cables have a diameter of $\sim$70 mm and weigh $\sim$10 kg/m.

We think that the BSSD should be composed of 0.05 mm-thick hydrophobic and impenetrable 3-D architecture glass-fiber, which does not corrode and offers an exceptionally slick surface difficult for any living marine organisms to colonize (bio-fouling). Currently, similar material is widely used in modern sails and for large yacht hulls. The structure, however, cannot be neglected or ignored and must be thoroughly inspected annually and, perhaps, each fiber-glass panel should be replaced after 5–10 years of service. An 85,000 m-long, 50 m-high and 0.05 mm-thick BSSD would total 212.5 $m^3$ in bulk. The BSSD's mass could total 303 metric tons. The BSSD would have a bulk ratio to Russian engineer P.M. Borisov's phenomenal Bering Strait Dam of merely 1/2,400,000. With a total BSSD facial area of at least 4,250,000 $m^2$, and the 2010 USD retail price of glass-fiber panels at

USD 50 per 10 $m^2$, the BSSD's curtain cost alone could sum at 2010 USD 425,000. The BSS Deflector would be hung from a sufficiently buoyant flexible pipe—perhaps a pliable air-pipe or aerogel-pipe that has the appearance of a linked chain of sausages—holding trapped stagnant air by ordinary vertically strung, depended in fact, stainless steel braided cables. Uniform-size simple concrete weights—like those employed in Halifax Harbor during World War II to anchor anti-submarine nets—would seal the glass-fiber drape's bottom/lower edge to the seafloor "permanently", and without causing excessive ecological harm because of its minimal spatial impact (area footprint). Basically, the BSSD's function is to partially compartmentalize the Arctic Ocean; its main tensile stress will be the impinging 800,000 $m^3/s$ Bering Sea current that the sturdy BSSD must divert successfully; surely, this current's constant impact will generate a non-destructive "spinnaker effect", with the BSSD slightly bulging northwards into the Bering Strait. Obviously, there must be some "play" in the several thousand dependent glass-fiber panels, some or all of them possibly infused with cement. There are other impermeable films available, derived from advanced material technologies, especially technical textiles exhibiting high-performance, purely functional and precisely woven or non-woven fabrics, which offer the BSSD's constructional and operational mega-engineers the attractive prospect of inexpensive seawater barriers that can safely be opened and closed often; some mega-engineers, indicating the use of expensive Kevlar, have proposed their use in the fabrication of reliable terrestrial canal lock-gates (Cullen 2005). The following image shows the most likely Alaska-Siberia layout for a BSSD structure (Figs. 10 and 11).

In 2007, the world's fleet of powered seagoing merchant marine ships was nearly 43,000. According to a reputable reporting agency, the Institute of Shipping Economics and Logistics (established in 1954), the 2007 fleet consisted of tankers (41%), bulk carriers (36%) and container ships (13%), with the remainder being general cargo and passenger vessel. It is noteworthy that tourist-carrying ships, some featuring luxury accommodates for wealthy travelers, have visited both the Southern Ocean where it laps Antarctica and the North Pole. We have not discovered anywhere in the available published literature any useful estimates of the number of ships that might eventually be routed through the Bering Strait on a published schedule if the voyage becomes feasible, desirable and profitable. During 1969, the 136,000 metric ton oil tanker *Manhattan*, protected by a specially-fitted, *ad hoc* steel-reinforced bow and hull belt, transited the Arctic Ocean, in both directions, to test the practicability of shipping Alaska's North Slope commercial oil product elsewhere. It is possible that mid- to late-twenty-first Century Arctic Ocean sea-conditions will permit commercial traffic to regularly utilize the route through the Bering Strait since that reduces drastically the voyage distance from Europe to Asia by ~11,000 km, and >19,000 km for those vessels that must pass South America's Cape Horn.

After the BSSD's completion, its successful commercial operation will require that people live and work "permanently" on the islands during all weather conditions because international navigational safety will demand a sea-lock control tower, residences and re-supply by shipments brought to the islands by vessels

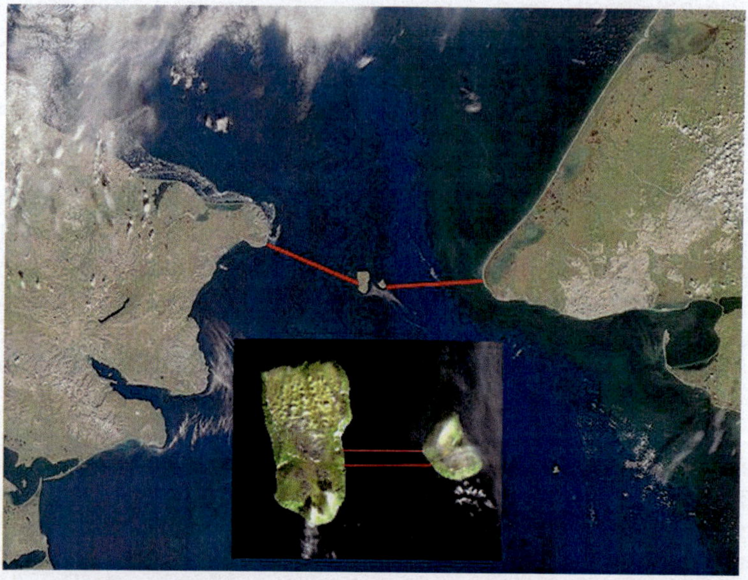

**Fig. 10** In the *biggest* image above, the summertime BSSD is indicated by the two straight red lines converging on the Diomede Islands. The *smaller inset* image shows the minimal useful shipping channel between the unequal length sea lock-gates stretching between the two Diomede Islands in the central part of the Bering Sea. We have dubbed this facility the "International Date Line Sea-lock"

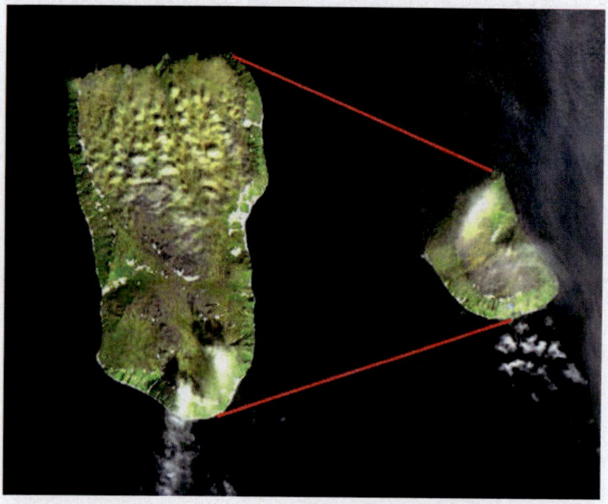

**Fig. 11** The image above shows the largest possible useful "International Date Line Sea lock" gates installable between the Diomede Islands which are necessary to limit the shipping channel seawater throughflow. The throughflow will be very slightly amplified by the BSSD because sea-level will be slightly higher south of the BSSD than north of it due to collision of the natural current with the barrier

**Fig. 12** Section of a recent USA-printed paper nautical chart showing the Bering Strait and vicinity. Obviously, more navigational marking facilities will have to be installed to adequately serve the commercial shipping and naval communities regularly utilizing the BSSD's "International Date Line Sea-lock" or simply passing over the BSSD

(Bolonkin and Cathcart 2007). Such a secondary dome facility enclosing all or a significant portion of an existing community is a large secondary envelope (dome) that conserves energy, although low-speed wind turbines could be inserted into the major ventilation ports to power generators for producing electricity. A domed facility's primary advantage as a near-total community envelopment would be in reducing infiltration of cold air above the Diomede Islands into the existing housing and offices and moderating the conditioned climate within the envelope. Authorities could also monitor the northward seawater flow through the Bering Strait in real-time (Cooper et al. 2006, Mizobata et al. 2010). And, accurate shipping charts will be necessary for commercial and naval vessels to navigate the new seaway. Perhaps, additional lighthouses, manned or automated, will be required (Fig. 12). During the last decade of the twentieth century, the use of computers for navigation of commercial vessels became standard. Shore-based facilities monitoring ship movements passing through straits such as Malacca and the English Channel have been established capitalizing on the ever-improving modalities of telecommunications (e.g., radar, radio and Internet). In other words, the BSSD will need to adopt the garb of the modern-day marine electronic seaway that would make the employment of all-weather marine pilots unnecessary. Since the whole BSSD, and the BSSD gated sea-lock, are at least 4 m below the ocean's surface that means various species of migratory whales ought to be able to swim presumably unimpeded through the sea-lock gates unmolested and without much physical or psychological hindrance.

Writing several decades before the twenty-first century's world-public concern on essentially unsettled Science-envisaged man-enhanced "Global Warming", the Canadian oceanographer Maxwell John Dunbar (1914–1995) was the only person to make any worthwhile attempt whatsoever to describe generally what might happen to the Pacific and Atlantic oceans, and to the Northern Hemisphere's climate, when a BSSD concept is realized. An accurate history of Arctic sea-ice, the remarkable reduction in area of which commenced "…in the late $19^{th}$ century", is just now being compiled (Polyak et al. 2010). We summarize Dunbar's conclusions by applying his thoughts to the BSSD. The BSSD alone would stop the seawater originating in the Bering Sea from entering the Arctic Ocean. Dunbar suggested in 1962 that such exclusive cessation would be felt mostly by the geographical fringes of North America, especially along Greenland's west coast and in Baffin Bay. The proportion of Arctic Ocean seawater in the Arctic Ocean/North Atlantic Ocean mixture, which composes the West Greenland Current, flowing northward, would be reduced by ~8–10%, causing a corresponding weakening of the Canadian Current on Baffin Island's east coast. The Labrador Current should become slightly warmer by ~0.5°C. In sum, M.J. Dunbar believed a BSSD-type structure would benefit climatically only Canada and Greenland because the present-day outflow of relatively cold Arctic Ocean seawater would be less when no BSSD existed (Dunbar 1962). If semi-anthropogenic "Global Warming" is becoming a reality (Masco 2010), then M. J. Dunbar's overall scientific thoughts on the subject are outdated and maybe even useless and strongly in need of update. Oceanography, like modern-day climatology, is not a settled profession purely in terms of geophysical certainty—facts. Factors are actually facts that stand in a causal relationship. To this day, such facts remain hard to obtain and sometimes very costly to understand (Greene et al. 2008). It seems warm water is already affecting the rate of land glacier melting of East Greenland when they are floating atop the intruding subtropical seawater (Straneo et al. 2010).

## 7 Environmental Impact

In comparison with other areas of construction engineering, hydraulic engineering risk problem has other dimensions. Through the hydrotechnical construction activity, water management works are carried out, which have great importance for all social and economic activity.

Damages caused by accidents in water harnesses can reach those caused by major natural disasters. Errors in hydraulic construction have serious repercussions, with large diffusion. This is why risk assessment must be made with utmost responsibility. In hydraulic engineering, the following forms of risk are involved (Nistreanu et al. 1999):

- Seismic risk, in connection to the occurrence of induced earthquakes;
- Seismic risk, on the occurrence of tsunami type, powerful waves;
- Risk of physical–chemical change in water quality;
- Biological risk by preventing free circulation of sea creatures;
- Risk of clogging by reducing the transport of silt;

- Environmental risk related to the influence the hydro facilities have on climate;
- Financial risk as a result of higher execution times of the work.

## 7.1 Induced Seismicity

It can be said that a connection was observed between the reservoirs, especially during their filling, and the local seismic activity. The ICOLD Committee report of 1978 refers, among other 20 cases, the case Tolbingo dam in Australia, where seismic recordings were made 13 years before lake filling. During this period a single earthquake was recorded, while in the 15 months that followed there were no fewer than 100.

These earthquakes (mainly superficial, with depths in the Earth-crust of 10 km) seem to be induced by large water reservoirs ($V > 1$ km$^3$, $H > 80$ m), but only under certain conditions. They are probably due to a phenomenon of release of existing tension, or by weakening the balance by efforts due to the relatively small weight of the stored water, or weakening of the ground due to extra infiltration.

However, increasing a reservoir is not a priori an element of generating earthquakes. Such large height dams, located in an area with high seismicity (California or Mexico) have not started, during lake filling, any seismic activity.

## 7.2 Tsunami Risk

The Great Alaskan Earthquake of March 28, 1964 generated a tsunami, which was also destructive in British Columbia, Canada, and in the U. S. States of Washington, California and Hawaii. The tsunami waves were particularly destructive at Vancouver Island. As recorded at tide gauges, the tsunami height was 1.4 m at Prince Rupert and 1.2 m at Tofino. Actual tsunami runup of 1964 was higher (White 1966).

Consequence of possibly the highest earthquake wave ever recorded, the town of Merak in Sumatra was destroyed in 1883 by the 46 m (151 ft) high tsunami, which accompanied the eruption from Krakatoa (Bishop 1885; Press 1956). The most recent alert of that kind took place in February 2010 after the Chile earthquake (Song 2010). This shows the huge extent of destruction upon the buildings emerged from the tsunami effect and the associated risk within the Bering water engineering. A special design must be associated to the bridge pillars of the gigantic construction supposed for the Bering Strait.

## 7.3 Physical–Chemical Quality of the Water

The stagnant water in the sea is quite different in physical–chemical quality to the water flowing free into the ocean before constraining. Qualitatively new processes appear in the region, as for example the water stratification. This occurs as a result of pressure differences that are caused by differences in density, due in turn to different temperatures, different salinity or to the different suspension content of water layers. Wind action in shallow water generally succeeds to mix the water well enough, and the process of stratification is unimportant or even nonexistent. For deep waters, however, water stratification is certainly entailed, and even large floods can not hinder this process. During the warm times the upper water (Epilimnion) is heated under the action of sunlight and warm water intake, while the lower zone (hipolimnion) remains at a temperature close to 4°C, corresponding to a maximum density of the water. These two areas are separated by a high thermal gradient zone (metalimnion). So the thermal variations due to wind and outdoor temperature differences are found only in Epilimnion. During the cold period, the low outdoor temperature, wind action and polar water intake get the Epilimnion at a lower temperature than the Hipolimnion, leading ultimately by vertical currents that are generated, to the disappearance of the Metalimnion and virtually to water uniformity in terms of temperature.

It results that, because of pressure variations due to differences in density, water movement occurs in different layers, so that some layers are to remain immobile, while others are moving in one direction or another. An example of horizontal currents is the movement of the surface layer of water, subject to the direct action of the sun, to areas shadowed by banks or clouds and vice versa. Density currents and stratification of seawater has a decisive influence on the physico-chemical and even biological properties of the seawater. The oxygen, the nutrients, the suspended matter concentration, the salinity and ecological processes, are different in the surface layers from those in the deep.

The lack of oxygen in the Hipolimnion causes anaerobic process producing ammonium and phosphorus dangerous at exchange between layers. The oxygen is necessary for aquatic life (breathing of fauna and flora) and to the self-purification capacity of the water by biodegradation of dissolved organic matter, in suspension or fallen to the bottom of the sea. The oxygen contained in water comes from a surface water penetration, catalyzed by wind action (the film due to oil or detergents pollution is an impediment) and from photosynthesis of aquatic flora (especially the phytoplankton). The content of the dissolved oxygen decreases with increasing temperature and pollution.

Deep reservoirs have a significant oxygen deficiency in the hipolimnion zone, which at discharges taken from this area (during the warm period) may give trouble in the downstream (failure to provide a minimum content of the dissolved oxygen). The oxygen content of the water from deep sea depend on: the importance of thermal stratification, water content in organic matter and the content in the products of decomposition.

Origin of deep seas oxygen deficiency is not clarified, the duration or extent of the deficit can not estimate. Downstream discharges lead to an overcapacity of water in nitrogen and oxygen. Nitrogen and oxygen content is directly proportional to the height of discharge and inversely proportional to water depth into which it discharges.

## 7.4 Biological Quality of the Water

The biological quality of the water means the totality of living beings (biocoenosis) and energetical, physical, chemical and biological conditions of the environment on a given area, relatively homogeneous. The biological water quality is essentially affected by the eutrophication phenomenon, favored by the achievement of hydro harnesses.

Eutrophication is the enrichment of water with nutrients above a limit from which on the worsening of the water quality begins (biological overproduction is reached). Depending on the trophic stages (trophos = food) the seas can be:

- Oligotrophic, characterized by the predominance of physical–chemical factors, where the nutrient is poor, with an minor biological production;
- Mesotrophic, characterized by the nutrient enrichment that leads to an increase in organic production and to a minor oxygen deficiency;
- Eutrophic, very rich in nutrients, causing rapid biological excessive growth and implicitly an aberrant extent of the oxygen deficit, especially in hipolimnion.

Eutrophication is accelerated by pollution with detergents, hydrocarbons and domestic and industrial effluents (phosphorus and nitrogen have a particular effect favoring eutrophication). Marine eutrophication effects are:

- Overdevelopment of plants and algae to shore, macrophytes and plankton at the bottom;
- Weight change of sea fish species in the economy by the occurrence of oxygen deficiency (salmonids to cyprinids replacement until the disappearance of the first);
- Reduced transparency and color change of the water;
- Loss of oxygen content in hipolimnion;
- Occurrence of hydrogen sulphide, free iron, magnesium and ammonia;
- Bottom deposit in large amounts of decomposing material;
- Occurrence of gas bubbles.

There are some bacteriological risks related to stagnant seawaters. Bacteriologically speaking one can say that the stagnant waters promote the development and transitation of pathogenic bacteria (Salmonella, Shigella, etc.).

## 7.5 Water Clogging

This effect is caused by silt regime change, warping being produced in the stagnant water (appearance of delta formation) and the concentration of nutritious but also harmful, substances, and downstream erosion of the bed occurs, change of the bed granulosity and a considerable diminishing of nutritious substances.

Water quality undergoes changes with respect to temperature, meaning that insolation increases and thermal stratification of the water occurs, water salinity is influenced by concentration, precipitation or dissolution. Turbidity reduction entails a reduced possibility for natural ventilation and oxygenation of the water and an accumulation of organic substances and chemicals. The appearance of a stagnant microclimate requires an increase of moisture and evaporation and provide a milder climate.

Another aspect stands in potential landslides in the sea, that may occur due to change in the flooding of the layers sensitive to softening, due to the erosion of the slopes base or due to the induced seismicity. Contour roads may be affected.

Underground hydrotechnical works can produce seepage, landslides and wetting of the slopes. Seeping water freezing may produce ice massives. On the surface, the transition from freshwater to the stalled water regime presents the shift to a new ecosystem, specifically in terms of physical, chemical and biological features. Flooding of large areas, which previously had an agricultural or forestry use, produces decommissioning of housing, eliminating communication paths, which are immediate and direct effects of the development projects.

In any stagnant sea occurs, over time, a complex process of aging (eutrophication), leading to the degradation of the water quality (low oxygen content in the appearance of a bad taste and smell, unavoidable even through the treatment etc.), to an accelerate clogging, to cancellation of fisheries and recreational utilizations.

The positive impact on the area remains clear, to develop fish farming, creating new communication channels with much lower travel times, and electricity production.

## *7.6 Potential Impact of Harnessing Upon the Climate*

Clearly the element of the marine harnessing that affects the climate is the emergence of stagnant waters. As the surface is higher, more important the effect on climate is. There are examples of facilities worldwide that significantly influence climate: Lake Volta in Ghana has changed the rainfall regime in the area, moving them from October in the months from July to August, or the great reservoir caused by the Aswan High Dam, which brought rain in a very dry area. Others have a relatively small influence, which is limited to the local microclimate.

The main cause of this specific microclimate is determined by the properties of the water, but the distance on land up to where the water basin influence is felt depends on its size and form, and on the weather condition (Fig. 13). In the case of a mega-engineering intrusion into the Bering Strait the influence of that change in water temperature distribution could be very important. The intention is to separate the cold, northern ocean from the wormier, southern Bering Sea and this may surely produce an atmospheric pressure imbalance that will generate high winds (Fig. 14).

The thermal regime of water surface and of the air from above is characterized by small annual and diurnal amplitudes, due to higher specific heat of the water,

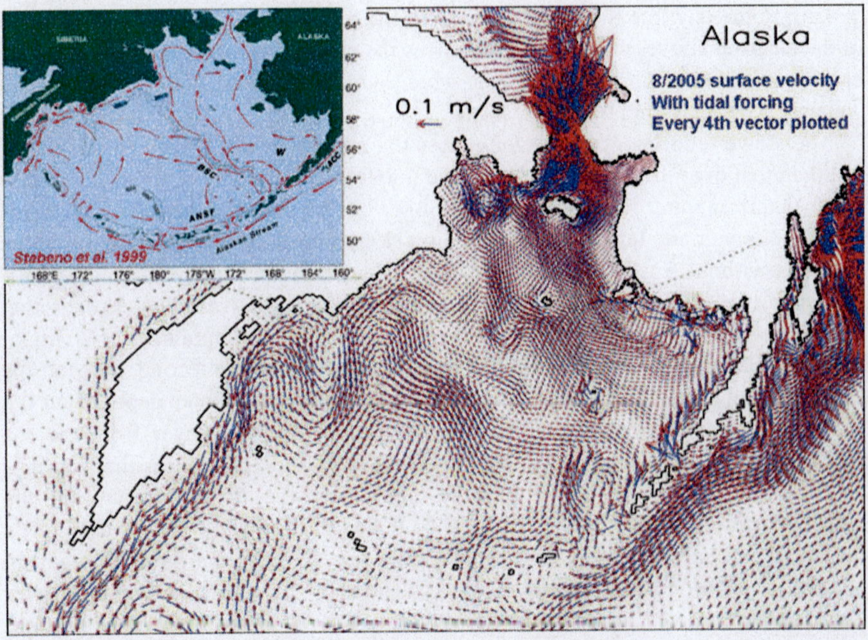


**Fig. 13** Tidal water currents within the Bering Sea and the surroundings (Greenemeier 2008)

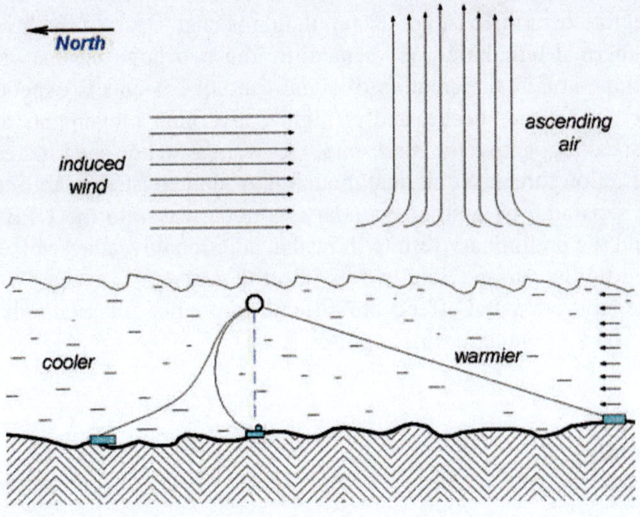


**Fig. 14** Presumable permanent induced wind structure above the BSSD which will affect shipping and human settled life-styles

due to higher water heat conductivity than that of the air, due to evaporation which maintains a constant heat exchange between the two media (water–air) and to resulting vertical currents that determine a turbulent mixing of the air.

During the daytime these causes prevent the excessive heating of the water and of the air from above, while at night reduces the cooling. Also, because of the same causes, in sunny summer afternoons, when land is at maximal temperatures, the water surface may remain much colder.

The slower cooling of the water causes the air over water bodies be warmer at fall than that over the land, producing real maximal differences at night. Raising of thermal minima and keeping the heat longer by the water makes the sea shores frosts be weaker and less frequent than a few km away. Usually, the duration of the intervals without frost exceeds 10–12 days in winter.

Regarding the air moisture over the water bodies, this usually exceeds that above the land, on account of a continuous process of evaporation. Maximum moisture differences are recorded during daylight, in the second part of the summer. Fog formation through evaporation over stagnant water depends on the air humidity, the temperature difference between the water layer from the sea surface and the atmospheric air and on the water salinity. To form that fog it is necessary that the water temperature be higher than the air temperature and the relative air humidity be greater than 90%. If this phenomenon occurs near roads (especially in case of major traffic), can give rise to accidents more common in winter, when in contact with the road surface ice is formed. Also in winter landscape in the area is covered by frost. Also, the fog changes the existing flora on the banks and could affect the health of the seaside residents. If the hydroelectric harness is in a windy area, these lead on pushing the mist into the surrounding areas, creating conditions for increased evaporation.

Wind regime over the basin of accumulation is characterized by the appearance of a local thermal circulation. By separating the two large basins with sensibly different temperatures, the creation of significant air currents is expected. During the day air divergences occur, and at night converging movements are formed. Generally speaking, above the reservoirs, the average wind speed increases due to decreased friction forces. Numerical modeling of this unusually large temperature imbalance, separated by a strait boundary, is underway into the UPB laboratory (Fig. 13) and the preliminary results show that additional weather worsening could be produced by the surface wind waves into the water.

All these environmental effects are difficult to predict quantitatively, however, they should be kept under a strict control.

## 8 Conclusions

We have proposed a removable and invisible Bering Strait Seawater Deflector that can be cheaply installed in a vital natural seaway. It is scalable. Unlike previous Bering Strait dam or bridge mega-project proposals, the BSSD would not contend with destructive aerial forces (storm winds), seawater waves or sea-ice jamming. The BSSD, a dependent membrane draped from a suspending cross-Bering Strait air-filled (un-pressurized) pipe that furnishes buoyancy to the BSSD, will float approximately 4 m below the seawater's surface in order to avoid damage from moving sea-ice that could pile against a sea-level floating boom. In other words, it will be placed in a, more or less, homogeneous environment. During the summertime, taking advantage of the curtain's play, the BSSD would be raised in sections, by mobile floating cranes, to the Bering Strait surface in order to facilitate regular maintenance work done by land-based crews housed nearby in docked or moored barges. Hence, the dependent draped barrier must have some give in it, which is operationally shown by the underwater "spinnaker effect". And, just like Japanese oceanographer Keiji Higuchi's Drake Passage sunken iceberg barrier in Antarctica's Southern Ocean, the Arctic BSSD will inevitably cause a slight increase in the sea-level on the side of the drape that receives the impacting seawater current. Submerged fiber-glass fabric sea-lock gates, placed at both ends of the shipping channel running between the Diomede Islands, will provide all large-draft commercial shipping with easy ingress and egress to the Arctic Ocean and Bering Sea, especially since the "International Date Line Sea-lock" end stoppers won't be required to open or close for moving ships with drafts <3 m. The sea-lock gates will be opened by a downward vertical movement instigated by seafloor-based electric winches contained in water-proof pods. The whole watery segment between the sea-gates, in fact, can be kept free of sea-ice by harnessing locally supplied geothermal energy intentionally harnessed to melt any seawater current/wind-trapped sea-ice pushed into the International Date Line Sea-lock" (van Huisen 1975).

We can offer only a reasoned estimate of the construction cost of the BSSD—we suspect it could be considerably less than the cost of a trans-Bering Strait railway/highway bridge, possibly with an financial expenditure of as little as 2010 USD 0.5 billion. One fact that especially bears crucially on our purely reasoned cost estimate is that no railroads or highways need be connected to the BSSD's Russia or Alaska termini. Only an overland fiber-optic telecommunications cable may need to be attached at the base of the curtain held in place by the flexible textile-constructed buoyancy pipe. Many BSSD mega-project parts and workers ought to be housed onboard floating barges safely sequestered in nearby seaports. Should adverse, unexpected climatic effects develop in the Arctic or in critical places elsewhere after emplacement of the BSSD, then the BSSD can be quickly removed, effectively alleviating any climatic adversities created by speedy direct elimination of the strictly man-induced mega-problem itself. However, it is to be hoped that the BSSD will perform successfully. By cooling the Chukchi Sea, the BSSD might, in some temporary way, help to retain the Arctic's on-the-verge-of-meltdown tundra in the same conditions of preservation as today—that is, stable, with no excessive methane greenhouse gas emissions causing undesirable regional or global climate change. Effectively, the BSSD constitutes perhaps the world's first unambiguously planned scalable geo-engineering experiment without enormous financial and biosphere costs and with reasonable human control during a societal-useful timeframe. The Fourth Intergovernmental Panel on Climate Change report issued in 2007, unconvincingly projected a smoothly rising global temperature increase as if Earth were Mars (Badescu 2010). The basic fiberglass panels forming our BSSD *in toto* might be considered, metaphorically, as a hinted, planned mega-engineering response that could temper some of the alarming facts that are sure to be publicized, circa 2014, in the world climate change report offered by the Fifth Intergovernmental Panel on Climate Change.

## References

Abbasi R et al (2010) Search for muon neutrinos from gamma-ray bursts with the IceCube neutrino telescope. Astron J 710:346–359

Anderson HH (1975) LUNAHEAT: warming the northern regions in the third millennium. Chart Mech Eng 23:43–45

Anon. (1889) The Poetsch freezing process in mining operations. Science 14:142–143

Anon. (1961) Bering Strait Dam. Nature 190:491

Arrigo KR (2008) Impact of a shrinking Arctic ice cover on marine primary production. Geophys Res Lett 35:GL035028

Badescu V (ed) (2010) Mars: prospective energy and material resources. Springer, NL, 700 pp

Behrman D, Isaacs JD (1992) John Isaacs and his oceans. American Geophysical Union ICSU Press, Washington DC, p 49

Bernshtein LB (1994) Power engineering in the 21st century: Penzhina tidal power station (Transmission of 190 TWhr of environmental clean energy to North America). Power Technol Eng 28:686–691

Bishop SE (1885) Krakatoa. Nature 31:288–289

Boe JH, Qu X (2009) September sea-ice cover in the Arctic Ocean projected to vanish by 2100. Nat Geosci 2:341–343

Boe J, Hall A, Qu X (2010) Sources of spread in simulations of Arctic sea ice loss over the twenty-first century. Clim Change 99:637–645

Bolonkin AA, Cathcart RB (2007) Inflatable 'Evergreen' dome settlements for Earth's Polar Regions. Clean Technol Environ Policy 7:125–132

Bolonkin AA, Cathcart RB (2008) Antarctica: a southern hemisphere wind power station? Int J Glob Environ Issues 8:262–273

Bolonkin AA, Cathcart RB (2009) Mega-projects: environment and technology. Nova Science Publishers, New York, 537 pp

Borgerson SG (2008) Arctic meltdown. Foreign Affairs 87:63–77

Borisov PM (1973) Can we control the arctic climate? Progress Publishers, Moscow, USSR, Glav 12:63–73

Bourassa R (1985) Power from the North. Prentice Hall, Ontario, 182 pp

Braun B, Nash WR (1985) Ground freezing for construction. ASCE Civil Eng 55:54–56

Brezhnev VA, Abramson VM, Zemelman AM, Vlasov SN, Koulaguin NI, Merkin VE, Razbeguin VN (2005) Russian underwater tunnels in the system of international transportation ways. Tunn Undergr Space Technol 20:595–599

Cameron RL (2005) The foundations of Antarctic glaciology. Arch Nat Hist 32:231–244

Cole T (1990) The bridge to tomorrow: visions of the Bering Strait Bridge. Alaska Hist 5:1–15

Collis C (2007) Cold colonies: Antarctic spatialities at Mawson and McMurdo stations. Cult Geogr 14:234–254

Conover T (2010) The routes of man: how roads are changing the World and the way we live today. Knopf, New York, 334 pp

Cooper LW, Codispoti LA, Kelly V, Sheffield GG, Grebmeier JM (2006) The potential for using Little Diomede Island as a platform for observing environmental conditions in Bering Strait. Arctic 59:129–141

Cullen MN (2005) Tension membrane water retaining structures. Trans Built Environ 79:217–220
Dunbar MJ (1962) Second report on the Bering Strait Dam. Northern co-ordination and Research Centre Department of Northern Affairs and National Resources (NCRC) 62-6, 23 pp
Elliott IH, Howard AJ (1994) Bored tunnels under deep waterways: what lies ahead? Tunn Undergr Space Technol 9:329–342
Etter L (2006) Feeling the heat; for Icy Greenland, global warming has a bright side. Wall Str J CCXLVIII:A1, A12
Foreman MGG (2006) Tidal energy in the Bering Sea. J Mar Res 64:797–818
Gautier DL et al (2009) Assessment of undiscovered oil and gas in the Arctic. Science 324:1175–1179
Gellert EP, Turley DM (1999) Seawater immersion ageing of glass-fiber reinforced polymer laminates for marine application. Composites A:1260–1264
Gordeev VV, Kravchishina MD (2010) Mechanisms of the recent sea ice decay in the Arctic Ocean related to the Pacific-to-Atlantic pathway. In: Nihoul JCJ, Kostianoy AG (eds) Influence of climate change on the changing Arctic and Sub-Arctic conditions. Springer, The Netherlands, p 161
Granier C (2006) Ozone pollution from future ship traffic in the Arctic northern passages. Geophys Res Lett 33:L13807
Greene CH, Pershing AJ, Cronin TM, Cecci N (2008) Arctic climate change and its impacts on the ecology of the North Atlantic. Ecology 89:524–538
Greenemeier L (2008) Turning the tide on harnessing the Ocean's abundant energy. Scientific American, October 20
Gualtieri L, Brigham-Grette J (2001) The age and origin of the Little Diomede Island upland surface. Arctic 54:12–21
Hansen K (2007) Building bridges with icy challenges. Geotimes 52:30–33
Heezen BC (1957) Whales entangled in deep-sea cables. Deep-Sea Res 1:105–115
Heezen BC, Johnson GL (1969) Alaskan submarine cables: a struggle with a harsh environment. Arctic 22:413–424
Heimann M (2010) How stable is the methane cycle? Science 327:1211–1212
Higuchi K (1971) A possibility of constructing a dam to change the general oceanic circulation. Collect Pap Sci Atmos Hydrosphere 9:355–363
Hornig JF (1999) Social and environmental impacts of the James Bay hydroelectric project. McGill-Queens University Press, Montreal, 169 pp
Howard R (2009) The Arctic Gold Rush: The New Race for Tomorrow's Resources. Continuum, London, 259 pp
Hu H, Wang J (2010) Modeling effects of tidal and wave mixing on circulation and thermohaline structures in the Bering Sea: process studies. J Geophys Res 115:C01006
Hu A, Meehl GA, Otto-Bliesner BL, Waelbroech C, Han W, Loutre M-F, Lambeck K, Mitrovica JX, Rosenbloom N (2010) Influence of Bering Strait flow and North Atlantic circulation on glacial sea-level changes. Nat Geosci 3:118–121
Husseiny AA (ed) (1978) Iceberg utilization. In: Proceedings of the first international conference held at Ames, Iowa. Pergamon, New York, 759 pp
Kitchen A, Miyamoto MM, Mulligan CJ (2008) A three-stage colonization model for the peopling of the Americas. PLoS One 3:1–7
Kudoyarov LI (1993) Problems of constructing an intercontinental railroad and tunnel under the Bering Strait. Gidrotekhnicheskoe Stroitel'stv 8:17–22
Kwok R (2010) Large sea ice outflow into the Nares Strait in 2007. Geophys Res Lett 37:L03502
Laulajainen R (2009) The Arctic sea route. Int J Shipp Transp Logist 1:55–73
Launder B, Thompson JMT (eds) (2009) Geo-engineering climate change. Cambridge, New York, 314 pp
Lawrence DM (2008) Accelerated Arctic land warming and permafrost degradation during rapid sea ice loss. Geophys Res Lett 35:L11506

MacAyeal DR (1984) Preventing a collapse of the West Antarctic Ice sheet: civil engineering on a continental scale. Ann Glaciol 4:302

Mackey KG, Fujita K, Gunbina LV, Kovalev VN, Imaev VS, Kozmin BM, Imaeva LP (1997) Seismicity of the Bering Strait region: evidence for a Bering block. Geology 25:979–982

Malecki EJ, Wei H (2009) A wired World: the evolving geography of submarine cables and the shift to Asia. Ann Assoc Am Geogr 99:360–382

Masco J (2010) Bad weather: on planetary crisis. Soc Stud Sci 40:7–40

McGuire AD (2009) Sensitivity of the carbon cycle in the Arctic to climate change. Ecol Monogr 79:523–555

Mizobata K, Shimada K, Woodgate R, Saitoh S-I, Wang J (2010) Estimation of heat flux through the Eastern Bering Strait. J Oceanogr 66:405–424

Molloy AE (1962) Arctic science and the nuclear submarine. Arctic 15:86–91

Nistreanu V, Nistreanu V (1999) Harnessing of water resources and its impact on the environment. ISBN 973-99065-8-3. BREN, Bucharest

Osafune S, Yasuda L (2010) Bidecadal variability in the Bering Sea and the relation with 18.6 year period nodal tidal cycle. J Geophys Res Ocean 115:C02014

Palmer AW (2002) Negotiation and resistance in global networks: the 1884 international Meridian conference. Mass Commun Soc 5:7–24

Partl R (1978a) Power from Greenland's glaciers. Water Power Dam Constr 30:42–50

Partl R (1978b) Power from glaciers: the hydropower potential of Greenland's glacial waters. Energy 3:543–573

Patterson WP, Dietrich KA, Holmden C, Andrews JT (2010) Two millennia of North Atlantic seasonality and implications for Norse colonies. Proc Am Acad Sci 107:5306–5310

Petroski H (2010) The essential engineer: why science alone will not solve our global problems. A.A. Knopf, New York, pp 169–172

Polyak L, Alley RB, Andrews JT, Brigham-Grette J, Cronin TM, Darby DA, Dyke AS, Fitzpatrick JJ, Funder S, Holland M, Jennings AE, Miller GH, O'Regan M, Savelle J, Serreze M, John KS, White JWC, Wolff E (2010) History of sea ice in the Arctic. Quart Sci Rev 29:1

Press F (1956) Volcanoes, ice, and destructive waves. Eng Sci 20(2):26–30

Richardson JG (1999) The bold effort in Siberia to complete the first global telegraphic circuit. Interdisc Sci Rev 24:307–313

Roye A, Gries T (2007) 3-D textiles for advanced cement based matrix reinforcement. J Ind Text 37:163–173

Sandal C, Nof D (2008) Laboratory experiments on the paleo-jamming of the Bering Strait. Deep-Sea Res I:1105–1117

Screen JA, Simmonds I (2010) The central role of diminishing sea ice in recent Arctic temperature amplification. Nature 464:1334

Sijp WP, England MH (2004) Effect of the Drake passage throughflow on global climate. J Phys Oceanogr 34:1254–1266

Smedsrud LH, Ingvaldsen R, Nilsen JEO, Skagseth O (2010) Heat in the Barents Sea: transport, storage, and the surface fluxes. Ocean Sci 6:219–234

Song J (2010) Tsunami that raced across Pacific delivers glancing blow to islands, Asian coasts. The Associated Press, Honolulu, Grand Forks Herald, 28 Feb 2010

Straneo F et al (2010) Rapid circulation of warm subtropical waters in a major glacial fjord in East Greenland. Nat Geosci 3:182–186

Swann AL et al (2010) Changes in Arctic vegetation amplify high-latitude warming through the greenhouse effect. Proc Natl Acad Sci 107:1295–1300

Thompson SR (1995) Sills of the global ocean: a compilation. Ocean Model 109:7–9

Timco GW, Weeks WF (2010) A review of the engineering properties of sea ice. Cold Reg Sci Technol 60:107–129

Turner J (2009) Record low surface air temperature at Vostok station, Antarctica. J Geophys Res Atmos 114:D24102

Van Huisen AT (1975) Geothermal channel and harbor ice control system. US Patent 3,889,473. Awarded 17 June 1975

Vermeij GJ, Roopnarine PD (2008) The coming Arctic invasion. Science 321:780–781
Wang M, Overland JE (2009) A sea ice free summer Arctic within 30 years? Geophys Res Lett 36:L07502
Westbrook GK (2009) Escape of methane gas from the seabed along the West Spitsbergen continental margin. Geophys Res Lett 36:L15608
White WRH (1966) The Alaska earthquake and its effects in Canada. Can Geogr J 72(6):210–219
Williams PD, Guilyardi E, Madec G, Gualdi S, Scoccimarro E (2010) The role of mean ocean salinity in climate. Dyn Atmos Oceans 49:108–123
Woodgate RA, Weingartner T, Lindsay R (2010) The 2007 Bering Strait oceanic heat flux and anomalous Arctic sea-ice retreat. Geophys Res Lett 37:L01602
Yusoff K (2005) Visualizing Antarctica as a place in time. Space Cult 8:381–398
Zaretskii YK, Korchevskii VF (2008) Railroad crossing of Nevel'skogo Strait from the Mainland to Sakhalin Island—alternate scheme with a dead-end Dike and navigation canal. Power Technol Eng 42:144–151

# Index

**D**